T0137157

Lecture Notes in Networks and Systems

Volume 227

The series "Lecture Notes in Networks and Systems" publishes the latest developments in Networks and Systems—quickly, informally and with high quality. Original research reported in proceedings and post-proceedings represents the core of LNNS.

Volumes published in LNNS embrace all aspects and subfields of, as well as new challenges in, Networks and Systems.

The series contains proceedings and edited volumes in systems and networks, spanning the areas of Cyber-Physical Systems, Autonomous Systems, Sensor Networks, Control Systems, Energy Systems, Automotive Systems, Biological Systems, Vehicular Networking and Connected Vehicles, Aerospace Systems, Automation, Manufacturing, Smart Grids, Nonlinear Systems, Power Systems, Robotics, Social Systems, Economic Systems and other. Of particular value to both the contributors and the readership are the short publication timeframe and the world-wide distribution and exposure which enable both a wide and rapid dissemination of research output.

The series covers the theory, applications, and perspectives on the state of the art and future developments relevant to systems and networks, decision making, control, complex processes and related areas, as embedded in the fields of interdisciplinary and applied sciences, engineering, computer science, physics, economics, social, and life sciences, as well as the paradigms and methodologies behind them.

Indexed by SCOPUS, INSPEC, WTI Frankfurt eG, zbMATH, SCImago.

All books published in the series are submitted for consideration in Web of Science.

More information about this series at http://www.springer.com/series/15179

Leonard Barolli · Isaac Woungang ·
Tomoya Enokido
Editors

Advanced Information Networking and Applications

Proceedings of the 35th International
Conference on Advanced Information
Networking and Applications (AINA-2021),
Volume 3

 Springer

Editors
Leonard Barolli
Department of Information
and Communication Engineering
Fukuoka Institute of Technology
Fukuoka, Japan

Isaac Woungang
Department of Computer Science
Ryerson University
Toronto, ON, Canada

Tomoya Enokido
Faculty of Business Administration
Rissho University
Tokyo, Japan

ISSN 2367-3370   ISSN 2367-3389   (electronic)
Lecture Notes in Networks and Systems
ISBN 978-3-030-75077-0   ISBN 978-3-030-75078-7   (eBook)
https://doi.org/10.1007/978-3-030-75078-7

This Springer imprint is published by the registered company Springer Nature Switzerland AG
The registered company address is: Gewerbestrasse 11, 6330 Cham, Switzerland

Welcome Message from AINA-2021 Organizers

Welcome to the 35th International Conference on Advanced Information Networking and Applications (AINA-2021). On behalf of AINA-2021 Organizing Committee, we would like to express to all participants our cordial welcome and high respect.

AINA is an international forum, where scientists and researchers from academia and industry working in various scientific and technical areas of networking and distributed computing systems can demonstrate new ideas and solutions in distributed computing systems. AINA was born in Asia, but it is now an international conference with high quality thanks to the great help and cooperation of many international friendly volunteers. AINA is a very open society and is always welcoming international volunteers from any country and any area in the world.

AINA international conference is a forum for sharing ideas and research work in the emerging areas of information networking and their applications. The area of advanced networking has grown very rapidly, and the applications have experienced an explosive growth especially in the areas of pervasive and mobile applications, wireless sensor networks, wireless ad hoc networks, vehicular networks, multimedia computing and social networking, semantic collaborative systems, as well as Grid, P2P, IoT, Big Data and Cloud Computing. This advanced networking revolution is transforming the way people live, work and interact with each other and is impacting the way business, education, entertainment and health care are operating. The papers included in the proceedings cover theory, design and application of computer networks, distributed computing and information systems.

Each year AINA receives a lot of paper submissions from all around the world. It has maintained high-quality accepted papers and is aspiring to be one of the main international conferences on the information networking in the world.

We are very proud and honored to have two distinguished keynote talks by Dr. Flora Amato, University of Naples "Federico II", Italy and Prof. Shahrokh Valaee, University of Toronto, Canada, who will present their recent work and will give new insights and ideas to the conference participants.

An international conference of this size requires the support and help of many people. A lot of people have helped and worked hard to produce a successful AINA-2021 technical program and conference proceedings. First, we would like to thank all authors for submitting their papers, the session chairs and distinguished keynote speakers. We are indebted to program track co-chairs, program committee members and reviewers, who carried out the most difficult work of carefully evaluating the submitted papers.

We would like to thank AINA-2021 general co-chairs, PC co-chairs, workshop co-chairs for their great efforts to make AINA-2021 a very successful event. We have special thanks to the finance chair and web administrator co-chairs.

We do hope that you will enjoy the conference proceedings and readings.

Leonard Barolli
Makoto Takizawa
AINA Steering Committee Co-chairs

Isaac Woungang
Markus Aleksy
Farookh Hussain
AINA-2021 General Co-chairs

Glaucio Carvalho
Tomoya Enokido
Flora Amato
AINA-2021 Program Committee Co-chairs

AINA-2021 Organizing Committee

General Co-chairs

Isaac Woungang	Ryerson University, Canada
Markus Aleksy	ABB Corporate Research Center, Germany
Farookh Hussain	University of Technology, Sydney, Australia

Program Committee Co-chairs

Glaucio Carvalho	Sheridan College, Canada
Tomoya Enokido	Rissho University, Japan
Flora Amato	University of Naples "Federico II", Italy

Workshops Co-chairs

Kin Fun Li	University of Victoria, Canada
Omid Ameri Sianaki	Victoria University, Australia
Yi-Jen Su	Shu-Te University, Taiwan

International Journals Special Issues Co-chairs

Fatos Xhafa	Technical University of Catalonia, Spain
David Taniar	Monash University, Australia

Award Co-chairs

Marek Ogiela	AGH University of Science and Technology, Poland
Arjan Durresi	Indiana University Purdue University in Indianapolis (IUPUI), USA
Fang-Yie Leu	Tunghai University, Taiwan

Publicity Co-chairs

Lidia Ogiela   Pedagogical University of Cracow, Poland
Minoru Uehara   Toyo University, Japan
Hsing-Chung Chen   Asia University, Taiwan

International Liaison Co-chairs

Akio Koyama   Yamagata University, Japan
Nadeem Javaid   COMSATS University Islamabad, Pakistan
Wenny Rahayu   La Trobe University, Australia

Local Arrangement Co-chairs

Mehrdad Tirandazian   Ryerson University, Canada
Glaucio Carvalho   Sheridan College, Canada

Finance Chair

Makoto Ikeda   Fukuoka Institute of Technology, Japan

Web Co-chairs

Phudit Ampririt   Fukuoka Institute of Technology, Japan
Kevin Bylykbashi   Fukuoka Institute of Technology, Japan
Ermioni Qafzezi   Fukuoka Institute of Technology, Japan

Steering Committee Chairs

Leonard Barolli   Fukuoka Institute of Technology, Japan
Makoto Takizawa   Hosei University, Japan

Tracks and Program Committee Members

1. Network Protocols and Applications

Track Co-chairs

Makoto Ikeda   Fukuoka Institute of Technology, Japan
Sanjay Kumar Dhurandher   Netaji Subhas University of Technology,
                                    New Delhi, India
Bhed Bahadur Bista   Iwate Prefectural University, Japan

TPC Members

Elis Kulla	Okayama University of Science, Japan
Keita Matsuo	Fukuoka Institute of Technology, Japan
Shinji Sakamoto	Seikei University, Japan
Akio Koyama	Yamagata University, Japan
Evjola Spaho	Polytechnic University of Tirana, Albania
Jiahong Wang	Iwate Prefectural University, Japan
Shigetomo Kimura	University of Tsukuba, Japan
Chotipat Pornavalai	King Mongkut's Institute of Technology Ladkrabang, Thailand
Danda B. Rawat	Howard University, USA
Akio Koyama	Yamagata University, Japan
Amita Malik	Deenbandhu Chhotu Ram University of Science and Technology, India
R. K. Pateriya	Maulana Azad National Institute of Technology, India
Vinesh Kumar	University of Delhi, India
Petros Nicopolitidis	Aristotle University of Thessaloniki, Greece
Satya Jyoti Borah	North Eastern Regional Institute of Science and Technology, India

2. Next Generation Wireless Networks

Track Co-chairs

Christos J. Bouras	University of Patras, Greece
Tales Heimfarth	Universidade Federal de Lavras, Brazil
Leonardo Mostarda	University of Camerino, Italy

TPC Members

Fadi Al-Turjman	Near East University, Nicosia, Cyprus
Alfredo Navarra	University of Perugia, Italy
Purav Shah	Middlesex University London, UK
Enver Ever	Middle East Technical University, Northern Cyprus Campus, Cyprus
Rosario Culmone	University of Camerino, Camerino, Italy
Antonio Alfredo F. Loureiro	Federal University of Minas Gerais, Brazil
Holger Karl	University of Paderborn, Germany
Daniel Ludovico Guidoni	Federal University of São João Del-Rei, Brazil
João Paulo Carvalho Lustosa da Costa	Hamm-Lippstadt University of Applied Sciences, Germany
Jorge Sá Silva	University of Coimbra, Portugal
Apostolos Gkamas	University Ecclesiastical Academy of Vella, Ioannina, Greece

Zoubir Mammeri	University Paul Sabatier, France
Eirini Eleni Tsiropoulou	University of New Mexico, USA
Raouf Hamzaoui	De Montfort University, UK
Miroslav Voznak	University of Ostrava, Czech Republic

3. Multimedia Systems and Applications

Track Co-chairs

Markus Aleksy	ABB Corporate Research Center, Germany
Francesco Orciuoli	University of Salerno, Italy
Tomoyuki Ishida	Fukuoka Institute of Technology, Japan

TPC Members

Tetsuro Ogi	Keio University, Japan
Yasuo Ebara	Osaka Electro-Communication University, Japan
Hideo Miyachi	Tokyo City University, Japan
Kaoru Sugita	Fukuoka Institute of Technology, Japan
Akio Doi	Iwate Prefectural University, Japan
Hadil Abukwaik	ABB Corporate Research Center, Germany
Monique Duengen	Robert Bosch GmbH, Germany
Thomas Preuss	Brandenburg University of Applied Sciences, Germany
Peter M. Rost	NOKIA Bell Labs, Germany
Lukasz Wisniewski	inIT, Germany
Hadil Abukwaik	ABB Corporate Research Center, Germany
Monique Duengen	Robert Bosch GmbH, Germany
Peter M. Rost	NOKIA Bell Labs, Germany
Angelo Gaeta	University of Salerno, Italy
Graziano Fuccio	University of Salerno, Italy
Giuseppe Fenza	University of Salerno, Italy
Maria Cristina	University of Salerno, Italy
Alberto Volpe	University of Salerno, Italy

4. Pervasive and Ubiquitous Computing

Track Co-chairs

Chih-Lin Hu	National Central University, Taiwan
Vamsi Paruchuri	University of Central Arkansas, USA
Winston Seah	Victoria University of Wellington, New Zealand

TPC Members

Hong Va Leong	Hong Kong Polytechnic University, Hong Kong
Ling-Jyh Chen	Academia Sinica, Taiwan
Jiun-Yu Tu	Southern Taiwan University of Science and Technology, Taiwan
Jiun-Long Huang	National Chiao Tung University, Taiwan
Thitinan Tantidham	Mahidol University, Thailand
Tanapat Anusas-amornkul	King Mongkut's University of Technology North Bangkok, Thailand
Xin-Mao Huang	Aletheia University, Taiwan
Hui Lin	Tamkang University, Taiwan
Eugen Dedu	Universite de Franche-Comte, France
Peng Huang	Sichuan Agricultural University, China
Wuyungerile Li	Inner Mongolia University, China
Adrian Pekar	Budapest University of Technology and Economics, Hungary
Jyoti Sahni	Victoria University of Technology, New Zealand
Normalia Samian	Universiti Putra Malaysia, Malaysia
Sriram Chellappan	University of South Florida, USA
Yu Sun	University of Central Arkansas, USA
Qiang Duan	Penn State University, USA
Han-Chieh Wei	Dallas Baptist University, USA

5. Web-Based and E-Learning Systems

Track Co-chairs

Santi Caballe	Open University of Catalonia, Spain
Kin Fun Li	University of Victoria, Canada
Nobuo Funabiki	Okayama University, Japan

TPC Members

Jordi Conesa	Open University of Catalonia, Spain
Joan Casas	Open University of Catalonia, Spain
David Gañán	Open University of Catalonia, Spain
Nicola Capuano	University of Basilicata, Italy
Antonio Sarasa	Complutense University of Madrid, Spain
Chih-Peng Fan	National Chung Hsing University, Taiwan
Nobuya Ishihara	Okayama University, Japan
Sho Yamamoto	Kindai University, Japan
Khin Khin Zaw	Yangon Technical University, Myanmar
Kaoru Fujioka	Fukuoka Women's University, Japan
Kosuke Takano	Kanagawa Institute of Technology, Japan

Shengrui Wang	University of Sherbrooke, Canada
Darshika Perera	University of Colorado at Colorado Spring, USA
Carson Leung	University of Manitoba, Canada

6. Distributed and Parallel Computing

Track Co-chairs

Naohiro Hayashibara	Kyoto Sangyo University, Japan
Minoru Uehara	Toyo University, Japan
Tomoya Enokido	Rissho University, Japan

TPC Members

Eric Pardede	La Trobe University, Australia
Lidia Ogiela	Pedagogical University of Cracow, Poland
Evjola Spaho	Polytechnic University of Tirana, Albania
Akio Koyama	Yamagata University, Japan
Omar Hussain	University of New South Wales, Australia
Hideharu Amano	Keio University, Japan
Ryuji Shioya	Toyo University, Japan
Ji Zhang	The University of Southern Queensland
Lucian Prodan	Universitatea Politehnica Timisoara, Romania
Ragib Hasan	The University of Alabama at Birmingham, USA
Young-Hoon Park	Sookmyung Women's University, Korea

7. Data Mining, Big Data Analytics and Social Networks

Track Co-chairs

Eric Pardede	La Trobe University, Australia
Alex Thomo	University of Victoria, Canada
Flora Amato	University of Naples "Frederico II", Italy

TPC Members

Ji Zhang	University of Southern Queensland, Australia
Salimur Choudhury	Lakehead University, Canada
Xiaofeng Ding	Huazhong University of Science and Technology, China
Ronaldo dos Santos Mello	Universidade Federal de Santa Catarina, Brasil
Irena Holubova	Charles University, Czech Republic
Lucian Prodan	Universitatea Politehnica Timisoara, Romania
Alex Tomy	La Trobe University, Australia
Dhomas Hatta Fudholi	Universitas Islam Indonesia, Indonesia
Saqib Ali	Sultan Qaboos University, Oman

Ahmad Alqarni	Al Baha University, Saudi Arabia
Alessandra Amato	University of Naples "Frederico II", Italy
Luigi Coppolino	Parthenope University, Italy
Giovanni Cozzolino	University of Naples "Frederico II", Italy
Giovanni Mazzeo	Parthenope University, Italy
Francesco Mercaldo	Italian National Research Council, Italy
Francesco Moscato	University of Salerno, Italy
Vincenzo Moscato	University of Naples "Frederico II", Italy
Francesco Piccialli	University of Naples "Frederico II", Italy

8. Internet of Things and Cyber-Physical Systems

Track Co-chairs

Euripides G. M. Petrakis	Technical University of Crete (TUC), Greece
Tomoki Yoshihisa	Osaka University, Japan
Mario Dantas	Federal University of Juiz de Fora (UFJF), Brazil

TPC Members

Akihiro Fujimoto	Wakayama University, Japan
Akimitsu Kanzaki	Shimane University, Japan
Kawakami Tomoya	University of Fukui, Japan
Lei Shu	University of Lincoln, UK
Naoyuki Morimoto	Mie University, Japan
Yusuke Gotoh	Okayama University, Japan
Vasilis Samolada	Technical University of Crete (TUC), Greece
Konstantinos Tsakos	Technical University of Crete (TUC), Greece
Aimilios Tzavaras	Technical University of Crete (TUC), Greece
Spanakis Manolis	Foundation for Research and Technology Hellas (FORTH), Greece
Katerina Doka	National Technical University of Athens (NTUA), Greece
Giorgos Vasiliadis	Foundation for Research and Technology Hellas (FORTH), Greece
Stefan Covaci	Technische Universität Berlin, Berlin (TUB), Germany
Stelios Sotiriadis	University of London, UK
Stefano Chessa	University of Pisa, Italy
Jean-Francois Méhaut	Université Grenoble Alpes, France
Michael Bauer	University of Western Ontario, Canada

9. Intelligent Computing and Machine Learning

Track Co-chairs

Takahiro Uchiya,	Nagoya Institute of Technology, Japan
Omar Hussain	UNSW, Australia
Nadeem Javaid	COMSATS University Islamabad, Pakistan

TPC Members

Morteza Saberi	University of Technology, Sydney, Australia
Abderrahmane Leshob	University of Quebec in Montreal, Canada
Adil Hammadi	Curtin University, Australia
Naeem Janjua	Edith Cowan University, Australia
Sazia Parvin	Melbourne Polytechnic, Australia
Kazuto Sasai	Ibaraki University, Japan
Shigeru Fujita	Chiba Institute of Technology, Japan
Yuki Kaeri	Mejiro University, Japan
Zahoor Ali Khan	HCT, UAE
Muhammad Imran	King Saud University, Saudi Arabia
Ashfaq Ahmad	The University of Newcastle, Australia
Syed Hassan Ahmad	JMA Wireless, USA
Safdar Hussain Bouk	Daegu Gyeongbuk Institute of Science and Technology, Korea
Jolanta Mizera-Pietraszko	Military University of Land Forces, Poland

10. Cloud and Services Computing

Track Co-chairs

Asm Kayes	La Trobe University, Australia
Salvatore Venticinque	University of Campania "Luigi Vanvitelli", Italy
Baojiang Cui	Beijing University of Posts and Telecommunications, China

TPC Members

Shahriar Badsha	University of Nevada, USA
Abdur Rahman Bin Shahid	Concord University, USA
Iqbal H. Sarker	Chittagong University of Engineering and Technology, Bangladesh
Jabed Morshed Chowdhury	La Trobe University, Australia
Alex Ng	La Trobe University, Australia
Indika Kumara	Jheronimus Academy of Data Science, Netherlands
Tarique Anwar	Macquarie University and CSIRO's Data61, Australia
Giancarlo Fortino	University of Calabria, Italy

Massimiliano Rak	University of Campania "Luigi Vanvitelli", Italy
Jason J. Jung	Chung-Ang University, Korea
Dimosthenis Kyriazis	University of Piraeus, Greece
Geir Horn	University of Oslo, Norway
Gang Wang	Nankai University, China
Shaozhang Niu	Beijing University of Posts and Telecommunications, China
Jianxin Wang	Beijing Forestry University, China
Jie Cheng	Shandong University, China
Shaoyin Cheng	University of Science and Technology of China, China

11. Security, Privacy and Trust Computing

Track Co-chairs

Hiroaki Kikuchi	Meiji University, Japan
Xu An Wang	Engineering University of PAP, China
Lidia Ogiela	Pedagogical University of Cracow, Poland

TPC Members

Takamichi Saito	Meiji University, Japan
Kouichi Sakurai	Kyushu University, Japan
Kazumasa Omote	University of Tsukuba, Japan
Shou-Hsuan Stephen Huang	University of Houston, USA
Masakatsu Nishigaki	Shizuoka University, Japan
Mingwu Zhang	Hubei University of Technology, China
Caiquan Xiong	Hubei University of Technology, China
Wei Ren	China University of Geosciences, China
Peng Li	Nanjing University of Posts and Telecommunications, China
Guangquan Xu	Tianjing University, China
Urszula Ogiela	Pedagogical University of Cracow, Poland
Hoon Ko	Chosun University, Korea
Goreti Marreiros	Institute of Engineering of Polytechnic of Porto, Portugal
Chang Choi	Gachon University, Korea
Libor Měsíček	J.E. Purkyně University, Czech Republic

12. Software-Defined Networking and Network Virtualization

Track Co-chairs

Flavio de Oliveira Silva	Federal University of Uberlândia, Brazil
Ashutosh Bhatia	Birla Institute of Technology and Science, Pilani, India
Alaa Allakany	Kyushu University, Japan

TPC Members

Yaokai Feng	Kyushu University, Japan
Chengming Li	Chinese Academy of Science (CAS), China
Othman Othman	An-Najah National University (ANNU), Palestine
Nor-masri Bin-sahri	University Technology of MARA, Malaysia
Sanouphab Phomkeona	National University of Laos, Laos
Haribabu K.	BITS Pilani, India
Shekhavat, Virendra	BITS Pilani, India
Makoto Ikeda	Fukuoka Institute of Technology, Japan
Farookh Hussain	University of Technology Sydney, Australia
Keita Matsuo	Fukuoka Institute of Technology, Japan

AINA-2021 Reviewers

Admir Barolli
Adrian Pekar
Ahmed Elmokashfi
Akihiro Fujihara
Akihiro Fujimoto
Akimitsu Kanzaki
Akio Koyama
Alaa Allakany
Alberto Volpe
Alex Ng
Alex Thomo
Alfredo Navarra
Aneta Poniszewska-Maranda
Angelo Gaeta
Anne Kayem
Antonio Loureiro
Apostolos Gkamas
Arjan Durresi
Ashfaq Ahmad
Ashutosh Bhatia
Asm Kayes
Baojiang Cui
Beniamino Di Martino
Bhed Bista
Carson Leung
Christos Bouras
Danda Rawat
Darshika Perera
David Taniar
Dimitris Apostolou
Dimosthenis Kyriazis
Eirini Eleni Tsiropoulou
Emmanouil Spanakis
Enver Ever
Eric Pardede
Ernst Gran
Eugen Dedu
Euripides Petrakis
Fadi Al-Turjman
Farhad Daneshgar
Farookh Hussain
Fatos Xhafa
Feilong Tang

Feroz Zahid
Flavio Silva
Flora Amato
Francesco Orciuoli
Francesco Piccialli
Gang Wang
Geir Horn
Giancarlo Fortino
Giorgos Vasiliadis
Giuseppe Fenza
Guangquan Xu
Hadil Abukwaik
Hideharu Amano
Hiroaki Kikuchi
Hiroshi Maeda
Hiroyuki Fujioka
Holger Karl
Hong Va Leong
Huey-Ing Liu
Hyunhee Park
Indika Kumara
Isaac Woungang
Jabed Chowdhury
Jana Nowaková
Jason Jung
Jawwad Shamsi
Jesús Escudero-Sahuquillo
Ji Zhang
Jiun-Long Huang
Jolanta Mizera-Pietraszko
Jordi Conesa
Jörg Domaschka
Jorge Sá Silva
Juggapong Natwichai
Jyoti Sahni
K Haribabu
Katerina Doka
Kazumasa Omote
Kazuto Sasai
Keita Matsuo
Kin Fun Li
Kiyotaka Fujisaki
Konstantinos Tsakos

Kyriakos Kritikos
Lei Shu
Leonard Barolli
Leonardo Mostarda
Libor Mesicek
Lidia Ogiela
Lin Hui
Ling-Jyh Chen
Lucian Prodan
Makoto Ikeda
Makoto Takizawa
Marek Ogiela
Mario Dantas
Markus Aleksy
Masakatsu Nishigaki
Masaki Kohana
Massimiliano Rak
Massimo Ficco
Michael Bauer
Minoru Uehara
Morteza Saberi
Nadeem Javaid
Naeem Janjua
Naohiro Hayashibara
Nicola Capuano
Nobuo Funabiki
Omar Hussain
Omid Ameri Sianaki
Paresh Saxena
Purav Shah
Qiang Duan
Quentin Jacquemart
Rajesh Pateriya
Ricardo Rodríguez Jorge
Ronaldo Mello
Rosario Culmone
Ryuji Shioya
Safdar Hussain Bouk

Salimur Choudhury
Salvatore Venticinque
Sanjay Dhurandher
Santi Caballé
Shahriar Badsha
Shigeru Fujita
Shigetomo Kimura
Sriram Chellappan
Stefan Covaci
Stefano Chessa
Stelios Sotiriadis
Stephane Maag
Takahiro Uchiya
Takamichi Saito
Tarique Anwar
Thitinan Tantidham
Thomas Dreibholz
Thomas Preuss
Tomoki Yoshihisa
Tomoya Enokido
Tomoyuki Ishida
Vamsi Paruchuri
Vasilis Samoladas
Vinesh Kumar
Virendra Shekhawat
Wang Xu An
Wei Ren
Wenny Rahayu
Wuyungerile Li
Xin-Mao Huang
Xing Zhou
Yaokai Feng
Yiannis Verginadis
Yoshihiro Okada
Yusuke Gotoh
Zahoor Khan
Zia Ullah
Zoubir Mammeri

AINA-2021 Keynote Talks

The Role of Artificial Intelligence in the Industry 4.0

Flora Amato

University of Naples "Federico II", Naples, Italy

Abstract. Artificial intelligence (AI) deals with the ability of machines to simulate human mental competences. The AI can effectively boost the manufacturing sector, changing the strategies used to implement and tune productive processes by exploiting information acquired at real time. Industry 4.0 integrates critical technologies of control and computing. In this talk is discussed the integration of knowledge representation, ontology modeling with deep learning technology with the aim of optimizing orchestration and dynamic management of resources. We review AI techniques used in Industry 4.0 and show an adaptable and extensible contextual model for creating context-aware computing infrastructures in Internet of Things (IoT). We also address deep learning techniques for optimizing manufacturing resources, assets management and dynamic scheduling. The application of this model ranges from small embedded devices to high-end service platforms. The presented deep learning techniques are designed to solve numerous critical challenges in industrial and IoT intelligence, such as application adaptation, interoperability, automatic code verification and generation of a device-specific intelligent interface.

Localization in 6G

Shahrokh Valaee

University of Toronto, Toronto, Canada

Abstract. The next generation of wireless systems will employ networking equipment mounted on mobile platforms, unmanned air vehicles (UAVs) and low-orbit satellites. As a result, the topology of the sixth-generation (6G) wireless technology will extend to the three-dimensional (3D) vertical networking. With its extended service, 6G will also give rise to new challenges which include the introduction of intelligent reflective surfaces (IRS), the mmWave spectrum, the employment of massive MIMO systems and the agility of networks. Along with the advancement in networking technology, the user devices are also evolving rapidly with the emergence of highly capable cellphones, smart IoT equipments and wearable devices. One of the key elements of 6G technology is the need for accurate positioning information. The accuracy of today's positioning systems is not acceptable for many applications of future, especially in smart environments. In this talk, we will discuss how positioning can be a key enabler of 6G and what challenges the next generation of localization technology will face when integrated within the new wireless networks.

Contents

An Improved DOA Estimation Algorithm of Neural Network Based on Interval Division

GuoBin Li[✉], XiaoOu Song, and Kun Shan

School of Information Engineering, Engineering University of PAP, Xi'an 710086, Shaanxi, China

Abstract. As neural network algorithm springs up, it is being applied to the DOA estimation more and more often. However, the robustness and real-time performance of the neural network algorithm under the condition of the multiple sources has always been a difficult problem. The DOA estimation algorithm of the neural network based on interval division has better robustness and real-time performance, which divides the signal into different angle intervals. However, the accuracy of signal division is difficult to be guaranteed, especially when the angle of signal is at interval edge, causing the large error. An improved interval division method is thus proposed in this paper. Firstly, by adjusting the weight of the antenna array, the beam of the antenna array is focused at the angle corresponding to each sub-region, therefore the spatial signal feature is improved, In the interval division, the edge overlapping division is adopted to improve the estimation accuracy of the interval edge angle. After experimental verification, the improved interval partition method can improve the performance of the algorithm.

1 Introduction

The array signal processing technology can effectively eliminate the interference and noise, enhance the useful signal and improve the system interference-resistant ability [1]. DOA estimation is an important research field of the array signal, which make use of phase difference between different antenna arrays to estimate the signal direction. In terms of the satellite communication, if we can estimate the DOA of interference signal accurately and in real time, it's convenient for us to carry out interference resistance. Particularly, in the face of the complicated and changeable dynamic interference, it is even more important to calculate the DOA of the interference signal.

The main problem in face of DOA estimation is how to estimate the direction of multiple sources in the space. The conventional DOA algorithms have MUSIC algorithm [2], Esprit Signal Parameter based on Rotation In variance Technology [3], Capon Algorithm [4], and some of their derivatives. The MUSIC algorithm has the higher estimation accuracy, which needs the number of known sources, and the computation brought by spectrum peak search is also huge. Capon algorithm does not need know the number of sources, but the resolution is too low under the condition of the small number of arrays. ESPRIT algorithm does not need spectral peak search, but the estimation error is large. In recent years, the rapid estimation DOA based on machine learning has gradually replaced traditional DOA algorithms in some cases. In 2018, Huang integrated MIMO and DOA estimation into the non-linear framework of the deep learning [5], which can simulate the situation of different channels well and obtain the good performance. In 2019, Wang et al. proposed a deep learning DOA algorithm based on time-frequency masking, which can be applied to various the antenna structures [6]. The DOA estimation algorithm based on neural network under the condition of multiple sources is also a key point of researches. The traditional neural network algorithm [7–9] finds the matching relationship between signal and angle by establishing the multi-signal model, but the robustness of the algorithm is not high, the network complexity is too large. The neural network algorithm based on interval division [10] divides the signal into different text intervals according to the angle. Then the estimation is performed in each interval, reducing the complexity of the network with the good robustness. However, it is difficult to guarantee the accuracy of the interval division, especially at the edge of the interval. Therefore, an overlapping division method of combining beam formation with the neural network is proposed in this paper. The accuracy of DOA estimation can be improved through improving the accuracy of interval division.

2   Signal Model of DOA Estimation

Considering that the signal is received by M linear arrays, the signal model structure is in Fig. 1 below, in which, M means the number of elements in the receiving array, d means the spacing between the elements and θ means the DOA of the signal to be solved. We can conclude that the time difference of the two adjacent elements in receiving the signal is:

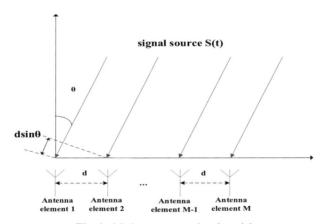

Fig. 1. M element array signal model

$$t = \frac{d \sin \theta}{c}. \tag{1}$$

Meanwhile, the phase difference between any element and the first element can be gotten as:

$$\varphi = \frac{2\pi f (i - 1)d \sin \theta}{c}, i = 1, \ldots, M. \tag{2}$$

in which f represents the frequency of the incident signal.

Assuming that the array receives N signals simultaneously (that is, the signal sparsity is N), we can obtain the receiving signal on the Kth antenna:

$$x_k(t) = \sum_{i=1}^{N} s_i(t) e^{-j\frac{2\pi f_i(k-1)d \sin \theta_i}{c}} + n_k(t). \tag{3}$$

Where, the $s_i(t)$ represents the amplitude of the ith interference signal, and the $n_k(t)$ represents the noise received by the kth element. The signal received by the whole antenna array can be expressed in the form of the matrix as:

$$\begin{bmatrix} x_1(t) \\ \cdots \\ x_M(t) \end{bmatrix} = \begin{bmatrix} 1 & \cdots & 1 \\ \cdots & \cdots & \cdots \\ e^{-j\frac{2\pi f_1(M-1)d \sin \theta_1}{c}} & \cdots & e^{-j\frac{2\pi f_N(M-1)d \sin \theta_N}{c}} \end{bmatrix} \begin{bmatrix} s_1(t) \\ \cdots \\ s_N(t) \end{bmatrix} + \begin{bmatrix} n_1(t) \\ \cdots \\ n_M(t) \end{bmatrix}. \tag{4}$$

It can also be shortened as:

$$X = AS + N \tag{5}$$

Where A is the direction matrix, S is the signal matrix, and N is the noise matrix. For different arrays, they have different directional matrixes. Meanwhile, the receiving matrix also has DOA information. Therefore, for different signal sources, direction matrixes and signal matrices are different, which provides a theoretical basis for our DOA estimation.

3 Improved DOA Estimation Algorithm of Neural Networks Based on Interval Division

3.1 DOA Estimation Algorithm of Neural Networks Based on Interval Division

The flow chart of the DOA estimation of neural networks based on space division can be divided into four steps, including pre-processing, rough classification (the division of signal intervals), sub division (angle estimation) and linear interpolation (Fig. 2).

The purpose of the pre-processing is to extract the features of the signal. The feature of the signal is taken as the output of the whole network, which is the same as the traditional neural network. The network also performs covariance processing (with the symmetry of the covariance matrix), takes the diagonal part of the signal, and then performs normalized processing to complete in extracting the features of the signal.

The main purpose of rough classification is to divide the signal into different intervals by encoder and decoder, and to make a rough classification on the signal DOA. The division method here is uniform non-overlapping type. The input of the encoder is the pre-processed feature matrix. If the angle responding to the feature matrix is within the corresponding interval of the encoder, the output of the decoder is the feature matrix, otherwise the output is the zero matrix with the same size. In this way, the signal features are divided into interval and rough classification is completed.

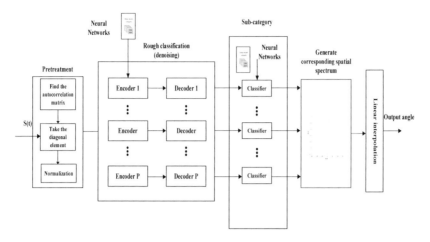

Fig. 2. Flow chart of DOA algorithm of neural networks based on interval division

Subdivision is a DOA estimation of the output sub-matrix in each decoder. Each classifier is independent to each other and only estimates the angle of the corresponding decoder in front of the classifier. The complexity of the network is greatly reduced, while the robustness of the classification effect under the condition of multiple sources is improved. The outputs of multiple classifiers are arranged in the order of interval to form spatial spectrum.

3.2   Improved Interval Division Method

In fact, the encoder and decoder only need to correspond to the signal feature of the interval. The signal feature of other intervals will influence the output of encoders. Although the training method of neural network can suppress the signal features of other intervals, it cannot suppress them completely, and still have some influence on the output. By adjusting the weight coefficient of each element, beam forming can align with the antenna's main beam at any direction, thus enhancing the signal gain in that direction, as shown in Fig. 3. After beam forming, the array gains the higher gain within $[0°, 30°]$. By using the feature of beam forming, each encoder can get the high gain signal corresponding to the angle in the interval, and improve the angle feature of the signal in the interval. The accuracy of classifier's classification is greatly improved.

In the interval division, if the uniform non-overlapping division method is adopted, the classification error of the angle of the edge at the interval will be very large, as shown in Fig. 4. Assuming the number of encoders is 6, the length of each sub interval is $30°$. When the signal of the source angle as an interval edge angle passes through the encoder, it is likely affected by other interval sources since the encoder does not perform well in classifying the edges. The signal is divided into the second interval, which has a greater effect on the DOA estimation of the later classifier.

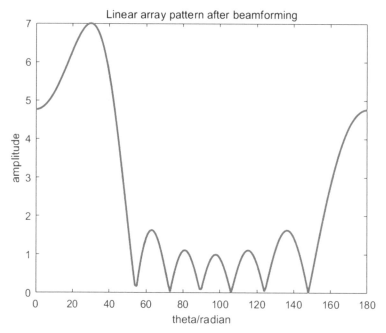

Fig. 3. Linear array pattern after beamforming

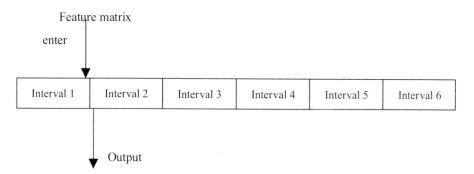

Fig. 4. Uniform non-overlapping method

The overlapping edges is adopted, as in Fig. 5, and it can solve this problem well. The angles of the edge adopt the overlapping method, namely, the angles of the overlapped regions are divided into two adjacent intervals, so the DOA estimation in two intervals are conducted. Finally, the estimation value of the two interval is weighted to get the final estimation, which avoids the DOA error resulted from the interval division.

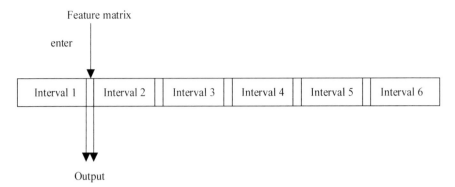

Fig. 5. Edge overlapping division method

The improved interval division mainly improves two parts, as Fig. 6. Beamforming is added to the input of encoders. In the interval division, the edge overlapping division is adopted. The beamforming technique is used to improve the feature of the sources in the intervals to some extent, which can improve the accuracy of the rough classification on sources. Not only does the edge overlapping classification improve the accuracy of interval edge estimation, but also play a great influence on the complexity of neural network algorithm.

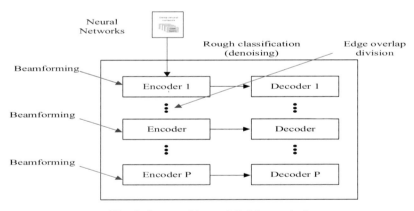

Fig. 6. Improved interval division method

3.3   Performance Analysis

The simulation parameters for the improved DOA estimation of the neural network algorithm based on interval division are shown in Table 1 as below:

Table 1. Simulation parameters for DOA estimation

Parameter	Value
Number of elements	7
Array spacing	0.5λ
Number of snapshots	1024
Learning rate	0.001
Batch size	32
Number of encoders	6
Number of iterations (encoder)	1000
Number of iterations (classifier)	300
Number of hidden layers(classifier)	3
Overlapping angle	$2°$
Signal-noise ratio	10 dB

The improved DOA estimation algorithm of the neural networks based on interval division is shown in Fig. 7. The experiment adopts two narrow-band signals with a 10-degree angle and an angle ranging from $[-60°, 60°]$. It can be seen that the improved DOA estimation algorithm of the neural networks based on interval division has a high accuracy in estimation.

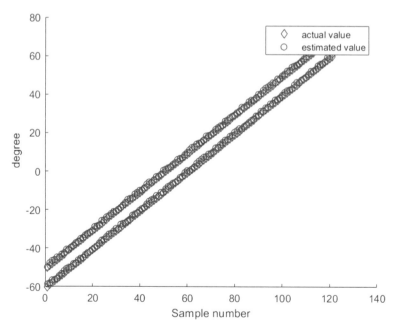

Fig. 7. Improved estimation results based on interval algorithm

Through counting the improved interval division method and the original method, the performance of the two methods is analyzed for error of the estimated value of the interval edge angle as shown in Table 2.

Table 2. Analysis of algorithm complexity

Mode	Test angle	Estimation error
Uniform non-overlapping method	[29°, 29.5°, 30°, 30.5°, 31°]	0.71
Improved interval division method	[29°, 29.5°, 30°, 30.5°, 31°]	0.68

After analysis, we can conclude that the improved DOA estimation algorithm based on interval division is more accurate in the estimation of interval edge angle than the previous.

4   Conclusion

In this paper, the DOA estimation algorithm of neural networks based on interval division is analyzed, and it is found that the accuracy of the estimated interval edge angle is not very high compared with other angles. Therefore, a new interval division is proposed. Firstly, the beamforming technology is used to improve the signal feature of the sources corresponding to subspace. Then, the edge overlapping division method is adopted to replace the original uniform non-overlapping division method. From the experimental data, the estimation error in this method is smaller than the original method, especially for the angle at the interval edge, the estimation accuracy is higher. Yet, the interval division of the encoder and decoder is only improved in this paper, and the improvement in the classifier needs further research.

References

1. Naidu, P.S.: Sensor Array Signal Processing. CRC Press, Inc. (2000)
2. Schmidt, R.O.: Multiple emitter location and signal parameter estimation. IEEE Trans. Antennas Propag. **34**(3), 276–280 (1986)
3. Roy, R., Kailath, T.: ESPRIT-estimation of signal parameters via rotational invariance techniques. IEEE Trans. Acoust. Speech Signal Process. **37**(7), 984–995 (1989)
4. Krim, H., Viberg, M.: Two decades of array signal processing research: the parametric approach. IEEE Signal Process. Mag. **13**(4), 67–94 (1996)
5. Huang, H., Yang, J., Huang, H., et al.: Deep learning for super-resolution channel estimation and DOA estimation based massive MIMO system. IEEE Trans. Veh. Technol. **67**(9), 8549–8560 (2018)
6. Wang, Z.Q., Zhang, X., Wang, D.L.: Robust speaker localization guided by deep learning-based time-frequency masking. IEEE/ACM Transactions on Audio, Speech, and Language Processing **27**(1), 178–188 (2019)
7. Zhu, C., Zhu, L.: A DOA estimation algorithm of satellite interference signals based on machine learning. Radio Commun. Technol. **45**(06), 586–590 + 585 (2019)

8. Advance, S., Politis, A., Virtanen, T.: Direction of arrival estimation for multiple sound sources using convolutional recurrent neural network. In: 2018 26th European Signal Processing Conference (EUSIPCO), Rome, pp. 1462–1466 (2018)
9. Nguyen, T.N.T., Gan, W.-S., Ranjan, R., Jones, D.L.: Robust source counting and DOA estimation using spatial pseudo-spectrum and convolutional neural network. IEEE/ACM Transactions on Audio, Speech, and Language Processing **28**, 2626–2637 (2020). https://doi.org/10.1109/TASLP.2020.3019646
10. Li, S.: Research of DOA estimation based on neural network. Harbin Institute of Technology (2019)

The Application in Handwritten Digit Recognition of Deep Belief Network Based on Improved Genetic Algorithm

Zhu Hong-Xia[1]([⊠]), Lu Xiu-Hua[1], and Zhu Hong-Ying[2]

[1] College of Science, Langfang Normal College, Langfang 065000, HeBei, China
[2] School of Mechanical and Precision Instrument Engineering, Xi'an University of Technology, Xi'an 710048, ShanXi, China

Abstract. The initial connection weight and threshold value of RBM in DBN have a certain influence on the recognition effect of the network. In this paper, the MNIST database is used as the data sample, and the improved genetic algorithm is applied to optimize the initial weight and threshold value in RBM, and the programming is realized by MATLAB language. The simulation results show that the improved DBN model is better than the traditional DBN network model and the traditional genetic algorithm improved DBN model in handwritten numeral recognition.

1 Introduction

As an important sub field of image recognition, handwritten numeral recognition technology has played an important role in more and more fields after years of development. It is often used in postal codes, bank bills and handwritten express documents and other fields. The most common research methods of this technology is neural network [1–3]. With the increase of the amount of data, the classical BP neural network is easy to fall into local optimum, which will affect the accuracy of recognition。In order to ensure the efficiency and accuracy of recognition in big data environment, more and more researchers are turning to use deep belief network (DBN) [3–5], The efficiency and accuracy of handwritten numeral recognition can be improved by using the advantages of this network in dealing with large-scale and complex data.

When only using deep belief network for recognition, the connection weights and threshold values of each node in unsupervised learning are randomly initialized, which makes the model easy to fall into local optimum in the training process. Intelligent evolutionary algorithm (IEA) is widely used in global optimization. In this paper, the improved genetic algorithm adaptive genetic algorithms (IAGA) in literature [6] and [7] are introduced into the deep belief network model to optimize the connection weights and threshold values of each node in the unsupervised learning part, so as to minimize the reconstruction error and enhance the recognition ability of the model.[1]

---

[1] This work was supported by Langfang Science and Technology Research and Development Self Fin- ancing Project (2019011008); the Natural Science Foundation of Hebei Province [no. F2018408040]; the Hebei Education Funds for Youth Project [no. QN2018047].

L. Barolli et al. (Eds.): AINA 2021, LNNS 227, pp. 10–16, 2021.
https://doi.org/10.1007/978-3-030-75078-7_2

2   The DBN Optimized by Adaptive Genetic Algorithm (IAGA-DBN)

2.1   DBN

DBN is a deep probabilistic digraph model with RBM as its basic unit. The structure is composed of multi-layer nodes. There is no connection between the nodes of each layer, and the nodes of two adjacent layers are fully connected. The lowest layer of the network is the observable variable, and the nodes of other layers are hidden variables. The connection between the top two layers is undirected, and the connection between the other layers is directed. The structure of deep belief network is shown in Fig. 1.

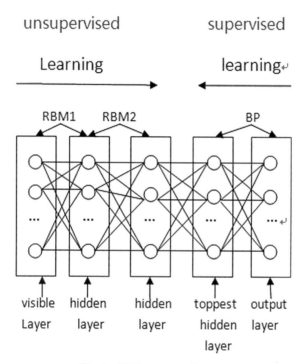

Fig. 1. DBN structure diagram

The training process of DBN is as follows:

(1) Bottom up learning process.
   The hidden layer neurons of the first RBM extract important features from the sample, and the trained features are activated as the input data of the second RBM. It can be regarded as "learning features from features" until all RBM hidden layers in DBN are trained. At this time, RBM obtains the connection weights and thresholds of each layer by unsupervised greedy learning.

(2) Top down parameter tuning process.

After the pre training process of RBM, DBN adjusts the weight of each layer through the method of supervised learning. The back propagation algorithm is used to fine tune the parameters, so that the output of each layer can restore the input of corresponding visual layer as much as possible. Because the weights of the network are obtained by learning from samples, it is close to the global optimal, so it can achieve good prediction effect.

2.2 IAGA

Genetic algorithm is a kind of swarm intelligence search algorithm. It imitates the phenomena of reproduction, hybridization and mutation in evolutionary biology, and utilizes genetic operators selection, crossover and mutation to solve problems. Although genetic algorithm has excellent global search ability, it has poor local optimization ability and is easy to fall into premature convergence phenomenon. This is because the crossover probability (PC) and mutation probability (PM) used in SGA algorithm are fixed in the whole evolution process, However, in the early stage of SGA, the population tends to gather around the individuals with low fitness. At this time, if the PC and PM are relatively large, it is easy to evolve better individuals; in the later stage of the algorithm, the population generally gathers around the individuals with higher fitness. If the smaller PC and PM are used, the better individuals in the population are not easy to be destroyed. Based on this, this paper adopts two methods that PC and PM can change with the population environment, that is adaptive genetic algorithm. one these methods, the calculation formula of PC and PM in literature [6] is as follows:

$$
pc = \begin{cases} \dfrac{pc_{max}-pc_{min}}{1+\exp\{\alpha[\frac{2(f_1-f_{avg})}{f_{rmmax}-f_{avg}}-1]\}} + pc_{min}, & f_1 \geq f_{avg} \\ pc_{max}, & f_1 < f_{avg} \end{cases} \tag{1}
$$

$$
pm = \begin{cases} \dfrac{pm_{max}-pm_{min}}{1+\exp\{\alpha[\frac{2(f_2-f_{avg})}{f_{max}-f_{avg}}-1]\}} + pm_{min}, & f_2 \geq f_{avg} \\ pm_{max}, & f_2 < f_{avg} \end{cases} \tag{2}
$$

The meanings of the symbols in formula (1) and formula (2) are as follows:

f_1——The maximum fitness of the cross individuals;

f_2——The fitness of the mutated individual;

f_{avg}——The average fitness of all individuals in the population;

f_{max}——The maximum fitness of all individuals in the population;

pc_{max}, pc_{min}, pm_{max}, pm_{min}——They are maximum and minimum crossover probability and maximum and minimum mutation probability respectively;

$\alpha = 0.903438$.

the calculation formula of PC and PM in literature [7] is as follows:

$$pc = \begin{cases} \dfrac{k_1}{1+\exp(\frac{k_2}{\phi})} \times \dfrac{f_{max}-f}{f_{max}-f_{avg}} + k_3 & f_1 \geq f_{avg} \\ k_4 & f_1 < f_{avg} \end{cases} \tag{3}$$

$$pm = \begin{cases} \dfrac{k_5}{1+\exp(\frac{k_6}{\phi})} \times \dfrac{f_{max}-f}{f_{max}-f_{avg}} + k_7 & f_2 \geq f_{avg} \\ k_8 & f_2 < f_{avg} \end{cases} \tag{4}$$

The meanings of the symbols in formula (3) and formula (4) are as follows:
$k_1 \sim k_8$: All of them are adaptive control parameters;
f_i is the fitness of individual i (i = 1, 2, …, m);

$$EX = f_{avg} = \frac{f_1 + f_2 + \cdots + f_m}{m};$$

$$DX = \frac{f_1^2 + f_2^2 + \cdots + f_m^2}{m} - EX^2;$$

$$\phi = \frac{EX + 1}{\sqrt{DX}}.$$

2.3   The Algorithm Fow of IAGA-DBN

The basic idea of optimizing DBN by IAGA algorithm is to optimize the initial weights and thresholds of restricted Boltzmann machines (RBMs) in DBN networks. By using the optimized weights and thresholds as the initial weights and thresholds of RBMs, the DBN network can improve the recognition rate and reduce the error to a certain extent. The optimization process is mainly divided into two steps: the first step is to determine the number of DBN network weights and thresholds and coding scheme; the second step is to use IAGA algorithm to complete the optimization.

The implementation steps of IAGA-DBN algorithm are as follows:

In the first step, the population size and DBN network structure are determined, and the population are randomly initialized by binary code;

In the second step, the fitness of the parent population is calculated, and the crossover probability and mutation probability are adaptively adjusted according to the fitness value, The value of probability is adjusted as shown in formula (1) and (2);

The third step is to select, cross and mutate the parent population to generate the offspring population. Single point crossing was used;

The fourth step is to repeat the second and third steps until the set accuracy requirements or the maximum number of iterations are reached;

In the fifth step, the weights and thresholds optimized by genetic algorithm are put into DBN to continue local optimization until the required error accuracy is achieved.

The algorithm flow chart of IAGA-DBN is shown in Fig. 2:

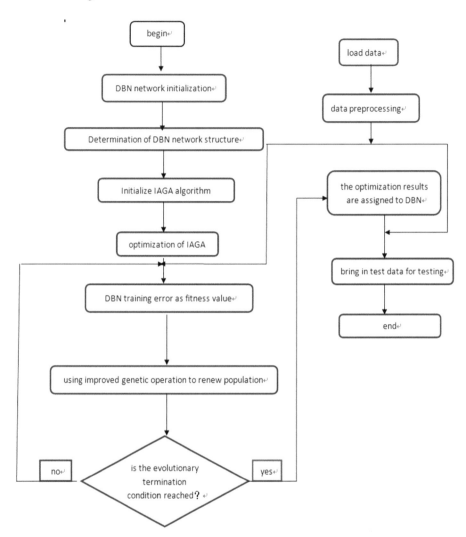

Fig. 2. Schematic diagram of algorithm structure of IAGA-DBN

3   Realization of Handwritten Digit Recognition Based on IAGA-DBN

3.1   Experimental Data

The Mnist database established by Yann Lecun and others in the world is the data used in this experiment. The data set contains 70000 images, 60000 images as training set and 10000 as test set, Each image is 28 × 28 pixels in size and has a corresponding label.

3.2 Test Comparison and Analysis of Each Model

Based on the above experimental data, the structure of DBN is as follows: there are 784 neurons in the input layer, The hidden layer has one layer. The number of neurons is 50 to 500, and the interval is 50. The excitation function of neurons is sigmoid function, and the output layer has 10 neurons. Softmax function is used for probability output, and the number of batches is 100. The parameters of basic genetic algorithm and improved genetic algorithm are as follows: the encoding mode is binary, the population size is 80, the individual length is 20, the decoding interval is $[-0.001, 0.001]$, and the genetic algebra is 10. In order to compare the performance of DBN, SGA-DBN, IAGA [6]-DBN and IAGA [7]-DBN, different numbers of neurons are taken from the hidden layer to compare and analyze the recognition error. The recognition error curve is shown in Fig. 3 below. It can be seen that the recognition error of IAGA [6]-DBN is less than that of DBN, SGA-DBN and IAGA [7]-DBN. When the number of hidden layer neurons is 400, the recognition accuracy of the three kinds of networks is shown in Table 1, and The recognition accuracy of IAGA [6]-DBN is 0.8%, 0.1% and 0.08% higher than the first three.

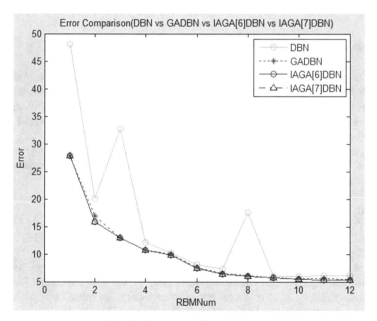

Fig. 3. The recognition error curve

Table 1. Comparison of experimental results of three different methods on Mnist data set

Model	Accuracy %
DBN	93.87
SGA-DBN	94.57
IAGA[7]-DBN	94.59
IAGA[6]-DBN	94.67

4   Conclusion

In order to improve the influence of DBN on handwritten numeral recognition due to different values of initial weights and thresholds, and the premature convergence of genetic algorithm, In this paper, an improved deep belief network based on IAGA is proposed. The experimental results show that the algorithm can give full play to the advantages of IAGA algorithm, effectively avoid local minima, and improve the recognition error of handwritten numeral set.

References

1. Murru, N., Rossini, R.: A Bayesian approach for initialization of weights in backpropagation neural net with application to character recognition. Neuroco mputing **193**, 192–205 (2016)
2. Hanx: Handwritten digital recognition based on GA-BP neural network. In: International Conference on Artificial Intelligence, Management Science and Electronic Commerce, pp. 5050–5052. IEEE (2011)
3. Zhang, X., Wu, L.: Handwritten digit recognition based on improved learning rate BP algorithm. In: 2nd International Conference on Information Engineering and Computer Scienc- e Proceedings, pp. 1–4. ICIECS (2010)
4. Song Xiaoru, W., Xue, G.S., et al.: Simulation study on handwritten numeral recognition based on deep neural network. Sci. Technol. Eng. **19**(5), 193–196 (2019)
5. Savich, A.W., Moussa, M.: Resource efficient arithmetic effects on RBM neural network solution quality using MNIST. In: 2011 International Conference on Reconfigurable Computing and FPGAs (ReConFig), pp. 35–40 (2011)
6. Guan Baolin, B.A., Lideng, B.P.: Neural network handwritten numerical recognition based on improved genetic algorithm. Control Instr. Chem. Ind. 40, 778–784 (2012). in Chinese
7. Chuang, C., Chellali, R., Yin, X.: Speech emotion recognition based on improved genetic algorithm optimized BP neural network. Appl. Res. Comput. **2**, 344–346 (2019)

A Fuzzy-Based Scheme for Admission Control in 5G Wireless Networks: Improvement of Slice QoS Considering Slice Reliability as a New Parameter

Phudit Ampririt[1(✉)], Ermioni Qafzezi[1], Kevin Bylykbashi[1], Makoto Ikeda[2], Keita Matsuo[2], and Leonard Barolli[2]

[1] Graduate School of Engineering, Fukuoka Institute of Technology, 3-30-1 Wajiro-Higashi, Higashi-Ku, Fukuoka 811-0295, Japan
[2] Department of Information and Communication Engineering, Fukuoka Institute of Technology, 3-30-1 Wajiro-Higashi, Higashi-Ku, Fukuoka 811-0295, Japan
{kt-matsuo,barolli}@fit.ac.jp, makoto.ikd@acm.org

Abstract. The Fifth Generation (5G) network is expected to be flexible to satisfy user requirements and the Software-Defined Network (SDN) with Network Slicing will be a good approach for admission control. The Quality of Service (QoS) is very important for 5G wireless networks. In this paper, we propose a Fuzzy-based scheme to evaluate the QoS considering 4 parameters: Slice Throughput (ST), Slice Delay (SD), Slice Loss (SL) and Slice Reliability (SR) as new parameter. We carried out simulations for evaluating the performance of our proposed scheme. From simulation results, we conclude that the considered parameters have different effects on the QoS performance. When ST and SR is increasing, the QoS parameter is increased. But, when SD and SL are increasing, the QoS is decreased.

1 Introduction

Recently, the growth of wireless technologies and user's demand of services are increasing rapidly. Especially in 5G networks, there will be billions of new devices with unpredictable traffic pattern which provide high data rates. With the appearance of Internet of Things (IoT), these devices will generate Big Data to the Internet, which will cause to congest and deteriorate the QoS [1].

The 5G network will provide users with new experiences such as Ultra High Definition Television (UHDT) on Internet and support a lot of IoT devices with long battery life and high data rate on hotspot areas with high user density. In the 5G technology, the routing and switching technologies aren't important anymore or coverage area is shorter than 4G because it uses high frequency for facing higher device's volume for high user density [2–4].

There are many research work that try to build systems which are suitable to 5G era. The SDN is one of them [5]. For example, the mobile handover mechanism with SDN is used for reducing the delay in handover processing and improve QoS. Also, by using SDN the QoS can be improved by applying Fuzzy Logic (FL) on SDN controller [6–8].

In our previous work [9], we proposed a fuzzy-based scheme for evaluation of QoS in 5G Wireless Networks considering three parameters: Slice Throughput (ST), Slice Delay (SD), Slice Loss (SL). In this paper, we consider Slice Reliability (SR) as a new parameter.

The rest of the paper is organized as follows. In Sect. 2 is presented an overview of SDN. In Sect. 3, we present application of Fuzzy Logic for admission control. In Sect. 4, we describe the proposed fuzzy-based system and its implementation. In Sect. 5, we explain the simulation results. Finally, conclusions and future work are presented in Sect. 6.

2   Software-Defined Networks (SDNs)

The SDN is a new networking paradigm that decouples the data plane from control plane in the network. In traditional networks, the whole network is controlled by each network device. However, the traditional networks are hard to manage and control since they rely on physical infrastructure. Network devices must stay connected all the time when user wants to connect other networks. Those processes must be based on the setting of each device, making controlling the operation of the network difficult. Therefore, they have to be set up one by one. In contrast, the SDN is easy to manage and provide network software based services from a centralised control plane. The SDN control plane is managed by SDN controller or cooperating group of SDN controllers. The SDN structure is shown in Fig. 1 [10,11].

- **Application Layer** builds an abstracted view of the network by collecting information from the controller for decision-making purposes. The types of applications are related to: network configuration and management, network monitoring, network troubleshooting, network policies and security.
- **Control Layer** receives instructions or requirements from the Application Layer and control the Infrastructure Layer by using intelligent logic.
- **Infrastructure Layer** receives orders from SDN controller and sends data among them.

The SDN can manage network systems while enabling new services. In congestion traffic situation, management system can be flexible, allowing users to easily control and adapt resources appropriately throughout the control plane. Mobility management is easier and quicker in forwarding across different wireless technologies (e.g. 5G, 4G, Wifi and Wimax). Also, the handover procedure is simple and the delay can be decreased.

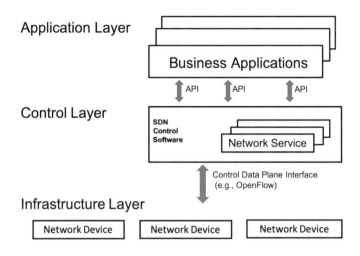

Fig. 1. Structure of SDN.

3   Outline of Fuzzy Logic

A Fuzzy Logic (FL) system is a nonlinear mapping of an input data vector into a scalar output, which is able to simultaneously handle numerical data and linguistic knowledge. The FL can deal with statements which may be true, false or intermediate truth-value. These statements are impossible to quantify using traditional mathematics. The FL system is used in many controlling applications such as aircraft control (Rockwell Corp.), Sendai subway operation (Hitachi), and TV picture adjustment (Sony) [12–14].

In Fig. 2 is shown Fuzzy Logic Controller (FLC) structure, which contains four components: fuzzifier, inference engine, fuzzy rule base and defuzzifier.

- **Fuzzifier** is needed for combining the crisp values with rules which are linguistic variables and have fuzzy sets associated with them.
- **The Rules** may be provided by expert or can be extracted from numerical data. In engineering case, the rules are expressed as a collection of IF-THEN statements.
- **The Inference Engine** infers fuzzy output by considering fuzzified input values and fuzzy rules.
- **The Defuzzifier** maps output set into crisp numbers.

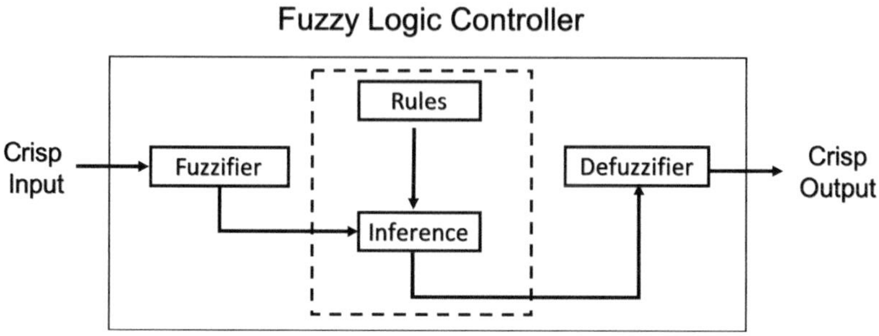

Fuzzy Logic Controller

Fig. 2. FLC structure.

3.1 Linguistic Variables

A concept that plays a central role in the application of FL is that of a linguistic variable. The linguistic variables may be viewed as a form of data compression. One linguistic variable may represent many numerical variables. It is suggestive to refer to this form of data compression as granulation.

The same effect can be achieved by conventional quantization, but in the case of quantization, the values are intervals, whereas in the case of granulation the values are overlapping fuzzy sets. The advantages of granulation over quantization are as follows:

- it is more general;
- it mimics the way in which humans interpret linguistic values;
- the transition from one linguistic value to a contiguous linguistic value is gradual rather than abrupt, resulting in continuity and robustness.

For example, let Temperature (T) be interpreted as a linguistic variable. It can be decomposed into a set of Terms: T (Temperature) = {Freezing, Cold, Warm, Hot, Blazing}. Each term is characterised by fuzzy sets which can be interpreted, for instance, "Freezing" as a temperature below 0 °C, "Cold" as a temperature close to 10 °C.

3.2 Fuzzy Control Rules

Rules are usually written in the form "IF x is S THEN y is T" where x and y are linguistic variables that are expressed by S and T, which are fuzzy sets. The x is a control (input) variable and y is the solution (output) variable. This rule is called Fuzzy control rule. The form "IF ... THEN" is called a conditional sentence. It consists of "IF" which is called the antecedent and "THEN" is called the consequent.

3.3 Defuzzificaion Method

There are many defuzzification methods, which are showing in following:

- The Centroid Method;
- Tsukamoto's Defuzzification Method;
- The Center of Are (COA) Method;
- The Mean of Maximum (MOM) Method;
- Defuzzification when Output of Rules are Function of Their Inputs.

4 Proposed Fuzzy-Based System

In this work, we use FL to implement the proposed system. In Fig. 3, we show the overview of our proposed system. Each evolve Base Station (eBS) will receive controlling order from SDN controller and they can communicate and send data with User Equipment (UE). On the other hand, the SDN controller will collect all the data about network traffic status and controlling eBS by using the proposed fuzzy-based system. The SDN controller will be a communicating bridge between eBS and 5G core network. The proposed system is called Integrated Fuzzy-based Admission Control System (IFACS) in 5G wireless networks. The structure of IFACS is shown in Fig. 4. For the implementation of our system, we consider four input parameters: Quality of Service (QoS), User Request Delay Time (URDT), Slice Priority (SP), Slice Overloading Cost (SOC) and the output parameter is Admission Decision (AD).

In this paper, we apply FL to evaluate the QoS parameter for the proposed system. For QoS evaluation, we consider 4 parameters: Slice Throughput (ST),

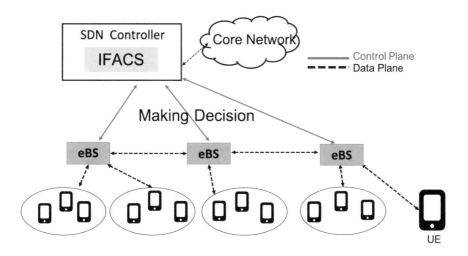

Fig. 3. Proposed system overview.

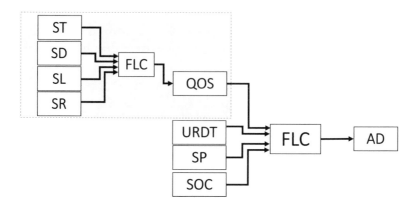

Fig. 4. Proposed system structure.

Slice Delay (SD), Slice Loss (SL) and Slice Reliability (SR) as new parameter. The output parameter is Quality of Service (QoS).

Slice Throughput (ST): The ST value is the abundance of data being sent and received within one second in a Slice. When SD value is high, the QoS performance is high.

Slice Delay (SD): The SD is a slice's data transmission time between user and the system. When SD value is high, the QoS performance is low.

Slice Loss (SL): The SL is the percentage of sending lost data in a slice. When the SL value is high, the QoS performance is low.

Slice Reliability (SR): The SR is the reliability of the slice. When the SR value is high, the QoS performance is high.

Quality of Service (QoS): The QoS is an important parameter for admission control. A user with good QoS will have high priority to be accepted in the network.

Table 1. Parameter and their term sets.

Parameters	Term set
Slice Throughput (ST)	Low (Lo), Moderate (Mr), Fast (Ft)
Slice Delay (SD)	Short (Sh), Intermediate (In), Long (Ln)
Slice Loss (SL)	Low (Lw), Medium (Md), High (Hg)
Slice Reliability (SR)	Poor (Po), Moderate (Mo), Good (Gd)
Quality of Service (QoS)	QoS1, QoS2, QoS3, QoS4, QoS5, QoS6, QoS7

Table 2. Fuzzy Rule base

Rule	ST	SD	SL	SR	QoS	Rule	ST	SD	SL	SR	QoS
1	Lo	Sh	Lw	Po	QoS3	41	Mr	In	Md	Mo	QoS4
2	Lo	Sh	Lw	Mo	QoS5	42	Mr	In	Md	Gd	QoS6
3	Lo	Sh	Lw	Gd	QoS7	43	Mr	In	Hg	Po	QoS1
4	Lo	Sh	Md	Po	QoS2	44	Mr	In	Hg	Mo	QoS2
5	Lo	Sh	Md	Mo	QoS3	45	Mr	In	Hg	Gd	QoS4
6	Lo	Sh	Md	Gd	QoS5	46	Mr	Ln	Lw	Po	QoS3
7	Lo	Sh	Hg	Po	QoS1	47	Mr	Ln	Lw	Mo	QoS4
8	Lo	Sh	Hg	Mo	QoS2	48	Mr	Ln	Lw	Gd	QoS6
9	Lo	Sh	Hg	Gd	QoS3	49	Mr	Ln	Md	Po	QoS1
10	Lo	In	Lw	Po	QoS2	50	Mr	Ln	Md	Mo	QoS2
11	Lo	In	Lw	Mo	QoS3	51	Mr	Ln	Md	Gd	QoS4
12	Lo	In	Lw	Gd	QoS5	52	Mr	Ln	Hg	Po	QoS1
13	Lo	In	Md	Po	QoS1	53	Mr	Ln	Hg	Mo	QoS1
14	Lo	In	Md	Mo	QoS2	54	Mr	Ln	Hg	Gd	QoS2
15	Lo	In	Md	Gd	QoS3	55	Ft	So	Lw	Po	QoS6
16	Lo	In	Hg	Po	QoS1	56	Ft	Sh	Lw	Mo	QoS7
17	Lo	In	Hg	Mo	QoS1	57	Ft	Sh	Lw	Gd	QoS7
18	Lo	In	Hg	Gd	QoS2	58	Ft	Sh	Md	Po	QoS5
19	Lo	Ln	Lw	Po	QoS1	59	Ft	Sh	Md	Mo	QoS6
20	Lo	Ln	Lw	Mo	QoS2	60	Ft	Sh	Md	Gd	QoS7
21	Lo	Ln	Lw	Gd	QoS4	61	Ft	Sh	Hg	Po	QoS3
22	Lo	Ln	Md	Po	QoS1	62	Ft	Sh	Hg	Mo	QoS5
23	Lo	Ln	Md	Mo	QoS1	63	Ft	Sh	Hg	Gd	QoS6
24	Lo	Ln	Md	Gd	QoS2	64	Ft	In	Lw	Po	QoS5
25	Lo	Ln	Hg	Po	QoS1	65	Ft	In	Lw	Mo	QoS7
26	Lo	Ln	Hg	Mo	QoS1	66	Ft	In	Lw	Gd	QoS7
27	Lo	Ln	Hg	Gd	QoS1	67	Ft	In	Md	Po	QoS3
28	Mr	Sh	Lw	Po	QoS5	68	Ft	In	Md	Mo	QoS5
29	Mr	Sh	Lw	Mo	QoS7	69	Ft	In	Md	Gd	QoS7
30	Mr	Sh	Lw	Gd	QoS7	70	Ft	In	Hg	Po	QoS2
31	Mr	Sh	Md	Po	QoS4	71	Ft	In	Hg	Mo	QoS3
32	Mr	Sh	Md	Mo	QoS5	72	Ft	In	Hg	Gd	QoS5
33	Mr	Sh	Md	Gd	QoS7	73	Ft	Ln	Lw	Po	QoS4
34	Mr	Sh	Hg	Po	QoS2	74	Ft	Ln	Lw	Mo	QoS6
35	Mr	Sh	Hg	Mo	QoS3	75	Ft	Ln	Lw	Gd	QoS7
36	Mr	Sh	Hg	Gd	QoS5	76	Ft	Ln	Md	Po	QoS2
37	Mr	In	Lw	Po	QoS4	77	Ft	Ln	Md	Mo	QoS4
38	Mr	In	Lw	Mo	QoS6	78	Ft	Ln	Md	Gd	QoS6
39	Mr	In	Lw	Gd	QoS7	79	Ft	Ln	Hg	Po	QoS1
40	Mr	In	Md	Po	QoS2	80	Ft	Ln	Hg	Mo	QoS2
						81	Ft	Ln	Hg	Gd	QoS4

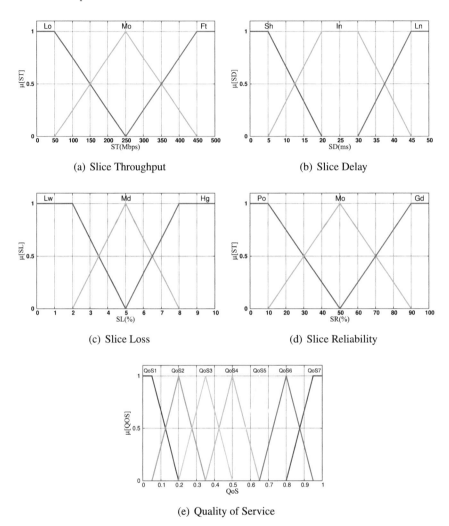

(a) Slice Throughput

(b) Slice Delay

(c) Slice Loss

(d) Slice Reliability

(e) Quality of Service

Fig. 5. Membership functions.

The membership functions are shown in Fig. 5. We use triangular and trapezoidal membership functions because they are more suitable for real-time operations [15–18]. We show parameters and their term sets in Table 1. The Fuzzy Rule Base (FRB) is shown in Table 2 and has 81 rules. The control rules have the form: IF "condition" THEN "control action". For example, for Rule 1: "IF ST is Lo, SD is Sh, SL is Lw and SR is Po THEN QoS is QoS3".

5   Simulation Results

In this section, we present the simulation result of our proposed scheme. The simulation results are shown in Fig. 6, Fig. 7 and Fig. 8. They show the relation of

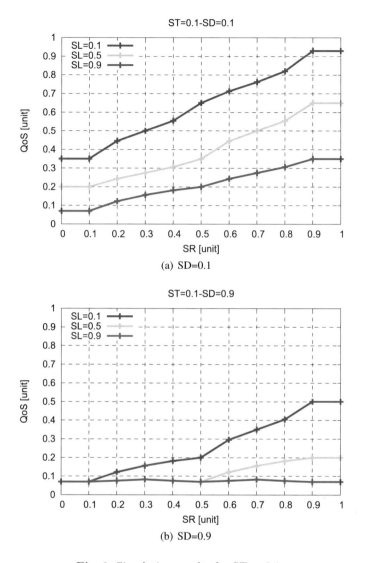

(a) SD=0.1

(b) SD=0.9

Fig. 6. Simulation results for ST = 0.1.

QoS with ST, SD, SL and SR. We consider ST and SD as constant parameters. We change the SL value from 0.1 to 0.9 and the SR from 0 to 1. In Fig. 6, we consider the ST value as 0.1. When SD increased form 0.1 to 0.9, we see that QoS is decreasing. When SD is 0.1, SL is 0.1, the QoS is increased by 32.02% when SR is increased form 0.3 to 0.8. This means that the system with high reliability can provide better QoS. When SD is 0.1 and SR is 0.7, the QoS is increased by 22.5% and 26.18% when SL is decreased form 0.9 to 0.5 and form 0.5 to 0.1, respectively. When the slice has a lot of data loss, the QoS performance will be decreased much more compared with the case of low data loss.

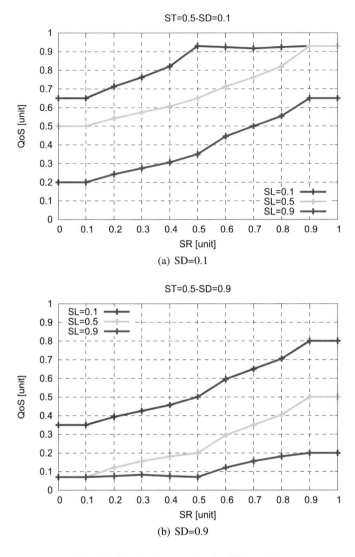

Fig. 7. Simulation results for ST = 0.5.

We compare Fig. 6 with Fig. 7 to see how ST has affected QoS. When SD is 0.1, SL is 0.5 and SR is 0.6, QoS is increased 26.74% by increasing ST from 0.1 to 0.5. This is because a higher ST value means the slice has more throughput and can provide better QoS.

In Fig. 8, we increase the value of ST to 0.9. We see that the QoS value is increased much more compared with the results of Fig. 6 and Fig. 7. We see that when the values of ST and SR parameters are increased, the QoS also increases.

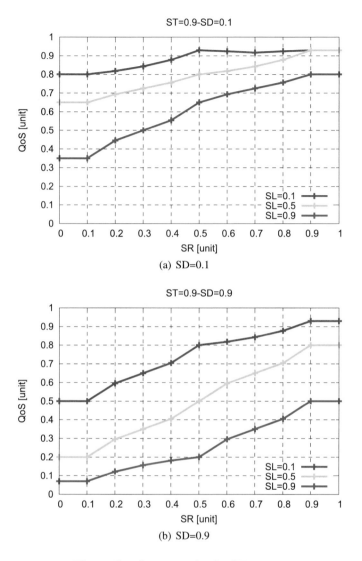

(a) SD=0.1

(b) SD=0.9

Fig. 8. Simulation results for ST = 0.9.

6   Conclusions and Future Work

In this paper, we proposed and implemented a Fuzzy-based scheme for evaluation of QoS. The evaluated QoS parameter will be used an input parameter for Admission Control in 5G Wireless Networks. We evaluated the proposed scheme by simulation. From simulation results, we conclude as follows.

- When ST and SR parameters are increased, the QoS value is increased. This means that QoS performance will be high.

- When SD and SL parameters are increased, the QoS parameter is decreased. Thus, the QoS performance will be low.

In the future, we would like to evaluate the Admission Control system by considering QoS parameter and other parameters.

References

1. Navarro-Ortiz, J., Romero-Diaz, P., Sendra, S., Ameigeiras, P., Ramos-Munoz, J.J., Lopez-Soler, J.M.: A survey on 5G usage scenarios and traffic models. IEEE Commun. Surv. Tutorials **22**, 1 (2020)
2. Hossain, S.: 5G wireless communication systems. Am. J. Eng. Res. (AJER) **2**(10), 344–353 (2013)
3. Giordani, M., Mezzavilla, M., Zorzi, M.: Initial access in 5G mmWave cellular networks. IEEE Commun. Mag. **54**(11), 40–47 (2016)
4. Kamil, I.A., Ogundoyin, S.O.: Lightweight privacy-preserving power injection and communication over vehicular networks and 5G smart grid slice with provable security. Internet of Things **8**, 100–116 (2019)
5. Hossain, E., Hasan, M.: 5G cellular: key enabling technologies and research challenges. IEEE Instrum. Meas. Mag. **18**(3), 11–21 (2015)
6. Yao, D., Su, X., Liu, B., Zeng, J.: A mobile handover mechanism based on fuzzy logic and MPTCP protocol under SDN architecture*. In: 18th International Symposium on Communications and Information Technologies (ISCIT-2018), pp. 141–146, September 2018
7. Lee, J., Yoo, Y.: Handover cell selection using user mobility information in a 5G SDN-based network. In: 2017 Ninth International Conference on Ubiquitous and Future Networks (ICUFN-2017), pp. 697–702, July 2017
8. Moravejosharieh, A., Ahmadi, K., Ahmad, S.: A fuzzy logic approach to increase quality of service in software defined networking. In: 2018 International Conference on Advances in Computing,Communication Control and Networking (ICACCCN-2018), pp. 68–73, October 2018
9. Ampririt, P., Ohara, S., Qafzezi, E., Ikeda, M., Barolli, L., Takizawa, M.: Integration of software-defined network and fuzzy logic approaches for admission control in 5G wireless networks: a fuzzy-based scheme for QoS evaluation. In: Barolli, L., Takizawa, M., Enokido, T., Chen, H.-C., Matsuo, K. (eds.) Advances on Broad-Band Wireless Computing, Communication and Applications, pp. 386–396. Springer, Cham (2021)
10. Li, L.E., Mao, Z.M., Rexford, J.: Toward software-defined cellular networks. In: 2012 European Workshop on Software Defined Networking, pp. 7–12, October 2012
11. Mousa, M., Bahaa-Eldin, A.M., Sobh, M.: Software defined networking concepts and challenges. In: 2016 11th International Conference on Computer Engineering & Systems (ICCES), pp. 79–90. IEEE (2016)
12. Jantzen, J.: Tutorial on fuzzy logic, Technical Report. Department of Automation, Technical University of Denmark (1998)
13. Mendel, J.M.: Fuzzy logic systems for engineering: a tutorial. Proc. IEEE **83**(3), 345–377 (1995)
14. Zadeh, L.A.: Fuzzy logic. Computer **21**, 83–93 (1988)
15. Norp, T.: 5G requirements and key performance indicators. J. ICT Standard. **6**(1), 15–30 (2018)

16. Parvez, I., Rahmati, A., Guvenc, I., Sarwat, A.I., Dai, H.: A survey on low latency towards 5G: ran, core network and caching solutions. IEEE Commun. Surv. Tut. **20**(4), 3098–3130 (2018)
17. Kim, Y., Park, J., Kwon, D., Lim, H.: Buffer management of virtualized network slices for quality-of-service satisfaction. In: 2018 IEEE Conference on Network Function Virtualization and Software Defined Networks (NFV-SDN), pp. 1–4 (2018)
18. Barolli, L., Koyama, A., Yamada, T., Yokoyama, S.: An integrated CAC and routing strategy for high-speed large-scale networks using cooperative agents. IPSJ J. **42**(2), 222–233 (2001)

Evaluation of a User Finger Movement Capturing Device for Control of Self-standing Omnidirectional Robot

Kenshiro Mitsugi[1], Keita Matsuo[2(✉)], and Leonard Barolli[2]

[1] Graduate School of Engineering, Fukuoka Institute of Technology (FIT),
3-30-1 Wajiro-Higashi, Higashi-Ku, Fukuoka 811-0295, Japan
mgm20108@bene.fit.ac.jp
[2] Department of Information and Communication Engineering, Fukuoka Institute
of Technology (FIT), 3-30-1 Wajiro-Higashi, Higashi-Ku, Fukuoka 811-0295, Japan
kt-matsuo@fit.ac.jp, barolli@fit.ac.jp

Abstract. Convenient systems and equipments to support humans are in great need for ever-growing populations of the elderly and those with disabilities caused by illness or injury. One of these system is the wheelchair, which can provide the user with many benefits such as maintaining mobility, continuing or broadening community and social activities, conserving strength and energy, and enhancing quality of life. However, when users use wheelchairs, they have to frequently stand and sit. This increases the physical burden on the user. For this reason, we proposed a self-standing omnidirectional robot. In order to support the user, the robot body must be able to flexibly make different movements and should be capable to deal with various control methods to meet diverse needs. In this paper, we present evaluation of a user finger movement capturing device for control of self-standing omnidirectional robot.

1 Introduction

Robots are being steadily introduced into modern everyday life and are expected to play a key role in the near future. Typically, the robots are deployed in situations where it is too dangerous, expensive, tedious, and complex for humans to operate.

One of the main features of world population in the 20th century was the increment of elderly people. According to WHO (World Health Organization) by 2025, the increase of population over aged 60 is predicted to reach 23% in North America, 17% in East Asia, 12% in Latin America and 10% in South Asia. There are 1 billion disabled persons in the world constituting nearly 14% of the global population.

In recent years, convenient facilities and equipments have been developed in order to satisfy the requirements of elderly people and disabled people. Among them, robot technologies are a common one which is used widely. They can provide the user with many benefits such as maintaining mobility, continuing

L. Barolli et al. (Eds.): AINA 2021, LNNS 227, pp. 30–40, 2021.
https://doi.org/10.1007/978-3-030-75078-7_4

or broadening community and social activities, conserving strength and energy, and enhancing quality of life.

In this paper, we present implementation of a user finger movement capturing device for control of self-standing omnidirectional robot.

The structure of this paper is as follows. In Sect. 2, we introduce the related work. In Sect. 3, we propose the self-standing omnidirectional robot and present our new control device. In Sect. 4, we describe the implemented of user finger capturing device. In Sect. 5, we evaluate the proposed system. Finally, conclusions and future work are given in Sect. 5.

2 Related Work

Most of the work for mobile robots has be done for improving the quality of life of disabled people. One of important research area is robotics. The persons having physical impairment often find it difficult to navigate the moving robot themselves. The reduced physical function associated with the age or disability make independent living more difficult. Many research works have been undertaken to reduce the problem of navigation faced by the physically and mentally challenged people and also older age persons. One of the suggestive measures is the development of a Brain Control Interface (BCI), that assist an impaired person to control the robot using his own brain signal. The research proposes a high-frequency SSVEP-based asynchronous BCI in order to control the navigation of a mobile object on the screen through a scenario and to reach its final destination [1]. This could help impaired people to navigate a robotic wheelchair. The BCIs are systems that allow to translate in real time the electrical activity of the brain in commands to control devices, provide communication and control for people with devastating neuromuscular disorders, such as the Amyotrophic Lateral Sclerosis (ALS), brainstem stroke, cerebral palsy, and spinal cord injury [2].

One of the key issue in designing wheelchairs is to reduce the caregiver load. Some of the research works deal with developing prototypes of moving robots that helps the caregiver by lifting function or which can move with a caregiver side by side [3,4]. The lifting function equipment facilitates easy and safe transfer from/to a bed and a toilet stool by virtue of the opposite allocation of wheels from that for a usual wheelchair. The use of lifting function and the folding of frames makes it more useful in indoor environments. Moving care Robots based on observations of people using integrated sensors can move with a caregiver side by side. This is achieved by a visual-laser tracking technique, where a laser range sensor and an omnidirectional camera are integrated to observe the caregiver.

Another important issue for the design of robot is the collision detection mechanism. The omnidirectional robots with collaborative controls ensures better safety against collisions. Such robots possess high level of ability when moving over a step, through a gap or over a slope [5,6]. To achieve omnidirectional motion, vehicles are generally equipped with an omniwheel consisting of a large number of free rollers or a spherical ball wheel. The development of such omniwheels attempts to replace the conventional wheel-type mechanism.

There are also other works which deal with vision design of robotic wheelchairs by equipping the wheelchair with camera for monitoring wheelchair movement and obstacle detection and pupil with gaze sensing [7,8]. Prototype for robots have been suggested in various research works, which are exclusively controlled by eye and are used by different users, while proving robust against vibration, illumination change, and user movement [9,10].

To enable older persons to communicate with other people the assisting devices have been developed. They can improve the quality of life for the elderly and disabled people by using robot technologies. The head gesture recognition is performed by means of real time face detection and tracking techniques. They developed a useful human-robot interface for RoboChair [11].

3   Proposed Self-standing Omnidirectional Robot

In this section, we describe the implementation of a self-standing Omnidirectional Robot. In our previous work, we implemented an omnidirectional wheelchair as shown in Fig. 1. This wheelchair can move omnidirectionly while keeping the direction, which is very good for different tasks. However, when users use the wheelchair, they have to frequently stand and sit. This increases the physical burden on the user. For this reason, we proposed a self-standing omnidirectional robot (see Fig. 2).

In order to confirm the movement of self-standing omnidirectional robot, we implemented a test model as shown in Fig. 3. This figure shows three omniwheels with small tires. With this structure, the wheel can rotate in front, back, right and left, so it is able to move in all directions.

3.1   Kinematics

For the control of the omnidirectional robot are needed the omniwheel speed, omnidirectional robot movement speed and direction.

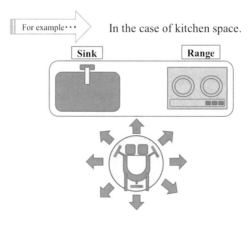

Fig. 1. Movement of omnidirectional wheelchair.

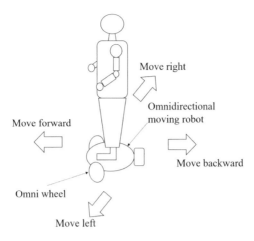

Fig. 2. Image of self-standing type omnidirectional robot.

Fig. 3. Test model of self-standing type omnidirectional robot.

Let us consider the movement of the omnidirectional robot in 2 dimensional space. In Fig. 4, we show the onmiwheel model. In this figure, there are 3 onmiwheels which are placed 120° with each other. The omniwheels are moving in clockwise direction as shown in the figure. We consider the speed for each omniwheel M1, M2 and M3, respectively.

As shown in Fig. 4, the axis of the omnidirectional robot are x and y and the speed is $v = (\dot{x}, \dot{y})$ and the rotating speed is $\dot{\theta}$. In this case, the moving speed of the omnidirectional robot can be expressed by Eq. (1).

$$V = (\dot{x}, \dot{y}, \dot{\theta}) \tag{1}$$

Based on the Eq. (1), the speed of each omniwheel can be decided. By considering the control value of the motor speed ratio of each omniwheel as linear and synthesising the vector speed of 3 omniwheels, we can get Eq. (2) by using

Reverse Kinematics, where (d) is the distance between the center and the omni-wheels. Then, from the rotating speed of each omniwheel based on Forward Kinematics, we get the omnidirectional robot moving speed. If we calculate the inverse matrix of Eq. (2), we get Eq. (3). Thus, when the omnidirectional robot moves in all directions (omnidirectional movement), the speed for each motor (theoretical values) is calculated as shown in Table 1.

$$
\begin{vmatrix} M_1 \\ M_2 \\ M_3 \end{vmatrix} = \begin{vmatrix} 1 & 0 & d \\ -\frac{1}{2} & -\frac{\sqrt{3}}{2} & d \\ -\frac{1}{2} & \frac{\sqrt{3}}{2} & d \end{vmatrix} \begin{vmatrix} \dot{x} \\ \dot{y} \\ \dot{\theta} \end{vmatrix} \tag{2}
$$

$$
\begin{vmatrix} \dot{x} \\ \dot{y} \\ \dot{\theta} \end{vmatrix} = \begin{vmatrix} \frac{2}{3} & -\frac{1}{3} & -\frac{1}{3} \\ 0 & -\frac{1}{\sqrt{3}} & \frac{1}{\sqrt{3}} \\ \frac{1}{3d} & \frac{1}{3d} & \frac{1}{3d} \end{vmatrix} \begin{vmatrix} M_1 \\ M_2 \\ M_3 \end{vmatrix} \tag{3}
$$

3.2 Control System of the Proposed Self-standing Omnidirectional Robot

For the control of the proposed self-standing omnidirectional robot, we considered motor control system. We used brushless motor (BLHM015K-50) and Raspberry Pi3 B+ as a controller. In Fig. 5 is shown the control system for the motor.

The implemented motor control system can connect to any control devices with TCP/IP and Bluetooth as shown in Fig. 6.

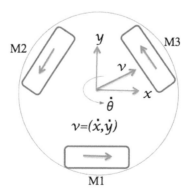

Fig. 4. Model of omniwheel.

Table 1. Motor speed ratio.

Direction (Degrees)	Motor speed ratio		
	Motor1	Motor2	Motor3
0	0.00	−0.87	0.87
30	0.50	−1.00	0.50
60	0.87	−0.87	0.00
90	1.00	−0.50	−0.50
120	0.87	0.00	−0.87
150	0.50	0.50	−1.00
180	0.00	0.87	−0.87
210	−0.50	−1.00	−0.50
240	−0.87	0.87	0.00
270	−1.00	0.50	0.50
300	−0.87	0.00	0.87
330	−0.50	−0.50	1.00
360	0.00	−0.87	0.87

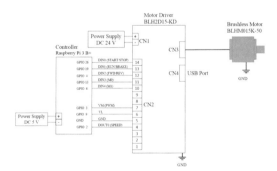

Fig. 5. Control system for the motor.

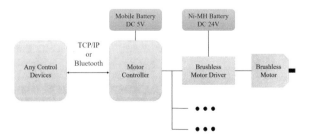

Fig. 6. Block diagram of control system.

4   Implementation of User Finger Capturing Device

In this section, we present our proposed control device using small ([mm] scale) movements of user finger. So far, we have been studying the wheelchair vision and IoT sensors [12,13]. Therefore, we tried to control the robot by following the human finger with a camera using our previous research. Some people with disabilities have a small range of their arm and fingers and can not operate devices like the joystick. Thus, the aim is to operate the robot by capturing small finger movements with a camera. In this way, the operation is easier than the joystick. By using video recognition as a controller, the physical burden for users can be reduced [14].

We show the image of the control device using small movement of user finger using machine learning in Fig. 7. In order to predict the user direction, we used Scikit-learn. Scikit-learn is an open source software for machine learning. It has a number of algorithms that support vector machine, random forest, k-means clustering and neural networks. We use neural networks for predicting the finger movement. In this work, we used ReLU (Rectified Linear Unit) as a activation function.

The proposed system can capture the movement of user finger top and track the orbit. We used this orbit to decide the directions for controlling the robot. However, it is difficult to get user finger directions, because each user has different habits of their finger movement.

In Fig. 8 is shown the captured finger top. Figure 9, (the left side of the figure) shows the finger movements (top, down, left and right), whereas the right side of the figure shows each directions (upper right, upper left, lower right and lower left). The system can track the orbit of user finger, which are showing by dots in the figure. We used these data to recognize the user directions.

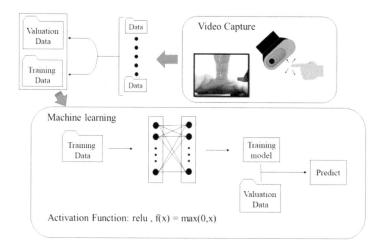

Fig. 7. Image of control device using small movements of user finger.

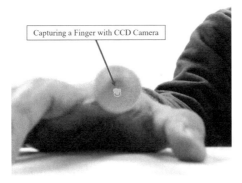

Fig. 8. Image of capturing finger at center position.

5   Experimental Results

In Fig. 10 are shown the user finger orbits for 4 directions (up, down, left and right) and in Fig. 11 are shown the user finger orbits for 8 directions (up, down, left, right, upper right, upper left, lower right and lower left). For 4 directions, the system can predict the user direction more easily than 8 directions. However, in this work we tried to detect for 8 directions. So, we tuned the system using ReLU activation function and 2 layer (400 400) neural network. Figure 11 shows complicated obits. However, our proposed control device with image sensor can track and predict small user finger movement correctly.

We show the experimented results of loss function in Fig. 12. When the Time Step is around 50, the Loss value is almost 0. The time step means the number of validations. According to our research, the predict rate of 98% is archived in case of 4 directions. The accuracy of the system for predicting the direction is over 99% in case of 8 directions.

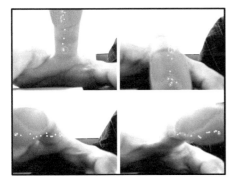

Fig. 9. Image of capturing a finger movement.

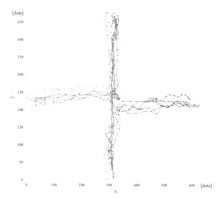

Fig. 10. Results of tracing small finger movement for 4 directions by camera.

We considered some active functions to use for our system. In order to decide an appropriate function, we compared accuracy rate of Rule, Logistic, Identify and Tanh. The results are shown Table 2. From the results, we decided to use Rule function, because the accuracy rate is higher than others.

Table 2. Compared accuracy rate.

Activation function	Rule	Logistic	Identity	Tanh
Number of learning time	400	400	400	400
Number of middle layer's neuron	400 × 400	400 × 400	400 × 400	400 × 40
Accuracy rate	99.72%	98.43%	97.78%	97.50%

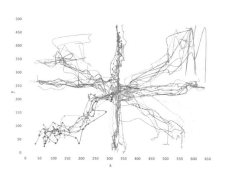

Fig. 11. Results of tracing small finger movement for 8 directions by camera.

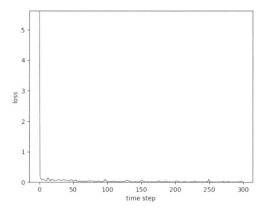

Fig. 12. Experiment results of loss function.

6   Conclusions and Future Work

In this paper, we proposed a self-standing omnidirectional robot and presented the implementation of user finger movement capturing device. We introduced some of the previous works and discussed their problems and issues. Then, we presented in details the kinematics and proposed self-standing robot. In addition, we have shown the evaluation of user finger movement capturing device for the control of self-standing omnidirectional robot. The evaluation results have shown that the proposed system has a good accuracy for detecting the user finger directions.

In the future work, we would like to enhance the accuracy and consider other control devices for users.

References

1. Diez, P.F., Mut, V.A., Perona, E.M.A., Leber, E.L.: Asynchronous BCI control using high-frequency SSVEP. J. NeuroEng. Rehabil. **8**(39), 8 (2011). https://doi.org/10.1186/1743-0003-8-39
2. Grigorescu, S.M., Luth, T., Fragkopoulos, C., Cyriacks, M., Graser, A.: A BCI-controlled robotic assistant for quadriplegic people in domestic and professional life. Robotica **30**(3), 419–431 (2012)
3. Mori, Y., Sakai, N., Katsumura, K.: Development of a wheelchair with a lifting function. Adv. Mech. Eng. **2012**, 9 (2012). https://doi.org/10.1155/2012/803014. Article ID 803014
4. Kobayashi, Y., Kinpara, Y., Shibusawa, T., Kuno, Y.: Robotic wheelchair based on observations of people using integrated sensors. In: Proceedings of IEEE/RSJ International Conference on Intelligent Robots and Systems, 11–15 October, USA (2009)
5. Ishida, S., Miyamoto, H.: Collision detecting device for omnidirectional electric wheelchair. Robotics **2013**, 8 (2013). Article ID 672826

6. Carlson, T., Demiris, Y.: Robotic wheelchair with collaborative control. In: Proceedings of IEEE International Conference on Robotics and Automation, pp. 5582–5587 (2010)
7. Jia, P., Hu, H.H., Lu, T., Yuan, K.: Head gesture recognition for hands-free control of an intelligent wheelchair. Ind. Robot Int. J. **34**(1), 60–68 (2007). https://doi.org/10.1108/01439910710718469
8. Arai, K., Mardiyanto, R.: Electric wheelchair controlled by eye-only for paralyzed user. J. Robot. Mechatron. **23**(1), 66–74 (2011)
9. Escobedo, A., Spalanzani, A., Laugier, C.: Multimodal control of a robotic wheelchair: using contextual information for usability improvement. In: Proceedings of IEEE/RSJ International Conference on Intelligent Robots and Systems (IROS-2013), pp. 4262–4267 (2013). https://doi.org/10.1109/IROS.2013.6696967
10. Gonzalez, J., Munoz, A.J., Galindo, C., Fernandez-Madrigal, J.A., Blanco, J.L.: A description of the SENA robotic wheelchair. In: Proceedings of IEEE Mediterranean Conference (MELECON-2006), pp. 437–440 (2006)
11. Wang, H., Grindle, G.G., Candiotti, J., Chung, C., Shino, M., Houston, E., Cooper, R.A.: The Personal Mobility and Manipulation Appliance (PerMMA): a robotic wheelchair with advanced mobility and manipulation. In: Proceedings of IEEE Engineering in Medicine and Biology Society, pp. 3324–3327 (2012). https://doi.org/10.1109/EMBC.2012.6346676
12. Matsuo, K., Barolli, L.: Implementation of omnidirectional wheelchair vision with small reflect mirrors: performance evaluation for tennis ball tracking considering different mirror angles. In: Proceedings of International Conference on Complex Intelligent and Software Intensive Systems (CISIS-2018), pp. 136–148. Springer, Cham (2018)
13. Matsuo, K., Kurita, T., Barolli, L.: A new system for management of IoT sensors considering Agile-Kanban. In: Barolli, L., Takizawa, M., Xhafa, F., Enokido, T. (eds.) Proceedings Workshops of the International Conference on Advanced Information Networking and Applications (WAINA-2019), pp. 604–612. Springer, Cham (2019)
14. Mitsugi, K., Matsuo, K., Barolli, L.: A comparison study of control devices for an omnidirectional wheelchair. In: Barolli, L., Amato, F., Moscato, F., Enokido, T., Takizawa, M. (eds.) Proceedings of the Workshops of the International Conference on Advanced Information Networking and Applications (WAINA-2020), pp. 651–661. Springer, Cham (2020)

Adaptive Architecture for Fault Diagnosis of Rotating Machinery

Ricardo Rodríguez-Jorge$^{(\boxtimes)}$, Laura Sánchez-Pérez, Jiří Bíla, and Jiří Škvor

Jan Evangelista Purkyně University, Pasteurova 3632/15,
Ústí nad Labem, Czech Republic
Ricardo.Rodriguez-Jorge@ujep.cz

Abstract. In this work, an adaptive architecture has been implemented for rotating machinery fault diagnoses of an industrial machine, where the input data used in this work has been taken from alternative current motors. The data sets for training and testing were recorded with a vibrometer. A dynamic neural architecture is previously trained with the training data set. Furthermore, an online monitoring system is implemented using the testing data set to detect abnormal behaviour of the monitored signal which can lead to a failure of the industrial machine. For the evaluation of the architecture, tests are performed using the obtained signal from different vibration tests.

1 Introduction

Monitoring the condition of a machine and its components is very important for avoiding unnecessary downtime and costs improving safety and enhancing the machine's lifetime by recognizing abnormal behavior in a machine component or a machine [1].

Many different methods have been developed for diagnoses in the detection of machine defects and can be classified as acoustic, temperature measurements, and vibration. Among these, vibration measurements in the time and frequency domain, which includes the shock pulse method, and acoustic emissions are the usual methods for measuring acoustic response and vibration [2].

In the literature, several diagnosis techniques are based on the analysis of electrical signatures, such as neutral voltage, rotor or stator current, and the analysis of electromagnetic size, such as the vibration signal or magnetic flux [3]. There are many "site factors" and variables that collectively form a complex vibration waveform [4].

Currently, fault diagnosis methods for rotating machinery include artificial intelligence (AI)-based and signal processing-based fault diagnosis methods. The AI-based fault diagnosis method can be used as a pattern recognition problem by using features extracted from previously collected signals. These methods include fuzzy inference, support vector machines (SVMs) and artificial neural networks. According to [5], AI-based methods exhibit three main disadvantages: (a) the selected features are only suitable for specific issues and require selection. In addition, the selected features require considerable manpower and time, because

L. Barolli et al. (Eds.): AINA 2021, LNNS 227, pp. 41–51, 2021.
https://doi.org/10.1007/978-3-030-75078-7_5

the feature quality determines the classification quality of an AI-based method; (b) the extraction of valid features relies on complex and advanced signal processing technologies and expertise in diagnosis; (c) artificial neural networks and SVMs have architectures that experience difficulty in learning complex and non-linear relationships in fault diagnosis. The design of higher-order neural unit architectures for fault diagnosis is encouraged.

From the above reviewed AI-based fault diagnosis, it is apparent that neural networks (NNs) have promising capabilities that can be implemented in rotating machinery fault diagnosis, and fault diagnosis with NNs is a subject of great interest in industrial maintenance due to the possibility of capturing the dynamic and structural aspects. Some authors are convinced that deep analysis is still needed [6].

In this research work, a pattern monitoring architecture is constantly fed new data obtained from vibrometer tests. The architecture has been implemented by a dynamic quadratic neural unit with sample-by-sample real time recurrent learning. In this sense, the system error function decreases considerable and the system algorithm is capable of efficiently modeling the vibration signal from the industrial machine. In addition, the architecture is able to show online the possible abnormal vibration behaviour.

The structure of this research paper is as follows: Sect. 2 illustrates different studies related to monitoring industrial machines; Sect. 3 presents a motivating example to illustrate the importance of the proposed architecture; Sect. 4 describes the architecture applied for rotating machinery fault diagnosis; the obtained results are shown and discussed in Sect. 5; finally, conclusions and future work are summarized in Sect. 6.

2   Related Work

In the literature, several types of measurements are used for the machine health monitoring of different components. Examples of this measurement types include power consumption, current, voltage, vibrations and acoustic emissions. However, there are multiple types of monitoring components, from which vibration signals very well represent the dynamics of machines, and they have been used frequently for machine health monitoring. The health managements and prognostics of rotating machines have been important research area for last decades [1,7].

For the rotary components, algorithms have been investigated for their health monitoring, such as EMD-, NN-, FT-based algorithms, and are frequently used. With regard to machine learning, [8] proposed a comparison between the artificial NN and support vector machine (SVM) methods for predicting bearing faults and obtained that the SVM method produced better results than the ANN method. Additionally, [9] developed a deep neural network (DNN)-based method for fault diagnoses, which is able to mine fault diagnoses from a given frequency spectrum for various diagnoses that can be classified into different health conditions. [10] presented an online fault diagnosis system for a rotating machine by using fractal theory to obtain the fractal dimension and lacunarity from the features of shaft orbital data; here, they used a backpropagation network learning algorithm to analyze the data. Table 1 presents a comparison of some related works to evaluate the importance of the proposed approach.

Table 1. Machine health monitoring comparison with rotary components.

Author	Monitoring components	Measurements	Applicable machines	Health assessment algorithms
[11]	Generator	Electric current	Wind turbine	EMD, ANN, PCA
[12]	Micro milling	Vision	Micro machining, machining center	PCA
[13]	Micro milling	Force, torque	Micro machining, machining center	WT, FLD, HMM
[14]	Micro turning	Deformation	Micro machining, machining center	Bayesian network
[15]	Micro milling, micro EDM	Power consumption, electric voltage, electric current	Machining center	Time domain analysis
[16]	Micro milling	Vision, wear	Micro machining	SPR
[17]	Micro grinding	Deformation, wear grains	Machining center	SPR
[18]	Micro EDM	Electric voltage, electric current, EDM pulse	Micro machining, micro EDM	Time domain analysis
[19]	Micro milling	Acoustic emission	Machining center	Short time Fourier transform (STFT), regression trees, NN
[20]	Micro drilling	Force, wear	Micro machining, machining center	ANN

In contrast to the related work, the presented fault diagnosis monitoring of rotating machines consists of firstly, a dynamic neural model which is trained with the training data set. Then, the adapted weights are saved and used during the online monitoring to test the testing data set. This makes the model able to identify patterns in the vibration signals to achieve the modeling of vibration values during the real time monitoring. Then, the fault diagnosis of a rotating machine consists of categorizing the state of the machine according to the severity categorization from the norm ISO 10816-1 standard guidelines. Currently, the unusual pattern detection of dynamic systems is a topic of scientific interest.

Fig. 1. Industrial machine A.C motor.

3   Motivating Example

The lack of an efficient rotating machinery fault diagnosis can be destructive for the industrial installation even though it can be seen as favorable economic effect. The process of performing corrective maintenance constantly affects diverse areas in a company, *e.g.*, the planned production, the useful life of equipment, man/hours (dead time), noncompliance with an established budget, metrics such as overall equipment effectiveness (OEE). All these factors directly affect the quality of the products made and the reliability perceived by the customer.

The aim when implementing fault diagnosis is to obtain a report that details the exact condition of the equipment. Most of the fault diagnosis techniques evaluate the evolution of a machine. Hence, carrying out or conducting a fault diagnosis inspection service for a single specific measurement is not convenient; however, the evolution over time can be studied, with this we can diagnose when intervention is necessary. The application of fault diagnosis to machines can reduce corrective maintenance and thereby reduce maintenance costs.

Accordingly, many machines are driven by electric motors, and it is essential to diagnose their health conditions with proper techniques. Mechanical vibration measurements of motors are one of the most popular techniques for this purpose. Another practical technique for electric motor health monitoring is to use electric signals such as power consumption, voltage or current. The Fig. 1 shows an industrial alternating current motor.

Rotary components are one of the most critical components of current and future manufacturing machines. Such machines are composed of large amounts of gears, bearings, shafts, and other types of rotary elements.

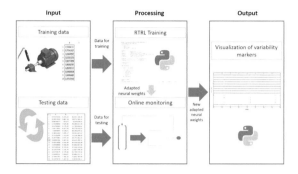

Fig. 2. Proposed adaptive architecture for fault diagnosis of rotatory machinery

4   Fault Diagnosis of Rotating Machinery

The proposed architecture for the automatic monitoring of industrial machines based on the vibration signals, can be seen in Fig. 2. The input data is obtained from a vibration tests that has been carried out in an alternating current motor, which belongs to an industrial furnace whose main function is to weld electronic components to a spline.

The study has been carried out by means of a vibrometer. Figure 3 shows the main screen of the vibrometer used and its main buttons. The sensor is positioned in the motor in three directions:

- *Axial:* sensor position that is in the direction of the axis line.
- *Radial:* sensor position perpendicular to the axis line.
- *Tangential:* 90° radial, tangent to the arrow.

Once the vibration measurement has been performed, two main characteristics, the general vibration and frequency, represented in revolutions per minute (*rpm*) of the motor in each position are obtained.

- *General vibration:* This is an oscillatory movement with a small amplitude. According to this, the machines present their own vibration signals which contain the information of each of their components. Therefore, a vibration signal captured from a machine is the vector sum of the vibration of each of its components. The measurement of this vibration is given in *mm/s*, which allows the recognition of most of the patterns of the primary failures and of the other components when they are in an evident state, such as imbalance, misalignment, mechanical play, and abrasive frictions.
- *Frequency:* It is defined as the number of complete cycles in a period of time. The characteristic unit is *cpm* (cycles per minute). There is an important relationship between the frequency and angular velocity of the rotating elements. The correspondence between *cpm* and *rpm* (cycles per minute - revolutions per minute) will identify the problem and the part responsible for the vibration. This relationship is due to the forces of the changing direction and

Fig. 3. Main screen of the vibrometer.

amplitude according to the rotational speed. Different problems are detected by the frequencies equal to the rotational speed or multiples thereof. Each type of problem shows a different vibration frequency. For this work, the unit of revolutions per minute (*rpm*) will be used.

4.1 RTRL Training

The DQNU uses an RTRL learning method and can be implemented in discrete and continuous real time. In Eq. 1 the adaptation of the weights system in the new adaptation time of the model is presented (*i.e.*, $w_{i,j}(k+1)$); where $w_{i,j}(k)$ stands for each individual weight, and $\Delta w_{i,j}(k)$ the neural weights increment of each discrete value from the vibration signal.

$$w_{i,j}(k+1) = w_{i,j}(k) + \Delta w_{i,j}(k) \tag{1}$$

Equation 2 describes the increment of the neural weights with RTRL, where $\tilde{y}(k+h)$ is the predicted neural output; $e(k+h)$, is the error in each discrete time $(k+h)$; μ, is the learning rate that determines the velocity of the learning process and in which proportion the neural weights are updated.

$$\Delta w_{i,j}(k) = \mu \cdot e(k+h) \cdot \frac{\partial \tilde{y}(k+h)}{\partial w_{i,j}} \tag{2}$$

where the DQNU output with RTRL is shown in Eq. 3, and the upper triangular matrix of weights $\mathbf{W}$ with neural bias $w_{0,0}$ is defined in Eq. 4

$$\tilde{y}(k+h) = \sum_{i=0}^{nx-1} \sum_{j=i}^{nx-1} x_j(k) \cdot w_{i,j} \cdot x_i(k) \tag{3}$$

$$\mathbf{W}(k) = \begin{bmatrix} w_{0,0} & w_{0,1} & \cdots & w_{0,nx-1} \\ 0 & w_{1,1} & \cdots & w_{1,nx-1} \\ \vdots & \vdots & \ddots & \vdots \\ 0 & \cdots & 0 & w_{nx-1,nx-1} \end{bmatrix} \tag{4}$$

Equation 5 is the column of the augmented vector $\mathbf{x}(k)$ of the neural input, $\mathbf{y}(k)$ stands for the real values, n is the number of real values that feed the neural input, k stands for the discrete time and h is the prediction horizon.

$$\mathbf{x}(k) = \begin{bmatrix} 1 \\ \tilde{y}(k+h-1) \\ \tilde{y}(k+h-2) \\ \vdots \\ \tilde{y}(k+1) \\ y(k) \\ y(k-1) \\ \vdots \\ y(k-n+1) \end{bmatrix} \tag{5}$$

4.2 Online Monitoring

The adaptive architecture contains an adaptive graphic monitor which allows displaying the vibration signal in real time. The architecture consists of detecting and visualizing unusual pattern that can indicate possible alert in the rotating machine according to the standards table of the vibration severity from the norm ISO 10816-1. The detection and visualization of unusual patterns is performed by adapting the neural weights during model adaptation. The architecture monitors online vibration signals obtained from vibration tests. This makes the system architecture able to identify patterns in vibration signals and to achieve the modeled values of the signal during the real-time monitoring. This monitoring is performed on the testing data set which is different data set from the training data set utilized during the training stage, which is illustrated on Fig. 2.

5 Experimental Results

The vibration recorded data stored in the testing data set were utilized to validate the adaptive monitoring architecture. It is important to point out that the adaptive model is capable of identifying variabilities of vibration signals showing different patterns for each case.

The adaptive model has been described in Sect. 4 for modeling vibration signals. The input configuration of the model has been $(n+h)$, and $(h-1)$ is the number of feedbacks to feed the input model, and the bias $\mathbf{x}(k=1)=1$. The number of training times has been set to $epochs = 100$, and the learning rate has been configured to $\mu = 0.001$. Besides, a window for the average of the neural weights increments has been considered as $window = 200$, the sensitivity parameter α will be used in the visualization of the variability markers.

Since the adaptive model is based on RTRL, it can be observed that the neural output models the behavior of the unstable signal of the real vibration signals, while markers indicate the relevant changes of the signal, allowing the patters formed in each type of vibration records to be displayed.

For the output obtained by the DQNU during the online monitoring is interpreted according to the ISO 10816-1 standard, which establishes the general conditions and procedures for the measurement and evaluation of vibrations. From this standard, the vibration severity table is derived, which is presented in Table 2, in which we relate the online monitoring of the adaptive evaluation with the *RMS* speed, and in this way, the value is located in the severity table whose values are associated with the state of the equipment.

Table 2. ISO 10816-1 standards table of vibration severity.

Vibration velocity		Velocity range limits and machine classes ISO standard 10816-1			
RMS (mm/s)	PEAK (in/s)	Small machines Class I	Medium machines Class II	Large machines Rigid supports Class III	Less rigid supports Class IV
0.28	0.02	Good	Good	Good	Good
0.45	0.03				
0.71	0.04				
1.12	0.06	Satisfactory			
1.80	0.10		Satisfactory		
2.80	0.16	Unsatisfactory (Alert)		Satisfactory	
4.50	0.25		Unsatisfactory (Alert)		Satisfactory
7.10	0.40	Unacceptable (Danger)		Unsatisfactory (Alert)	
11.20	0.62		Unacceptable (Danger)		Unsatisfactory (Alert)
18.00	1.00			Unacceptable (Danger)	
28.00	1.56				Unacceptable (Danger)
45.00	2.51				

According to the severity table, the machines are divided into four types of classes and speed range limits, which are described below:

Machine Classes

Class I: individual parts that connect to a machine during normal operation. (Electric motors that do not exceed 15 kW are typical examples of parts in this category).

Class II: medium-sized machines (generally motors with 15 to 75 kW output), without special foundations, or rigid machines (above 300 kW) mounted on special foundations.

Class III: large motors and other machines with large rotating masses mounted on stiff and heavy foundations, which are relatively hard in the direction of vibration measurement.

Class IV: large engines and other machines with large rotating masses mounted on relatively flexible foundations in the direction of the vibration measurement (for example, a turbo generator, especially those with lightweight substructures).

Speed Range Limits

Good: value indicating a good machine condition. Vibration from new or recently reconditioned machines should be within this zone.

Satisfactory: machines with values within this zone are normally considered acceptable for long-term, unrestricted operation.

Unsatisfactory (alert): machines with values within this zone are normally considered unsatisfactory for long-term continuous operation. Generally, the machine can be operated for a limited period of time in this condition until an appropriate opportunity for repair arrives.

Unacceptable (danger): vibration values within this zone are normally considered severe enough to cause damage to the machine.

5.1   Visualization of Variability Markers

The architecture contains a visualization of variability markers, in this stage unusual adaptive neural weights are evaluated and visualized by a graphical interface design. For the detection of changes in the vibration signal, a sensitivity parameter is initialized experimentally as $\alpha = 2.92$. This sensitivity parameter is considered in the online monitoring as shown in Fig. 2. The online monitoring influences a condition that allows drawing markers according to the evaluation of the increment parameters. To allow drawing of the markers, a condition is considered as in Eq. 6.

$$|\Delta \mathbf{W}(k)| \geq \alpha \cdot \overline{\Delta \mathbf{W}(k)} \tag{6}$$

where $\overline{\Delta \mathbf{W}(k)}$ stands for the average of the weight increments, and $|\cdot|$ stands for absolute value. Accordingly, a window of a set of values has been considered for the average weight increment.

6   Conclusions

In this paper, a fault diagnosis monitoring architecture is proposed for rotating machinery based on the vibration signals obtained from vibration tests. The monitoring architecture is presented by a dynamic quadratic neural unit which is able to adapt itself with the real-time recurrent learning training. In its current stage, the adaptive architecture plots the variability markers, and the patterns of variability are visualized. Therefore, this paper concludes that the health condition monitoring of an industrial machine using the DQNU with RTRL optimization is highly efficient.

In future work, vibration data from an alternating current motor in real time using Raspberry Pi will be recorded. Additionally, the signals obtained from the vibration that present some values outside the optimal range will be sent to an alert in the machine through an Andon system, which indicates that maintenance is required through assigned colors.

Acknowledgements. This project is supported by Jan Evangelista Purkyně University. Title of the project - Predictive maintenance of an industrial machine using neural networks.

References

1. Janssens, O., Van de Walle, R., Loccufier, M., Van Hoecke, S.: Deep learning for infrared thermal image based machine health monitoring. IEEE/ASME Trans. Mechatron. **23**(1), 151–159 (2018)
2. Unal, M., Onat, M., Demetgul, M., Kucuk, H.: Fault diagnosis of rolling bearings using a genetic algorithm optimized neural network. Measurement **58**, 187–196 (2014)
3. Djamila, B., Tahar, B., Hichem, M.: Vibration for detection and diagnosis bearing faults using adaptive neurofuzzy inference system. J. Electr. Syst. **14**(1), 95–104 (2018)
4. Sohaib, M., Kim, C.-H., Kim, J.-M.: A hybrid feature model and deep-learning-based bearing fault diagnosis. Sensors **17**(12), 2876 (2017)
5. Tang, S., Shen, C., Wang, D., Li, S., Huang, W., Zhu, Z.: Adaptive deep feature learning network with Nesterov momentum and its application to rotating machinery fault diagnosis. Neurocomputing **305**, 1–14 (2018)
6. Rodriguez Jorge, R.: Artificial neural networks: challenges in science and engineering applications. Front. Artif. Intell. Appl. **295**, 25–35 (2017)
7. Lee, G.Y., Kim, M., Quan, Y.J., et al.: Machine health management in smart factory: a review. Mech. Sci. Technol. **32**, 987–1009 (2018)
8. Patel, J., Upadhyay, S.: Comparison between artificial neural network and support vector method for a fault diagnostics in rolling element bearings. Procedia Eng. **144**, 390–397 (2016)
9. Jia, F., Lei, Y., Lin, J., Zhou, X., Lu, N.: Deep neural networks: a promising tool for fault characteristic mining and intelligent diagnosis of rotating machinery with massive data. Mech. Syst. Signal Process. **72**, 303–315 (2016)
10. Chang, H.-C., Lin, S.-C., Kuo, C.-C., Lin, C.-Y., Hsieh, C.-F.: Using neural network based on the shaft orbit feature for online rotating machinery fault diagnosis. In: 2016 International Conference on System Science and Engineering (ICSSE), pp. 1–4 (2016)
11. Malik, H., Mishra, S.: Artificial neural network and empirical mode decomposition based imbalance fault diagnosis of wind turbine using TurbSim, FAST and Simulink. IET Renew. Power Gener. **11**(6), 889–902 (2017)
12. Zhu, K., Yu, X.: The monitoring of micro milling tool wear conditions by wear area estimation. Mech. Syst. Signal Process. **93**(1), 80–91 (2017)
13. Hong, Y., Yoon, H., Moon, J., Cho, Y.-M., Ahn, S.-H.: Tool-wear monitoring during micro-end milling using wavelet packet transform and Fisher's linear discriminant. Int. J. Precis. Eng. Manuf. **17**, 845–855 (2016)
14. Mandal, S., Sharma, V.K., Pal, A.: Tool strain- based wear estimation in micro turning using Bayesian networks. Proc. Inst. Mech. Eng., Part B: J. Eng. Manuf. **230**(10), 1952–1960 (2016)
15. Tristo, G., Bissacco, G., Lebar, A., Valentinčič, J.: Real time power consumption monitoring for energy efficiency analysis in micro EDM milling. Int. J. Adv. Manuf. Technol. **78**, 1511–1521 (2015)

16. Szydlowsk, M., Powałka, B., Matuszak, M., Kochmański, P.: Machine vision micro-milling tool wear inspection by image reconstruction and light reflectance. Precis. Eng. **44**, 236–244 (2016)
17. Wen, X., Gong, Y.: Modeling and prediction research on wear of electroplated diamond micro - grinding tool in soda lime glass grinding. Int. J. Adv. Manuf. Technol. **91**, 3467–3479 (2017)
18. Wang, J., Qian, J., Ferraris, E., Reynaerts, D.: In-situ process monitoring and adaptive control for precision micro-EDM cavity milling. Precis. Eng. **47**, 261–275 (2017)
19. Griffin, J.M., Diaz, F., Geerling, E., Clasing, M., Ponce, V., Taylor, C., Turner, S., Michael, E.A., Mena, F.P., Bronfman, L.: Control of deviations and prediction of surface roughness from micro machining of THz waveguides using acoustic emission signals. Mech. Syst. Signal Process. **85**(15), 1020–1034 (2017)
20. Patra, K., Jha, A., Szalay, T., Ranjan, J., Monostori, L.: Artificial neural network based tool condition monitoring in micro mechanical peck drilling using thrust force signals. Precis. Eng. **48**, 279–291 (2017)

Unsupervised Classification of Medical Documents Through Hybrid MWEs Discovery

Alessandro Maisto[✉]

University of Salerno, Via Giovanni Paolo II, 132, 84084 Fisciano, (SA), Italy
amaisto@unisa.it

Abstract. The automatic processing of medical language represents a clue for computational linguists due to intrinsic feature of these sub-codes: its lexicon comprises a vast number of terms that appear infrequently in texts. In addition, the presence of many sub-domains that can coincide in a single text complicates the collection of the training set for a supervised classification task. This paper will tackle the problem of unsupervised classification of medical scientific papers based on a hybrid Multiword Expression Discovery. We apply a morpho-semantic approach to extract medical domain terms and their semantic tags in addition to the classic MWEs discovery strategies. The collected MWEs will be used to vectorize texts and generate a network of similarities among corpus documents. With this approach, we try to solve both problems caused by the medical domain features. The presence of a vast lexicon of low-frequency terms is dealt with by extracting many semantic tags with a small dictionary; the issues of co-occurring sub-domains are solved by generating clusters of similarity values instead of a rigid classification.

1 Introduction

Text classification is a common task in Computational Linguistics and Machine Learning. In the last few years, supervised classification models have reached high accuracy levels in classifying text belonging to many different domains. Nevertheless, medicine's technical-scientific language represents an exception for many statistical and machine learning algorithms as much in text classification as in other computational linguistics tasks. In fact, the medical domain lexicon is characterised by a number of technical lemmas larger than that of other domains. In addition, those lemmas, which can be organised in taxonomies, sub-domains, and notional fields, appear in technical-scientific texts with low frequency values [1]. A significant number of words from the medical domain lexicon could be defined as "rare event" [2]. Free linguistic resources for the medical domain such as taxonomies, ontologies, dictionaries, or *vade mecum* are uncommon and limited. Furthermore, scientific medical texts often mix many sub-domains, making difficult a classification based on pre-defined classes.

L. Barolli et al. (Eds.): AINA 2021, LNNS 227, pp. 52–61, 2021.
https://doi.org/10.1007/978-3-030-75078-7_6

In this work, we propose a hybrid approach for the unsupervised classification of medical scientific documents. The idea is to address the large number of medical terms, starting from a restricted number of morphemes that, if combined, allow the recognition of a considerable number of terms. This approach, called "morpho-semantics", allows the identification of those words that belong to the same sub-domain (e.g., *mio-cardio* and *cardio-pathy* belong to the sub-domain of "cardiology"). The morphemes included in our dictionary represent a set of formative elements that possess semantic self-sufficiency [3]. Thanks to a set of combination rules, we employ morphemes to build a list of tagged medical terms. Finally, we use syntactic patterns to recognise domain Multiword expressions [4,5].

We combined the Rule-based approach with a statistical Multiword extraction in order to vectorize the documents. We subsequently compared vectorized documents to generate correlation values and build a network of texts to identify clusters or document communities. Section 2 presents a brief state of the art about unsupervised classification, Multiword extraction, and Morpho-semantics; in Sect. 3, we present our methodology; Sect. 4 will illustrate our experimentation and the obtained results; Sect. 5 contains the conclusions.

2 State of Art

The unsupervised algorithms make it possible to identify the structure of unlabelled data and to explore text patterns perhaps understudied or unknown [6]. Unsupervised approaches are based on the inferences that are carried out by clustering data into different clusters without labeled responses [7]. [8], for example, uses the Pointwise Mutual Information of the document's words with a positive or negative reference word to assign a polarity label to unlabelled reviews. [9] use sets of keywords for each target category and search them into the text to assign a label. [10] and [11] model the unsupervised classification task as a similarity problem between two sets of words, a set of keywords for each category, and the text. [12] build networks of similarity between texts to utilize network classification algorithms to classify texts without a set of pre-edited categories. The use of a specialized lexicon to give structure to unstructured texts with classification purposes could improve the classification task [13]. In particular, the presence of Multiword Expressions (MWE) helps the semantic analysis of specialistic documents [14,15]. They could represent 90% of words' characteristic of a specialized language [16]. MWEs make it possible to the summarise the meaning of the text in which they occur, representing a list of text's keywords [17]. We employ two approaches for MWEs extraction in this work: a morpho-semantic approach and a statistical approach [18]. Morpho-semantic approaches have already been applied to the medical domain. [19] employed a Morpho-Semantic approach to identify and transform terminal morphemes in the English medical dictionary. [20] proposed a classification of the medical lexicon based on formative elements of Latin and Greek origin; [21] classify the terms of diseases ending in -itis; [22] propose the same approach on the terminology of surgical operations ending in -ectomy or -stomy; [23] on the suffix -osis. [24], propose a morpho-semantic strategy to build medic taxonomy in the Italian language. [1] present an automatic translation of terms of rare diseases based on a multi language dictionary of medical morphemes. Statistic MWEs discovery is traditionally based on the extraction of candidates among the n-grams of a text and the classification of those n-grams in MWEs and non-MWEs. The first step, the extraction of candidates, could be performed both with pattern-based and frequency-based approaches. Pattern-based approaches are used when the MWE POS tags are restricted [25]. Frequency-based techniques extract all the n-grams that exceed a given threshold [26,27]. The classification of the extracted n-grams is performed using Association Measures [28,29] such as Log-Likelihood ratio [30], T-score [31], or Pointwise Mutual Information [32], also for n-grams with more than two components [33].

3   Classification Methodology

This paper presents an unsupervised classification methodology that takes advantage of a hybrid MWEs Discovery method to vectorize documents. We employed a Morpho-semantic strategy for the extraction of medical-specialised MWEs and their sub-domain tags. In addition, we supplement the MWEs database with a MWEs discovery methodology based on the PMI Association Measure. After extracting the MWEs, we build a vectorized version of each document based on MWEs frequency in texts. We also generate a Cosine Similarity score for each text vectors pairs to build a network of similarities. The final classification has been performed through the application of a Modularity Class algorithm over the generated network. The pipeline of our methodology is the following:

1. Syntactic Parsing of Texts
2. Morpho-semantic extraction of medical domain lexicon
3. Morpho-semantic-based MWEs discovery
4. PMI MWEs discovery
5. Text vectors similarity extraction
6. Classification of documents

3.1   Hybrid MWEs Discovery

The medical domain lexicon includes a large number of specialistic terms and an even more significant number of specialistic MWEs. Because of this feature, the extraction of semantically tagged MWEs from medical texts should require extensive electronic dictionaries.

The morpho-semantic approach provides reasonably good results with the use of a restricted dictionary. It allows the analytical description of the meaning of the words included in the same "Morphological Family" [34]. With this app-roach, we can infer the sub-domain of the extracted word by extending a single morpheme's tags to the complete word. For example, the morphemes *acusia-*, *-oma*, *myco-* or *gastro-* can make up a vast number of terms such as "hypera-cusia", "carcinoma", "gastritis", "mycosis". Those terms can be easily related to the medical sub-domains "otolaryngology", "oncology", "gastroenterology", and "dermatology".

The choice of a morpho-semantic strategy is suitable for two main aspects: first, we can assign tags to a vast number of terms starting from a small dictio-nary. Those tags can be used for label texts or categories once the classification has been performed. In addition, this approach ensures that single words and compound words related to them belong to the medical domain without the need for probabilistic calculations.

The first step of the morpho-semantic MWEs discovery is the extraction of single domain words and their tags. To extract the single words, we use a dictionary of about 1000 morphemes subdivided into four classes:

- Confixes (CFX): neoclassical formative elements with a full semantic value (i.e. *mammo-*, *-cephalia*);
- Pre-suffixes Confixes (CFXS): includes the same elements as Confixes but, in some cases, without the last letter(s) (i.e. *mamm-*, *cephali-*);
- Prefixes (PFX): morphemes that appear in the first part of the word and can connote it with a specific meaning (i.e. *-hypo*, *-hyper*);
- Suffixes (SFX): morphemes that appear in the final part of the word and can connote a specific meaning (i.e. *-oma*, *-ite*) or syntactic class (i.e. *-ary*, *-able*).

A set of morphological patterns combines the morphemes and searches for words that belong to the medical domain. The identified terms must include two or more dictionary morphemes but no other morphological items. Patterns search through the lemmas generated by a previous parsing stage, performed by *Core-nlp* parser [35]. The list of the morphological patterns is the following:

- CFX-CFXS-SFX (i.e. *Athero-scler-osis*)
- CFX-CFX (i.e. *Cephalo-pathy*)
- CFX-CFX-CFXS-SFX (i.e. *Bronco-pneumo-path-ic*)
- PFX-CFX (i.e. *Hyper-acusia*)
- PFX-CFX-CFX (i.e. *Peri-cardio-pathy*)
- PFX-CFX-CFXS-SFX (i.e. *Hemi-megal-enceph-aly*)
- CFXS-SFX (i.e. *cephal-itis*)

Using the extracted terms as markers, we started a medical pattern-based MWEs discovery that includes Nouns and Adjectives principally. The discovery was led by the medical terms, which act as a marker. The patterns used, applied to the parsed text are depicted in Fig. 1.

Fig. 1. Patter used in the medical MWEs discovery

The first Pattern catches all Nouns and Adjectives that points to the marker (i.e. "interventional cardiology"); the second Pattern captures the sequence Noun-Preposition-Noun in which the marker did not represent the head of the phrase (i.e. "treatment of cardiopathy"); the third Pattern identifies all the nouns and adjectives that point to the same Noun pointed by the marker (i.e. "coronary artery bypass").

Morpho-semantic MWEs discovery could, in some documents produce a small number of results. In order to improve the dimensionality of the MWEs vectors, we performed, in parallel, a MWEs discovery based on n-grams extraction. We extracted bigrams and trigrams that didn't include punctuation symbols and calculated the *Association Ratio* [32], or Pointwise Mutual Information, for each bigram (1), extending the formula to the trigrams (2) [36]:

$$I(x, y) = \log_2 \frac{P(x, y)}{P(x)P(y)} \quad (1) \qquad I(x, y, z) = \log_2 \frac{P(x, y, z)}{P(x)P(y)P(z)} \quad (2)$$

We extracted single words, bigrams, and trigrams frequency from the entire corpus with no punctuation. The threshold for bigrams has been set to 7, while the threshold for trigrams to 15.

We evaluated the task of MWEs discovery on a sample of 10 texts. The overall precision achieved is 0.607. The MWEs extraction based on PMI obtains precision values ranging between 0.48 to 0.60, while the Morpho-semantic-based Discovery reaches values between 0.85 to 0.95.

3.2 Document Classification

Once the MWEs have been extracted, we generate a vector for each text. The vectors' dimension corresponds to the total number of MWEs extracted from the corpus, and the values represent the number of occurrences of each MWE in the selected text. We also included into the vectors all the medical tags extracted. The classification starts comparing vectors to generate similarity values among texts. As similarity measure, we used the Cosine Similarity. The result was a document to document-similarity table that could be read as a weighted edge graph. The final classification has been generated applying the *Modularity Class* algorithm [37] to the graph. With this approach, we generate a class attribution for each document based on the distance among all the corpus documents. Moreover, we can use the similarity values to select texts that are halfway between two classes.

In the next section, we will show the methodology's results on a corpus of medical papers.

4   Experiment

To test the proposed methodology, we collect a Corpus of Medical Papers extracted from the *British National Corpus* (BNC) [38] or manually extracted from the free-to-read papers proposed by Google Scholar. In particular, the manually extracted papers refer to particularly 'hot' medical subject, the COVID-19 Pandemic. From the BNC, we select a set of newspaper articles about AIDS, Smoking and general health and from the Health Pages of The Independent, as well as scientific papers from the British Medical Journal. The topics selected range from COVID-19, AIDS, cancer, cardiology, diabetes, National Health System, dementia, and hepatitis. The total number of selected documents is 74. The paper's bibliographies have been excluded from our analysis.

The Modularity Class algorithm has been launched over the network with a Resolution parameter of 0.4 and produces 15 communities or categories. The network shown in Fig. 2 shows the generated graph: the node's colour represents that node's affiliation with a specific class. The titles of the paper have been shortened and they retaining their original meaning. The distance has been calculated with the application of a *ForceAtlas2* Layout Algorithm [39]. ForceAtlas2 is a force-direct layout that uses weights to simulate physical attraction to spatialise a network. Weighted edges give their nodes an attraction force while, in general, nodes repel each other.

We analyse the generated communities in order to evaluate the precision of the classification. We studied the general sense of each paper of the corpus and tried to find commonalities among the same class documents. Table 1 contains a list of the classes with the proposed common thread and the number of document they include.

The evaluation of the results has been based on the presence of consistent documents in the classes. With this evaluation technique, our classifier has a precision of 0.89, with ten classes composed of only compatible texts.

Fig. 2. Document's network and identified communities

Table 1. Classes composition and common thread

Classes	Common thread	No. Documents
0	Sexually transmitted disease	5
1	Infants + Glue ear	6
2	Covid-19	8
3	Cardiac and Pulmonary disease	4
4	Covid-19 Diffusion	3
5	Virus disease therapies	3
6	Family Inheritance	3
7	Cardiac disease + Health structures	5
8	Inheritance	5
9	Cancer	5
10	Diagnosis + mortality rate	4
11	Diabetes	6
12	Respiratory disease	3
13	Health structure	3
14	Smoking	9

5   Conclusion and Future Work

In this paper, we tested an unsupervised classification methodology based on MWEs extracted with a hybrid strategy. We employ a standard MWEs discovery based on PMI and a MWEs extraction based on morpho-semantics. Due to the feature of the medical domain, our approach is suitable for two reasons. The first being the domain lexicon, composed of many terms that appear with low frequencies. Morpho-semantics allows for the treatment of a significant number of specific words with a small dictionary and with excellent results (the MWEs discovery based on morpho-semantics boasts a 93% precision rate). The second reason being the unsupervised classification based on Modularity Class. By generating similarity values between texts and considering the results as a network, we can automatically calculate communities of similar documents. In addition, we keep track of documents that are partially related to a class or are in the middle between two or more categories. For example, this is a big area in the graph in which COVID papers are concentrated. The paper about "Covid and Obesity" appears near an article about "Nutrition and Diabetes". Instead, a paper about "Renal failure" is far from the centre because it differs from all other documents. In future works, we will implement a more effective association ratio on MWEs discovery and develop a labelling system for automatically generated classes[1].

References

1. Maisto, A., Guarasci, R.: Morpheme-based recognition and translation of medical terms. In: Okrut, T., Hetsevich, Y., Silberztein, M., Stanislavenka, H. (eds.) International Conference on Automatic Processing of Natural-Language Electronic Texts with NooJ, pp. 172–181. Springer, Cham (2015)
2. Möbius, B.: Rare events and closed domains: two delicate concepts in speech synthesis. Int. J. of Speech Technol. **6**(1), 57–71 (2003)
3. Iacobini, C.: Composizione con elementi neoclassici. La formazione delle parole in italiano. Tübingen: Max Niemeyer Verlag, pp. 69–95 (2004)
4. Balzano, W., Del Sorbo, M.R.: Genomic comparison using data mining techniques based on a possibilistic fuzzy sets model. Biosystems **88**(3), 343–349 (2007)
5. Amato, A., Balzano, W., Cozzolino, G., Moscato, F.: Analysis of consumers perceptions of food safety risk in social networks. In: Barolli, L., Takizawa, M., Xhafa, F., Enokido, T. (eds.) International Conference on Advanced Information Networking and Applications, pp. 1217–1227. Springer, Cham (2019)
6. Grimmer, J., Stewart, B.: Text as data: the promise and pitfalls of automatic content analysis methods for political texts. Polit. Anal. **21**(3), 267–297 (2013)
7. Thangaraj, M., Sivakami, M.: Text classification techniques: a literature review. Interdisc. J. Inf. Knowl. Manag. **13**, 117–135 (2018)
8. Turney, P.D.: Thumbs up or thumbs down? semantic orientation applied to unsupervised classification of reviews. arXiv:preprint cs/0212032 (2002)

[1] This paper has been produced with the financial support of the Project financed by Campania Region of Italy 'REMIAM - Rete Musei intelligenti ad avanzata Multimedialita'. CUP B63D18000360007.

9. Ko, Y., Seo, J.: Automatic text categorization by unsupervised learning. In: COL-ING 2000 Volume 1: The 18th International Conference on Computational Linguistics (2000)
10. Yang, L., Li, C., Ding, L., Li, Q.: Combining lexical and semantic features for short text classification. Procedia Comput. Sci. **22**, 78–86 (2013)
11. Miller, T., Dligach, D., Savova, G.: Unsupervised document classification with informed topic models. In: Proceedings of the 15th Workshop on Biomedical Natural Language Processing, pp. 83–91 (2016)
12. Maisto, A., Pelosi, S., Stingo, M., Guarasci, R.: A hybrid method for the extraction and classification of product features from user generated contents. Lingue e Linguaggi **22**, 137–168 (2017)
13. Catone, M.C., Falco, M., Maisto, A., Pelosi, S., Siano, A.: Automatic text classification through point of cultural interest digital identifiers. In: Barolli, L., Hellinckx, P., Natwichai, J. (eds.) International Conference on P2P, Parallel, Grid, Cloud and Internet Computing, pp. 211–220. Springer, Cham (2019)
14. Amato, F., Boselli, R., Cesarini, M., Mercorio, F., Mezzanzanica, M., Moscato, V., Persia, F., Picariello, A.: Challenge: processing web texts for classifying job offers. In: Proceedings of the 2015 IEEE 9th International Conference on Semantic Computing (IEEE ICSC 2015), pp. 460–463. IEEE (2015)
15. Amato, F., Casola, V., Mazzocca, N., Romano, S.: A semantic approach for fine-grain access control of e-health documents. Log. J. IGPL **21**(4), 692–701 (2013)
16. Elia, A., Cardona, G.R.: Discorso scientifico e linguaggio settoriale. un esempio di analisi lessico-grammaticale di un testo neuro-biologico. Quaderni del Dipartimento di Scienze della Comunicazione–Università di Salerno, Cicalese A., Landi A., Simboli, linguaggi e contesti (2) (2002)
17. Bolasco, S., et al.: Statistica testuale e text mining: alcuni paradigmi applicativi. Quaderni di Statistica **7**, 17–53 (2005)
18. Amato, F., Casola, V., Mazzeo, A., Romano, S.: A semantic based methodology to classify and protect sensitive data in medical records. In: 2010 Sixth International Conference on Information Assurance and Security, pp. 240–246. IEEE (2010)
19. Pratt, A.W., Pacak, M.: Identification and transformation of terminal morphemes in medical English. Methods Inf. Med. **8**(2), 84–90 (1969)
20. Wolff, S.: The use of morphosemantic regularities in the medical vocabulary for automatic lexical coding. Methods Inf. Med. **23**(4), 195–203 (1984)
21. Pacak, M.G., Norton, L.M., Dunham, G.S.: Morphosemantic analysis of-ITIS forms in medical language. Methods Inf. Med. **19**(2), 99–105 (1980)
22. Norton, L.M., Pacak, M.G.: Morphosemantic analysis of compound word forms denoting surgical procedures. Methods Inf. Med. **22**(1), 29–36 (1983)
23. Dujols, P., Aubas, P., Baylon, C., Grémy, F.: Morpho-semantic analysis and translation of medical compound terms. Methods Inf. Med. **30**(1), 30 (1991)
24. Amato, F., Mazzeo, A., Elia, A., Maisto, A., Pelosi, S.: Morphosemantic strategies for the automatic enrichment of Italian lexical databases in the medical domain. Int. J. Grid Util. Comput. **8**(4), 312–320 (2017)
25. Baldwin, T.: Deep lexical acquisition of verb-particle constructions. Comput. Speech Lang. **19**(4), 398–414 (2005)
26. Biber, D., Conrad, S., Cortes, V.: If you look at...: lexical bundles in university teaching and textbooks. Appl. Linguist. **25**(3), 371–405 (2004)
27. Brooke, J., et al.: Unsupervised multiword segmentation of large corpora using prediction-driven decomposition of n-grams. In: COLING 2014, the 25th International Conference on Computational Linguistics: Technical Papers (2014)

28. Manning, C., Schutze, H.: Foundations of Statistical Natural Language Processing. MIT Press, Cambridge (1999)
29. Evert, S., et al.: E-view-affilation–a large-scale evaluation study of association measures for collocation identification. In: Proceedings of eLex 2017–Electronic Lexicography in the 21st Century: Lexicography from Scratch (2017)
30. Dunning, T.E.: Accurate methods for the statistics of surprise and coincidence. Comput. Linguist. **19**(1), 61–74 (1993)
31. Church, K., Gale, W., Hanks, P., Hindle, D.: Using statistics in lexical analysis. In: Lexical Acquisition: Exploiting On-line Resources to Build a Lexicon, pp. 115–164 (1991)
32. Church, K., Hanks, P.: Word association norms, mutual information, and lexicography. Comput. Linguist. **16**(1), 22–29 (1990)
33. Petrović, S.: Collocation extraction measures for text mining applications (2007)
34. Jacquemin, C., Tzoukermann, E.: NLP for term variant extraction: synergy between morphology, lexicon, and syntax. In: Strzalkowski, T. (ed.) Natural Language Information Retrieval, pp. 25–74. Springer, Dordrecht (1999)
35. Manning, C.D., Surdeanu, M., Bauer, J., Finkel, J., Bethard, S.J., McClosky, D.: The Stanford CoreNLP natural language processing toolkit. In: Association for Computational Linguistics (ACL) System Demonstrations, pp. 55–60 (2014)
36. Lyse, G.I., Andersen, G.: Collocations and statistical analysis of n-grams. Exploring Newspaper Language: Using the Web to Create and Investigate a Large Corpus of Modern Norwegian, ser. Studies in Corpus Linguistics, pp. 79–109. John Benjamins Publishing, Amsterdam (2012)
37. Blondel, V.D., Guillaume, J.-L., Lambiotte, R., Lefebvre, E.: Fast unfolding of communities in large networks. J. Stat. Mech: Theory Exp. **2008**(10), P10008 (2008)
38. Leech, G.N.: 100 million words of English: the British National Corpus (BNC) (1992)
39. Jacomy, M., Venturini, T., Heymann, S., Bastian, M.: ForceAtlas2, a continuous graph layout algorithm for handy network visualization designed for the Gephi software. PLoS ONE **9**(6), e98679 (2014)

Building a Pos Tagger and Lemmatizer for the Italian Language

Alessandro Maisto[1(✉)] and Walter Balzano[2]

[1] University of Salerno, Via Giovanni Paolo II, 132, 84084 Fisciano, (SA), Italy
amaisto@unisa.it
[2] University of Naples Federico II, Via Claudio, 21, 80125 Napoli, (NA), Italy
walter.balzano@unina.it

Abstract. In this work, we present two modules for a python open-source library for the analysis of the Italian language. The modules include a Pos tagger based on Averaged Perceptron Tagger and a Lemmatizer, based on the vast collection of linguistic data held by the Department of Politics and Communication Science of the University of Salerno. While the Averaged Perceptron Tagger algorithm is mostly used for the the English language from famous python libraries such as NLTK or Spacy, the Lemmatizer represents an entirely original module that relies on a vast electronic dictionary characterized by the presence of syntactic, morphological, and semantic tags. We present our approach and a preliminary experiment in which we compare our module results with the results of another widely used Pos-tagger and Lemmatizer as Tree-Tagger.

1 Introduction

In this work, we present two new components of an open-source Natural Language module (NLP) for the analysis of the Italian language: a Part-Of-Speech Tagger and a Lemmatizer, two of the core tools for the analysis of natural language.

Part-Of-Speech (POS) tagging is one of the most important and used pre-processing steps in Natural Language Processing task; it is considered the necessary baseline for every further linguistic analysis [1]. It consists of attributing to each wordform its grammatical category and disambiguates that terms which could be considered ambiguous in the selected language. Although there are several available English language resources, the number of tools currently available for the Italian language is drastically reduced. The situation is even more reduced when we consider only open-source tools. Due to this deficiency of open-source resources, we propose an averaged perceptron POS Tagger, based on an algorithm widely used in most common python libraries for English language (NLTK[1], Spacy[2]). As it regards the Lemmatizer, instead, we assume that

---

A. Maisto edited Sects. 1, 2, 3, 4, 5; W. Balzano collaborated in the project.

[1] http://www.nltk.org/.

[2] https://spacy.io/.

© The Author(s), under exclusive license to Springer Nature Switzerland AG 2021
L. Barolli et al. (Eds.): AINA 2021, LNNS 227, pp. 62–71, 2021.
https://doi.org/10.1007/978-3-030-75078-7_7

a morphologically rich language like Italian requires a lexicon-based approach to cope with the richness of word forms and to provide the best precision. So we propose a Lemmatization module that takes advantage of the considerable amount of linguistic data provided by the Italian Electronic Dictionary developed in the Department of Political and Communication Sciences of the University of Salerno. In the Sect. 2 we briefly present some "state-of-art" Postaggers and Lemmatizers; in Sect. 3 our modules are presented; in Sect. 4 is shown a comparison between our modules and two of most used free resources for Italian language Pos Tagging and Lemmatization. Finally, in Sect. 5 we present some possible future researches.

2   Related Works

Since the '70s automatic tag of Part-Of-Speech was considered an essential application for future computational linguistics research.

[2] propose a rule-based approach in the TAGGIT program and reach the 77% of precision in disambiguated the Brown Corpus [3].

[4] presents a program that uses a linear-time dynamic programming algorithm to find POS assignment to words based on lexical and contextual probability with an estimated precision of 95–99%. The stochastic method first calculates Lexical Probability by dividing the frequency of a word for the same word's frequency with a specific POS tag. The stochastic method proposed in the paper makes use of lexical disambiguation rules. Then, the model calculated the contextual probability: *The probability of observing part of speech X given the following two parts of speech Y and Z, is estimated by dividing the trigram frequency XYZ by the bigram frequency YZ.*

In 1992, [5] present a POS tagger based on Hidden Markov Model (HMM). He defines five features that a Part-Of-Speech tagger requires. A POS Tagger must be:

- **Robust** because it must work with text corpora that contain ungrammatical constructions, isolated phrases, unknown words, or non-linguistic data.
- **Efficient** because it must tag the largest number of words in the shortest possible time.
- **Accurate** because it should assign the correct POS tag to every word encountered.
- **Tunable** as meaning that it should take advantage of linguistic insights.
- **Reusable** as the effort required to retarget the tagger to new corpora, tagsets, or languages should be minimal.

Authors use HMM because this kind of model permits complete flexibility in the choice of training corpora, tagsets, or languages, reducing time and complexity. This model reaches 96% of precision on Brown Corpus.

An example of a Rule-based POS tagger was presented by [6]. In this method, the algorithm automatically acquires the rules and reaches an accuracy comparable to stochastic taggers. Once the algorithm assigns to each word the most likely tag, estimated by examining a large tagged corpus without regards to context, it improves its performances using two procedures:

a. unknown words capitalized are considered as Proper Nouns;
b. a procedure attempts to assign the tag most common for words ending in the same three letters.

In a second step, the tagger acquires patches to improve performance. Patch templates are based on context, lexical properties, and distribution region.

The importance of context in POS tagging is also underlined by [7], which uses probability and some contextual features parallelly. Non-rare words (words that occur more than five times in the corpus) are tagged using the simple probability model. Rare and unknown words, on the contrary, are tagged as words with similar prefixes or suffixes. Tested on the Wall St. Journal corpus, the model obtains a total accuracy of about 95%.

Another statistical model could be found in [8–10].

Currently, Pos Tagging is essentially considered a "solved task", with state-of-the-art taggers achieving a precision of 97%–98% [11,12].

The best algorithms are considered the Stanford Tagger (version 2.0), which uses the maximum entropy cyclic dependency network [13]; BI-LSTM-CRF [14], and NLP4J [15] that use Dynamic Feature Induction. In [16], a new perceptron algorithm was presented for training tagging models as an alternative to maximum entropy models. The Averaged Perceptron Tagger uses a discriminative, feature-rich model. Features are modeled using *feature functions*: $\phi(h_i, t_i)$ when h_i represents the history and t_i the tag. History h_i is a complex object modeling different aspects of the sequence being tagged. It contains previously assigned tags and other contextual features such as the form of the current word [17–19].

For a sequence of words w of length n in a model with d feature functions, the scoring function is defined as:

$$score(w, t) = \sum_{i=1}^{n} \sum_{s=1}^{d} \alpha_s \phi_s(h_i, t_i)$$

with α_s as feature weights paired with a feature function ϕ_s. The Viterbi algorithm is used to compute the highest scoring sequence of tags.

[20] proposes a new implementation of Averaged Perceptron Tagger highlighting some behaviors as the algorithm's aversion to the excess of information or too complex features and establishing that the maximum accuracy usually came between the 4th and the 8th interaction. In this implementation, a set of morphological features are added to the standard perceptron tagger implementation in order to reflect the complexity of Czech and, in general, Slavic languages. Based on [20], [21] propose a different set of features for the English and Czech language due to the typological difference between the two idioms.

For what concern Lemmatization task, The lemmatization task consists of converting a token into its common base form the lemma. Most algorithms use external lexical information, as *Morfette* [22], or *SEM* [23] for French texts.

In particular, Morfette uses a probabilistic system to joint morphological tagging and lemmatization from morphologically annotated corpora. Morfette system is composed of two learning modules, one of that for lemmatization. A decoding module searches for the best sequence of pairs of morphological tags and lemmas for an input sequence of word forms. The class assigned to a wordform - lemma pair is the corresponding shortest edit script (SES) between the two reversed strings. SES represents the shortest sequence of instructions which transforms a string w into a string w'. The precision of Morfette on the lemmatization task is 93% for Romanian, 96% for Spanish, and 88% for the Polish language.

[24] use a training Dictionary to search the longest common substring of the full form and the lemma. The system generates lemmatization rules determining the derivation rules that lead to flexed forms using the dictionary.

Speaking about Italian resources, we can cite an Italian version of TreeTagger [25], an Italian model for OpenNLP [26], TagPro [27], CORISTagger [28], Tanl POS tagger [29], ensemble-based taggers [30] and Pattern tagger [31].

3   Proposed Approach

As we mentioned above, our work includes two modules: a Postagger Module and a Lemmatization Module. Both are written in Python 2.7 and are composed of two Python Classes, some dictionary files, and a ".pickle" in which semi-supervised generated data are stored.

The Pos Tagger module, called *Mr. Tag*, is based on an implementation of an Averaged Perceptron Tagger[3] for the English language. We start expanding the set of features used by the original model to make them more suitable for the Italian language. Then, we perform a semi-supervised training phase on a gold-standard corpus: we train the pos tagger both on a ten million-word as on a one million-word tagged corpus extracted from the "Paisá Corpus" [32], a recollection of over 380.000 different documents belonging to over 1.000 different web pages. The introduced features are focused on morphosyntactical differences between English and Italian. Our Feature Set is shown in Table 1.

---

[3] https://github.com/sloria/textblob-aptagger.

Table 1. Set of features used for *Mr Tag* PosTagger

Type	Features
Morphology	First character of present word
	Last 4 characters of present word
Syntax	Previous word tag
	Second previous word tag
	Two previous word tags
	Next word tag
	Second next word tag
	Two next word tags
	Previous and next word tags
	Two previous and two next word tags
Context	Present word
	Next word
	Previous word
	Two previous words
	Two next words
	Next and previous word
	Suffix of previous word
	Prefix of previous word
	Second previous word
	Second next word

Since perceptron-based taggers are mainly based on language-dependent features, we started looking for features effective for the Italian language. To select representative features, we performed a morphological analysis and notice that, on average, Italian words are longer than English ones [16]. So we choose to take into account the last 4 characters of each word (i.e. Italian adverbs, normally end with the suffix *mente* and in verbs declinations we could have suffix as *asse* or *ebbe* in construction like *mangi-a-sse* or *dorm-i-rebbe*), and one character in the case of prefixation phenomena.

Instead, syntactic and context features have been expanded by considering a bigger context for each word, including one or two words before and after the target word; this is due to the extensive use of determiners and distant dependencies that characterize the Italian language.

The POS Tagger, enriched with this set of features, has been trained on a silver standard corpus [33], the "Paisá Corpus", a large Italian Corpus (250 million tokens) of text extracted from web pages and annotated in CoNLL[4], with ten iterations. To verify how the pos tagger's performance change when the corpus

---

[4] CONLL.

dimension change, we use two different portions of the corpus of respectively 1 and 10 million-words. We used the "DELA" Tag Set [34,35] that includes:

- Names: N
- Verbs: V
- Adjectives: A
- Adverbs: AVV
- Determiners: DET
- Prepositions: PREP
- Pronouns: PRON
- Conjunctions: CONG
- Numerics: NUM
- Interjections: INTER
- Others: X

The Lemmatization module, called *Mr. Lemmi*, is based on a set of dictionaries annotated in "DELA", particularly on DELAF, the Italian Electronic Dictionary of Flexed Forms. DELAF includes over 1 million flexed Italian forms, and it has been divided into six sub-dictionaries to improve the algorithm's performance. *Mr. Lemmi* uses a different dictionary for each part of speech (except for Adverbs). Lemmatizer works searching the pair token-tag through a set of dictionaries. The POS Tag helps to select the correct dictionary where to find the token. For invariable part-of-speech such as adverbs, interjection, or conjunctions, the system chose as lemma directly the token. Dictionary entries are composed of four parts: the token, the lemma, the POS or grammatical category and a list of syntactic and semantic properties separated by the "+" symbol. An example of a dictionary entry is the following:

- Token: *poliziotti* - "policemen"
- Lemma: *poliziotto* - "policeman"
- POS: *N* - Noun
- Properties: $m+p+Human$ - masculine, plural, human noun

To avoid Lemmatization errors and correct Pos Tagging errors, if a word is not present on a variable of its specific tag, a new iteration searches the word in all variables relative to other tags. In this way, the Lemmatizer can correct some pos tagging errors, such as when an unambiguous token is wrongly associated with a POS. In this way, if the word *perché*, "why", was wrongly tagged as a Noun, and the first iteration is unable to find it in the Noun variable, the second iteration search it in all other tag variables and can find it in Conjunction variable and then, correct the tag. If a word is not present in any other variable, then the algorithm assigns it the original tag from the pos tagger and uses the token as the lemma. If the word were *hobbit*, which is not present in the DELAF dictionary, after the second iteration, the resulting lemma would be *hobbit*. In the case of ambiguous tokens such as *pesca* which in Italian can belong to the Verbs (*to fish*; *Antonio pesca ogni domenica* - "Antonio fishes every Sunday") or to Nouns (*peach*; *ho mangiato una succosa pesca* - "I ate a juicy peach") the Lemmatizer is not able to find the error.

4 Experiment and Evaluation

Both POS Tagger and the Lemmatizator have been tested on a gold standard of 10.000 tokens extracted by the ItWak corpus and manually verified.

To evaluate *Mr. Tag* and *Mr. Lemmi*, we decided to compare their performances with Tree-Tagger[5] [25], one of the most well-known tools for annotating text with part-of-speech and lemma information.

We compared the tools evaluating the results obtained from the analysis on the same data portion. An overview of the comparison is presented in Tables 2 and 3.

Table 2. Comparison of POS tagging task

Algorithms	Precision
Mr.Tag 1M	0,923
Mr.Tag 10M	0,969
Tree-Tagger	0,907

Table 3. Comparison of lemmatization task

Algorithms	Precision	Time
Mr. Lemmi	0,973	20.52 s
Tree-Tagger	0,961	2,54 s

As shown by Tables 2 and 3, our modules reach excellent results in terms of precision for both tasks, overcoming Tree-Tagger. Conversely, for what concerns the time, we reach a good result with the POS Tagger module, but the Lemmatizator still requires many adjustments. Tree-Tagger performs both tasks concurrently, so the time expressed by Table 3 is the summed time of POS Tagging and Lemmatization.

Since the POS Tagger trained on one million-word corpus reached the lowest results, we did not apply lemmatization. Finally, we analyze in a qualitative way the results of the Lemmatization phase. The vast majority of errors derives from an incorrect part-of-speech tagging in ambiguity cases (i.e. *regola,N*, lemmatized as *regolare*, "to rule", instead of *regola,N* that could be lemmatized as *regole,N*, "rule"), or not (i.e. *seconda,NUM* that must be lemmatized as *secondo,NUM* "second").

Both for Tree-Tagger and our system, the pos tagging phase results were lower than the lemmatization phase results. In many cases, the lemma with the wrong POS Tag coincides with the correct one. In some cases, the mechanism of cross-checking implemented in our Lemmatizer corrects some gross errors. For example, the word *onnipresenti*, "omnipresent" tagged as Verb, has been corrected as Adjective. The same happens with *arboree*, "arboreal" initially tagged as Noun and then corrected as Adjective.

As concerns the Mr. Tag pos tagger trained on one million-word corpus, it reaches the lowest values with many Adjective, Noun, or Verb classification errors. Some example of these errors includes the verb *è*, "is", tagged as Noun, adjective *francese*, "french" or *basso*, "short" considered ad Nouns, or the Noun *pianoforte* tagged as Adjective.

---

[5] http://www.cis.uni-muenchen.de/~schmid/tools/TreeTagger/.

5   Conclusions and Future Work

In this work, we presented the development of a novel Italian Pos Tagger and Lemmatization modules able to compete with other freely available systems. The POS Tagging module is based on Averaged Perceptron Tagger and is trained on the silver-standard corpus Paisà. We train the POS Tagger with one million-words and ten million-word portions of the corpus to study how the results change when the training set dimension changes. The system reaches precision values comparable to other famous models as Tree-Tagger but with more computation time. In addition, since the Lemmatization module is based on a large set of dictionaries, we implement a cross-check mechanism to correct the POS tagging errors related to unambiguous words.

As a next step, we plan to improve the algorithm's performances by converting dictionaries of flexed forms directly in a more convenient data structure to minimize the time for retrieving data.

References

1. Amato, A., Balzano, W., Cozzolino, G., Moscato, F.: Analysis of consumers perceptions of food safety risk in social networks. In: Barolli, L., Takizawa, M., Xhafa, F., Enokido, T. (eds.) International Conference on Advanced Information Networking and Applications, pp. 1217–1227. Springer, Cham (2019)
2. Greene, B.B., Rubin, G.M.: Automatic grammatical tagging of English. Department of Linguistics. Brown University (1971)
3. Francis, W., Kucera, H.: Frequency analysis of English usage (1982)
4. Church, K.W.: A stochastic parts program and noun phrase parser for unrestricted text. In: Second Conference on Applied Natural Language Processing, pp. 136–143. Association for Computational Linguistics (1988)
5. Cutting, D., Kupiec, J., Pedersen, J., Sibun, P.: A practical part-of-speech tagger. In: Proceedings of the Third Conference on Applied Natural Language Processing, pp. 133–140. Association for Computational Linguistics (1992)
6. Brill, E.: A simple rule-based part of speech tagger. In: Proceedings of the Workshop on Speech and Natural Language, pp. 112–116. Association for Computational Linguistics (1992)
7. Ratnaparkhi, A., et al.: A maximum entropy model for part-of-speech tagging. In: Proceedings of the Conference on Empirical Methods in Natural Language Processing, Philadelphia, USA, vol. 1, pp. 133–142 (1996)
8. Toutanova, K., Manning, C.D.: Enriching the knowledge sources used in a maximum entropy part-of-speech tagger. In: Proceedings of the 2000 Joint SIGDAT Conference on Empirical Methods in Natural Language Processing and Very Large Corpora: Held in Conjunction With the 38th Annual Meeting of the Association for Computational Linguistics, vol. 13, pp. 63–70. Association for Computational Linguistics (2000)
9. Giménez, J., Marquez, L.: SVMTool: a general POS tagger generator based on support vector machines. In: Proceedings of the 4th International Conference on Language Resources and Evaluation. Citeseer (2004)
10. Denis, P., Sagot, B., et al.: Coupling an annotated corpus and a morphosyntactic Lexicon for state-of-the-art POS tagging with less human effort. In: PACLIC, pp. 110–119 (2009)

11. Toutanova, K., Klein, D., Manning, C.D., Singer, Y.: Feature-rich part-of-speech tagging with a cyclic dependency network. In: 2003 Conference of the North American Chapter of the Association for Computational Linguistics on Human Language Technology, vol. 1, pp. 173–180 (2003)
12. Shen, L., Satta, G., Joshi, A.: Guided learning for bidirectional sequence classification. In: ACL, vol. 7, pp. 760–767. Citeseer (2007)
13. Manning, C.D.: Part-of-speech tagging from 97% to 100%: is it time for some linguistics? In: Gelbukh, A.F. (ed.) International Conference on Intelligent Text Processing and Computational Linguistics, pp. 171–189. Springer, Berlin (2011)
14. Huang, Z., Xu, W., Yu, K.: Bidirectional LSTM-CRF models for sequence tagging. arXiv preprint arXiv:1508.01991 (2015)
15. Choi, J.D.: Dynamic feature induction: the last gist to the state-of-the-art. In: Proceedings of NAACL-HLT, pp. 271–281 (2016)
16. Collins, M.: Discriminative training methods for hidden Markov models: theory and experiments with perceptron algorithms. In: Proceedings of the ACL-02 Conference on Empirical Methods in Natural Language Processing, vol. 10, pp. 1–8. Association for Computational Linguistics (2002)
17. Amato, F., Casola, V., Mazzocca, N., Romano, S.: A semantic approach for fine-grain access control of e-health documents. Log. J. IGPL **21**(4), 692–701 (2013)
18. Amato, F., Boselli, R., Cesarini, M., Mercorio, F., Mezzanzanica, M., Moscato,V., Persia, F., Picariello, A.: Challenge: processing web texts for classifying job offers. In: Proceedings of the 2015 IEEE 9th International Conference on Semantic Computing (IEEE ICSC 2015), pp. 460–463. IEEE (2015)
19. Amato, F., Casola, V., Mazzeo, A., Romano, S.: A semantic based methodology to classify and protect sensitive data in medical records. In: 2010 Sixth International Conference on Information Assurance and Security, pp. 240–246. IEEE (2010)
20. Votrubec, J.: Morphological tagging based on averaged perceptron. In: WDS 2006 Proceedings of Contributed Papers, pp. 191–195 (2006)
21. Hajič, J., Raab, J., Spousta, M., et al.: Semi-supervised training for the averaged perceptron POS tagger. In: Proceedings of the 12th Conference of the European Chapter of the Association for Computational Linguistics, pp. 763–771. Association for Computational Linguistics (2009)
22. Chrupała, G., Dinu, G., Van Genabith, J.: Learning morphology with Morfette (2008)
23. Constant, M.,Tellier, I., Duchier, D., Dupont, Y., Sigogne, A., Billot, S.: Intégrer des connaissances linguistiques dans un crf: application à l'apprentissage d'un segmenteur-étiqueteur du français. In: TALN, vol. 1, p. 321 (2011)
24. Kanis, J., Müller, L.: Automatic lemmatizer construction with focus on OOV words lemmatization. In: Matoušek, V., Mautner, P., Pavelka, T., (eds.) International Conference on Text, Speech and Dialogue, pp. 132–139. Springer, Berlin (2005)
25. Schmid, H.: Treetagger—a language independent part-of-speech tagger. Institut für Maschinelle Sprachverarbeitung, Universität Stuttgart **43**, 28 (1995)
26. Morton, T., Kottmann, J., Baldridge, J., Bierner, G.: Opennlp: a Java-based NLP toolkit (2005)
27. Pianta, E., Zanoli, R.: TagPro: a system for Italian PoS tagging based on SVM. Intelligenza Artificiale **4**(2), 8–9 (2007)
28. Favretti, R.R., Tamburini, F., De Santis, C.: CORIS/CODIS: a corpus of written Italian based on a defined and a dynamic model. A Rainbow of Corpora: Corpus Linguistics and the Languages of the World. Lincom-Europa, Munich (2002)

29. Attardi, G., Fuschetto, A., Tamberi, F., Simi, M., Vecchi, E.M.: Experiments in tagger combination: arbitrating, guessing, correcting, suggesting. In: Proceedings of Workshop Evalita, p. 10 (2009)
30. Dell'Orletta, F.: Ensemble system for part-of-speech tagging. In: Proceedings of EVALITA, vol. 9, pp. 1–8 (2009)
31. De Smedt, T., Daelemans, W.: Pattern for Python. J. Mach. Learn. Res. **13**, 2063–2067 (2012)
32. Lyding, V., Stemle, E., Borghetti, C., Brunello, M., Castagnoli, S., Dell'Orletta, F., Dittmann, H., Lenci, A., Pirrelli, V.: The paisa corpus of Italian web texts. In: Proceedings of the 9th Web as Corpus Workshop (WaC-9), pp. 36–43 (2014)
33. Hahn, U., Tomanek, K., Beisswanger, E., Faessler, E.: A proposal for a configurable silver standard. In: Proceedings of the Fourth Linguistic Annotation Workshop, pp. 235–242 (2010)
34. Elia, A.: Dizionari elettronici e applicazioni informatiche. In: JADT (1995)
35. Elia, A., Marano, F., Monteleone, M., Sabatino, S., Vellutino, D.: Strutture lessicali delle informazioni comunitarie all'interno di domini specialistici. In: Statistical Analysis of Textual Data, Proceedings of 10th International Conference "Journées D'Analyse Statistique des Données Textuelles", pp. 9–11. Università" La Sapienza, Roma (2010)

Extracting Information from Food-Related Textual Sources

Alessandra Amato[(✉)], Francesco Bonavolontà, and Giovanni Cozzolino

Università degli Studi di Napoli Federico II, Ed. 3 - Via Claudio, 21,
80125 Napoli, Italy
{alessandra.amato,francesco.bonavolonta,giovanni.cozzolino}@unina.it

Abstract. Often when we think of the world of analysis, we always think of numerical experimental data that will then have to be somehow processed and then visualized on a graph. In reality, analysis has no prejudice regarding the type of "data to be analyzed". This can be a simple number and an image, a sound, or even a text. And it is precisely the analysis of texts, and in particular of the language used, that we will discuss in this article. This article aims to exploit Python libraries that allow Language Processing and text analysis in general to extract the information of interest.

1 Introduction

The World Wide Web consists of billions of interconnected documents, otherwise known as websites. The source code of websites is written in HTML (Hypertext Markup Language). HTML source code is a mixture of human-readable information and machine-readable code, known as tags. The web browser (e.g. Chrome, Firefox, Safari or Edge) processes the source code, interprets the tags and presents the information contained therein to the user.

In order to extract from the source code only information of interest to the user, special software is used. These programs, known as "web scrapers", "crawlers", "spiders" or simply "bots", search the source code of websites for predefined patterns and extract the information they contain. The information obtained from web scraping is summarised, combined, evaluated and stored for further use.

The popular Python programming language is particularly suitable for creating web scraping software. As websites are constantly updated, web content also changes. For example, the design or content of pages can change. A web scraper is written for the specific structure of a page. If the structure of the page changes, the web scraper must be adapted, a task that is particularly easy with Python.

Python is also very well suited for word processing and web resource retrieval, both technical foundations for web scraping. Python also represents an established standard for data analysis and processing. In addition to the general suitability of the language, Python stands out with a thriving programming

L. Barolli et al. (Eds.): AINA 2021, LNNS 227, pp. 72–80, 2021.
https://doi.org/10.1007/978-3-030-75078-7_8

ecosystem that includes libraries, open source projects, documentation and language references, as well as forum posts, bug reports and blog articles.

Specifically, there are several sophisticated tools for performing web scraping with Python, such as *Scrapy, Selenium,* and *BeautifulSoup* [1]. In this article we propose a comprehensive overview of *BeautifulSoup* module.

2   Motivation

The number of users and applications of the Social Web is constantly increasing. The main architects of these processes are the users themselves, able to generate contents and traffic in the network through the creation, connection, comment, tag, vote, retransmission, upload or download of resources in a participatory logic, where the interaction between users is the added value [2–4].

What stimulates users the most is their involvement in a variety of social activities, such as creating social relationships, suggesting and sharing resources with friends or not, creating groups or communities, commenting on other people's activities and profiles; what fuels this market is that it is not only users who benefit from this process, but also the social applications themselves, which increase their range of action and their ability to act as the amount of information they share increases. On the other hand, the semantic web (or intelligent web) has as its main purpose the description of online resources through a representation that is understandable and processable by a computer.

Natural language processing or web scraping techniques provide a new way to process and share data on the web in a standardised way [5–7]. This concur to make the information addressed to the final user to be comprehensible also from a computer [8,9].

The real challenge in many contexts is to be able to further formalise this information by creating domain ontologies [10], or even higher ontologies, by which to conceptualise a particular piece of knowledge uniquely, improve search accuracy, overcome differences in terminology, facilitate interoperability of information systems and infer relationships that, in some cases, may prove to bring new information.

3   WEB Scraping

The basic scheme of the web scraping process is quite simple [11]. Firstly, the web scraper developer analyses the HTML source code of the relevant page. Usually there are unique patterns based on which the desired information can be extracted [12]. The scraper is programmed according to these patterns. The rest of the work is done automatically by the scraper [13,14]:

- Accessing the website via the URL;
- Automated extraction of structured data according to the templates.
- Summarise, store, evaluate, combine, etc. the extracted information.

In this section we focus on a set of Python libraries that allow us to parse html documents fetched from the Web and to extract specific section from them [15, 16].

We recall *urllib.request* module, useful to extract information from website [17, 18], like shown in the following code block.

```python
# we recall the modules useful to extract information␣
 ↪from website
from urllib.request import urlopen
# Page that we want to open
url="http://www.columbia.edu/~fdc/sample.html"
# This baseurl will be needed to extract a picture
baseurl="http://www.columbia.edu/~fdc/"
# webpage Handler
page=urlopen(url)
#read the content of the web page
html_bytes= page.read()
print(html_bytes)
#store the text taken from the webpage with the␣
 ↪proper codify
html=html_bytes.decode("utf-8")
```

Then, we used *BeautifulSoup*, in particular "html.parser", to parse the html code and the method ".get_text()" to access the text without the html commands [19].

```python
from bs4 import BeautifulSoup
# Handler of the html code
soup = BeautifulSoup(html, "html.parser")
# Getting the text
soup2 = soup.get_text()
```

Here we used string spliting and character indexes to extract the table of contents

```python
import re
re.findall("CONTENTS",soup2)

# the method .split divides the string into 2 parts
# the first containing the word "CONTENTS" and the
 ↪second right after it
splitted = soup2.split("CONTENTS")

# The method .splitlines returns a list made of lines
splitted2 = splitted[1].splitlines()


indexes = []
# With this cycle we obtained the indexes for the
 ↪empty lines
for i,x in enumerate(splitted2):
  # if a line is empty
  if x == "":
      indexes.append(i)

indexes2 = []
app=0
# In this cycle we obtain the indexes of 2
 ↪consecutive "\n"
for i in indexes:
  # This if checks if 2 consecutive lines are empty
  if app == i-1:
    indexes2.append(i)
  app = i
print (indexes2)
# The Last print gave us the indexes where the table
 ↪of contents starts and ends,
# in particular we had to add 1 to the starting and
 ↪subtract 1 to the ending
stringa = ""
for i in range(2,13):
  stringa += splitted2[i]+"\n"
print(stringa)
```

The result of the previous code block is the Table of Content of the website.

3.1   Extracting Meta-data

```python
import bs4
# hr are horizontal lines
listt = soup3.find_all('hr')
# Since the page ends with an hr we want to discard␣
↪the last
text = ["" for i in range(len(listt)-1)]

i = 0
el = listt[0]
# This while is a cycle that checks if we find an hr␣
↪before starting
while(el):
    # next_sibling gives us the next tag
    el = el.next_sibling
    # this if is needed to avoid "list out of index"␣
↪error
    if i<len(listt)-1:
        # This if checks if we reached another hr and␣
↪if so it goes to next chapter
        if el == listt[i+1]:
            i += 1
        # Since the webpage ends with an hr we need to␣
↪stop at the penultimate hr
        if i == len(listt)-1:
            break

        # isinstace checks the type of el, and here is␣
↪needed to get the text properly
        if isinstance(el, bs4.element.Tag):
            text[i] += el.get_text()
        elif isinstance(el, bs4.element.NavigableString):
            text[i] += el

print(text[-1])
```

We figured out that the website that we are working on is written in htlm5 and this causes some issues to html.parser, so we decided to use *html5lib* to make the parsing easier.

In the following block code we show how to extract all the paragraph of a web document page [20, 21].

```
soup3 = BeautifulSoup(html, "html5lib")
# p stands for paragraph
paragraph = soup3.find_all("p")
print(paragraph[2].string)
```

The result is the following:

```
You can still create Web pages on your own computer␣
␣and look at them with your computer's Web browser,␣
␣but for other people to see them, you have to upload␣
␣them to the "big" computer that has the Web browser.␣
␣ The rest of this document is about how to create␣
␣your first Web page.
```

4   Extracting Images

For the extraction of images, we parsed the html source from the website in order to detect the *img* tags, then we created the url and through the *requests* package we got access to the image. Finally, we used the packages *"PIL"* and *"io"* to encode the image and *"matplotlib"* to display the picture.

In the following we show the code block that implement the pipeline:

```
import requests
from PIL import Image
from io import BytesIO
from matplotlib import pyplot as plt
# picture filename
image = soup3.find_all("img")[0]
# url of the picture
imageurl = baseurl+image["src"]
# handler of the picture
response = requests.get(imageurl)
# conversion into bytes
img = Image.open(BytesIO(response.content))
# displaying the picture with matplotlib
plt.imshow(img)
plt.axis("off")
plt.show()
```

5   Ontology Building

Data gathered from NLP pipeline can be exploited for ontology's construction. In this section we show how we used information parsed from medical reports to build an ontology that describe risks related to food consumption [22].

The following listing reports ontology's main classes and relationships.

```python
from owlready2 import *
onto = get_ontology("http://test.org/onto.owl")
with onto:
    class Dietary_Survey(Comprehensive_Database):pass
    class Survey_Methodologies(Dietary_Survey): pass
    class Food_Categories(FoodEx): pass
    class Chemical_Occurrence(Dataset): pass
    class Food_Consumption(Dataset): pass
    class Hazard(Thing): pass
    class Chemical(Hazard): pass
    ### Properties
    class Disaggregate_In(ObjectProperty):
        domain = [Data_Provider]
        range = [Food_Categories]
    class Used_For(ObjectProperty):
        domain = [EPIC_Soft]
        range = [Dietary_Survey]
```

In the following listing main individuals are described.

```python
### Individuals

    EuroFIR = Classification_System("EuroFIR")
    COST99 = Classification_System("COST99")
    Eurocode2 = Classification_System("Eurocode-2")
    EFG = Classification_System("EFG")
### FoodCategories
    Grains = One("Grains_and_grain-based_products")
    Vegetable =_
 ↪One("Vegetables_and_vegetable_products")
    Starchy = One("Starchy_roots_and_tubers")
    Legumes = One("Legumes_nuts_and_oilseeds")
    Fruits = One("Fruit_and_fruit_product")
    Meat = One("Meat_and_meat_products")
    Fish = One("Fish_and_other_seafood")
    Milk = One("Milk_and_dairy_products")
    Eggs = One("Eggs_and_egg_products")
    Sugar = One("Sugar_and_confectionary")
    Oils = One("Animal_and_vegetable_fats_and_oils")
    Juices = One("Fruit_and_vegetable_juices")
    Non_alcoholic = One("Non-alcoholic_beverages")
    Alcoholic = One("Alcoholic_beverages")
    Water = One("Drinking_water")
    Spices = One("Herbs_spices_and_condiments")
    Composite = One("Composite_food")
    Snacks = One("Snacks_desserts_and_other_foods")
onto.save(file="file1", format="rdfxml")
```

In Fig. 1 we show the result of the building process.

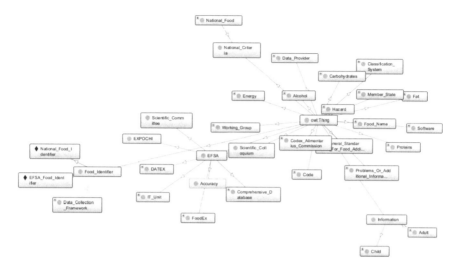

Fig. 1. Ontology of Food's consumption risks

6   Conclusions

BeautifulSoup is the oldest of the Python web scraping tools presented. Like Scrapy, it is an HTML parser. Its process looks like the following:

URL → HTTP Request → HTML → BeautifulSoup.

Unlike Scrapy, scraper development with BeautifulSoup does not require object-oriented programming. Instead, the scraper is written as a simple script. BeautifulSoup is, therefore, probably the easiest way to derive targeted information from the "tag soup".

Thanks to BeautifulSoup it is relatively easy and powerful to develop NLP applications that can extract and derive knowledge from web pages. In future work we foresee to integrate this techniques into a knowledge base information system that uses extracted information to derive further knowledge.

Acknowledgement. This paper has been produced with the financial support of the Project financed by Campania Region of Italy 'REMIAM - Rete Musei intelligenti ad avanzata Multimedialita'. CUP B63D18000360007.

References

1. BeautifulSoup.   https://www.crummy.com/software/BeautifulSoup/bs4/doc/ (2020). (Accessed 28 Jan 2021)
2. Angrisani, L., Bonavolontà, F., Tocchi, A., Moriello, R.S.L.: Frequency domain measurement node based on compressive sampling for sensors networks. In: 2015 IEEE International Workshop on Measurements & Networking (M&N), pp. 1–5. IEEE (2015)

3. Angrisani, L., Arpaia, P., Bonavolonta, F., Moriello, R.S.L.: Academic FabLabs for industry 4.0: experience at university of Naples Federico II. IEEE Instrum. Measur. Mag. **21**(1), 6–13 (2018)
4. Angrisani, L., Bonavolontà, F., Liccardo, A., Moriello, R.S.L., Ferrigno, L., Laracca, M., Miele, G.: Multi-channel simultaneous data acquisition through a compressive sampling-based approach. Measurement **52**, 156–172 (2014)
5. Thomas, D.M., Mathur, S.: Data analysis by web scraping using Python. In: 2019 3rd International conference on Electronics, Communication and Aerospace Technology (ICECA), pp. 450–454. IEEE (2019)
6. Mahto, D.K., Singh, L.: A dive into web scraper world. In: 2016 3rd International Conference on Computing for Sustainable Global Development (INDIACom), pp. 689–693. IEEE (2016)
7. Chaulagain, R.S., Pandey, S., Basnet, S.R., Shakya, S.: Cloud based web scraping for big data applications. In: 2017 IEEE International Conference on Smart Cloud (SmartCloud), pp. 138–143. IEEE (2017)
8. Balzano, W., Sorbo, M.R.D.: Genomic comparison using data mining techniques based on a possibilistic fuzzy sets model. Biosystems **88**(3), 343–349 (2007)
9. Gisolfi, A., Balzano, W.: Constructing and consulting the knowledge base of an expert systems shell. Expert. Syst. **10**(1), 29–35 (1993)
10. Cozzolino, G.: Using semantic tools to represent data extracted from mobile devices. In: 2018 IEEE International Conference on Information Reuse and Integration (IRI), pp. 530–536 (2018)
11. Mitchell, R.: Web Scraping with Python: Collecting More Data from the Modern Web. O'Reilly Media Inc., Newton (2018)
12. Amato, F., Castiglione, A., Cozzolino, G., Narducci, F.: A semantic-based methodology for digital forensics analysis. J. Parallel Distrib. Comput. **138**, 172–177 (2020)
13. Lawson, R.: Web Scraping with Python. Packt Publishing Ltd., Birmingham (2015)
14. Jarmul, K., Lawson, R.: Python Web Scraping. Packt Publishing Ltd., Birmingham (2017)
15. Amato, F., Casola, V., Mazzeo, A., Romano, S.: A semantic based methodology to classify and protect sensitive data in medical records. In: 2010 Sixth International Conference on Information Assurance and Security, pp. 240–246. IEEE (2010)
16. Popovski, G., Seljak, B.K., Eftimov, T.: A survey of named-entity recognition methods for food information extraction. IEEE Access **8**, 31586–31594 (2020)
17. Kushmerick, N., Weld, D., Doorenbos, R.: Wrapper induction for information extraction. Citeseer (1997)
18. Amato, F., De Pietro, G., Esposito, M., Mazzocca, N.: An integrated framework for securing semi-structured health records. Knowl.-Based Syst. **79**, 99–117 (2015)
19. Zheng, C., He, G., Peng, Z.: A study of web information extraction technology based on beautiful soup. JCP **10**(6), 381–387 (2015)
20. Amato, A., Cozzolino, G., Giacalone, M.: Opinion mining in consumers food choice and quality perception. In: Barolli, L., Hellinckx, P., Natwichai, J. (eds.) International Conference on P2P, Parallel, Grid, Cloud and Internet Computing, pp. 310–317. Springer, Cham (2019)
21. Vidal, L., Ares, G., Machín, L., Jaeger, S.R.: Using Twitter data for food-related consumer research: a case study on "what people say when tweeting about different eating situations". Food Qual. Prefer. **45**, 58–69 (2015)
22. Amato, A., Balzano, W., Cozzolino, G., Moscato, F.: Analysis of consumers perceptions of food safety risk in social networks. In: Barolli, L., Takizawa, M., Xhafa, F., Enokido, T. (eds.) International Conference on Advanced Information Networking and Applications, pp. 1217–1227. Springer, Cham (2019)

PaSy - Management of a Smart-Parking System Based on Priority Queues

Walter Balzano[✉], Walter Galiano, and Silvia Stranieri

University of Naples Federico II, Naples, Italy
silvia.stranieri@unina.it

Abstract. One of the most tedious problems affecting smart cities is traffic congestion. Several studies demonstrate that the main cause of this phenomenon is drivers looking for a free parking slot, operation that requires a slow driving and, hence, slows down the entire traffic trend. For this reason, many researchers are proposing innovative techniques to speed up the parking process and improve the traffic configuration. In this work, we design an innovative smart-parking system based on a consortium-linke mechanism, in such a way that both consortium-members and drivers can take advantage from the deal. Specifically, if the user wants to reserve a slot, the system first performs a user profiling, then proposes several choices of available parking areas, in an opportune order with respect to the user's preferences. Reservation requests can be made through a mobile app and they are dealt with priority queues. From one hand, each driver optimizes his driving time while maximizing his preferences, on the other hand consortium-members obtain an important visibility.

1 Introduction

Vehicular ad Hoc Networks (VANETs) constitute an emerging research field due to their numerous potential applications. These networks are made of several vehicles communicating between them through broadcast messaging. Each vehicle is equipped with on board side units (OBUs), and the communication is performed by relying on road side units (RSUs), placed all over the streets [9]. Through such a communication scheme, vehicles can transmit road information to each other, by using road infrastructure items as a bridge, when the communication range does not cover the distance between them. One of the most interesting applications of VANETs is about parking strategies. Indeed, several studies demonstrate that drivers looking for a free parking slot seriously contribute to traffic congestion. According to [17], the average time spent hunting for parking is 17 h in USA, for an average cost of 72.7 billion dollars. It is pretty clear that this phenomenon should be handled to reduce traffic congestion, and hence improve road security, and VANETs provide an interesting instrument to address such a problem for researchers in this field. In this work, we face the parking wasting time problem, by introducing a consortium-like model. Specifically,

L. Barolli et al. (Eds.): AINA 2021, LNNS 227, pp. 81–90, 2021.
https://doi.org/10.1007/978-3-030-75078-7_9

several parking areas are supposed to be part of a consortium, which constitutes a guarantee for both consortium-members and drivers. The main actors involved in our study are drivers, parking-owners, and the environment. Drivers are the ones that are affected the most by the traffic congestion: indeed, according to an English study [2], the time wasted by them for the parking process is approximately 2549 h in the lifetime. The possibility of a quick parking plays a main role to improve the life of all citizens, because it increases access within communities and the result can be an upgrade of economic growth. Thanks to the development of new smart-parking systems, the drivers stress is lightened and the driving experience becomes more enjoyable. These are all significant factors, especially from the social point of view: just think of how many times we gave up on a restaurant dinner for "fear" of spending more time in the car than at the table. Parking-owners are supposed to be updated with IoT technologies. During the years the idea of parking is changing and, if we look forward, the scenario is becoming ever more smarter than the actual one, especially with the introduction of IOT technologies and modern sensors that support both customers and owners. First, the car park must be equipped with an Internet connection so to update information on available parking spaces and occupancy rate; it must also be provided with one or more RFID (Radio-Frequency Identification) readers, which verify the driver's identity through RFID tags or cards and, once verified, the driver can access or exit the car park. The other component which need to be taken into account is the environmental one. Nowadays, the development of technologies, especially related to smart-mobility, is based on "green" issues. The problem of finding a parking space, on the street or in a car park, represents a drawback for drivers but also for the environment, due to the constant emission of pollutants in the atmosphere from the exhaust gas of the cars, and it makes the living conditions worse in one or more parts of the city. The consequences of what has just been explained are: air pollution, noise pollution, and road connection difficulties. According to ISTAT analyses in 2016 [1], the problems felt the most by families in relation to the area in which they live were: air pollution (38.0%), traffic (37.9%) and parking difficulties (37.2%). On the Italian podium of the most "stressed parking" regions we find Lazio (50,1%), Liguria (50%), and Campania (43,5%).

2   Related Work

Traffic congestion problem contributes to an increment of drivers stress level, a decrease of road security conditions, but it has also in important impact on pollution levels. With no doubts, drivers searching for a free parking slot constitute one of the main causes of this phenomenon. For these reasons, researchers coming from different fields are proposing several solutions to the parking problem. Authors in [11], for instance, propose an algorithm for smart parking based on a known optimization problem, which is the ant colony: the approach is bio-inspired, since the algorithm imitates the way the ants chose a path rather than another, by following the pheromone released by the other ants. With the same spirit, the propose an optimization algorithm that maximizes the number of free parking slot met during the chosen path. Similarly, author of [10] proposes a variant of the known distance geometry problem which, through a logic approach, applies an inverse procedure with respect to the classical one: starting from a network in which any node position is known, the build a corresponding graph containing useful information about parking slot. Another interesting approach to the parking problem is presented in [8], where they propose an algorithm to allocate vehicles in parking areas, by varying known memory allocation algorithms, in a destination-based key: they modify First, Best, and Worst-fit algorithms in such a way that the distance from the selected slot and the destination of the driver is taken into account. Also authors of [16] focus on parking problem by using IoT (Internet of Things). Indeed, they develop an IoT module to keep track of parking availability on one side, and an application allowing users to verify such an availability on the other [4,6]. An interesting critical point is raised in [3], where authors use blockchains to preserve drivers privacy. A survey on VANET trust is proposed in [13]. The innovative aspect of our work is the integration of a consortium-like mechanism in the parking process. For a survey on smart parking challenges and solutions, you can consult [15].

3   Integration Model

3.1   Profiling

In this section, we define the parameters specifying the preferences entered, during the profiling, by users and parking-owners. To the best of our knowledge, all researches on smart parking systems consider only a few parameters, such as *cost*, that is the amount a user intends to spend for parking in a given parking area, *distance*, between the driver's starting point and his car park or garage destination, and *time*, that represents the number of hours or minutes needed to reach the parking area. We enrich by adding new parameters, through the user profiling, as well as the profiling of parking-owners joining the consortium. In this sense, from one hand, each driver optimizes his driving time while maximizing his preferences, on the other hand consortium-members obtain an important visibility[5].

3.1.1   Driver Parameters

We consider the following parameters for drivers: *Max cost*, maximum amount the user is willing to spend in order to park his car; *Max distance*, represents the maximum distance allowed between the target parking area and user's destination; *Min stop*, value representing the minimum time during which the driver expects to occupy the parking area. This is needed to discard those garage whose closing hour does not match with driver needs; *Key release*, a binary value indicating if the parking-owner requires the key release from the driver; *Departure time*, parameter needed to establish a priority in the waiting queue, but this aspect will be discussed in the following sections of the paper; *Subscription*, as in the previous case, is needed as part of the priority evaluation.

3.1.2   Parking-Owner Parameters

For owners, instead, we take into account: *Min cost*: if there is no hourly cost, the minimum cost overlaps the fixed cost of a parking; *Hourly cost*, hourly parking rate; *Capacity*, the actual number of parking slots; *Working timetable*, if the car park is open 24 h a day, no timetable should be provided; otherwise, the parking-owner must specify the opening and closing time; *Key release*, as in the case of drivers, also the owners can set this parameter, for reasons of space and potential displacements of cars.

3.1.3   Preference Tables

In the example tables, we have an anticipation of an hypothetical screen of profiling on the mobile app. The users and the owners can specify if a field should be a constraint or not (the first column of each sub-table contains this information); in the first case, the system does not propose an incompatible car park to the user; in the second case, instead, if there are no constraints (for example a parking-owner sets the key release only as preference) the system may propose the incompatible car parks as the last in the list of possible user choices (Table 1).

Table 1. Driver and parking-owner preferences

	X	Preference		X	Preference
MaxCost		5\euro	MinCost		3\euro
MaxDistance	X	500 m	HourlyCost	X	2\euro/hour
MaxStop		45 min	Capacity		300
KeyRelease	X	no	Timetable	X	06:00–24:00
DepartureTime	X	18	KeyRelease	X	yes

3.2   Mathematical Model

In order to formalize our proposal, we build the mathematical model that helps the system to propose several choices of available parking areas, in an opportune order with respect to the user's preferences.

3.2.1   The Transportation Model

The focus will be on one of the linear programming models, *the transportation model*, that is an optimization problem with a linear objective function and linear constraints, that goes well with the current study. It consists in a number of localities, *origins*, that produce a fixed quantity of available goods and a certain number of clients, *destinations*, who require precise quantities of goods. Now, there is a company, that produces goods, who wants to transport it from m

supply centers to n demand centers, trying to minimize the shipping costs. The cost of shipping one unit from origin i to destination j is c_{ij} and if you shipped x_{ij} units, the cost would be $c_{ij} \, x_{ij}$. So the result is the following objective function:

$$z = \sum_{i=1}^{m} \sum_{j=1}^{n} c_{ij} x_j; \tag{1}$$

3.2.2   Our Model

After introducing the *transportation model* with the general formulation, now we apply it to our specific case. We can assume that the *m origins* are the cars looking for a parking space, and the *n destinations* are the parking areas that can accept the cars; the amount that can be carried by each origin, instead, is always equal to one, since we consider as "goods to transport" the car itself. Anyway, let's define in detail the formulation:

1. *Variables*

$$x_{ij} = \begin{cases} 1 \; if \; the \; car \; i \; parks \; in \; the \; car \; park \; j \\ 0 \qquad\qquad otherwise \end{cases}$$

   If the driver i parks in the car park j then $x_{ij} = 1$; otherwise $x_{ij} = 0$;

2. *Constraints*
   a.

$$0 \leq x_{ij} \leq 1; \qquad i = 1, ..., n; \quad j = 1, ..., m; \tag{2}$$

   b.

$$\sum_{j=1}^{n} x_{ij} \leq 1; \tag{3}$$

   The car i can park at most in one of the n active car parks;
   c.

$$\sum_{i=1}^{m} x_{ij} \leq k_j; \tag{4}$$

   Each car park j can contain at most k_j, that is the capacity of a car park, vehicles of m total cars looking for parking;
3. *Objective function*

$$min \; z = \sum_{i=1}^{m} \sum_{j=1}^{n} c_{ij} x_{ij}, \; x_{ij} \in [0, 1]; \tag{5}$$

As one can see from the objective function, we want to minimize the cost c_{ij} for every car i parking in the garage j. In the preference tables we have shown how, with this study, the user can choose a car park more suitable to his needs. To apply this, we replace, in the formula, the variable c_{ij} with a function that we define *general cost function*.

3.2.3   General Cost Function

We start by defining three classes of parameters useful for the enunciation:

- *Preference parameters*: are parameters that express the degree of preference. In this case α and β represent the preferences with respect to the price and the time for parking the car i in the garage j. The two values are chosen such that their sum is equal to one; for example if α is equal to 0.8 (80%), then β is equal to 0.2 (20%); in this case, parking areas less expensive, but possibly more distant than others, will be offered to the user.

$$\pi \equiv (\alpha, \beta, ...); \tag{6}$$

- *Real parameters*: in R is represented the value of all the real parameters useful to the computation of the *general cost function*

$$R \equiv (\frac{c_j}{c_{max}}, \frac{T_{ij}}{T_{max}}, ...); \tag{7}$$

  – c_j: is the cost of parking area j; it depends on the parking formula, hourly or fixed cost;
  – c_{max}: is the maximum cost that a member of the consortium can reach;
  – T_{ij}: is the time needed by the driver i to reach the parking area j;
  – T_{max}: is the maximum time needed by a driver to reach to complete travel and parking;
- *Binary parameters*: in B is represented the value of all binary constraints; there is a logical AND between the various constraints; all $b_1...b_n$ must be equal to one:

$$B = b_1 \ AND \ b_2 \ AND \ ... \ b_n; \tag{8}$$

In the following, we propose a simple example:

$$b_1 = \begin{cases} 1 \ if \ the \ user \ i \ and \ car \ park \ j \ are \ compatible \ on \ key \ release \\ 0 \qquad\qquad\qquad\qquad otherwise \end{cases}$$

If b_1 is equal to zero, for example, the function doesn't take a value and the car park j not be a proposal considered by the system for the user i;
- *Formulation*

$$G_{ij} = B\pi R = B \left(\alpha \frac{c_j}{c_{max}} + \beta \frac{T_{ij}}{T_{max}} + ... \right). \tag{9}$$

This is the final function that summarizes all three classes of parameters. As we can see, the model is easily expandable by adding any extra parameters to the respective categories;
- *Final objective function*

$$min \ z = \sum_{i=1}^{m} \sum_{j=1}^{n} c_{ij} x_{ij} = \sum_{i=1}^{m} \sum_{j=1}^{n} G_{ij} x_{ij}, \quad dove \ x_{ij} \in [0, 1]. \tag{10}$$

3.3 Priority Queues

As said before, the registered users also have the possibility to join a consortium with subscription, but this is not mandatory. Depending on the status of the user, the system will assign a different priority to each one. Clearly, the subscribers have higher priority than non-subscribers. Moreover, also among non-subscribers one can make a distinction: in fact, the mechanism is based on the creation of several different queues of reservation, based on the priority.

3.3.1 Subscription Categories

(a) *Fidelity*: is a monthly subscription; the main advantage is to have a higher priority than non-subscribers at the time of reservation and after the departure;

(b) *Gold*: is an annual subscription; one of the important aspects regarding the parking problem is also represented by a user group that has problems in finding a parking space nearby home, since the slots on the street are few and always occupied. Thanks to this type of subscription, the user will have a parking space, in the nearest car park, during the entire year;

(c) *Prime*: Prime: is an annual subscription, the most expensive; Prime owners have a parking slot reserved in each parking area member of the consortium. They do not need to reserve a slot through the app, because it is always available.

3.3.2 Non-subscriber Users

Among non-subscribers, we recognize two categories of users:

(a) *Batch users*: those who make their reservation a long time before departure, for example those who book in the morning to reserve a parking space for dinner in the evening;

(b) *Real time users*: those who book at the moment of departure.

Clearly, they are handled differently.

3.4 Queues Categories

As previously said, we use six different queues with diverse semantics to handle any user according with the corresponding priority, according to his kind of subscription.

We now introduce all the various types of queues used in this model:

(a) *Batch queue*: non-interactive; the Batch users have a higher priority than the Real time users in the Final queue;

(b) *Real time queue*: interactive; not convenient in terms of priority;

(c) *Fidelity queue*: Fidelity subscribers have top priority in the Final queue, regardless of the moment of booking;

(d) *Gold queue*: populated by Gold subscribers;

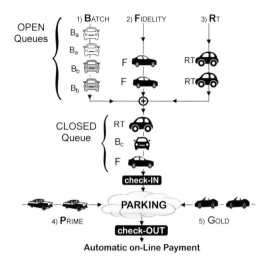

Fig. 1. Priority queues - categories

(e) *Prime queue*: populated by Prime subscribers;
(f) *Final queue*: is the final departure queue, we have all the reservations ordered by priority.

Figure 1 summarizes queues types and their operation.

3.4.1   Batch Moments

As you can see from the Fig. 2, there are three different moments concerning Batch users, ranging from reservation to parking. The following timeline will help us.

Fig. 2. Timeline of batch moments

1. B_a: requests entered in the reservation queue;
2. B_b: requests are virtually allocated, being updated continuously to guarantee the best solution to the user; payment is done at one hour from the departure;
3. B_c: requests, at the time of departure, are physically allocated to assigned parking places.

4   Conclusions

This work proposes an innovative consortium-like smart parking strategy. The solution provided takes into account all drivers preferences, by also benefit the consortium-members, which earn visibility and several guarantees. Numerous are the possible hints for future work. Indeed, not only our model can be easily enriched with other parameters taking into account new relevant aspects for road security, but also we aim at exploiting formal techiques based on the strategic reasoning for multi-agent systems. In this setting, one can consider cars as intelligent agents and the desired solution may come as an equilibrium among the agents, in a way similar as it has been done in [7, 12, 14, 18][1].

References

1. Annuario statistico italiano 2016 (2016)
2. Motorists spend 106 days looking for parking spots, May 2013
3. Al Amiri, W., Baza, M., Banawan, K., Mahmoud, M., Alasmary, W., Akkaya, K.: Privacy-preserving smart parking system using blockchain and private information retrieval. In: 2019 International Conference on Smart Applications, Communications and Networking (SmartNets), pp. 1–6. IEEE (2019)
4. Amato, F., Casola, V., Gaglione, A., Mazzeo, A.: A semantic enriched data model for sensor network interoperability. Simul. Model. Pract. Theory **19**(8), 1745–1757 (2011)
5. Amato, F., Moscato, V., Picariello, A., Colace, F., De Santo, M., Schreiber, F.A., Tanca, L.: Big data meets digital cultural heritage: design and implementation of scrabs, a smart context-aware browsing assistant for cultural environments. J. Comput. Cult. Herit. (JOCCH) **10**(1), 1–23 (2017)
6. Amato, F., Moscato, V., Picariello, A., Sperlì, G.: Extreme events management using multimedia social networks. Futur. Gener. Comput. Syst. **94**, 444–452 (2019)
7. Aminof, B., Kwiatkowska, M., Maubert, B., Murano, A., Rubin, S.: Probabilistic strategy logic. In: Kraus, S. (ed.) Proceedings of Twenty-Eighth International Joint Conference on Artificial Intelligence, IJCAI 2019, pp. 32–38 (2019). ijcai.org
8. Balzano, M., Balzano, W., Sorrentino, L., Stranieri, S.: Smart destination-based parking for the optimization of waiting time. In: Workshop of the International Conference on Advanced Information Networking and Applications, pp. 1019–1027. Springer, Berlin (2020)
9. Balzano, W., Del Sorbo, M.R., Stranieri, S.: A logic framework for C2C network management. In: 2016 30th International Conference on Advanced Information Networking and Applications Workshops (WAINA), pp. 52–57. IEEE (2016)
10. Balzano, W., Stranieri, S.: LoDGP: a framework for support traffic information systems based on logic paradigm. In: International Conference on P2P, Parallel, Grid, Cloud and Internet Computing, pp. 700–708. Springer, Berlin (2017)
11. Balzano, W., Stranieri, S.: ACOp: an algorithm based on ant colony optimization for parking slot detection. In: Workshop of the International Conference on Advanced Information Networking and Applications, pp. 833–840. Springer, Berlin (2019)

---

[1] This paper has been produced with the financial support of the Project financed by Campania Region of Italy 'REMIAM - Rete Musei intelligenti ad avanzata Multimedialita'. CUP B63D18000360007.

12. Cermák, P., Lomuscio, A., Mogavero, F., Murano, A.: Practical verification of multi-agent systems against SLK specifications. Inf. Comput. **261**, 588–614 (2018)
13. Hussain, R., Lee, J., Zeadally, S.: Trust in VANET: a survey of current solutions and future research opportunities. IEEE Trans. Intell. Transp. Syst. (2020)
14. Jamroga, W., Malvone, V., Murano, A.: Natural strategic ability. Artif. Intell. **277** (2019)
15. Khalid, M., Wang, K., Aslam, N., Cao, Y., Ahmad, N., Khan, M.K.: From smart parking towards autonomous valet parking: a survey, challenges and future works. J. Netw. Comput. Appl. 102935 (2020)
16. Khanna, A., Anand, R.: IoT based smart parking system. In: 2016 International Conference on Internet of Things and Applications (IOTA), pp. 266–270. IEEE (2016)
17. McCoy, K.: Drivers spend an average of 17 hours a year searching for parking spots
18. Rubin, S., Zuleger, F., Murano, A., Aminof, B.: Verification of asynchronous mobile-robots in partially-known environments. In: Chen, Q., Torroni, P., Villata, S., Hsu, J.Y., Omicini, A. (eds.) PRIMA 2015: Principles and Practice of Multi-agent Systems - 18th International ConF, 2015. LNCS, vol. 9387, pp. 185–200. Springer, Berlin (2015)

An Ontological Model to Represent Food Consumption Risks

Alessandra Amato[(✉)], Francesco Bonavolontà, and Giovanni Cozzolino

Università degli Studi di Napoli Federico II, Ed. 3 - Via Claudio,
21 - 80125 Naples, Italy
{alessandra.amato,francesco.bonavolonta,giovanni.cozzolino}@unina.it

Abstract. Digital technology innovation is vividly affecting the health and economic aspects of food production, distribution, and consumption. Goods in transit, trade, exotic products from countries where food and agricultural legislation is not necessarily as stringent as in Europe. But also, mass catering and large supermarket chains with distribution of products over long distances. And finally, great use of preserved products, which must be kept within the cold chain and which a blackout can put at risk. In this work we present an ontology model to describe the food supply chain and the risks related to it.

1 Introduction

The topic of food safety is of great importance because more and more attention is paid to the health of citizens and to what they eat. Italians have proven to be a very attentive people to what they put in their plates, to the quality of ingredients, to the freshness of food, but also to its origin.

In an increasingly globalised food market, in fact, it is even more important to have knowledge of the food chain, to know the place of origin of a product but also the various intermediate steps that have led up to the counter of the supermarket, the greengrocer or the favourite restaurant [1, 2].

"We are what we eat" is no longer just a saying, but a way of life now shared: we are increasingly attentive to what we buy and consume, we are increasingly sensitive to the type of cultivation, the organic origin of food, the types of farming and the way food is treated and stored. Our health is at stake and we are well aware of this.

For this reason, the issue of food traceability has now taken a leading role in guiding our consumption choices. The enormous changes that have affected the food system, which is no longer characterised by a close relationship between production and consumption but in which food preservation plays a fundamental role, today pose new problems and critical points to be resolved in order to guarantee food safety.

Thanks to the Internet of Things, for example, network-enabled sensors located in agricultural environments, shipping containers, factories, stores, and

L. Barolli et al. (Eds.): AINA 2021, LNNS 227, pp. 91–96, 2021.
https://doi.org/10.1007/978-3-030-75078-7_10

kitchens are generating data that can be used to ensure food traceability, while reducing resource consumption, costs, and waste.

Building high-capacity data-sharing networks, however, has exposed a key obstacle: the content of existing Internet-connected information remains isolated due to a multiplicity of local languages and data dictionaries [3–5].

Addressing data harmonisation issues that span multiple domains [6–8], from food safety to consumer health and convenience, would require more than a well-defined hierarchical vocabulary, an ontology would be needed.

One of the goals of this work is to describe an ontology about Safety Food scenario. "Food safety" refers to the conditions and practices that preserve the quality of food to prevent contamination and food-borne illness [9–11].

2   Case Study

The case study of interest refers to a Safety Food scenario [12]. In this scenario, the Data Collection and Exposure Unit (DATEX) of the European Food Safety Authority (EFSA) developed a new system called FoodEx, following an EFSA's Scientific Committee's opinion on exposure assessment, recommending the urgent collection of available consumption data at an aggregated level followed by an expanded collection of data at a detailed level. The main objective of FoodEx is to facilitate the assessment of dietary exposure to hazardous chemicals and microbiological by allowing accurate matching of the datasets on chemical occurrence and food consumption. It provides for different food descriptions and classification systems and is a hierarchical system based on 20 main food categories that are further divided into subgroups up to a maximum of 4 levels. FoodEx codify all foods and beverages present in the food consumption database provided by 20 Member States and collected from 22 different national dietary surveys.

The requested data should be representative at national level for at least the adult population and was collected at individual level by means of a 24 h recall or dietary record. Data providers were asked to codify all foods and beverages present in the food consumption database according to the draft FoodEx classification system, at the most disaggregated level possible. The transmission of food consumption data was accomplished through the Data Collection Framework (DCF), an application designed by the IT Unit of EFSA.

The FoodEx system is user friendly and flexible about the food classification of the different Nations. Although there are some differences in the classifications systems of the national dietary surveys, has been demonstrated that all data providers were able to classify correctly the large majority of their food items at least at the 2nd level of the FoodEx. To support FoodEx has been created a Working Group that develops the standardisation of the food classification and the description system. One of the main task of FoodEx is to develop an harmonised classification of the food consumption data included in the Comprehensive Database whose main goal is to improve the accuracy of EFSA's exposure assessment calculations.

3 The Represented Information

We will now describe how we have modelled the structure and interaction of the various entities involved in the Food Safety activity carried on by the European Food Safety Authority (EFSA).

In Fig. 1 we report the ontology which describes the taxonomy of the most relevant components.

Fig. 1. Taxonomy. The bold text are classes, the non-bold are instances that have a property so protege displays them among the classes.

The Classification System entity is the focal point of the EFSA food safety project. As we can see in Fig. 2 It is actually composed of several classification systems (in particular EFG and Eurocode-2) created within different research projects (such as COST99 and EuroFIR) in order to obtain an harmonised classification.

It relies on the Comprehensive Database and on the Code as it provides a discriminant method to identify Food Categories. As we can see from the Fig. 3, the codification has been made by the Data Providers along with the Food Consumption dataset construction, based on the FoodEx classification system mainly developed by the DATEX unit and the Working Group.

Detailed previously available information differentiated for Adults and Children were also used by the EFSA alongside with Dietary Surveys data collected on international scale. Proposals focused on children were also made by EFSA with the EXPOCHI with data from 13 Member States. With respect to the identification of food, two identifier were transmitted and stored into the database, an EFSA Food Identifier and a national one.

Food Consumption data were then used also to evaluate the exposure to Hazardous Chemicals. Identification and Use of Other Sources are also key aspects to complement the Comprehensive Database. SAS Enterprise Software was used

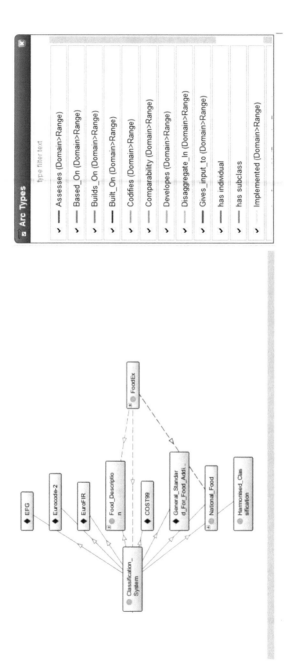

Fig. 2. Classification system properties.

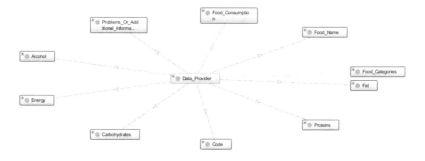

Fig. 3. Data providers

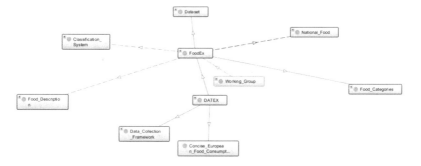

Fig. 4. DATEX as pivot of the FoodEx-classification system connection.

to extract information from Food Consumption data, while Belgium, Germany and Netherlands used EPIC-soft for data crawling on Dietary Surveys (Fig. 4).

4   Conclusions

In this work we presented an ontology model to describe and formalise the Safety Food scenario. "Food safety" refers to the conditions and practices that preserve the quality of food to prevent contamination and food-borne illness. We foresee to exploit this representation in order to develop a system that infers knowledge from data gathered from food supply chains, like sensors located in agricultural environments, shipping containers, factories, stores, and kitchens.

Acknowledgements. This paper has been produced with the financial support of the Project financed by Campania Region of Italy 'REMIAM - Rete Musei intelligenti ad avanzata Multimedialita'. CUP B63D18000360007.

References

1. Van Kleef, E., Frewer, L.J., Chryssochoidis, G.M., Houghton, J.R., Korzen-Bohr, S., Krystallis, T., Lassen, J., Pfenning, U., Rowe, G.: Perceptions of food risk management among key stakeholders: results from a cross-European study. Appetite **47**(1), 46–63 (2006)
2. Eiser, J.R., Miles, S., Frewer, L.J.: Trust, perceived risk, and attitudes toward food technologies 1. J. Appl. Soc. Psychol. **32**(11), 2423–2433 (2002)
3. Amato, F., De Pietro, G., Esposito, M., Mazzocca, N.: An integrated framework for securing semi-structured health records. Knowl.-Based Syst. **79**, 99–117 (2015)
4. Amato, F., Moscato, V., Picariello, A., Sperli'ì, G.: Extreme events management using multimedia social networks. Futur. Gener. Comput. Syst. **94**, 444–452 (2019)
5. Amato, F., Moscato, V., Picariello, A., Sperlí, G.: Kira: a system for knowledge-based access to multimedia art collections. In: 2017 IEEE 11th International Conference on Semantic Computing (ICSC), pp. 338–343. IEEE (2017)
6. Angrisani, L., Bonavolontà, F., Tocchi, A., Lo Moriello, R.S.: Frequency domain measurement node based on compressive sampling for sensors networks. In: 2015 IEEE International Workshop on Measurements and Networking (M&N), pp. 1–5. IEEE (2015)
7. Angrisani, L., Bonavolontà, F., Liccardo, A., Moriello, R.S.L., Ferrigno, L., Laracca, M., Miele, G.: Multi-channel simultaneous data acquisition through a compressive sampling-based approach. Measurement **52**, 156–172 (2014)
8. Angrisani, L., Arpaia, P., Bonavolonta, F., Lo Moriello, R.S.: Academic fablabs for industry 4.0: experience at university of Naples Federico II. IEEE Instrum. Meas. Mag. **21**(1), 6–13 (2018)
9. de la Calle, S.d.P., Moreno, E.R., Gaspar, T.V., Alonso, P.R., Torres, J.M.: Sources of information on food consumption in Spain and Europe. Nutricion hospitalaria **31**(3), 29–37 (2015)
10. Martinez-Victoria, E., de Victoria, I.M., Martinez-Burgos, M.A.: Intake of energy and nutrients; harmonization of food composition databases. Nutricion hospitalaria **31**(3), 168–176 (2015)
11. World Health Organization et al.: Gems/food programme: report of the WHO working group on collection of food consumption data (COFOCO), 30 January 2012 (2012)
12. European Food Safety Authority: Evaluation of the FoodEX, the food classification system applied to the development of the EFSA comprehensive European food consumption database. EFSA J. **9**(3), 1970 (2011)

Intelligent Cloud Agents in Multi-participant Conversations for Cyber-Physical Exploitation of Cultural Heritage

Angelo Ambrisi, Rocco Aversa, Massimo Ficco, Danilo Cacace, and Salvatore Venticinque$^{(\boxtimes)}$

Department of Engineering, University of Campania, Aversa, Italy
{angelo.ambrisi,rocco.aversa,massimo.ficco,
salvatore.venticinque}@unicampania.it,
danilo.cacace@studenti.unicampania.it

Abstract. Most of all existing solutions for developing conversational systems have been designed for dyadic and one-on-one communication with users. In this work, we present the design of mechanisms and interaction models for enabling dyadic agents to converse in multi-participants chat. We aim at complementing the visit of archaeological sites and distributed museums with a social dimension fostered by intelligent agents in one cyber-physical world with live and remote users who enjoy the assets of the cultural environment. A cloud-based prototype implementation is also provided.

1 Introduction

Current research advances in Artificial Intelligence (AI) have exploited the development of cyber-physical systems, in which context aware software agents support users in their daily activities. Conversational agents represent a class of software, which can interact through the same communication platforms, and supports communication between human users. They have been used to implement social network, messaging apps, chats and voice calls, speech assistant. The development of intelligent conversational agents, which are capable of natural language processing, have been investigated since the beginning of computer science. Nowadays the challenge does not deal with developing agents that exhibit intelligent behaviour indistinguishable from that of a human, rather than to assist the user in the exploitation of digital services in the smartest way. The result is a cyber-physical world, where human users converse with software agents as they usually do with other users in several application contexts. Users continuously receive alerts, advertising, recommendations, contact the customer care, ask for information or for executing tasks by a voice or text message, sometimes neglecting if on the other side there is someone or something smart. Despite technologies for developing conversational systems have made a great stride [8],

L. Barolli et al. (Eds.): AINA 2021, LNNS 227, pp. 97–106, 2021.
https://doi.org/10.1007/978-3-030-75078-7_11

nearly all existing solutions have been designed for dyadic, one-on-one communication with users.

Extending communication mechanisms and modeling advanced behavior for enabling dyadic agents to converse in multi-participants chat is the research problem investigated in this work. In order to improve the visitors' experience, research effort has been also spent for the enhancement of cultural heritage by using intelligent virtual guides, both in physical and virtual environments [1, 2, 6]. In the Cleopatra[1] project, we aim at complementing the visit of archaeological sites and distributed museums with a social dimension in a cyber-physical world, where cloud-based software agents animate users' interaction with the assets of the cultural environment [7]. The rest of the paper is organized as follows: Sect. 2 presents related work. In Sect. 3, the Cleopatra Project is presented. Analysis of requirements of the agents based platform is introduced in Sect. 4. Section 5 presents the main issues addressed by the design of a software bridge that works enabling the collaboration of dyadic agents to participate in a multi-users conversations. A prototypical implementation of the proposed bridge is described in Sect. 6. Finally, conclusions and future work are summarized in Sect. 7.

2   Related Work

The sector of Artificial Intelligence has been explored for decades and the interest in the development of Chatbots and Intelligent Virtual Assistants is still growing today. In [11] author discuss about the importance of data transfers using natural language platforms, such as Dialogflow [9], as a key process for chatbot development. In [10] a comprehensive review of research on chatbots is presented. It is supplemented by a review of commercial and independent technology. The paper highlights that current chatbots are designed primarily for dyadic interactions with chat-oriented and/or task-oriented roles. According to a statistic made by the authors, about 90% of the articles in the literature is focused on dyadic chatbots. The most popular commercial platforms for developing NLP conversational agents include Google's Dialogflow, Amazon Lex, IBM Watson, Microsoft LUIS and RASA [3]. An example of multiparty interaction shows a novel experience, where users interact with multiple, text-based conversational systems as if they were sitting around a table [5]. Another survey is presented in [13], where a classification of different tasks is performed regarding the analysis of the multi-participant chat and the research areas that motivated these tasks. Authors highlight a lack of widely-deployed techniques for automated analysis of multiparticipant chat, which if available could aid a large variety of users and data analysts. The virtual heritage concept has the potential to incorporate some of the knowledge aspects through the use of avatars. Through the use of AI, the virtual agents can absorb the relevant knowledge and become the knowledge carriers. Few attempts that use rich individual models of virtual agents for capturing some of the cultural attributes are mostly focused on building culturally

---

[1] http://cleopatra-project.cloud.

adaptive agents [4]. Virtual guides are another popular direction in deploying avatars in virtual heritage. In [12] authors presented an experimental semantics-driven conversational interface design for chatbots in museum settings, targeting visitors to converse about exhibits and learn information about their style, the artists, the era, and other aspects related to them. The results about different scenarios are compared to see how the semantics considered for the design is transferred to the implementation and to the user perception.

3   The Cleopatra Project

Cleopatra (CoLlaborative ExploratiOn of cyber-PhysicAl culTuRal lAndscapes) is a research project aiming at designing a new generation of cyber physical systems, where human users, devices and intelligent software agents are enabled for multi-modal collaboration within a hybrid (physical, augmented and virtual) cultural environment. Cleopatra aims at exploiting a new kind of social dimension for leveraging the interest and augmenting the cultural experience. Research and technological advances will be evaluated in two relevant case studies located in Campania, in order to integrate various points of view and different communication needs. The first scenario will be set in the historical-artistic field, with the creation of the "MudiR-Museo diffuso del Risorgimento" in the Terra di Lavoro (Caserta district), identifying and coordinating different cultural resources of the local system. The second scenario, in the archæological field, will develop routes for visiting the fortified sites of the Samnites in Northern Campania, thanks to a virtual agent and digital reconstructions, to the discovery of a territory rich of history, nature and traditions.

A cyber-physical system will be dynamically built as a temporary P2P overlay of users, who are visiting the same or different distributed cultural assets. The user's smartphone will be a gate to join the network to access a wider cultural space. The overlay will be built over a standard P2P protocol. It will be extended with advanced functionalities, intelligent software agents and services implemented as members of the same network. Visitors, by their avatars, represent channels, which allow to reach and open a window with different view of the same site or on remote sites. Visitor's profiles, characterized by location, cultural interest, device capability and authorization policies will be automatically updated and used to support the dynamic composition of thematic environments, which users can join or leave according their own interest. Intelligent software agents will join the community as moderators, who are in charge of delivering certified information answering questions or intervening in a dialog, to collect feedback, recommend physical or virtual itinerary. They will responsible to guide the collaborative fruition raising the interest, recovering the cultural coherence of the visit and avoiding that users are loss in a meaningless interaction with the applications or with other users.

4   Architecture Design

In this section, the design of a software platform is presented. It implements a bridge between a multi-participant chat and a not dyadic agent developed as multiple collaborating NLP dyadic agents deployed in the cloud. In particular, we focus on the support of multiple communication patterns. The bridge has been designed as an agent itself, that forwards messages to three dyadic NLP agents, who exploit a common storage as a service to share the knowledge extracted from their conversations. In Fig. 1, the UML use case diagram for the agent bridge is shown. The agent bridge receives from the Multi User Chat (MUC) three kinds of messages:

- *Private messages*, which are explicitly addressed to the agent (it cannot be read from the other). They belong to an independent conversation that can be or cannot be semantically related to the conversation ongoing in the public chat.
- *Tagged messages*, which are public messages. Its explicit receiver is the agent itself. They can start a thread in the public conversation, but this thread is readable and open to the others.
- *Untagged messages*, which are public message. Its explicit receivers is not the agent or any explicit receiver.

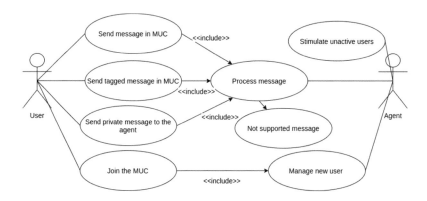

Fig. 1. Use case UML diagram of the agent bridge.

In addition, the agent bridge is able to detect some events, such as the joining of a new user to the MUC, a long idle time of a user in the conversation, a long global idle time in the conversation, and the disconnection of a user.

In Fig. 2, an high level UML component diagram of the software platform is shown. Each conversational agent is an instance of NLP dyadic agents. The service storage is used to share users' data and the relevant information about the common conversation, which are relevant to the specific application. Two different interfaces are implemented by the MUC agent. The first one supports the communication with the MUC, the other one allows for the interaction with the NLP agents.

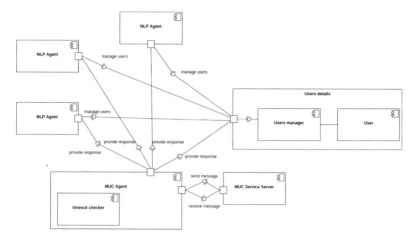

Fig. 2. High level UML components diagram of the not dyadic agent.

5 Bridging Multi Users Conversation with Dyadic Agents

In order to exploit the relevant resources that are available for the development of NLP dyadic agents, a bridge is designed for enabling the collaboration of dyadic agents to participate in a multi-users conversations, and appearing as one participant. We focus here on the main issues which have been addressed to support the participation to a conversation with multiple users, which is a difficult task even for a human.

5.1 Parallel Conversations versus Multiple Threads

Usually, a dyadic agents can handle many conversations in parallel, each one with one user, but each conversation proceeds independently from the others. They are isolated communications, and the first message from a user represents the beginning of a new conversation. A dyadic agents needs to identify and manage multiple conversations at the same time. Messages can be received from any senders on a public channel, where everything can be read by other participants. When the messages is sent from a different user, the agent cannot assume that a new conversation is starting from scratch, because previous messages have been eventually read in the public chat by the user himself. A conversation includes messages from different senders, which start or participate to some threads at any time. In each thread, the agent cannot neglect what it has been discussed before. Moreover, messages can be received on a private channel by the same senders. In this case, the conversation must be handled as a thread of the same conversation, because the agent cannot neglect what the sender said on the public channel and must care about the private conversation in order to decide what to post in the public channel. Moreover the agent must take into account that the private messages are not automatically disclosed to the others.

5.2   Own Messages versus Multiple Receivers

A dyadic agent is the receiver of each incoming message. In a multi-users context the receiver of the message can be undefined (addressed to all joined users), can be a public message tagged with a specific receiver (the agent itself or another user who joined the conversation), or can be a private message to the agent. The receivers could be also implicit, in the sense that the body of the message specifies that the it is addressed to all the participants who satisfy some conditions. Obviously, this issue requires the capability to distinguish multiple communications patterns within the conversation or within one thread. In particular, the last one requires to understand the semantic of the message body.

5.3   Replies versus Posts

A dyadic agent always tries to reply to the received message. In a multi-users conversation, when the agent is not the explicit receiver of a message, it needs to understand when it should or it could respond. In particular, the response could be not required or not expected from the agent, or could have been already provided by others before the agent is ready to respond. A response could depend on a sequence of messages, for example, the choice among exclusive options or a vote. A response could not be timely, as a bid not valid anymore in an auction. We can say that when the agent is not the explicit receiver of a message, it will not reply, but it will make a decision about posting a message. Of course, it could send messages proactively, but usually such an initiative can be programmed as a reaction to some events (long silence, connection of a new users, disconnection of a user, situations awareness, goal activation).

6   Prototype Development

The developed prototype has been implemented in Python. It supports the interaction of the agent bridge with a Multi User Chat (MUC) by the XMPP protocol. The XMPP protocol stands for Extensible Messaging and Presence Protocol. It is a set of open technologies for instant messaging, presence, and multi-party chat. It supports voice and video calls, collaboration and other functionalities by a number of standard extensions. Prosody[2] is the XMPP server technology which has been used to hosts the MUC. The agent bridge implements an XMPP client interface to the MUC and a REST interface to three dyadic conversational agents. The slixmpp library has been used to develop the XMPP interface. The REST interface implementation is based on the Flask library. Conversational NLP agents have been implemented using the Open Source RASA platform. Rasa Open Source is a machine learning framework to automate text- and voice-based assistants. Each NLP agent handles one kind of messages: private messages, public tagged message and public untagged messages. The interface between the agent bridge and the conversational agents uses a REST RASA

---

[2] https://prosody.im/.

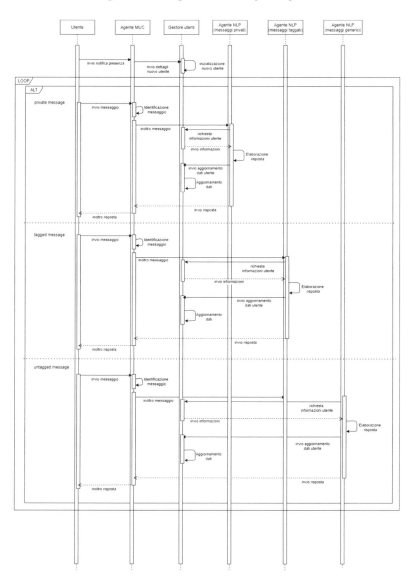

Fig. 3. UML sequence diagram illustrating the three different mechanisms supported by the not dyadic agents.

channel. The message exchange between the agent bridge and a NLP agent is implemented as an HTTP POST request with a json body. Each message from the MUC, forwarded to a NLP agent, includes a sender and a text:

```
{ "sender": "id_user",
  "message": "Hi there!"}
```

Each response returned from a NLP include a text and eventually a link (e.g., to an image):

```
{ "text": "Hey Rasa!",
  "image": "http://example.com/image.jpg" }
```

The storage service also exposes a REST interface that provides methods for handling information about all users, such as their identity, their nick name in the MUC, and eventually, to put and read relevant data about the conversation.

In Fig. 3, we detailed the sequence diagram of the three different communication mechanisms. After a first phase of initialization, a message is received by the agent bridge. According to the kind of message, it will be forwarded to the related NLP agent. The NLP agent, with eventually the support of the storage service, produces a response that is returned to the agent bridge. The NLP agent eventually update the common storage. In the case of untagged messages a response is not mandatory.

As simple case study, we provide in Fig. 4 an excerpt from a conversation between a salesclerk agent and two visitors. We suppose the visitors automatically join the chat-room once they pass trough a defined area of the museum or the salesclerk can welcome them at a specific time of the visit. In Fig. 4a, the salesclerk offers some gadgets for sale and asks to contact him directly for getting additional information or for buying some stuff. When the interested user tags the salesclerk in a message, he is invited to contact the agents privately for organizing the shipment. In Fig. 4b, a private conversation is used to collect personal information such as visitor's address.

(a) Multi-participants conversation.   (b) Private conversation.

Fig. 4. Salesclerk's public and private conversation.

7   Conclusion

In this work, we proposed a communication mechanism that extends the capability of NLP dyadic agents to join multi-participant conversations. We designed and implemented a software bridge that allows developer for combining three NLP agents, which can collaborate to handle public, private and tagged messages in one multi-participant conversation. The software platform, in the context of the Cleopatra project, will be the baseline for developing multi-agents applications, which will complement the visitors' experience in an archaeological site, or in a geographically distributed museum, with a social dimension. It will support the interaction of virtual avatars with visitors of the cultural site, either live or from remote, to foster their engagement with cultural assets and with other users.

Acknowledgements. The Cleopatra Project has been funded by the University of Campania "Luigi Vanvitelli" through the VALERE 2019 research program.

References

1. Aversa, R., Di Martino, B., Ficco, M., Venticinque, S.: A simulation model for localization of pervasive objects using heterogeneous wireless networks. J. Simul. Model. Pract. Theory, 1758–1772 (2011)
2. Bickmore, T., Vardoulakis, L., Schulman, D.: Tinker: a relational agent museum guide. Auton. Agents Multi-agent Syst. **27** (2013)
3. Bocklisch, T., Faulkner, J., Pawlowski, N., Nichol, A.: Rasa: open source language understanding and dialogue management. In: NIPS Workshop on Conversational AI (2017)
4. Bogdanovych, A., Ijaz, K., Simoff, S.: The city of Uruk: teaching ancient history in a virtual world. In: Intelligent Virtual Agents. Springer, Heidelberg (2012)
5. Candello, H., Pinhanez, C., Pichiliani, M.C., Guerra, M.A., de Bayser, M.G.: Having an animated coffee with a group of chatbots from the 19th century. In: Extended Abstracts of the 2018 CHI Conference on Human Factors in Computing Systems, CHI EA 2018, pp. 1–4 (2018)
6. Kopp, S., Gesellensetter, L., Krämer, N., Wachsmuth, I.: A conversational agent as museum guide - design and evaluation of a real-world application. Intell. Virtual Agents **3661**, 329–343 (2005)
7. Linrun, Q., Kangshun, L.: The research of intelligent agent system architecture based on cloud computing. In: Proceedings of the 12th International Conference on Computational Intelligence and Security (CIS), pp. 1–13 (2016)
8. Masche, J., Le, N.T.: A review of technologies for conversational systems. In: Advanced Computational Methods for Knowledge Engineering, pp. 212–225 (2018)
9. Navin, S., Agrawal, A.: Introduction to Google Dialogflow, pp. 13–54. Apress, Berkeley (2020)
10. Seering, J., Luria, M., Kaufman, G., Hammer, J.: Beyond dyadic interactions: considering chatbots as community members. In: Proceedings of the 2019 CHI Conference on Human Factors in Computing Systems, CHI 2019, pp. 1–13. Association for Computing Machinery, New York (2019)

11. Singh, A., Ramasubramanian, K., Shivam, S.: Introduction to Microsoft Bot, RASA, and Google Dialogflow, pp. 281–302. Apress, Berkeley (2019)
12. Spiliotopoulos, D., Kotis, K., Vassilakis, C., Margaris, D.: Semantics-driven conversational interfaces for museum chatbots. In: Rauterberg, M. (ed.) Culture and Computing, pp. 255–266. Springer, Cham (2020)
13. Uthus, D.C., Aha, D.W.: Multiparticipant chat analysis: a survey. Artif. Intell. **199–200**, 106–121 (2013)

E-government and Italian Local Government: Managerial Approach in Two Macro Areas to Improve Manager's Culture and Services

A. Marino[1], P. Pariso[1], and M. Picariello[2(✉)]

[1] Dipartimento di Ingegneria, Università degli Studi della Campania "L. Vanvitelli",
Via Roma, 29, Aversa, CE, Italy
{alfonso.marino,paolo.pariso}@unicampania.it
[2] Aversa, Italy
michele.picariello@unicampania.it

Abstract. The purpose of this paper is, theoretically and empirically, explore how public e-government policies are translated into practice. The study intends to analyze the change in management culture within the Campania (South Italy) and Liguria (North Italy) Local Public Government (CPG and LPG, respectively) in the context of the Italian Public Administration (IPA). The aim of the study is to understand, through a SWOT and factor analysis, the elements that are relevant to change the process logic from bureaucratic to competitive. The main data of the sector will be presented. The analysis of the data underlines the strategic role of management in the sector. Furthermore, items of questionnaire Likert statements, using factor analysis and varimax rotation, underline different management typologies. We also analyze how the Italian managers characterize organizational structures and provision of services. The conclusions indicate possible actions to be taken to create efficiency and productivity within the CPG and TPG and to effect change in the picture of the complex IPA.

Keywords: E-government · E-bureaucracy · ICT and public sector reforms

1 Introduction

The adoption of E-Government (Egov) in public sector organizations has been often associated with Reform programs aiming at reducing the inefficiencies generated by bureaucratic culture. Egov is quite difficult to define with many different definitions in the public domain. Defining Egov too narrowly as electronic service delivery only can result in exercises that are overly complex and costly. Such a definition can also miss the transformative potential of Egov to speed-up decision-making, streamline or reduce processes, or reduce costs of engagement. However, it is important to have a common working definition to guide public bodies in developing focus and priorities. It is expected that any working definition will change over time as opportunities and priorities develop. In Italian Local Government (ILG) Egov experience, as an information system enabled innovation in government, is transforming technological platforms and the way

L. Barolli et al. (Eds.): AINA 2021, LNNS 227, pp. 107–116, 2021.
https://doi.org/10.1007/978-3-030-75078-7_12

governments in procure services and engage with suppliers. Furthermore, the organizational bottlenecks and the high incidence of political decision makers constitute the sector's negative features. A competitiveness-enhancing reform based on professional skills, management capabilities and training needs to be implemented but the process is still ongoing. Egov can be a driving force for local development. As an expanding sector, it dynamizes traditional economic activities and enhances local manager specificities, offering people new opportunities for a better service quality. The term Egov, as used by the Italian Government Project, applies to the use of ICT as a tool to achieve better government. Moreover, numerous studies [3] have shown that a timely analysis may be supportive to overcoming obstacles such as the Egov context. External barriers and internal challenges to Egov implementation. The study intends to analyze two Italian regions: Campania and Liguria. The two regions contribute to the creation of wealth and economic growth due to the number of users and Italian regions.

2 Literature Review

2.1 Italian Public Administration

Egov does not operate alone. The context in which Egov is taking place and the ability of governments to respond to these external pressures are determinant for the ultimate success of Egov. In particular, the broader information society of which Egov is one component plays a role in professionality, management and cultural assets. All these items within IPA somewhat are outstanding. All of these factors affect the willingness of businesses and citizens to use, or take up, Egov services. The failure to respond to an ever-changing environment and expectations can result in barriers to Egov implementation. The local level consists of regions, provinces, towns and other less relevant local bodies. All of them enjoy political and administrative autonomy. The former, in particular, is based on the principle that all administrative functions. Provinces and towns are also developing ICT autonomy, in the framework of a process that should enhance their independence from the state resources. This is the case of provinces and towns, while the regions still depend on the state resources. Towns have an analogous structure and the relevant legislation provides the direct use of sources, introducing an element of democracy at this level of the Italian legal order. The local administration has to be distinguished by the peripheral articulation of the central administration, meaning all bodies depending on the ministries but peripherally located. Moreover, at local level, particularly in South Italy there is an important barrier to develop Egov: digital divide. The digital divide is an important barrier to Egov in that people who do not have access to the Internet will be unable to benefit from online services. While Egov can also improve services to citizens through other channels, the inability to provide online services to all citizens can hold back Egov projects.

2.2 Case Study Context

Technological advancements and the search by suppliers for new markets have resulted in a bewildering array of technical solutions in search of problems to fix. Italian Government face the challenge of fostering the development of Egov while there is still great uncertainty regarding fast moving technological change, and it is difficult to anticipate future policy impacts. New technologies are tempting because they often promise better solutions and enticing possibilities for business change. More often, however, they promise solutions that purport to enable an organization to implement IT without changing its services processes. It is therefore not surprising that public sector organizations keep trying to develop systems based on new technologies. Experience shows, however, that systems built on emerging and unknown technologies are very susceptible to failure. In some instances, the potential benefits might warrant taking such huge risks; most often, this is not the case. IPA, have recognized a potential for adding to service supply in areas that were previously not considered attractive [1] for public services. Additionally, public services can enable public authorities to achieve a variety of objectives, such as improving employment [2] and the physical environment of an area. Campania and Liguria managers' experiences are within this track. It is thus important, on Egov grounds alone, for Italian Governments to continue policies and specific interventions to reduce the digital divide. Campania and Liguria for e.g., have large central government proxies: urban transportation; public support for business; health care and social services [7]; education and research; housing. All these proxies are complemented by quality services, which are also the result of management, professional skills and staff training.

2.3   Northern vs Southern IPA

A change in the relationship between the northern and southern IPA is necessary. Many regions, local government are developing services that are no customer-focused. As governments are developing more and more services, they are also coming to the realization that they often do not know what kind of Egov citizens want. Egov tools have provided governments with new ways to provide information and to consult with citizens. However, determining the preferences of citizens and businesses, with regard to the structure and content of services, rests a true challenge for Italian Government. In part because many people would hard-pressed to be able to articulate their expectations of government, even with a full understanding of the technological possibilities. These changes can affect the consequences of service development on social and economic environments [4]. New logic and cultural challenges are linked to the application of management theory in the CPA and LPA. Such issues include the characteristics of the tasks that the public is supposed to carry out; the normative foundation of their work and; in this context, some strategic organizational and managerial elements are considered in order to improve management and managers. These three elements are referred to as the external and internal task context, the normative context and the organizational context.

2.3.1   The External and Internal Task Context

Two circuits typically produce the service's experience. The first relates to quality patterns and motivations. The aim is low external barriers to entry. External Egov barriers often concern breakdowns, missing components or lack of flexibility in the government-wide frameworks that enable Egov. The result can be an inability to achieve a whole-of-government perspective in services implementation. This is particularly true when Egov is treated as a merely technical issue rather than one that concerns the basic service delivery mandate of government, or when regions ignore how additional value can be created by better collaborating with other regions providing related services. In this sense, the barriers are not external to local government itself, but rather concern responsibilities that are broader than the sphere of activity of any region. The second is more diffuse and complicated. It concerns the local managers' goal for which the specific services can be seen as a means and an end. In this latter approach, public service is not about an individual's concerns, but rather about the reproduction and development of their country's culture. In this way, IPA carries out both aggregative and integrative functions. On the one hand, it must take as a point of departure the citizens' needs; on the other, it socializes and regulates the citizens' behavior.

2.3.2   The Normative Context

The normative context contains the considerations, principles and demands to which the managers and organization must generally relate. Therefore, there are varying elements, all of which can be seen as restrictions on internal processes and the way in which services are produced and distributed. The success of Egov initiatives and processes are highly dependent on government's role in ensuring a proper legal framework for their operation. The introduction and uptake of Egov services and processes will remain minimal without a legal equivalence between digital and paper processes. As a first step, identifying these areas would help indicate where redundant or overlapping normative were in place: an agreed process of regular examination would provide an opportunity to get rid of requirements that have outlived their usefulness. Moreover, legislation enacted in order to protect the privacy and security of citizens' data can impede data sharing across government. The issues are related to resources use, productivity, efficiency and quality of services.

2.3.3   The Organizational Context

Starting from 24 privileged witnesses from each three areas, that make up the sector some strategic organizational and managerial variables have been considered. Motivational factors and working organization are essential, knowing how and why to motivate employees is an important managerial skill. Furthmore, IPA legislation is strategic in order to formulate a legal and regulatory framework for the sustainable development and management of public service. Job stability [6] is also an important element. Furthermore, investment in information technology (ICT), the role of private organization [8], and managerial culture are major human factors at the organization level. These variables will be investigated in order to better understand mutual conditioning and possible innovation of the managers.

3   Methodology

A sample of Italian public managers was selected in October 2017; the interviews were conducted beginning in November 2017. 800 managers were interviewed from each region, Campania and Liguria. A large part of interviews (70%) was conducted inside the organization and the remainder (30%) by skype interview. It can be said that the IPA is characterized by the presence of a few large units; the presence of small units with at most ten employees; the presence of staff with high service life. The managers have been investigated by responding to a questionnaire. Starting from privileged witnesses, three key points emerge from the analysis: social background variables of managers, productivity variables and efficiency variables. The questionnaire comprised 30 pre - developed, 15 for each part, Likert statements, designed to measure the five different areas of the questionnaire. Specifically, respondents were asked to indicate the level of criticism on a seven-point scale, ranging from "strongly criticizes" (7) to "low criticizes" (1) on different items. The 30 Likert statements were explored by principal components factor analysis and varimax rotation, which resulted in a four - factor solution, two for each region. The purpose of the factor analysis [5] was to combine the statements into a set of factors that were deemed to represent a first example of organizational types linked to the interviews of managers into different regions. The internal consistency of each factor was examined by Cronbach's alpha tests. All the alpha coefficients were above 0.5, which means that high correlation existed between the items.

4   Empirical Results

The results were obtained by questioning a sample of managers from Campania and Liguria. The sample distribution by geographical and education factors, is show in Tables 1 and 2. Table 1 shows the general lack of tertiary sector educational qualifications among the staff.

Table 1. Managers by region

Campania public managers (South Italy)	Liguria public managers (North Italy)
33% on the coast	33% on the coast
33% in the middle	33% in the middle
33% inland	33% inland
Total managers 800	Total managers 800

The decision to divide the region into three main areas is dictated by the different orientation of the management culture concerning the privileged witnesses. Within the three-macro areas, culture and service orientation is homogeneous in relation to the school curriculum and previous experience in the field. Table 2, shows the number of graduates for the two regions. "No educational level" in Campania, is the item with the greatest number of responses. In Liguria a degree in management, law, engineering, cultural heritage (also these prevail for Campania) and Secondary school diploma represent 58% of the responses.

Table 2. Managers education level by region

Campania managers	Liguria managers
University 10%	University 30%
Secondary high school 25%	Secondary high school 28%
Junior high school 20%	Junior high school 31%
No education level 45%	No education level 19%
Total 100% (70% male)	Total 100% (80% male)

The purpose of the factor analysis, which resulted in a three factors solution, was to combine the statements into a set of factors that were deemed to represent the organizational types linked to the interviews of managers. Specifically, items with higher loadings, 16 factors, (see Table 3) were considered (alpha coefficients above 0.5) as more important and as having a greater influence on organizational types.

Table 3. Factor analysis Italian public managers

Factor names and items	Mean	S.D.	Factor loading	Alpha
Improving productivity Campania managers	**5.0**			**0.83**
Motivational factor to entry	5.25	1.57	0.78	
The role of public management	5.14	1.59	0.65	
Control of public sector	5.19	1.36	0.58	
Coordination of public sector	5.61	1.43	0.61	
Improving productivity Liguria managers	**5.73**			**0.90**
Relations customers – allotment of duties	5.19	1.37	0.53	
Low level of information technology	5.15	1.37	0.49	
Private control	6.15	1.11	0.81	
Private coordination	6.25	1.23	0.88	
Improving efficiency Campania managers	**5.26**			**0.80**
Work organization as problem	5.19	1.54	0.75	
Mutual help relation with other managers	5.00	1.04	0.62	
Job stability	5.17	1.13	0.52	
Salary	5.13	1.11	0.69	
Improving efficiency Liguria managers	**5.73**			**0.81**
Managerial culture	5.71	1.24	0.75	
Quality of service	5.16	1.51	0.72	
Credit and information by banks	5.23	1.37	0.54	
Public legislation	5.33	1.47	0.64	

Managers have been asked what the critical points to improve productivity. The results are categorized in Table 4. Southern Italian managers emphasized motivational factors to entry, especially incentives for productivity and training, and then factors described as 'the form of management' and 'coordination of the tourism sector'. Northern Italian public managers considered the first element to be 'coordination' and 'control of public sector'. Both groups of managers underline the necessity of a new normative context and strongly criticize the role of the Italian Government.

Table 4. Factor loading items linked to the critical points for improving productivity

Campania managers	Liguria managers
Motivational factor to entry (0.78)	Relation customer – allotment of duties (0.53)
The role of public management (0.65)	Low level of information technology (0.49)
Control of public sector (0.58)	Private control (0.81)
Coordination of public sector (0.61)	Private coordination (0.88)

The critical factors to efficiency items (Campania, see Table 5) are work organization, mutual help relationships with other agencies and salary. Managers underline the absence of hierarchical influences and 'professionality' into the service supply. They also point out the need to improve mutual relationships with other managers. The managers underline the modalities by which the different members of the organization undertake their specific tasks, professional functions and roles. The salary is the last element; managers argue that individual economic reward should be taken into account to improve productivity and efficiency. Northern managers, from Liguria, (see Table 5) underline the importance of 'managerial culture' and 'quality of service' in terms of paying more attention to the specific managerial culture of the sector and the needs of its users. One important bottleneck is quality of service in the relations between customers: with reference to the need for a quick response about the coordination and control of information flows.

Starting from Tables 4 and 5, the data shows four different organizational types: insensitive organization, sensitive organization, participated organization and proactive organization. In the insensitive organization, the main characteristics refer to low attention to the identification of the user's needs, and to productivity and efficiency. This configuration is present in a large part of Campania, particularly inland. This type of organization takes into account the following factors to improve productivity: motivational factors to entry (0.78) and the role of public management (0.65). In addition, improving efficiency: work organization as problem (0.75) and salary (0.69). The sensitive organization shows interest in the knowledge of the user's needs, productivity and efficiency. This configuration is present in Campania on the coast and inland in Liguria. In these areas, the organization takes into account the following factors to improve productivity: control of public sector (0.58) and coordination of public sector (0.61). Improving efficiency: mutual help in relations with other managers (0.62) and job stability (0.52). Participated organization in the middle of Liguria takes into account the

Table 5. Factor loading items linked to the critical points for improving efficiency

Campania managers	Liguria managers
Working organization as a problem (0.75)	Managerial culture (0.75)
Mutual help relation with other manager (0.62)	Quality of service (0.72)
Job stability (0.52)	Credit and information by banks (0.54)
Salary (0.69)	Public legislation (0.64)

following factors for improving productivity: relations with customers - allotment of duties (0.53), low level of information technology (0.49). Improving efficiency: public legislation (0.64), credit and information by bank (0.54).

The proactive organization shows great interest in the user's requests, productivity and efficiency. This configuration is present in a large part of the Liguria coast. This type of organization takes into account the following factors for improving productivity: private control (0.81) and private coordination (0.88). Improving efficiency: managerial culture (0.75) and quality of service (0.72). From four different organizational types, do not emerge a common vision of Italian Egov. The establishment of overall frameworks is an important step in meeting common Egov challenges. While taking a common or shared approach promotes the consistency of service systems, it should not, however, be construed as shifting the responsibility away from the managers that are responsible for everyday implementation of services. Indeed, a number of serious challenges to Egov implementation remain even once all of the appropriate frameworks are put in place. These challenges involve ensuring that a common understanding and sense of mission is shared across all levels of government and ensuring the necessary leadership to accomplish this. It also requires improving coordination and collaboration, clarifying public strategy, ensuring that government officials have the necessary skills and tools to carry out their mission and to monitor and evaluate success. A common vision is not a goal in itself, but a means to achieve policy priorities.

5   Discussion

The four organizational typologies, the insensitive, sensitive, participated and proactive organization, highlighted in the empirical results how managers have strategic influence on the services productivity of the CPA and LPA linked to different areas investigated in the country. On the other hand, in Campania and Liguria, two different visions emerge. In recent years, there has been an increasing interest in the use of management theory within the IPA. Namely, in two aspects: first, an interest in the application of management theory in the Local Public Administration, CPA and LPA. This takes the form of importing ideas and methods developed in and for the private sector. The assumption is that the private sector is better than the public sector in some specific ways: private sector organizations are more cost conscious, more inclined to implement modern personnel management and more capable of developing corporate culture as a steering instrument. Such a debate considers the incentives for productivity and particularly the necessity to create some reliable measures of management efficiency in the local administration. The second aspect is an interest in the use of management theory in the study of the public sector. Here the aim is somewhat different. In Liguria, the main variables are those of the private control and coordination, the importance of ICT and allotment of duties. Campania instead, always improving its productivity, underlines the importance of control and coordination of public sector, the motivational factor and the role of public management. In order to improve efficiency the two regions underline a different variables set that for Liguria is the managerial culture, quality of service, role of bank and public legislation. In Campania instead such a different variables set is working organization, mutual help, job stability and salary. The difference between the regions within the same country highlights that despite the profound differences in managerial approach, both attract a growing number of services. Operative and theoretical action are important to improve the public sector. At operative level, some organizational actions are strategic for managers in different regions. Particularly a cross contamination between managers in the different regions could be an interesting operative approach to improve efficiency and productivity in Campania by large part of Liguria managers from the coast.

6 Conclusions

Making a change means taking into account the different experiences and culture of managers, in each of these broad areas in which there are significant differences in relation to the location and culture. It is important to modify operational values and decisions. The skills required for Egov are not simply formal, as general managers need broad skills to engage in Egov decision making. Managers must be able to lead the organization's ICT department and outside partners, and they must be able to integrate the organization's strategy with the broader goals of the organization. The need for greater investment in ICT represents a strategic variable to change. The role of the IPA is an open question and therefore, in the diagnosis of reform there is no "one best way". Managerial approach and managers roles are strategic variables to improve public sector and its performance.

References

1. Bertot, J., Estevez, E., Janowski, T.: Universal and contextualized public services: digital public service innovation frame work. Gov. Inf. Q. **33**(2), 211–222 (2016)
2. Commission of the European Communities: The Public Administration of the Community: A Study of Concentration, Competition and Competitiveness. Bruxelles, Office for Official Pubblications of the European Communities (2016)
3. Di Martino, B., Marino, A., Rak, M., Pariso, P.: Complex, Intelligent, and Software Intensive Systems (2020)
4. Dekker, R., Bekkers, V.: The contingency of governments' responsiveness to the virtual public sphere: a systematic literature review and meta-synthesis. Gov. Inf. Q. **32**(4), 496–505 (2015)
5. Hair, J.F., Anderson, R.E., Tatham, R.L., Black, W.C.: Multivariate Data Analysis with Readings. Prentice-Hall, Hemel Hempstead (1995)
6. Lee, G., Kwak, Y.H.: An open government maturity model for social media based public engagement. Gov. Inf. Q. **29**(4), 492–503 (2012)

7. Marino, A., Pariso, P., Picariello, M.: Digital Health: The Italian Experience in the European Context (2021)
8. Porumbescu, G.A.: Linking public sector social media and e-government website use to trust in government. Gov. Inf. Q. **33**(2), 291–304 (2016)

Container Based Simulation of Electric Vehicles Charge Optimization

Rocco Aversa, Dario Branco, Beniamino Di Martino,
and Salvatore Venticinque[✉]

Department of Engineering, University of Campania,
via Roma 29, 81031 Aversa, Italy
{rocco.aversa,beniamino.dimartino,salvatore.venticinque}@unicampania.it,
dario.branco@studenti.unicampania.it

Abstract. This paper proposes the exploitation of simulation techniques to evaluate energy optimization strategies in smart micro-grids. In particular, a container based deployment approach allows for running simulations in the Cloud, evaluating multiple scenarios and optimization algorithms. Here we present both the simulator technology and an original two-phases optimization algorithm that computes a sub-optimal solution in real time. We introduce a simple scenario with real data.

1 Introduction

Two of the greatest environmental concerns are pollution and $CO2$ Emissions due to vehicles. Use electrical vehicles could the solution of both these problems. However, when EVs are charged at home and, even sometimes, when the charging takes place at public or commercial Charging stations, they start to stress the electrical distribution network. On the other hand, Smart Grids have the potential to consider EVs not only as a load, but also as a flexible power source. Smart meters can provide information to carry out an optimal schedule to optimize the available power in the grid. A comparable research study, performed in Portugal, reveals a positive correlation between charging of electric vehicle and solar power [6].

The spread of IoT technologies provides real time data that can be exploited for developing smart solutions, to improve energy utilization in micro-grids. The GreenCharge project provides innovative cities with technological solutions and business models for cost-effective implementation and management of charging infrastructures for electric vehicles in smart micro-grids [7]. GreenCharge is testing its innovative solutions in practical trials in Barcelona, Bremen and Oslo. Simulation will be used in the GreenCharge Evaluation Loop to operate the measures in a virtual environment where the Pilots can be extended, overcoming real limitations, and the measures can be easily complemented with missing functionalities [4].

L. Barolli et al. (Eds.): AINA 2021, LNNS 227, pp. 117–126, 2021.
https://doi.org/10.1007/978-3-030-75078-7_13

This contribution focuses on the design and implementation of an optimal loads scheduler that exploits charging flexibility of electric vehicles and the potential vehicle to grid capability. The objective of the optimal schedule is to maximize the utilization of energy production by decentralized renewable energy sources and to reduce the power peak. The optimization techniques exploits the load shifting and the control of power level of charged EVs. A software simulator, developed within the activities of the GreenCharge H2020 research project, is used for the experimental activities. Here we present the container based architecture of the simulator and a simple case study that uses real data to reproduce a realistic scenario.

2 The Greencharge Simulator

The GreenCharge simulator reproduces in a virtual environment the events that occur in a real pilot using a collection of real misured data [3]. It is based on the original CoSSMic simulator [1], and allows to extend the evaluation capability in real pilots, which are limited in the heterogeneity and number of devices and in the duration of operating trials. The simulator is based on the discrete-event simulation (DES) model where the system appears as a discrete sequence of events in time [2]. In Fig. 1 the conceptual model of the container based deployment configuration of the GreenCharge simulator is shown. Using both a virtual or a real network many containerized components interoperate trough a loosely coupled integration. The blue boxes represent the simulation engine and its Graphical User Interface (GUI). They use a volume to access simulation input and output data such as the configuration of scenarios, input time-series and output results. The XMPP server provides a peer-to-peer communication overlay for multi-agents distributed implementation. A volume is used to save user-credentials, since the simulator can be used by multiple users who can run their simulations in parallel, in one or in multiple containers. An optimization model can be integrated as Energy Management Systems (EMS) that runs in its own container and uses the Simulator interface to receive simulation events and to return the optimal energy schedule. The GreenCharge project will evaluate two different EMS innovative technological solutions, developed by the University of Oslo and by the Eurecat partner. Here we investigate an alternative solution that is used to demonstrate how the simulation platform works. The user can access the Simulator GUI by the web interface of the hosting container.

In Fig. 2 the GreenCharge Simulator Graphic User Interface (GUI) is shown. The Control Panel represents a kind of dashboard of the tool to set and overlook a simulation session. In this panel, after completing the configuration phase, we can set the day and the starting time of the simulation. Pressing the start button activates all the simulation agents and starts the scheduling process. As the simulation progresses over time, the actual simulation time is updated, allowing the user to keep track of its evolution. The *simulation scenario* is described by two XML files. The *neighborhood.xml* file describes the static configuration of the micro-grid, that means the list of device with their parameters and the

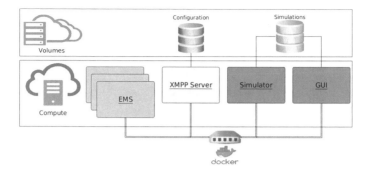

Fig. 1. Container based deployment configuration of GreenCharge Simulator

Fig. 2. Simulator GUI

topology of their connections. The *loads.xml* file defines the events which will occur during the simulation, such as the start of an EV charge session, the booking of a charge point, the update of prediction of PV energy production, or the planned utilization of a heating-cooling device. The *inputs* include a list of time-series which correspond to the energy profile of devices such as washing-machines or production profile of photo-voltaic panels. The output include a log of the messages exchanged between the EMS and the simulation engine. The remaining results consist of time-series directly uploaded by the EMS or computed by the simulator according to the schedule received from EMS.

3   Formulating the Energy Management Problem

Self-consumption can be defined as the share of total photovoltaic production consumed directly by the owner of the plant [5]. In Fig. 3, areas A and B correspond to the interaction with the electricity grid in terms of demand and generation, respectively. Area A corresponds to the power absorbed from the electricity network by a building. Area B corresponds to the injection into the electricity grid of the surplus power produced by the photovoltaic system. The overlapping

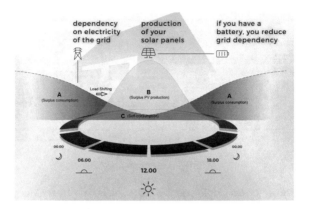

Fig. 3. Self consumption

parts, i.e. area C, correspond to the power used directly inside the building. This area is sometimes referred to with the term of absolute self-consumption, but what is usually meant by the term self-consumption is the self-consumed part relating to total production. The increase in self-consumption may provide greater profits of the plants and may decrease the pressure on the electricity distribution network (Grid).

Therefore, the goal of an energy management system here is to find an optimal schedule of energy loads, that maximizes the self-consumption, without violating constraints set by the users about the earliest start time and the latest start time of the appliances. Example of shiftable loads are the ones generated by a washing-machine or a dishwasher.

The formulation of the complete optimization problem, taking into account energy producers, energy consumers and available stotage batteries, aims to maximize the ratio between EC_{res} and EP_{res}, where EC_{res} is the renewable energy consumed in the neighborhood and EP_{res} represents the renewable energy produced. Maximizing this ratio is equivalent to minimizing the energy drained from the grid and the energy exported from the panel to the grid.

Moreover, assuming that storage can either provide or drain energy at one time, and that max charging power is equal to max discharging power, the discrete minimization problem is formulated in Eq. 1.

$$min\left(\sum_{n=0}^{N}\left(\left| \sum_{i=0}^{I} P_{res}^{i}(n\Delta t) - \sum_{j=0}^{J} C_{con}^{j}(n\Delta t) \right. \right. \right.$$
$$\left. \left. \left. + \sum_{k=0}^{K} x_k * PMAX_{storage}^{k} \right| \right) \right) : x_{n,k} \in [-1,1] \forall k \in K, \forall n \in N \quad (1)$$

This formula is valid if and only if the storage can only accumulate the energy produced by the local renewable energy sources. In particular:

- N i the number of discrete time intervals
- I is the number of Producers
- J is the number of Appliances
- K is the number of energy Storage
- $P_{res}^i(n\Delta t)$: is the power produced by i-th renewable source at time $n\Delta t$
- $C_{con}^j(n\Delta t)$: is the power consumed by j-th appliance at time $n\Delta t$
- $x_{n,k} * PMAX_{storage}^k$: power provided by k-th energy storage at time $n\Delta t$.

We also suppose that $J1$ appliance generates shiftable loads, while loads of $J2$ appliance cannot be shifted, such as lights, tv and any other devices that are not monitored and controlled. All these devices represent a background load that can be subtracted to the available renewable production. In Eq. 2 we split the contribution of shiftable and not shiftable loads.

$$\sum_{j=0}^{J} C_{con}^j(n\Delta t) =$$

$$\sum_{j_1=0}^{J_1} C_{Back}^{j_1}(n\Delta t) + \sum_{j_2=0}^{J_2} C_{Sh}^{j_2}(t_0^{j_2}, n\Delta t) \qquad (2)$$

with $J1 + J2 = J$. In order to simplify the problem, looking for a sub-optimal solution, we propose to address the optimization problem in two steps. First we find the schedule for shiftable loads, than we try to shift and modulate the charging of energy storages.

The first optimization problem is formulated in Eq. 3.

$$min\left(\sum_{n=0}^{N}\left(\left|\sum_{i=0}^{I} P_{res}^i(n\Delta t) - \sum_{j_1=0}^{J_1} C_{Back}^{j_1}(n\Delta t))\right.\right.\right.$$

$$\left.\left.\left. - \sum_{j_2=0}^{J_2} C_{Sh}^{j_2}(t_0^{j_2}, n\Delta t)\right|\right)\right) : t_0^{j_2} \in [t_{est}^{j_2}, t_{lst}^{j_2}] \forall j_2 \qquad (3)$$

Hence, here we aim at finding the best set of starting times (t_0) of the shiftable loads that maximizes the self-consumption with $t_0^{j_2}$ between the earliest start time $t_{est}^{j_2}$ and the latest start time $t_{lst}^{j_2}$. Once the start time of shiftable loads has been assigned we deal with the optimal charging/discharging of the storages. In particular, we address this problem as a linear minimization problem every Δt seconds. In Eq. 4, $R_{res}(n\Delta t)$ represents the *Residual power* in the n_{th} time interval, while the $k_t h$ x is the real decision variable for each storage. If $x_k > 0$ in the $n^t h$ interval, it means that the storage k will charge in that in interval, it will discharge otherwise.

$$min\left(\left|R_{res}(n\Delta t) - \sum_{k=0}^{K} x_{n,k} * PMAX_{storage}^k\right|\right) : x_k \in [-1,1] \forall k \qquad (4)$$

4   Optimization Algorithm

The optimization algorithm works in two phases. In a first phase a Genetic Algorithm is used to find the start time for the load shift. In a second phase a many linear optimization problems are solved, one for each time slot within which the power value of charging EVs is kept constant.

4.1   GA Based Load Shifting

To provide results in a defined and certain time, the optimal schedule will be computed using evolutionary optimization techniques, with the addition of specialized genetic operators studied for the type of addressed problem.

In our case, we models an individual as a list of start-time of the loads that must be scheduled. Therefore, the number of genes is variable and depends on the user's real-time needs. The same applies to optimization constraints in terms of load start-time preferences. The less flexible they are, the less there will be room for optimization, making the response time lower.

The optimum is computed without limits of iterations, but using the quadratic norm of the population as a stopping criterion. The quadratic norm (qn) represents an index that measures how much the population is scattered in the research space.

4.2   Linear EV Charge Optimization

The linear minimization problem of Eq. 4 is solved in 144 time interval of 10 min each. EVs are modeled as stationary storage. Basically it is possible to use the energy stored in car's battery whenever possible to supply power to the loads present within the neighborhood if necessary. Any intervals where there is not enough self-production to meet the energy demands would be balanced by the battery energy, thus increasing self-consumption On the other hand, if an EV k does not support the V2G capability, the $x_k > 0$ constraint will be set.

Other constraints limit the maximum amount of energy that can be drained from and to the EV energy storage. Such constraints are defined in Eq. 5.

$$\begin{cases} \frac{x_k * PMAX_{storage}^k}{6} \leq E_t - E_n & \text{if } \frac{x_k * PMAX_{storage}^k}{6} > 0 \\ \frac{x_k * PMAX_{storage}^k}{6} \leq E_n & \text{if } \frac{x_k * PMAX_{storage}^k}{6} < 0 \end{cases} \tag{5}$$

Where E_n is the Energy stored in the EV battery at the n_{th} interval and E_t is the total battery capacity. Consequently, $E_t - E_n$ is the energy that can still be charged. The previous constraint states that the energy charged within a ten-minute interval (the sampling step chosen) cannot be greater than the energy necessary for the complete charge of the battery, on the other hand, the Energy discharged within a 10 min interval cannot be greater than the amount of energy already present in the battery. Moreover, the last constraint is set to avoid that the power peak in each time interval does not overcome the physical threshold.

Finally, in order to satisfy the EV energy demand before the departure time, the algorithm dynamically set some x_k values before the linear solver is run. The idea is to leave the car battery free to charge and discharge freely until the last available timeslot after which it would no longer be possible to charge the car up to the necessary level in the desired time even with the maximum charge power. The algorithm, in each time slot, computes the necessary power value to satisfy the charging demand. If the required value, divided by the maximum charging power, belongs to $[-1, 1]$, it means that the achievement of the target SoC is still feasible. On the other hand, to avoid that in the next slot is too late to charge at the desired level even if the maximum power is used, then a lower threshold is set for deciding to charge at maximum power in the current interval. Such a threshold value is set equal to 0.9 in the following experiments. A special case of this algorithm is when V2G is enabled. In this case the algorithm prevents the usage of V2G in the i time slot, if it could cause in the $i + 1$ slot that the required power to achieve the target SoC exceeds the maximum value. Logically this procedure also affects the optimal result, because we have inhibited the use of V2G in an slot in advance. Howeve, since the slot duration is only 10 min, then the impact is limited.

5   Experimental Results

The simulation scenario includes 2 photo-voltaic plants and 16 dishwasher, 2 washing-machines and three EVs. The power profiles of all devices have been extracted from real measured data, but, in order to configure a larger workload, the same device is replicated in the proposed experiment with random EST and LST constraints, whose difference is no more than one hour. The PV plants produces 24.9 kWh from 08:00 to 18:15. The stopping criterion for computing the theoretical optimum is a value of the quadratic norm equals to 10^{-6}. It is reached on average in 100 iterations. The average value of self-consumption has been 59%, that corresponds to an green energy consumption of 14.7 kWh. Figure 4a shows the optimal schedule compute by the Genetic Algorithm in a specific run that converged after 81 iterations. In Fig. 4b we see in green the self-consumed energy. It is straightforward to observe that, because of the constraints, some loads consume from the grid before the PV plants start to produce. On the other hand the power peak exceed at the PV power more than once. The blue line corresponds to the PV power consumption In a second phase the linear optimization computes the optimal charge of the three EVs. We considered the real brand and models which have been monitored in trials: two instance of a VW e-Golf with a 24 kWh battery and a Peugeot iOn with a 16 kWh battery. The maximum charge power for both was limited by the charging point. All the required parameters, including the arrival and departure time, and the status of

Table 1. Input parameters for the EVs charge optimization.

EV	Capacity	Max power	Arrival (soc, time)	Departure
EV1	24 KWh	3.6 kW	50%; 10:40	70%; 20:40
EV2	24 KWh	3.1 kW	25%; 09:20	90%; 17:40
EV3	12 KWh	1.8 kW	25%; 08:15	85%; 16:16

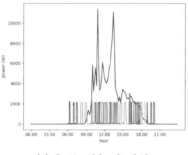

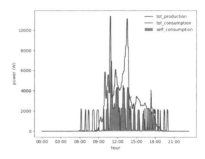

(a) Optimal loads shift   (b) Self Consumption

Fig. 4. Optimal schedule after the loads shift.

charge on arrival and the target one, are listed in Table 1. In Fig. 5 it is shown a comparison of results, in terms of self-consumption, with and without V2G support. In Fig. 5a it is shown that the algorithm is able to consume all the energy produced by the PV plants, but it cannot reduce the power peak when it needs to charge the EV to comply with the desired energy level at the departure time. The stacked power of charging EVs in Fig. 5b try to saturate the PV panel in the beginning, but especially EV2 must charge at maximum power before leaving. This behaviour causes a power peak that is partially compensated by the PV production. In Fig. 5c the self-consumption is still 100%, but the algorithm exploits the V2G support to minimize the energy exchange with grid using the available energy stored in the EV batteries. It can be observed that there is not power consumed from the grid while the PV is producing. On the other hand, a higher power peak respect to the previous case is due to the necessity to charge EV1. In fact EV1 is the last one to leave and the one that provides to the grid most of its energy.

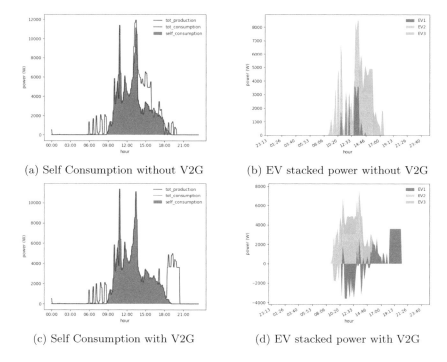

(a) Self Consumption without V2G

(b) EV stacked power without V2G

(c) Self Consumption with V2G

(d) EV stacked power with V2G

Fig. 5. Effects of EV charge on self-consumption with and without V2G.

6   Conclusion

We presented a container based deployment solution for the evaluation of energy management strategies in smart micro-grid scenarios based on simulation. The container based approach allows to speed up the evaluation activities deploying instances of the simulator in a distributed systems, or in Cloud, and running multiple optimization strategies working on different scenarios. We focused on the evaluation of an original optimization algorithm that aims at maximizing the self-consumption of decentralized PV energy production in a smart-microgrid, exploiting the flexibility of EV charging, with and without the support of V2G capability. The experimental results demonstrates the feasibility of the evaluation approach. Further improvements are required to take into account conflicting goals, such as power peak minimization and realistic battery models. Investigation on high performance and scalability issues of the proposed deployment configuration is needed.

Aknowledgements. Authors of this paper, on behalf of GreenCharge consortium, acknowledge the European Union and the Horizon 2020 Research and Innovation Framework Programme for funding the project (grant agreement no. 769016).

References

1. Amato, A., Aversa, R., Di Martino, B., Scialdone, M., Venticinque, S.: A simulation approach for the optimization of solar powered smart migro-grids. Adv. Intell. Syst. Comput. **611**, 844–853 (2018)
2. Amato, A., Venticinque, S.: Big data for effective management of smart grids. In: Data Science and Big Data: An Environment of Computational Intelligence. Springer, Heidelberg (2017)
3. Aversa, R., Branco, D., Di Martino, B., Venticinque, S.: Greencharge simulation tool. In: Advances in Intelligent Systems and Computing (AISC), vol. 1150, pp. 1343–1351 (2020)
4. Di Martino, B., Colucci Cante, L., Venticinque, S.: An ontology framework for evaluating e-mobility innovation. In: Complex, Intelligent and Software Intensive Systems, pp. 520–529. Springer, Cham (2021) International Publishing
5. Luthander, R., Widén, J., Nilsson, D., Palm, J.: Photovoltaic self-consumption in buildings: a review. Appl. Energy **142**, 80–94 (2015)
6. Nunes, P., Farias, T., Brito, M.C.: Enabling solar electricity with electric vehicles smart charging. Energy **87**, 10–20 (2015)
7. Venticinque, S., Di Martino, B., Aversa, R., Natvig, M., Jiang, S., Sard, R.E.: Evaluating technology innovation for e-mobility. In: 2019 IEEE 28th International Conference on Enabling Technologies: Infrastructure for Collaborative Enterprises (WETICE), pp. 76–81. IEEE (2019)

Cloud Computing Projects:
A Bibliometric Overview

Teodor-Florin Fortiş[(✉)] and Alexandra-Emilia Fortiş

Faculty of Mathematics and Informatics, Department of Computer Science,
West University of Timişoara, Timişoara, Romania
{florin.fortis,alexandra.fortis}@e-uvt.ro

Abstract. Several projects were selected for funding in 2009, through
the European Commission's FP7 ICT call 5, covering various topics associated with Cloud Computing. Through this paper we intend to investigate the impact these projects have generated at a distance of 10 years
from their initial activities, evaluating at the same time the dimensions
and strength of the corresponding consortia, the typical structure of
collaboration, or the evolution of the corresponding research trends.
The findings associated with this research are usually processed via
VOSViewer.

Keywords: Bibliometrics · Cloud computing projects · VOSViewer

1 Introduction

Several projects dealing with various cloud computing topics were selected for
funding, following the FP7 ICT call 5 from 2009, under the specific objective:
Internet of Services, Software and Virtualisation. While cloud computing was
not clearly identified in the text of this call, the growing interest around this
novel subject was at the core of the development of several projects under the
"open, scalable, dependable service platforms, architectures, and specific platform components" specific objective outcome.

Among the declared challenges and objectives of these projects we can mention the identification of "innovative distributed programming models that simplify the development of Cloud-based services, allowing for ordinary programmers to take full advantage of the seemingly unbounded amount of computational
power and storage available on demand in large scale Cloud infrastructures"[1], the
development of an "advanced model-driven methodology and tools for REuse and
Migration of legacy applications to Interoperable Cloud Services"[2], the development of an open-source API and "platform that enables applications to negotiate

---

[1] Cloud-TM https://cordis.europa.eu/project/id/257784.
[2] REMICS https://cordis.europa.eu/project/id/257793.

L. Barolli et al. (Eds.): AINA 2021, LNNS 227, pp. 127–138, 2021.
https://doi.org/10.1007/978-3-030-75078-7_14

Cloud services as requested by their users"[3], or "to create a dependable ecosystem of providers and consumers that will be the foundation of an efficacious operation of services and infrastructures"[4].

These projects, among others from the same FP7 call, generated a large interest during their implementation and proved to be influential in the following years. In current paper we investigate, from a bibliometric perspective and ten years after, some aspects related with the influence that the research outcomes of these projects have generated.

2   Background Information

2.1   Bibliometric Investigations

Different bibliometric investigations have been employed to provide different perspectives related with the evaluation of research or of specific scientific outlets. For the case of scientific outlets, there are investigations usually oriented towards the identification of the leading trends for a specific period of time [6,8,20], analysis of their impact [8], a citation and publication landscape [20], as a means of comparison between similar scientific outlets or analysis of a scientific domain [6].

For carrying out such investigations, miscellaneous tools for bibliometric analysis, statistical approach and data visualization have to be used on the large amount of metadata associated to scientific outputs, such those provided in [2,3,6,20]. One can exploit the results arising from these investigations in management strategies, to improve visibility and accessibility of research data, to establish and consolidate research groups and to better share experiences and expertise on specific domains.

In our analysis, we use some of the capabilities included in VOSViewer, such as co-authorship, bibliographic coupling, co-occurrences of keywords. [2,3] For the investigation on collaborative aspects, we use the different metrics, as explained in [17]: the collaborative index (CI), degree of collaboration (DC), and collaborative coefficient (CC).

2.2   The Projects Under Analysis

For our investigations we selected four cloud computing projects that received EU funding through the FP7 ICT call 5 from 2009 (FP7-ICT-2009-5). With EU contributions ranging from 1.7 million euros to 7.1 million euros and a size of consortia from 4 to 16 organizations, the projects offered important advances in specific cloud computing challenges, formulated in the context of the call.

---

[3] mOSAIC https://cordis.europa.eu/project/id/256910.
[4] OPTIMIS https://cordis.europa.eu/project/id/257115.

Cloud-TM: A Novel Programming Paradigm for Cloud Computing. The project considered "the development of a self-optimizing middleware aimed at simplifying the development and administration of applications deployed on large scale Cloud Computing infrastructures", as a response to "the lack of programming paradigms and abstractions capable of bringing the power of parallel programming into the hands of ordinary programmers". The outcomes of the project include the Cloud-TM Autonomic Manager, the Cloud-TM Data Platform or the Autoplacer. [13]

mOSAIC: Open-Source API and Platform for Multiple Clouds. One of mOSAIC's goals was to "build of an open-source and portable platform for using Cloud services based on a proposed API and Cloud usage patterns", as a response to some identified weaknesses, such as the "lack of common programming model for Cloud-oriented applications;" or the "platform dependability and non-portability due to different APIs for different types of resources", among others. [7] Also, the project proposed a reference cloud ontology (the mOSAIC Cloud ontology), or a set of tools addressing "user-centric SLA management and dynamic negotiation". [12,19]

REMICS: REuse and Migration of Legacy Applications to Interoperable Cloud Services. The main objective was "to specify, develop and evaluate a tool-supported model-driven methodology for migrating legacy applications to interoperable service cloud platforms", with particular attention paid to the migration process by supporting "recovery process with the BLU AGE tool and the use of SoaML and forward engineering with Modelio tool". [10] The cloud modelling framework CLOUDMF "includes a set of tools that aims at facilitating the provisioning, deployment and adaptation of multi-cloud systems by leveraging upon model-driven engineering techniques and methods". [5]

OPTIMIS: Optimized Infrastructure Services. Characterized by innovation in the field of sustainable IT services, the project "focuses on open, scalable and dependable service platforms and architectures that allow flexible and dynamic provision of advanced services". [4] The outcome of the project – the OPTIMIS Toolkit – incorporates research performed in the direction of "optimizing the whole service life cycle, including service construction, deployment and operation, on a basis of aspects such as trust, risk, eco-efficiency and cost". [18]

3   Data Collection and Methodology

We are now going to describe the data collection methodology that was used for our analysis. As our research is based on the scientific outputs of several FP7 research projects, we included in our data store all valid entries which were declared in the context of the corresponding research projects, and are relevant for our investigations, including 'conference papers' and 'book chapters'.

3.1 Data Sources

In order to retrieve the relevant information, our first search was for project information[5] and corresponding projects' websites. As the four projects under analysis were active more than six years before our investigations, finding their websites was not a trivial task. In the case of two projects, the corresponding websites are still available, while for the other two projects, the domains associated with project's website are no longer available. Consequently, we searched on http://web.archive.org/ for the most recent version of project's websites.

Next, based on the list of scientific outputs available on the archived websites, for each project we constructed lists of papers which were indexed on Scopus, which was used as the primary source for meta-information. Scopus was considered as the primary source of information for our research, as we found that this database was offering the best coverage for projects' scientific outputs. The summary of retrieved information is presented in Table 1.

3.2 Data Preparation

For each of the selected projects two different lists were prepared: the list of scientific outputs indexed in Scopus, and the lists of references to project's scientific outputs. In order to support the different bibliometric investigations, a full export as comma separated values (CSV) files was considered. Further processing of the exported data sets was required, in order to get uniform representation of the various meta-information used in our investigations, including

Table 1. Summary of retrieved project information

Project	OPTIMIS	REMICS	mOSAIC	Cloud-TM
Generated impact				
h-index	22	9	17	11
i-index	42	12	44	19
Document structure				
Documents	71	27	65	38
Journal papers	13	4	12	3
Conference papers	56	21	51	32
Book chapters	1	2	2	1
Open access	29	9	8	20
Citations				
Citations	1,877	290	1,109	274
Citations per paper	26.43	10.74	17.06	7.21

---

[5] Information available on Cordis website, https://cordis.europa.eu/.

Table 2. Highly cited documents with project affiliation

Document	Citations
A survey of intrusion detection techniques in cloud, [9, OPTIMIS]	465
OPTIMIS: a holistic approach to cloud service provisioning, [4, OPTIMIS]	253
An adaptive hybrid elasticity controller for cloud infrastructures,[1, OPTIMIS]	171
An analysis of mOSAIC ontology for cloud resources annotation, [12, mOSAIC]	137
Portable cloud applications - from theory to practice, [16, mOSAIC]	114
Portability and interoperability between clouds: challenges and case study, [15, mOSAIC]	102
An empirical study of the state of the practice and acceptance of model-driven engineering in four industrial cases, [11, REMICS]	81
When scalability meets consistency: genuine multiversion update-serializable partial data replication, [14, Cloud-TM]	61

author names, author keywords, index keywords, or author affiliations. This processing was performed with OpenRefine[6], by using its clustering capabilities to fix minor differences in the aforementioned meta-information.

4 Results

There are various facets that can be explored by using bibliometric investigations. With an analysis based on the scientific output generated by several research projects, we are limiting our research at: a) identification of collaboration patterns and inter-project collaborations; b) keywords analysis and alignment with main objectives of the projects; c) analysis of research impact in citing papers, based on co-occurrences of keywords.

4.1 Impact Overview

Starting from the refined data, we can extract a set of basic results, strictly related with the raw data we have obtained. Thus, we can offer an overview of the collaborative size of the corresponding projects and an initial estimation of their impact.

An initial estimation of the generated impact for each of the four projects can be offered by using the h-index (self-citations excluded) and i-index values. Information about those two metrics are included in Table 1. We may notice that more than half of the total number of papers registered for the four projects received at least 10 citations, each. In Table 2 we present the top performing papers for each project, most of them with at least 100 citations.

---

[6] https://openrefine.org/.

A bibliographic coupling analysis can be used to discover a strong relation for some of the highly cited papers, as depicted in Fig. 1b, even if there is little information about effective collaborations between the different projects. For this analysis we considered only highly cited papers, with at least 50 citations. The bibliographic coupling is even stronger when we focus on authors (at least 5 documents and 65 citations), instead of countries, suggesting a high level of common research interest between the corresponding projects, as shown in Fig. 2. The identified clusters are in relation with the mOSAIC and OPTIMIS projects for the country analysis, and also include REMICS researchers for the authors-oriented analysis.

Moreover, the co-authorship analysis from Fig. 1a, based on country affiliations, show that there are strong links between the countries involved in the four projects. The co-authorship analysis was realized for countries with at least 4 documents and 100 citations, with clusters related with the OPTIMIS, mOSAIC and REMICS projects.

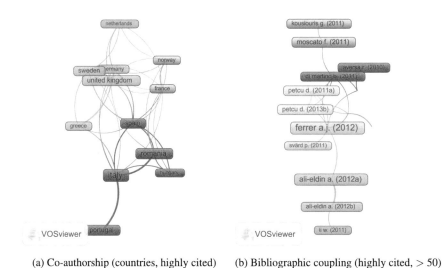

(a) Co-authorship (countries, highly cited)   (b) Bibliographic coupling (highly cited, > 50)

Fig. 1. Two perspectives on project's impact

4.2   Distribution of Papers and Collaboration

Starting from the numbers that suggest the size of the research effort involved in these projects (with 203 – total number of publications; 263 – total number of unique authors), one can extract some additional information linked with some collaborative aspects. By using the distribution of papers, starting from the number of authors, we can compute the CC (collaborative coefficient), DC (degree of collaboration), or CI (collaborative index), to emphasize some of the collaboration patterns, as exposed in Table 4. [17]

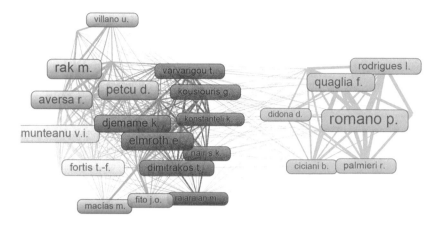

VOSviewer

Fig. 2. Bibliographic coupling for authors, all projects, at least 65 citations per author

The values of DC indicate that most of the articles are with at least 2 authors, with more single-authored papers for mOSAIC and none for REMICS. The value of CI is just the average number of authors, which must be analyzed in conjunction with the value of CC, showing that while the average number of authors is close to 4, there is a majority of papers with at 3 authors for the set of papers from all projects (the value of CC should be 0.6667).

The typical number of authors range from 2.8942 (mOSAIC, 56.71% of papers with 2 or 3 authors) to 3.2927 (REMICS, with 62.96% of papers with 3 or 4 authors). For the set of documents from all projects, there is a slightly higher value for papers with 3 or 4 authors, compared with papers with 2 or 3 authors (49.26% vs 48.27%), with a large number of documents with 3 authors (29.04%).

Table 3. Computed collaborative metrics

Project	OPTIMIS	REMICS	mOSAIC	Cloud-TM	ALL
DC	0.9718	1.0000	0.9552	0.9737	0.9803
CI	4.4648	3.5926	3.6269	3.7895	3.9458
CC	0.6937	0.6963	0.6545	0.6802	0.6788

While we can discover important collaborative work inside each of these projects, one can investigate the potential common research interest of these projects, by employing 'co-authorship' and 'bibliographic coupling' analyses between the various couples of projects. It is worth to notice that we can discover a weak 'bibliographic coupling' between the different projects, with one notable exception. In Fig. 3b, the mOSAIC-OPTIMIS diagram for 'bibliographic coupling' shows that there exist enough common research interests between the two projects, even if the 'co-authorship' analysis from Fig. 3a reveals a relatively low level of direct collaboration between these projects.

4.3   Co-occurrences of Keywords Analysis for Research Areas

Next, we use the co-occurrences of all keywords as the main instrument for this section, in order to discover how the four projects aligned with their major research questions, and the interrelations that exist between these research areas. In order to use the co-occurrences of all keywords, first we have to refine even more the input data for the projects under analysis. The thesaurus file is used to remove the obvious or generic keywords, such that only the most relevant ones will show in the report. Once this filter is in action, we can characterize the projects in terms of relevant keywords as described in Table 4.

The overall picture emphasizes the strong interrelations between these projects, based on the co-occurrences analysis performed over the full bibliometric database, as show in Fig. 4.

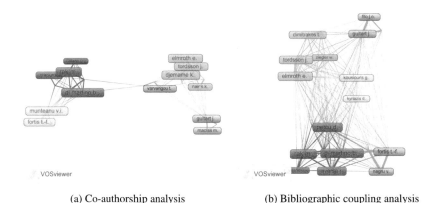

(a) Co-authorship analysis   (b) Bibliographic coupling analysis

Fig. 3. The OPTIMIS-mOSAIC comparison

Table 4. Keyword characterization of projects

OPTIMIS	mOSAIC	REMICS	Cloud-TM
computing resource, quality of service, service level agreements, business model, data privacy, access control, risk assessment, computer simulation, elasticity, and optimization	application requirements, software portability, interoperability, resource provisioning, service level agreements, API, programming models, semantics, multi-agent systems, and cloud governance	legacy systems, legacy applications, interoperability, model-driven, model-driven methodology, migration process, service oriented architecture, information services, and software engineering	experimental studies, learning systems, optimization, transactional memory, transactional systems, replication protocol, transactional data, software transactional memory, storage allocation, and concurrency control

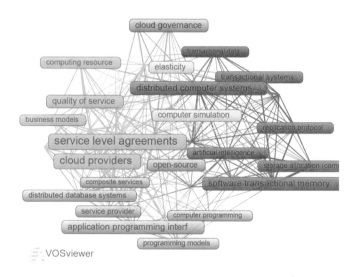

Fig. 4. Co-occurrences of keywords in all analyzed projects

A similar investigation performed on the set of papers citing documents from the four projects identify a good overlap with the main research areas, together with hints on additional research areas influenced by the outcomes of the four projects. The relevant keywords (at least 120 occurrences in citing documents) for these new directions of research, as shown in Fig. 5, include: big data, internet of things, intrusion detection, computer crime, energy efficiency or energy utilization.

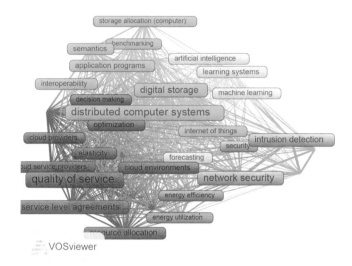

Fig. 5. Co-occurrences of keywords in citing documents

5   Conclusions

In this paper we offered a bibliometric overview of four cloud computing projects which were implemented in the period 2010–2013. The analysis shows that there are similar collaborative patterns in the four projects, even if the size of their scientific outputs is different, and a similar structure in terms of authorship. Our investigations also emphasize that, based on a strong bibliographic coupling, there were some common research interests in these projects. Moreover, the co-authorship analysis from the set of projects' documents and the set of citing papers shows a wide country coverage, as an indication of the intense project collaborations wide interests in projects' outcomes.

References

1. Ali-Eldin, A., Tordsson, J., Elmroth, E.: An adaptive hybrid elasticity controller for cloud infrastructures. In: 2012 IEEE Network Operations and Management Symposium. IEEE (2012). https://doi.org/10.1109/noms.2012.6211900
2. van Eck, N.J., Waltman, L.: Software survey: VOSviewer, a computer program for bibliometric mapping. Scientometrics **84**(2), 523–538 (2009). https://doi.org/10. 1007/s11192-009-0146-3
3. van Eck, N.J., Waltman, L.: Visualizing bibliometric networks. In: Measuring Scholarly Impact, pp. 285–320. Springer, Heidelberg (2014). https://doi.org/10. 1007/978-3-319-10377-8_13

4. Ferrer, A.J., Hernández, F., Tordsson, J., Elmroth, E., Ali-Eldin, A., Zsigri, C., Sirvent, R., Guitart, J., Badia, R.M., Djemame, K., Ziegler, W., Dimitrakos, T., Nair, S.K., Kousiouris, G., Konstanteli, K., Varvarigou, T., Hudzia, B., Kipp, A., Wesner, S., Corrales, M., Forgó, N., Sharif, T., Sheridan, C.: OPTIMIS: a holistic approach to cloud service provisioning. Futur. Gener. Comput. Syst. **28**(1), 66–77 (2012). https://doi.org/10.1016/j.future.2011.05.022
5. Ferry, N., Chauvel, F., Rossini, A., Morin, B., Solberg, A.: Managing multi-cloud systems with CloudMF. In: Proceedings of the Second Nordic Symposium on Cloud Computing and Internet Technologies - NordiCloud 2013. ACM Press (2013). https://doi.org/10.1145/2513534.2513542
6. Iqbal, W., Qadir, J., Tyson, G., Mian, A.N., ul Hassan, S., Crowcroft, J.: A bibliometric analysis of publications in computer networking research. Scientometrics **119**(2), 1121–1155 (2019). https://doi.org/10.1007/s11192-019-03086-z
7. Martino, B.D., Petcu, D., Cossu, R., Goncalves, P., Máhr, T., Loichate, M.: Building a mosaic of clouds. In: Euro-Par 2010 Parallel Processing Workshops, pp. 571–578. Springer, Heidelberg (2011). https://doi.org/10.1007/978-3-642-21878-1_70
8. Merigó, J.M., Mas-Tur, A., Roig-Tierno, N., Ribeiro-Soriano, D.: A bibliometric overview of the journal of business research between 1973 and 2014. J. Bus. Res. **68**(12), 2645–2653 (2015). https://doi.org/10.1016/j.jbusres.2015.04.006
9. Modi, C., Patel, D., Borisaniya, B., Patel, H., Patel, A., Rajarajan, M.: A survey of intrusion detection techniques in cloud. J. Netw. Comput. Appl. **36**(1), 42–57 (2013). https://doi.org/10.1016/j.jnca.2012.05.003
10. Mohagheghi, P., Berre, A.J., Henry, A., Barbier, F., Sadovykh, A.: REMICS-REuse and migration of legacy applications to interoperable cloud services. In: Towards a Service-Based Internet, pp. 195–196. Springer, Heidelberg (2010). https://doi.org/10.1007/978-3-642-17694-4_20
11. Mohagheghi, P., Gilani, W., Stefanescu, A., Fernandez, M.A.: An empirical study of the state of the practice and acceptance of model-driven engineering in four industrial cases. Empir. Softw. Eng. **18**(1), 89–116 (2012). https://doi.org/10.1007/s10664-012-9196-x
12. Moscato, F., Aversa, R., Di Martino, B., Fortiş, T., Munteanu, V.: An analysis of mosaic ontology for cloud resources annotation. In: 2011 Federated Conference on Computer Science and Information Systems (FedCSIS), pp. 973–980 (2011)
13. Paiva, J., Ruivo, P., Romano, P., Rodrigues, L.E.T.: AutoPlacer: scalable self-tuning data placement in distributed key-value stores. ACM Trans. Auton. Adapt. Syst. **9**(4), 19:1–19:30 (2014). https://doi.org/10.1145/2641573
14. Peluso, S., Ruivo, P., Romano, P., Quaglia, F., Rodrigues, L.: When scalability meets consistency: genuine multiversion update-serializable partial data replication. In: 2012 IEEE 32nd International Conference on Distributed Computing Systems. IEEE (2012). https://doi.org/10.1109/icdcs.2012.55
15. Petcu, D.: Portability and interoperability between clouds: challenges and case study. In: Towards a Service-Based Internet, pp. 62–74. Springer, Heidelberg (2011). https://doi.org/10.1007/978-3-642-24755-2_6
16. Petcu, D., Macariu, G., Panica, S., Crăciun, C.: Portable cloud applications—from theory to practice. Futur. Gener. Comput. Syst. **29**(6), 1417–1430 (2013). https://doi.org/10.1016/j.future.2012.01.009
17. Savanur, K., Srikanth, R.: Modified collaborative coefficient: a new measure for quantifying the degree of research collaboration. Scientometrics **84**(2), 365–371 (2009). https://doi.org/10.1007/s11192-009-0100-4

18. Tordsson, J., Montero, R.S., Moreno-Vozmediano, R., Llorente, I.M.: Cloud brokering mechanisms for optimized placement of virtual machines across multiple providers. Futur. Gener. Comput. Syst. **28**(2), 358–367 (2012). https://doi.org/10.1016/j.future.2011.07.003
19. Venticinque, S., Aversa, R., Martino, B.D., Rak, M., Petcu, D.: A cloud agency for SLA negotiation and management. In: Euro-Par 2010 Parallel Processing Workshops, pp. 587–594. Springer, Heidelberg (2011). https://doi.org/10.1007/978-3-642-21878-1_72
20. Zurita, G., Shukla, A.K., Pino, J.A., Merigó, J.M., Lobos-Ossandón, V., Muhuri, P.K.: A bibliometric overview of the journal of network and computer applications between 1997 and 2019. J. Netw. Comput. Appl. **165**, 102, 695 (2020). https://doi.org/10.1016/j.jnca.2020.102695

Applying Patterns to Support Deployment in Cloud-Edge Environments: A Case Study

Beniamino Di Martino[1,2] and Antonio Esposito[1(✉)]

[1] Department of Engineering, University of Campania "Luigi Vanvitelli",
Aversa, Italy
beniamino.dimartino@unina.it, antonio.esposito@unicampania.it
[2] Dept Computer Science and Information Engineering, Asia University, Taichung,
Taiwan

Abstract. A major trend followed by IT experts and Software developers in recent years is represented by the "Cloudification" of existing applications, with a strong shift of computations and data from local and centralized servers to remote, distributed data-centers. Indeed, using Cloud resources has reduced, for most SMEs, both the initial investments in hardware and software assets and maintenance costs, making it a viable choice in many situations. On the other hand, Cloud Computing requires to store consistent volumes of data on remote databases, with a series of consequences on data privacy that need to be carefully addressed. Moreover, the advent of the Internet of Things, with the huge quantity of data that smart devices continuously produce and consume, often in real time, renders the transfer of information to and from remote servers too cumbersome, as it relies on network speed and continuous availability. New programming paradigms have thus emerged, such as Cloud-Edge, which tries to combine benefits deriving from the exploitation of the resources offered by Cloud architecture and the need to consume data locally. The Cloud-Edge paradigm requires a careful design of the integration between Cloud and Edge architectures, in order to avoid bottlenecks and efficiently exploit both local and remote resources. In this paper a methodology based on Architectural, Computational and Deployment Patterns will be presented to support the deployment of applications in Cloud-Edge environments, starting from pre-existing software solutions.

1 Introduction

Recently a great interest has been shown by IT developers and experts in migrating existing software applications to the Cloud, in order to exploit the almost limitless amount of resources made available by Cloud platforms and the "pay-as-you-go" economic model used to manage them. This resulted in the movement and storage of huge amounts of data from local, private data-centers to large remote servers. Although the Cloud model offers tangible advantages and benefits, especially in terms of revenues, return on investment and better use of

L. Barolli et al. (Eds.): AINA 2021, LNNS 227, pp. 139–148, 2021.
https://doi.org/10.1007/978-3-030-75078-7_15

existing hardware structures, it still shows weaknesses. First, since the data is not in the direct possession of the customer, as most of the time it is stored remotely, security problems can arise. Second, but not least, the simple fact that a remote server must be reached to initiate a computation and receive a result can hinder actual applications. Real-time applications need to provide quick and immediate responses, which cloud platforms can't always guarantee. Furthermore, the Cloud is highly dependent on the Internet connection to function: if there is a network failure, the services simply cannot be reached.

This is one of the biggest challenges when dealing with real-time and potentially critical applications. The Internet of Things relies heavily on real time to deliver results. Imagine smart robots in factories: they need to immediately analyze sensor data, to react accordingly to the environment. If all calculations were done in the Cloud, delays in communications could slow down work or lead to potential security threats. Furthermore, in a more general perspective, the enormous amount of data to be transferred using the current Internet networks could further aggravate local congestion and cause communication problems.

The term Edge Computing, has bee used to indicate the processing of information at the edge of the network, where the data is produced. It appears to have an architectural model that is opposed to Cloud Computing: indeed, while in Cloud Computing the supply of IT solutions take place through a centralized infrastructure located far from the actual user, in Edge computing we have a decentralized and distributed IT infrastructure, made up of minor data processing centers, the so-called micro-datacenters, located close to the user.

Having a local computation infrastructure to locally manage critical calculations, especially when real time results are expected, is a huge advantage of Edge Computing, but it is not enough: it is also necessary to carefully restructure the calculation to exploit the local infrastructure, and in particular to balance the weighting of the computational load on the computation nodes. The application of the calculation distribution methodology allows the exploitation of the computational resources available at the edges of the software network, and the balancing of the computational loads, which will be distributed in such a way as to minimize the need for a Cloud-based central server, are at the base of Edge-Cloud platforms.

The advantage of Cloud Edge Computing therefore consists in performing inexpensive computations on edge devices in order to reduce the amount of data to be sent to the Cloud. The delay caused by the forwarding of data to the Cloud is compensated by the ability to perform the most expensive operations with high computing resources, which are not available at the edge.

In this paper a Pattern based approach, involving the use of Architectural, Computational and Deployment Patterns, will be exploited to support the design and deployment of analytics for Cloud Edge environments. Well known algorithms and analytics will be taken in consideration to define the mapping between existing solutions describable by Patterns and the Cloud-Edge scenario. Selection of Patterns, their mapping and subsequent refinement will be guided by performance metrics.

2   State of the Art

This section will introduce Architectural, Deployment and Computational Patterns referring to the Cloud Edge paradigm, which will be than used as a knowledge base to identify the ones best fitting our test case. A Pattern is defined as a general and reusable solution to a common and recurring problem, within a given context in software design. The purpose of a Pattern is to avoid new implementations prone to errors and bugs, both in design and in development, since they can be seen as a set of best practices that a programmer should follow when building an application or software system.

Cloud Patterns can be considered as a particular category of Patterns, focusing on describing problems and solutions related to Cloud Computing. Cloud models describe the commonalities of cloud computing environments and the design of applications for cloud computing. Some models can be helpful in understanding the proper organization of the software stacks on which applications depend. Patterns can also be helpful in understanding what changes may be required to application code for a successful migration to the cloud computing environment, especially when coupled with semantic technologies [5]. The same considerations can be made with regard to the Edge Computing environment.

Computational Patterns describe how the computations should be distributed among processing nodes, in order to efficiently implement different kinds of algorithms. Of course, choosing a specific Computational Pattern depends on the characteristics of the algorithm and of the computational nodes involved. The European project Toreador[1] has exploited a series of major Computational Patterns, such as **MapReduce** [4], **Producer/Consumer**, **Bag of Tasks**, **Pipeline** and **Tree Computation**, investigating on their differences and fields of application [6].

Another interesting source of information regarding Computational Patterns is represented by [7], where a series of Patterns have been described and distinguished according to the their specific objective. Remarkable examples are the **Backtrack Branch and Bound** and **Circuits** Patterns.

An **Architectural Pattern** is a general and reusable solution to a problem that commonly occurs in software architecture in a given context. Architectural models address various problems in software engineering, such as computer hardware performance limits, high availability, and minimization of business risks. Architectural Patterns generally provide solutions based either on the composition of Microservices, Model Driven Control concepts or Cluster-based organizations [9]. N-Tier Patterns follow the Model Driven Control theory, and organize functionalities into three main layers:

- A **Presentation Layer** that manages communication with external actors or clients, and is responsible to present information.

---

[1] http://www.toreador-project.eu/.

– An **Application Logic Layer** that deals with the data processing necessary to produce the results to be forwarded to the presentation level.
– A **Resource Management Level** that manages the data consumed for the operation of the entire system.

More layers can be added, or existing ones can be customized, according to the specific objective of the Pattern.

Finally, **Deployment Patterns** offer a high level view of how software components can be actually deployed on a target architecture, following a series of recommendation aiming at improving performances. Many provider-specific Deployment Patterns exist, like Canary, Blue-Green and Red-Black deployments defined by Octopus [8], or Multi-region, Multi-tenant and Geodes Patterns from Azure [1].

3   The Methodology

The proposed methodology aims at defining a Pattern-based approach to seamlessly design analytics for Cloud-Edge environments, starting from existing and well known algorithms. The overall procedure can be divided into four main phases, as also described in Fig. 1.

– The first step of the procedure presented here involves the application of Computational Patterns to the source algorithm. This phase involves the analysis of the Computational Patterns identified in the state of the art, in order to determine the ones that best fit the chosen algorithm. The selection of the Computational Pattern is therefore based on the algorithm and on performance and efficiency metrics.
– The second step involves the decomposition in components of the previously identified Computational Patterns according to suitable Architectural Patterns. This phase of the procedure requires the recognition of the specific architectural layer to which each component of the Computational Patterns must be connected.
– The third phase of the procedure requires to perform a deployment on a specific scenario, which can be a Cloud, Edge or Cloud-Edge environment. Deployment Patterns are applied to specify how the components are mapped to the physical nodes in the adopted scenario. Furthermore, performance aspects must be taken into consideration: deploying components that often interact with each other on distant physical nodes can introduce unacceptable and unnecessary delays that degrade performance. The components that are tightly coupled are generally part of the same architectural level. Therefore, by carrying out a correct decomposition of the Computational Pattern components following architectural patterns, the most efficient type of deployment to be performed becomes more evident. Another aspect to be considered is the heterogeneity of the available physical devices, and therefore the different computing resources they offer.

– The last step of the procedure requires to apply analytical models and/or to apply simulation techniques to assess the performance of the designed solution, in order to eventually identify bottlenecks and potential optimization strategies before actual deployment.

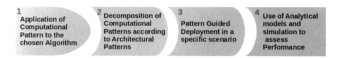

Fig. 1. Steps of the proposed procedure

The following sections are focused on the first three steps of the procedure, leaving the use of analytical models and simulation techniques to future works.

However, in order to apply the described procedure, it is necessary to select an algorithm to map and deploy to a Cloud-Edge scenario. In this paper, the **Ķ-means** clustering algorithm has been applied. The goal of the algorithm is to identify groups of homogeneous elements within a set of data. In particular, the K-Means tries to establish a fixed number of clusters which represent groups that divide the input data on the basis of a certain similarity between them. Two elements are at the core of the algorithm: the **centroids** and the **data points**. Assuming that each analysed object can be can be represented as a numerical feature vector, spatially mapped to a data-point, a single centroid characterizes a certain cluster and, at the end of each iteration, it will represent the midpoint of the objects included in that specific cluster. The K-Means algorithm follows an iterative procedure and can be divided into three phases:

– **Start:** establish the number of data points and the number of centroids by dividing the space of iterations into k partitions in a random manner. These partitions must be disjoint and the sum of them must contain all the starting elements. This phase is initiated by a generic Client, as seen in Fig. 2.
– **Assign Prototype:** with regard to this phase, the algorithm takes into account each data point and associates it with the cluster whose centroid is closest. In order to support the distribution of calculations on sparse devices, the algorithm takes into consideration each data point and only associates it with the cluster whose centroid is closest. Thus, a partial average is made between all the data points belonging to the same cluster. The partial average operation allows not to broadcast all the starting data points onto the network, since only a small number of information that depends on the number of clusters k chosen during the initialization phase will be sent.
– **Position Update:** once the previous phase is finished, the next step of the algorithm is to update the position of the centroids. The new position will be characterized by being the midpoint of all the data points belonging to this cluster: as, during the previous phase, there are possibilities that a data point can pass from one group to another, it is necessary to update centroid

so that it represents the midpoint of the data points belonging to that cluster at that particular moment. The component in charge of this phase is referred to as Updater.

The K-Means algorithm iterates through the second and third phases until a condition of convergence is reached.

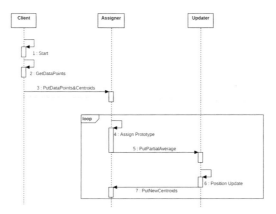

Fig. 2. Sequence diagram describing the K-mean algorithm

3.1   Selection of the Computational Pattern

Since the K-means algorithm works iteratively, by producing new centroids at each iteration and elaborating/consuming them in order to update the clusters, the **Producer/Consumer Pattern** seems suitable to describe its computations. Producer/Consumer represents the circumstance in which we have a computational node who produces a task or a series of tasks, that will be then processed by the so-called consumer. This pattern often requires a data buffer within which the producer inserts the tasks to be executed. Such a structure turns out to be particularly useful when there is a strong difference between the computational abilities of the producer and of the consumer.

Figure 3 shows how the components of the K-means and the interactions among them can be mapped on a Producer Consumer Pattern. The main operations are structured as follows.

- The **Start** operation, consisting in establishing the number of data points and the number of centroids by randomly dividing the space of iterations into k partitions.
- The **DoConsuming** operation, in which Consumers will consider each data point separately and associate it with the cluster whose centroid is the closest. The single consumer makes a partial average of the data points associated with each cluster. Since each element is represented by a vector, the methodology used to measure the concept of proximity is that of Euclidean distance.

– **DoProducing** operation in which a single producer updates the position of the centroids. The new position will be characterized by being the midpoint of all the data points belonging to this cluster. The new centroids will be forwarded, through the Queue, back to consumers to allow a new iteration of the algorithm if necessary.

Figure 3 highlights how the initialization phase is performed by one or more Master components, which receives objects from external sources. Consumers communicate with producers only through the Queue.

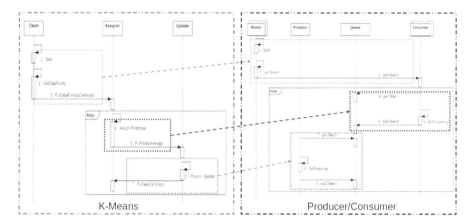

Fig. 3. Mapping of the K-means to the Producer Consumer Pattern, described through sequence diagrams

3.2 Mapping to an Agnostic Architectural Pattern

Once the Producer Consumer pattern has been mapped to the K-Means algorithm, it is possible to map the computational components on an N-Tier Architecture, using one of the Architectural Patterns approaches introduced in Sect. 2. The chosen Producer Consumer Pattern has been thus mapped to a four-layer architecture:

– **Data Ingestion Layer:** the Detector component, which detects the data, and the Master component which receives data from the Detector and carries out the algorithm initialization phase are part of the data acquisition level.
– **Application Logic Layer:** from the analysis of the state of the art this represents the level of the system that deals with the processing of the data necessary to produce the results. Consumers are thus part of this level.
– **Resource Management Layer:** it is the level that manages the data necessary for the functionality of the entire system. The Queue belongs to this layer.
– **Application Logic Layer:** a second application layer is needed to separate Consumers and Producers, which only interact through the Queue.

Only the first Data Ingestion Layer has not been already presented, as the others have been previously mentioned in Sect. 2. This layers has the specific objective to acquire data from external sources, and it is generally required as an additional tier in situations where data input represent a fundamental part of the architecture, while data presentation is not relevant. In our case, the software will be interfaced with sensors that will provide data, but that are not further involved and that require no specific feedback.

Figure 4 shows the application of the Producer Consumer Pattern on the selected N-tier architecture.

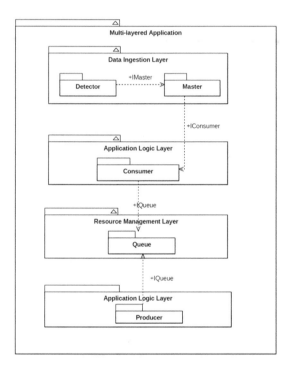

Fig. 4. Mapping to a generic 4-tier architecture pattern

3.3 Mapping to an Agnostic Deployment Pattern

In a generic Cloud-Edge scenario, where no specific platform has been taken in account, we consider only sensors whose goal is to detect information, or data. Therefore, they do not have any calculation resources at their disposal, that is the reason why data points are forwarded across the network to other edge devices. In this simplified deployment scenario, each sensor will act as a simple Detector and is associated with a specific device and can only forward information to it. This is reflected in the 4-tier Architecture described in Sect. 3.3, where the Presentation Layer has been substituted by a Data Ingestion Layer.

In the target deployment Pattern, each device will host a Master component for the reception of data points by the Detectors, which will start the algorithm initialization phase. Similarly, each edge node will contain a Consumer to associate each data point to the cluster having the minimum distance from it, and to perform a partial average of the clusters. The Queue turns out to be a component to which both consumers and producers can freely access. Having a centralized structure such as the Cloud, the Queue was deployed right on it.

Finally, in this Cloud-Edge scenario, only a single Producer has been considered. It receives the partial averages from the edge devices through the Queue and updates the position of the centroids. The Deployment Diagram of the Producer Consumer computational pattern on an n-tier architecture for a Cloud Edge scenario is shown in Fig. 5. In this example, devices hosting Masters and Consumers have a minimum computation capability, and are thus represented by Raspberry devices which provide both computation resources and configuration flexibility for eventual experiments. The deployment shown here has been obtained through a generalization of the Geode Pattern defined in [2].

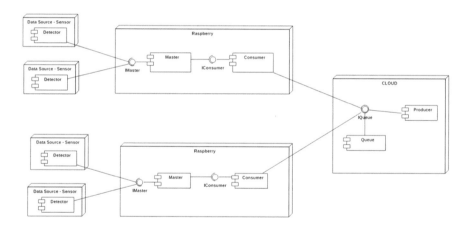

Fig. 5. Generic deployment of the architecture described in Sect. 3.3

4 Conclusion and Future Work

In this paper a methodology has been proposed for the deployment of applications to Cloud-Edge scenarios, employing Patterns to enable the mapping of algorithms to distributed environments. The K-Means algorithm has been used as a test case to describe how, by using a Producer/Consumer Computational Pattern, and an N-tier Architectural pattern, it is possible to distribute the computations in a Cloud-Edge scenario and to support the deployment by following a standard approach.

In the future both analytical and simulation based models will be employed to assess the performances of the deployed solution, so to compare different strategies deriving from the application of different Patterns. As of now, LogP [3] and Bulk Synchronous Parallel (BSP) [11] models are being taken in consideration as analytical models, while simulations are being carried out on the EdgeCloudSim environment [10].

References

1. Azure: Cloud design patterns (2021). https://docs.microsoft.com/it-it/azure/architecture/patterns/. Accessed 08 Feb 2021
2. Azure: Cloud design patterns - geode (2021). https://docs.microsoft.com/it-it/azure/architecture/patterns/geode. Accessed 08 Feb 2021
3. Culler, D., Karp, R., Patterson, D., Sahay, A., Schauser, K.E., Santos, E., Subramonian, R., Von Eicken, T.: LogP: towards a realistic model of parallel computation. In: Proceedings of the Fourth ACM SIGPLAN Symposium on Principles and Practice of Parallel Programming, pp. 1–12 (1993)
4. Dean, J., Ghemawat, S.: MapReduce: simplified data processing on large clusters. Commun. ACM **51**(1), 107–113 (2008)
5. Di Martino, B., Esposito, A., Cretella, G.: Semantic representation of cloud patterns and services with automated reasoning to support cloud application portability. IEEE Trans. Cloud Comput. **5**(4), 765–779 (2015)
6. Di Martino, B., Venticinque, S., Esposito, A., D'Angelo, S.: A methodology based on computational patterns for offloading of big data applications on cloud-edge platforms. Future Internet **12**(2), 28 (2020)
7. Keutzer, K., Mattson, T.: Our pattern language (OPL): A design pattern language for engineering (parallel) software. In: ParaPLoP Workshop on Parallel Programming Patterns. vol. 14, pp. 10–1145. Citeseer (2009)
8. Octopus: Patterns and practices (2021). https://octopus.com/docs/deployments/patterns. Accessed 08 Feb 2021
9. Pahl, C., Fronza, I., El Ioini, N., Barzegar, H.R.: A review of architectural principles and patterns for distributed mobile information systems. In: WEBIST. pp. 9–20 (2019)
10. Sonmez, C., Ozgovde, A., Ersoy, C.: Edgecloudsim: An environment for performance evaluation of edge computing systems. Trans. Emerg. Telecommun. Technol. **29**(11), e3493 (2018)
11. Williams, T.L., Parsons, R.J.: The heterogeneous bulk synchronous parallel model. In: International Parallel and Distributed Processing Symposium. pp. 102–108. Springer (2000)

Obserbot: A Totally Automated Watcher to Monitor Essential Services

Gregorio D'Agostino$^{(\boxtimes)}$ and Alberto Tofani

ENEA, CR Casaccia, Via Anguillarese 301, Rome, Italy
{gregorio.dagostino,alberto.tofani}@enea.it

Abstract. "Obserbot" is the name of a long term project resulting from the crasis of the terms "observe" and "bot" (abridged form for robot). In fact it represents a framework to monitor internet and extract some knowledge without any human supervision. The service is available 24 h a day and 365 days a year. Obserbot targets are presently two: official news media and twitter. A rough mass of textual data are collected form those two sources. A in-line semantic analysis of the collected stream allows to extract information of a specific Domain of Interest (DoI). The hardware and software architecture allowing collection is rather versatile and can be employed for several different purposes, however the semantic analytics is strictly DoI dependent. In its present form, Obserbot can handle information related to all essential services. Essential services are those activities performed by network of utilities that allow good provision: water supply, energy supply (e.g. gas and electricity), fuel supply, fresh food supply etc. and responding to other fundamental human needs such as transports, mobility and social connectivity.

1 Introduction

The obserbot project is a transversal activity supported by different projects sharing the need to watch at citizens' reactions to specific activities or social initiatives. It provides a barometer of the citizens' perception of those activities, mostly the provision of essential services. It allows real time collection of people reactions to service outages or malfunctions by twitter and related articles published on main internet media. Digested knowledge from those sources is deployed on a web-server to be accessed by authorised users. At the moment access is granted to analysers only, however statistics are made available to others such as water or electricity providers.

Poorness of real data is one of the most important obstacles to validate models of Critical Infrastructures (CI) and the Systems of Systems (SyoSy) resulting from their interdependencies [3]. There are several means to validate a model that depend on the level of abstraction of modelling (especially the system representation) and availability of data. Thereby the importance of the acquisition of precious real data on undesired events [6]. Historical data possibly represent the elective reference for testing model; however, they are often covered by non disclosure constraints imposed by the confidential policy of the operators

L. Barolli et al. (Eds.): AINA 2021, LNNS 227, pp. 149–158, 2021.
https://doi.org/10.1007/978-3-030-75078-7_16

owning and managing the CI's. Moreover, since historical data very frequently do not cover the different infrastructures involved in critical events but are limited to the one originating it, the systemic view and the interdependence effects are missed.

Human based acquisition of information is rather accurate, yet very expensive in terms of allocated human time. Therefore automated mechanisms (and software platforms) for acquisition and creation of databases (DB) for real events repository are extremely useful. Due to the proliferation of infrastructures and consequently the number of their failures, the human acquisition is not sustainable in the long run for scientific purposes. This problem is specially important in Europe as legal Regulation of the sectors do not compel Operators to report on their outages. Furthermore, collection of data from media may in principle provide a final user-based (possibly unbiased) perspective of events based on impact and consequences on population.

In this scenario we have developed a framework and a related platform to perform data acquisition. We have named this tool "Obserbot" making the crasis between observer and robot. Next section will describe the different software components of the Obserbot and the algorithms behind them.

The present work (nor the "obserbot" itself) does not pretend to be competitive with large human assisted and very expensive projects (such as [7,8]), nor to replace historical data acquisition as reported by the owners of the different infrastructures by means of their SCADA systems and related logs. It represents a non expensive tool to monitor the impact of undesired events as perceived by cytizens. Detailed analysis of critical events require knowledge of the infrastructures, their interdependence and details of managements that only the owners have access to. However, systemic analysis, may largely benefit of the repository. Moreover, it allows quantitative sentiment analysis of media with respect to CIs management and their effectiveness.

Beside the former ongoing activity, Obserbot is now expanding to watch at some specific human activities such as the Local Energy Communities (LEC's). A LEC is a group of people exchanging goods and services without explicit money exchange or bank transaction, yet resorting to "tokens" handled by a block-chain like infrastructure. Each token corresponds to an equivalent amount of KWh (or Joule) in the electricity bill. Most of people belonging to the LEC share the same DSO (Distribution System Operator) and normally they also share the substation providing the electric energy. ECListener (Electricity Community Listener) is the name associated with that specific activity.

2   The Obserbot Framework

Our framework and the software platform that implements it basically consist of the following parts: acquisition modules that *scrape* web sources; the analytical codes to analyse and classify the news; the event reconstructor (which gathers information in the same event coming from different media sources); the event repository; and a web server interface that will allow the access to the repository in a regulatory manner. Figure 1 shows the different modules of the framework and the potential end users as for example decision makers, scientists analysers, critical infrastructure operators and other software platforms such as for example the CIPCast Decision Support System. Section 4 provides some further information on the news analyser.

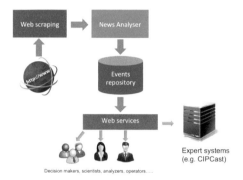

Fig. 1. The Obserbot platform

3   Platform Description

We built up on bare metal a dependable platform to perform obserbot activities. The hardware consists of a Dell rack endowed with the following hardware:

- a UPS unit: Dell Smart-UPS SRT 3000 VA
- a switch of 10 Gb/s bandwidth capacity by both copper ethernet and optical fiber: EMC N3024ET
- a PowerEdge R640
- Two workstations Precision 7920

The platform is located in a room endowed with fire alarm, differential and thermal electric switches to prevent excessive temperature increase or voltage instabilities. The UPS is capable to feed the whole platform for about 30 min to allow human intervention. Automated switch off of the whole platform, in case of longer outages, is planned yet not implemented.

The server is endowed with a redundant power supply, not the workstations, yet. The workstations and the server are endowed with the "idrac" interface which allows full remote access also upon accidental malfunctionings. Due to epidemic access policy restrictions, most of the configurations have been performed remotely using the idrac interfaces. However the system is also endowed with an analogical switch (kvm) that allows local console access and maintenance activities. It is worth mentioning that apart from the very first deployment (physical set up), practically all software installation was performed remotely.

To build up the Obserbot platform from scratch, only open source software was employed. CentOS8 Stream was selected as operative system for the server and the workstations.

The different interoperable services (micro-services), required to set up the framework, have been all implemented using Docker containers. All services are managed by a Kubernetes cluster including the server and the workstations. This makes the system resilient with respect to loss of one or even two of the physical platforms (server and workstation). All disks are set up in raid1 configurations mode to provide data availability robustness. The stream of data is handled by Kafka, which is also installed under Kubernetes. Its persistent storage layer is managed by a CEPH cluster. The CEPH cluster allocates three disks of 2 TB each distributed on the server and the two workstations. The required repositories (hosted by Mongo or Mysql data base management systems) are also implemented under Kubernetes rand CEPH.

The web server which provide access to the digested information was developed by a NGINX Ingress controller, always deployed under Kubernetes. This allows a resilient webservice.

4   The Semantic Analysis

Currently, the platform automatically analyses and stores news regarding the Italian territory and the language of the news is Italian. In the present version, the platform relies on the Google web alert service [5] that performs the scraper functions, providing one feeds it with a suited set of keywords. It is worth noting that this dependence on an external engine will be removed in the next releases, by developing the module *web scraping* for the direct net inspection and acquisition of CI news from specific web sources. The present acquisition mechanism is based on a set of "keywords" that trigger rough data acquisition by Google (Fig. 2). The user can configure the service by selecting a set of keywords and temporal frequency for receiving the news discovered by the service on the web related to each of the specified keyword. In general the user can configure several "web alerts". Typically, each day the service sends an email to the user containing a report with the web links to the discovered news.

The selection of such set of keywords is crucial to achieve adequate performances. The present set of keywords was assessed in two steps which represent a general method extendible to any domain of interest: a preliminary educated guess and an "Amplification" by Natural Language Processing [4] analysis.

In the preliminary phase an initial (human given) set of keywords is selected to perform a first harvesting. In our case this corresponds to configure the Google Alerts service according to the preliminary set. This initial set should be very selective. That is keywords should be absolutely specific of the critical events of CIs and not general words to indicate malfunctioning or unpleasant situations. As a result of the preliminary harvesting, an alert corpus is build up by gathering the web contents of the news resulting from the Google Alert daily reports. This set of web-page contents, form the initial rough "alert corpus", is inspected

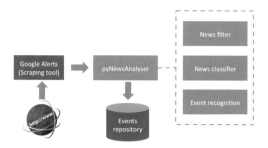

Fig. 2. The pyNewsAnalyser tool

using the Python NTLK tool [2] and further analysed by our developed tools. The analysis takes into account the following features:

- Words frequency distribution, to find the most common words used in the context of CI failures and extreme natural events web news. Ultimately this can lead to a preliminary ontology.
- Comparison with a reference Corpus and Frequency Lexicon of Written Italian ([1] reference corpus hereafter). The objective of this step is to discover words with a frequency higher than in the usual written language. Such words are used to discriminate news related to the desired domain and spurious ones e.g. ...*the 10 mins blackout of the football team caused....* The most effective selection of topic words was achieved by attributing each word a "temperature" when compared to the reference corpus along the line of paper [9]: the higher the temperature the more inherent the word to the domain of interest.
- The context of the most frequent words. In this case the objective is to identify common key phrases to be used in the scraper that is ultimately in the present approach changing the input for the Google Alert service. Typical phrases are also employed to receive or select only pertinent news.

Figure 2 shows the main modules of the *pyNewsAnalyser* tool. The *news filter* module, leveraging also the dictionary produced in the preliminary phase, filters the rough web texts to locate those portions of text in the web page (*web text entries* in the following) that are inherent to CI related information. During this phase the procedure eliminates all web text that does not contain information related to CI mis-functioning and/or extreme natural events. For each web news, the output of the *news filter* phase consists of a set of *web text entries* that are passed to the further analyses. All the filtered texts are classified and stored in the events repository.

Given the set of pertinent texts, the *news classifier* tries to understand which are the infrastructures involved in the events and to assign some web entries categories. At the present stage, the considered technological infrastructures are the following: electrical network, water network, telecommunication network, viability infrastructure. On the other hand, the *news classifier* uses the following possible categories: natural events, ordinary maintenance, technological infrastructure failure. In both cases the algorithm can assign the *none* tag when it is not possible to iidentify the involved technological infrastructure and/or the news category. It is worth stressing that the analyser looks for each potential infrastructure involved by a non exclusive approach; that is once an infrastructure is found to be involved, the search continues to check if there are others involved as well. The same strategy is employed for the categories. The final classification does not provide a single category, but the set of all of those describing the event. As atypical example, during a *planned maintenance* a *fault* may occur; in this case both *maintenance* and *fault*, are returned as output categories. Similarly for involved infrastructures.

During the Natural Language Processing analysis and refinement a basic set of key words have been chosen. In particular, the analysis has allowed the assessment of three set of keywords or dictionaries:

- *Filter dictionary* - 250 keywords that are more frequent in web news related to CI networks failures and/or extreme natural events. These set of keywords are used by the *news filter* module to filter web text.
- *Infrastructure dictionary* - 55 keywords used to identify the involved technological infrastructures.
- *Category dictionary* - 45 keywords used to identify the web entry category.

The following procedure summarises the *analyzeWebNews()* main steps of the algorithm.

analyzeWebNews()
begin
$alert\_list = getAlerts()$
$F = loadFilterDictionary$
$I = loadInfrastructureDictionary$
$C = loadCategoryDictionary$
for all $alert \in alert\_list$ **do**
  $alert_c = htmlCleanup(alert.html\_code)$
  $(web\_content, score) = filterText(alert_c, F)$
  if $score \geq THRESHOLD$ **then**
    $i\_list = getInfrastructures(web\_content, I)$
    $c\_list = getCategories(web\_content, C)$
    $saveAlert(alert, i\_list, c\_list)$
  end if
end for
end

The procedure is executed each day and analyses the web news as returned by the Google Alert Service. As first step the procedure analyse each alert contained in the Google Alert Service report. The procedure *getAlerts()* returns a list of alert objects such as $alert = (id, description, html\_code, web\_text_content, link, date)$ containing the alert id (that is automatically generated by the procedure), the short alert description contained in the report, the html code of the web page, the entire alert web content, the link to the original web page and the current date. Then, after loading the three dictionaries the procedure analyses each alert. The procedure *htmlCleanup()* eliminates all html code (e.g. scripts, images, styling code, advertisements) not related to the CI news and returns the html code containing relevant information. The *filterText()* procedures takes in put the relevant html code, the filter dictionary and returns the tuple (*webcontent*, *score*). The score value is used to discard not relevant web news. For relevant web news the procedure discovers the involved technological infrastructures and the web news categories by means of the *getInfrastructures()*, *getCategories()* procedures.

To enrich the database attributes for each alert other modules are applied to geo-localise the event. This step was performed resorting to the Italian census data referred to local geographical entities (regions, provinces, municipalities) which also contains the (lat,long) data. The geo-localisation also allows to represent the different recognised events in a GIS environment. This feature allows integration with other ongoing simulation activities. The ultimate objective of "Obserbot" will be the automatic collection and classification of all news related to CI failures and extreme natural events. The default classification schema for CI failure events is described in the following:

- *Description*. The filtered visible text of the page will be stored as the description text of a given event;
- *Web text*. The database stores the source html code of the page;
- *Link*. The link of the scraped web page;
- *Date*. The date as reported in the web news of the failure event;
- *Scraping Date*. The date pyCINews has scraped the page;
- *Extension*. If possible, the tool should be able to assign an extension to the CI outage. Possible values for this attribute are: town, rural area, city district, city, provence, administrative region, region area (a specific portion of a territory), national.
- *Duration*. If possible, the tool should be able to assign a duration attribute to the CI outage. Possible values for this attribute are: very brief (less than 1 h), brief (1 h–2 h), medium (2 h–8 h), long(8 h–24 h), very long (more than 24 h);
- *Hazards*. The accidental or naturally occurring phenomenon causing the CI failures event. The tool will consider natural hazards (e.g. heavy rain, heavy snow, ice, strong wind, flash flood, landslide, earthquake), technical hazards (technical failures) and anthropic hazards
- *Geo-localization*. This attribute will store the toponymy names contained in the web page event description

5   Results

This section shows the results of the pyCINews tool applied to 6508 Google Alerts collected from June 2016 to May 2017. Figure 3 shows the distribution of alerts according to the different categories. It is worthwhile to note how the majority of collected alerts are related to faults and/or maintenance (about 60%). Moreover, the most frequent type of event involves both the categories (30%). This can be

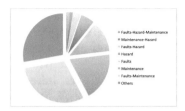

Fig. 3. The distribution of alerts according to the different categories

interpreted considering that during maintenance the resilience of the technological networks is lowered and hence natural events have an higher probability to induce outages or malfunctioning. In fact, hazards out of the maintenance periods are more rare. There are approximately 27% of alerts that do not follow under the previous categories or the tool has not be able to categorise. Figure 4 reports the classification of alerts with respect to the involved infrastructures. In this case the most affected sector is the hydric one (~34%). This can be at least partly explained by the ageing of the infrastructures. Most of them have been buildup or renewed in the fifties or earlier and hence they need severe restoration or total replacement. The second most affected sector is the electric one (~24%). In this case, as in the rest of Europe, the consumption pro capite has been growing during time whereas the infrastructure has been not developed accordingly. Apparently, the telecom sector seems to be less prone to contingencies. This is consistent with the known large resilience of the infrastructure due to huge number of redundancies. The other sectors (gas, oil, etc.) contributes for the ~28% of the total alerts. In future release of the tools these group will be further classified. Concerning the geo-localisation the analytical module was able to attribute toponyms to about 80% of alerts. The main reason why the tool is not able to perform better is the existence of roads, avenues (e.g. *Via Roma*), squares (e.g. *Piazza Bologna*) and other local toponymies which coincide with those of the municipalities. Currently the tool discards such ambiguous cases. However, the future releases will overcome the problem. There are also ambiguous toponymies due to municipalities sharing their names. In some cases proximity to other known locations can be used to disambiguate. In other cases this ambiguity remains and more sophisticated analysis are required.

Finally, the analytical module was validated using a sample set of alerts (V_A in the following). In particular, each alert $v_a \in V_A$ was manually analysed to assign the correct categories and identify the involved infrastructures. We compared the human results with those achieved by the analytical module. To perform a validation of the analytical module performance ad hoc assessment of the general concepts *precision* and *recall* (used in information retrieval) was required. In particular, let us denote with $C = f, m, h$ the set of categories to be assigned to each alert. The result of each category attribution to each alert will be indicated by a triple containing for each category either its symbol (f, m, h) or its negation $\bar{f}, \bar{m}, \bar{h}$. For example the triple $f, \bar{m}, h$ indicates that the alert refers to a

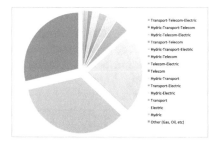

Fig. 4. The distribution of alerts related to the different infrastructures

fault and hazard event. For each alert the precision was measured by the number of correct algorithm responses divided by the number of questions (three in this case). If the previous example represent the human response and $f, \bar{m}, \bar{h}$ is the automated response the precision is $2/3$. To estimate the recall one may employ the number of correct categories found by the analytical module divided by the number of actual categories as found by the human. In the previous example the recall would be $1/2$. Similar definitions were employed for the infrastructures recognition.

Table 1 reports the resulting precision and recall for the performed validation against a random sample of 51 alerts.

Table 1. Analytical module validation results

	Categories	Infrastructures
Precision	0.79	0.94
Recall	0.91	0.94

6 Conclusions and Further Developments

In this paper we report on an ongoing activity to perform a totally unsupervised monitoring of news media. Present focus is on critical infrastructure Quality of Service (QoS) as perceived by their users. However the framework is highly versatile. We presented the state of the art of an expert system capable to observe the news in the network and to create automatically a repository of contingency events for critical infrastructures. *Obserbot* is the name we coined making a crasis between observer and robot. The idea is not to substitute well assessed projects for critical events recognition (which require a lot of funds and people to be sustained), but to provide an automated tool to monitor media. We provided a taxonomy of the events (as they appear on the media) and a tool to index news accordingly. We also estimated precision and recall of the automated recognition process compared to the human one.

At the present stage the system still resorts to the google scraper and performances are not excellent (around 90% success compared to human inspection); nevertheless we are able to provide significant statistics on the principal contingencies in Italy for a given observation period. Similar results are also achieved by direct web crawling of the principal newspapers' certified websites. However our private crawling is more time consuming. A very promising development of the Obserbot project is the inspection of twitter and other social sources to the purpose of providing early warning for infrastructures inoperabilities. Preliminary results on twitter are encouraging, thus revealing that the reference corpus and related hot words are significantly different from articles on the newspapers.

Acknowledgements. The obserbot project is hosted at ENEA, however it benefits of several collaborations among which is worth mentioning the University of Rome I "Sapienza" - Information Department - and University of Campania "Vanvitelli". Some activities are also performed in collaboration with ENEA CRESCO center in Portici. This work was supported by Project 1.7 "Technologies for the efficient penetration of the electric vector in the final uses" within the "Electrical System Research" Programme Agreements 19-21 between ENEA and the Ministry of Economic Development. The authors wish to thank Beniamino Di Martino, Mario Navarra (Univ. di Campania "Vanvitelli"), Emiliano Casalicchio (Univ. Roma I "Sapienza") and Giuseppe Santomauro (ENEA-TERIN/ICT) who are actively collaborating in the further developments of the project. Thanks are also due to Luigi La Porta for his continuous help in the platform development and management.

References

1. CNR - Institute of Cognitive Sciences and Technologies Searching - Corpus and Frequency Lexicon of Written Italian. http://www.istc.cnr.it/grouppage/colfisEng
2. Natural Language Toolkit. http://www.nltk.org
3. D'Agostino, G., Scala, A.: Networks of Networks: The Last Frontier of Complexity. Understanding Complex Systems. Springer, Cham (2014)
4. Jurafsky, D., Martin, J.H.: Speech and Language Processing: An Introduction to Natural Language Processing, Computational Linguistics, and Speech Recognition, 1st edn. Prentice Hall PTR, Upper Saddle River (2000)
5. https://www.google.it/alerts
6. Luiijf, E., Nieuwenhuijs, A., Klaver, M., van Eeten, M., Cruz, E.: Empirical findings on critical infrastructure dependencies in Europe - critical information infrastructure security. In: Third International Workshop, CRITIS 2008, Rome, Italy, 13–15 October 2008. Revised Papers
7. http://www.globalincidentmap.com/
8. http://www.risidata.com/
9. Rêgo, H.H.A., Braunstein, L.A., D'Agostino, G., Stanley, H.E., Miyazima, S.: When a text is translated does the complexity of its vocabulary change? Plos ONE **9** (2014). https://doi.org/10.1371/journal.pone.0110213

Saving Electronic Health Record (EHR) in private and secure mHealth system with blockchain Smart contracts

Arij Alfaidi[(⊠)] and Edward Chow

Department of Computer Science, University of
Colorado at Colorado Springs, Colorado Springs, USA
{aalfaidi,cchow}@uccs.edu

Abstract. Electronic healthcare records (EHRs) have become popular because they facilitate sharing of information, and they take small space compared to paper chart systems. Significant improvements continue to be seen in the EHR field because of technological advancement. The current study introduces Health Record Chain (HRC), a mobile application that facilitates communication between patients and caregivers. The technology eliminates the need for physical examination, which encourages social distancing, especially during the COVID-19 pandemic. However, it raises concerns about user's privacy and adherence to relevant regulations, For instance, Health Insurance Portability and Accountability Act and General Data Protection Regulation GDPR. The suggested system integrates blockchains to control access to user data, ensuring privacy and confidentiality.

1 Introduction

Mobile applications have become integral tools in the healthcare institution because they promote communication between the patients and the caregivers, improving the services' quality. Besides, the technology enables tracking and monitoring of patients' conditions in real-time [1]. Despite the numerous benefits of mobile health applications (mHealth apps), several concerns arise about privacy and cybersecurity issues. Hackers take advantage of these systems' susceptibility to acquiring confidential and private details of the users with malicious intends [2]. Hence, the security features of the mHealth apps should be enhanced to protect users from exploitation by cybercriminals. In this regard, the proposed system integrates blockchain technology to hash the health record encrypt data shared in mHealth apps to ensure privacy and prevent access from unauthorized quarters.

2 Previous Work

Different improvements and inventions are being undertaken to enhance EHR. Some studies have highlighted the use of blockchain in communication processes. For example, Zyskind et al. have illustrated that blockchains can be utilized for consent to the executives by eliminating encoded information [2]. Lastly, Azaria and Asaph propose

L. Barolli et al. (Eds.): AINA 2021, LNNS 227, pp. 159–164, 2021.
https://doi.org/10.1007/978-3-030-75078-7_17

developing a blockchain technology based on Ethereum to link patients to healthcare providers regardless of the geographical separation [5]. The current study uses these insights to recommend a system that facilitates record recovery, information sharing, flexibility, and permission request using blockchain mining technology.

3 Blockchain

Blockchains have evolved from keeping financial ledgers to implementing numerous computer applications, including decentralized resources, transmitted between multiple states using cryptographically secured pathways [6]. Nodes are employed to encode information into valid forms uploaded into a blockchain [4]. Applications of blockchains have expanded, making them suitable for communication purposes to promote cybersecurity and protect users' privacy.

In blockchain, the security of a computer and stability of logic transmission is dependent on the peer-to-peer protocol because they share information with the available and relevant nodes. The system can pose queries and get acceptable outcomes, promoting the whole network's certainty [2]. Hence, blockchains can ensure the security of communication systems and can be controlled by smart contracts. The idea was fully implemented into Ethereum, which involves creating smart contracts, which are commands within a blockchain, allowing storage of on-chain state. The technology also accommodates a wide array of programming languages, making it flexible for managing healthcare records [7]. Besides, blockchains' ability to allow advanced functionalities, multiparty arbitration, and mining enables the systems to conform to users' needs and regulatory systems. Programmers are considering blockchains in numerous applications due to the mentioned benefits, such as security and flexibility.

3.1 Consortium Blockchain

Different blockchain systems have been developed in the recent past. Consortium blockchains are the most utilized. The consensus of these systems is regulated by pre-selected nodes, which validate a transaction. For instance, if there are six nodes on a chain and the addition of a block requires validation by four entities, then the upload can be restricted to the selected participants. This type differs from private blockchains

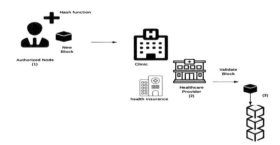

Fig. 1. Using consortium blockchain in a mHealth application. When the patient adds a new record all the authorized nodes should validate adding the record to blockchain.

because it is partially centralized. The proposed system will use a consortium-type blockchain because the data will require more than one node for verification. Figure 1 illustrates the use of consortium blockchain in a mHealth application. According to the diagram, different nodes (insurance, health care provider) must validate a change before implementation.

4   System Implementation

The proposed system HRC aims to improve user's privacy. It is an integration of the existing electronic medical records (EMR) with blockchain nodes. HRC features include Ethereum client, database administrator, back-end library, and EHR manager. The software will be operated with the SQLite database and accessed via mobile application, web doctor interface and privacy officer web page. Besides, Ethereum explicit Web3 API of the JSON RPC implemented using Swift Xcode will be employed to offer connectivity. A smart contract with user information and Ether address will be utilized. When 'Authorization Contract' is implemented, it will save users' EHR information securely in an external database DB.

Fig. 2. HRC system Architecture. Number 1; when the user registers, the webserver will send it to the W3 node server that communicates with blockchain to create Ether (wallet) address. This will give the user a hash address to encrypt his/her data. Numbers 2 and 3 in the graph will show how the system will create this wallet address for the user, and it will be sent to the webserver to allow registering the user shown in number 4 in the graph. Number 5 shows gathering health data from Apple watch to help to make a history of health data for the user to check. After collecting the data, it will be sent to the doctor's website to make the graph's diagnosis shown in number 6. After the doctor makes the diagnosis, it will be sent to the health app to accept adding it or reject it. The user can delete and edit his/her EHR data, shown in number 7 in the graph. If the user accepts adding the health data to history, it will be sent to the webserver, demonstrated in number 8 in the diagram. These data will add to the DB encrypted by a hash address, shown in number 9 in the graph.

The Ether address will be hashed to the users' data, which will be encrypted using the Proxy re-encryption method. Therefore, the system will allow users to edit or delete their data on request since the data is saved on external DB.

We implemented a second smart contract to address the relationship between the patient and the health provider. "PatieonHRAccess" controls the access to the patient data. It will ensure that the patient will approve to share his EHR data with the doctor web server. Another smart contract, "DataControl," implemented for the health provider to control updating or deleting data from user DB. Privacy controller webpage is created that will allow the authorized one after the DataControl smart contract achieved. Privacy controller webpage make the update on patient health available after getting patient approval. Figures 2 provide a schematic representation of HRC.

5   Discussion

Our HRC implementation improved data quality and quantity for medical research by encouraging the developer to use blockchain in the mHealth system. HRC will save the medical data off-chain for addressing blockchain's scalability issue at the same time to protect the user's right to privacy. GDPR gives the patient the Right to remove all his/her medical records, so when saving data off-chain, we can remove the hash value, and the wallet address will be discarded. The patient controls his/her EHR data, the patient will know where it will be saved and who will have access to it, and he/she has to permit to update the data. At the same time, the patient will have the medical care needed and communicate with health providers at any time. The proposed system enhances both data quality and quantities by incorporating blockchain technology in a mHealth application. The technology will gather patients' data and then send it to the doctors for diagnosis. Patients will have the right to remove their information from the system and choose their preferred doctors. Moreover, the user interface is easy to operate. Thus, it can be concluded that the system promotes privacy and is easy to use. Figure 3 and 4 show the interface of mobile health applications and the doctor website.

Fig. 3. HRC Mobile Application interface. It shows the user profile and his health data record. These data were collected from wearable health sensors. Users can communicate with the doctor anytime. Users can delete\Update their health data anytime.

Ethereum's exchanges require Ether, a system cash unit, to be prepared by the system. Ether can be earned by mining, granting an adequate measure to a node that solves the computational puzzle. Therefore, care suppliers are boosted to take an interest in mining to support the continuation of their services. In like manner, when patients wish to share their data, they will be required to spend Ether or host the goal get-together

Fig. 4. HRC doctor webpage. The doctor can check patient health data anytime. Doctor can approve communication requests from patients. Doctor can communicate with a patient.

reserve them. Giving patients Ether or having them pay for it tends to be controlled by healthcare provider system owners. Performance evaluation has been done using Xcode simulations. The registration of a new patient is fast and efficient.

6 Conclusion and Future Work

Electronic health records are becoming popular, but cyber threats and concerns about privacy limit them. The proposed system integrates blockchain technology to ensure security and confidentiality. However, this field research is still in rudimentary stages, necessitating future research to identify the most appropriate technologies to employ. Future works will involve using chameleon hash functions to investigate the effects of redactable blockchain then use it in mHealth applications.

We will include our HRC system's installation and configuration steps and deposit them to a GitHub site for broader distribution to have a more overall impact. We hope to improve the future mobile healthcare system application. We will create more sets of representative operational scenarios as a benchmark for evaluating the performance of the prototype system we have developed.

Acknowledgments. I want to thank Tabuk University in Tabuk KSA, for their support in making this system evaluated and published.

References

1. Arora, S., Yttri, J., Nilsen, W.: Privacy and security in mobile health (mHealth) research. Alcohol Res. Curr. Rev. **36**(1), 143 (2014)
2. Zyskind, G., et al.: Decentralizing privacy: using blockchain to protect personal data. In: 2015 IEEE Security and Privacy Workshops (SPW), pp. 180–184 IEEE (2015)
3. Cernian, A., et al.: PatientDataChain: a blockchain-based approach to integrate personal health records. Sensors **20**(22), 6538 (2020)
4. Alfaidi, A., Chow, E.: Redactable Blockchain in Mobile Healthcare System
5. Azaria, A., et al.: Medrec: using blockchain for medical data access and permission management. In: 2016 2nd International Conference on Open and Big Data (OBD). IEEE (2016)

6. Esposito, C., et al.: Blockchain: a panacea for healthcare cloud-based data security and privacy? IEEE Cloud Comput. **5**(1), 31–37 (2018)
7. Lin, I.C., Liao, T.C.: A survey of blockchain security issues and challenges. IJ Netw. Secur. **19**(5), 653–659 (2017)

iNote: A Low Cost Banknote Recognition System for Visually Impaired Persons

Md. Shanjinur Islam and Ashikur Rahman[(✉)]

Department of Computer Science and Engineering, Bangladesh University
of Engineering and Technology, Dhaka, Bangladesh
1505066.msi@ugrad.cse.buet.ac.bd, ashikur@cse.buet.ac.bd

Abstract. Paper currency recognition is an important area of pattern recognition and plays a vital role in day to day life of visually impaired persons. It has various potential applications including electronic banking, currency monitoring systems, money exchange machines, etc. Although there are many potential solutions for paper currency recognition system, those are either logically inefficient or economically infeasible. This paper describes a low cost automated hardware system for currency recognition using color sensors and unsupervised neural networks designed for visually impaired people. This device specifically can be used where paper currency notes have different colors for each class of notes. As a low cost solution, we have used minimal hardware components required to make the system workable. This device reads inputs from banknotes using color sensors and sends data to a trained neural network to predict the category of the note. The experimental results show that the built system is highly accurate.

Keywords: Color sensor · Image processing · Feature extraction · Neural network · Kohonen Self Organising Map · Arduino · Currency recognition · Visually impaired

1 Introduction

Visual impairment, also known as vision impairment or vision loss, is a decreased ability to see to a degree that causes problems not fixable by usual means, such as wearing glasses [7]. In 2015, it was estimated that the number of visually impaired at the global level is about 253 million people [9]. Among them 217 million people have visual impairment and 36 million people are totally blind [1]. In Bangladesh, more than 750,000 people are blind among 30+ age group population, of which 80% are due to cataract. Approximately 120,000 cataract patients are added every year [16]. Over 6 million people need vision correction by spectacles and other means.

Now-a-days, it is quite common that visually impaired people ride the bus, raise kids or shop for groceries. While making the purchase, often they need to pay the bills using banknotes. Usually, the banknotes come into two flavours– i) paper banknotes, and ii) plastic banknotes. Unlike many other countries, all

L. Barolli et al. (Eds.): AINA 2021, LNNS 227, pp. 165–176, 2021.
https://doi.org/10.1007/978-3-030-75078-7_18

banknotes are paper based in Bangladesh. The visually impaired people are incapable of recognizing the denominations of paper currencies due to their sameness in size, and texture. Therefore, (sometimes) it becomes extremely difficult for them to locate the proper banknotes while making a purchase without assistance from others. Sometimes the assistance is also required to ensure that they have been given the correct change amount by the sales clerk. Even people with partial sight may have trouble distinguishing a 50 taka banknote from a 500 taka banknote due to their sameness in color, especially if the banknote is old and worn.

Being a developing country, Bangladesh never took any legal initiative to make its bank currency accessible to its blind and visually impaired residents. Many of these people can not even bear the high expenses to recover from visual impairment. Moreover, the state-of-the-art currency recognition techniques either is scanner based or image-processing based both of which require equipment that are not low in cost and sometimes very difficult to operate, especially for visually impaired persons.

In this paper, we present a light, portable, easy to use and most importantly, cheap currency reader, dubbed as *iNote*, that identify and announce the denomination of a Bangladesh banknote inserted into a (color) sensor on the device. The significant characteristic of Bangladesh Banknotes is that each paper currency is of different color. So, in our solution we used color sensors to detect the color of banknotes so that it can be easily classified/recognised. The iNote is small enough to be fit into a side pocket. It can identify all currency of Bangladesh, assuming the banknotes are not too old, crumpled or otherwise damaged.

2   Literature Review

Several research works [2–6,8,10,12,13] have been conducted on currency recognition in the past few years. Majority of these works have been targeted a particular country's currency system. For example, in [5], the authors present their work for Korean banknotes, in [10] for Pakistani currency, in [2] for Egyptian banknotes and finally, the works in [3,4], and [8] for Bangladesh banknotes. All of these researches can be classified into mainly of two categories a) Template Matching based, and b) Deep Learning based. Both of these methods require heavy computation and significant hardware performance.

In [2], the authors propose a system where a simple image processing utilities has been used to ensure fast processing and robustness as much as possible. The basic techniques utilized in the proposed system include image foreground segmentation, histogram enhancement, region of interest (ROI) extraction and finally template matching based on the cross-correlation between the captured image and the data set. The experimental results demonstrate that the proposed method can recognize Egyptian paper money with high quality that reaches 89% and can process in a very short time. However, any image capturing equipment is expensive.

In [3], a camera-based automatic currency recognizer for Bangladeshi bank notes has been developed that assists visually impaired people in Bangladesh.

They have exploited the deep learning architecture for classification of bank notes and have evaluated the performance of their model using a novel dataset consisting of nearly 8000 images of Bangladeshi bank notes. To verify the effectiveness and efficacy of the proposed solution, they have developed a mobile Android application, and evaluated and validated the application with the users from a blind community. Although the use of deep neural network might improve the detection accuracy, it requires machines with high processing capacity which is very costly.

3   System Overview and Problem Formulation

In this section we provide system architecture and formal description of the problem that we solve in this paper.

3.1   System Overview

We use Semi-supervised approach to solve the low cost currency recognition problem. The process has two core phases. 1) Scanning Phase 2) Prediction phase.

Scanning Phase. The scanning phase provides us with a digital copy of paper currency. In this phase we use our hardware's color sensors to collect data from paper currencies and these data are passed to the prediction phase.

Prediction Phase. Prediction phase is where it is determined in which category the paper currency belongs to. Sensor data is passed to the trained model and the model predicts the class by comparing with the trained labels.

In the following section we provide a formal description of these two phases.

3.2   Problem Formulation

Scanning Phase: Suppose after scanning **n** data points are collected from a particular paper currency. Let us assume that the array of sensor data is denoted by, $V = \{\langle v_1^r, v_1^g, v_1^b \rangle, \langle v_2^r, v_2^g, v_2^b \rangle, ... \langle v_n^r, v_n^g, v_n^b \rangle\}$. Here i^{th} data is the tuple $\langle v_2^r, v_2^g, v_2^b \rangle$, and v_i^r represents **red** filter value, v_i^g represents **green** filter value and v_i^b represents **blue** filter value of the ith data point respectively.

Training Phase: Suppose, there are k classes. For each class there are p bank notes. From each of these banknotes n sensor data points are collected. These sensor data are used to train **class representation vector** for each class. As there are k classes, so there will be k trained representation vectors.

Prediction Phase: V sensor data array is scanned from the test note and passed through the trained model and results representational vector z. If i^{th} representational vector is closest to the representational vector z, then i class is selected as the predicted class of the banknote.

4   Methodology

Paper currencies of Bangladesh have an unique feature. Each paper currency has a different color. So our idea of the solution is what if we use color sensor to detect color values from the currencies, process them using an algorithm and predict their classes. Figure 1 shows all banknotes of Bangladesh.

Fig. 1. Bangladeshi paper-currencies

4.1   Algorithm

At first the color vector of each banknote is determined using the following procedure. Each RGB pixel of a banknote consists of three values. Red, green and blue. Now if we interpret three color quantities as three individual dimensions of a three-dimensional space then each pixel may be viewed as a vector lying on a three-dimensional space. Figure 2(a) shows this diagrammatically. The angle θ between two vectors corresponding to two RGB values can be used for defining similarities between colors.

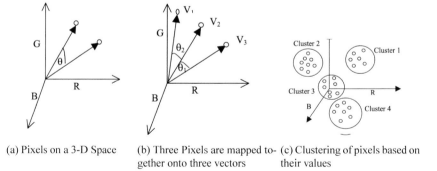

(a) Pixels on a 3-D Space   (b) Three Pixels are mapped to- (c) Clustering of pixels based on
                              gether onto three vectors   their values

Fig. 2. Describing the process of determining similarities and clustering pictorially

As for example let us consider three pixels p_1, p_2, p_3 which are mapped onto three individual vectors V_1, V_2, V_3 respectively as shown in Fig. 2(b). Now, let $\theta_1 > \theta_2$. Then obviously the vector V_2 is much closer to the vector V_3 then to the vector V_1. So pixel p_2 will be much similar to the pixel p_3 then to the pixel p_1 with respect to their color content. In this way we can easily map each and every pixel of an image onto this three-dimensional space and try to find the similarities between them. The pixels will be grouped together based on their similarities and number of groups will depend on the required accuracy level. This phenomenon has been shown pictorially in Fig. 2(c).

4.1.1   Network Structure

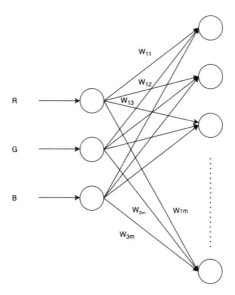

Fig. 3. Kohonen layer

The first block of our solution is a clustering layer. The clustering is needed in order to establish a solution. We choose Kohonen Self Organising Map [11] as our clustering algorithm. And training Kohonen Layer is an integral part of the solution. So, we have two choices while training the Kohonen Layer. Firstly, using RGB values from random images or using sensor Values from the sensors. Kohonen Layer consists of m number of neurons where m is the number of clusters we want. The accuracy depends on more cluster we create. There are three input neurons as we only of three Red, Green, and Blue quantities of a color pixel and obviously, the number of neurons has to be equal to the size of input vector (Fig. 3).

Algorithm 1: TRAINING THE KOHONEN LAYER

Input: $V = [[v_1^r, v_1^g, v_1^b], [v_2^r, v_2^g, v_2^b], ... [v_n^r, v_n^g, v_n^b]]$
Output: Trained Kohonen Layer
$i \leftarrow 1$
while $i \leq n$ **do**
    for $k \leftarrow 1$ **to** 3 **do**
        for $j \leftarrow 1$ **to** m **do**
            $VAL_j = \sum_k v_{ik} * w_{kj}$
    Assign the input vector to cluster L_j for which the value of VAL_j is maximum
    for $i \leftarrow 1$ **to** n **do**
        $w_{ij} \leftarrow w_{ij} + \beta * (v_j - w_{ij})$
    $i \leftarrow i + 1$

Training the Kohonen Layer: In the training phase we train the network so that the weight vectors are adjusted based on the training vectors and act as a learned color value. The weight vectors are initialized randomly. As learning progresses the weight vector ultimately gets learned with the values of all samples collected at the time of training. During training process the input vectors are color samples applied one by one and for an input vector, v_i^r, v_i^g, v_i^b the weight adjustment is made as follows. For each neuron $Lj, j = 1, 2, 3..., m$ of Kohonen layer, the output VAL_j is calculated using the equation:

$$VAL_j = \sum_k v_{ik} * w_{kj}$$

After calculating all the values, the neuron having the maxi-mum value is declared as the winning neuron for the current input vector. The weights connected with the winning neuron Lj is updated using the following formula:

$$[w_{ij}(new) = w_{ij}(old) + (v_j w_{ij}(old))]$$

where, $i = 1, 2, ..., n$. When a vector having the similar type of color vector is applied to the input layer, the same neuron is going to fire and the same weights are updated. But for a vector of different color, a different neuron gets fired and different weights get updated.

Representative Vector Generation: The second step in our solution is to generate *representative vectors*. There are k category of banknotes we are classifying, so there would be k representative vectors, one for each class. Firstly, for each category, a total of p banknotes are collected for training and for each note of a specific category, n color data are collected. So the shape of vector would be a vector of $p * n * 3$

Firstly, q^{th} banknote from p banknotes is selected. Each color value from n color points of the qth note is applied to the Kohonen Layer. For each point, a specific neuron from m neurons will be the winner. We will keep track of which neuron wins how many times of those n color points. After passing all of these n data points, we have a record of which neuron won how many times for these data. From this winning count we create a vector, which has a dimension equal to m. This vector constitutes the representative vector for the specific banknote of the specific category.

Now, there will p number of representative vectors for each of the banknotes for a specific category. Category representative vector is calculated by averaging all of these p vectors. For each of k classes we will have k representative vectors, one for each category.

Algorithm 2: REPRESENTATIONAL VECTOR GENERATION

Input: $P =$
$$[[[v_{11}^r, v_{11}^g, v_{11}^b], [v_{12}^r, v_{12}^g, v_{12}^b], ...[v_{1n}^r, v_{1n}^g, v_{1n}^b]], ..., [[v_{p1}^r, v_{p1}^g, v_{p1}^b], [v_{p2}^r, v_{p2}^g,$$
$$v_{p2}^b], ...[v_{pn}^r, v_{pn}^g, v_{pn}^b]]]$$

Output: Vector of size m

$sum \leftarrow$ Zero Array of length m

for *each* v *in* P **do**

    $rv \leftarrow$ Zero Array of length m

    for $i \leftarrow 1$ *to* n **do**

        for $k \leftarrow 1$ *to* 3 **do**

            for $j \leftarrow 1$ *to* m **do**

                $VAL_j = \sum_k v_{ik} * w_{kj}$

        $j \leftarrow$ cluster L_j for which the value of VAL_j is maximum

        $rv[j] \leftarrow rv[j] + 1$

    $sum \leftarrow sum + rv$

$sum \leftarrow (1/p) * sum$

return sum

Prediction: The prediction activity has two phases. First representative vector generation and second, comparison with trained representative vectors. After collection of n sensor data points of the test banknote, we pass these points to our trained Kohonen Layer. For each point, a specific neuron will be the winner and after n points we will have a count which neuron has won how many times. This is basically the representative vector whose length is m where ith value indicate number of times neuron i won for a particular instance. Now, we move to the second network as shown in Fig. 5. Here number of input neurons is equals to the number of neurons of Kohonen Layer which is m and output layer has k neurons which are defined as, O_p, $p = 1, 2, 3..., k$. The weights, W of each neuron is equal to the trained representative vector of each class. We calculate Euclidean distance with each trained representative vectors. If the distance between representative

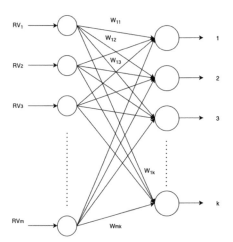

Fig. 4. Prediction layer

vector of a class p and vector of the test banknote RV is minimum, we classify our sample as of class p. Total complexity of this method is equals $O(k * m + p)$ where k is number of samples provided, m is size of Kohonen Layer network layer and p is number of classes trained (Fig. 4).

Algorithm 3: PREDICTING PAPER-CURRENCY

Input: $V = [[v_1^r, v_1^g, v_1^b], [v_2^r, v_2^g, v_2^b], ... [v_n^r, v_n^g, v_n^b]]$
Output: Predicted Class
$RV \leftarrow$ Zero Array of length m
for $i \leftarrow 1$ **to** n **do**
    for $k \leftarrow 1$ **to** 3 **do**
        for $j \leftarrow 1$ **to** m **do**
            $VAL_j = \sum_k v_{ik} * w_{kj}$
    $j \leftarrow$ cluster L_j for which the value of VAL_j is maximum
    $RV[j] \leftarrow RV[j] + 1$
for each class p **do**
    for $j \leftarrow 1$ **to** m **do**
        $dist_p = \sum_i RV[j] * w_{ij}$
$p \leftarrow$ neuron O_p for which the value of $dist_p$ is minimum
return p

---

5   Results

In this section we present the experimental results.

5.1   Hardware

As a low cost solution, we use only three components all of which cost less than ten dollars together. They are:

- Arduino Nano [15]
- TCS3200 Color Sensor [14]
- Push Button

5.1.1   Device Details

Arduino Nano (Fig. 5(a)) is used because we needed a device that can be be handled easily and at the same time can be easy to carry around. Being *nano* it takes less space consequently it is more easier to carry and roam around.

The second component TCS3200 senses color light with the help of an 88 array of photo-diodes. Then using a Current-to-Frequency Converter the readings from the photo-diodes are converted into a square wave with a frequency directly proportional to the light intensity. Finally, using the Arduino Board we can read the square wave output and get the results for the color. The photo-diodes have three different color filters. Sixteen of them have red filters, another 16 have green filters, another 16 have blue filters and the last 16 photo-diodes are clear with no filters. Each 16 photo-diodes are connected in parallel, so using the two control pins S_2 and S_3 we can select which of them will be read. The sensor has two more control pins, S_0 and S_1 which are used for scaling the output frequency. The frequency can be scaled to three different preset values of 100 %, 20 % or 2%. This frequency-scaling function allows the output of the sensor to be optimized for various frequency counters or micro-controllers.

5.1.2   Implementation Problems and Solutions

During the hardware design, we had to face many problems. Firstly, The task was to find the ideal hardware device type which is hustle-free to travel with. We worked weeks before finalizing to a pen-shaped device. It is easy to hold, it is designed thus way so that any note can be scanned through horizontally.

Secondly, sensor values change drastically with the change of distance between itself and the target object. So we had to fix this by setting a fixed height from target object to sensor so that distance always remain same between each target object and sensor.

Next problem we faced is environmental condition. As color sensor values change due to any kind of environmental light changes, we had to isolate the whole sensor into a single chamber where external lights can not enter. Only LEDs built inside of the sensors are used.

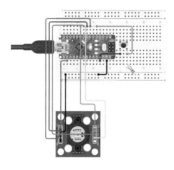

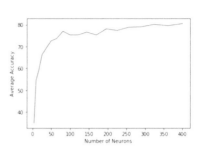

(a) Circuit Diagram of Proposed Solution   (b) Number of Neurons vs Accuracy

Fig. 5. The device and its accuracy

5.2 Experimental Results

First, we collect sensor data from 30 different paper currency notes of 6 different classes both front and backside. And then tested our algorithm with this data set. We trained our model with varying number of neurons, learning rate, neighborhood factor and batch size. Firstly we varied number of neurons while setting learning rate = 0.5, sigma = 0.7 and iterations = 30000. We also calculated class-wise accuracy while varying number of neurons. The result is shown in Fig. 6

After these results, we found that, Numbers of neurons = 64, Learning Rate = 0.5 , Sigma = 0.7 and Iterations = 30000 suits the best while considering computational cost. After setting these hyper parameters, only considering 30 notes per class gave us an extraordinary average accuracy of 93.33% for 6 classes of paper currencies. The confusion matrix the confusion matrix for this approach is shown in Table 1(a).

Then we used data augmentation, in our case, the data augmentation is done by adding some noise to the sensor data mimicking real life *sensor data read errors* that generally occurs in hardware devices. So, we mapped all of these 60 currency notes to augmented 3600 notes where we used 70 percent of total data set for training and rest for testing. With this large data set our algorithm generates an average of 74.7% accuracy for six different classes. Table 1(b) shows the confusion matrix for this large data set. From this confusion matrix it is clearly evident that, firstly, there are a significant portion of 50 taka notes that are miss classified as 500 taka notes and vice-versa. Same phenomenon has also been observed between 100 taka and 1000 taka. The reason for these miss-classifications lie on the fact that 50 taka and 500 taka bank not pair and 100 taka and 1000 taka banknote pair are very similar in color. Consequently, in some cases, sensor data from these pairs overlap with each other which results in more mis-classifications.

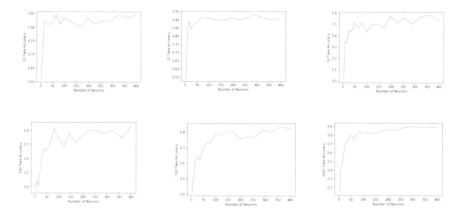

Fig. 6. Number of neurons vs class-wise accuracy

Table 1. Confusion matrices

(a) Confusion matrix with small data set

	10	20	50	100	500	1000
10	11	1	0	0	0	0
20	0	12	0	0	0	0
50	0	0	11	0	1	0
100	0	0	0	10	2	0
500	0	0	0	0	12	0
1000	0	0	0	0	0	12

(b) Confusion matrix with large data set

	10	20	50	100	500	1000
10	158	4	12	0	0	6
20	0	159	21	0	0	0
50	11	29	94	14	28	4
100	0	0	12	118	25	25
500	1	2	9	21	144	3
1000	8	0	10	27	10	125

Finally, we varied the hyper parameters to see their effects on results. The following table contains detailed results with subsequent values of hyper parameters being used (Table 2).

Table 2. Effect of Hyper parameters on Accuracy

Data size	Map size	Sigma	LR	SS	Accuracy
60	49	0.7	0.5	15	93.3
3600	64	0.7	0.5	20	74.7

LR = Learning Rate, SS = Sample Size

6   Conclusions and Future Work

In this paper, we presented a low-cost, neural network based paper currency recognition system. We also provide a detailed hardware design and enlist all its components which cost under ten dollars all together. The accuracy of the system is also measured. For reasonable size data sets, the accuracy reaches up to

93.3%. Thus the proposed system is a low cost viable solution with a very good accuracy measure. In future, we plan to apply our system for plastic banknotes and find the efficacy of the system.

References

1. Ackland, P., Resnikoff, S., Bourne, R.: World blindness and visual impairment: despite many successes, the problem is growing. Community Eye Health **30**(100), 71–73 (2017)
2. Semary, N., Fadl, S., Eissa, M., Gad, A.: Currency recognition system for visually impaired: egyptian banknote as a study case (2015). https://doi.org/10.1109/ICTA.2015.7426896
3. Hasan, M., Tripto, N.. Ali, M.: Developing a Bangla currency recognizer for visually impaired people. In: ICTD 2019 (2019)
4. Rahman, M.M., Bruce Poon, M., Amin, A., Yan, H.: Recognizing Bangladeshi Currency for Visually Impaired. Machine Learning and Cybernetics, pp. 129–135. Springer, Heidelberg (2014)
5. Chae, S.-H., Kim, J.K., Pan, S.B.: A study on the Korean banknote recognition using RGB and UV information. In: Communication and Networking, pp. 477–484. Springer (2009)
6. Pathrabe, A.M., Bawane, B.D.: Paper currency recognition system using characteristics extraction and negatively correlated NN ensemble. Int. J. Latest Trends in Comput. **1**(2), 121–124 (2010). (E-ISSN: 2045-5364)
7. Manduchi, R., Coughlan, J.: (Computer) vision without sight. Commun. ACM **55**(1), 96–104 (2012)
8. Jahangir, N., Chowdhury, A.R.: Bangladeshi banknote recognition by neural network with axis symmetrical masks. In: 2007 10th International Conference on Computer and Information Technology, pp. 1–5 (2007). https://doi.org/10.1109/ICCITECHN.2007.4579423
9. Pascolini, D., Mariotti, S.P.: Global estimates of visual impairment. Br. J. Ophthalmol. **96**, 614–618 (2011)
10. Ali, A., Manzoor, M.: Recognition system for Pakistani paper currency. World Appl. Sci. J. **28**(12), 2069–2075 (2013)
11. Kohonen, T.: The self-organizing map. Proc. IEEE **78**(9), 1464–1480 (1990). https://doi.org/10.1109/5.58325
12. Abburu, V., Gupta, S., Rimitha, S.R., Mulimani, M., Koolagudi, S.G.: Currency recognition system using image processing. In: 2017 Tenth International Conference on Contemporary Computing (IC3), Noida, pp. 1–6 (2017). https://doi.org/10.1109/IC3.2017.8284300.
13. Sarfraz, M.: An Intelligent paper currency recognition system. Procedia Comput. Sci. **65**, 538–545 (2015). https://doi.org/10.1016/j.procs.2015.09.128. ISSN 1877-0509
14. https://components101.com/tcs3200-color-sensor-module
15. Louis, L.: Working Principle of arduino and using it as a tool for study and research. Int. J. Control Autom. Commun. Syst. **1** (2018). https://doi.org/10.5121/ijcacs.2016.1203
16. Daily Star news. https://www.thedailystar.net/news-detail-206301. Accessed 13 Jan 2021

Design of a Diagnostic Support Method Utilizing Interrogation Information in Traditional Chinese Medicine

Ryo Nakagawa[1(✉)], Yuka Komuro[1(✉)], Yui Takahashi[1(✉)], Takashi Seki[2(✉)],
Yoshiaki Rikitake[1(✉)], and Akiko Takahashi[1,3(✉)]

[1] National Institute of Technology Sendai College, 4-16-1 Ayashi-Chuo, Aoba-ku, Sendai,
Miyagi 989-3128, Japan
nakagawa.ryo.r2@dc.tohoku.ac.jp, {a1911511,
a2011518}@sendai-nct.jp, {yoshiaki,akiko}@sendai-nct.ac.jp
[2] Fuji Toranomon Orthopedic Hospital, 1067-1, Kawashimata, Gotemba,
Shizuoka 412-0045, Japan
[3] Graduate School of Information Sciences, Tohoku University, 6-3-09 Aramaki-aza, Aoba,
Aoba-ku, Sendai, Miyagi 980-8579, Japan

Abstract. Traditional Chinese medicine (TCM) features diagnosis of a "pattern" that reflects the mental and physical condition of the patient and suggests an appropriate treatment. However, diagnosis requires enormous knowledge and experience, and it can be difficult to determine an appropriate treatment method regardless of clinical experience. TCM doctors need a framework for sharing knowledge and diagnostic methods and supporting treatment determinations. Here a diagnostic support method is proposed that extracts relevant patterns using the TCM database that stores the patient's interrogation information as well as knowledge about TCM. The method classifies patterns based on the relationships among them and provides the pattern extraction to the doctor as a model diagram.

1 Introduction

In recent years, the importance of Traditional Chinese medicine (TCM) using Kampo prescriptions has been reaffirmed due to the growing desire of modern people to return to nature. This desire is a response to changes in eating habits and increase in the prevalence of modern diseases which are difficult to cure with Western medicine [1]. According to the "Kampo Prescriptions Survey 2011," 89% of doctors prescribe Kampo with usual diagnosis. In addition, 59% of doctors answered that treatment by TCM was effective for diseases that are difficult to cure with Western medicine. In this way, the role of TCM has been established as complementary medicine [2].

The concept of "pattern" is a feature of diagnosis in TCM. Pattern refers to the pathological condition such as the individual's constitution and the degree of progression of symptoms in TCM; it is similar to the disease name in Western medicine. A TCM doctor diagnoses the relevant pattern from the patient's symptoms and living environment and determines the appropriate treatment method. However, unlike the disease names in Western medicine, patients can exhibit multiple patterns those are intricately related. In addition, since few diagnostic methods in TCM are based on quantitative data, it is difficult to make an appropriate diagnosis without considerable knowledge and clinical experience. In another result from the above survey [2], it was found that many doctors prescribed Kampo to patients according to the diagnostic criteria of Western medicine, and since they rarely used the diagnostic criteria of TCM, there were cases in which their treatment was consequently insufficient. It has been reported that many doctors canceled Kampo prescriptions because of its insufficient treatment effects. Therefore, it is necessary to provide a quantitative, objective framework that supports the determination of the appropriate treatment method regardless of the doctor's level of experience with TCM while sharing a wealth of knowledge about TCM.

In this paper, for the purpose of supporting the diagnosis of doctors in TCM, we use a method of extracting pattern candidates using patient interrogation information and the relationship between patterns based on the knowledge of TCM, for each patient's symptoms and mental and physical condition. We propose a method for creating a relational model among patterns to be provided to doctors as a model diagram.

2   Related Research and Issues

2.1   Traditional Chinese Medicine (TCM)

Traditional medicine is the treatment that examines the entire body and mind of patients in addition to disease, and the World Health Organization (WHO) International Classification of Diseases 11th Edition (ICD-11) includes the disease classification of East Asian traditional medicine [3]. TCM is a traditional medicine that was originally developed in ancient China based on traditional Chinese philosophy and spread over the East Asian countries. Rather than seeing humans as a collection of parts such as internal organs, TCM considers these parts individuals that are cohesively related to each other. TCM characteristically analyzes the particular constitution of each person and the cause and course of disease onset [4]. TCM doctors diagnose using four examinations: observation (to look), auscultation (to hear and to smell), interrogation (to ask), and palpation (to touch) [5]. The doctor decides the appropriate treatment for the patient based on the four examinations.

2.2   Related Research

Many kinds of research have been conducted to present patterns that correspond to patients from information on TCM. The tongue diagnosis support system of Watsuji et al. [7], which diagnoses the nature and progress of pathological conditions, by observing tongue quality (in particular tongue mucosa) and tongue coat. They created a fuzzy rule-based on these variables, and the doctor was asked to input the condition of the patient's tongue using the fuzzy scale to automatically determine the corresponding patterns. Comparison with tongue diagnosis by a doctor was used to verify the tongue diagnosis support system. Seventy percent of the diagnosis results were in agreement, showing that rule-based results that matched the patients' clinical diagnoses could be obtained.

Furthermore, Igarashi et al. proposed an inter-pattern relational modeling method [8] that analyzed complex relationships among patterns for the purpose of supporting diagnosis by TCM. This method objectively analyzed the causes that affected other patterns and the results that were affected by other patterns using the DEMATEL method. The positional relationships among patterns were objectively determined using ISM analysis, which clarifies the hierarchical structure. Specifically, "relationship in which the symptoms of one pattern worsen and cause another pattern" was extracted from TCM knowledge as described in the literature and was used as the direct or indirect relationship basis for the DEMATEL method. The proposed method created a total-influence matrix whose elements were the total-influence strengths of the relationships between patterns. This matrix made it possible to analyze the strength of the relationship among a given patient's patterns. This method further created a reachable matrix with binary elements expressing the presence or absence of a relationship based on the total-influence matrix. In the ISM analysis, the most downstream evidence was extracted using a reachable matrix, in which the strength of the influence of the total-influence matrix was expressed by 0, 1. The most downstream pattern is a pattern that does not affect any pattern other than the pattern that is located in or below it. By repeating the extraction of the most downstream pattern except for the pattern extracted last time, the pattern can be divided into layers. An experiment to verify the effectiveness of the proposed method demonstrated the strengths of relationships among the patterns widely recognized in TCM. It was confirmed that the patterns classified into each level by ISM analysis occupied their own level or had a relationship with the pattern located in the underlying level of the hierarchy. The proposed approach thus made it possible to objectively grasp the overall picture of the complex relationships among the patterns of a given patient.

2.3 Technical Issues and Proposals

As mentioned above, many kinds of research have been conducted to support TCM diagnosis, but they are not enough to diagnose correctly. It is necessary to solve the following technical issues.

(P1) It is difficult to identify the pattern based on a simple medical examination.

Diagnosis in TCM had been determined by four examinations. But observation, auscultation, and palpation are not easy to acquire by themselves for the doctors without enough knowledge and experiences. So, information obtained from inter-rogation may be the best for the diagnosis support system. Therefore, a method for identifying the pattern by simple diagnosis using interrogation is required.

(P2) It is difficult to classify the multiple patterns presented based on the patient's physical and mental symptoms.

Patients often exhibit multiple patterns. However, the conventional diagnostic sup-port method supports only a limited pattern or grasps the whole picture of the pattern exhibited. In such a method, it is not possible to define which physical or mental symptoms are responsible for the pattern presented by the support method. Therefore, a doctor using such a method may make an incorrect diagnosis, leading to an inappropriate treatment decision.

In the present research, the above problem is solved by a diagnostic support method that realizes the following proposals.

(S1) Realization of the pattern candidate extraction method based on the interrogation information.

Based on the information obtained from the interrogation, we realize a function to provide the doctor with patterns that may show the physical and mental conditions of the patient. This function is made possible that probable patterns can be correctly identified based on the interrogation alone. By linking with this function and an electronic TCM interrogation sheet, the relevant patterns are extracted from the subjective symptoms of the patient using the TCM database that stores the patient's interrogation information described in the sheet as well as TCM knowledge. At the same time, this function provides a pattern search and pattern display to illustrate patterns similar to the extracted pattern. This function allows the doctor to know, at the start of the medical examination, the pattern that may correspond to the patient.

(S2) Realization of an interpattern relational model creation method that classifies patterns by hierarchical cluster analysis.

Hierarchical cluster analysis is a method for grouping data groups according to similarity coordinate. By applying hierarchical cluster analysis based on the relationships among patterns obtained from the TCM knowledge descriptions in the literature, it is possible to classify patient patterns according to the degree and progress of their physical and mental symptoms. In addition, a model diagram showing the classified patterns and the relationships among the patterns is created and presented to the doctor. As a result, doctors can determine appropriate treatment methods based on the patterns associated with the physical and mental symptoms of the patient. Thus, this method makes an accurate diagnosis regardless of the doctor's diagnosis experience.

3 Diagnostic Support Method

3.1 Outline of the Diagnostic Support Method

The diagnostic support method proposed in this study automatically extracts the relevant patterns using information based on pattern descriptions in TCM literature. Classification is performed using hierarchical cluster analysis based on relationships among patterns. By providing the classified pattern to the doctor with a model diagram, the method allows the pattern corresponding to the patient to be visually grasped, thus assisting in an appropriate treatment decision. Thus, the diagnostic support method consists of the pattern candidate extraction method that extracts patterns based on interrogation information and the interpattern relational model creation method that classifies patient patterns based on relationships among the patterns and provides these patterns to the doctor as a model diagram.

3.2   Pattern Candidate Extraction Method

Outline of Pattern Candidate Extraction Method. Figure 1 provides an outline of the pattern candidate extraction method. The pattern candidate extraction method acquires the patient's information using an electronic interrogation sheet. The method then automatically obtains the patient's pattern by and collating the patient's information with pattern information in the TCM database. These functions are realized by two mechanisms: the interrogation information management mechanism and the pattern candidate extraction mechanism. The flow of diagnosis based on this method is shown below.

Step 1. The patient uses an electronic terminal, such as a personal computer or tablet, to enter symptoms in the electronic interrogation sheet.
Step 2. The information from the electronic interrogation sheet is obtained to the interrogation information management mechanism, and after data transformation, it is saved in the TCM database.
Step 3. The patient's interrogation information and pattern information stored in the TCM database is sent to the pattern candidate extraction mechanism.
Step 4. The pattern candidate extraction mechanism extracts patterns and displays them to the doctor.

The TCM database consists of the following databases.

Interrogation Item Database. Stores interrogation items, answer formats and answer options for data fields in the electronic interrogation sheet.

Interrogation Information Database. Stores the information entered in the electronic interrogation sheet.

Patient Information Database. Stores information about the patient's personal information and patterns. Interrogation information is managed for each patient by linking with the interrogation information database.

Pattern Information Database. Stores pattern information, such as the name of the pattern, the outline of the pattern, the pattern's typical symptoms, and the treatment method, based on the TCM literature.

Interpattern Relational Database. Stores the patterns that occur when the symptoms of each pattern progress, based on the TCM literature.

The role of each mechanism is as follows.

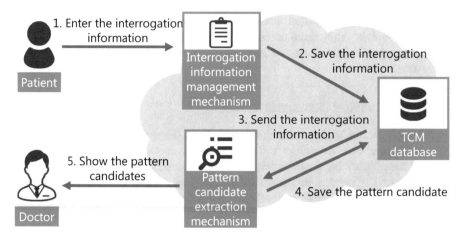

Fig. 1. Outline of the pattern candidate extraction method

Interrogation Information Management Mechanism. An outline of the interrogation information management mechanism is shown in Fig. 2. This mechanism consists of three functions: an electronic interrogation sheet creation function, an interrogation information registration function, and an interrogation information acquisition function. The electronic interrogation sheet creation function creates an electronic interrogation sheet to digitize and manage the patient's physical and mental symptoms information as interrogation information. The interrogation information registration function registers the interrogation information obtained by the electronic interrogation sheet for each patient. The interrogation information acquisition function displays the interrogation information specified by the doctor in a form corresponding to the electronic interrogation sheet. These functions make it possible to provide necessary interrogation information to doctors.

Pattern Candidate Extraction Mechanism. Figure 3 shows an outline of the pattern candidate extraction function. This mechanism consists of two functions: a match rate calculation function and a severity determination function. The pattern candidates are automatically extracted from the interrogation information acquired by the interrogation information management mechanism, and the match rate and severity of the pattern symptoms are analyzed. The match rate calculation function extracts candidate patterns by comparing the symptoms reported in the interrogation information as obtained by the interrogation information management mechanism with the symptoms of each pattern stored in the pattern information database. The match rate of the patterns corresponding to the patient is then calculated by the fraction of symptoms in the pattern information database that matches those reported by the interrogation sheet. The severity determination function determines the severity of a patient's symptoms in the interrogation information database based on the adverbs contained in the interrogation text stored in the interrogation item database. Specifically, the severity of the adverb is determined based on the verbal rating scale used in clinical practice. The verbal rating scale, used in clinical practice, is used as an index to determine severity according to adverbs. If the term "every day, often" is included, the severity is considered to be high, whereas if the term "sometimes, a little" is included, the severity is considered to be low.

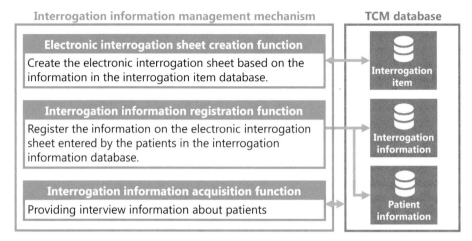

Fig. 2. Outline of the interrogation information management mechanism

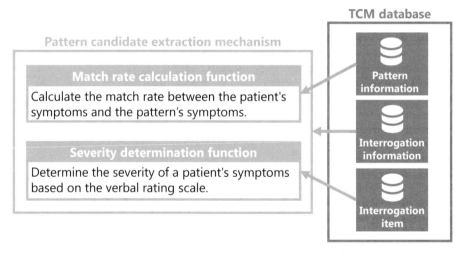

Fig. 3. Outline of the pattern candidate extraction mechanism

3.3 Interpattern Relational Model Creation Method

Outline of Interpattern Relational Model Creation Method.
Hierarchical cluster analysis is a type of multivariate analysis that groups data according

	Liver depression	Liver blood deficiency	Kidney yang deficiency	Dampness	Fluid retention
Liver depression	0	0	0.25	0	0.75
Liver blood deficiency	0	0	0	0	0.5
Kidney yang deficiency	0	0	0	0	0
Dampness	0.75	0.25	0.19	0	0.69
Fluid retention	0	0	0	0	0

Fig. 4. Total- influence matrix

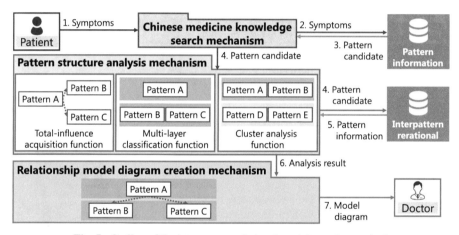

Fig. 5. Outline of the interpattern relational model creation method

to a similarity criterion [9]. When performing hierarchical cluster analysis, we define a distance function and merger method for determining the combination to be clustered. In the interpattern relational model creation method, the Euclidean distance is used as the distance function, and the Ward method is used as the merger method. The Euclidean distance is the distance of a straight line connecting two points in multivariate space. The Euclidean distance $d(p, q)$ is calculated from two data where $p = (p_1, p_2, \ldots, p_n)$ and $q = (q_1, q_2, \ldots, q_n)$ exist in the n-dimensional space.

$$d(p, q) = \sqrt{\sum_{i=1}^{n} (q_i - p_i)^2} \tag{1}$$

In Ward's method, the sum of squares between the data is determined as the distance between clusters. The sum of squares $D(C_1, C_2)$ in the potential cluster is given by the following formula using the two clusters, C_1 and C_2, and the centroid gc of each cluster.

$$D(C_1, C_2) = \sum_{x \in C_1 \cup C_2} d\left(x, g_{C_1 \cup C_2}\right) - \sum_{x \in C_1} d\left(x, g_{C_1}\right) - \sum_{x \in C_2} d\left(x, g_{C_2}\right) \quad (2)$$

$$g_c(p, q) = \frac{1}{2}(p_1 + q_1, p_2 + q_2, \ldots, p_n + q_n) \quad (3)$$

Classification is performed by combining the data with the smallest values obtained by and the merger method as a new cluster.

To provide the coordinates for distance calculation in the hierarchical cluster analysis, we use a total-influence matrix calculated by the interpattern relational modeling method [8]. Figure 4 shows the total-influence matrix when analyzing the relationships between pairs of each of five patterns. Each element in the total-influence matrix represents the strength of the relationship between the patterns in row i and the pattern in column j. The sum of rows (R) indicates the strength of relationship that pattern i exerts on other patterns. Furthermore, the sum of columns (C) indicates the strength of relationship exerted upon pattern j by other patterns. $R + C$ when $i = j$ can be regarded as the degree of the patient's physical and mental symptoms of pattern i. Moreover, $R - C$ can be regarded as the progress of the patient's physical and mental symptoms of pattern i. The more positive the degree of cause, the more fundamental a case of the pattern the patient exhibits, and the more negative the degree of cause, the more likely the extracted patterns are to represent an additive result. This method is possible to classify the pattern by using $R + C$ indicating the degree of the patient's physical and mental symptoms and $R - C$ indicating the progress of them as the coordinates for distance calculation in the hierarchical cluster analysis.

Outline of Interpattern Relational Model Creation Method. Figure 5 shows an outline of the interpattern relational model creation method. This method analyzes the relationships among patterns based on patient symptom input and provides it as a model diagram. The flow of diagnosis based on this method is as follows.

Step 1: The patient enters the symptoms using the interrogation sheet. The TCM knowledge search mechanism searches for pattern candidates that include these symptoms.
Step 2: The pattern structure analysis mechanism is applied to the pattern candidates to obtain the results for the DEMATEL method, the ISM analysis, and the hierarchical cluster analysis.
Step 3: Based on the results in Step 2, the relational model creation mechanism creates a model diagram showing the relationships among the patterns and provides it to the doctor. Based on the created model diagram, the doctor determines the appropriate treatment method.

The role of each mechanism is as follows.

The TCM Knowledge Search Mechanism. This mechanism extracts patterns including patient symptoms obtained from the interrogation, from the pattern information database. The extracted patterns are sent to the pattern structure analysis mechanism.

The Pattern Structure Analysis Mechanism. This mechanism analyzes the interpattern relationship by the DEMATEL method, the ISM analysis, and the hierarchical cluster analysis. Thus, this mechanism consists of three functions: a total-influence acquisition function, a multilayer classification function, and a cluster analysis function. The total-influence acquisition function extracts the inter-pattern relationship corresponding to the patient's symptom from the interpattern relational database and calculates the total-influence matrix based on it. The multilayer classification function calculates a reachable matrix in which the presence or absence of each element of the total-influence matrix is represented by 0, 1, and grasps the entire structure of the patient's patterns by extracting the most downstream patterns. The cluster analysis performs hierarchical cluster analysis based on the total-influence matrix and classifies patterns according to the degree and progress of patient's physical and mental symptoms. The analysis results obtained by these functions are sent to the relational model diagram creation mechanism.

The Relational Model Diagram Creation Mechanism. This mechanism creates a model diagram showing the patient's interpattern relationship, based on the results obtained by the pattern structure analysis mechanism. The interpattern relationships are illustrated via arrows, place the affecting pattern at the root of the arrow and affected pattern at the tip of the arrow. Also, by selecting either multilayer classification or hierarchical cluster analysis, the position of the pattern is reflected in the model diagram based on the results of each analysis. The doctor grasps the entire structure of the patient's pattern by the model diagram reflecting the multilayer classification, and decides the appropriate treatment method by the model diagram reflecting the hierarchical cluster analysis.

4   Conclusion

To support the TCM diagnosis for doctors, we proposed a diagnostic support method consisting of a pattern candidate extraction method and an interpattern relational model creation method. The pattern candidate extraction method that utilizes the interrogation information consisting of the interrogation information management mechanism and the pattern candidate extraction mechanism extracts the patient's pattern candidates from the electronic interrogation sheet. Also, the interpattern relational model creation method consisting of a TCM knowledge search mechanism, a pattern structure analysis mechanism, and a relational model diagram creation mechanism provides a model diagram to support the patient's appropriate treatment.

In the future, we will integrate the pattern candidate extraction method and the pattern relationship model creation method. Furthermore, we will implement and evaluate this integrated method as a diagnostic support system that doctors can use when making a TCM diagnosis.

References

1. Dewer, T., et al.: Health from the Viewpoint of TCM in University Students, pp. 73–75. Bulletin of Tokai Gakuin University (2014)

2. Japan Kampo Pharmaceutical Preparation Association, Kampo Prescription Survey (2011). https://www.nikkankyo.org/serv/pdf/jittaichousa2011.pdf. Accessed 20 Dec 2020
3. WHO: ICD-11 for Mortality and Morbidity Statistics, WHO. https://icd.who.int/browse11/l-m/en. Accessed 28 Apr 2020
4. Matsumoto, K., et al.: Development of oriental medicine expert system-aiming at universalization of oriental medicine, Hira Autonomy No. 6, 7, 132–135 (1987)
5. Maciocia, C.: The Foundations of Chinese Medicine, 2nd edn. Churchill Livingstone, London (2005)
6. Katayama, K., et al.: Analysis of Questionnaire for Traditional Medicine and Development of Decision Support System, Evidence-Based Complementary and Alternative Medicine (2014)
7. Watsuji, T., et al.: Construction of a diagnostic support system for traditional TCM. J. Soc. Biomed. Fuzzy Syst. 5(1), 23–30 (2003)
8. Igarashi, A., et al.: A method for modeling relationships between patterns to support information sharing. J. Inf. Process. Soc. Jpn. 61(3), 667–675 (2020)
9. Yoshihara, K., et al.: Overview of cluster analysis. J. Surf. Anal. 21(1), 10–17 (2014)

Software-Defined Networking: Open-Source Alternatives for Small to Medium Sized Enterprises

Paul Thornley$^{(\boxtimes)}$ and Maryam Bagheri

Sheffield Hallam University, Sheffield, UK
b9017293@my.shu.ac.uk, maryam.bagheri@shu.ac.uk

Abstract. SDN Networking is a new and emerging technology and is receiving significant considerations within organizations. Previous research predominantly emphases on SDN deployment with large Local Area Networks (LAN's) and its use in corporate Wider Area Networks. SDN adoption within small networks is an area of research that has not been explored in any detail. This paper aims to analyze a market-leading SDN solution – Cisco DNA Centre- and identify key technologies for utilization with SME-based networks. Furthermore, this research aims to provide open-source SDN solutions suitable for Small to Medium-sized networks and show how those identified technologies could be developed and deployed within Open-source solutions. This research shows that Open-source SDN controllers are a viable and deployable solution within small to medium-sized business networks.

1 Introduction

The demands made on today's business computer networks is arguably at its highest, more than it has ever been. Key stakeholders and decision-makers within organizations continually face challenges when considering the technologies and concepts available in delivering efficiency and effectiveness within IT infrastructure. Organizational budgets and technological complexities are often in competition within key stakeholder considerations, where any introduction of technologies with increased complexity, often have a significant impact on available budgets [18]. Additionally the move towards concepts such as the Internet of Things (IoT), borderless networks, and cloud computing has increased the need for the underlying infrastructure to adapt and evolve in response to business demand. Whilst there are many techniques and guidelines outlining design methodologies to provide scalability and expansion of networks [3], these often do not consider how networks respond to demand dynamically. This inability to evolve its physical infrastructure poses challenges when introducing new technologies into infrastructure and is referred to as 'Internet ossification' [15]. As a response, organizations have moved towards network programmability and automation to address the issue of

---

The original version of this chapter was revised: the chapter title has been changed. The correction to this chapter is available at https://doi.org/10.1007/978-3-030-75078-7_71

© The Author(s), under exclusive license to Springer Nature Switzerland AG 2021, corrected publication 2021
L. Barolli et al. (Eds.): AINA 2021, LNNS 227, pp. 188–199, 2021.
https://doi.org/10.1007/978-3-030-75078-7_20

'Internet Ossification' and provide network infrastructure that responds dynamically to network needs and demands. One such concept is the paradigm of Software Defined Networking and its key objective of introducing network automation and programmability within IT infrastructure.

According to current figures, the Software Defined Networking market was worth $8 Billion (USD) in 2019 and is expected to grow 40% and be worth an estimated $100 Billion (USD) by 2025 [2]. The operational benefits of introducing SDN based technologies within a networked infrastructure has been widely suggested and supported by the projected increase in the market value of SDN technologies. Several studies suggest that the implementation of SDN technologies provides significant improvements in network scalability, elasticity, network management, and response to demand in comparison to using traditional networking hardware and concepts [9, 11]. This makes SDN technologies a significant consideration for organizational key stakeholders when considering the re-development or implementation of networked infrastructure. However, market-leading SDN solutions such as Cisco DNA Centre are significantly costly and often complex to implement. With smaller profit margins and lower IT budgets, this could result in SME businesses being unable to introduce SDN technologies within their infrastructure due to SDN solution costs.

Currently, there is diminutive research and documentation available that provides affordable suggestions and guidance for the development and deployment of SDN based technologies that are purposely targeted at SME businesses.

To address this, this paper aims to ascertain how SDN technologies are applied to smaller networks typically found within SME IT infrastructure. We will analyze and identify key technologies within a market-leading Software Defined Networking solution and compared them against open-source or lower cost SDN alternatives. Prototype networks to demonstrate the viability of SDN within smaller SME networks will be created in EVE-NG simulator.

2 Literature Review

SDN solutions today are becoming a major consideration for organizations and key stakeholders within the business. Limitations within traditional networking infrastructure are one of the driving factors that has contributed to the shift towards Software-Defined Networking concepts. Lack of options within the programmability and automation of networking nodes often results in un-responsive networks, poor load balancing, and require configuration to be conducted through different vendor-specific configuration interfaces [11]. In the mid-2000s researchers at Stanford University sought to address this issue by developing a method for logically centralizing the control and data planes of networking devices [5].

Figure 1 demonstrates the abstraction of the control plane and links to a centralized SDN controller. The data plane of networking nodes remains within the device itself, resulting in the device being only concerned with the operation of forwarding data. This centralization of the control plane within the SDN controller provides opportunities to create a holistic perspective of network performance and control plane-based configurations [11, 17]. To provide network software and hardware abstraction, the

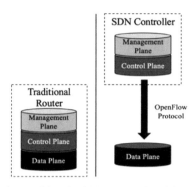

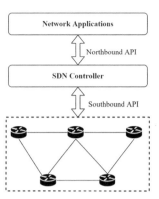

Fig. 1. Traditional v SDN - control and data planes

Fig. 2. SDN architectural

centralized SDN controller uses Northbound (NBI) and Southbound (SBI) Application Programmable Interfaces [11]. Figure 2 shows the placement of the controller with the NBI providing the link to software and the SBI providing communication to network devices. By creating this abstraction of software and hardware within the paradigm of networking, could arguably increase the ease of management and configuration of networking devices, even under network growth and expansion of the infrastructure. Since SDN initial conception, several research projects were conducted aiming to analyze and identify the key benefits of introducing SDN into an organizations network. A study conducted in 2014, introduced SDN networking gradually into a large network and operated SDN alongside legacy networking devices [12]. In their study, it was found that using a centralized controller achieved faster network convergence times after link failures identified through the Spanning Tree Protocol (STP). Also, faster download speeds through improved multi-path forwarding of data as compared to the traditional legacy devices. These benefits are further supported in a study conducted in 2014, where five key SDN deployment benefits were identified. The author in [10] argues the dynamic nature of SDN through its response to networking data and the ability to make changes in real-time in response to returned data. Furthermore, they argue the elasticity of SDN technologies where additional nodes, links and routing decisions can be either be created or removed in response to networking needs.

2.1 SDN Architecture in Smaller Networks

A large amount of research has been focused on the design, deployment, and implementation of SDN technologies within bigger networking architectures such as large corporations and datacenters. Studies such as the development of a large-scale SDN testbed in [8] and Google's B4 project [9] provide evidence in support of the benefits inherited through the deployment of SDN across multiple large campus networks and datacenters. But they fail to suggest how these concepts could be introduced into smaller network architectures where skills, resources, and budgets are significantly limited. Without the research of SDN technologies focused on the deployment within smaller networks could provide difficulties in organizations deciding to move towards SDN architectures.

Research performed in 2014 suggested a methodology for the deployment of low-cost hardware and open-source software to create a small test-bed network. This was achieved by introducing the Floodlight SDN Controller to a centrally located Raspberry Pi device and operating as the main SDN controller. Besides, the switching fabric of the network consisted of further Raspberry Pi devices operating as Open Virtual Switches (OVS). This SDN deployment on Raspberry Pi devices provided systems such as Quality of Service (QoS), Load Balancing, and topology overviews at a very low cost in comparison to traditional networking hardware [10]. The deployment of SDN based technologies within this study evidenced the possibility of applying this concept of networking within smaller SME-based topologies.

The author in [6] developed the 'Neto-App' to address the shortfalls found within SDN technologies in supporting smaller academic and business networks. The 'Neto App' applied the concept of an underlay network consisting of layer 3 routing and on overlay network providing network segmentation by applying a technology called Virtual Extensible Local Area Networks (VXLAN). They suggested by applying the underlay/overlay concept, an SME organization can deploy an SDN based solution on limited networking hardware and introduce benefits inherited using SDN technologies.

There is strong evidence in support of applying SDN technologies within SME networks. This can bring efficiency, management, development, and scalability improvements throughout all areas of the network. However, SMEs seeking to adopt new technologies within existing infrastructure often experiences barriers such as technical support, poor understanding of new systems, and overall costs [16]. With the increasing costs of ICT based OPEX and limited profit margins within SME's, the ability to fund specialist expertise or staff for SDN development may be limited. Arguably, this could effectively 'price-out' SME businesses from deploying SDN technologies. This could prevent SMEs from experiencing benefits inherited through the introduction of SDN concepts and fall behind competitors that have access to new technologies and larger IT budgets. This provides research opportunities in analyzing key aspects of SDN technologies and identifying solutions that could be potentially implemented within an SME network at little or zero cost.

3   Methodology

This research intends to establish methods of deploying key SDN features identified from a market-leading solution and applying open-source alternatives with an SME. From this, we can formulate the main question: how can open-source alternative SDN technologies provide free or low-cost solutions for small to medium-sized businesses? To provide a solution to this question, first, the key technologies of the Cisco DNA Centre will be identified. By determining the key features of Cisco DNAC which provides the core functionality, we will use previous literature and implement the same features in an open-source alternative suitable for SMEs.

3.1   Cisco DNA Centre

Cisco DNA Centre (DNAC) is an SDN solution for the enterprise networks released by Cisco, the world's largest networking company. Cisco DNAC is a highly featured network controller and management dashboard that automate networks, deploy group-based secure access and network segmentation, provide assurance and can manage hetero-geneous network devices by integrating Cisco solutions with third-party technologies. Case studies from Cisco do show that the Cisco DNAC solution can be implemented and introduce benefits within smaller organizations. However, this solution is considerably expensive and SME organizations may find this solution exceeds financial constraints and maintain underperforming networks consisting of traditional networking devices. This would suggest SME's seeking to introduce SDN technologies within their infrastructure would need to consider alternatives due to limited budgets and other financial constraints.

3.2 Key Technologies in Cisco DNA Centre

In Cisco SD-Access architecture, to promote network segmentation across both under-lay and overlay fabrics, Cisco DNAC uses the Location Identifier Separation Protocols (LISP) within the underlay fabric and VXLAN technologies within the overlay fabric. By combining the two technologies allows for the distribution of segmented layer 2 LANs networks across a layer 3 routed underlay, increasing network performance by removing issues inherited by spanning tree and link redundancy [7]. Arguably, the use of layer 3 routed underlay devices could provide SME organizations with the ability to easily expand their networks across multiple geographical locations. Furthermore, LISP provides a holistic approach to the domain and network segmentation as opposed to individual devices performing this functionality. This reduces configuration tasks performed by network engineers and developers seeking to implement network segmentation potentially reducing costs within the development.

Whilst it is recognized that Cisco DNAC contains many more functions and scope for SDN development, the purpose of this research is to establish the viability of VXLAN and LISP configurations for use within an SME. A virtualized platform was used to create prototype concept networks consisting of the technologies and the identified configuration methods.

3.3 Open Source SDN Controller for SME-Based Network

The controller plays a critical role within the functionality of an SDN-based network and serious considerations should be made on the choice of the controller and its placement. Misconfiguration and deployment of an SDN controller can distribute errors and performance issues throughout an organization's infrastructure [19]. To prevent this, key stakeholders and network engineers must fully consider how SDN is structured and how some of the key programming languages and protocols are applied within the SDN architecture. As such, these considerations may be a deciding factor when choosing an SDN controller within an SME-based network. Table 1 shows the most prominent SDN controllers in use today and suggests the wide variety of programming languages and APIs used within its operation. Of note, the REST API can be utilized within the SBI of most identified controllers and can be supported by any programming language that includes a REST API library. This provides software developers with the opportunity to create network-based applications to configure control plane logic using a variety of programming languages, irrespective of the programming language used to create the controller itself.

Table 1. SDN Controller and programming languages [19, 20]

Controller	Programming language	Northbound APIs
Pox	Python	ad-hoc
Ryu	Python	REST
Nox	C++	ad-hoc
Floodlight	Java	Rest, Java, RPC, and Quantum
Beacon	Java	ad-hoc
OpenDaylight	Java	REST, RESTCONF, XMPP and NETCONF
ONOS	Java	REST and Neutron

Of note is the OpenDaylight SDN controller that provides the largest scope for API development within SDN-based developments. RESTCONF and NETCONF are network-based APIs that are not new to the world of networking. These APIs are often utilized within network programmability applications with many networking device operating systems supporting the use of these APIs [4]. Since these APIs have been widely available outside the paradigm of SDN architecture, many network engineers and developers will already possess knowledge of these APIs. Arguably, since developers would potentially hold the relevant skill sets using APIs such as RESTCONF and NETCONF, this makes the OpenDaylight controller a significant consideration within SDN architecture within SME's.

3.4 OpenDaylight SDN Controller

In 2013, the open network foundation with the Linux Foundation created the Open-Daylight (ODL) SDN controller. The motivational factor for the development of ODL was the prevalence of issues within previously released SDN controllers. As a result, a new SDN controller was established and involved the collaboration of multiple vendors creating a more efficient and stable SDN controller. This has made ODL one of the most utilized open-source SDN controllers and has had a significant influence on commercially available SDN solutions [1].

One such utilization of ODL was within a research project conducted in 2018 [6] and utilized the ODL controller within a small topology designed for the deployment within an SME and utilized the underlay and overlay fabrics. They argued the importance of introducing network orchestration and automation to promote scalability and simple operation. Within this study, they successfully deployed ODL as a viable SDN controller within a small academic or SME topology [6] and utilized various APIs available within ODL to facilitate network automation. This study and the collaborative approach to the development of ODL show this as a potentially viable SDN controller within an SME network. However, the study failed to discuss how the underlay and overlay fabrics were introduced into the topology in which the 'NetO-App' was introduced. Furthermore, it is recognized there are many other SDN controllers available, its impact within the paradigm of SDN development and research and resultant improvements within this sector makes ODL a significant choice when considering the implementation of SDN but the deployment method would need to be explored.

4   Deploying LISP, VxLAN Within OpenDaylight SDN Controller

For the deployment of an SDN-based underlay and overlay fabric utilizing LISP, VXLAN and an OpenDaylight SDN controller with smaller infrastructures, we show several examples in the form of prototype networks using the EVE-NG network virtualization tool to establish the viability of these technologies within an SME network.

4.1   Location Identifier Separation Protocol (LISP) Within OpenDaylight

To facilitate LISP based activities within an SDN architecture, ODL provides the LISP Flow Mapping Services containing a Mapping Server (MS) and Mapping Resolver (MR) [16] (Fig. 3). This architecture operates similarly to LISP within Cisco DNAC, this provides a centralized location where EIDs can be mapped to their respective RLOC. Similarly, the Mapping Resolver, analogous to the Domain Name Service (DNS), resolves requests and updates sent from RLOCs within the SDN network. This is achieved by using the LISP protocol through the SBI and allows the register and request of RLOCs from any device that supports this protocol. The RESTCONF API can be used to explicitly map EID to RLOC within the mapping server through the LISP Flow Mapping Service NBI [16] providing programmatic opportunities to introduce automation into EID and RLOC mapping. Whilst the LISP protocol is a Cisco proprietary technology, it is available as an open-source technology and could provide the basis of underlay fabrics within an open-source SDN solution such as OpenDaylight.

Figure 4 shows LISP deployment within a network, we created a core consisting of five CSR1000v routers, of which three have been selected to operate as the Routing Locators (RLOC) and function as both ingress and egress routers (xTR) into the core network. To provide connectivity between each of the xTR devices, OSPF has been configured to provide routing within the network Core. Each of the campus routers has been introduced to mimic host devices connected to the inside interface of the xTR and example LISP over a geographical separation. Within the Core-MS-MR router, two roles are performed, the first is the Map-Server and allows each of the RLOC routers to register the connected devices' Endpoint Identifiers (EID) and form an RLOC-to-EID database within the core router.

The second role is the Map-Resolver which handles Map-Requests (MR) from xTR devices when the device requires the location of an EID within the network. The MR performs a lookup to match the destination EID with the RLOC it is connected to and informs the xTR which RLOC the data should be forwarded on to.

To demonstrate the viability of LISP deployment with a small network, we removed LISP services from the router and utilized it as a Core edge access router with an Ubuntu

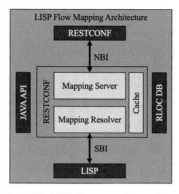

Fig. 3. ODL LISP flow mapping architecture

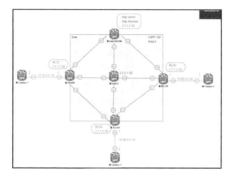

Fig. 4. LISP network diagram

Server operating behind the Core edge device. Figure 5 shows the network diagram and deployment within EVE-NG virtualization software with CSR3 and CSR2 providing xTR operations. Similar to the previous demonstration, OSPF has been used for Core routing and the LISP Mapping services moved to a centralized OpenDaylight SDN Controller which deployed on an ubuntu server.

The key benefit of introducing the OpenDaylight SDN controller to services such as LISP is the programmable opportunities within the network. This enables organizations to create solutions and programs that a unique and fully tailored towards the organization's network requirements. Whilst controllers such as OpenDaylight can be utilized to operate and manage many networking aspects, Rest APIs within the controller can be used to abstract information which can be later used to make programmatic decisions and network changes.

4.2   Virtual Extensible Local Area Networks and OpenDaylight

As previously discussed with the Cisco DNAC-based literature, the deployment of under-lay and overlay networks could be achieved through the implementation of VXLAN technologies. The VXLAN technology can provide layer 2 extensions of segmented

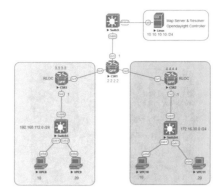

Fig. 5. LISP and OpenDaylight SDN network topology

networks across different domains and increasing the number of available LANs significantly compared to traditional VLAN segmentation [14]. In a study of virtual machine
(VM) migration between distributed datacenters, the OpenDaylight SDN controller was
successfully utilized to orchestrate the efficient migration of the VMs both inter-domain
and inter-LAN through the utilization of VXLAN. Applying this approach to network
segmentation would allow both the migration and movement of physical devices and
virtual devices.

Typically, VXLAN uses the 'Flood and Learn' approach to distribute broadcast,
unknown unicast and multicast (BUM) traffic to every VTEP device within the same
multicast group. This reduces the need to flood broadcast traffic throughout the entire
network and limiting network device utilization. This can be achieved within the data
plane using either multicast or the 'Ingress Replication' method. An advantage of using
multicast VXLAN is the scalability through network growth, as long as the switch resides
within the same multicast group and the VXLAN process, it will receive BUM traffic
for the configured networks. As a solution, we have demonstrated how VXLAN can
be implemented within a small network using the 'Flood and Learn' approach using
firstly the unicast ingress replication method (Fig. 6) and secondly, the multicast method
(Fig. 7).

In some scenarios, data transfers may require windowing and error-checking mechanisms as found within TCP. As such, using a multicast VXLAN would be unsuitable
as this method of messaging uses the less reliable but more efficient UDP transport
protocol [13]. However, within a scenario where the IP address is known but the MAC
address is not, then ARP requests would need to be broadcast throughout the Layer 2
network to establish the destination MAC. Since we are operating L2 over L3 overlay
mechanisms need to be introduced to ensure replication of broadcast messages within
the underlay. A solution is the manual replication of unknown unicast traffic through
Protocol-Independent Multicast (PIM). Figure 6 shows a network diagram of a small network where PIM is introduced to manually replicate (multicast) unknown unicast traffic
and sent to pre-configured VTEP destinations. However, organizations using this method
would not benefit from automatic device discovery as provided within Cisco DNAC. In
response to this, the Any-source Multicast Rendezvous Point utilizing the 'Flood and

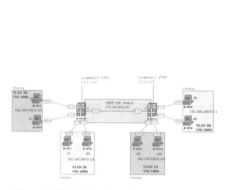

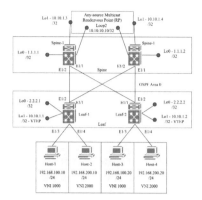

Fig. 6. VXLAN unicast ingress replication   **Fig. 7.** VXLAN any-source multicast

Learn' concept was exampled to enable automatic discovery of VTEP devices within the same multicast group (Fig. 7). This method of VXLAN deployment provides small business scalability opportunities through expansion of layer 2 over layer 3 campus sites whilst simplifying the configuring new devices with network growth.

5 Conclusion

The key motivation of this study was to examine how SDN concepts could be introduced into small to medium organisations using open-source alternatives to minimize financial costs in comparison to commercially available systems. We examined some of the key features within both commercially available controllers and open-source alternatives with some features drawing similarities. Of significance is the professional and comprehensive solution provided by Cisco DNAC and its end-user interface. This would be a good option for organizations seeking to deploy SDN within their network but considering budgets, it may not be a financially viable option. However, through our research, several key features of Cisco DNAC such as LISP and VXLAN were matched with services that could be introduced using an open-source SDN controller such as OpenDaylight (ODL). The functionalities provided by the ODL SDN controller to support services such as LISP and VXLAN drawing significant similarities to the Cisco SD-Access architecture. This would allow for an open-source alternative to introduce underlay and overlay fabrics within SME infrastructure with zero-cost compared to Cisco DNAC. Organizations would still need to establish whether current networking hardware can support technologies such as LISP and VXLAN.

Although we have shown open-source alternatives to commercial solutions, further quantitative research would provide opportunities to analyze metrics such as performance and effectiveness between open source and commercial solutions.

References

1. Badotra, S., Panda, S.N.: Evaluation and comparison of OpenDayLight and open networking operating system in software-defined networking. Clust. Comput. **23**(2), 1281–1291 (2020). https://doi.org/10.1007/s10586-019-02996-0

2. Bhutani, A., Wadhwani, P.: Software Defined Networking (SDN) Market Size by Component (2019). https://www.gminsights.com/industry-analysis/software-defined-networking-sdn-market
3. Cisco Systems Inc. Cisco Validated Design Program (2020). https://www.cisco.com/c/en/us/solutions/enterprise/validated-design-program/networking_solutions_products_genericcontent0900aecd80601e22.html
4. Cisco Systems. Programmability Configuration Guide, Cisco IOS XE Amsterdam, 17. 1. x (2019b). https://www.cisco.com/c/en/us/td/docs/ios-xml/ios/prog/configuration/171/b_171_programmability_cg/restconf_protocol.html
5. Godanj, I., Nenadić, K., Romić, K.: Simple example of software defined network. In: Proceedings of 2016 International Conference on Smart Systems and Technologies, SST 2016, pp. 231–238 (2016). https://doi.org/10.1109/SST.2016.7765665
6. Gedia, D., Perigo, L.: Neto-App: a network orchestration application for centralized network management in small business networks. IT Ind. 6(3), 61–72 (2018)
7. Gooley, J., Schuemann, D., Yanch, D., Curran, J.: Cisco Software-Defined Wide Area Networks: Designing, Deploying and Securing Your Next Generation WAN with Cisco SD-WAN. Cisco Press, UK
8. Huang, T., Yu, F. R., Zhang, C., Liu, J., Zhang, J., Liu, Y.: A survey on Large-Scale Software Defined Networking (SDN) testbeds: approaches and challenges. In: IEEE Communications Surveys and Tutorials, vol. 19, Issue 2, pp. 891–917. Institute of Electrical and Electronics Engineers Inc. (2017). https://doi.org/10.1109/COMST.2016.2630047
9. Jain, S., Kumar, A., Mandal, S., Ong, J., Poutievski, L., Singh, A., Venkata, S., Wanderer, J., Zhou, J., Zhu, M., Zolla, J., Hölzle, U., Stuart, S., Vahdat, A.: B4: Experience with a globally-deployed software defined WAN. Comput. Commun. Rev. 43(4), 3–14 (2013). https://doi.org/10.1145/2534169.2486019
10. Jarschel, M., Zinner, T., Hossfeld, T., Tran-Gia, P., Kellerer, W.: Interfaces, attributes, and use cases: a compass for SDN. IEEE Commun. Mag. 52(6), 210–217 (2014). https://doi.org/10.1109/MCOM.2014.6829966
11. Kreutz, D., Ramos, F.M.V., Verissimo, P.E., Rothenberg, C.E., Azodolmolky, S., Uhlig, S.: Software-defined networking: a comprehensive survey. Proc. IEEE 103(1), 14–76 (2015). https://doi.org/10.1109/JPROC.2014.2371999
12. Levin, D., Schmid, S., Canini, M., Schaffert, F., Feldmann, A.: Panopticon: Reaping the Benefits of Incremental SDN Deployment in Enterprise Networks. In: 2014 USENIX Annual Technical Conference, p. 333 (2014)
13. Naranjo, E.F., Salazar Ch, G.D.: Underlay and overlay networks: the approach to solve addressing and segmentation problems in the new networking era: VXLAN encapsulation with Cisco and open-source networks. In: 2017 IEEE 2nd Ecuador Technical Chapters Meeting, ETCM 2017, 1–6 January 2017 (2018)
14. Noghani, K.A., Kassler, A., Gopannan, P.S.: EVPN/SDN assisted live VM migration between geo-distributed data centers. In: 2018 4th IEEE Conference on Network Softwarization and Workshops, NetSoft 2018, pp. 182–186 (2018)
15. Nunes, B.A.A., Mendonca, M., Nguyen, X.N., Obraczka, K., Turletti, T.: A survey of software-defined networking: past, present, and future of programmable networks. IEEE Commun. Surv. Tutor. 16(3), 1617–1634 (2014)
16. OpenDaylight Project: LISP Flow Mapping User Guide—LISP Flow Mapping master documentation (2018a)
17. Prajapati, A., Sakadasariya, A., Patel, J.: Software defined network: future of networking. In: Proceedings of the 2nd International Conference on Inventive Systems and Control, ICISC 2018, pp. 1351–1354 (2018)
18. Tarutė, A., Gatautis, R.: ICT impact on SMEs performance. Procedia Soc. Behav. Sci. 110, 1218–1225 (2014). https://doi.org/10.1016/j.sbspro.2013.12.968

19. Zhao, Y., Iannone, L., Riguidel, M.: On the performance of SDN controllers: a reality check. In: 2015 IEEE Conference on Network Function Virtualization and Software Defined Network, NFV-SDN 2015, pp. 79–85 (2016)
20. Zhu, L., Karim, M. M., Sharif, K., Li, F., Du, X., Guizani, M.: SDN Controllers: Benchmarking and Performance Evaluation (2019). ArXiv. https://arxiv.org/abs/1902.04491

A Survey of Description Methods
for Fieldbus Systems

Darshit Pandya[(✉)]

University of Rostock, Rostock, Germany
darshit.pandya@uni-rostock.de

Abstract. Fieldbus systems have been around in the manufacturing industries for quite a while now, and with the increasing trends in Building Automation Systems it has also entered into other domains. In order to closely monitor, visualise and analyse these systems for enabling and aiding services such as energy management, predictive maintenance and strengthening security, it is essential to describe these system in an appropriate manner. This paper discusses different methods available for describing these systems, and how those methods vary from each other.

1 Introduction

In the automation pyramid, the fieldbus systems work at the Field level and quite often at the Control level. As the name suggests, it is a bus communication system that connects several terminal devices such as sensors, actuators and control devices such as PLCs with the help of different kinds of physical media. Fieldbus system, as described in [2], is *the digital way to integrate devices to the control system*. The stern requirements for communication at field level is what distinguish fieldbus systems. Traditional industrial systems were connected in a point-to-point manner, which dramatically increases the need of cabling, and restricts the communication scope.

In order to monitor, visualise or analyse a system, there is a need of system description in an appropriate format. A concrete example for this would be *Security Analysis*, where it is vital to have an apt system description, so as to perform the analysis better to strengthen the security of the corresponding entity. Naturally the content and the depth of the system description will depend on its further application. As discussed earlier fieldbus systems are used in critical environment where several devices communicate with each other using compact messages and tiny overhead. Thus, when we talk about the description of such distributed networked systems, we are essentially looking at describing two core elements of the system.

- Describing the network
- Describing the devices

The network description might include details of how the network is laid out, what physical media is used, the logical topologies, the functional topologies,

L. Barolli et al. (Eds.): AINA 2021, LNNS 227, pp. 200–210, 2021.
https://doi.org/10.1007/978-3-030-75078-7_21

etc. Of course, working at a certain abstraction level might make more details obscure. On the other hand, device description might include details about individual devices, about their available interfaces, manufacturing details, type of device, etc. Combining these details might give us an overview of what devices are being used and how and what devices connected to each other.

Usually all the existing fieldbus technologies use these two primary descriptions in order to describe the entire system. The data format that most of these technologies use is Extensible Markup Language (XML). The justification, as explained in [1], is reliability. Because XML-based formats are based on their Schema Definitions (XSDs), it is possible to check these models for their reliability and consistency of both semantics and syntax (which arguably are the most important elements of a data model).

The standard IEC 61158 (*Industrial Communication Networks - Fieldbus specifications*) describes 79 existing technologies, and a possible method on how these technologies could be amalgamated in a single foundation and representation [1]. The standard originally discussed eight *'types'* of protocols (and its related profile families), also specifying fieldbus systems' *generic concepts* [11]. The systems use different approaches while describing devices and networks. The device descriptions are more often used for correctly configuring the devices in the design phase and analysing each device against its utility in test phase. Usually the device descriptions (mostly type information about the device) are provided by the device manufacturers, and the individual parameters according to the device's implementation could be adjusted using proprietary engineering tools.

This device description, along with *Network descriptions* (such as network topologies, which are not only limited to nodes and edges' information anymore, but also has logical and functional variants which provides information on how the communication is laid out, and which edges on the network are communicating amongst themselves), are enough for general system description. Apart from these distinct information sets, a separate set of information might also be derived from the conjunction of these descriptions, and that is how a system, and its characteristics could be described.

2   Description Methods

2.1   EDDL

As discussed earlier, this is one of the more widely used (text based) device description language, used mainly in the automation industry [1]. It is a standardised way for exchanging and accessing device description, provided by it's manufacturer [8]. The EDDL standard is described in *IEC 61804-3 (Devices and Integration in enterprise systems)*. The description (majorly) consists of two elements, the information/data that a device can provide, and what *communication commands* do we need in order to gain that information data [8]. These two features are important since it makes insertion and management of a device easier. In case of feature updates (which are rare in critical infrastructures), we just need to amend the EDDL file to make the integration uncomplicated [8]. Thus, a field device can be appended into the system without hindering or disturbing the runtime functionality of the system. The standard IEC 61804-3 identifies EDDL as a *generic* language for describing [3],

- *Device parameters and its functions*
- *Graphical representation of the device*
- *Device's interactions with control devices*

These files are predesigned by manufacturers, and while adding them in the system, the relevant parameters are adjusted. Furthermore, there are several tools (developed by the supporting companies), that helps in configuring the system, debugging, run simulations, etc. [3]. Also, because it is a text-based description, it becomes easier to work with it. The image below gives us a broader view of the information EDDL covers (Fig. 1).

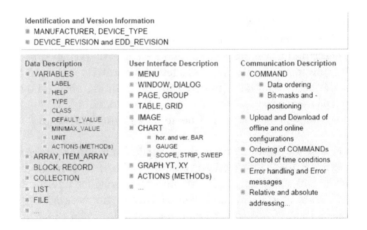

Fig. 1. EDDL Structure and keywords [3]

2.2 ESI and ENI

With the increasing trends of remotely controlling and monitoring industrial processes, in order to efficiently and smoothly conduct business processes, there has been also an increase in Ethernet and IP based technologies at Fieldbus layer. *EtherCAT* is one such technology. Originally developed by Beckhoff Automation, EtherCAT uses traditional computer network technologies without risking industrial requirements [9]. EtherCAT has been internationally standardised in *IEC 61158* since 2007 [9]. EtherCAT is used in several applications with hard and soft real time requirements, in automation technology.

There are numerous solutions available in order to describe traditional IT networks (which could be helpful here too). For instance, *Network Description Language* or *Network Markup Language*. But EtherCAT uses its indigenous description methods [4,5] to describe device and its network respectively. These are

EtherCAT Slave Information (ESI) and *EtherCAT Network Information (ENI)*. Both these standards/descriptions are based on XML (Fig. 2).

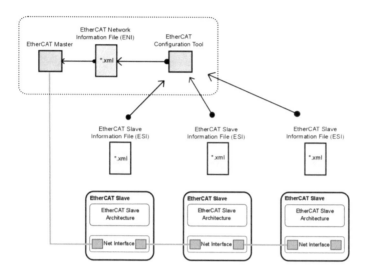

Fig. 2. EtherCAT description methodology [10]

- Just like EDDL, the ESI files come with the device with all its network-accessible attributes/properties and their mapping options [9]. This information could be further used by the configuration tool to configure the network online or offline.
- The ENI files detail the network topology, the cyclical commands sent periodically and how commands could be initiated for each device. The EtherCAT master has these ENI Files and it sends commands according to its content [1].

2.3   EDS

Another (non XML based) approach is *Electronic Data Sheets*, which work with EtherNet/IP networks and describe how devices can be used on the network [1]. The ODVA organisation controls three different technologies (namely, *Common Industrial Protocol, EtherNet/IP and DeviceNet*) which all use EDS for device description. The description contains objects, attributes and the services that are accessible in the device [1]. It contains ASCII representations of devices' object parameters and how the object could be addressed. It does not have hierarchical structures, rather it follows *Index/Subindex description* method. This description method is not limited to ODVA organisation's technologies, but other systems such as CANOpen systems also used EDS in order to describe system components, before migrating to its own XML based specification (CiA 311) which is discussed in the following section. As is the case with many other description

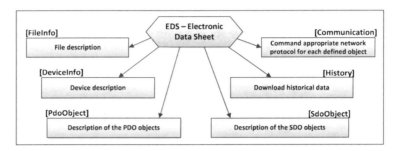

Fig. 3. EDS file structure [13]

methods, EDS are provided by the device manufacturer with necessary information to communicate with the device and to configure it using proprietary tools during the system setup phase. As discussed earlier it contains (in ASCII format) two key elements [12],

- *Communication parameters*
- *Objects of the device*

Thus in essence, each device in a system will have its own manufacturer provided EDS, with the relevant information. The ultimate motivation behind having/creating this file is to make device integration easy. Figure 3 shows the file structure of EDS [13].

As can be observed from the Fig. 3, it is not only possible to describe the device and the appropriate network command for each defined object, but it is also possible to describe and categorise,

- *Process Data Objects (PDO)*
- *Service Data Objects (SDO)*

Both objects and their services have different goals. *Process Data Objects* (in general) deal with the real time communication between devices whereas *Service Data Objects* helps in read/write operations in the object dictionary of a device [14].

2.4   CiA 311

A CANOpen FD (Flexible Data-Rate) device's specific method for device description is CiA 311, which is in XML format. [6] describes its advantages and use cases in detail. It also describes how CiA 311 trumps over the outdated electronic data sheets CiA 306 that were used by CANOpen Field devices. The major motivation for the update was to integrate application information, device description and communication information in the same description [6]. Although EDS was sufficient to describe this information, as it was responsible for exchanging information between CANOpen applications, network and device configuration tools, etc. in CANOpen systems for years [6], the major disadvantage was EDS' structural capabilities. The structure of EDS only allowed for *sections and key-value-pairs* but not any additional *structuring or grouping of parameters* (since it uses the Windows INI format) [6]. Also, as discussed earlier it did not allow a hierarchical structure, rather it followed an Index/Subindex description.

The new XML-based format was designed based on the ISO standard 15745 *Industrial automation systems and integration - Open systems integration framework* [6], which extensively describes different protocol-independent device profiles and communication profiles for system integration and smoother interoperability. As discussed, [6] details how CiA 311 is based upon the meticulously categorised structure of ISO 15745. The structure of ISO 15745 Profile Container suggests a primary division between

- *Device Profile*: Profile Header and Profile Body
- *Communication Network Profile*: Profile Header and Profile Body

Device Profile Body	Communication Profile Body
• Device Manager device monitoring information • Device Identity vendor, part number, … • Device Function key characteristics • Application Process description of application, parameters, functions, ..	• Application Layers list of CANopen objects, dummy mapping, .. • Transport Layers CAN bit rates • Network Management Layers general CANopen features

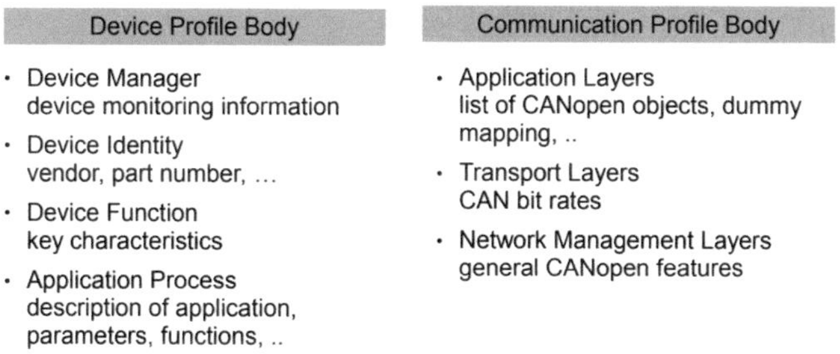

Fig. 4. CANOpen FD XML device descriptions [15]

Most of the information are in the respective Profile Bodies [15]. Following the structure in the standard, the *Device Profile Body* further contains information about device monitoring, application descriptions, parameters templates, device identity, etc. [15] and the *Communication Profile Body* has the device interface information, and other communication relevant details as shown in the Fig. 4. The Application Process section of the Device Profile Body is usually the biggest section [15]. The core idea behind this structurisation/categorisation is to have links or references between two profiles [15]. (For instance we can have a link between an application process from the Device Profile Body to a CANopen object in the Communication Profile Body).

2.5 FIBEX

A seemingly large issue that prevailed in the automotive and measuring industry in the past has been the incompatibility across tools and not having a uniform description language, which made data exchange and conversion between tools difficult [7]. Using this issue as a motivation, *ASAM (Association for Standardisation of Automation and Measuring)* came up with an XML based cross-company standard schema called *FIBEX (Fieldbus Exchange Format)*.

According to [7] the FIBEX standard details a multi-protocol- capable format for the exchange of data in *message-oriented* communication systems. A FIBEX file is divided into several areas, and each area details its unique functionalities. Using objects from these areas, we can describe generic network topologies (with routing options), we can also describe interfaces, and further append the description using technology-specific extensions (for FlexRay, MOST, CAN, TTCAN, LIN and Ethernet). With all the mentioned features about FIBEX, there is no question about its ability of designing and managing large heterogeneous networks [16]. The strength of FIBEX lies in *compatibility across tools, data exchange and its integrity* [16]. [16] details FIBEX's use cases (which includes *Bus Monitoring, Simulation, Communication Scheduling, Bus System Definition, etc.* that lead to the core specification of FIBEX representation.

Using XML design features, FIBEX decomposes a communication system into several elements. A few of the major elements are as follows,

- *ECU (Electronic Control Unit)*
- *Function*
- *Signal*
- *Cluster*
- *Topology*

The elements listed here are just a few of them, but this shows the depth of FIBEX's description [16]. As shown in the diagram, the entities belonging to these elements are related to each other with logical cardinality and hence describe a system. As described in [16] FIBEX consists of a basic definition (fibex.xsd) that is a protocol-independent definition, which could be further extended to protocol specific descriptions using additional definitions.

2.6 AutomationML

Two concerns are addressed in [1] regarding device description languages and topology description. This includes

- the detached type specific information for a device and the individual configurable parameter of the same device,
- and the unavailability of the network topology information outside the proprietary engineering tool (across other tools).

The paper [1] further tries to overcome both the issues using *AutomationML* with an example. AutomationML is an XML-based data format, which follows an object-oriented paradigm [1]. The AutomationML community has developed a technology independent proposal on modelling the communication model of a system. It is not only limited to describing network features in terms of topologies, but it also has the capabilities of describing the *geometrical features* (using COLLADA) and *functional features* (using PLCopen XML) of a device.

However the key idea modelling the communication systems (also discussed in [1]) is the strict separation of the physical aspects of the network (dealing with the physical media, cables and wires of the network infrastructure), and the logical topology (dealing with the logical addresses and the logical interaction between devices). In order to make it technology independent, AutomationML has communication white papers [17] that includes a set of *Role Classes* and *Interface classes*. Using these, one can model technology specific communication systems [1].

With its object oriented design, AutomationML provides the capabilities to initiate object instances for manufacturer provided *Device type*, and modify individual parameters of these instances, link them with other instances, create hierarchies, consequently integrate device information with system specific configuration parameters and network information [18]. Furthermore, it is also possible to model the behavior and sequencing of a system using the extended PLCopen XML features as discussed earlier [18].

As can be seen in the Fig. 5, a typical AutomationML file consists of four divisions.

- The first one being *InstanceHierarchy*, which as the name suggests lists the object instances as discussed earlier [18]. It is the most important modelling element in AutomationML, with its *InternalElements* hierarchy.

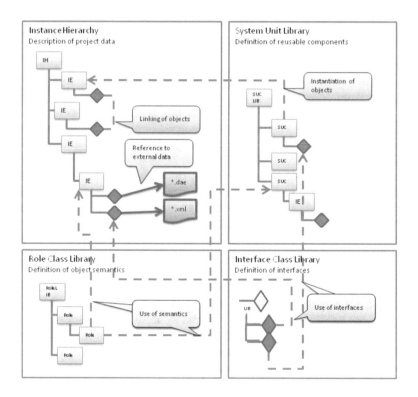

Fig. 5. AutomationML architecture [18]

- The second one is the *Role Class Library* where an abstract *'Role'* (or a functionality) could be assigned to an instance (e.g. MechanicalPart or CommunicationRoleClassLib), and hence it defines the contextual semantics for the object instance [18].
- Similarly there is an *Interface Class Library* where an abstract relation could be defined between elements or between elements and external information not covered by the model (for instance, SignalInterface, PhysicalEndPoint, etc.) [18].
- And the final division in the architecture is *Unit classes*. These are the reusable modular elements, which could be used as templates for system modelling [18].

Every division apart from InstanceHierarchy has a set of basic classes already defined by AutomationML. Using these system-independent elements, and establishing links between them, it is possible to create a system-specific (communication and functional) model.

3 Evaluation

There are naturally multiple factors on which these description methods could be evaluated. For instance, the possibility to describe systems from multiple domains, the information depth or the abstraction layer at which these descriptions work, how detailed these descriptions are, etc. Each of these description methods were designed with a clear initial motivation, and usage, which has been mentioned briefly for each method.

Another factor could be the information coverage. Is it primarily a device description or a network description, or can it describe both. Can the description be extended to contain interface information. Furthermore, in device description, does it only include manufacturer provided *device type information* or does it also include system specific configurable parameters. Similarly for network description, how many different network topologies (*physical, logical, functional*) can it describe, can it describe the routing?, etc.

We can take a look at the encoding of information in each description. In what format/s can we describe the system. This is an important factor since the available output format/s dictates its usage across other tools. Also, with formats based on a pre-defined schema (like XML), it is also possible to validate and check the description for any kind of consistency errors. Encoding in traditional formats such as Windows INI and others, might limit the description structurally, and its capability to contain more information.

As discussed in the previous points, there can be multiple individual factors on which these description methods could be evaluated. In order to decide which description method is suitable for a particular application, the user must curate a subset of application context dependant factors, and decide upon the favorable description method. As seen in Table 1 we have listed down the description methods discussed here with its ability to describe a system, till what lengths can it describe it, in what format can it describe, and usually in what industries are these formats used. This table basically works as a summary for the paper.

Table 1. Overview of description methods

Description methods	Domain	Device/Network description	Encoding formats	Information depth
EDDL	Automation	Device	Text based	- Device parameters - Graphical representation - Communication commands
ESI and ENI	Automation	ESI for device ENI for network	XML	- ESI : Manufacturer provided file Network-accessible attributes/properties, - ENI : Details the network topology
EDS	Automation spacecraft engineering, etc.	Device	ASCII	Communication parameters objects of device
CiA 311	Automation robotics	Both	XML	Device profile communication profile
FIBEX	Automotive and others	Both	XML	Huge inter-related data model for a system
AutomationML	Technology independent	Both	XML	Object oriented modular design, possible to create system specific descriptions

4 Summary

This paper discusses several description methods available for describing fieldbus systems (in several cases, other systems too). The methods discussed here are not limited to single domain, rather they are chosen from different domains to show the versatility of some of these descriptions. Some of these descriptions are not only limited to a single entity description (like device description or network description) but they cover the entire general system description in a system-independent manner. The goal of this paper was to exhibit the more widely used system description methods, how are they different from others, what were their initial design purpose, and how it has evolved.

Acknowledgements. This research is financed by grant from the German Ministry of Economics and Energy.

References

1. Rentschler, M., Drath, R.: Vendor-Independent modeling and exchange of fieldbus topologies with AutomationML. In: 2018 IEEE 23rd International Conference on Emerging Technologies and Factory Automation (ETFA), Turin, pp. 956–963 (2018). https://doi.org/10.1109/ETFA.2018.8502630
2. Eren, H.: Fieldbus Systems. Handbook of Measuring System Design, 15 July 2015. https://doi.org/10.1002/0471497398.mm933
3. Winkel, L., Diedrich, C., Berge, J., Blevins, T.: Keeping Systems and Communication Up-to-date using EDDL. https://pdfs.semanticscholar.org/6a23/191ea7fc9a8c02890589c583d98c7b1c21b9.pdf. Accessed 25 Jan 2021
4. EtherCAT Technology Group. ETG.2000 EtherCAT Slave Information (ESI) Specification

5. EtherCAT Technology Group. ETG.2100 EtherCAT Network Information (ENI) Specification

6. Gedenk, T.: emtas GmbH, 'Use cases and advantages of the XML device description format for CANopen FD devices' (2017). https://www.can-cia.org/fileadmin/resources/documents/conferences/2017_gedenk.pdf. Accessed 25 Jan 2021

7. Bachmann, T.: 'FIBEX XML format and AUTOSAR development' via EETimes (2009). https://www.eetimes.com/FIBEX-XML-format-and-AUTOSAR-development/. Accessed 25 Jan 2021

8. Banerjee, S., Großmann, I.D.: An electronic device description language based approach for communication with DBMS and file system in an industrial automation scenario. In: 2016 IEEE 21st International Conference on Emerging Technologies and Factory Automation (ETFA), Berlin, pp. 1–4 (2016). https://doi.org/10.1109/ETFA.2016.7733682.

9. EtherCAT Technology Group. 'EtherCAT - the Ethernet Fieldbus'. https://www.ethercat.org/pdf/english/ETG_Brochure_EN.pdf. Accessed 25 Jan 2021

10. Langlois, K., et al.: EtherCAT tutorial: an introduction for real-time hardware communication on windows [tutorial]. IEEE Robot. Autom. Mag. **25**(1), 22–122 (2018). https://www.researchgate.net/publication/323718831

11. ELAND Cables. 'IEC 61158 Cable'. https://www.elandcables.com/electrical-cable-and-accessories/cables-by-standard/iec-61158-cable. Accessed 25 Jan 2021

12. JVL intelligent motors. 'Ethernet/IP - Read More'. https://www.jvl.dk/810/industrial-protocol-cip. Accessed 25 Jan 2021

13. Gaitan, N.C.: An uniform description solution for devices based on DCON ASCII protocol. In: 2014 18th International Conference on System Theory, Control and Computing (ICSTCC), Sinaia, pp. 821–825 (2014). https://doi.org/10.1109/ICSTCC.2014.6982520

14. CSS Electronics. 'CANopen Explained - A Simple Intro (2020). https://www.csselectronics.com/screen/page/canopen-tutorial-simple-intro/language/en. Accessed 25 Jan 2021

15. Gedenk, T.: emtas GmbH. 'Use Cases and Advantages of the new XML Device Description for CANopen FD Devices.' https://www.can-cia.org/fileadmin/resources/documents/slides/icc_2017_slides_gedenk.pdf. Accessed 25 Jan 2021

16. Krammer, J., Bornat, P., Dengel, G., Kuttenkeuler, S., Lukas, R., et al.: FIBEX – an exchange format for networks based on field busses. In: 2nd Embedded Real Time Software Congress (ERTS 2004), Toulouse, France. hal-02275436 (2004)

17. AutomationML. 'AutomationML Whitepaper Communication', September 2014. https://www.automationml.org/o.red/uploads/dateien/1459418220-Automation ML%20Whitepaper%20-%20AutomationML%20Communication%20v1_Sept 2014.pdf. Accessed 25 Jan 2021

18. Schmidt, N., Lueder, A.: AutomationML in a Nutshell, November 2015. https://www.automationml.org/o.red/uploads/dateien/1447420977-AutomationML %20in%20a%20Nutshell_151104.pdf. Accessed 25 Jan 2021

A Hybrid Message Delivery Method for Vehicular DTN Considering Impact of Shuttle Buses and Roadside Units

Masaya Azuma[1], Shota Uchimura[1], Yoshiki Tada[1], Makoto Ikeda[2(✉)],
and Leonard Barolli[2]

[1] Graduate School of Engineering, Fukuoka Institute of Technology,
3-30-1 Wajiro-higashi, Higashi-ku, Fukuoka 811-0295, Japan
[2] Department of Information and Communication Engineering,
Fukuoka Institute of Technology, 3-30-1 Wajiro-higashi,
Higashi-ku, Fukuoka 811-0295, Japan
makoto.ikd@acm.org, barolli@fit.ac.jp

Abstract. In recent years, hybrid approaches have become common in a wide-range of fields, such as vehicle engines, communication system and improved crop varieties. By combining the advantages of hybrid approaches better methods can be proposed and implemented. In this paper, we propose a hybrid message delivery method for Vehicular Delay Tolerant Networking (DTN). We focus on the impact of shuttle buses and road-side units with communication capabilities for delivering the messages. We use Epidemic, Spray and Wait (SpW), and the proposed hybrid method as the message delivery protocols. From the simulation results, we observed that the delivery ratio of hybrid method considering Epidemic-based routing for regular vehicles and SpW-based routing for road-side units is good for the considered scenario.

Keywords: Vehiuclar DTN · Hybrid · Epidemic · SpW

1 Introduction

Mobile Ad-hoc Networks (MANETs) have been considered an effective communication method for low-speed mobile terminals. For high speed mobile terminals, the message delivery method needs to consider sparse and dense situations. In a dense environment, there are many neighboring nodes and the delivery ratio to the destination is generally high. On the other hand, in a sparse network, the conventional protocol will time out and the message will not reach the destination in short time [11]. In order to address this problem, Delay-/Disruption-/Disconnection-Tolerant Networking (DTN) [6] has attracted attention. In [13], we have evaluated a message delivery method in Vehicular DTN considering Enhanced Dynamic Timer (EDT) [7]. The EDT method is one of the recovery function for reducing the storage usage of each vehicle. However, the EDT method communication performance is decreased in sparse networks. For this

L. Barolli et al. (Eds.): AINA 2021, LNNS 227, pp. 211–218, 2021.
https://doi.org/10.1007/978-3-030-75078-7_22

reason, some hybrid methods have been proposed to address the overhead and delay problems [4,5].

In [5], the authors have proposed a hybrid method, which combines MANET and DTN to decrease the number of transmissions in the network. But, this work does not consider the evaluation of roadside units or other applications.

In [4], the authors have proposed a hybrid DTN routing method which considers the Epidemic-based routing method and the SpW-based routing method to improve the storage usage of the vehicles.

In this paper, we propose a hybrid message delivery method which consider the Epidemic-based routing method and Spray and Wait (SpW) based routing method. We focus on the impact of shuttle buses and road-side units with communication capabilities for delivering the messages.

The rest of the paper is structured as follows. In Sect. 2, we give the overview of DTN protocols for inter-vehicle communications. In Sect. 3 is described the hybrid message delivery method for Vehicular DTN. In Sect. 4, we provide the evaluation system and the simulation results. Finally, conclusions and future work are given in Sect. 5.

2   Overview of DTN Protocols

DTN can provide a reliable internet-working for space tasks [3,9,15]. The space networks have possibly large latency, frequent link disruption and frequent disconnection. In Vehicular DTN, the intermediate vehicles store messages in their storage and then transmit to others. The network architecture is specified in RFC 4838 [2].

The famous DTN protocols are Epidemic [8,14] and SpW [12]. Epidemic routing is performed by using two control messages to replicate a bundle message. Each vehicle periodically broadcasts a Summary Vector (SV) in the network. The SV contains a list of stored messages of each vehicle. When the vehicles receive the SV, they compare received SV to their SV. The vehicle sends a REQUEST message if the received SV contains an unknown message. In Epidemic routing, consumption of network resources and storage state become critical problems, because the vehicles replicate messages to neighbors in their communication range. Then, the received messages remain in the storage and the messages are continuously replicated even if the end-point receives the messages. Therefore, the recovery schemes such as timer and anti-packet may delete the replicated messages in the network. For the sparse network, the anti-packet deletes the replicated messages too late due to the large delay. In the case of the conventional anti-packet, the end-point broadcasts the anti-packet, which contains the list of messages that are received by the end-point. Vehicles delete the messages according to the anti-packet and replicate the anti-packet to other vehicles. In the case of the conventional timer, the messages have a lifetime and they are punctually deleted when the lifetime of the messages is expired. The challenge is to formulate an optimal lifetime that applies to various environments.

The SpW achieves resource efficiency by setting a strict upper bound on the number of copies per bundle message allowed in the network. The protocol is composed of two phases: the spray phase and the wait phase. Spray phase is terminated when the number of replications reaches the upper limit. When a relay vehicle receives the replica, it enters the wait phase, where the relay simply holds that particular message until the destination is encountered directly. It is possible to suppress the communication cost compared with conventional Epidemic.

3   Proposed Hybrid Message Delivery Method

In this section, we describe the hybrid message delivery method considering regular vehicles, shuttle buses and road-side units as message relaying terminals. Our method uses Epidemic or SpW protocol depending on the terminal type. In this paper, we evaluate the hybrid method in two cases. In Table 1, we shown the list of protocols for different terminals. We use Epidemic for Case 1 and SpW for Case 2 as a message delivery protocol. Sequence numbers 3 and 4 use the hybrid method. For the sequence numbers 2 and 6, the road-side units are not activated.

Table 1. List of protocols for different terminals.

Seq	Case ID	Vehicle and shuttle bus	Road-side unit	Type
1	Case 1	Epidemic	Epidemic	Conventional
2	Case 1	Epidemic	Not activated	Conventional
3	Case 1	Epidemic	SpW	Hybrid
4	Case 2	SpW	Epidemic	Hybrid
5	Case 2	SpW	SpW	Conventional
6	Case 2	SpW	not activated	Conventional

4   Evaluation of Proposed Method

In this section, we evaluate the hybrid method in Vehicular DTN on the Scenargie simulator [10].

4.1   Evaluation Setting

We consider a mesh grid road model considering regular vehicles, shuttle buses and road-side units (see Fig. 1). In this evaluation, we deploy a maximum of 100 vehicles/km^2. The following two cases are prepared for this evaluation.

- Case 1: Regular vehicles use Epidemic routing as a message delivery method.
- Case 2: Regular vehicles use SpW as a message delivery method.
- The message start-point and road-side units are fixed, the destination is mobile. Regular vehicles continue to move on the roads based on the map-based random way-point mobility model. Shuttle buses continue to move only on the main road.

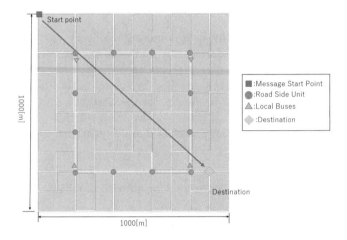

Fig. 1. Road model.

For this evaluation, we consider four vehicular densities: 25, 50, 75 and 100 vehicles/km^2. Table 2 shows the simulation parameters used for the network simulator. For evaluation, the start-point replicates 40 kinds of bundle messages to the relay regular vehicles and shuttle buses. Then the replicated messages deliver to the destination. The simulation time is 2,000 s. The ITU-R P.1411 propagation model is used in this simulation [1]. We consider the interference from obstacles at 5.9 GHz radio channel.

We evaluate the performance of storage usage, delivery ratio and delay for different densities of vehicles. The storage usage indicates the average of the storage state of each vehicle. The delivery ratio indicates the value of the generated bundle messages divided by the delivered bundle messages to the destination. The delay indicates the transmission delay of the bundle to reach the destination.

Table 2. Simulation parameters.

Parameter	Value
Simulation time ($T_{\max}$)	$2,000$ [seconds]
Area dimensions	$1,000$ [m] $\times$ $1,000$ [m]
Number of regular vehicles	25, 50, 75, 100 [vehicles/km^2]
Number of shuttle buses	4
Number of road-side units	0 or 12
Minimum speed	8.333 [m/s]
Maximum speed	16.666 [m/s]
Storage size of vehicles	10 [Mbytes]
Message start and end time	10–400 [seconds]
Message generation interval	10 [seconds]
SpW: Hop threshold	1–17 [seconds]
SV transmission interval	1 [second]
Message size	$1,000$ [bytes]
PHY model	IEEE 802.11p
Propagation model	ITU-R P.1411
Radio frequency	5.9 [GHz]
Antenna model	Omni-directional

4.2 Simulation Results

We evaluate the hybrid method considering different terminals (regular vehicle, shuttle bus and road-side unit) compared to conventional method for Case 1 and Case 2.

We present the simulation results of storage usage for different vehicles in Fig. 2. For 25 and 50 vehicles, the storage usage increases with increasing the simulation time. We have also observed that the storage usage drops after $1,800$ seconds for Case 1. We observed that hybrid method for Case 1 and Case 2 use less storage usage compared with conventional method.

We present the simulation results of the delivery ratio in Fig. 3. The delivery ratio for Case 1 reached 100%, while for Case 2 finally reached 58% after 75 vehicles due to the effect of maximum number of replications for SpW.

We present the simulation results of delay in Fig. 4. For Case 1, Epidemic considering road-side units are better regardless of the number of vehicles. For Case 2, the delay is almost the same. The delay results in Case 1 are shorter than Case 2. However, we can notice that the delay in this scenario is quite large.

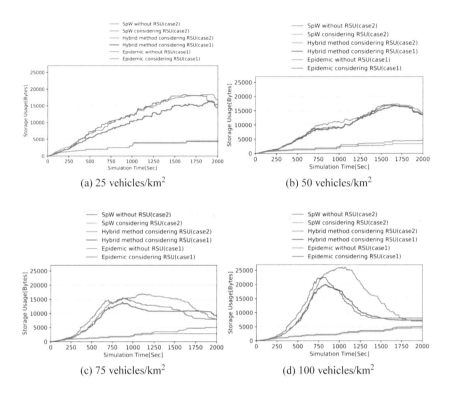

Fig. 2. Storage usage for different vehicles.

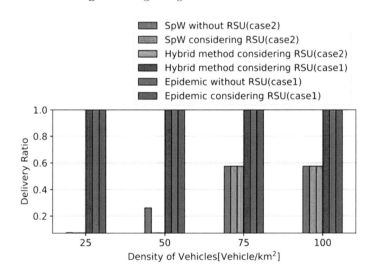

Fig. 3. Delivery ratio.

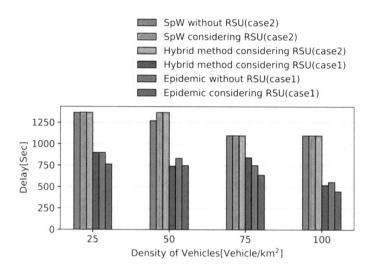

Fig. 4. Delay.

5   Conclusions

In this paper, we proposed a hybrid message delivery method which consider the Epidemic-based routing method and Spray and Wait (SpW) based routing method. We evaluated the network performance of shuttle buses and road-side units with communication capabilities for delivering the messages. From the simulation results, we found that the delivery ratio of hybrid method considering Epidemic-based routing for regular vehicles and SpW-based routing for road-side units is good for the considered scenario. Also, all bundle messages reached the destination even if for sparse network.

In future work, we would like to reduce the delay time and formulate the maximum number of replications that can be adapted to various scenarios.

Acknowledgment. This work has been partially funded by the research project from Comprehensive Research Organization at Fukuoka Institute of Technology (FIT), Japan.

References

1. Rec. ITU-R P.1411-7: Propagation data and prediction methods for the planning of short-range outdoor radiocommunication systems and radio local area networks in the frequency range 300 MHz to 100 GHz. ITU (2013)
2. Cerf, V., Burleigh, S., Hooke, A., Torgerson, L., Durst, R., Scott, K., Fall, K., Weiss, H.: Delay-tolerant networking architecture. IETF RFC 4838 (Informational), April 2007

3. Fall, K.: A delay-tolerant network architecture for challenged Internets. In: Proceedings of the International Conference on Applications, Technologies, Architectures, and Protocols for Computer Communications, SIGCOMM 2003, pp. 27–34 (2003)
4. Henmi, K., Koyama, A.: Hybrid type DTN routing protocol considering storage capacity. In: Proceedings of the 8th International Conference on Emerging Internet, Data and Web Technologies (EIDWT 2020), pp. 491–502, February 2020
5. Ito, M., Nishiyama, H., Kato, N.: A novel routing method for improving message delivery delay in hybrid DTN-MANET networks. In: Proceedings of the IEEE Global Communications Conference (GLOBECOM 2013), pp. 72–77 (2013)
6. Kawabata, N., Yamasaki, Y., Ohsaki, H.: Hybrid cellular-DTN for vehicle volume data collection in rural areas. In: Proceedings of the IEEE 43rd Annual Computer Software and Applications Conference (COMPSAC 2019), vol. 2, pp. 276–284, July 2019
7. Nakasaki, S., Ikeda, M., Barolli, L.: A message relaying method with enhanced dynamic timer considering decrease rate of neighboring nodes for Vehicular-DTN. In: Proceedings of the 14th International Conference on Broad-Band Wireless Computing, Communication and Applications (BWCCA 2019), pp. 711–720, November 2019
8. Ramanathan, R., Hansen, R., Basu, P., Hain, R.R., Krishnan, R.: Prioritized epidemic routing for opportunistic networks. In: Proceedings of the 1st International MobiSys Workshop on Mobile Opportunistic Networking (MobiOpp 2007), pp. 62–66 (2007)
9. Rüsch, S., Schürmann, D., Kapitza, R., Wolf, L.: Forward secure delay-tolerant networking. In: Proceedings of the 12th Workshop on Challenged Networks (CHANTS 2017), pp. 7–12, October 2017
10. Scenargie: Space-time engineering, LLC. http://www.spacetime-eng.com/
11. Solpico, D., Tan, M.I., Manalansan, E.J., Zagala, F.A., Leceta, J.A., Lanuza, D.F., Bernal, J., Ramos, R.D., Villareal, R.J., Cruz, X.M., dela Cruz, J.A., Lagazo, D.J., Honrado, J.L., Abrajano, G., Libatique, N.J., Tangonan, G.: Application of the V-HUB standard using LoRa beacons, mobile cloud, UAVs, and DTN for disaster-resilient communications. In: Proceedings of the IEEE Global Humanitarian Technology Conference (GHTC 2019), pp. 1–8, October 2019
12. Spyropoulos, T., Psounis, K., Raghavendra, C.S.: Spray and wait: an efficient routing scheme for intermittently connected mobile networks. In: Proceedings of the ACM SIGCOMM Workshop on Delay-Tolerant Networking 2005 (WDTN 2005), pp. 252–259 (2005)
13. Tada, Y., Ikeda, M., Barolli, L.: Performance evaluation of a message relaying method for resilient disaster networks. In: Proceedings of the 15th International Conference on Broadband and Wireless Computing, Communication and Applications (BWCCA 2020), pp. 1–10, October 2020
14. Vahdat, A., Becker, D.: Epidemic routing for partially-connected ad hoc networks. Technical report, Duke University (2000)
15. Wyatt, J., Burleigh, S., Jones, R., Torgerson, L., Wissler, S.: Disruption tolerant networking flight validation experiment on NASA's EPOXI mission. In: Proceedings of the 1st International Conference on Advances in Satellite and Space Communications (SPACOMM 2009), pp. 187–196, July 2009

Optimum Path Finding Framework for Drone Assisted Boat Rescue Missions

Kemal Ihsan Kilic[(✉)] and Leonardo Mostarda

Computer Science Division, University of Camerino, Camerino, Italy
{kemal.kemal,leonardo.mostarda}@unicam.it

Abstract. In boat rescue operations, Unmanned Aerial Vehicles (UAVs or drones) can travel long distances by utilizing the grid of floating charging stations (CSs) on the sea. To respond quickly and/or in an economic way to rescue calls, the "optimum path" from the Base Station (BS) to the boat and back to the BS should be estimated for the missions. Generally the optimum path is the shortest path involving hops from the BS via CSs to the boat and back to the BS. However, multiple objectives can be considered for two parties, drones and boats, like priority of boats, the number of chargings for the UAV, and the average waiting time for the boats. We proposed an heuristic extension called "red-gray path" which provides savings for the flight distance, depending on the boat position in the CS grid. The "drone range", which is the maximum flight range that UAV can fly with the battery, is the fundamental parameter for the design of the rescue infrastructure. The choices for the geometry of the CS grid plays important role in the effectiveness of the heuristic we proposed. We presented mathematical analysis on the effectiveness of two different deployment strategies for the CSs based on the degree of benefiting from the heuristic we proposed. Namely the triangular and square CS grids. While the square grid provides better savings for the red-gray path heuristic, the triangular grid offers better coverage for the proposed heuristic with less number of CSs for the same mission area.

1   Introduction

With the diminishing size of electronic chips, many interesting application areas have emerged. One of such rising technologies of recent years is the utilization of UAVs for various tasks. Today, drones, like quadcopters or hexacopters are used in many monitoring, search and rescue, industrial inspection, communication, and delivery applications. An extensive survey on the application areas of drones can be found in [12]. In recent years there is a growing interest in the drone assisted off-shore search and rescue applications [8, 13]. Drones can reach quickly and economically to the desired location as they fly in the air and they need little energy due to their small size. They can collect information and send it back making two-way communication possible. They can carry important medical supplies or other necessary items. Since they fly in the air, whether the mission is over the land or over the sea, their path is direct and almost without any

L. Barolli et al. (Eds.): AINA 2021, LNNS 227, pp. 219–231, 2021.
https://doi.org/10.1007/978-3-030-75078-7_23

obstacle. In this context, our work proposes a design framework for such search and rescue missions assisted by drones by considering the static configuration of CSs and boats. Our current goal was to design a dispatcher system for the drones (currently for a single drone) by considering static boat configuration that later can be used as a part of a dynamic optimization framework.

The proposed framework includes CS grid deployment strategies, heuristic path-finding with multi-objective optimizations, mathematical analysis of the CS deployment in connection with the proposed heuristic. For the framework the drone range is the crucial parameter as it determines the CS grid configuration and is the base of the heuristic, the "red-gray path", we proposed. For this heuristic, we assumed that the drones can not be charged on the boat. However, if the boats have a charging facility for drones this does not nullify the usage of the heuristic we proposed. The "red edge" is the edge that the drone can fly from any CS to the boat but can not return on the same edge. Whereas the "gray edge" is the edge that UAV can fly to and come back without depleting the onboard battery. From this perspective, the "red-gray path" represents the pair of red and gray edges where the total distance of them is less than or equal to the drone range. The arbitrary spacing of the CSs can create "blind spots" in the mission region where boats can not be reached. Currently, in our framework we considered the case in which there is a single drone starting the mission from the single BS, traversing the boats that asked to be rescued in an "optimum way" and returning to the BS back again. The dispatcher system optimizes the traversal of the drone according to the objectives integrated into the optimization scheme. In our framework, not only we are considering multiple-objectives in addition to the shortest tour, but also we are optimizing for multiple parties. The drones and the boats to be rescued are the two parties considered in the optimization scheme. While the rescue party aims to use less energy for the flight and less number of chargings, the boat party desires less waiting time. These goals for two parties can be in conflicting states. Sometimes the shortest tour for the drone can yield a worse average waiting time for the boats compared to the longer tours. Such conflicts should be resolved with proper multi-objective optimization methods. Generally, the Pareto optimization method is used for such cases. We are proposing a priority-based scheme. Especially for urgent missions, the shortest tour for the drone can be completely overridden in favor of the least average waiting time for the boats. While the CS grid deployment properties and the profit from the proposed heuristic are elaborated with analyses, the algorithm part is still in the process of experimentation and verification. For this reason, in the discussions related to the optimum tour finding algorithms, prototype runs, and results will be presented.

Although the problem is closely related to the TSP (single drone) and the generic Vehicular Routing Problem (VRP) (multiple drones), the fact that offshore flights are required on a regular geometric CS grid makes it a special and novel variant. In this sense, our study offers novel contributions to mainstream TSP and VRP research. On the other hand, many concepts and methods can be borrowed from TSP and VRP research can help to solve the boat rescue problem.

Both TSP tour [4] and VRP [5] are known as NP-Hard problems. The family of VRP problems is related to two fundamental combinatorial NP-Hard problems. Namely TSP and bin-packing [2]. For this reason, approximation algorithms are proposed, in which the sub-optimum solution is reached in a polynomial time.

For the rest of the paper, in Sect. 2 we included a summary of the previous related work. The proposed framework is presented and case studies are analyzed in Sect. 3. We shared our conclusions and ideas for future work in Sect. 4.

2   Related Work

Drone assisted search and rescue operations are one of the important application areas of the drones that emerged in recent years. In several studies, the advantages of using drones in such operations are presented [1,9,10]. Drones provide economic ways for such operations. They can be fully automated and can reach places that are dangerous for human health.

Delivery with truck-drone systems is another related work we should mention. Traveling Salesman Problem with Drone (TSP-D) is a very widely studied topic. Especially in drone assisted delivery systems, the delivery truck is paired with drones that can serve the customer request. While the truck goes on its way the drone or drones assist the delivery, helping the truck in the TSP tour. The review of such delivery systems and the variants of them can be found in [11,14]

The fact that we are currently considering a single drone from a single static BS (depot) going and serving the static boat requests, turns the problem into classical TSP. However, the necessity of charging for the drone and the configuration of the CS grid makes the problem a variant of TSP. In our framework, there is no truck but the drone. The path of the drone depends on the CS grid connections. On the other hand, the configuration of CSs should be adjusted according to the range of the drone that is utilized. In this sense, while the boats can be regarded as "the clients", the CSs on the way can be regarded as special clients or depots that should be visited. In addition to these differences, we aim to evaluate TSP tour by considering multi-objectives. Further details and methods on the Multi-Objective TSP (MOTSP) can be found in [3,6,7]

3   Proposed Framework

The proposed framework for the dispatcher we described in Sect. 1 consists of a static CS grid, BS at a static location, and the rescue drone. The mission is controlled by the BS centrally. Once the rescue request or requests arrive at BS, the optimum path depending on the objectives is estimated and the drone is deployed for the mission. Currently, we do not consider dynamic rescue requests that can be arrived during the mission for the drone in mission. However, the dispatcher can deploy another drone for the new request or requests by estimating the optimum path once again for the new requests. Figure 1 shows a representative configuration of the problem.

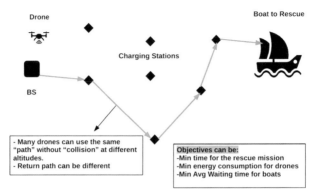

Fig. 1. Boat rescue operation with drones.

To study the proposed framework we developed a prototype user interface (UI) in which the user can configure the mission region and select boat positions on a real-world map. The software converts the geographical structure to a graph structure and runs algorithms on the graph structure. In Fig. 2 a screenshot can be seen for this software.

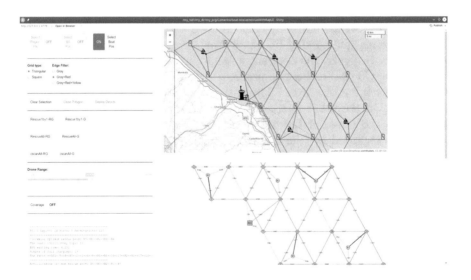

Fig. 2. Prototype UI developed for the rescue system.

The first step in the mission infrastructure is CS deployment. For the deployment of CSs so far we studied two different deployment configurations. Namely square grid and triangular grid. In order to prevent "blind spots" in the mission region, CSs should be spaced in a special geometry and according to the range of the drones that are selected for the rescue missions. For each case, blind spots should be avoided by estimating the optimum (min number of CSs and no blind spot) inter CS spacing. For any selected geometry for the CS grid, if the inter CS spacing is less than the optimum spacing there will be no blind spots. But more than necessary CSs will be required. If the inter CS spacing is greater than the optimum spacing, then there will be blind spots in the mission region. In Figs. 3a and 3b, these situations are depicted.

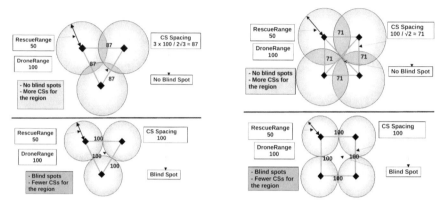

(a) Triangular grid CS deployment properties.

(b) Square grid CS deployment properties.

Fig. 3. CS grid deployment properties.

When the drone departs the BS it should reach at least one of the CSs with the initial amount of fuel which can be regarded as "full capacity". After recharging on any CS the drone reaches full capacity. The drone after arriving at the boat should have enough fuel to visit at least one of the CSs on its way back. The next CS can be the same or different one. If the drone can safely go, rescue the boat, and come back without depleting all the fuel on any path, we call this path "gray edge" (length less than or equal to half of the drone range) on the graph. Assuming that boats do not have any charging facility for the drone if the boat is in the range of the drone but there is no way to turn back we call this path "red edge" (length greater than half of the drone range) on the graph. These type of edges are relative to the boat position and also depends on the range of the drones. If boats can provide a charging facility then any edge that can be reachable from CSs is a gray edge. The combination of "red-gray" edges sometimes can make the rescue operation possible if the sum of their lengths is less than or equal to the drone range. We call such an edge pair "good red-gray path". In Fig. 4 such a condition is depicted. The red-gray path edges are dynamically added to the graph structure as the boats make rescue calls. This red-gray path heuristic is a general heuristic that can be integrated into any optimum path-finding algorithm like Shortest Path and A^*. After the good red and gray edges added to the graph structure, the path-finding algorithm can just find the optimum path to the boat. However, whether the final destination of the path is a boat or the CS after the gray edge should be considered for the next traversal.

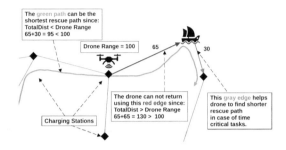

Fig. 4. The good red-gray path making the rescue possible.

Assuming that the boats can ask for rescue at any point with equal probability, we can present an analysis on the frequency of having a such red-gray edge pair. This analysis can justify the importance of using such a heuristic. Figures 5a and 5b helps us to understand the analysis focused on the triangular and square regions among different CS grid deployment. The CSs are at the vertices of the triangles and squares. The circles represent the rescue region of the drone from the CSs. Any point in the circle of the associated charging stations has a gray edge leading to it from the station. Outside of the circle points are connected with the "red" edge to the associated CS. The green areas represent the locus of the rescue points with "good red-gray paths" (the sum of distance is less than or equal to drone range) pairs. The red regions represent the locus of the "bad red-gray" edges (when a total distance greater than the drone range). The probability of having a good red-gray path is simply the ratio between the green area and the total area of the triangle. A point sampling is carried out in the triangular and square regions. Then for each sampled point, the type of the point and the edge distances to vertices are analyzed.

On Tables 1 and 2 point statistics are presented for triangular and square grids respectively. Small letters "b" and "g" represent bad and good respectively, whereas capital "R" is for Red and "G" is for Gray. "1gR+1G+1bR" means the point that is connected to vertices with 1 good Red edge, 1 Gy edge, and 1 bad Red edge. The total number of points marked with "(*)" gives the total number of points with at least one good red-gray edge pair. The probability of having at least one good red-gray path is the ratio between the total number of points with at least one good red-gray edge pair and the total number of points. For the triangular CS grid, sampling with 69483 points gives us 0.78 and for the square CS grid, sampling with 160801 points gives us 0.82. This means for each rescue call, about 78% of the time for the triangular grid and about 82% of the time for the square grid, we can have a good red-gray path pair that can give us a better shortest path for the mission. Because of the rounding and choice

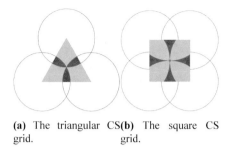

(a) The triangular CS grid. **(b)** The square CS grid.

Fig. 5. The prob. of having a "good red-gray path" (the green region) in different CS grid types.

Table 1. Point statistics for triangular grid. 69483 points are sampled.

Point Type	N	Prob.
3bR	0	0
1G+2bR	0	0
2G+1bR	15171	0.218
3G	5	0.00007
(*)1gR+2G	68	0.00098
(*)1gR+1G+1bR	14428	0.208
(*)2gR+1G	39811	0.573

Table 2. Point statistics for square grid. 160801 points are sampled.

Point Type	N	Prob.
4bR	0	0
1G+3bR	0	0
2G+2bR	29452	0.183
3G+1bR	64	0.0004
4G	5	0.00003
(*)1G+3gR	736	0.0046
(*)2G+2gR	0	0
(*)1G+2gR+1bR	66752	0.415
(*)3G+1gR	64	0.0004
(*)2G+1gR+1bR	63728	0.396
(*)1G+1gR+2bR	0	0

for considering boundary conditions with equality or without equality (when comparing the distances), the numbers in Tables 1 and 2 contains some errors.

After finding the probability of having a red-gray edge pair we need to find out the probability of benefiting from this heuristic. For this, the direction of the drone and the direction of the red-gray path should be aligned in a special way in order to benefit from this heuristic. If the drone meets the gray edge first, it will just save the boat and return to BS via that gray edge. But, if the drone meets the red edge first, it will utilize the red-gray path with savings and return to BS. Figure 6 summarizes all possible directions drone can approach the red-gray edge pair and the outcome of these directions for the triangular grid. A similar analysis can be made for the square grid. Now we can estimate the probability of benefiting from the red-gray path heuristic by using data from Tables 1 and 2. Firstly, we should have a "good direction" (gD in Eqs. 1 and 2) for the boat in which there is a good red-gray path. For the triangular grid Fig. 6 suggests that we can have $\frac{1}{3}$ probability to choose a direction for the drone in the case of the triangular grid when we have a single good Red edge (1gR). In Eq. 1 the probability of benefiting from the red-gray path is given as 0.45. Similar to the triangular grid case, for the square grid the drone can approach the good red-gray edge pair from one of the 4 directions. In Eq. 2 the probability of benefiting from the red-gray path is given as 0.31.

$$P(Benefit_{Tri}) = P(1gR) \times P(gD) + P(2gR) \times P(gD) \tag{1}$$
$$= 0.21 \times 0.333 + 0.57 \times 0.666 = 0.45$$
$$P(Benefit_{Sq}) = P(1gR) \times P(gD) + P(2gR) \times P(gD) + P(3gR) \times P(gD) \tag{2}$$
$$= 0.3964 \times 0.25 + 0.415 \times 0.5 + 0.0046 \times 0.75 = 0.31$$

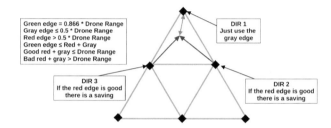

Fig. 6. The coming direction of the drone towards a "good red-gray path" in the triangular CS grid.

Finally, we need to know the actual savings from this heuristic. For this we need to consider the edge statistics given on Tables 3 and 4. The edge length metrics on these tables are normalized according to the drone range. On tables, "gG" (good Gray) refers to the gray edges that can be part of any red-gray edge pair and "bG" (bad Gray) edge type refers to the gray edges that can not be part of any red-gray edge pair. In short, if a pairing (total edge distance less than nor equal to Drone Range) is possible, both red and gray edges becomes "good". The actual savings depends on the path length, the number of boats that can be saved in a single tour, and other factors. However, for the actual savings, we can analyze the last part of the rescue path and present actual figures. The first column, gRG (good Red-Gray Pair), of Table 3 suggests that when we have a good red-gray edge pair, the average total length is about 0.914 units (times the Drone Range in the case of a triangular grid). This edge pair can save us from using an edge between CSs which has a distance of 0.866 unit and a gray edge which has an average length of 0.334 units. In total the length of this path is $0.866 + 0.334 = 1.2$ unit. Assuming that we use a "good gray" edge which has an average length of 0.271 unit, the total length with an edge between CSs becomes $0.866 + 0.271 = 1.137$ units. So we need 1.137 units to reach the boat without using any red edge. With the red-gray edge pair, we can reach the boat in 0.652 units by using the red edge. Here the saving in reaching the boat is about 43%. If we consider the fact that we need to take the gray edge every time we take

Table 3. Edge statistics for triangular grid. 208449 (3 edges * 69483 points) edges are sampled.

Edges	N	Prob.	Min	Max	Avg
gRG	94118	0.68	0.866	1	0.914
bRG	44838	0.32	1	1.253	1
All-RG	138956	1	0.866	1.253	0.968
gR	94118	0.45	0.502	0.866	0.652
bR	29599	0.14	0.502	0.779	0.677
All-R	123717	0.59	0.502	0.866	0.658
gG	54307	0.26	0	0.496	0.271
bG	30425	0.15	0.364	0.502	0.448
All-G	84732	0.41	0	0.502	0.334
All	208449	1.00	0	0.866	0.527

Table 4. Edge statistics for square grid. 643204 (4 edges * 160801 points) edges are sampled.

Edges	N	Prob.	Min	Max	Avg
gRG	199504	0.35	0.707	1	0.839
bRG	376064	0.65	1	1.366	1.075
All-RG	575568	1	0.707	1.366	0.993
gR	199504	0.45	0.502	1	0.608
bR	189448	0.14	0.502	0.999	0.749
All-R	388952	0.59	0.502	1	0.677
gG	131280	0.26	0	0.498	0.254
bG	122972	0.15	0.251	0.502	0.419
All-G	254252	0.41	0	0.502	0.334
All	643204	1.00	0	1	0.541

the red edge then the saving becomes about 19.3% with 0.914 unit path. This saving is for the last edges of the path. This case is shown in Fig. 7a.

To find out the actual saving amount for the square grid configuration, Fig. 7b shows an example scenario in which the comparison can be seen between the regular shortest path and the shortest path with the red-gray path heuristics. The average length of the "Good Red-Gray Pair" (gRG) is listed as 0.839 units on Table 4. This edge pair can save us from using two edges between CSs which has a distance of 0.71 units and a gray edge which has an average length of 0.334 units. In total the length of this path is $2 \times 0.71 + 0.334 = 1.754$ units. Assuming that we use a "good gray" edge which has an average length of 0.254 units, the total length of the path becomes $2 \times 0.71 + 0.254 = 1.674$ units. So we need 1.674 units to reach the boat without using any red edge. With the red-gray edge pair, we can reach the boat in 0.608 unit by using the red edge. Here the saving in reaching the boat is about 64%. If we consider the fact that we need to take the gray edge every time we take the red edge then the saving becomes about 50% with 0.839 unit path. This saving is for the last edges of the path. This case is shown in Fig. 7b.

The other comparison that should be made is the coverage effectiveness of the CS grid configurations. For this, we investigated the area per CS ratios and presented analysis in Figs. 8a and Fig. 8b by taking the limits when the number of vertices approaches infinity. The triangle grid limit approaches 2 triangular areas per CS while the square grid approach value of 1 square area per CS. Considering the analysis we have presented above and assuming drone range as a unit of measurement for distances, triangular grid without "blind spots" (Fig. 3a) gives 2 *times the area of the triangle* $= \sqrt{3} \times 0.866^2 = 1.30 \ unit^2$ per CS. On the other hand, the square grid without "blind spots" (Fig. 3b) gives *Area of the square* $= 0.71^2 = 0.5 \ unit^2$ per CS. This means a triangular grid

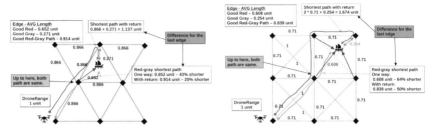

(a) Savings from red-gray edge heuristic in trian-(b) Savings from red-gray edge heuristic in gular grid.   square grid.

Fig. 7. Savings from red-gray edge heuristic for different CS grid configurations.

can cover 2.6 times more area with the same number of CSs. In other words, a triangular grid needs fewer CSs for the same mission region.

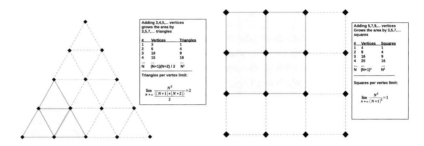

(a) The ratio of triangular area per CS for the(b) The ratio of squares per CS for the square triangular grid.   grid.

Fig. 8. The ratio of unit area per CS for the triangular and square grid.

On Table 5 we have summarized our findings for both CS deployment configurations. While the square grid deployment can provide a better probability of savings and a better amount of savings from the red-gray path heuristics, the number of CSs necessary for the mission is higher compared to the triangular grid. This trade-off should be considered in CS deployment phase of the rescue mission.

The optimum tour finding algorithm (fast approximate) we are proposing is similar to the hard disk scheduling algorithms. However, instead of a rotating plate, we have a rotating drone. The boats are clustered according to sector angles and for each sector, they are sorted according to their geographical distances from the BS. The tour cost is estimated from clock-wise and anti-clockwise. The optimum direction is then selected for the mission. However, for each sector, the optimum direction (towards BS or away from BS) of visiting boats should be decided. In Fig. 9 these issues are summarized.

Table 5. Comparison of triangular and square grid CS configuration.

Grid type	Prob. of having a Good Red-Gray Path	Prob. of using a Good Red-Gray Path	Savings 1-way	Savings Return	Mission Area per CS
Triangular	0.78	0.45	43%	20%	1.30
Square	0.82	0.31	64%	50%	0.5

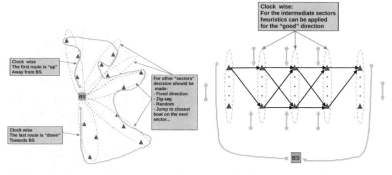

(a) Overall direction of the proposed algo.   **(b)** Choice for sector direction.

Fig. 9. The summary of the algorithm we are developing for the optimum path.

4   Conclusions and Future Works

Although the boat rescue problem we presented can be regarded as an instance of the classical TSP, still there are peculiarities that make it a novel problem and they need to be addressed. The regular geometry of the CS grid provides substantial positional information and homogeneous edge distances between CSs. These properties can be exploited in many ways for designing simple and efficient path-finding algorithms, even to a degree to utilize "Manhattan Distances" to come up with a simple formula. The red-gray path heuristic is a simple graph augmentation technique that can provide further savings for the path length and can be integrated into any path-finding algorithm. We presented an analysis of the effectiveness of this technique. Triangular and square CS grid configurations are analyzed and compared. The triangular grid provides more effective coverage and a higher probability of benefiting from the red-gray path heuristic. The square grid provides more savings from the red-gray path heuristic although the probability of benefiting from the heuristic is lower than the triangular grid Currently, we are working on the single-drone version of the problem which is like a TSP tour. In the future, once we verify the algorithms we are developing, we will extend our work by considering multiple drones serving multiple boats. This version of the problem is similar to VRP. The BS in our case can be regarded as a depot. The boats are the clients that request service and the drones are the vehicles. In this regard, the variant of VRP for our work can be named as "single depot, fuel-constrained, routing problem with static service request". However, for the routes, we have CSs that should be visited for the drones to go further in replying to the service (rescue) request of the boats.

References

1. Claesson, A., Schierbeck, S., Hollenberg, J., Forsberg, S., Nordberg, P., Ringh, M., Olausson, M., Jansson, A., Nord, A.: The use of drones and a machine-learning model for recognition of simulated drowning victims–a feasibility study. Resuscitation **156**, 196–201 (2020). https://doi.org/10.1016/j.resuscitation.2020.09.022
2. Fenton, A.: The bees algorithm for the vehicle routing problem. Master's thesis, Department of Computer Science, University of Auckland (2016). http://arxiv.org/abs/1605.05448
3. Guo, X., Ji, M., Zhao, Z., Wen, D., Zhang, W.: Global path planning and multi-objective path control for unmanned surface vehicle based on modified particle swarm optimization (PSO) algorithm. Ocean Eng. **216**, 107693 (2020). https://doi.org/10.1016/j.oceaneng.2020.107693
4. Karp, R.M.: Reducibility among combinatorial problems. In: Miller, R.E., Thatcher, J.W., Bohlinger, J.D. (eds.) Complexity of Computer Computations: Proceedings of a symposium on the Complexity of Computer Computations, held March 20–22, 1972, at the IBM Thomas J. Watson Research Center, Yorktown Heights, New York, and sponsored by the Office of Naval Research, Mathematics Program, IBM World Trade Corporation, and the IBM Research Mathematical Sciences Department, pp. 85–103. Springer, Boston (1972). https://doi.org/10.1007/978-1-4684-2001-2_9
5. Khuller, S., Malekian, A., Mestre, J.: To fill or not to fill: the gas station problem. ACM Trans. Algorithms **7**(3), 36:1–36:16 (2011). https://doi.org/10.1145/1978782.1978791
6. Lianshuan, S., Zengyan, L.: An improved pareto genetic algorithm for multi-objective TSP. In: 2009 Fifth International Conference on Natural Computation, vol. 4, pp. 585–588 (2009). https://doi.org/10.1109/ICNC.2009.510
7. Lust, T., Teghem, J.: The multiobjective traveling salesman problem: a survey and a new approach. In: Coello Coello, C.A., Dhaenens, C., Jourdan, L. (eds.) Advances in Multi-Objective Nature Inspired Computing, pp. 119–141. Springer, Heidelberg (2010). https://doi.org/10.1007/978-3-642-11218-8_6
8. Marinei: The growing role for aerial drones in the maritime industry. Marinei news. https://www.marine-i.co.uk/news/article/80/the-growing-role-for-aerial-drones-in-the-maritime-industry. Accessed 29 Dec 2020
9. McRae, J.N., Gay, J.C., Nielsen, B.M., Hunt, A.P.: Using an unmanned aircraft system (drone) to conduct a complex high altitude search and rescue operation: a case study. Wilderness Environ. Med. **30**(3), 287–290 (2019). https://doi.org/10.1016/j.wem.2019.03.004
10. Mishra, B., Garg, D., Narang, P., Mishra, V.: Drone-surveillance for search and rescue in natural disaster. Comput. Commun. **156**, 1–10 (2020). https://doi.org/10.1016/j.comcom.2020.03.012
11. Roberti, R., Ruthmair, M.: Exact methods for the traveling salesman problem with drone. Optimization Online (2019). http://www.optimization-online.org/DB_FILE/2019/11/7457.pdf

12. Shakhatreh, H., Sawalmeh, A.H., Al-Fuqaha, A., Dou, Z., Almaita, E., Khalil, I., Othman, N.S., Khreishah, A., Guizani, M.: Unmanned aerial vehicles (UAVs): a survey on civil applications and key research challenges. IEEE Access **7**, 48572–48634 (2019). https://doi.org/10.1109/ACCESS.2019.2909530
13. Silvagni, M., Tonoli, A., Zenerino, E., Chiaberge, M.: Multipurpose UAV for search and rescue operations in mountain avalanche events. Geomatics Natural Hazards Risk **8**(1), 18–33 (2017). https://doi.org/10.1080/19475705.2016.1238852
14. Vásquez, S.A., Angulo, G., Klapp, M.A.: An exact solution method for the TSP with Drone based on decomposition. Comput. Oper. Res. **127**, 105127 (2021). https://doi.org/10.1016/j.cor.2020.105127

Algorithmic Aspects of the Maximum 2-edge-colorable Subgraph Problem

Alessandro Aloisio[1,2(✉)] and Vahan Mkrtchyan[1]

[1] Gran Sasso Science Institute, School of Advanced Studies, L'Aquila, Italy
{alessandro.aloisio,vahan.mkrtchyan}@gssi.it
[2] Department of Information Engineering, Computer Science and Maths,
University of L'Aquila, L'Aquila, Italy

Abstract. A k-edge-coloring of a graph is an assignment of colors from a set of k colors to edges of the graph such that adjacent edges receive different colors. A maximum k-edge-colorable subgraph in a graph is a k-edge-colorable subgraph containing a maximum possible number of edges. In the maximum k-edge-colorable subgraph problem we are given a graph and an integer k, the goal is to find a maximum k-edge-colorable subgraph together with its k-edge-coloring. In this paper, we consider the maximum 2-edge-colorable subgraph problem and present some results that deal with the fixed-parameter tractability of this problem. Our main results state that the problem is fixed-parameter tractable with respect to carvingwidth and pathwidth.

1 Introduction

In this paper, we consider the maximum 2-edge-colorable subgraph from the perspective of parameterized complexity theory. The goal in this problem is to find a 2-edge-colorable subgraph in the input graph such that it contains the largest number of edges. It is easy to see that it amounts to finding two edge-disjoint matchings such that the total number of taken edges is maximized. One can easily show that this problem is equivalent to taking vertex-disjoint paths and even cycles in the input graph such that the resulting subgraph contains maximum number of edges. In this paper, we present some algorithmic results for this problem that deal with the fixed-parameter tractability of this problem.

Parameterized complexity theory has been raising in popularity in the last few years because it looks at giving a finer running time analysis with respect to the classical complexity theory. This allows us to better understand where the difficulties of a problem lie. The main idea behind this notion is the following: for a problem under consideration the goal is to find exact algorithms whose running times are based not only on the input size but also on one or more parameters. An FPT algorithm requires that the parameter does not appear in the exponent part of the input size, but just as a multiplicative (non constant)

This work has been partially supported by the Italian MIUR PRIN 2017 Project ALGADIMAR "Algorithms, Games, and Digital Markets".

factor of it. This leads to a good time function with a constant exponent on the input size. We will not go into the detail of parameterized complexity theory that can be found in [17].

The rest of this paper is organized as follows. In Sect. 2, a formal definition of the problem is provided. The related work is given in Sect. 3. Section 4 discusses our results on the carvingwidth. Then we present our results on pathwidth in Sect. 5. Finally, we conclude the paper in Sect. 6, where we summarize our results and outline the directions for future work.

2 Problem Statement

In the following we assume that graphs are undirected, finite, with no loops and no multiple edges. We also consider multigraphs that may contain multiple edges but no loops. We use $V(G)$ and $E(G)$ to denote the vertices and the edges of a graph G; $d_G(u)$ for the degree of a vertex u; $\delta(G)$ and $\Delta(G)$ for the minimum and maximum degree of the vertices in G.

A subset of $E(G)$ is a matching of G if no vertex of G is incident to two edges from it. A maximum matching is a matching that contains the largest possible number of edges. A subgraph K of a graph G is spanning, if $V(K) = V(G)$.

A graph G is k-edge colorable if, for $k \geq 0$, each of its edges can be colored with a color in $[k] = \{1, \ldots, k\}$, such that there are no two adjacent edges of the same color. The *chromatic index* of G, denoted with $\chi'(G)$, is the smallest k for which G is k-edge-colorable. The classical theorem of Shannon states that, for any multi-graph G, $\Delta(G) \leq \chi'(G) \leq \left\lfloor \frac{3\Delta(G)}{2} \right\rfloor$ holds, see [31,34]. Moreover, the classical theorem of Vizing states that, for any multi-graph G, $\Delta(G) \leq \chi'(G) \leq \Delta(G) + \mu(G)$ holds, see [34,36]. Here $\mu(G)$ denotes the maximum multiplicity of an edge of G. A multi-graph G is class I, if $\chi'(G) = \Delta(G)$, otherwise it is class II.

Clearly, if $k < \chi'(G)$, we cannot color all the edges of G with k colors. Therefore, a natural question is finding the maximum number of edges in $E(G)$ that can be colored with k colors. A subgraph H of G is called maximum k-edge-colorable, if H is k-edge-colorable and contains the maximum number of edges among all k-edge-colorable subgraphs of G. For $k \geq 0$ and a graph G, let $\nu_k(G) = \max\{|E(H)| : H \text{ is a } k\text{-edge-colorable subgraph of } G\}$. Clearly, a k-edge-colorable subgraph is maximum if it contains exactly $\nu_k(G)$ edges. Please note that $\nu_1(G)$ is the size of a maximum matching of G, which we will simply denote with $\nu(G)$. As already said, the paper focuses on the exact solvability of the maximum k-edge-colorable subgraph problem. Its precise formulation is the following:

Maximum k-edge-colorable subgraph
Input: A graph $G = (V(G), E(G))$, and an integer $k \geq 1$.
Solution: A subgraph H of G that is k-edge-colorable.
Goal: Maximize the number of colored edges.

It can be easily proven that the decision version of the maximum k-edge-colorable subgraph is NP-complete for every fixed $k \geq 2$. In fact, when G is cubic and $k = 2$, we have that $\nu_2(G) = |V|$ if and only if G contains two edge-disjoint perfect matchings. The latter condition is equivalent to saying that G is 3-edge-colorable, which is an NP-complete problem as Holyer has demonstrated in [23].

We study the maximum k-edge-colorable subgraph problem from the perspective of parameterized complexity theory, which provides a better complexity analysis than the classical theory of NP-completeness.

The following sections are mainly about the case $k = 2$ of the maximum k-edge-colorable subgraph problem. We decided to tackle this case because, even if the 2-edge coloring problem is polynomially solvable, the maximum 2-edge-colorable subgraph problem is hard, as mentioned before. We present some results that deal with the fixed-parameter tractability of this problem with respect to two graph-theoretic parameters. In particular, we showed the fixed-parameter tractability of our problem with respect to carvingwidth and pathwidth. For further details on theoretical aspects, the reader is invited to see [17,37].

3   Related Work

Graph colorings represent a class of very famous classical combinatorial problems, which have been investigated in depth by many researches in the last decades. In general, coloring problems can be formulated in a simple way, but they are very hard to solve.

One of the interesting problems considered in this area is the estimation of the size of the maximum k-edge-colorable subgraph. There are many papers where the ratio $\frac{\nu_k(G)}{|E(G)|}$ has been investigated. [22,30] prove lower bounds for this ratio in case of regular graphs and $k = 1$. For regular graphs of high girth the bounds are improved in [19]. Albertson and Haas investigated the problem in [1,2] when G is a cubic graph. See also [27], where it is shown that for every cubic multigraph G, $\nu_2(G) \geq \frac{4}{5}|V(G)|$ and $\nu_3(G) \geq \frac{7}{6}|V(G)|$. Moreover, [12] proves that for any cubic multigraph G, $\nu_2(G) + \nu_3(G) \geq 2|V(G)|$, and in [27,28] Mkrtchyan et al. showed that for any cubic multigraph G, $\nu_2(G) \leq \frac{|V(G)|+2\nu_3(G)}{4}$. Finally, in [24], it is shown that the sequence ν_k is convex in the class of bipartite multigraphs.

Bridgeless cubic graphs that are not 3-edge-colorable are called snarks [16], and the ratio for snarks is investigated by Steffen in [32,33]. This lower bound has also been investigated in the case when the graphs need not be cubic in [21,25]. Kosowski has investigated the problem from the algorithmic perspective [26]. As we have mentioned previously, the problem of finding a maximum k-edge-colorable graph in an input graph is NP-complete. Thus, it is natural to investigate the (polynomial) approximability of the problem. In [18] for each $k \geq 2$ an approximation algorithm for the problem is presented. There for each fixed value of $k \geq 2$, algorithms are proved to have certain approximation ratios and these ratios are tending to 1 as k goes to infinity. In [26], two approximation algorithms for the maximum 2-edge-colorable subgraph and maximum 3-edge-colorable subgraph problems are presented whose performance ratios are $\frac{5}{6}$ and $\frac{4}{5}$, respectively. Finally, note that the results of [18] are improved for $k = 3, ..., 7$ in [25].

Some structural properties of maximum k-edge-colorable subgraphs of graphs are proved in [12,29].

4   Graphs with Bounded Carving-Width

In the following, we will see that the maximum 2-edge-colorable subgraph problem is fixed-parameter tractable with respect to the *carving-width* [13,20] of a graph.

Let G be a graph, and let S be a subset of $V(G)$. As usual, let $(S, V(G) \setminus S)$ be the set of edges between S and $V(G) \setminus S$, that is $(S, V(G) \setminus S)$ is an edge cut of G. Following the definition given in [20], we consider now a decomposition (T, ϕ) of G, where T is a rooted binary tree with exactly $|V(G)|$ leaves, and all internal nodes are of degree three, except for the root, which has degree two; and ϕ is a one-to-one relationship between $V(G)$ and the leaves of T. Removing an edge e of T returns a partition of the leaves of T into two subsets, so it also partitions $V(G)$ in two subsets S_1 and S_2. The width of an edge e is the number of edges with one vertex in S_1 and the other in S_2, that is $w(e) = |(S_1, S_2)|$. Please note that in [13,35] carving-width and carving decomposition are defined on unrooted sub-cubic graphs.

The tree T is called a *carving* of G, and (T, ϕ) is called a *carving decomposition* of G. The width of (T, ϕ) is the maximum weight $w(e)$ over all $e \in E(T)$. The *carving-width* of G, denoted by $cw(G)$, is the minimum width over all carving decompositions of G. We set $cw(G) = 0$ if $|V| = 1$.

Given a graph G with fixed carving-width, it is possible to find a carving decompositiong of G in linear time, see [14,35]. This result allows to state the next theorem.

Theorem 1. *The maximum 2-edge-colorable subgraph problem is FPT tractable with respect to the carving-width h, and, given a carving decomposition of width h, it can be solved in $O(3^{2h}(2|V(G)| - 1))$ time.*

Proof. Let G be a connected graph with a carving decomposition (T, ϕ) of width h, where *root* is the root of T. We can assume w.l.o.g. that G is connected because coloring two disjoint graphs can be done separately. Denote by $\{0, 1, 2\}$ the set of colors, where 1, and 2 are called true colors, while 0 is dummy and means 'not colored'. Let $c[u, v]$ be the pair $(c, (u, v))$, where $c \in \{0, 1, 2\}$, and $(u, v) \in E$. Let also p be a function that takes in input a pair $c[u, v]$, and returns 1 if $c \in \{1, 2\}$, i.e., it is a true color, and returns 0, otherwise. We use $c[u, v]$ instead of $(c, (u, v))$ to avoid cluttered notation. Please note that only the dummy color 0 can be used on two or more adjacent edges, because it essentially says that an edge is not colored. We have extended the set of colors in order to make the proof more readable.

Now, we describe a dynamic programming algorithm to find a maximum 2-edge-colorable subgraph of G, which exploits the structure and the properties of the carving decomposition. To avoid confusion between the vertices of the graph, and the ones in the tree, we will call the vertices of T nodes.

For a node i of T, denote with $T(i)$ the subtree with i and all its descendants. Let S_i be the set of vertices corresponding to the leaves in $T(i)$, and let S_{i-} be the set $V(G) \backslash S_i$. Define $G(i)$ as the subgraph of G induced by S_i, and let $G(i, i^-)$ the subgraph of G induced by S_i and the edges in (S_i, S_{i-}). Let $f(i, A)$ be the optimum value for the maximum 2-edge-colorable subgraph problem restricted to $G(i, i^-)$, where A is a set of $|(S_i, S_{i-})|$ edge-color pairs, one for each edge in (S_i, S_{i-}), i.e. $A = \bigcup_{(u,v) \in (S_i, S_{i-})} \{c[u, v]\}$, which satisfies the constraint: "The edge $(u, v) \in (S_i, S_{i-})$ is colored with the color c in the edge-color pair $c[u, v] \in A$".

If i is a leaf corresponding to a vertex $u \in V$, then one of the following conditions holds: a) $f(i, A) = \sum_{(u,v) \in (S_i, S_{i-})} p(c[u, v])$ if there are no two edges in (S_i, S_{i-}) with the same true color assigned; b) $f(i, A) = -\infty$ otherwise.

If i is an interior node of T with two sons j and k, then we set $f(i, A) = -\infty$ if there exist two adjacent edges in (S_i, S_{i-}) colored with a same true color. If it is not the case, we compute $f(i, A)$ by solving the following maximisation problem, where $A = (B \cup C) \backslash (B \cap C) = B \triangle C$, i.e., the symmetric difference between B and C.

$$\max \quad f(j, B) + f(k, C) - \sum_{(u,v) \in ((S_j, S_{j-}) \cap (S_k, S_{k-}))} p(c[u, v]) \tag{1}$$

$$\text{s.t.} \quad c[u, v] \in (B \cap C) \quad \forall (u, v) \in ((S_j, S_{j-}) \cap (S_k, S_{k-}))$$

In fact, by definition of carving decomposition, we have that $S_i \subseteq S_j \cup S_k$, $S_j \cap S_k = \emptyset$, and that the edges in (S_i, S_{i-}) are the ones in $(S_j, S_{j-}) \cup (S_k, S_{k-})$ minus the ones in $(S_j, S_{j-}) \cap (S_k, S_{k-})$, i.e., $(S_j, S_{j-}) \triangle (S_k, S_{k-})$. This means that, for every B and C where the edges in (S_i, S_{i-}) have the colors given in A, and each edge in $(S_j, S_{j-}) \cap (S_k, S_{k-})$ has a same color both in B and in C, $f(i, A)$ is equal to the best value $f(j, B)$ plus the best value $f(k, C)$, minus the profit given by the the edges in $(S_j, S_{j-}) \cap (S_k, S_{k-})$, because it is counted twice. Clearly, for a specific A, the variables of the Problem (1) are the colors to be assigned to the edges in $(S_j, S_{j-}) \cap (S_k, S_{k-})$, that is the pairs of edge-colors in $B \cap C$.

Clearly, the optimum of the maximum 2-edge-colorable subgraph problem is the maximum value $f(root, \emptyset)$, because there are no vertices in S_{root-}. The time complexity at each leaf i is $O(3^h)$, because there are at most h edges in (S_i, S_{i-}), and there are 3 possible colors, $\{0, 1, 2\}$. The time complexity at each internal node i of T is given by all the possible colors assigned to every edge in $(S_j, S_{j-}) \cup (S_k, S_{k-})$. Since the maximum number of edges in each partition is h, the carving width, there are at most $2h$ edges in $(S_j, S_{j-}) \cup (S_k, S_{k-})$. Moreover, since the number of colors is 3, the time complexity at each internal node of T is $O(3^{2h})$. In conclusion, since each internal node has degree three (two sons), and in a perfect binary tree there are $2|V(G)| - 1$ nodes, then T has at most $2|V(G)| - 1$ nodes. This means that the time complexity of the dynamic programming algorithm is $O(3^{2h}(2|V(G)| - 1))$.

5 Graphs with Bounded Pathwidth

Here we show that the maximum 2-edge-colorable subgraph problem is fixed-parameter tractable with respect to the *pathwidth* of a graph. We first recall some basic definitions that we use in Theorem 2. Informally, a *path decomposition* of a graph G is a way of representing G as a path-like structure.

Definition 1 ([20]). A path decomposition of a graph G is a set $\mathscr{P} = (X_1, \ldots, X_r)$ of subsets of $V(G)$, that is $X_i \subseteq V(G)$ for each $i \in \{1, \ldots, r\}$, called *bags*, such that

- (i) for every $u \in V(G)$ there exists $i \in \{1, \ldots, r\}$ with $u \in X_i$;
- (ii) for every $(u, v) \in E(G)$, there exists $i \in \{1, \ldots, r\}$ with $u, v \in X_i$;
- (iii) for every three bags X_i, X_j, and X_k, with $i \leq j \leq k$, it holds that $X_i \cap X_k \subseteq X_j$.

The *width* of a path decomposition equals $\max_{i \in \{1, \ldots, r\}} |X_i| - 1$, and the *pathwidth* of a graph G, is the minimum width of a path decomposition of G. To avoid confusion between the vertices of the graph, and the ones in the path P, we will call nodes the vertices X_i in the path decomposition. A property of path decompositions [20], called here *pathwidth separator* property, is that for every three nodes X_i, X_j, and X_k, with $i < j < k$, each path that connects a vertex in $X_i \setminus X_j$ with a vertex in $X_k \setminus X_j$ contains a vertex in X_j. Thus, node X_j separates the vertices in $X_i \setminus X_j$ from the ones in $X_k \setminus X_j$. Here we use the *nice path decomposition*, which is a specific type of path decomposition.

Definition 2 ([20]). A path decomposition of a graph G is *nice* if $|X_1| = |X_r| = 1$, and for every $i \in \{1, 2, \ldots, r-1\}$ there is a vertex $v \in V(G)$, such that either $X_{i+1} = X_i \cup \{v\}$ (introduce node), or $X_{i+1} = X_i \setminus \{v\}$ (forget node).

It can be easily shown that, starting from a path decomposition $\mathscr{P}$ of width h, it is always possible to build a nice path decomposition of width h in linear time, see [20]. Moreover, in [15] it is presented a linear time algorithm that finds a path decomposition for graphs of small pathwidth. Since property (iii) in Definition 1 says that every vertex $v \in V(G)$ belongs to a consecutive set of bags, the number of nodes in a nice path decomposition is at most twice the number of vertices in $V(G)$.

Theorem 2. *The maximum 2-edge-colorable subgraph problem is FPT tractable with respect to the pathwidth h, and, given a path decomposition of width h, can be solved in $O((h^2 - h + 1)7^{(h+1)}|V(G)|)$ time via a dynamic programming algorithm.*

Proof. Let G be a graph with a path decomposition of width h, which can be computed in linear time [15]. Let $\{0, 1, 2\}$ be the set of available colors where 1, and 2 are true colors, while 0 is dummy and means 'not colored'. Let $p\colon \{0, 1, 2\} \to \{0, 1\}$ be a function, which is equal to 1 if and only if the input is a true color, 0 otherwise (i.e., $p(0) = 0$, $p(1) = 1$, and $p(2) = 1$). Let also $c\colon E(G) \to \{0, 1, 2\}$ be a function which takes an edge (u, v) as input, and returns the color $c((u, v))$ assigned to it. In order to avoid cluttered notation, in the following we will write $c(u, v)$ instead of $c((u, v))$.

The first step of the algorithm is to compute a nice path decomposition $\mathscr{P} = (X_1, \ldots, X_r)$ with the same width h, which can be done in linear time [20]. Denote by $G(X_i)$ the subgraph induced by the vertices in $\bigcup_{j=1}^{i} X_j$. Let $f(X_i, \mathscr{A})$ be the maximum value of the maximum 2-edge-colorable subgraph problem on $G(X_i)$, where $\mathscr{A}$ is a collection of $|X_i|$ non empty subsets of colors $A(v)$, with $v \in X_i$, that satisfies the constraint: "The colors *incident* to the vertex v are those in $A(v)$, for every $v \in X_i$".

Here, as usual, a color is *incident* to a vertex v if it is used at least in one edge incident to v. Please note that only the dummy color 0 can be incident more than once to a vertex, because it means that an edge is not colored. At each node X_i of $\mathscr{P}$, we compute the values $f(X_i, \mathscr{A})$, for every possible collection $\mathscr{A}$, via a dynamic programming algorithm that starts at X_1, ends at X_r, and exploits the pathwidth separator property. If there is no solution of the constrained problem, then we set $f(X_i, \mathscr{A}) = -\infty$. Since in X_1 there is only one vertex v, and in $G(X_1)$ there are no edges, one of the two conditions holds: a) $f(X_1, \mathscr{A}) = 0$, if $\mathscr{A} = \{A(v)\} = \{\emptyset\}$; b) $f(X_1, \mathscr{A}) = -\infty$, otherwise. In fact, there cannot be colors in $G(X_1)$, so there is only one solution with $\mathscr{A} = \{\emptyset\}$, and optimum value 0.

If X_{i+1} is an *introduce node*, with $X_{i+1} = X_i \cup \{v\}$, we can calculate the value $f(X_{i+1}, \mathscr{A})$, for a given collection $\mathscr{A}$ of color sets, solving the maximisation problem below.

$$\max \quad f(X_i, \mathscr{B}) + \sum_{u \in (X_i \cap V(v))} p(c(u, v))$$

$$\text{s.t.} \quad B(u) = A(u) \quad \forall u \in X_i \setminus V(v)$$
$$\qquad B(u) = A(u) \quad \forall u \in (X_i \cap V(v)) \mid c(u, v) = 0 \qquad (2)$$
$$\qquad B(u) = A(u) \setminus \{c(u, v)\} \quad \forall u \in (X_i \cap V(v)) \mid c(u, v) \neq 0$$
$$\qquad A(v) = \bigcup_{u \in X_i} c(u, v)$$

where $V(v)$ is the set of the vertices incident to v, that is the *neighbourhood* of v. The variables of Problem (2) are $\mathscr{B}$ and $c(u, v)$, for every $u \in X_i$. The fourth constraint makes explicit how the colors incident to v are used on the edges (u, v), with $u \in X_i$. The first three constraints, instead, are used to properly define $\mathscr{B}$, that is a collection of color subsets for X_i derived from $\mathscr{A}$. In particular, the first one states that $B(u)$ is equal to $A(u)$ for every vertex $u \in X_i$ non adjacent to v; the second states that $B(u)$ is equal to $A(u)$ if the color used for the edge (u, v)

is *0*; and the third states that $B(u)$ is equal to $A(u)$ minus the true color used for the edge (u, v), because it cannot be used again for the edges in $G(X_i)$ incident to u. The objective function sums the already computed optimum $f(X_i, \mathscr{B})$ with the number of edges in $G(X_i)$ incident to v that receive a true color.

For any *forget node* $X_{i+1} = X_i \setminus \{v\}$, the value $f(X_{i+1}, \mathscr{A})$ for a specific collection of incident color sets $A(u)$, with $u \in X_i$, is computed by solving the following maximisation problem.

$$\begin{aligned} \max \quad & f(X_i, \mathscr{B}) \\ \text{s.t.} \quad & B(u) = A(u) \quad \forall u \in X_i \setminus \{v\} \\ & B(v) \subseteq \{0, 1, 2\} \end{aligned} \tag{3}$$

In fact, since $G(X_{i+1}) = G(X_i)$, the value $f(X_{i+1}, \mathscr{A})$ essentially equals the maximum value $f(X_i, \mathscr{B})$ for every possible collection of color sets $\mathscr{B}$ compatible with $\mathscr{A}$, that is $B(u) = A(u)$ for every $u \in X_i \setminus \{v\}$. The variable of Problem (3) is the collection $\mathscr{B}$.

Once this dynamic programming algorithm stops, the optimum of the 2-edge-colorable subgraph problem is the maximum value $f(X_r, \mathscr{A})$, for every possible collection of color subsets $A(u) \subseteq \{0, 1, 2\}$ of the incident colors in each vertex $u \in X_r$. Please note that there are exactly 7 non empty subsets of $\{0, 1, 2\}$. Now, we compute the complexity of the algorithm. At each *introduce node*, we solve at most $7^{(h+1)}$ Problems (2), one for every possible subset of incident colors $A(u)$, and for every $u \in X_{i+1}$, which are at most $h+1$. Moreover, for each specific $\mathscr{A}$ in Problem (2), the maximum number of edges (u, v) with $u \in (X_i \cap V(v))$ is h, because $|X_i \cap V(v)| \leq |X_i| \leq h$. If $h \geq 3$, we can color these edges only if $0 \in A(v)$, and we can do it in at most $h(h - 1)$ different ways. In fact, if $A(v)$ contains one true color, there are h possibilities to select the edge associated with it; while, if $A(v)$ contains both the true colors, the possibilities are $h(h-1)$. Clearly, if $h \leq 2$, there can be few ways of coloring these edges. This means that, for a specific $\mathscr{A}$, Problem (2) has at most $h(h - 1)$ solutions, and the complexity for an *introduce node* is $O(7^{(h+1)}h(h - 1)) = O(7^{(h+1)}(h^2 - h))$.

At any *forget node*, the complexity is at most $7^{(h+1)}$, given by all the subsets of colors $A(u)$, with $u \in X_{i+1}$, and all the possible subsets of colors $B(v)$ for the forget vertex v. In conclusion, since there are at most $|V(G)|$ *introduce nodes* and $|V(G)|$ *forget nodes*, the time complexity of the dynamic programming algorithm is $O((h^2 - h + 1)7^{(h+1)}|V(G)|)$.

6   Conclusion and Future Work

In this paper, we considered the maximum 2-edge-colorable subgraph problem. The problem belongs to the class of classical graph coloring problems, which have been of great interests in graph theory and algorithms. We addressed it from the perspective of the fixed-parameter tractability theory. Our results state that this problem is fixed-parameter tractable with respect to the carving-width and the pathwidth. We believe that our line of research is worth for further investigation. In particular, it is interesting to check whether it is possible to write both FPT and polynomial algorithms for other classes of graphs, as it is done in [3–5,7,9]. Some other well known parameters that could be analysed are the treewidth, and the cliquewidth. Another line of research that seems interesting is the generalization of the problem to the weighted case, where each color can have different weights even in combination with specific edges, similar to what it is proposed in [4,6,8,10,11].

References

1. Albertson, M., Haas, R.: Parsimonious edge colouring. Discrete Math. **148**, 1–7 (1996)
2. Albertson, M., Haas, R.: The edge chromatic difference sequence of a cubic graph. Discrete Math. **177**, 1–8 (1997)
3. Aloisio, A., Navarra, A.: Constrained connectivity in bounded X-width multi-interface networks. Algorithms **13**(2), 31 (2020)
4. Aloisio, A., Navarra, A., Mostarda, L.: Energy consumption balancing in multi-interface networks. J. Ambient. Intell. Humaniz. Comput. **11**(8), 3209–3219 (2019). ISSN 1868-5145
5. Aloisio, A., Arbib, C., Marinelli, F.: Cutting stock with no three parts per pattern: work-in-process and pattern minimization. Discrete Optim. **8**(2), 315–332 (2011)
6. Aloisio, A., Arbib, C., Marinelli, F.: On LP relaxations for the pattern minimization problem. Networks **57**(3), 247–253 (2011)
7. Aloisio, A., Navarra, A.: Balancing energy consumption for the establishment of multi-interface networks. In: SOFSEM 2015: Theory and Practice of Computer Science, vol. 8939, pp. 102–114. Springer, Heidelberg (2015)
8. Aloisio, A.: Coverage, subject to a budget on multi-interface networks with bounded carving-width. In: WAINA, Advances in Intelligent Systems and Computing, vol. 1150, pp. 937–946. Springer, Cham (2020)
9. Aloisio, A., Navarra, A.: Budgeted constrained coverage on bounded carving-width and series-parallel multi-interface networks. Internet Things **11**, 100259 (2020)
10. Aloisio, A., Budgeted constrained coverage on series-parallel multi-interface networks. In: AINA, Advances in Intelligent Systems and Computing, vol. 1151. Springer, Cham (2020)
11. Aloisio, A., Navarra, A., Mostarda, L.: Distributing energy consumption in multi-interface series-parallel networks. In: WAINA 2019, Advances in Intelligent Systems and Computing, vol. 927, pp. 734–744. Springer, Cham (2019)
12. Aslanyan, D., Mkrtchyan, V., Petrosyan, S., Vardanyan, G.: On disjoint matchings in cubic graphs: maximum 2-edge-colourable and maximum 3-edge-colourable subgraphs. Discrete Appl. Math. **172**, 12–27 (2014)
13. Belmonte, R., Hof, P., Kamiński, M., Paulusma, D., Thilikos, D.M.: Characterizing graphs of small carving-width. Discrete Appl. Math. **161**, 1888–1893 (2013)
14. Bodlaender, H.L.: A linear time algorithm for finding tree-decompositions of small treewidth. IAM J. Comput. **25**, 1305–1317 (1996)
15. Cattell, K., Dinneen, M.J., Fellows, M.R.: A simple linear-time algorithm for finding path-decompositions of small width. Inf. Process. Lett. **57**(4), 197–203 (1996)
16. Cavicchioli, A., Meschiari, M., Ruini, B., Spaggiari, F.: A survey on snarks and new results: products, reducibility and a computer search. Discrete Math. **28**(2), 57–86 (1998)

17. Cygan, M., Fomin, F.V., Kowalik, L., Lokshtanov, D., Marx, D., Pilipczuk, M., Pilipczuk, M., Saurabh, S.: Parameterized Algorithms, pp. 3–555. Springer, Cham (2015). ISBN 978-3-319-21274-6
18. Feige, U., Ofek, E., Wieder, U.: Approximating maximum edge colouring in multigraphs. Lecture Notes in Computer Science, vol. 2462, pp. 108–121 (2002)
19. Flaxman, A.D., Hoory, S.: Maximum matchings in regular graphs of high girth. Electron. J. Comb. **14**(1), 1–4 (2007)
20. Fomin, F.V., Kratsch, D.: Exact Exponential Algorithms. Springer, Heidelberg (2010)
21. Fouquet, J.L., Vanherpe, J.M.: On parsimonious edge-colouring of graphs with maximum degree three. Graphs Comb. **29**(3), 475–487 (2013)
22. Henning, M.A., Yeo, A.: Tight lower bounds on the size of a maximum matching in a regular graph. Graphs Comb. **23**(6), 647–657 (2007)
23. Holyer, I.: The NP-completeness of edge-colouring. SIAM J. Comput. **10**(4), 718–720 (1981)
24. Karapetyan, L., Mkrtchyan, V.: On maximum k-edge-colourable subgraphs of bipartite graphs. Disc. Appl. Math. **257**, 226–232 (2019)
25. Kamiński, M.J., Kowalik, L.: Beyond the Vizing's bound for at most seven colours. SIAM J. Discrete Math. **28**(3), 1334–1362 (2014)
26. Kosowski, A.: Approximating the maximum 2- and 3-edge-colourable problems. Discrete Appl. Math. **157**, 3593–3600 (2009)
27. Mkrtchyan, V., Petrosyan, S., Vardanyan, G.: On disjoint matchings in cubic graphs. Discrete Math. **310**, 1588–1613 (2010)
28. Mkrtchyan, V., Petrosyan, S., Vardanyan, G.: Corrigendum to "On disjoint matchings in cubic graphs". Discrete Math. **313**(21), 2381 (2013)
29. Mkrtchyan, V., Steffen, E.: Maximum Δ-edge-colourable subgraphs of class II graphs. J. Graph Theory **70**(4), 473–482 (2012)
30. Nishizeki, T.: On the maximum matchings of regular multigraphs. Discrete Math. **37**, 105–114 (1981)
31. Shannon, C.E.: A theorem on colouring the lines of a network. J. Math. Phys. **28**, 148–151 (1949)
32. Steffen, E.: Classifications and characterizations of snarks. Discrete Math. **188**, 183–203 (1998)
33. Steffen, E.: Measurements of edge-uncolourability. Discrete Math. **280**, 191–214 (2004)
34. Stiebitz, M., Scheide, D., Toft, B., Favrholdt, L.M.: Graph Edge Colouring. Wiley, Hoboken (2012)
35. Thilikos, D., Serna, M., Bodlaender, H.: Constructive linear time algorithms for small cutwidth and carving-width. In: Proceedings of the 11th International Conference on Algorithms and Computation (ISAAC), Taipei, Taiwan, 18–20 December 2000, pp. 192–203 (2000)
36. Vizing, V.: On an estimate of the chromatic class of a p-graph. Diskret Analiz **3**, 25–30 (1964)
37. West, D.: Introduction to Graph Theory. Prentice-Hall, Englewood Cliffs (1996)

Effective Use of Low Power Heterogeneous Wireless Multimedia Sensor Networks for Surveillance Applications Using IEEE 802.15.4 Protocol

Nurdaulet Kenges[1]([✉]), Enver Ever[2], and Adnan Yazici[1]

[1] Department of Computer Science, School of Engineering and Digital Sciences,
Nazarbayev University, Nur-Sultan 010000, Kazakhstan
{nurdaulet.kenges,adnan.yazici}@nu.edu.kz
[2] Computer Engineering Program, Middle East Technical University, Northern
Cyprus Campus, 99738 Mersin 10, Güzelyurt, Turkey
eever@metu.edu.tr

Abstract. With the recent technological advancements in video sensors, digital signal processing units and digital radio interfaces, low-cost devices such as Wireless Multimedia Sensor Networks (WMSNs) became an important part of surveillance applications. The scarcity of resources for these devices introduces new challenges in terms of energy consumption, and Quality of Service (QoS) guarantees. One of the challenges is processing and transmission of large volumes of multimedia data. In this paper, the efficiency of WMSN applications where the computations are performed at the sensor nodes is considered. A real-life testbed is employed as well as simulation models in Castalia. The network using proposed architecture showed better results in terms of Quality of Service and Quality of Experience, while still preserving application requirements.

1 Introduction

In recent years, wireless multimedia sensor networks (WMSNs) have received lots of attention from the research community, resulting in various types of contributions such as physical layer enhancements, medium access control mechanisms, routing and clustering algorithms and different kinds of internet of things (IoT) applications [1,2]. Unlike the wireless sensor networks (WSNs), which usually collect scalar data, WMSNs focus more on providing and delivering multimedia data. Recent technological advances in embedded systems and communication protocols allow WMSNs to be used in a variety of applications, such as real-time environmental monitoring, health-care and video surveillance [1–3,10].

In sensor networks, a sensor node is a small electronic device. It has a limited energy source, and its battery cannot be recharged in some applications. Unlike other wireless networks, it is generally difficult or impractical to replace or charge batteries, which are exhaustible. Limitations of WMSNs in terms of battery lifetime, processing capacity and communication bandwidth create new challenges to the research community. Multimedia sensors generate a significant amount of data to be processed and delivered, which can be a challenging task with such limitations. Li et al. [10] suggest that the problem of a large volume of data can be dealt with by reducing the data redundancy or increasing the possible bandwidth in the network. The first approach tries to reduce the amount of the data to be transmitted and the latter approach can be achieved by using multi-channel MAC protocols, where data can be transferred by several nodes in parallel, utilizing the bandwidth more efficiently. However, the transmission rate of most of the the standards such as the IEEE 802.15.4 protocol is low (i.e. 250 Kbit/s), and may not be sufficient to stream large numbers of video frames timely.

In this study, the resources of the WMSN nodes are used for local processing to extract useful information from the images and videos using image analysis and machine learning techniques. The extracted information is expected to be smaller in size compared to the multimedia data, which means that less bandwidth and less energy are required to send the information to the sink. The results obtained show the effectiveness of using local processing techniques in WMSNs in terms of energy efficiency, QoS and QoE. The paper shows a detailed analysis of the performance of real-life testbed and a simulation.

The rest of the paper is organized as follows. In Sect. 2, existing solutions for data redundancy in WMSNs are considered. Heterogeneous wireless multimedia sensor networks under study are considered in Sect. 3. The methods used for the evaluation of the proposed architectures are discussed in Sect. 4. Numerical results are presented with discussions in Sect. 5 and finally, in Sect. 6 the conclusions are discussed.

2   Related Works

Various solutions are available in the existing literature to reduce data redundancy in WMSNs [8,10,15]. Multimedia In-network Processing is one of the methods that can be used, where a group of nodes in the network collaborate to reduce data redundancy and possibly improve the level of inference. Distributed Source Coding is another solution where the computation burden is shifted from encoders (camera sensors) to decoders (sink), allowing camera sensors nodes to spend less energy on computation [9]. Another method is to use incorporate data storage and query processing methods which allows the network to query the data only when it is needed, spending less energy on the communication compared to the normal case where all the data generated transmitted to the sink for further processing [4,10].

There are two main approaches to implement multimedia in-network processing which are multimedia data fusion and online multi-view video summarization. In the first method, head nodes collect information from neighbouring nodes and combine it to make a summarized report. Thus possibility of having similar data is reduced and less information is sent since the nodes do not try to send everything directly to the sink. Usually in WMSNs, multimedia data fusion is multimodal because of the sensor heterogeneity [15]. Therefore, in WMSNs, the head node can fuse several types of sensors such as audio, video and scalar data. For example, in [8] a tracking algorithm is presented where camera sensors are mounted between two microphones. This allows the system to perform target localization using colour-based change detection in video frames and on time difference of arrival estimation between two microphones. Distributed source coding inverts one-to-many paradigm, so that the computation cost is shifted from encoders part (camera sensors nodes) to decoders (usually a sink) unlike most of the video coding paradigms which use a one-to-many approach in video encoders and decoders. For example in [14], the PRISM which is a video coding paradigm, allows to transfer the computationally expensive video encoder motion-search module to the video decoder. For WMSNs, this may substantially decrease energy consumption in the network, as most of the decoding work will be performed on the sink side [11].

Although in most of the applications using WMSNs the data generated and sensed from nodes go to the sink for further analysis and processing, studies such as [5] show that with recent technological developments, it is possible to equip some devices with a greater amount of flash storage and with more effective processors. This allows a network to process and store the information within the network (In-network Data Storage and Query Processing). Only the final result is transferred to the sink. With data storage functionality we may also add query processing techniques, so that information is delivered upon request. This would significantly reduce the communication cost between the network and the sink. In-network Data Storage and Query Processing could be the best alternative to local processing in terms of communication cost. However, these approaches require huge storage capacity and high processing power of nodes which may not be feasible for some applications using WMSNs. Additional database management methods should also be implemented in the network.

In this study, local processing techniques are employed for surveillance applications. The proposed system is evaluated in terms of energy efficiency of the network, and overall performance of the network (throughput, transmission delay and packet loss). Real time test beds and simulation results are employed for this purpose.

3   The Heterogeneous Wireless Multimedia Sensor Network Under Study

In this section the details of the test bed established are presented with three different architectures. In our experiments, we use two main architectures. Architecture 1a and 1b are based on wireless Xbee communication between nodes, whereas architecture two uses the Bluetooth. Architectures 1a, 1b, and 2 are illustrated in Fig. 1, Fig. 2 and Fig. 3 respectively.

Fig. 1. Architecture 1a: connection through Xbee modules

Fig. 2. Architecture 1b: MySignals with Bluetooth connection

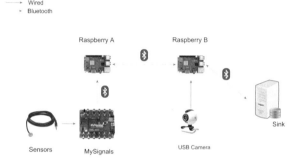

Fig. 3. Architecture 2: connection through Bluetooth modules

In the experiments, we defined two main scenarios. The main difference between them is whether or not nodes use local processing on the edge. In the first scenario, the network sends all the generated data to the sink without adding any intelligence. In the second scenario, one or all of the nodes process generated scalar/visual data to extract the piece of information needed for the application. The first scenario is illustrated inf Fig. 4. In the first phase, the whole process starts by getting sensor data from MySignals board [13]. The temperature sensor is employed for this scenario for acquiring scalar information. In the second phase, all the sensor data which was recorded goes to "Raspberry Pi A" where data is checked for a potential anomaly. For the experiment, the threshold was taken as 30 °C. The regular temperature of the room is about 20–22 °C, therefore the recording temperature higher than 30 °C is considered as a signal for the network that a person started to interact with the system. In the third phase, the system triggers the camera sensor and captures the frame. All the data collected from the sensors, name of the person and image are in turn sent to the sink. The second scenario is presented in Fig. 5. The first and the second phases are the same as the first scenario. In the third phase, the system triggers the camera and takes an image of the person. The image is then sent to face recognition library located in the "Raspberry Pi B" to recognize a person. If the person is in the database of the system, the system will define the name of the person, otherwise, the name will be set as "unknown". The sink will only receive scalar sensor and textual information of the image which is the name of the person in the image.

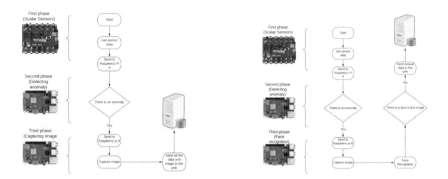

Fig. 4. Scenario 1: sending Everything   **Fig. 5.** Scenario 2: sending Knowledge

For the local processing in the "Raspberry Pi B" node, we used face recognition tools which are based on Dlib C++ library. Since the library is based on C++ language, it is compatible with the Raspbian System. The library allows us to recognize and manipulate faces from the command line. The Model in the library was built using dlib's state-of-the-art face recognition tools [6]. It uses Network Architecture based on ResNet-34 [7] with a few layers removed and the number of filters per layer reduced by half. The model is able to achieve up to 99.38% of accuracy on the Labeled Faces in the Wild benchmark.

The applications described are motivated by a case study where a smart house system is employed for elderly people. The system can be considered as an early stage prototype. Developing the system further by including additional MySignals sensors (ECG, heart rate, blood pressure, body position etc.), it is possible to monitor the health of the elderly people in the smart house environment. Furthermore, the face recognition function with the database of trusted people can be used as a security system. Several cameras may be installed on the territory. If the person approaching the house is not in the list of trusted people, the homeowner can be notified by the system.

4   Performance Evaluation

Performance evaluation is performed using a real-life testbed and simulation runs are employed in Castalia, which uses Omnet++ platform.

The experiments are conducted based on the architectures presented and the scenarios discussed in Sect. 3. The network consists of 2 Raspberry Pi 3B+ devices. Each device represents a node in the WMSN. The network has one sink node connected to Raspberry Pi B node, which has a camera to collect multimedia data. For communication, Xbee and Bluetooth modules are used for each node. MySignals board feeds the network with scalar data. Energy consumption is recorded using the RuiDeng AT34.

The simulation results are obtained using Omnet++ framework, which is an object-oriented modular discrete-event network simulator based on C++ language [12]. The edge processing energy consumption is defined as "baselineNodePower", which is the constant minimum amount of power drawn. The initial simulation runs consider the scenario of the testbed and is identical to the real-life version of the experiment. Once the accuracy of the simulation setup is validated using the results obtained from the testbed, the size of the network is expanded for evaluating larger scale configurations.

5 Numerical Results and Discussions

The initial experiments consider an extreme case where all the temperature values from MySignals are higher than 30°, which means that all of the incoming readings are treated as anomaly. This scenario is used to evaluate the maximum possible energy consumption of the network. In Table 1 the energy consumption levels of the MySignals is given for one hour in Architecture 1a. The average value is presented for energy consumed for the case where scalar data is sent to the network every second.

Table 1. Wired MySignals results, Architecture 1a, both scenarios

	Voltage, V	Current, A	Consumed energy
Attempt 1	5.01 V	0.171 A	3070 J
Attempt 2	4.99 V	0.171 A	3078 J
Attempt 3	5.01 V	0.171 A	3074 J
Average			3074 J

Table 2. MySignals with Bluetooth module results, Architecture 1b and 2, both scenarios

	Voltage, V	Current, A	Consumed Energy
Attempt 1	5.01 V	0.196 A	3420 J
Attempt 2	5.01 V	0.195 A	3409 J
Attempt 3	5.01 V	0.195 A	3412 J
Average			3414 J

The same configuration is used to present results of MySignals this time for Architecture 1b in Table 2. The results show that this time the MySignals board consume more energy since the Bluetooth module is not as efficient as the wired connection, however, the wireless connection introduces flexibility in terms of mobility of the sensors employed.

Table 3 summarizes the energy consumption results of the whole network in detail for every Architecture and Scenario described.

Table 4 shows the number of times Sink receives information from the camera In scenario "Knowledge", the information about the user is shown as the name of the person, while in "Everything" this is raw images of the person. We can

Table 3. Comparison of energy consumption

Scenario	Architecture 1a	Architecture 1b	Architecture 2
Knowledge	23 849 J	24 189 J	22 285 J
Everything	23 786 J	24 126 J	21 853 J

Table 4. Information received by Sink

Scenario	Received by sink	Consumed energy
Knowledge	1936 names	23 849 J
Everything	63 frames	23 786 J

see that the system that is configured for the scenario "Knowledge" outperforms the other scenario in terms of amount of information delivered while spending about the same amount of energy.

Similarly in Table 5, the time required for each event (anomaly detection and face recognition) to do a full cycle is presented. The results show that for the end-user, the scenario "Knowledge" is the better option, since in the other scenario customer should wait around 57 s to receive an image/frame of the video.

The maximum data rate of XBee modules is around 250 kbps in an ideal case. In Table 5 the energy and time required to send an image in an ideal case using XBee is calculated. The time required to send an image in such conditions is still twice more than in scenario "Knowledge". Therefore making local processing and sending information to the sink is significantly faster than sending raw video frame in the network with a connection through XBee modules.

Table 5. Energy and time required for each event to do full cycle

	Knowledge	Everything	Everything ideal case
Raspberry pi A	4 J	297 J	9 J
Raspberry pi B	3 J	345 J	7 J
Total	7 J	642 J	16 J
Time	1.86 s	57 s	2.8 s

5.1   Simulation

The parameters used for the simulation are same as the ones used for the real-life testbed. Simulation time is one hour. The number of nodes considered is four. CC2420 is standard radio parameter used in the radio chips. For the Application, we created "customApplication" class, where the routing of the network resembles scenarios described in Sect. 3. StaticGTS802154 was used as MAC protocol since it is the protocol used in ZigBee devices. Table 6 is used to define baseline node power for every node in the Simulation parameters. The baseline power consumption of a node is the constant minimum amount of power drawn (processor base consumption, other electronic circuits consumption, etc.). Duration of the experiment which uses the testbeds is also one hour, therefore the value of

Table 6. Converting joules to power

Device	Consumed, J	Consumed, mWh	Power, mW
Raspberry pi A	8 406 J	2 335 mWh	2 335 mW
Raspberry pi B (with Camera)	12 369 J	3 435 mWh	3 435 mW
MySignals	3 074 J	853 mWh	853 mW

Power in mW will be the same as Energy consumed in mWh. The packet rates are defined as one packet per second as described in scenario 1. The summary of the parameters used for simulation can be given as:

sim-time-limit = 3600 s

SN.numNodes = 4

SN.node[*].Communication.Radio.RadioParametersFile = "../Parameters/ Radio/CC2420.txt"

SN.node[*].ApplicationName = "CustomApplication"

SN.node[*].Communication.MACProtocolName = "StaticGTS802154"

SN.node[0].ResourceManager.baselineNodePower = 0 # Sink Node

SN.node[1].ResourceManager.baselineNodePower = 3435 # Raspberry Pi B

SN.node[2].ResourceManager.baselineNodePower = 2335 # Raspberry Pi A

SN.node[3].ResourceManager.baselineNodePower = 853 # MySignals

SN.node[0].Application.packet.rate = 0 #pkt/sec

SN.node[*].Application.packet.rate = 1 #pkt/sec

* - means for every node

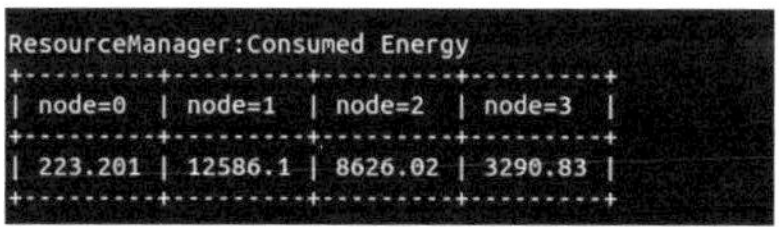

Fig. 6. Energy consumed per node ($node0 \Rightarrow Sink, node1 \Rightarrow Raspberry PiB, node2 \Rightarrow RaspberryPiA, node3 \Rightarrow Mysignals$)

Fig. 7. Packets transmitted per node ($node1 \Rightarrow RaspberryPiB, node2 \Rightarrow RaspberryPiA, node3 \Rightarrow Mysignals$)

After running the experiment, the value for energy consumption for the whole network is calculated as 24995 J. Which is close to the results we obtained from the testbed in Architecture 1a and scenario "Knowledge", measured as 23 849 J. The discrepancy is less than 4.8%.

When we consider the results presented in Fig. 6, we can see that the energy consumption provided in the simulation is very similar to the real one. In the figure, the Sink node (with index 0) has a small amount of energy consumed, this is the energy consumed on the reception of the packets.

Figure 7 shows the number of packets transmitted from each node, and Fig. 8 shows the reception per node is higher than it can be expected. These values show the received packets in Communication level.

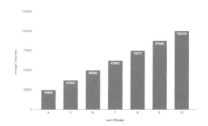

Fig. 8. Packets received per node ($node0 \Rightarrow Sink, node1 \Rightarrow Raspberry$ $PiB, node2 \Rightarrow RaspberryPiA, node3 \Rightarrow$ $Mysignals$)

Fig. 9. Energy consumption vs Number of nodes

Once the simulation configuration is validated through comparisons with the testbed it is possible to analyse the systems proposed for different scales. This is motivated by the smart house case study, where several camera sensors are needed to monitor the space inside/outside of the house. Therefore we introduced additional nodes with Raspberry Pi B parameters (the node with camera). In Fig. 9 it is possible to see that the energy consumption grows linearly as the number of nodes increases.

6   Conclusion

In this study various architectures and scenarios are considered for a typical smart house surveillance system. For the applications considered, the reduction of data redundancy using the edge processing did not affect the results in terms of energy efficiency, however, it the reduction strategy was useful in terms of Quality of Service. This approach dramatically reduced the size of the data being transmitted in the network and the sink received more information for the same given time and amount of energy spent. Both scenarios showed the same result in terms of energy consumption.

For the future studies, it is expected that with the increasing distance between nodes, the energy consumption to transfer the data will also rise. In such conditions using local processing in the network could be effective in terms of energy efficiency as well. Therefore, it is desirable to consider the scaling of the network to the degree of larger smart house applications with several surveillance cameras and multiple health monitoring sensors.

It is also possible to use the prototype in development of stat of the art systems for monitoring the temperature of the people crossing borderlines between countries. This can be particularly useful as a fast and automated detection tool of people with infections.

Acknowledgement. This work was funded by NU Faculty-development competitive research grants program, Nazarbayev University, GrantNumber-110119FD4543.

References

1. Akyildiz, I.F., Melodia, T., Chowdhury, K.R.: A survey on wireless multimedia sensor networks. Comput. Netw. **51**(4), 921–960 (2007)
2. Akyildiz, I.F., Melodia, T., Chowdury, K.R.: Wireless multimedia sensor networks: a survey. IEEE Wirel. Commun. **14**(6), 32–39 (2007)
3. Akyildiz, I.F., Melodia, T., Chowdhury, K.R.: Wireless multimedia sensor networks: applications and testbeds. Proc. IEEE **96**(10), 1588–1605 (2008)
4. Diallo, O., Rodrigues, J.J.P.C., Sene, M., Lloret, J.: Distributed database management techniques for wireless sensor networks. IEEE Trans. Parallel Distrib. Syst. **26**(2), 604–620 (2013)
5. Diao, Y., Ganesan, D., Mathur, G., Shenoy, P.J.: Rethinking data management for storage-centric sensor networks. In: CIDR, vol. 7, pp. 22–31 (2007)
6. DLib C++ library. http://dlib.net/. Accessed 04 Jan 2020
7. He, K., Zhang, X., Ren, S., Sun, J.: Deep residual learning for image recognition. In: Proceedings of the IEEE Conference on Computer Vision and Pattern Recognition, pp. 770–778 (2016)
8. Huiyu, Z., Taj, M., Cavallaro, A.: Target detection and tracking with heterogeneous sensors. IEEE J. Sel. Top. Sign. Process. **2**(4), 503–513 (2008)
9. Izhar, M.A.M., Aljohani, A.J., Ng, S.X., Hanzo, L.: Distributed joint source coding and trellis coded modulation for symbol-based Markov sources. IEEE Trans. Veh. Technol. **6**(5), 4031–4041 (2017)
10. Li, S., Kim, J.G., Han, D.H., Lee, K.S.: A survey of energy-efficient communication protocols with QoS guarantees in wireless multimedia sensor networks. Sensors **19**(1), 199 (2019)
11. Lu, Q., Luo, W., Wang, J., Chen, B.: Low-complexity and energy efficient image compression scheme for wireless sensor networks. Comput. Netw. **52**(13), 2594–2603 (2008)
12. OMNeT++ Community Site. https://omnetpp.org/. Accessed 10 Jan 2020
13. Pinto, A., Zhang, Z., Dong, X., Velipasalar, S., Vuran, M.C., Gursoy, M.C.: Energy consumption and latency analysis for wireless multimedia sensor networks. In: 2010 IEEE Global Telecommunications Conference GLOBECOM 2010, pp. 1–5. IEEE (2010)
14. Puri, R., Majumdar, A., Ramchandran, K.: PRISM: a video coding paradigm with motion estimation at the decoder. IEEE Trans. Image Process. **16**(10), 2436–2448 (2007)
15. Wu, Y., Chang, E.Y., Chang, K.C.-C., Smith, J.R.: Optimal multimodal fusion for multimedia data analysis. In: Proceedings of the 12th Annual ACM International Conference on Multimedia, pp. 572–579 (2004)

A Weak Signal Capture Algorithm Considering DBZP and Half-bit Difference

Kun Shan[✉], XiaoOu Song, and GuoBin Li

School of Information Engineering, Engineering University of PAP,
Xi'an 710086, Shaanxi, China

Abstract. This paper analyzed the shortcomings of the traditional half bit alternation algorithm in weak signal capture sensitivity and capture efficiency, and proposed and verified a DBZP-half-bit differential algorithm. This alternate algorithm not only has a high signal-to-noise ratio gain, but also ensures the capture speed. This algorithm mainly performs the differential coherence operation of adjacent blocks after dividing the signal into half-bit blocks, and then implements the DBZP algorithm with the FFT module added in the block to reduce the complexity of the frequency search unit. The improved algorithm is discussed and verified by simulation, which can achieve good capture performance in actual conditions.

1 Introduction

Under environments such as urban canyons, tunnels, high-rise buildings, etc. due to weak signal strength, general receivers and acquisition algorithms often cannot capture satellite signals normally, which affects the normal operation of the navigation system. In an open environment with a strong signal, the signal power is above –160 dBW, and under the influence of environmental changes, the reduction of the signal direction elevation angle or multipath interference, the signal power is generally attenuated by 10–30 dBW. The receiver cannot work normally, which affects the positioning accuracy.

The most direct and effective way to capture weak signals is to extend the integration time through coherent integration and incoherent integration, increase the signal-to-noise ratio gain, and solve the impact of the corresponding bit hop and square loss at the same time. Normally, receiver acquisition algorithms include half-bit alternate algorithm, full-bit algorithm, maximum likelihood bit synchronization algorithm, circular correlation estimation algorithm, etc. [1, 2]. The main purpose of the above-mentioned algorithms is to eliminate the influence of the possible jumps of each bit of the navigation data bit. Increase the coherent accumulation time as much as possible and improve the signal-to-noise ratio, so that the signal can be captured smoothly.

Based on the principle and gain of coherent integration and non-coherent integration, this essay analyzes the principle of half-bit alternate algorithm, and combines half-bit alternate algorithm and differential coherence to propose an improved half-bit differential algorithm, then combines the characteristics of half-bit differential algorithm and DBZP algorithm, achieve the final DBZP-half-bit differential capture algorithm, and compare the half-bit alternate algorithm with final DBZP-half-bit differential capture algorithm through simulation [3, 4].

L. Barolli et al. (Eds.): AINA 2021, LNNS 227, pp. 252–259, 2021.
https://doi.org/10.1007/978-3-030-75078-7_26

2   Weak Signal Capture Algorithm

For weak signals that cannot be captured, coherent integration and non-coherent integration are usually performed by accumulating the correlation values or the square of the correlation of adjacent C/A code periods to increase the signal-to-noise ratio gain and detect weak signals. The half-bit alternating algorithm uses both coherent integration and incoherent integration, which has obvious gains for weak signal capture, but due to the square loss of incoherent integration and only half the length of the signal provides effective information, the capture efficiency is reduced.

2.1   Half-bit Alternation Algorithm

For ordinary satellite signals, the navigation data bit is one bit every 20 ms, and there may be a jump, which limits the length of the coherent integration. Take N segments of 20 ms signal data, and divide each segment of signal into two blocks with a length of 10 ms, divided into n = 2N blocks, as shown in Fig. 1:

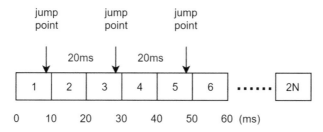

Fig. 1. Schematic diagram of half-bit alternate algorithm

Perform a 10 ms coherent integration in each block, and the correlation result is recorded as V_n, and then the odd-numbered blocks V_1, V_3,...V_{2N-1} and the even-numbered blocks V_2, V_4,...V_{2N} are respectively non-coherently accumulated to obtain:

$$\begin{cases} \Lambda_A = V_1^2 + V_3^2 + ... + V_{2N-1}^2 \\ \Lambda_B = V_2^2 + V_4^2 + ... + V_{2N}^2 \end{cases}. \tag{1}$$

Apparently, the jump points must all exist in the odd or even block groups, and the group without the jump points has an obvious peak in the incoherent integration result, thereby achieving capture. For a 200 ms signal, $N = 10$, the signal-to-noise ratio gain obtained by coherent integration is:

$$G_L = 10 \lg L = 10 \lg 10 = 10 \ (\text{dB}). \tag{2}$$

The signal-to-noise ratio gain obtained incoherently is:

$$G_N = 10 \lg N - L(N) = 10 - L(N) \ (\text{dB}). \tag{3}$$

Where L(N) is the square loss.

$$L(N) = 10\log\left[\frac{1 + \sqrt{1 + \frac{9.2N}{D_c(1)}}}{1 + \sqrt{1 + \frac{9.2}{D_c(1)}}}\right] \text{(dB)} . \tag{4}$$

$$D_c(1) = \left[erf^{-1}(1 - 2P_{fa}) - erf^{-1}(1 - 2P_d)\right]^2 . \tag{5}$$

Among them P_{fa} is the false alarm probability and P_d is the detection probability.

It can be seen from Eqs. (4) and (5) that the square loss depends on the false alarm probability and the detection probability, and will increase with the increase of N, so the improvement effect of the signal-to-noise ratio is not good. Although the half-bit alternate algorithm avoids bit jumps, the square loss and the need for twice the length of the signal affect the signal-to-noise ratio gain and real-time capture [5, 6].

2.2 Differential Coherent Integral

In order to solve the square loss problem of the non-coherent integration algorithm, propose a differential coherent integration algorithm. The coherent integration results of two adjacent sections are conjugate multiplied, and the results of the N times differential multiplication are accumulated:

$$S = \sum_{i=1}^{N} V_i V_{i-1}^* . \tag{6}$$

As two adjacent integration sections are in the same data bit, the influence of the positive and negative data bits can be eliminated, so squaring is not required before accumulation. Meanwhile, since the noises of different integration sections are independent of each other, the mean value of noises is still 0 after multiplication, avoiding the appearance of square loss [7]. Therefore, the gain can reach close levels theoretically to coherent integral:

$$G_S \approx 10\lg N \text{ (dB)} . \tag{7}$$

3 Improved Capture Algorithm Process

In the half-bit alternate algorithm, when the integration time is longer, the square loss caused by incoherent integration will also increase significantly, and even the signal-to-noise ratio gain of incoherent integration is basically offset. Therefore, the differential algorithm is combined with the half-bit alternate algorithm. Avoid incoherent accumulation and solve square loss. The specific process is as follows:

At the receiver end with 1 ms as the delay, take 10 sets of data that are long enough, each set of data is divided into blocks of 10 ms, and coherent integration is performed in the block, and the correlation result of the mth group and nth block is marked as $V_{m,n}$(m = 1, 2, 3, …, 10, n = 1, 2, 3, … 2N), each group of odd-numbered blocks and adjacent blocks are conjugate multiplied, and finally the multiplied result After summing, the modulus value is taken, and the threshold judgment is performed. As shown in Fig. 2:

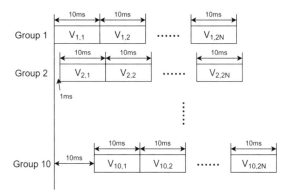

Fig. 2. Schematic diagram of half-bit differential algorithm

The relevant values at the time of each group's judgment are:

$$\begin{cases} \Lambda_1 = \left| \sum_{i=1}^{N} V_{1,2i-1} \times V_{1,2i}^* \right| \\ \Lambda_2 = \left| \sum_{i=1}^{N} V_{2,2i-1} \times V_{2,2i}^* \right| \\ \cdots \\ \Lambda_{10} = \left| \sum_{i=1}^{N} V_{10,2i-1} \times V_{10,2i}^* \right| \end{cases} \qquad (8)$$

There is one group of data in 10 groups, and the transition edge is within 1 ms after the block edge. During the capture process, when the local code phase is aligned, the transition edge and the block edge are also aligned at the same time, and the correlation value is significantly higher than other groups., Thereby capturing the signal. This eliminates the impact of bit jumps. At the same time, because the two blocks of differentiate coherent data are within the same data bit, and after conjugate multiplication, the positive and negative effects of the data bit are eliminated, so that the multiplication result can be directly without squaring. The addition avoids the influence of the square loss. And the half-bit differential algorithm does not discard the odd-numbered or even-numbered blocks. It makes full use of all signal data, and the signal-to-noise ratio gain is also significantly increased.

In order to improve the capture speed while ensuring the capture sensitivity, in the process of obtaining the correlation results $V_{m,n}$ by coherent integration in the half-bit differential algorithm, the parallel code phase capture based on DBZP is used, and an additional FFT module is added after the DBZP module as a frequency search module. M-point FFT operation is performed on the correlation results of M blocks of the same code phase to realize two-dimensional fast search of code phase and frequency at the same time, reducing the amount of calculation (Figs. 3 and 4). The principle flow chart is:

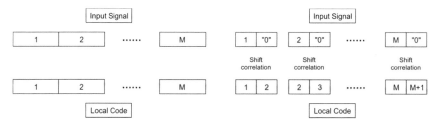

Fig. 3. "Block", "Double Block", "Zero Padding"

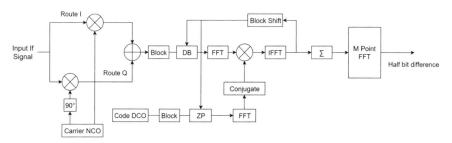

Fig. 4. DBZP-FFT flowchart

In DBZP, the number of blocks and the number of block shifts have a strict meaning and cannot be arbitrarily selected. The number of blocks determines the final number of FFT operation points for the same code phase, and also determines the search range of the Doppler frequency that can be obtained. Assuming that the number of sampling points in each block is Sb, the correlation after partitioning reduces the sampling rate of the data to f_s/S_b, and the number of data points contained in each block decreases from Sb to 1, so the resolution remains unchanged, In order to ensure that the search range includes the entire Doppler frequency shift range, the number of blocks M should satisfy:

$$M \geq \left[\frac{f_{\max} - f_{\min}}{\Delta f} \right] + 1 . \tag{9}$$

The block shift operation is because the ranging code in each block is incomplete when the block operation is performed. The cyclic convolution can only obtain the code phase correlation value corresponding to the length of the block. The shift is required to make the convolution of each block The length meets one period of the ranging code, so the number of block shifts Z should meet:

$$Z \geq \frac{N_S}{S_b} = \frac{f_S}{1000} / \frac{f_s}{f_{max} - f_{min}} = \left[\frac{f_{max} - f_{min}}{1000} \right] + 1 . \tag{10}$$

By the usage of DBZP algorithm in the block to perform related operations, which effectively improves the capture efficiency and capture performance. Combined with the half-bit differential algorithm, the capture of weak signals can be completed more efficiently and reliably.

4 Simulation Results and Analysis

The simulation uses the B1I signal of the Beidou satellite navigation system, which is divided into two types of navigation messages, one is the D1 navigation message broadcast by IGSO/MEO satellites, and the other is the D2 navigation message broadcast by GEO satellites, both of which are modulated by QPSK and carrier frequency It is 1561.098 MHz, using C/A code spreading with a code length of 2046 and a code rate of 2.046 MHz.

The bit period of the D1 navigation message is 20 ms, and the data rate is 50 bps; the bit period of the D2 navigation message is 2 ms, and the data rate is 500 bps. The Doppler frequency shift range of the B1I signal is ±10 KHz, so the number of shifts Z is 21; the frequency resolution is set at 500 Hz, so the number of blocks M is 41. Use the DBZP-half-bit differential algorithm and the traditional half-bit alternate algorithm to capture the 200 ms D1 navigation message. Figures 5 and 6 show that the two algorithms have a carrier-to-noise ratio of 20 dB. The capture effect of Hz Beidou B1I signal.

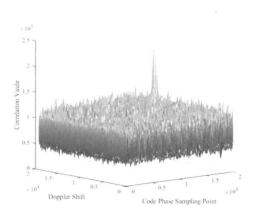

Fig. 5. Half-bit alternate algorithm (20 dB·Hz)

As the carrier-to-noise ratio is 20 dB·Hz, both algorithms can capture the signal, but the acquisition effect of the half-bit alternate algorithm is obviously inferior to the DBZP-half-bit differential algorithm, and the noise suppression effect is not obvious. Figures 7 and 8 show the capture effects of the two algorithms on signals with a carrier-to-noise ratio of 15 dB·Hz.

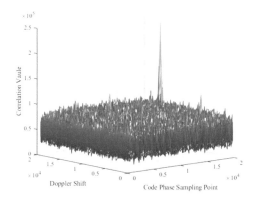

Fig. 6. DBZP-half-bit differential algorithm (20 dB·Hz)

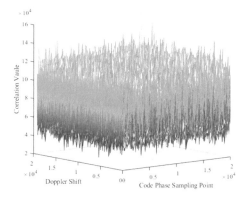

Fig. 7. Half-bit alternate algorithm (15 dB·Hz)

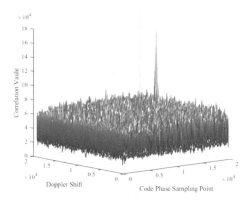

Fig. 8. DBZP-half-bit differential algorithm (15 dB·Hz)

As the carrier-to-noise ratio is 15 dB·Hz, the half-bit alternate algorithm cannot capture the signal normally, while the DBZP-half-bit differential algorithm can still capture the signal smoothly. The detection probability of the two algorithms varies with the carrier-to-noise ratio as shown in Fig. 9. The DBZP-half-bit differential algorithm is optimized by about 6–7 dB.

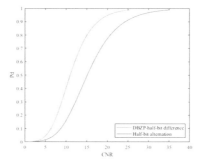

Fig. 9. Schematic diagram of carrier-to-noise ratio-detection probability

5   Conclusion

This essay made a proposal of a capture algorithm based on DBZP-half-bit difference. On the basis of half-bit block and difference operation, combined with DBZP algorithm and FFT operation module, it reduces the amount of capture operation and increases the capture gain while increasing the capture speed. Through algorithm analysis and simulation, the feasibility and effectiveness of the algorithm are verified. The signal with a carrier-to-noise ratio of 15 dB·Hz can be successfully captured, and it can be applied well in satellite systems such as GPS and BeiDou.

References

1. Zhang, W., Ghogho, M.: Computational efficiency improvement for unaided weak GPS signal acquisition. J. Navig. **65**(2), 363–375 (2012)
2. Kovář, P., Kačmařík, P., Vejražka, F.: Interoperable GPS, GLONASS and GALILEO software receiver. Gyroscopy Navig. **2**(2), 69–74 (2011)
3. Wang, X., Ji, X., Feng, S.: A scheme for weak GPS signal acquisition aided by SINS information. GPS Solutions **18**(2), 243–252 (2014)
4. Lin, Z., Aimeng, L., Jicheng, D., et al.: BeiDou signal acquisition with Neumann-Hoffman code modulation in a degraded channel. Sensors **17**(2), 322 (2017)
5. Hao, F., Yu, B., Gan, X., et al.: Unambiguous acquisition/tracking technique based on sub-correlation functions for GNSS Sine-BOC signals. Sensors **20**(2), 485 (2020)
6. Jianing, W., Baowang, L., Zhe, X.: Weak GPS signal acquisition method based on DBZP. J. Syst. Eng. Electron. **29**(02), 236–243 (2018)
7. Cui, H., Li, Z., Dou. Z.: Fast acquisition method of GPS signal based on FFT cyclic correlation. In: Proceedings of the 13th International Conference on Wireless Communications, Networking and Mobile Computing (WiCOM 2017), vol. 9. Scientific Research Publishing (2017)

Scalable Leader Election Considering Load Balancing

Radha Rani[1(✉)], Saurabh Rashpa[1(✉)], Dharmendra Prasad Mahato[2,3(✉)], and Van Huy Pham[4(✉)]

[1] Department of Computer Science and Engineering,
National Institute of Technology Hamirpur, Hamirpur, India
radharani_phdcse@nith.ac.in
[2] Department of Computer Science and Engineering,
National Institute of Technology Hamirpur, Hamirpur, Himachal Pradesh, India
dpm@nith.ac.in
[3] Faculty of Information Technology, Ton Duc Thang University,
Ho Chi Minh City, Vietnam
dharmendra@tdtu.edu.vn
[4] Faculty of Information Technology, Ho Chi Minh City, Vietnam
phamvanhuy@tdtu.edu.vn

Abstract. Distributed computing consists of a model where multiple parts of a system are accessed by different computing machines for the betterment of work efficiency. To get better results, electing a leader is one of the most critical tasks. Leader election is the process of assigning one of the processes as the leader whose work is to organize all the provided jobs that are distributed between the different computer nodes and to provide them with the required resources. The obstacle that is dealt with here is to elect a processor that can act like a leader from among the set of multiple processors using distributed protocols. To explain this in further detail, let us consider that we have n number of processors, and among these n processors, we have n number of processors, and among these n processors we have cn number of processors which can be considered as bad or corrupt and *(1-c)n* number of processors among them can be considered as good or not corrupt. Here the value of c is a fraction value and is fixed. The problem that is to be dealt with is to select a processor with a probability that has to be constant, a single processor from the given n number of processors which can act as their leader, no matter which set of the given cn processors are bad here. The scalability that is mentioned hereof leader election being scalable refers to the fact that every good or non-corrupt processor that is available amongst the total n number of processors sends and also processes several bits. And these number of bits that are being sent and processed by the various processors are polylogarithmic in n. Here we can say that the number of bits that are sent over or processed by a node belongs to a function that is polynomial in the logarithm of n.

L. Barolli et al. (Eds.): AINA 2021, LNNS 227, pp. 260–269, 2021.
https://doi.org/10.1007/978-3-030-75078-7_27

1   Introduction

Leader election in distributed computing is considered as one of the major problems. Let us consider that we have n number of processors at our disposal. From among these n processors, we know cn of them are corrupt or bad and the remaining $(1-c)n$ processors are good where the value of c is fixed. Here we assume that before the beginning of the algorithm a rival that is all-knowing and is not bounded by any computational power picks up bad processors. The purpose of this rival is to manage the actions of the bad processors. This is done to greatly increase the likelihood of electing a bad processor as the leader.

Our objective is to come up with an algorithm that makes sure that a good processor is named as the leader. Here the probability is constant irrespective of whichever set of cn processors are found to be bad. Furthermore, the issue of load balancing also has to be resolved. Over the years many research works have been done to devise leader election algorithms that have been effective in reducing the number of rounds needed to elect a leader and on variable c i.e. the fraction of bad processes that can be permitted.

2   Brief Introduction to Leader Election in Distributed Systems

In distributed systems, the process of assigning a node as the organizer of equally distributed tasks among multiple nodes or computers is known as leader election [1–3]. The election of a leader is a very crucial task in distributed computing where the flow of data is among nodes separated by physical distance. Assigning a node as the organizer of tasks in a distributed network is an issue that is challenging in nature and which requires appropriate leader election algorithms. The communication between different nodes in a distributed computing network is carried out with the help of message passing or by using a shared memory resource. Coordination among the different nodes is an important requirement for the efficient execution of any task in a distributed network. For a distributed network to be called a purely distributed system the decisions made in the network should not be controlled by a single central node. Instead, all of the nodes present in the system should interact with each other to come to an appropriate mutual decision.

It is noticed that when a decision is made all the nodes do not come to a common conclusion, this results in communication that is time-consuming and inconsistent [4]. This inconsistency leads to difficulty in coordination between the nodes when all the nodes are needed to be consistent. In distributed systems, leader election can be used as a technique for breaking the symmetry of distributed networks. So, to decide upon a node that can act as a central node to control the distributed system, a node is elected amongst all the nodes which can act as their leader. The leader serves the purpose of acting as a central control in the decentralized network. So basically, the main purpose of a leader election algorithm is to select a node that can manage the tasks of the system.

The leader is elected based on some criteria such as selecting a node with the greatest identifier. Upon the election of the leader, a specific state is attained by the nodes which are known as the terminated state. These states are divided into two categories; the elected and the non-elected states. Upon entering any one of the states, the node remains always in that state [5]. For the correct execution of any leader election algorithm, it must satisfy the condition of safety and liveness. The condition of safety for the algorithm states that at any point in time only one node can become the leader by entering into the elected state. The condition of liveness says that at last all of the nodes will enter any one of the two states, elected or non-elected state [6]. The exchange of information between the nodes is carried out by sending messages to each other until a common agreement is achieved. And upon reaching a decision, one of the nodes is assigned as the leader and every other node will accept it as the leader.

3   Brief Introduction to Load Balancing

The ability of a system to process things at high speed is highly needed. Since the very beginning of developing computers, the focus has always been on the performance of the system. Work has been done on how to increase the performance of a system that ushered in the age of supercomputers. In fields like defense, business, and scientific communities, the need for high-performing computers has always been in demand. Since supercomputers are expensive so a new concept of parallel distributed computing emerged. In today's times, parallel distributed computers are available for common use. In it, multiple processors are connected over a large number of systems which has made it very popular when large computing resources are required. The system having hundreds of powerful processors has also been developed recently [7].

In a distributed network, some of the nodes will be computing at a greater speed than others while some will be computing at a slower speed [8]. If the processing and communication speed of the network is not accounted for, the performance of the entire distributed system will be dependent on the node having the slowest processing speed. Thus, in a nutshell, load balancing techniques handle the idleness of nodes by averting them from being idle and other nodes from being overloaded.

Now the load balancing algorithm is applied in two situations. First is static load balancing which is applied during compilation time in which task assignment is done before the runtime [9]. And once the jobs are assigned, there is no change in the distribution of tasks. The second one is dynamic load balancing which is applied at execution time. In this category, dynamic distribution of tasks is done among the nodes and tasks can be redistributed again. In SLB (Static Load Balancing), for efficient distribution of tasks to improve the performance, the knowledge about the task and the system has to be known beforehand. And the work can be assigned only once so there is almost negligible communication overhead. Where in DLB (Dynamic Load Balancing), reassignment of tasks is possible during runtime so a considerable amount of communication overheads are present and their number increases upon increasing the number of processors. The overall quality of any load balancing algorithm depends on the number of steps taken to reach a balanced state and the amount of load that is moved.

3.1 Need of Load Balancing

The load balancing problem is the distribution of loads among the processing elements. In a system, when there are multiple nodes, the probability of some nodes being idle and some being overloaded with work is very likely. So the goal of the algorithms used for load balancing in these systems is to distribute the load on each element so that none of the nodes becomes either overloaded or idle. This suggests that every processing node has an equal amount of load at any point of a given time. Thus the performance of a system can improve to a significant amount if the design of the load balancing algorithm is up to the mark.

3.2 Issues in Load Balancing

There are numerous issues when load balancing is taken into concern:

- In a distributed computing network, the bandwidth of communication channels is finite, and since the processing units are physically far from each other so the decision has to be made by the load balancing algorithm whether the migration of tasks can be allowed.
- Division of a job on an arbitrary basis might not be possible which will further add constraints in dividing of tasks.
- The processors used can vary from one another in their architecture, operating system, processing speed, memory, and disk capacity.
- Based on the workload provided by the users, the load on processors and the network can be varying in time.
- Every job is further divided into smaller jobs and all these can have varying execution times which can add to the complexity.

3.3 Advantages of Load Balancing

The advantages of load balancing are as follows:

- Waiting time for any task is reduced.
- System resources are utilized to maximum capacity.
- The reliability and stability of the system are improved.
- Small jobs are not starved.
- Response time of the task is minimized.
- System throughput is maximized.
- The possibility of any future modification is possible.
- The overall performance of the entire system is improved.

Due to the given advantages, the field of load balancing techniques has become a research-intensive field.

4   Network Model and Assumptions

The leader election problem is solved with the help of using a communication model which is known as the full information model. In this model, the processor nodes in the network communicate with one another by using a broadcast channel. Thus all of the processors are aware of what every other processor is communicating. Each of the processors in this model has an identity that is unique to itself only. Also, every processor can know the identity of any processor which sends any message. Rounds of communication take place and in every single round, any processor can interact with any other processor in the network. It is taken into the assumption that each of the corrupt processors has acquired the message sent by any non-corrupt processor. A further assumption is made that due to the synchronization of all the nodes during each round, the messages sent in that round are received before the transmission of any other message of the next round. Private arbitrary bits are assigned to each node which is known to only that node in the network.

Before the commencement of the said algorithm, the rival selects the bad processor nodes. This rival then manages what these nodes do to increase the probability of making them the leader. The rival has no bound in its computational power so any cryptographic assumption is ruled off.

But this model has a major drawback associated with it. Using the full information model, one cannot design an algorithm in which each node sends only a sublinear number of bits. This is because communication is done in this model by broadcasting the messages. To overcome this problem, another model called to point to point full information model is taken into use. In this model, the communication takes place between just two processors where one is at the transmitting end while the other is at the receiving end. In this model, the corrupt processors can read every message while the good ones can only see the messages meant for them. Rest everything is similar to the full information model. Where every node has an identity (id) unique to them. This id is also known by any node that receives a message from the other node. In this model too, communication takes place in rounds.

The difference which makes this model harder than the previous model is that in this model multiple processors can receive a different set of messages from the same corrupt processor. Whereas in the full information model, the corrupt processor could only send one message at a time. This was due to the fact of it following a broadcast way of communication. So the challenge is to reduce the count of how many bits a good processor can send and process. Another assumption we make is that a node can ignore a message being sent to it and still incur no extra cost. Also, a corrupt node can only send a limited amount of messages.

4.1 Assumptions

It is assumed that:

- Every corrupt processor can acquire all the messages sent by all the non-corrupt processors.
- Due to the synchronization of all the nodes during each round, the messages sent in that round are received before the transmission of any other message of the next round.
- The rival has no bound in its computational power.
- A node can ignore a message being sent to it and still incur no extra cost.
- And a corrupt node can only send a limited amount of messages.

5 Related Work

In the following, we have surveyed works that are directly related to the topic. Mostly the focus is on leader election in distributed computing and various load balancing techniques. Both static and dynamic techniques [10–12] to balance load have been discussed in this paper.

Ben-Or and Linial [13, 14] gave One of the first techniques to elect a leader in a full information model. Their protocol was a one-round protocol and it could handle up to $1/(ln\ n)$ fraction of corrupt processors. These results were further worked upon [15–17] and improved to handle more corrupt processors. Saks [16] in his research paper presented that there does not exist a protocol that can handle $\frac{n}{2}$ fraction of corrupt processors.

Alon and Naor [18] gave a protocol that had an optimal capability to recover from corrupt nodes but the number of rounds it required for that was linear. Danny Dolev et al. [19] improved upon previous algorithms for Byzantine agreement to achieve agreement between several processes. This agreement depends upon a network of processes that are unreliable but can conduct multiple synchronized rounds of exchanging information. Uriel Feige [20] in his paper stated some new selection protocols for full information models. He used the concept of the lightest bin principle for leader election. This technique had the advantage that it can reduce the number of players from the total count and can also maintain the fraction of good players. Rafail Ostrovsky et al. [21] proposed in their paper a solution to the problem of leader election in the full information model. They studied the problem in two different settings. The first is the cryptographic setting and the second being the information-theoretic setting.

6 Proposed Method

First, we take into account the lightest bin protocol which is given by Feige [20] in his paper. his protocol can be explained as a game where players throw balls into random bins. In this game, there are multiple bins in front of the players. They randomly throw balls at any bin and the balls of the players inside the bin containing the least quantity of balls are moved to the next round. This bin is known as the lightest bin. The players which are left out of the lightest bin are discarded. The players then again throw these balls into the bins and recursion of the protocol is done. This protocol elects a good leader from the n number of processors, among which more than $\frac{2}{3}rd$ of processors are good and suitable for being a leader. We can see from his results that this is possible in $ln\ n$ rounds where each node has to broadcast once in every round. Now this protocol is for the full information model and we adapt it for the point to point information model by removing each broadcast operation and replacing it with byzantine agreement algorithm (Fig. 1).

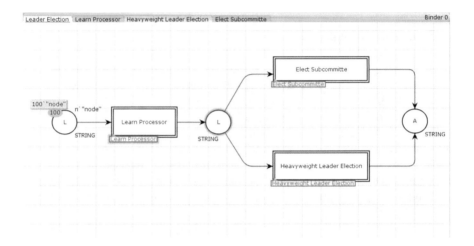

Fig. 1. Leader election model using Colored Petri Nets

7   Proposed Algorithm

In our algorithm, first, we make a subset of processors. This subset is known as a committee and this concept was given by Bracha [22]. Secondly, our algorithm puts into use the lightest bin protocol given by Feige [20] which is adapted to work in point to point full information model. Lastly, the Byzantine algorithm of agreement given by Dolev [19] is used which elects a good leader with a positive constant probability.

Basically in our algorithm, we first group the processors. The size of these groups is of polylogarithmic number. Every processor in our network is allotted to numerous groups at a time. Every group then selects among themselves good processors which can participate in further rounds. This selection of processors

[23] occurs simultaneously in all the groups. This step is then repeated on all the selected processors until only a polylogarithmic number of nodes are present at last. Finally, these processors again run the algorithm for the final time to elect a leader.

Algorithm 1. Algorithm: To Elect a Leader

1: $j \leftarrow 0$
2: **while** $j \leq i$ **do**
3:     For all nodes A of the committee on the layer j, every processor belonging to the node A finds about all of the remaining members of A by calling Learn Processor function on each node.
4:     If $j \leq i$, run subcommittee protocol on every node A and elect a subcommittee.
5:     If $j = i$, elect leader using lightest bin protocol.
6:     If processor elected to two or more subcommittee stop their participation in further rounds.
7:     $j \leftarrow j+1$
8: **end while**
9: If processor p is assigned to a number of nodes greater than 8 it is stopped from participating in future election.
10: To elect a leader every processor makes a random poll. And a processor gives a response to the one it thinks of as leader. Thus if most of the queries of processor gets a responses it thinks of itself as a leader

---

The step 3 of this Leader Election Algorithm is used to provide the capability to balance the load on the network.

8   Conclusion

Our leader election algorithm assumes that there are n processors and among them only less than $\frac{1}{3}rd$ of processors are corrupt.

Following is a list of properties that the algorithm has:

- Precisely one of the good processors is elected as the leader.
- $A(1) - o(1)$ number of processors are aware of their leader.
- Each of the good processors can handle only a fixed number of bits which is a polylogarithmic function in n.

Thus in this paper, we have worked upon the implementation of a leader election algorithm by modeling it using Colored Petri Nets. The algorithm processes a polylogarithmic number of bits and is thus scalable for large-scale networks like peer-to-peer networks. More work can be done in further reducing the number of rounds in this algorithm. Furthermore, research can be done to analyze if the number of good processors required to elect a leader can be reduced to any more extent.

References

1. Burman, J., Doty, D., Nowak, T., Severson, E.E., Xu, C.: Efficient self-stabilizing leader election in population protocols. CoRR, vol. abs/1907.06068 (2019). http://arxiv.org/abs/1907.06068

2. Gasieniec, L., Stachowiak, G.: Fast space optimal leader election in population protocols. CoRR, vol. abs/1704.07649 (2017). http://arxiv.org/abs/1704.07649

3. Berenbrink, P., Kaaser, D., Kling, P., Otterbach, L.: Simple and efficient leader election. In: Seidel, R. (ed.) 1st Symposium on Simplicity in Algorithms (SOSA 2018), ser. OpenAccess Series in Informatics (OASIcs), vol. 61, pp. 9:1–9:11. Schloss Dagstuhl–Leibniz-Zentrum fuer Informatik, Dagstuhl, Germany (2018). http://drops.dagstuhl.de/opus/volltexte/2018/8302

4. Kim, T.W., Kim, E.H., Kim, J.K., Kim, T.Y.: A leader election algorithm in a distributed computing system. In: Proceedings of the Fifth IEEE Computer Society Workshop on Future Trends of Distributed Computing Systems, pp. 481–485 (1995)

5. Singh, G.: Leader election in complete networks. In: Proceedings of the Eleventh Annual ACM Symposium on Principles of Distributed Computing, ser. PODC 1992, pp. 179–190. Association for Computing Machinery, New York (1992). https://doi.org/10.1145/135419.135457

6. Chow, Y., Luo, K.C.K., Newman-Wolfe, R.: An optimal distributed algorithm for failure-driven leader election in bounded-degree networks. In: Proceedings of the Third Workshop on Future Trends of Distributed Computing Systems, pp. 136–141 (1992)

7. Luling, R., Monien, B., Ramme, F.: Load balancing in large networks: a comparative study. In: Proceedings of the Third IEEE Symposium on Parallel and Distributed Processing, pp. 686–689 (1991)

8. Sanchez-Rodriguez, D., Macías, E., Suarez, A.: Effective load balancing on a lan-wlan cluster, pp. 473–479 (2003)

9. Haddad, E.: Dynamic optimization of load distribution in heterogeneous systems. In: Proceedings Heterogeneous Computing Workshop, pp. 29–34 (1994)

10. Ingram, R., Shields, P., Walter, J.E., Welch, J.L.: An asynchronous leader election algorithm for dynamic networks. In: IEEE International Symposium on Parallel & Distributed Processing, pp. 1–12. IEEE (2009)

11. Schiper, N., Toueg, S.: A robust and lightweight stable leader election service for dynamic systems. In: 2008 IEEE International Conference on Dependable Systems and Networks With FTCS and DCC (DSN), pp. 207–216. IEEE (2008)

12. Vasudevan, S., Kurose, J., Towsley, D.: Design and analysis of a leader election algorithm for mobile ad hoc networks. In: Proceedings of the 12th IEEE International Conference on Network Protocols, 2004. ICNP 2004, pp. 350–360. IEEE (2004)

13. Ben-Or, M., Linial, N.: Collective coin flipping, robust voting schemes and minima of banzhaf values. In: 26th Annual Symposium on Foundations of Computer Science (SFCS 1985), pp. 408–416 (1985)

14. Linial, N., Saks, M.: Collective coin flipping and other models of imperfect randomness (1988)

15. Kahn, J., Kalai, G., Linial, N.: The influence of variables on boolean functions. In: Proceedings of the 29th Annual Symposium on Foundations of Computer Science, ser. SFCS 1988, pp. 68–80. IEEE Computer Society, USA (1988). https://doi.org/10.1109/SFCS.1988.21923

16. Saks, M.: A robust noncrytographic protocol for collective coin flipping. SIAM J. Discret. Math. **2**(2), 240–244 (1989). https://doi.org/10.1137/0402020
17. Ajtai, M., Linial, N.: The influence of large coalitions. Combinatorica **13**, 129–145 (1993)
18. Alon, N., Naor, M.: Coin-flipping games immune against linear-sized coalitions. In: Proceedings of the 31st Annual Symposium on Foundations of Computer Science, ser. SFCS 1990, pp. 46–54. IEEE Computer Society, USA (1990). https://doi.org/10.1109/FSCS.1990.89523
19. Dolev, D., Fischer, M., Fowler, R., Lynch, N., Strong, H.: An efficient algorithm for byzantine agreement without authentication. Inf. Control **52**, 257–274 (1982)
20. Feige, U.: Noncryptographic selection protocols. In: Proceedings of 40th IEEE Foundations of Computer Science (FOCS) (1999)
21. Ostrovsky, R., Rajagopalan, S., Vazirani, U.: Simple and efficient leader election in the full information model. In: Proceedings of the Twenty-Sixth Annual ACM Symposium on Theory of Computing, ser. STOC 1994, pp. 234–242. Association for Computing Machinery, New York (1994). https://doi.org/10.1145/195058.195141
22. Bracha, G.: An o(log n) expected rounds randomized byzantine generals protocol. J. ACM **34**, 910–920 (1987)
23. King, V., Saia, J., Sanwalani, V., Vee, E.: Scalable leader election, pp. 990–999 (2006)

Formal Methods to Analyze Energy Efficiency and Security for IoT: A Systematic Review

Luciana Pereira Oliveira[1]([⊠]), Arley Willer Neves da Silva[2],
Lucas Pereira de Azevedo[2], and Maria Vitória Lima da Silva[2]

[1] IFPB Campus João Pessoa,
Av. Primeiro de Maio, 720 - Jaguaribe, João Pessoa, PB, Brazil
`luciana.oliveira@ifpb.edu.br`
[2] IFPB Campus Itaporanga,
R. Projetada, SN - Vila Moco, Itaporanga, PB, Brazil
`arley.silva@ifpb.edu.br`,
`{lucas.azevedo,maria.silva.7}@academico.ifpb.edu.br`

Abstract. Typically, third parties must not capture the data transmitted in IoT (Internet of Things) constituted of the devices with limited capacities and different configurations. Hence, proposals for safety communications are analyzed to identify if it consumes lower energy through formal methods, simulations and measurements. However, the two last limit the evaluation with a finite set of scenarios. For this reason, this study used a systematic methodology to investigate formal methods used by articles in the context of the low-power security solution for IoT. The results correspond to a diversity of mathematics tools, research topics, analytical strategies, and other information to motivate the use of formal methods to generalize the evaluation of new research with an unambiguous description in all potential scenarios.

1 Introduction

The sensor's characteristics in WSN (Wireless Sensor Networks) and Internet of Things (IoT) have similarities and require security mechanisms with energy constraint. The data captured and analyzed by devices are considered as critical for several WSN and IoT scenarios that should contain low power and secure mechanism to protect the network and devices against attacks [6]. For example, the sensor should be authenticated before to start the transmit the information that should not be understood and intercepted by third parties [3]. Therefore, it is important to go beyond simulations and measurements that evaluate solutions considering a finite set of scenarios. For this, the formal methods provide mechanisms to generalize the study with precise and unambiguous description in all potential scenarios.

This work found 370 secondary technical papers (surveys and reviews) and only 49 contained information about the IoT, energy and security. However, none

L. Barolli et al. (Eds.): AINA 2021, LNNS 227, pp. 270–279, 2021.
https://doi.org/10.1007/978-3-030-75078-7_28

of the secondary work have been focused to comprehend analytical researches used with mathematical conceptual to assume various conditions and to provide a generic results. Although it is possible to find 474 papers that contain results from simulations and 110 papers that present results of the evaluations based in real experiments, only 72 works described the results based on mathematical analysis. Moreover almost 100% are paper with mathematical notation were published in journals classified at higher index in Scopus[1].

Therefore, the purpose of this research based on Systematic Literature Review (SLR) method is to comprehend the diversity of formal methods to evaluate security mechanisms with low energy consumption for IoT (including WSN). The main commitments of this study are highlighted as follows:

- Presenting information extracted from papers that used the proof methods or model checking to analize energy, security and sensors (IoT or WSN) in analytical approach.
- Designing recommendations to be addressing future research with energy, IoT, security and formal methods.

The organization of this research based on systematic review is considered as follows. Section 2 presents the background about formal methods. The Sect. 3 describes related works. The research methodology is clarified in Sect. 4. The discussion of the papers selected by SRL is described in Sect. 5, including the recommendations to be addressing future research with energy, iot, security and analytical approach. Finally, the conclusion and future works are in Sect. 6.

2   Formal Methods

The formal methods allow the description of the properties (requirements) of a solution (system, algorithms, process and procedures) in a generic and abstract way at a high level with the use of mathematical concepts. It is also possible to describe the desired properties of the solution. In this context, this study found that researchers use mathematical definitions to obtain results through the following modalities:

a) Simulation - the research defines preliminaries information (function, expression and other mathematical notation) that will be implemented and used on simulation environment where there are restrictions for limited inputs. Some studies implements the simulator and others use known simulator, for example network simulator like NS3, OMNet ++, NS2, OpeNet and others.

B) Model checking - in first step, a formal description (known as formal specification) is specified with a precise mathematical model (generally modeled by automata [1] - the set of all possible states and transitions). In second step, the formal properties (states required in temporal logic, for example) is executed by checker model that automatically determines whether the property satisfies

---

[1] Scopus - https://www.scopus.com/.

or not. A variant of this is the probabilistic model checking that allows the formal automatic verification of the system involving stochastic behavior like as Markov [2].

C) Manual proof - Usually, it contains a logic sequence manually described by three blocks (preliminaries, problem statement and proof development) to deduce a conclusion [4]. The preliminaries contains the axioms (definitions that are considered necessarily evident and true), vocabulary, mathematical notation. The problem statement is the goal description with one o more following works - theorem (a main result that will be proved), lemma (used by complex proof as an intermediate result), proposition (theorem or lemma), conjecture (property that remains unproven) and corollary (a conclusion of a previous theorem or lemma)). The proof development is the sequence of logical deductions [4].

There are several types of the proof development. The intersection of strategies to design proof from authors [4] and [5] are five strategies that are described in following (direct, contrapositive, contradiction, counterexample and induction). For example, [4] describe the proof by cases that was not presented by [5]. The proof by cases is usually used by simulation when the preliminary block has not the enough information and a set of finite cases is possible to be used with additional information to reach the conclusion, for example, $(PVQ) \rightarrow R \equiv (P \rightarrow R) \wedge (Q \rightarrow R)$.

The direct proof is when each step of the strategy is an axiom or conclusion of previous step, so the provide enough information to prove the result directly. Each step, the preliminary block of the proof provide enough data to execute the proof directly, for example, $P \rightarrow C$.

The contrapositive proof is when the negation of preliminary block results a negation of the conclusion, for example, $P \rightarrow C \equiv \neg C \rightarrow \neg P$.

The proof by contradiction is used when it is hard to prove preliminary block as true, e.g. about infinity. Initially to assume $P \rightarrow C$ is false, for example the proof will demonstrate that $\neg C$ is false in order to find the conclusion (C) as true and by contradiction $P \rightarrow C \equiv true$.

The counterexample proof is used to find an example where the sentence is not true. So, this can be used to disprove the property and the lack of counterexamples will keep sentence as a conjecture.

The induction proof is commonly used when there is a recursive function, a ordered set and a universal property over an infinite set. This is constituted of two main steps: base and induction. For example, in base step, $P \rightarrow C$ is valid for element k in a set S and, in inductive hypothesis, $P \rightarrow C$ for some element n in S. The proof will demonstrate that $P \rightarrow C$ for the element n + 1 in S and to conclude that $P \rightarrow C$ for all elements greater than or equal to k in S. Another example, is to prove by induction using Nash equilibrium where the same function will repeat for each of the players and if to find a moment when all player keep the fixed strategy and no player has a better strategy to play than his current set of the strategy, then the end of the game was identified.

3   Related Work

The research carried out by [6] was the first and unique study, using the SLR method, that reviews the experimental evaluation considering security and energy in the IoT. They identified 40 secondary technical papers (surveys or reviews) between 2011 and 2019 containing information about the IoT, energy, and security.

Instead of covering all 40 articles, they chose the 8 most relevant papers and compared the articles as to the type of review, year of the article, paper selection process, the evaluation through practical experiments, main topic, range of the year analyzed, the intensity of the terms security and energy addressed.

They did not identify any systematic reviews considering IoT, energy, and security and the range of the year analyzed since 2003 and completed in 2018.

The authors from [6] worked with experimental measurement. This paper extends [6] to extract other information in terms of evaluations: formal method which is another technique for analyzing new security solutions.

Additional information is that it was possible to find [9] as only a secondary work that partially addressed the subject of formal methods. This work in [9] analyzed papers about security and energy in data aggregation, considering articles between 2006 and 2011. So the paper analyzed did not contain information about IoT explicitly, because the database from [6] has papers from 2004 to 2014 that described only about WSN and, from 2015 to 2019, it was possible to identify papers with explicit IoT.

Therefore, this paper is the first SLR study in terms of the formal methods to evaluate security and energy in the IoT (including WSN). According to the existing secondary papers, the existing deficiencies propose that is important to do a comprehensive literature review to address these weaknesses as follows:

- The present studies do not provide an in-depth study of formal methods used by papers containing three areas (security, IoT, and energy).
- Only one study evaluated the association among three areas.
- The structure of the presented studies usually does not have a systematic review.

4   Methodology

A systematic review is a reproducibility type of study, because it is conducted through a set of well-defined steps to select and analyze the most relevant studies. This work followed the same method executed by [6] that used three steps: data planning with research questions; execution; and analyzes that generate the results of the graphics, tables, and descriptions.

4.1   Research Questions (RQs)

In order to understand the evaluations of studies that merger energy, security, and IoT, this SLR paper formulated 5 questions: i) what are the publication trends of selected papers?; ii) which formal methods are in the selected studies?; iii) what are the theoretical models used and are there some reused models?; iv) how are energy constraints in theoretical models?; and v) what recommendations for future work that will conduct formal analysis in the field of IoT, energy, and security?

4.2   Search Process

This study analyzed the database from [6] with 5671 papers extracted from the following electronic databases: IEEE Xplore, ACM digital library, and Science Direct. They collected articles using variations of the words IoT, energy, and security that were described in [6].

After obtaining the database, this SLR applied inclusion (IC) and exclusion (EC) criteria to identify a viable number of papers to analyze:

- EC1: Works not written in English.
- EC2: Papers does not contain security words (security or secure or malicious or crypt or shared key or confidential or attack) and energy (energy or power consumption) in the title or the abstract.
- EC3: Secondary studies (abstract or title contains the word review or survey).
- EC4: Duplicate articles.
- EC5: Secondary work identified after reading the full article.
- EC6: Incomplete work, because the paper has no value related to formal methods.
- EC7: Papers without analytical data in terms of energy consumption and safety.
- EC8: Low-quality papers (indexed below 50 by Scopus)
- IC1: Papers that were not excluded by EC1, EC2, EC3, and EC4
- IC2: Papers did not exclude by any of the exclusion criteria.

4.3   Studies Selection

Each paper was analyzed by two stages in the execution phase. In Stage 1, title and summary of all articles were confronted with EC1, EC2, EC3, EC4, and EC8 to select them by IC1. In stage 2, all papers were read in order to exclude works by CE5, EC6, and EC7. The selected paper was classified by IC1 and IC2. As a result, a total of 5633 studies was excluded and 38 primary studies from the 3 electronic bases were selected and included to extract information related to RQs. All selected papers are in https://sourceforge.net/projects/aina2020/files/base_metodo_formais.xlsx.

5   Search Results

5.1   RQ1: What Are the Publication Trends of Selected Papers?

The studies found in this review are in a time period between 2006 and 2019. The Fig. 1 (A) shows the number of included papers by year of publication. From

2004 to 2014, the selected paper described only about WSN and, from 2015 to 2019, it was possible to identify a total of thirteen papers with explicit IoT. This indicates that the area has become more mature from 2015.

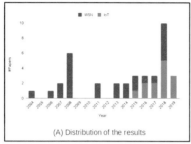
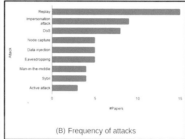

(A) Distribution of the results   (B) Frequency of attacks

Fig. 1. Selected papers

The quality of the papers were obtained by Scopus that provides an indication of the journal impact using numerical value. Two papers were removed the SLR, because had quality less than 25 in Scopus index and 90% of the papers has high index by scopus. This SRL found 9 distinct conferences and 19 journals. The major number of publications were identified in Computer Networks (5 papers), Ad Hoc Networks (4 papers), Computer Communications (3 papers), Information Sciences (2 papers) and IEEE International Conference on Communications (2 papers).

A total of 24 distinct attacks were identified. The Fig. 1 (B) and the most frequency referenced by papers were Replay, Impersonation, and Denial-of-Service (DoS) attacks. It is an important highlight that some papers describe a solution for more than one kind of attack.

A total of the 3 papers did not present performance data, because they described manual proof or by model checking analysis. The energy was evaluated by papers using the explicit and implicit function or mathematical expression. A total of the 25 papers evaluated the energy in terms of consumption, overhead, or saving and 10 papers performed the energy cost implicitly by computation and communication cost. It is an important highlight that some papers evaluated the solution for more than one kind of performance analysis. For example, [7] evaluated the energy cost explicitly, the computational and communication cost. Figure 2 (A) presents the most frequently referenced by papers were general metrics energy cost, computational cost (including a number of computations, storage, time execution, memory) and communication cost (including bits transmitted, lifetime, reliability, packet size, delay, communication overhead bandwidth, and delivery ratio).

A total of 15 research topics were identified and 5 scenarios (smartgrid, smart homes, Underwater, and health-care) were referenced. Figure 2 (B) presents authentication scheme and key establishment as the most frequently research topics.

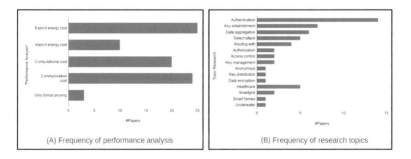

(A) Frequency of performance analysis

(B) Frequency of research topics

Fig. 2. Frequency

5.2 RQ2: Which Formal Methods Are in the Selected Studies?

This study found more than 474 articles with mathematical definitions to obtain results through simulation. Besides, another set of articles consisting of 38 elements was identified with formal methods, 19% with model and 81% with manual verification. The simulated articles were discarded from this SRL due to the large volume and the analytical modalities were analyzed in depth. A total of 38 selected papers was found: 34 used manual strategy, 8 using model checking and 4 used both.

Figure 3 (A) presents five kinds of manual proof: induction, contradiction, contrapositive, direct, and by cases. The last two strategies were more used. Four formal methods were identified: Nash Equilibrium, Bilinear Pairings (commonly in analysis tools of cryptographic protocols), BAN-logic (usually in security analysis), and the LP (used in the problem of maximizing or minimizing a linear function). Some papers used more than one strategy to execute manual proof.

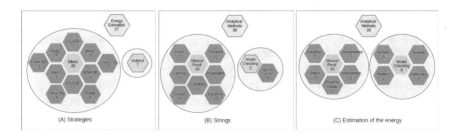

(A) Strategies

(B) Strings

(C) Estimation of the energy

Fig. 3. Strategies, Strings and Estimation of the energy in selected papers

The papers that used the model checking as an formal method selected AVISPA, scyther, or Markov (probabilistic model checking) as the tool and automata (with state and transitions). The Stochastic Automata Networks (SAN) was classified as a subtype of automata strategy, but SAN is also a derivation of Markov. Among them, the AVISPA was the most referenced.

5.3 RQ3: What Are the Theoretical Models Used and Are There Some Reused Models?

The manual proofs and model checking were analyzed in terms of the strings used to describe the problem statement. Figure 3 (B) shows the strings used by papers. The paper with Markov or Automata strategy used ST (states and transitions) to describe the problem. The studies like as [8] with manual and model checking did not choose ST to describe the article goal, because the problem was presented by a string from manual proof and analyzed by model checking. The proof, theorem, and equation were the most referenced strings in manual proof. Only six papers used complex proof containing lemma string as an intermediate result of the proof.

Several articles contained the following theoretical models: network, security, attack, and energy. A total of 14 papers referenced the attack model (been nine using adversary as a subtype), four works associated an energy model, and 14 other distinct models presented by studies were the following: automata, threat, LP, oracle, heuristic, mathematical programming, framework, energy, system, communication, delay, framework, traffic, and game-theoretic model. Moreover, as can be seen in Fig. 3 (B), 12 papers reused model defined by other authors. The Dolev-Yao Attack model was reused by 5 distinct papers.

5.4 RQ4: How Are Energy Constraints in Theoretical Models?

Although only six paper contains the theoretical energy model, several papers analyzed the energy consumption by estimation as presented in Fig. 3 (C). The articles about energy constraints used estimation of the energy directly by metric in Joules or Watts and indirectly by the computation or computation cost. All articles that analyzed energy directly used data published by other documents or articles. However, the majorly of the papers contain energy data from MICA2 device. In another hand, the papers with energy estimation indirectly used data from their paper. As highlight some papers described solutions for more than one kind of device and some papers described only chip radio (CC2420 or CC2520), others informed only the CPU or microcontroller (ATMEGA128L, Intel i7, MSP430 or 32-bit Cortex-M3). For example, Mica2 has ATmega128L and the CC1000.

5.5 RQ5: What Recommendations for Future Work that Will Conduct Formal Analysis in the Field of IoT, Energy, and Security?

In terms of gaps, future researches should propose new energy models that can be used by formal analysis of the security in IoT, because this SRL found only one energy model and model checking is the formal method with a low number of studies. This SLR found 21 attacks type, existing some gaps in this context. For example, Mirai (a malware to remotely controlled IoT devices like IP cameras) exemplified one attack that was not studied by formal method.

A higher rate of acceptance of articles with formal methods in journals was identified. There may be a low rate for sending conferences.

This SRL recommends manual proof as formal analysis, induction, or direct proof as a strategy to demonstrate the problem, reusing Dolev-Yao as an attack model and estimating energy by reference that contains the energy consumption of the MICA2. This recommendation is because new studies to be easily compared to existing results from older works. On the other hand, if model checking is chosen as a formal method, then the AVISPA tool should be used because it was more referenced. Moreover, energy cost, computational cost, and communication cost should be chosen to execute the performance analysis.

New papers should use at least network, attack, and security as analytical models to describe the problem. Besides, when new analytical studies consider reusing recommendation models by this SRL, then they will provide information in a way that will be more easily compared with other articles that used the same model.

6   Conclusion

The presented results of this SLR aimed at investigating the characteristics of formal methods in context of IoT, energy and security. This review has helped to understand the current trends and state-of-the-art with mathematical methodologies and to identify research gaps and future directions. After analyzing each selected paper, they are classified according to their aims and the way they work. Thus, this SLR studied two formal methods: (i) model checking and (ii) manual proof. Also, it has identified the main theoretical models, tools, and performance analysis used.

The results of this SLR suggest that there is a predominance of formal methods that manual proof with direct or induction strategies, using the strings proof and theorem to describe the problems. However, if the researchers choose model checking as an formal method, they should use AVISPA as a tool, because it was more selected among work with model checking.

This SLR encourages researchers to continually reuse existing mathematical models like Dolev-Yao that was an attack model used by three papers, to choose data from MICA2 device to estimate energy. Moreover, it was possible to find gaps in the energy model and attacks analyzed by existing papers.

Therefore, the results achieved in this study can be used as a guide for researchers to identify which strategies, attacks, theoretical models, tools, and energy estimation best fit for new articles can be more broadly compared to existing publications.

Acknowledgment. The authors would like to thank the Federal Institute of Paraíba(IFPB)/Campus João Pessoa for financially supporting the presentation of this research and, especially thank you, to the IFPB Interconnect Notice - No. 01/2020.

References

1. Almeida, J.B., Frade, M.J., Pinto, J.S., Melo de Sousa, S.: An overview of formal methods tools and techniques. In: Rigorous Software Development. Undergraduate Topics in Computer Science. Springer, London (2011). https://doi.org/10.1007/978-0-85729-018-2_2
2. Hermanns, H., Katoen, J., Meyer-Kayser, J., et al.: A tool for model-checking Markov chains. STTT **4**, 153–172 (2003). https://doi.org/10.1007/s100090100072
3. Chavan, A.A., Nighot, M.K.: Secure and cost-effective application layer protocol with authentication interoperability for IoT. In: International Conference on Information Security, Privacy (ICISP 2015), 11–12 December 2015, Nagpur, India (2015)
4. Jimenez, D.M., Viladrosa, R.C., Sanchez, M.A.H.: Formal Proof: Understanding, writing and evaluating proofs. http://openaccess.uoc.edu/proofs.pdf
5. Hopcroft, J.E., Ullman, J.D., Motwani, R.: Introdução à Teoria de Autômatos, Linguagens e Computação. Translation 2a. edition. Editora Campus (2001)
6. Oliveira, L.P., Vieira, M.N., Leite, G.B., de Almeida, E.L.V.: Evaluating energy efficiency and security for internet of things: a systematic review. In: Barolli, L., Amato, F., Moscato, F., Enokido, T., Takizawa, M. (eds.) Advanced Information Networking and Applications. AINA 2020. Advances in Intelligent Systems and Computing, vol. 1151. Springer, Cham (2020). https://doi.org/10.1007/978-3-030-44041-1_20
7. Boudia, O.R.M., Senouci, S.M., Feham, M.: A novel secure aggregation scheme for wireless sensor networks using stateful public key cryptography. Ad Hoc Netw. **32**, 98–113 (2015). https://doi.org/10.1016/j.adhoc.2015.01.002
8. Sani, A.S., Yuan, D., Jin, J., Gao, L., Yu, S., Dong, Z.Y.: Cyber security framework for internet of things-based energy internet. Future Gener. Comput. Syst. **93**, 849–859 (2018). https://doi.org/10.1016/j.future.2018.01.029
9. Dhasian, H.R., Balasubramanian, P.: Survey of data aggregation techniques using soft computing in wireless sensor networks. IET Inf. Secur. **7**(4), 336–342 (2013). https://doi.org/10.1049/iet-ifs.2012.0292

Multi-focus Image Fusion Algorithm Based on Grey Relation of Similarity in Contourlet Domain

Lili Jiao[✉], Jie Deng, Lili Zhang, and Qiutong Lin

Police Officer College of the Chinese People's Armed Police Force, Chengdu 610213, China

Abstract. Aiming at solving the problem of multi-focus image fusion, this paper proposes a visible-light images fusion method which is based on grey relation of similarity in Contourlet domain. Firstly, the source images are decomposed by multi-scale Contourlet transform, then the high-frequency coefficients of each fused image are divided into small blocks. The different fusion strategies of high-frequency coefficients are formulated based on the similarities of these sub-blocks which are determined by the Grey Euclid Relational Degrees of grey theory, while the low-frequency coefficients are fused by arithmetic mean method; Finally, the fused image is reconstructed by the inverse Contourlet transform. The experimental results show that the proposed method is superior to pyramid-based methods and wavelet-transform-based methods in terms of information entropy, standard deviation and definition.

1 Introduction

Image fusion is an important part of image processing, which plays an important role in computer vision, target recognition, remote sensing and medical image processing. Multi-focus image fusion refers to the organic combination of image information from different focus areas to obtain more abundant, accurate and reliable information for the same scene. The fusion result is more in line with the visual characteristics of human or machine, which is convenient for subsequent processing [1].

At present, multi-focus image fusion methods are mainly based on spatial domain and transform domain. Spatial domain fusion methods mainly include pixel based method and block based method. The advantages of these methods which are easy to implement, but the fusion scheme is more complex [2]. In the transform domain, wavelet transform is widely used in image fusion due to its good time-frequency local analysis characteristics [3, 4]. However, wavelet transform has only horizontal, vertical and diagonal information, which is isotropic and mainly reflects the location and characteristics of singular points. However, it is difficult to express the singularity of lines and planes such as edges and linear features in two-dimensional images [5]. Curvelet transform [6] and Contourlet transform [7] proposed in recent years are more suitable for depicting curves or lines in two-dimensional images than wavelet transform due to their anisotropy and other characteristics, and are rapidly introduced into visible image fusion [8, 9].

When observing an image, the human eye usually pays less attention to the gray value of pixels in the image, and pays more attention to the similarity of edge, contour, region and other features between the image to be evaluated and the reference image, which is a process similar to comparison and classification [8]. With the deepening of human research, structural similarity, regional similarity and other factors have become an important index to objectively evaluate image quality and fusion effect [10–12]. Therefore, in the process of designing the image fusion method, we should make full use of and retain the similarity information in the image to ensure that the fused image is more suitable for human observation and machine analysis. In this paper, the Contourlet transform is used as a multi-scale analysis tool to detect the similarity between the source images from different scales. Based on this, this paper is to discuss the fusion technology in multi-focus visible images.

2 Grey Relational Analysis of Contourlet Coefficient

2.1 Grey Relational Analysis

Grey theory, put forward by Professor Deng Julong, is a new theory on the problems of small sample, poor information and uncertainty. It has a relatively perfect method system, among which grey relational analysis is an important basis of data analysis system. The basic idea of grey relational analysis is to analyze and compare the geometric shapes of the corresponding curves of the data series. The closer the geometric shapes are, the greater the correlation degree is; conversely, the smaller the correlation degree is [13].

The unique advantage of grey relational analysis technology in processing small sample data is also applicable to the data sequence formed by small neighborhood pixels in the image, and has produced a lot of research results in image fusion, image denoising and other fields. For example, the edge points and non-edge points of panchromatic image are calculated by grey relational analysis and fuzzy reasoning, and then the multi-spectral image is transformed by IHS which based on the brightness component carried out the histogram matching of panchromatic image. Finally, the fused image is obtained by inverse IHS transform [14].

Differently, we use Grey Euclid Relational Degree to calculate the similarity between the Contourlet domain corresponding to each scale of high-frequency coefficient sub-graphs of the image to be fused, and formulate fusion strategies from different levels with the combination of regional energy. The main steps of calculating Grey Euclid Relational Degree are as follows [13]:

1. Abstracting reference sequence X_0 and comparison sequence X_j from grey system. Where:

$$x_0 = \{x_0(k)|k = 1, 2, \cdots, N \in \text{integer}\}$$

$$x_j = \{x_j(k)|j = 1, 2, \cdots, I \in \text{integer}, k = 1, 2, \cdots, N \in \text{integer}\}$$

2. The grey correlation coefficient ξ_{0j} between the reference series x_0 and the comparison series x_j is calculated according to the following formula (1).

$$\xi_{0j}(k) = (\Delta \min + \zeta \Delta \max)/(\Delta 0j(k) + \zeta \Delta \max). \tag{1}$$

Where: $\Delta \min = \min_{\forall j,k} |x_0(k) - x_j(k)|$; $\Delta \max = \max_{\forall j,k} |x_0(k) - x_j(k)|$;

$$\Delta 0j(k) = |x_0(k) - x_j(k)| \cdot j = 1, 2, \cdots, I; k = 1, 2 \cdots, N$$

The resolution coefficient is a constant ζ ($\zeta = 0.5$), $\Delta \max$ and $\Delta \min$ represent the maximum absolute difference and the minimum absolute difference between the reference sequences and comparison sequences, respectively, $\Delta 0j(k)$ represents the absolute difference between $x_0(k)$ and $x_j(k)$.

3. Computing the Grey Euclidean Integral Relevance Degree R_{0j} between the reference sequence X_0 and the comparison sequence X_j.

$$R_{0j} = 1 - (1/\sqrt{N}) \times \left[N(\overline{R}_{0j} - 1)^2 + \sum_{k=1}^{N} \varepsilon_{0j}^2(k) \right]^{1/2}. \tag{2}$$

Where: $\overline{R}_{0j} = (1/N) \times \sum_{k=1}^{N} \xi_{0j}(k)$; $\varepsilon_{0j}(k) = \xi_{0j}(k) - \overline{R}_{0j}$; $\sum_{k=1}^{N} \varepsilon_{0j}(k) = 0$.

Obviously, according to formula (2), the Grey Euclid Relational Degree takes account of not only the influence of the average correlation coefficient $\overline{R}_{0j}$ of reference sequence and comparison sequence at each point on the correlation degree, but also the influence of the fluctuation value $\varepsilon_{0j}(k)$ of the correlation coefficient at each point on the correlation degree. When $\overline{R}_{0j}$ is constant, if $\varepsilon_{0j}(k)$ increases, the correlation degree decreases; when $\varepsilon_{0j}(k)$ is constant, if $\overline{R}_{0j}$ increases, then the correlation degree increases. This effectively avoids the problem of inaccurate calculation of Deng's correlation degree when dealing with large local fluctuation data.

2.2 Analyzing the Similarity Between Contourlet Domains of the Fused Source Images Through the Grey Euclid Relational Degree

Classically, the arithmetic average method is suitable for fusion in multi-focus visible image fusion because the average brightness of each source image is basically the same, so the low-frequency coefficients in the transform domain are close. However, the fusion strategy of high-frequency coefficients with different clarity or fuzziness in the same region of different source images is the key factor to determine the quality of the fused image. Assuming that the visible images to be fused are A and B, the high-frequency Contourlet coefficients of different scales and directions are obtained by multi-level Contourlet decomposition, which contain the main feature information such as image details, texture and structure. The main steps to analyze the similarity between the high frequency coefficient subgraphs at the corresponding scale and from the corresponding direction of the two source images by using Grey Euclid Relational Degree are given below.

1. The size of the high-frequency coefficient subgraphs at the corresponding scale and from the corresponding direction obtained by Contourlet transform of source images

A and *B* are the same. Therefore, the high-frequency coefficient matrixes at each scale and from each direction can be one-dimensionalized, and the one-dimensional data sequences *FA* and *FB* of *i*-scale *j*-direction subgraphs can be obtained.

2. *FA* and *FB* are divided into *n* subsequences of equal length, denoted as *FA*(*t*) and *FB*(*t*) ($t = 1, 2, ...$); Obviously, *n* here is the average number of segments of one-dimensional data sequence of *i*-scale *j*-direction subgraph, let $n = 4$.

3. The mean and standard deviation of *n* subsequences *FA*(*t*) were calculated: $\mu_{FA(1)}$, $\mu_{FA(2)}$, ...,$\mu_{FA(n)}$ and $\sigma_{FA(1)}$, $\sigma_{FA(2)}$, ...,$\sigma_{FA(n)}$. Similarly, the mean and standard deviation of *n* subsequences *FB*(*t*) are obtained as follows: $\mu_{FB(1)}$, $\mu_{FB(2)}$, ..., $\mu_{FB(n)}$ and $\sigma_{FB(1)}$, $\sigma_{FB(2)}$, ..., $\sigma_{FB(n)}$.

4. With $\{\mu_{FA(1)}, \mu_{FA(2)}, ..., \mu_{FA(n)}, \sigma_{FA(1)}, \sigma_{FA(2)}, ..., \sigma_{FA(n)}\}$ is the reference sequence of *i*-scale *j*-direction subgraph features in image *A*, $\{\mu_{FB(1)}, \mu_{FB(2)}, ..., \mu_{FB(n)}, \sigma_{FB(1)}, \sigma_{FB(2)}, ..., \sigma_{FB(n)}\}$ is the comparison sequence of *i*-scale *j*-direction subgraphs in image *B*.

5. According to (1)–(2), the similarity $R_{AB}^{i,j}$ between *A* and *B* in *i*-scale *j*-direction subgraphs is calculated.

6. Let $\theta \in (0, 1]$ be the threshold of correlation degree. When $R_{AB}^{i,J} \geq \theta$, it is considered that the subgraphs of *i*-scale *j*-direction of *A* and *B* are very similar; otherwise, there are differences. Next, according to the similarity and local energy difference, we choose different high-frequency coefficient fusion strategies to effectively improve the overall detail clarity of the fused image. The new image fusion method in Contourlet domain is to be presented as follows.

3 Multi-focus Image Fusion Algorithm in Contourlet Domain Based on the Grey Relation of Similarity

Basic thought of this paper is to use Contourlet transform to decompose multi-focus source images *A* and *B* to obtain low-frequency coefficients and high-frequency coefficients. Then, the low-frequency coefficients are taken as the arithmetic average value. The high-frequency coefficients are determined by the similarity of Grey Euclid Relational Degree and local energy. Finally, the fused image is obtained by inverse Contourlet transform.

The main flow chart is shown in Fig. 1.

The main steps are as follows:

Step 1 Source images *A* and *B* are decomposed by Contourlet domain to obtain low-frequency coefficients in Contourlet domain and high-frequency coefficients at all scales and from all directions.

Step 2 Fuse the low-frequency coefficients of Contoulet domain. The specific fusion strategy is to obtain the fused low-frequency coefficients by arithmetic average method for the low-frequency coefficients of *A* and *B*.

Step 3 Fuse the high frequency coefficients of Contoulet domain. The specific fusion strategy can be divided into the following steps:

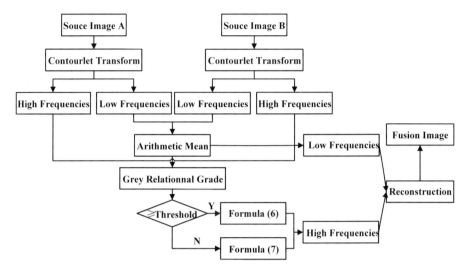

Fig. 1. Schematic diagram of multi-focus image fusion algorithm in Contourlet domain based on grey relation of similarity

① The high frequency coefficients at i scale and from j direction in Contourlet domain are divided into 4×4 blocks, and N_B blocks are obtained. The local energy of k block is obtained by the following formula:

$$E_A^{i,j,k} = \sum_{x=1}^{4}\sum_{y=1}^{4}\left[I_A^{i,j,k}(x,y)\right]^2; E_B^{ij,k} = \sum_{x=1}^{4}\sum_{y=1}^{4}\left[I_B^{i,j,k}(x,y)\right]^2. \quad (3)$$

Where: $i = 1, 2, \ldots, m$ (m is the number of decomposition levels); $j = 1, 2, \ldots, 2^l$ (l is the decomposition series of each layer); and $I^{i,j,k}$ represents the high-frequency coefficient of Contourlet domain in the k-th block of i-scale j-direction subgraph.

② According to the local energy sum of the k block ($k \in [1, N_B]$) at i scale and from j direction, 4×4 high energy area coefficient block $MaxE^{i,j}$, and low energy region coefficient block $MinE^{i,j}$ are generated.

$$\underset{\forall x,y \in [1,4]}{Max\ E^{i,j,k}}(x,y) = \begin{cases} I_A^{i,j,k}(x,y), & if\ E_A^{i,j,k} > E_B^{i,j,k} \\ I_B^{i,j,k}(x,y), & if\ E_A^{i,j,k} \le E_B^{i,j,k} \end{cases}. \quad (4)$$

$$\underset{\forall x,y \in [1,4]}{Min\ E^{i,j,k}}(x,y) = \begin{cases} I_A^{i,j,k}(x,y), & if\ E_A^{i,j,k} < E_B^{i,j,k} \\ I_B^{i,j,k}(x,y), & if\ E_A^{i,j,k} \ge E_B^{i,j,k} \end{cases}. \quad (5)$$

Based on the results of N_B block processing, the high-energy matrix $MaxE^{i,j}$ and the low-energy high-frequency coefficient matrix $MinE^{i,j}$ with the same size as the original i-scale j-direction subgraphs are synthesized.

③ According to Sect. 2.1, Grey Euclid Relational Degree is used to compare the similarity between i-scale j-direction subgraphs, so as to determine the high-frequency coefficients of i-scale j-direction subgraphs after fusion.

$$I_F^{i,j} = R_{A,B}^{i,j} \times MaxE^{i,j} + \left(1 - R_{A,B}^{i,j}\right) \times MinE^{i,j}, \ \ if \ \ R_{A,B}^{ij} \geq \theta. \tag{6}$$

$$I_F^{i,j} = MaxE^{i,j} \times k_1 + MinE^{i,j} \times k_2, \ \ if \ \ R_{A,B}^{i,j} < \theta. \tag{7}$$

Step 4: Inverse the low and high frequency coefficients in Contourlet domain to obtain fused image.

It can be seen from the above steps that when the information of each scale subgraph of the source image is very similar, formula (6) not only emphasizes the detailed information in the high-energy region coefficient, but also takes account of the detailed information in the low-energy region coefficient; on the contrary, when the information between the scale subgraphs of the source image is greatly different, formula (7) improves the detailed information of the high-energy region and on the other hand increases the low-energy region characteristic information of the area. Therefore, this method not only synthesizes the high-frequency information of each focused image to a large extent, but also effectively improves the high-frequency coefficient. That is to say, on the basis of obtaining the clear details of different focused images, high-quality fusion images can be obtained.

4 Simulation Experiment

Using information entropy, standard deviation and definition as evaluation indexes, this method is compared with contrast pyramid method (DBD), ratio pyramid method (BL), Laplace pyramid method (LP), filter subtract decimate (FSD) pyramid method, Harr wavelet transform method and DBSS wavelet transform method. Taking the 512×512 test image clock image as an example, the fusion results are shown in Fig. 2, and the corresponding fusion indexes are shown in Table 1.

It can be seen from Fig. 2 that the fusion effect of pyramid transform, wavelet transform and the CG method in this paper takes advantage of multi-scale decomposition to propose detailed information from different focused images, and effectively retains the texture and edge information that the human eye is concerned about. The CG method' result on the information entropy, standard deviation and definition of the fused image are the largest, which reflects the advantage of Contourlet transform to represent the information of two-dimensional image showed on the Table 1. It also further shows that the proposed method integrates the detailed information of the clear area in the source image, effectively improves the image resolution, so we get a point that the CG method is superior to the previous fusion methods.

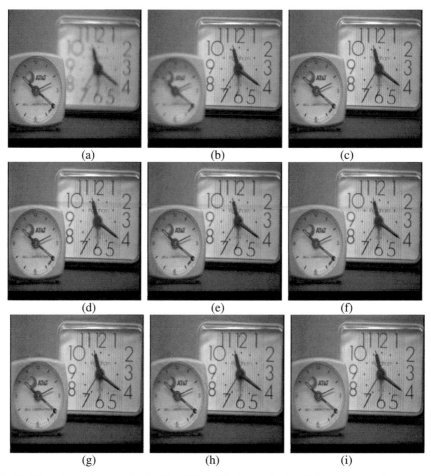

Fig. 2. Clock fusion images obtained by different fusion methods: (a) Source image A, (b) Source image B, (c) DBD method, (d) BL method, (e) LP method, (f) FSD method, (g) Harr method, (h) DBSS method, (i) CG method

Table 1. Quality comparison of different clock fusion images

Fusion method	Information entropy	Standard deviation	Definition
DBD	7.2985	51.9812	3.9904
BL	7.3367	51.1835	2.8242
LP	7.3613	51.8365	3.9940
FSD	7.3676	47.8822	3.0772
Haar	7.3464	50.5332	3.9243
DBSS	7.3670	51.0544	4.1000
CG	7.3756	52.3952	4.2060

In order to test the generality of the algorithm, this paper takes the registered 512 × 512 test image Pepsi image as an example, the three evaluation indexes of the fused image information: entropy, standard deviation and definition is showing on the Fig. 3.

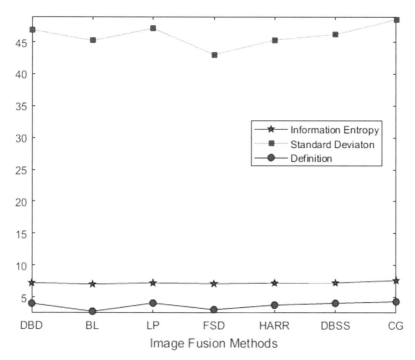

Fig. 3. The curves were the quality comparison of different Pepsi fused images. The larger the median value of the image, the better the effect of image fusion, that is to say, the fusion method which has the highest position in the same curve is the one to get the best effect of image fusion. It can be seen from Fig. 3 that this method is best.

5   Conclusion

In this paper, a multi-focus image fusion method in Contourlet domain based on similarity grey correlation is proposed by using Contourlet transform, grey relational analysis and regional energy information. In this method, the arithmetic average method is used to fuse the low-frequency coefficients in Contourlet domain, and the high-frequency coefficients fusion strategy is formulated by using the similarity of transformation domain subgraphs and local area energy based on grey relational analysis. The high-frequency coefficients with large local energy are strengthened when the similarity between fusion subgraphs is strong, as well as the high-frequency coefficients are enhanced when the similarity is weak. Finally, the fused image is reconstructed by inverse Contourlet transform. Experimental results show that this method is superior to other fusion methods such as wavelet transform in Multi-focus visible image fusion.

References

1. Wan, R., Ma, M.: A multi-focus image fusion method in curvelet domain. J. Shaanxi Normal Univ. (Nat. Sci. Ed.) **40**(5), 18–22 (2012)
2. Zhu, S., Qu, P.: Multi-focus image fusion based on wavelet transform and guided filtering. Meas. Control Technol. **39**(9), 103–107 (2020)
3. Wang, X., Niu, S., Zhang, K., et al.: Infrared dim target image fusion based on wavelet transform and feature extraction. J. Northwest. Polytechnical Univ. **38**(4), 723–731 (2020)
4. Wang, S.: Research on wavelet image fusion algorithm based on region energy and contrast. Ship Electron. Eng. **40**(5), 48–51 (2020)
5. Jiang, T., Zhou, J.: Research on image fusion based on combining curvelet transform. Informatiz. Res. **46**(3), 23–27, 45 (2020)
6. Candes, E.J, Demanet, L., Donoho, D., et al.: Fast discrete curvelet transform. Appl. Comput. Math. 1–43 (2005)
7. Do, M.N., Vetterli, M.: Contourlets: a new directional multiresolution image representation. In: Proceedings of IEEE International Conference on Image Processing, Rochester, NY, pp. 497–501. IEEE (2002)
8. Miao, A., Wan, R., Yin, Y.-L.: Multi-focus image fusion based on grey relation of similarity in curvelet domain. Acta Electronica Sin. **40**(10), 1984–1988 (2012)
9. Liu, K., Li, H.: Image fusion algorithm based on correlation among contourlet coefficients and PCA. Comput. Eng. Appl. **48**(16), 10–14, 62 (2012)
10. Luo, X., Wu, X.: An evaluation method of image fusion based on region similarity. Acta Electronica Sin. **38**(5), 1152–1155 (2010). (in Chinese)
11. Zhang, Y., Jin, W.: Image fusion assessment method based on structural similarity and region of interest. Acta Photonica Sin. **40**(2), 311–315 (2011). (in Chinese)
12. Zhang, L., Zeng, G., Wei, J., et al.: Multi-focus image fusion based on self-similarity. Comput. Eng. Des. **39**(9), 2805–2810 (2018)
13. Deng, J.: The Basic Method of Grey System. Huazhong University of Science and Technology Press, Wuhan (1996)
14. Zhang, H., Gu, Y., Guo, Q.: Image fusion based edge detection of grey relational analysis and fuzzy inference. Remote Sens. Inf. **35**(1), 15–27 (2020)

NEA: An SDN Switch Architecture Suitable for Application-Oriented MAC

Diego Nunes Molinos[1(✉)], Romerson Deiny Oliveira[2], Marcelo Silva Freitas[3], Marcelo Barros de Almeida[4], Pedro Frosi Rosa[1], and Flavio de Oliveira Silva[1]

[1] Faculty of Computing, Federal University of Uberlandia (UFU),
Joao Naves de Avila Avenue, 2121, Santa Monica, Uberlandia, Brazil
`{diego.molinos,pfrosi,flavio}@ufu.br`
[2] Department of Computer Science, Unimontes, Montes Claros, Brazil
`romerson.oliveira@unimontes.br`
[3] Department of Exact Sciences, Federal University of Jataí (UFJ), Jataí, Brazil
`msfreitas@ufg.br`
[4] Electrical Engineering Faculty (FEELT), Federal University of Uberlandia (UFU),
Joao Naves de Avila Avenue, 2121, Santa Monica, Uberlandia, Brazil
`marcelo.barros@ufu.br`

Abstract. After several decades most of the initial Internet design remains the same. The most important changes just happened on the application and physical layers while network and transport layers remain the same. Internet architecture has been turning into a limiting factor for its own evolution, and currently, several researchers have been working on new approaches that aim at redesigning the architecture. Software-Defined Networking (SDN) has been standing out for delivering more flexibility to the network. Based on the SDN paradigm, the Title Model represents a revolutionary approach to reaching out to the applications' new requirements. The Entity Title Architecture (ETArch) is the materialization of this model, offering improvements concerning the network's addressing and routing aspects. The programmability at the data link level is essential for the efficient use of network resources, and the ETArch is not restricted only to the software elements of the architecture. The network element responsible for the physical interconnection inside the network must have the capacity to guarantee Quality of Service (QoS) requirements driven by the applications, providing support to the communication space.

1 Introduction

Several types of communication have been making massive use of the Internet. The applications are demanding more and more resources from the network. Although those applications require mobility, security, energy efficiency, bandwidth, and Quality of Service (QoS), they usually rely on the network to satisfy their requirements. Software-Defined Networking (SDN) offers better and efficient control over networking devices from a centralized controller to improve network management.

L. Barolli et al. (Eds.): AINA 2021, LNNS 227, pp. 289–297, 2021.
https://doi.org/10.1007/978-3-030-75078-7_30

Due to the decoupling of the data plane and control plane, as expected in SDN, a distance from the functions implemented in hardware and software is created, making the hardware's behavior a single granular network fabric [4]. This contributed so that all flexibility and programmability were carried over to the network control level, practically unexplored at the data plane level. The fact is that relying entirely on the software layer to provide flexibility and programmability can result in loss of quality of service (QoS).

The Entity Title Architecture (ETArch) is based on the SDN paradigm, which aims to offer an architecture with features and support to context-oriented algorithms that allow changes and adjustments according to the network requirements [2,3].

ETArch enables requirements defined by the applications to permeate through the architecture layers down to the data plane level. At this level, the networking device must comprehend those requirements and modify their operation's context to guarantee, for example, mobility [3].

It is important to note that several proposes offer flexibility and programmability at the lower level of the network. Some of them aim to explore the parallel processing feature from reconfigurable platforms like Field Programmable Gate Array (FPGA) and the capacity of dynamic reconfiguration to enable programmability at the Link level [5].

The fact is, there is no easy way to develop a programmable solution using reconfigurability. Reconfigurable platforms that offer dynamic reconfiguration require a great effort from the designer to design hardware blocks, data structures, orchestration, and control of all elements. On the other hand, some proposals aim to explore the network flexibility through the software solutions, for instance, [11] and [10]. It is important to note that these platforms' programmability is directed to their application field, and they tend to be limited.

This work aims to present a new switch architecture for SDN networks with MAC driven by the application called NEA Network Element. This architecture is Linux-based, and it uses some native Linux tools such as Bridges, VLANs, and Ebtables to enable flexibility at the lowest level.

The remainder of the document is structured as follows: Sect. 2 brings fundamental concepts of ETArch, presents the state-of-the-art related to SDN network elements, and the general design of ETArch Network Element. Section 3 presents the details for the architecture. Finally, Sect. 4 offers a discussion about this proposal, some concluding remarks, and suggestions for future work.

2   Background

This section provides a desirable background about the ETArch environment and important concepts, Networks Elements and a brief about ETArch switch.

2.1   Entity Title Architecture - ETArch

To approximate the Application layer semantically to the lower layers, allowing the requirements defined by the applications to permeate through the architecture layers, the Entity Title Architecture (ETArch) has been designed.

ETArch does not have a fixed layer structure as in the TCP/IP architecture [2]. In place of transport and network layers, ETArch defines the Communication layer. As shown in Fig. 1, this layer differs from traditional ones by being flexible and shaping according to application-specific communications requirements such as, for example, QoS, low power consumption, and bandwidth.

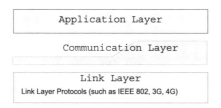

Fig. 1. ETArch layer architecture pattern. Figure adapted from [8]

From the ETArch perspective, it is important to comprehend the concepts of Entity, Workspace, and the control agent to facilitate the understanding of how the architecture works.

2.1.1   Entity and Title

Entities can be defined as any element that wants to communicate in a distributed environment [1]. Entities can be applications, devices, computers, a network element, or even a sensor. Entities are only identified by a title and a set of capabilities.

2.1.2   Workspaces

The Workspaces are logical multi-end communication buses driven by the applications and carries a set of requirements and they don't exist in the initial network configuration. They are created when an entity wants to offer content. When another entity is willing for this content, DTS must register it and attach it to the Workspace that offers that content.

ETArch specifies two types of Workspaces, namely, Data and Control Workspaces [8]. In the Fig. 2 is observed as an example of Data and control Workspace. In this figure, the DTS domain consists of two agents, DTSA1 and DTSA2, each controlling a Network Elements (NEs) subset. To this end, each agent abstracts the topology of the NEs it controls.

2.1.3   Domain Title Service - DTS

A DTS represents a set of ETArch controllers in a distributed system and their goal is to separate the network into manageable subnets within ETArch [2]. A Domain Title Service Agent (DTSA) acts in the control plane, managing the network elements. DTSA is responsible for creating, managing, and dropping Workspaces on network elements. It is important to note that each DTSA exercises control over a limited number of network elements [5]. A typical ETArch environment is illustrated in Fig. 2.

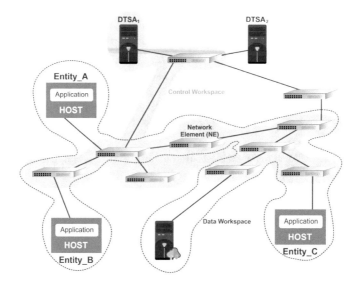

Fig. 2. Typical ETArch environment with data workspace, control workspace, DTSA and entities

It is important to note that each DTSA's act on the network control plane, and all communication between DTSA's and network elements is performed through a Control Workspace. On the other hand, the network elements act on the network data plane level.

2.2   Network Element in ETArch Architecture

Usually, SDN implementations use OpenFlow-based switches or make use of some virtualization technique [6,7]. The ETArch architecture requires some flexible support to the programmability in the lowest level of the network. The flexibility of SDN networks is coupled to the capacity of programmability present in their layers. It is observed that several implementations assume that there is connectivity at the link level of the network.

In this sense, ETArch needs a flexible network element that supports several Workspaces per physical port. As previously presented, Workspace is a logical path that connects all entities that are part of the same communication domain in the ETArch architecture. Workspaces by nature are a multicast communication domain segregated. The same entity can be part of several Workspaces, which means that a given Workspace does not have direct contact with another Workspace. In addition to its intrinsic multicast nature, it can also be emphasized that QoS's concept is also intrinsic to its nature, just because DTSAs on network elements deploy a Workspace according to the requirements driven by the application [8]. It is appropriate to think that Workspaces can offer finer-grained resource management.

3   NEA: Network Element Architecture for SDN Networks

Communication reliability has become a visible challenge for SDN networks. The separation of the control plane from the network data plane allows achieving a certain level of flexibility in the network management. However, it is observed that, regardless of the model or architecture used, every communication has different goals and requirements. Applications such as banking transactions, streaming services, and download/upload services have different communications requirements.

Naturally, communications require an end-to-end data routing infrastructure, but what is needed for a banking application in terms of security is quite different from what is required when downloading/uploading images on social networks.

In the ETArch, the main issue is directed towards multicast communications. Basically, ETArch institutes communications through Workspaces, where, for example, multicast are naturally implemented. The purpose is to have reliable transmissions regardless of the number of connected entities.

Focusing on the network element, the adoption of the Linux-based system to improve network solutions has gained attention in recent years. Projects such as Open vSwitch [9], Rocker Switch [12], and SwitchDev [13] are examples of L2/L3 data plane forwarding that make use of the Linux operating system to compose their solutions for data forwarding.

In this scenario, a dedicated solution offers more performance. However, it suffers from upgrades and the insertion of additional modules. Linux-based solutions offer a lower cost. Updating a new feature or even an upgrade is not a critical task, all of this, not to mention the ease of building a scalable solution. The NEA architecture overview is illustrated in Fig. 3.

The NEA architecture was designed to be modular, making it easy to upgrade or update. The NEA core was based on natives modules of Linux kernel such as Bridges, 802.1P/Q, VLANs, and Ebtables tools to support lowest level forwarding.

3.1   VLAN Filtering

In addition to manipulating the routing table and flow control, Linux Bridges can offer VLAN-oriented flow filtering, which contributes to QoS's improvement.

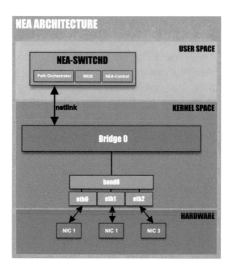

Fig. 3. NEA - Network Element Architecture overview

The protocols used in VLANs provide filtering services that support dynamic definition and establishment of communication groups in a network environment. The frames are filtered and addressed only to a particular group of recipients.

To make it possible to materialize the Workspace concept in this proposal and the fact that Linux Bridge is critical to handle all interface traffic, it is necessary to use IEEE802.1Q tagging mechanisms to identify within Bridge which frames are relative to certain Workspace for orchestration policies. It is important to note that many network interface cards do not have support for tagging frames. Therefore, all the tags will be inserted on the ingress flow and removed after the egress flow. This scenario is illustrated in Fig. 4.

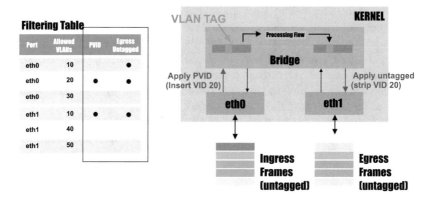

Fig. 4. VLAN-bridge routing architecture - untagged packets - Figure adapted from [14]

NEA offers support to the ETArch architecture through the NEA-SWICTHD. It is a user-space module that enables the ETArch controller interface (DTSA) and acts with the Workspace Orchestrator.

3.2 NEA-Switchd

As shown in Fig. 5, it can be seen that the NEA-Switchd module is composed of the modules: Path Orchestrator, NEA-Control, and WorkspaceDataBase (WDB).

The main sub-module is the Path Orchestrator, which is responsible for the orchestration of Workspaces. The Path Orchestrator sub-module through the Netlink interface will manipulate the Bridge routing structures. This includes the creation, removal, update, definition of flow priority levels. As already explained in the previous topic, the structure of a Workspace is associated with a VLAN.

The NEA-Control allows primitives from other network elements through the ETArch protocols and messages from the network controller (DTSA).

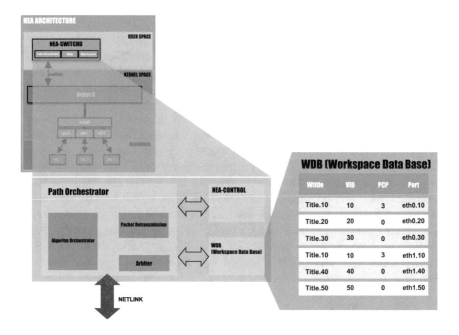

Fig. 5. Network Element Architecture modules

3.2.1 Arbiter

The Arbiter module is responsible for receiving data traffic from Bridge and making the forwarding decision. Possible decisions are: (i) Forward the ETArch primitives to the NEA-Control module, these primitives are for communication between network elements and constitute what the Network Control Workspace, (ii) Forward the primitives from the network controller for NEA-Control, and (iii) Provide inputs for the Algorithm Orchestrator module to perform the orchestration.

3.2.2 Workspace Data Base (WDB)

Besides the entire data structure for L2 forwarding is already inserted at the Kernel level and managed by Bridge, including the FDB (Forwarding Data Base), it is necessary to establish a database to store information from the Workspaces, the WDB table is formed by the following fields: (i) Workspace Title, it is the identification of the Workspace as provided for in the title model already explained earlier in this document, (ii) VID, the field used to relate a given Workspace to a VLAN framework, (iii) PCP, used to define the priority level of that VLAN and (iv) Port, to identify the Workspace-linked port VLAN.

4   Concluding Remarks

Exploring flexibility and programmability at all network architecture layers has become the key to improving applications' QoS. One problem with this is that all implementations assume there is connectivity at the network link level, and they end up, in turn, dedicating efforts to build solutions that explore the highest levels of network architecture.

This work presents a network element for SDN networks with MAC support driven by applications with a focus on the ETArch architecture. The flexibility is achieved by using VLAN techniques associated with the ability to program high-level functions for managing and materializing the Workspace concept.

It is important to note that the concept of Workspace can assume different interpretations based on the reference element, that is, having the network element as a reference, Workspace is materialized through the construction and maintenance of routing tables, using a buffer space for the temporary storage of primitives to guarantee the aggregation of multicast traffic. From the network controller's point of view, it is a logical bus, free of topology and geographic dimension.

Through state-of-the-art, there is a need to offer support to flexible network elements with fine granularity. The flexibility offered by techniques such as OpenFlow offers limited flexibility and coarse granularity compared to this proposal.

The future work will address issues such as the complete switch engine solution proposed in this work and a scenario to compare the NEA performance concerning Openflow.

Acknowledgments. This project has been built-up with the support of MEHAR team. We want to thank all who contributed to our research. Brazilian agency CAPES has partially funded this work.

References

1. de Souza Pereira, J.H., Kofuji, S.T., Rosa, P.F.: Horizontal address ontology in internet architecture. In: 2009 3rd International Conference on New Technologies, Mobility and Security, pp. 1–6. IEEE (2009)
2. de Oliveira Silva, F., Gonçalves, M.A., de Souza Pereira, J.H., Pasquini, R., Rosa, P.F., Kofuji, S.T.: On the analysis of multicast traffic over the entity title architecture. In: 2012 18th IEEE International Conference on Networks (ICON), pp. 30–35. IEEE (2012)
3. Silva, F., Corujo, D., Guimaraes, C., Pereira, J., Rosa, P., Kofuji, S., Neto, A., Aguiar, R.: Enabling network mobility by using IEEE 802.21 integrated with the entity title architecture. In: Proceedings of the IV WPEIF: SBRC 2013 Workshops, pp. 29–34 (2013)
4. de Oliveira Silva, F., de Souza Pereira, J.H., Rosa, P.F., Kofuji, S.T.: Enabling future internet architecture research and experimentation by using software defined networking. In: 2012 European Workshop on Software Defined Networking, pp. 73–78. IEEE (2012)
5. Kalyaev, A., Melnik, E.: FPGA-based approach for organization of SDN switch. In: 2015 9th International Conference on Application of Information and Communication Technologies (AICT), pp. 363–366. IEEE (2015)
6. McKeown, N., Anderson, T., Balakrishnan, H., Parulkar, G., Peterson, L., Rexford, J., Turner, J.: OpenFlow: enabling innovation in campus networks. ACM SIGCOMM Comput. Commun. Rev. **38**(2), 69–74 (2008)
7. Bifulco, R., Rétvári, G.: A survey on the programmable data plane: abstractions, architectures, and open problems. In: 2018 IEEE 19th International Conference on High Performance Switching and Routing (HPSR), pp. 1–7. IEEE (2018)
8. Oliveira, R.D., Molinos, D.N., Freitas, M.S., Rosa, P.F., de Oliveira Silva, F.: Workspace-based virtual networks: a clean slate approach to slicing cloud networks. In: CLOSER, pp. 464–470 (2019)
9. Pfaff, B., Pettit, J., Koponen, T., Jackson, E., Zhou, A., Rajahalme, J., Amidon, K.: The design and implementation of open vswitch. In: 12th USENIX Symposium on Networked Systems Design and Implementation (NSDI 2015), pp. 117–130 (2015)
10. Bosshart, P., Daly, D., Gibb, G., Izzard, M., McKeown, N., Rexford, J., Schlesinger, C., Talayco, D., Vahdat, A., Varghese, G., Walker, D.: P4: programming protocol-independent packet processors. ACM SIGCOMM Comput. Commun. Rev. **44**(3), 87–95 (2014). https://doi.org/10.1145/2656877.2656890
11. OF-CONFIG 1.2. OpenFlow Management and Configuration Protocol. Open Networking Foundation. Available via DIALOG (2014). https://opennetworking.org/wp-content/uploads/2013/02/of-config-1.2.pdf
12. Feldman, S.: Rocker: switchdev prototyping vehicle. In: Proceedings of Netdev 0.1. Available via DIALOG (2015). https://people.netfilter.org/pablo/netdev0.1/papers/Rocker-switchdev-prototyping-vehicle.pdf
13. Pirko, J., Feldman, S.: Ethernet switch device driver model (switchdev). Available via DIALOG (2014). https://www.kernel.org/doc/Documentation/networking/switchdev.txt
14. Makita, T.: Virtual switching technologies and Linux bridge. Available via DIALOG (2014). https://studylib.net/doc/18879676/virtual-switching-technologies-and-linux-bridge

A Survey on Data Science Techniques for Predicting Software Defects

Farah Atif, Manuel Rodriguez[(✉)], Luiz J. P. Araújo, Utih Amartiwi, Barakat J. Akinsanya, and Manuel Mazzara

Innopolis University, Republic of Tatarstan, Innopolis, Russia
{f.atif,m.rodriguez.osuna,l.araujo,u.amartiwi,
b.akinsanyaa}@innopolis.university,
m.mazzara@innopolis.ru

Abstract. In recent years, data science has been used extensively to solve several problems and its application has been extended to several domains. This paper summarises the literature on the synergistic use of Software Engineering and Data Science techniques (e.g. descriptive statistics, inferential statistics, machine learning, and deep learning models) for predicting defects in software. It shows that there is a variation in the use of data science techniques and limited reasoning behind the choice of certain machine learning models but also, in the evaluation of the obtained results. The contribution of this paper has to be intended as a categorization of the literature according to the most used data science concepts and techniques from the perspectives of descriptive and inferential statistics, machine learning, and deep learning. Furthermore, challenges in software defect prediction and comments on future research are discussed and forwarded.

1 Introduction

Software products and platforms have become, today, omnipresent in every domain and an important asset for every company. Despite the evolution of programming languages, coding environments, modern design patterns, and management approaches that aim to facilitate building software with minimum ambiguity and higher reliability. The presence of defects is imminent and can be discovered during the coding or testing phase or later after deployment.

In the literature, the term defect can be found under different naming such us error, fault, and failure. In the work of Mantyla and Lassenius [42] they grouped defect definitions into three categories according to the works in this field. The defects can be perceived as failures when they cause the crashing of the system or yield faulty results [32,42,54]. From the IEEE Standard failure is defined as "the inability of a function to meet the expected requirements". Other defined defects as faults, which are errors in the code but not necessarily lead to system failure [1,8] [48]. IEEE defines faults as "incorrect decision taken while understanding the given information, to solve problems or in implementation of process". Also, defects can be considered as a deviation from quality [17,25,33,49]. From the PSP perspective, the defect is counted when a change in the program occurs.

L. Barolli et al. (Eds.): AINA 2021, LNNS 227, pp. 298–309, 2021.
https://doi.org/10.1007/978-3-030-75078-7_31

Data Science (DS) is a scientific field and paradigm that is driving a research evolution in statistics, computer science, and Artificial Intelligence (AI). It emphasizes the use of general methods without changing its application, irrespective of the domain. DS aims to meet the challenges of the increasing demand and sustainable future while ensuring the best solutions. Over the years, DS has proved its ability to improve the software engineering process by following methodologies to retrieve useful insights from data [26].

Software Defects Prediction (SDP) in early stages is essential for software reliability, and quality [4]. Detecting defects after the release of the software is an exhausting, and time-consuming process. Hence, the ability to make earlier predictions will help to reduce the effort, and cost of resolving these defects. This survey aims to investigate different Data Science approaches and applications for predicting defects in software.

In this paper, we continue the work initiated in [3] about the investigation of synergistic approaches between Software Engineering and Data Science. Here we do not intend to provide a full Systematic Literature Review, but instead to summarise an overview of the most commonly used techniques. The collection of references and papers have been conducted querying the major research database such as Google Scholar, DBLP, ACM Digital Library, and IEEE explore using generic keywords such as *software defects, descriptive statistics, statistical inference, defects prediction in software, regression models, software quality*. We left a systematic and more comprehensive approach for a future extension of this work.

The remaining of this paper is summarised as follows. Section 2 describes defects, their impact and their nature. Section 3 presents methods employed for predicting defects . Subsections 1 and 2 show the descriptive and statistic inference methods. Section 4 Regression models, Sect. 5 presents Classification models for predicting defects. A discussion of the main findings from studies is presented in Sect. 6. Suggestions for future work is shown in Sect. 7.

2   Predicting Defects

Software defects can be attributed to many factors, the most common one is human error [24], these types of errors can be classified according to their nature, the severity of the impact they can cause and the priority they have in being fixed.

The nature of the defects can be grouped into main categories that include performance errors which are bound to the speed, resource consumption response time and stability of the software [37]. Compatibility defects affect the performance of the software on particular types of devices, operating systems, and hardware [46]. Security defects are related to weaknesses that allow a potential security attack [43]. The severity is measured with a number from one to four. A severity error number one is the highest and corresponds to an outage in the system and considerable impact to the user. The severity error number two also implies damage. However, a temporary solution or a circumvention to this issue is available. The remaining severity errors three and four are representative of lesser damage that can range from a touch and feel problem to an annoyance [52].

As aforementioned, predicting defects at early stages contribute to guarantee the reliability of the software and reduce the costs of maintenance. The process of predicting defects aims to classify software modules into fault-prone and none fault-prone. In the following, we demonstrate data science techniques used for predicting defects in software.

3   Descriptive and Inferential Statistical Methods in Defect Prediction

Statistical methods are important for understanding the data and make the right use of it. In software defect prediction different techniques derived from statistical concepts have been used such as descriptive statistics, and inferential statistics. In this section, we provide the literature review of the techniques used in software DP.

3.1   Descriptive Statistics

Descriptive statistics is a summary statistic used to organize, present, and analyze data. It relies on numerical and graphical techniques [26]. In software defect prediction, descriptive statistic has been used widely to describe the metrics obtained from the software [6,21,40,51,57] by calculating the mean, the variance, the median, the maximum, and the minimum. The most used metrics in this area are Chidamber and Kemerer (CK) set of object-oriented metrics, lines of code (LOC) metric, McCabe Cyclomatic complexity metric, and Halstead's metrics. Descriptive statistics is also used to understand the dataset in this work [10] the mean of lines of code and the mean of vulnerabilities were calculated for each application in the dataset.

In the works of [7,11], the mean and the standard deviation have been used for measuring the uncertainty of the predictions. Which is caused by the flipping between cross-validation runs by the classifiers. Metrics used for model evaluation are namely the accuracy, precision, recall, and F-score. This work [7] suggests that the oscillations in predictions can be caused by the fold generation. Also, non-stratified folding may cause class imbalance, which can significantly affect the performance of the model.

3.2   Statistical Inference Methods

Statistical Inference (SI) is the process of inferring something about a population based on what is measured in the sample. It is divided into two types, parametric and non-parametric. The parametric statistics requires some assumptions about the population parameters. The most used assumption about data is the normal distribution, some papers also considered the homogeneity of variance. In the work of [38], the Shapiro-Wilk test for normality has been used, and Levene's test for homogeneity. If the data does not meet the assumption, then non-parametric statistics should be applied.

Some researchers do such a preliminary process to identify important features and metrics. In this work [58], correlation analysis has been used before building the model to check whether there is a relationship between the number of defects and the metrics. [2] selected k-best features using Chi-square for three datasets. To handle the problem of the non-normal distribution of data, density kernels such as Gaussian, Epanechnikov, Biweight, and Triweight are used in [29]. Furthermore, correlation analysis both parametric (Pearson) and non-parametric (Spearman) supports the evaluation of relationships between metrics, the existence of col-linearity, and evaluate the prediction [39]. For instance, Pearson and Spearman correlations are used to measure the goodness of fit of regression model in [6, 23, 28], and [21]. In [53] correlation between independent values has been used to check col-linearity. However, in real data, the correlation between metrics exists and can affect the prediction quality. To cover the multi-collinearity, dimensionality reduction technique called principal component analysis (PCA) was applied for this problem in [58].

Comparative analysis in defect prediction is used to compare models [20, 36], defect metrics, process metrics, and evaluation metrics [38]. The parametric statistics used to compare between two groups are z-test [20] and T-test [36] while Analysis of Variance (ANOVA) used to compare more than 2 groups. In [38] Wilcoxon matched pair, a non-parametric comparative analysis is also used to identify which process metrics improve defect prediction models based on product metrics. In [36], Analysis of Variance (ANOVA) has been used with Post-hoc analysis and Bayesian approach to compare the result of defect prediction between models. To evaluate the model, Area Under ROC Curve (AUC) and an alternative named H measures have been used for model evaluation. These evaluations are compared by using non-parametric statistics called the Friedman test and got no significant difference between them. Moreover, the review analysis in [31] also suggests using effect size for better evaluation.

4   Regression Models

Regression techniques are powerful statistical methods that have been exploited for predictions and in classification. In this section, we present the use of linear regression, logistic regression, and Poisson regression in DP.

4.1   Linear Regression

(LR) Regression methods allow predicting unknown variable based on one or more known variables [21]. It has been used in DP for different purposes and in different ways. In [58] LR was used to predict defects on a dependency graph to identify dependencies between pieces of code, it was also used by [23] to predict defects from defect history in SAP Java code.

4.2 Logistic Regression

Logistic regression is a generalisation of the linear regression to binary classification [45]. In the works of [6,21,51] it was used to study the relationship between the metrics and fault-proneness of classes.

It also has been used as a classifier for LSTMs [9,56] and deep belief network (DBN) [58]. Many works employed logistic regression, which predicts the probability of being a defect. If the predicted probability is greater than a threshold -usually 0.5- they label it as a defect, if it is less than the threshold, then it is labeled as non-defect [30,45].

The work of [51] held an empirical validation of object-oriented metrics for predicting the fault proneness in the classes on a large dataset. It compared the performance of logistic regression with two machine-learning methods (decision trees (DT) and artificial neural networks (ANN)).

With a different approach in [9] Random Forests (RF) outperformed significantly logistic regression on Samsung dataset. However, logistic regression performed better than RF on smaller dataset Promise.

In the work of [21] both linear and logistic regressions were used, univariate and multivariate on CK metrics. Univariate regression analysis is used to examine the effect of each metric separately, while multivariate regression analysis examines the common effects of the metrics in fault-proneness detection [21]. They used both regressions because logistic regression only classifies a class into a defect or not. However, it is necessary to predict the number of bugs in a class, that is why linear regression has been used to predict the number of bugs.

4.3 Poisson Regression

Poisson regression is another type of regression that is used to model counted data and contingency tables. In DP, it can be used for small components of software [12]. It was observed that Poisson provides an estimate of several defects in a small component unlike techniques such as discriminant analysis which identifies components that are likely to have defects with a limited number of defects. The methodology consists of calculating the probability of defects regarding an expected number of defects using a defined set of metrics.

5 Classifying Software Defects Using Machine Learning and Deep Learning

Classification models aim to classify software modules into defect-prone and non-defect-prone either using software metrics or analyzing log files. In the following section, we demonstrate the use of machine learning and deep learning models for software DP.

5.1   Machine Learning Models

The result of recent research using machine learning and deep learning models for DP tasks has shown promising results. The most commonly used ML models for DP tasks are Random Forest (RF), Support Vector Machine (SVM), Naive Bayes (NB), Logistic Regression, K Nearest Neighbourhood (KNN), Bayesian Networks (BN), Radial basis function (RBF), Decision Tree (DT) and the most commonly used DL models for DP tasks are Artificial Neural Network (ANN), Convolutional Neural Network (CNN), Long short-term memory (LSTM), Multi-Layer Perceptron (MLP).

The work of [14] used weighted SVM to handle the problem of class imbalance.

Several researchers have used SVM with different software metrics. In the work of [41], SVM was used on CK metrics and on McCabe Halstead (basic and derived), line count, branch count, misc, error count and density metrics in [11,19]. Another way SVM was used to predict defects was in [14] using log files. SVM was applied on context vectors obtained with random indexing which is a text analysis technique. The context vectors represent an entry in a log file which represents an event in the system. [18] held a comparative study between the performance of SVM and ANN, The result of the research showed that SVM outperformed ANN when viewing fault-proneness prediction as a binary classification task with the NASA dataset.

In the work of [7,27,55], RF models were applied for DP tasks. RF model is an ML model that consists of a set of decision trees that classify input data. The output of RF is chosen from the majority predicted value from decision trees. [20] and [34] studied the effectiveness of random forests comparing to some other classification models (logistic regression, Naive Bayes, Discriminant analysis) on NASA datasets. On the same dataset, [11] showed that SVM, RF, and MLP outperformed KNN, Logistic Regression, RBF, and DT. For CM1dataset, SVM slightly outperformed RF and MLP, RF outperformed SVM and MLP for PC1 dataset and MLP outperformed RF and SVM for KC1 and KC3 dataset. In [7], a comparative study of the performance of four classifiers SVM, RF, and NB on NASA datasets. RF got the best result, however, when measuring the stability of the model for the DP task, NB showed more stability compared to other classifiers, which revealed that the model had a low level of flipping in different runs which in opposite SVM is the less stable in KC4 dataset due to its sensitivity to the training data.

Causal methods came to offer better decision making compared to regression models. BN is a directed graph where the nodes represent uncertain variables and arcs that represent a causal relationship between variables [13,44]. According to [13] the power of BN is its flexibility since the probability values can be assigned to variables and hence the joint probability distribution can be recalculated and conditioned using Bayesian propagation. However, Bayesian models suffer from subjectivity when introducing expert judgements, but their main advantages are using a limited number of metrics and coping with missing data. In [7] NB showed more stability in predicting defects on many datasets compared to SVM and RF.

5.2 Deep Learning Methods

Recent studies showed that neural networks are being applied for classifying modules into defect-prone and non-defect-prone. Feedforward neural network allows a model to learn by propagating the adjusted weights on neurons calculated from the output. The input layer inputs the calculated software metrics which are fed forward into the neural network and the output layer outputs the result of the classification of the module. Through the experiments of [11, 15] the accuracy obtained with neural networks in overall is as good as other ML models like SVM and RF.

CNN provided a new approach for DP tasks than other ML models built using hand-crafted features [35] that correlate with defects but don't capture the semantic of the code source [9]. The result of their work showed an improvement of 12% and 16% in terms of F-measure, comparing to DBN-based and traditional features-based respectively. In [35] CNN was used for DP task as a feature generator to generate semantic and structural features of the source code which was later combined with logistic regression as classifier.

Software predictions using deep generative models were investigated by [22]. The result of their work shows that Stack Sparse Auto-Encoder (SSAE) and Deep Belief Network (DBN) achieve better performance. In the work of [5], different types of neural networks have been used for DP, which are namely Feedforward Neural Network (FNN), Recurrent Neural Network (RNN), Artificial Neural Network (ANN), and Deep Neural Network (DNN). In the experiment Genetic Algorithm (GA) was used to select the features and Particle Swarm Optimization (PSO) for building the cluster. The models were trained on different datasets from NASA promise software engineering repository. The result of the study reveals that DNN gave the best accuracy.

6 Discussion

The overview of the literature shows that in DP, statistical techniques are used in the pre-modeling step and model evaluation, while machine learning models are utilized for the predictions. Descriptive statistics are used to understand the distribution of the data of each variable or samples and identify the possibility of outliers. Inferential statistics helps in identifying the correlation between variables, normalizing data, proceed hypothesis testing for identifying the relevant features for a model. However, sometimes the features are not related directly to defect so that their relation is not significant statistically. Moreover, a machine learning algorithm can cover this problem so that it can predict the defect more accurately.

Many papers studied the performance of ML models on specific datasets such as SVM, Random Forests, Bayes Networks, MLP, NN, CNN, and LSTM.

When outliers are present in the dataset SVM's tend to be more effective since they can control them using the margin parameter C, but also they are good when dealing with higher dimensional data, the density functions allow to map to higher dimensions in an optimized way [11]. Random Forest is too robust to outliers and noisy data which make their accuracy high and more significant for large datasets [20]. Logistic regression is also as good as many ML models when features and outputs are linearly related but it doesn't allow to combine multiple features at the same time [56]. Therefore, techniques of multiple classifications as one vs one and one vs all must be used for multiple classifications. However, these machine learning models suffer from different limitations that are hard to reach a consensus about which model is better. We remark many researchers claiming that some models are better than others for some specific datasets and specific metrics. The generalization of results is still questionable till today due to many factors that influence experiments such as metrics chosen and datasets, because the performance of a model is highly dependent on the dataset and metrics [44]. [16] study affirms that classifiers perform differently in different datasets and the focus should be on studying different classifiers instead of proving the performance of one. Ensemble learning in this case can be adequate to shift the focus from the model to the quality of dataset and metrics to use.

Moreover, there here has been little reasoning about the choice and size of software metrics. Appropriate features are very important for the model, it's essential to have a set of effective metrics that truly contribute to DP to reduce the parameter size of the model, but also to understand the relationship between them and their effect on the prediction. The work of [44] highlighted the relationship between different software metrics and defects tested on different datasets. It would be interesting to explore the effectiveness of obtained metrics, first, in predicting defects and, second, on datasets used in the experiments adding the NASA datasets to clarify assumptions made about the influence of metrics, dataset, and the model. We also recommend the use of techniques of feature selection like cross-validation or inferential statistic techniques before conducting any experiment.

The NASA datasets are the most used by researchers in this area, it allows comparing the performance of methods proposed in the literature, but they do not help to generalize results to other software. Moreover, NASA datasets suffer from duplicated cases [50] especially in MC1, PC2 and PC5 datasets the rate of repeated instances are estimated 79%, 75% and 90% respectively [19]. In ML duplicated cases prevent the model from learning useful patterns since data that are in training are seen in the validation. Also, reducing the size of the dataset can bias the model and hence less accurate results. For this reason, data preprocessing should be conducted properly before any experiment.

Another problem that is encountered in DP is the imbalanced data, which means that samples of some classes are frequent than others. This phenomenon influences the classifiers generally by over-predicting majority classes. The survey [30] shows that using the resampling technique can help to overcome this data imbalance problem.

7 Conclusion and Future Research

Software bugs, errors, defects, or different names can be assigned to them, but they remain a quality problem in software engineering. Experts claim that 3 or 4 defects per 1000 lines of code are enough to make a drastic degradation of the quality of the software. Moreover, the impact of defects are negative on economic, business and human sides [47]. Defect prediction aims to automate the prediction of defects and help testers in spotting and correcting the right defective components for reliable software.

Throughout this paper, we surveyed the application of Data Science in DP, whose contribution to this domain is remarkable. The literature is abundant of works that interest in this area of software engineering. Among the sub-fields of Data science, ML models got the largest part of the interest of researchers. Many ML models have been applied to different datasets with different techniques, and many empirical studies were conducted to measure the effectiveness and power of models against others. The inferential and descriptive statistic concepts were not addressed specifically in DP. However, they were used and present in most of the experiments to measure the goodness of fit and the goodness of estimate of models, and in other cases to evaluate the capability of models in different conditions.

References

1. Ieee standard glossary of software engineering terminology: IEEE Std 610(12–1990), 1–84 (1990). https://doi.org/10.1109/IEEESTD.1990.101064
2. Agarwal, S., Gupta, S., Aggarwal, R., Maheshwari, S., Goel, L., Gupta, S.: Substantiation of software defect prediction using statistical learning: an empirical study. In: 2019 4th International Conference on Internet of Things: Smart Innovation and Usages (IoT-SIU), pp. 1–6. IEEE (2019)
3. Akinsanya, B., Araújo, L.J., Charikova, M., Gimaeva, S., Grichshenko, A., Khan, A., Mazzara, M., Shilintsev, D., et al.: Machine learning and value generation in software development: a survey. In: Proceeding of International Conference on Software Testing, Machine Learning and Complex Process Analysis (TMPA-2019) (2019)
4. Alsawalqah, H., Faris, H., Aljarah, I., Alnemer, L., Alhindawi, N.: Hybrid smote-ensemble approach for software defect prediction. In: Computer Science On-line Conference, pp. 355–366. Springer (2017)
5. Ayon, S.I.: Neural network based software defect prediction using genetic algorithm and particle swarm optimization. In: 2019 1st International Conference on Advances in Science, Engineering and Robotics Technology (ICASERT), pp. 1–4. IEEE (2019)
6. Basili, V.R., Briand, L.C., Melo, W.L.: A validation of object-oriented design metrics as quality indicators. IEEE Trans. Software Eng. **22**(10), 751–761 (1996)
7. Bowes, D., Hall, T., Petrić, J.: Software defect prediction: do different classifiers find the same defects? Software Qual. J. **26**(2), 525–552 (2018)
8. Burnstein, I.: Practical software testing: a process-oriented approach. Springer (2006)

9. Dam, H.K., Pham, T., Ng, S.W., Tran, T., Grundy, J., Ghose, A., Kim, T., Kim, C.J.: A deep tree-based model for software defect prediction. arXiv preprint arXiv:1802.00921 (2018)
10. Dam, H.K., Tran, T., Pham, T.T.M., Ng, S.W., Grundy, J., Ghose, A.: Automatic feature learning for predicting vulnerable software components. IEEE Trans. Softw. Eng. **47**, 67–85 (2018)
11. Elish, K.O., Elish, M.O.: Predicting defect-prone software modules using support vector machines. J. Syst. Softw. **81**(5), 649–660 (2008)
12. Evanco, W.M.: Poisson analyses of defects for small software components. J. Syst. Softw. **38**(1), 27–35 (1997)
13. Fenton, N., Neil, M., Marsh, W., Hearty, P., Marquez, D., Krause, P., Mishra, R.: Predicting software defects in varying development lifecycles using Bayesian nets. Inf. Softw. Technol. **49**(1), 32–43 (2007)
14. Fronza, I., Sillitti, A., Succi, G., Terho, M., Vlasenko, J.: Failure prediction based on log files using random indexing and support vector machines. J. Syst. Softw. **86**(1), 2–11 (2013)
15. Gayathri, M., Sudha, A.: Software defect prediction system using multilayer perceptron neural network with data mining. Int. J. Recent Technol. Eng. **3**(2), 54–59 (2014)
16. Ghotra, B., McIntosh, S., Hassan, A.E.: Revisiting the impact of classification techniques on the performance of defect prediction models. In: Proceedings of the 37th International Conference on Software Engineering-Volume 1, pp. 789–800. IEEE Press (2015)
17. Gilb, T., Graham, D.: Software Inspections. Addison-Wesley, Reading (1993)
18. Gondra, I.: Applying machine learning to software fault-proneness prediction. J. Syst. Softw. **81**(2), 186–195 (2008)
19. Gray, D., Bowes, D., Davey, N., Sun, Y., Christianson, B.: Using the support vector machine as a classification method for software defect prediction with static code metrics. In: International Conference on Engineering Applications of Neural Networks, pp. 223–234. Springer (2009)
20. Guo, L., Ma, Y., Cukic, B., Singh, H.: Robust prediction of fault-proneness by random forests. In: 15th International Symposium on Software Reliability Engineering, pp. 417–428. IEEE (2004)
21. Gyimothy, T., Ferenc, R., Siket, I.: Empirical validation of object-oriented metrics on open source software for fault prediction. IEEE Trans. Software Eng. **31**(10), 897–910 (2005)
22. Hasanpour, A., Farzi, P., Tehrani, A., Akbari, R.: Software defect prediction based on deep learning models: Performance study. arXiv preprint arXiv:2004.02589 (2020)
23. Holschuh, T., Pauser, M., Herzig, K., Zimmermann, T., Premraj, R., Zeller, A.: Predicting defects in sap java code: an experience report. In: 2009 31st International Conference on Software Engineering-Companion Volume, pp. 172–181. IEEE (2009)
24. Huang, F., Strigini, L.: Predicting software defects based on cognitive error theories. In: 2018 IEEE International Symposium on Software Reliability Engineering Workshops (ISSREW), pp. 134–135. IEEE (2018)
25. Humphrey, W.S.: A discipline for software engineering. Pearson Education India (1995)
26. Igual, L., Seguí, S.: Introduction to data science. In: Introduction to Data Science, pp. 1–4. Springer (2017)

27. Jacob, S.G., et al.: Improved random forest algorithm for software defect prediction through data mining techniques. Int. J. Comput. Appl. **117**(23), 18–22 (2015)
28. Janes, A., Scotto, M., Pedrycz, W., Russo, B., Stefanovic, M., Succi, G.: Identification of defect-prone classes in telecommunication software systems using design metrics. Information sciences **176**(24), 3711–3734 (2006). https://doi.org/10.1016/j.ins.2005.12.002
29. Ji, H., Huang, S., Lv, X., Wu, Y., Feng, Y.: Empirical studies of a kernel density estimation based Naive Bayes method for software defect prediction. IEICE Trans. Inf. Syst. **102**(1), 75–84 (2019)
30. Kamei, Y., Shihab, E., Adams, B., Hassan, A.E., Mockus, A., Sinha, A., Ubayashi, N.: A large-scale empirical study of just-in-time quality assurance. IEEE Trans. Software Eng. **39**(6), 757–773 (2012)
31. Kampenes, V.B., Dybå, T., Hannay, J.E., Sjøberg, D.I.: A systematic review of effect size in software engineering experiments. Inf. Softw. Technol. **49**(11–12), 1073–1086 (2007)
32. Laitenberger, O.: Studying the effects of code inspection and structural testing on software quality. In: Proceedings Ninth International Symposium on Software Reliability Engineering (Cat. No. 98TB100257), pp. 237–246. IEEE (1998)
33. Laitenberger, O., DeBaud, J.M.: An encompassing life cycle centric survey of software inspection. J. Syst. Softw. **50**(1), 5–31 (2000)
34. Lessmann, S., Baesens, B., Mues, C., Pietsch, S.: Benchmarking classification models for software defect prediction: a proposed framework and novel findings. IEEE Trans. Software Eng. **34**(4), 485–496 (2008)
35. Li, J., He, P., Zhu, J., Lyu, M.R.: Software defect prediction via convolutional neural network. In: 2017 IEEE International Conference on Software Quality, Reliability and Security (QRS), pp. 318–328. IEEE (2017)
36. Li, L., Lessmann, S., Baesens, B.: Evaluating software defect prediction performance: an updated benchmarking study. arXiv preprint arXiv:1901.01726 (2019)
37. Liu, H.H.: Software Performance and Scalability: A Quantitative Approach, vol. 7. Wiley, New York (2011)
38. Madeyski, L., Jureczko, M.: Which process metrics can significantly improve defect prediction models? an empirical study. Software Qual. J. **23**(3), 393–422 (2015)
39. Malhotra, R.: Comparative analysis of statistical and machine learning methods for predicting faulty modules. Appl. Soft Comput. **21**, 286–297 (2014)
40. Malhotra, R., Jain, A.: Fault prediction using statistical and machine learning methods for improving software quality. J. Inf. Process. Syst. **8**(2), 241–262 (2012)
41. Malhotra, R., Kaur, A., Singh, Y.: Empirical validation of object-oriented metrics for predicting fault proneness at different severity levels using support vector machines. Int. J. Syst. Assur. Eng. Manag. **1**(3), 269–281 (2010)
42. Mäntylä, M.V., Lassenius, C.: What types of defects are really discovered in code reviews? IEEE Trans. Software Eng. **35**(3), 430–448 (2008)
43. McGraw, G.: Software security. IEEE Secur. Priv. **2**(2), 80–83 (2004)
44. Okutan, A., Yıldız, O.T.: Software defect prediction using Bayesian networks. Empir. Softw. Eng. **19**(1), 154–181 (2014)
45. Panichella, A., Oliveto, R., De Lucia, A.: Cross-project defect prediction models: L'union fait la force. In: 2014 Software Evolution Week-IEEE Conference on Software Maintenance, Reengineering, and Reverse Engineering (CSMR-WCRE), pp. 164–173. IEEE (2014)
46. Pobereżnik, L.: A method for selecting environments for software compatibility testing. In: 2013 Federated Conference on Computer Science and Information Systems, pp. 1355–1360. IEEE (2013)

47. Pressman, R.S.: Software Engineering: A Practitioner's Approach. Palgrave Macmillan, London (2005)
48. Runeson, P., Andersson, C., Thelin, T., Andrews, A., Berling, T.: What do we know about defect detection methods? [software testing]. IEEE Softw. **23**(3), 82–90 (2006)
49. Runeson, P., Wohlin, C.: An experimental evaluation of an experience-based capture-recapture method in software code inspections. Empir. Softw. Eng. **3**(4), 381–406 (1998)
50. Shepperd, M., Song, Q., Sun, Z., Mair, C.: Data quality: some comments on the NASA software defect datasets. IEEE Trans. Software Eng. **39**(9), 1208–1215 (2013)
51. Singh, Y., Kaur, A., Malhotra, R.: Empirical validation of object-oriented metrics for predicting fault proneness models. Software Qual. J. **18**(1), 3 (2010)
52. Sullivan, M., Chillarege, R.: Software defects and their impact on system availability: a study of field failures in operating systems. FTCS **21**, 2–9 (1991)
53. Thapaliyal, M., Verma, G.: Software defects and object oriented metrics-an empirical analysis. Int. J. Comput. Appl. **9**(5), 41–44 (2010)
54. Wiegers, K.E.: Peer Reviews in Software: A Practical Guide. Addison-Wesley, Boston (2002)
55. Yang, X., Lo, D., Xia, X., Sun, J.: TLEL: a two-layer ensemble learning approach for just-in-time defect prediction. Inf. Softw. Technol. **87**, 206–220 (2017)
56. Yang, X., Lo, D., Xia, X., Zhang, Y., Sun, J.: Deep learning for just-in-time defect prediction. In: 2015 IEEE International Conference on Software Quality, Reliability and Security, pp. 17–26. IEEE (2015)
57. Zhou, Y., Leung, H.: Empirical analysis of object-oriented design metrics for predicting high and low severity faults. IEEE Trans. Software Eng. **32**(10), 771–789 (2006)
58. Zimmermann, T., Nagappan, N.: Predicting defects using network analysis on dependency graphs. In: 2008 ACM/IEEE 30th International Conference on Software Engineering. pp. 531–540. IEEE (2008)

Survey on Blockchain Applications for Healthcare: Reflections and Challenges

Swati Megha[✉], Hamza Salem, Enes Ayan, Manuel Mazzara, Hamna Aslam, Mirko Farina, Mohammad Reza Bahrami, and Muhammad Ahmad

Lab of Software and Service Engineering, Innopolis University, Innopolis, Russia

Abstract. Blockchain is a decentralized, distributed, public, and digital ledger that is used to record transactions across many computers. At the top of Blockchain, Ethereum introduces smart contracts to support on-chain storage and enable Decentralized Apps (*DApps*) to interact with the Blockchain programmatically. In the healthcare domain, programmable Blockchains provide different solutions to resolve some key challenges, such as immutability, security, data provenance, robustness, availability, and privacy. This paper aims to categorize and evaluate different solutions in the healthcare domain that makes use of programmable Blockchain or part of it. We use a software engineering approach to organize the existing 23 papers in multiple categories such as challenges addressed and quality attributes promoted. We will further categorize solutions based on how much they have used Blockchain and show the other research which Blockchain is not recommended to use. This work is intended to provide researchers in the field with a well-defined and structured categorization, plus insights into the existing literature.

1 Introduction

Healthcare, the organized provision of medical care in the form of preventive, curative, rehabilitative or palliative health services to individuals within a community, is universally recognised as a fundamental human right; a natural and inalienable entitlement that belongs to all humans living on this planet [37].

Blockchain is a distributed [29] system recording and storing transaction records. More precisely, it is a shared, immutable record of peer-to-peer transactions built from linked transaction blocks and stored in a digital ledger. This technology, based on a peer-to-peer (P2P) topology, was originally developed in 1991 by Stuart Haber and Scott Stornetta, who wanted to realise a system where any document timestamps could not be tampered with [16].

Blockchain depends on established Crypto-graphic techniques that allow each participant in a network to securely interact with other members of the same network. Crucially, in such networks there is no central authority governing the interactions. This typically allows data to be stored on several thousands

L. Barolli et al. (Eds.): AINA 2021, LNNS 227, pp. 310–322, 2021.
https://doi.org/10.1007/978-3-030-75078-7_32

of servers and information to be recorded digitally while being processed distributively in an immutable audit trail that subsequently becomes available to everyone in the network and beyond it [11].

Blockchain is said to offer flexible and innovative solutions for a number of different domains, including but not limited to: Data Security, Management Access Control, Integrity, and Interoperability. In recent years, many practitioners emphasised the potential beneficial impact of Blockchain technology in healthcare [6,14,21,38]. Medical doctors and other researchers thus investigated the ways in which this revolutionary technology can be put into service by placing the patient at the center of the medical ecosystem, thus increasing their security, privacy, and even the interoperability of their health data.

For example, "MeDShare" is a system that uses smart contracts for data flow [38]; "Modum.io" is a startup that combines IoT (Internet of Things) with Blockchain technology [6] and several smart contract-based solutions mentioned in [14,21,38]. Gcoin is a blockchain based supply surveillance system developed to stop the production and the procurement of counterfeit drugs [35]. Med-Chain brings the benefit of blockchain, digest chain, and P2P network techniques together, addressing the issue of data-sharing in wearable technology [4].

In this review paper we surveyed 23 state of art blockchain in healthcare system proposals and then taxonomized them into several categories based on the healthcare requirements and quality attributes they addressed. In Sect. 2, we look at the requirements and quality attributes in the healthcare industry. Section 3 analyses the challenges of implementing Blockchain in healthcare. Section 4 offers ethical insights related to the management of healthcare data. Section 5 presents a categorization of existing Blockchain-based solutions into five different quality attributes, based on Sect. 2 and 3. Section 6 and 7 conclude by summarising what we have achieved and by pointing out possible research directions.

2   Requirements of Healthcare Industry

Healthcare systems are fragmented and have a nested hierarchy [19]. Because of these characters, several subsystems are created. Subsystems differ on several grounds, such as different techniques for financing, different processes for facilitating services, medicines, and equipment, and different types of treatment of patients. These subsystems, most of the time, work independently, and therefore they create significant gaps in the provision of healthcare.

Therefore, to solve the abovementioned problems a generic solution, fostering integration is needed. It is very likely to think of using digital ledger to bind these subsystems together. It is compelling to use Blockchain to ensure security, privacy, and traceability within fragmented healthcare systems.

This work analyzed and concluded whether or not Blockchain's solution could help the healthcare systems. If Blockchain's solution could help, then can we use it to transform the healthcare system as a whole, or it would be suitable only in specific areas. If not, then what makes Blockchain not suitable for the healthcare systems. For the above-stated reasons, this work analyzed different research works in this area that supports and provide solutions to transform the healthcare system through Blockchain and others that hold a slightly negative perspective towards introducing Blockchain solutions into the healthcare systems. Further, we concluded this analysis. Like any other industry, the healthcare industry is facing challenges that need to be addressed to ensure better service to patients. This section discusses several challenges being faced by the healthcare industry at current times.

1. **Supply chain tracking and tracing to prevent Counterfeit drug**
   World health organization defines counterfeit drugs as "Products deliberately and fraudulently produced and/or mislabelled with respect to identity and/or source to make it appear to be a genuine product" [24]. The Drugs life-cycle starts from a fundamental study followed by a non-clinical study, clinical trial, then granting permit application followed by production and manufacturing and at the end marketing and release. Regulation is required at every stage of this life-cycle to avoid production and supply of counterfeit drugs. According to world health organization data counterfeit drug has a market of worth 21 billion dollars which is makes up approximately 10% of the global medicine market. And this is due to the loopholes or incapability to collect and analyze information about the drug supply chain [8]. The counterfeit drug causes economic loss to governments and pharmaceutical injury, but also give rise to crimes while it is being manufactured and procured. Whereas the major risk here is public health because the counterfeit drug could also contain ingredients in the wrong quantities that means it might be toxic or might have no effect on the patients. Toxic treatment of ineffective treatment both can lead to a dangerous effect on a patient's health. A very recent case happened in Nigeria where reputedly 60k–80 children were counterfeit malaria vaccine which generally is a combination of several medications but in this case contained only chloramphenicol [24]. One of the major reasons for the Manufacturing and procurement of counterfeit drugs insufficient and/or complex supply chain surveillance. You might be surprised to know that counterfeit drugs are available through the legal supply chain small part of which consist of online pharmacy. This happens because stages of the supply chain starting manufacturing to marketing and consumption are vulnerable. After all, it is being handled by several stakeholders until it reaches patients and can provide entry to counterfeit drugs through any one of them [5]. Reducing the links in the supply chain can reduce the supply of counterfeit drugs. One way to achieve is through license restrictions and another is through technology that can track the production and procurement of drugs.

2. **Solutions to Care coordination Issue:** Care coordination is defined as organizing patient care activities and bringing all participants involved in the care on the same page by sharing patient information among them [22]. Care coordination is done purposefully to achieve safe and effective healthcare for the patient. The care coordination problem is a well-known problem in the healthcare industry and is related to the fundamental problems of medicine. Patients receive care from multiple healthcare organizations, the sharing of

patient data among these organizations is critical for care coordination, and accurate and timely sharing is equally important for effective care. Healthcare data are highly sensitive and Here comes the conflict of data privacy and accessibility. Patient data accessibility by these organization is critical for effective healthcare at the same time patents needs to have control over their data to harness their data privacy rights [9]. At current times providing access to sensitive data, avoiding its misuse at the same time ensuring personal privacy and anonymity is one of the major issues in the healthcare industry. Another major problem in the sphere of care coordination is the heterogeneous data format being used by these healthcare organizations that require conversion of shared data from one format to another thereby consuming time hence makes data sharing among organizations difficult and at the same time reduces the tempo of healthcare services.

3. **Secure Electronic Health Records (EHRs) sharing:** Even though the dilemma of data privacy and accessibility i.e. the care coordination issues is being taken care of and the patient and healthcare organization agrees upon EHRs sharing with the mutual concept which will rule out the data accessibility concern at the institutional side and data privacy concern at patients side still EHRs sharing is challenging because the usual data-sharing network used at the current time cannot guarantee the data will be inaccessible to a third party other than the patient or the institutions which generates a data security threat while the share happens. Major security challenges in healthcare systems are related to authorization, access control, and non-repudiation of records. These attributes are crucial for promoting quality attributes such as availability of information and date, integrity, and confidentiality. In today's world of healthcare gadgets, medical records could be of several types, ones that are conventional patient files and others, data from healthcare gadgets and sensors. Data is being saved into the database and must be retrieved only by authorized individuals, thereby requires strict access control to eliminate tampering and copying [27].

4. **Smart healthcare data Management:** This is the age of information, and like any other industry, the healthcare industry is also growing day by day. There has been an increase in the number of wearable and other IoT devises for health monitoring in the last decade. Medical data in today's age comes from several smart devices and sensors connected through the internet and mobile applications. IoT has also given a new dimension to the way we create, share, and access such data. We are producing health-related data with ease daily, and this generates the requirement of ease of access, secure transfer, and retrieval of this data [27]. But our traditional healthcare system doesn't have efficient ways to share and manage the continuously generated data from sensors and other IoT devices. Also, traditional approaches s of sharing over a network do not provide flexibility with metadata changes. The wearable and IoT health devise produce health data every second hence also required continuous and efficient storage as well as secured sharing to the medical institutions on patient's demand. A patient-driven data storing and sharing scheme is required especially in the case of wearable technologies.

3 Fundamental Challenges of Implementing Blockchain Architecture in Healthcare Systems

Many researchers conclude that specific barriers are preventing the introduction of Blockchain into healthcare systems. In this section, we analyze the research works that have a negative perspective towards the introduction of Blockchain into the healthcare system. Several researchers believe that interoperability, scalability, storage, social acceptance, and standardization are the challenges that prevent the introduction of Blockchain into the healthcare system.

1. **Security and Privacy of Data:** Blockchain-based application architecture will remove the need of third parties to carry out transactions; an entire community will be responsible for transaction verification instead of any third party on Blockchain architecture. This means that all nodes can have access to data transmitted by any on a node in Blockchain-based architecture, thereby creating a risk of compromise with data security and privacy. As an example, because of the absence of a third party, a patient needs to provide access to their data to another user in the chain for emergency cases. This emergency contact in the chain can further provide access to their data to set of other users in the chain, thereby creating massive security and privacy risk [1].
   The obvious solution here is to increase the security to restrict such data sharing. Researchers analyze that increasing security could hinder data transfer from one block to another, thereby the receiver could only have limited or incomplete access to data, which in turn will not serve the purpose of ease of data sharing.
   One of the security attacks prevalent in the Blockchain network is called a 51% attack. This attack can occur when a miner or team of miners owes 51% of the blocks in the network. They can have the authority to prevent any transaction by not providing the required consent, which seems very risky as compared to other conventional attacks in centralized systems and especially in case of medical records, which can lead to life and death scenarios.

2. **Managing Storage Capacity, Scalability, and IoT overhead:** The underlying architecture of Blockchain is designed in such a way that it does not require massive storage. The healthcare industry generates a considerable amount of data daily and requires its processing and transferring as the Blockchain network will grow in the healthcare system. It will create an issue with the storage of data. For example, a single patient could have several data such as test reports, scans such as x-rays, CT scan, MRI scan, history of medication, tests, results, surgeries, and many more such records will be available to every node in the Blockchain environment and such data from all patients available over all nodes will require a tremendous amount of storage space. Also, every transaction in the Blockchain environment is recorded, and this will only contribute to the growth of data, which will reduce the searching and access speed of records. Hence Blockchain network in the healthcare scenario will need massive storage and seems not fit for purpose to many researchers [1].

In this age of the IoT, users are generating data at massive scale from smart devices, which in the future would require integration with the healthcare system and thereby increase the size of data to be stored and retrieved. Researchers can foresee that such type of data will be generated daily and will increase the overhead [27].

3. **Standardization Challenges:** An International standardization will be required for authentication and certification in the Blockchain environment. Standardization is also crucial for scrutiny and safety of records within the Blockchain network. Standardization of data is essential for seamless and effective data transfer. However, ironically healthcare system is fragmented, and even in a nation, no universal standard is prevalent. In this case, standardization of all kinds of medical records at the global level seems impractical to researchers at least for now [1,27].

4   Brief Ethical Reflections on Blockchain Technology

Blockchain can deliver prime data security and full privacy, especially if complemented by secure computation. It represents a significant technological breakthrough, which -as we have seen above- promises to help overcome existing potential ethical barriers (such as transparency and accountability) in the implementation of learning healthcare system in society. Nevertheless, some important ethical considerations underlie the usage of this technology. Such considerations seem to threaten its scalability and potentially affect its desirability [23]. We briefly discuss some of these considerations below.

It is unquestionable that a digital identity system (a blockchain) can provide an immutable and secure identity that can be uniquely tied to a person's biometrics (such as fingerprints or iris scans). Such a digital system may indeed come in handy and be even extremely helpful in several different situations (it could, for instance, allow a member of a disenfranchised group -say a political refugee- to cross a border and get access to vital medical care which could be precluded on their country – [36]). However, how secure the personal information recorded on that blockchain could be? Who would be able to access to it, and who could guarantee that such a sensitive information would never be used against that person, to expose or exploit her in the future [15,26,39].

Otherwise stated, how a blockchain system is coded, who has access to it, who finances it, to whom the custodians of the information (e.g., healthcare organizations) must report and which rules govern the system (for instance, a reliable international legal framework for dealing with situations where the data subject is deceased), have fundamental intentional and unintentional social and ethical consequences [7].

In this short contribution, owning it to lack of space, we can't explore the moral and socio-cultural ramifications of these important concerns; however, we would like to point out that understanding the ethical and moral impact of these consequences matters profoundly for both the individual and the society in which she lives. To ensure that the best outcomes for individuals and communities are achieved, we would like to emphasise that any blockchain system should be intentionally designed with people in mind and should be driven by an ethical approach that takes the citizen and its community at heart. Thus, in implementing blockchain technology in the wider society, significant attention should be put on issues surrounding governance, personal identity, state rules, verification/authentication, access, data transfer, data ownership, security, trust, manipulation, and accountability.

5   Possible Blockchain Use Case in Health Industry

Table 1. Quality Attributes mapping to research papers

Data security/privacy	Decentralized management	Access control	Integrity	Interoperability
[4, 18, 33, 34, 41]		[13, 31]	[25, 41]	[18, 33]
[3, 6, 9, 20, 35, 43]	[9, 28]	[6, 9, 28]	[9, 35]	[9]
[10, 14, 40]	[17, 21]	[2, 12, 14, 38]		[2, 14, 32, 42]

This section describes the categorization of existing Blockchain-based solutions into five different quality attributes. We choose 23 different Blockchain-based proposed and existing solutions. Some solutions focus on a single quality attribute like integrity or interoperability, and others promote more than one quality attribute. Data management and data sharing, security, and privacy are some of the top requirements of the healthcare industry. Therefore, most of the proposed solutions are promoting the related quality attribute to address these requirements. There are five quality attribute categories, namely Data Security and Privacy, Decentralized Management, Access Control, Integrity, and Interoperability. Each paper in this survey is categorized in one or more of these categories in Table 1.

Some papers listed in Table 1 focus on cross-institutional sharing of data in the health industry. Cross-institutional sharing is a challenge because of different data structure, privacy, and security concerns involved in this process, the value of data exchange within the healthcare system is significant and hence requires a solution.

Paper [35] address the issue of complex and inefficient supply chain surveillance system to prevent the production and procurement of counterfeit drugs. The paper proposes a Gcoin blockchain as the base of the data flow of drugs. Also, suggest changing the regulation module for inspection and examination of the only model to the surveillance net model. The proposed scheme ensures Data transparency and surveillance at every stage of the supply chain. Addressing the same issue paper [34] presents Blockchain benefits and use cases to advance biomedical/healthcare data ledger. Five different used cases promoting five different quality attributes are presented in the paper. It proposes to use Blockchain as a data backbone of the healthcare industry, for an audit of clinical research proposals, in the supply chain to ensure original manufacturer and ownership transfer and to improve robustness for counterfeit drug prevention/detection systems and the last use case to improve patient confidence in consent recording systems

Paper [9] directly addresses the care coordination problem addresses in the challenge section, it aims for cross-institutional sharing of patient's data while maintaining the privacy and anonymity of the patient. Balance between. The paper suggests applying blockchain and smart contracts to Electronic Health Records (EHRs) to remove the conflict between data privacy and accessibility in the healthcare domain. Based on what the author of this paper is suggesting they propose a general BC based architecture for a global scale EHR followed by the scenario of usage. The Paper also discusses the ethical issues of heath domain data. The authors have not implemented the EHR but suggest that the system could be implemented using the Ethereum platform and its smart contract solution. They look forward to implementing a functional prototype as their future work. Paper [10] proposes a largescale information architecture to access Electronic Health Records (EHRs) based on Smart Contracts as information mediators. Paper [42] addresses the interoperability challenge through a Blockchain-based healthcare apps.

Clinical research is important for medical advancement the parties involved in clinical research are the participants and the clinical research institute. Clinical research requires the collection of participant's sensitive health data, Paper [3] address the care coordination issue in the case of clinical research. Mostly participants of such research are will to be the subject of such research and share their data on certain agreement to harness their definition of data privacy at the same time research institution try to ensure that the data collected from the participant are authentic. The paper provides proof-of-concept implementation to protect the interest of both the parties which allows participants to keep their data private until they reach on mutual agreement with e institution on the other hand ensures that institutions are receiving authentic data from the patient after the agreement. Paper removes the conflict of data security and privacy in this case. The paper provides use cases, reference architecture, and implemented a Proof-of-Concept (PoC) using RiotOS and Ethereum. In the end users real-world data set to evaluate the system performance. As future work, they look forward to implementing IoT in other stages of the clinical trial.

Paper [40] addresses the EHRs data security issue. The paper proposes a blockchain-based decentralized EHRs sharing network called MedRec. The authors presented the main structure of the whole network and contracts to be used in the network and further explains the whole mechanism of the working of the network. The proposed system claims to address eight main security issues namely confidentiality, privacy, access control, integrity, data authenticity, user authentication, audibility, and transparency. Paper has no future work insight. On the similar ground Paper [14] proposes a framework and did a comparative performance analysis with paper [40]. The framework is being called Ancile that uses 6 unique smart contracts namely Consensus, Classification, Service History, Ownership, Permissions, and Re-encryption in an Ethereum-based blockchain.

The proposed solution ensures data privacy and integrity. The authors look forward to continuing to researching blockchain technology to meet legislative standards for medical data and protection of patient privacy.

Paper [33] also addresses the EHRs secure sharing and Care coordination problem as paper [9] and 3 with two main focuses of this paper are security and to solve the issue of heterogeneous format/structure and meaning of data and the security. Proposes a blockchain bases network that ensures cones over the data syntax. Then harnesses network-wide keys and smart contracts to ensure security while sharing. The paper also presents 3 algorithms for different modules of the proposed network. The proposed solution ensures data consistency and security. The future work presented is very ambitious as they discuss the possibility of nested blocking and AI diagnosis output back to the blockchain

Paper [21] focuses on securing patient data, dissemination of patient's medical data is a risk to patient's privacy, and the healthcare industry needs active management to protect these records. The paper [38] proposes a system called MeDShare the system to ensure medical data sharing among medical big data custodians in a trust-less environment. MeDShare uses smart contracts for the flow of data. Paper [12] focused on solving the issue of authority, access control is a significant part of data security, especially in healthcare. The solution proposes a distributed ledger, a consortium Blockchain that stores operations as transactions that ensure that different facilities know all the parties that can act over the e-Health resources. The solution from [32] has its focus on creating Blockchain-based health ecosystems that are patient-centric in terms of interactions to enable trust, balancing the pluralistic morality of identity. Paper [4] addresses the data management issue and focuses especially on wearable technology data-sharing issues. The paper proposes a scheme called Med Chain, MedChain brings the benefit of blockchain, digest chain, and P2P network techniques. The paper presents detailed working of the scheme with module architecture and efficiency analysis and their evaluation result shows that the scheme can achieve data security and flexibility, as the future work authors are willing to overcome the limitation of MedChain.

On the similar ground paper [18] presents another blockchain-based platform called BlocHIE for information exchange for both personal healthcare data and electronic health record EHRs. The paper presents a minimal viable product and have extensively evaluated the algorithms. The minimum viable product ensures efficient data sharing and transparency among users. Paper [28] address data management issues with the perspective of identity and access management using blockchain. Hyperledger Fabric an Open source blockchain framework is being used to develop a prototype. The paper proposes the concept with only EHRs use case and focuses on system efficiency and scalability.

Paper [13] proposes a solution for restricted use in terms of area where surgical instruments associated with an RFID tag can be used and not allowed for use outside the permissible zone. Paper [17] proposes permissioned Blockchain for implementing Medical Administration Records (MAR) whereas [31] uses Ethereum Blockchain for the same purpose. Paper [20] proposes an access control manager to securely store and access data required by the image classification application (Stacked Denoising Auto-encoders (SDA)) during retraining in Real-time from an external data storage. Paper [43] has the proposal of a body sensor network to design a lightweight backup and efficient recovery scheme for keys of health Blockchain. In paper [6] a start-up idea using IoT (Internet of Things) sensor devices is being proposed called modum.io. The start-up uses blockchain technology to reduce operational costs in the pharmaceutical supply-chain and in to assert data immutability public accessibility. Paper [25] addresses data management and secure sharing issue by proposing a decentralized permission Blockchain for data privacy and a mobile application to collect health data from patients' health gadgets, which are further shared with the healthcare provider and insurance companies over a secured network.

Although several paper addressed above presents schemes to improve the interoperability issues in the healthcare domain there almost no scheme or guideline is being proposed to evaluate healthcare applications based on blockchain. One very interesting paper [41] provides evaluation metrics to assess Blockchain-based DApps in terms of their feasibility, intended capability, and compliance in the healthcare domain.

6   Conclusion

In this paper, we have investigated the current use of Blockchain in the healthcare domain. Blockchain technology presents a decentralized network and because of the sensitive nature of data in this domain has potential for use in healthcare. The study aims to categorize and evaluate different solutions in the healthcare domain that makes use of programmable Blockchain or part of it. We have used a software engineering approach to organize the existing work in multiple categories, such as challenges addressed and quality attributes promoted.

This work is intended to provide researchers in the field with a well-defined and structured categorization, plus insights into the existing literature. To achieve this objective, we have surveyed 23 solution based on Blockchain. Some of them can be called decentralized applications (Dapps) and the others used parts of the Blockchain but not entirely. These solutions are proposed to solve some challenges inside the healthcare domain, such as data integrity, data security, data privacy (authentication and authorization), and data management. On the other hand, there is still considerable uncertainty concerning the use of Blockchain in healthcare. Several researchers believe that interoperability, scalability, storage, social acceptance, standardization are challenges that prevent the introduction of Blockchain into the healthcare domain.

These results have highlighted the importance of knowing the use case before making any suggestion for a solution, whether the technology is Blockchain or not, and the importance of categorization based on quality attribute and the challenges addressed. In our research, we propose to analyze any technology through software engineering methods. Despite the hype and tending for a particular technology, potential challenges must be defined as well as if the technology can provide the solution. Our work concludes that Blockchain can be used and provide multiple solutions in the healthcare domain depending on the use case and the challenges in the project.

7   Future Work

For future work, we propose to examine customized Blockchain settings and specifications for different application domains. A single-solution approach seems inadequate in consideration of various concerns regarding ethics, privacy, authority, and efficient data storage and access. We look into examining the customized settings for Blockchain technology according to the services it needs to address. There is always a trade-off between variables. As an example, seamless data access compromises privacy, and decentralization raises the issues of authenticity. Service and functional priorities differ from one organization to the other. In respect to such concerns, adopting the Blockchain for customizable options is indispensable.

Blockchain has a variety of exciting potential for transforming the healthcare industry, many of which have not yet been implemented. Some areas in the healthcare industry (insurance, pharmacology, smart hospitals, etc.) can use Blockchain more.

One of the highest costs of loading healthcare processes and the medical system is the constant monitoring of the flow of money and services provided. Unfortunately, well-equipped and advanced medical centers with more specialized departments have longer accounting processes. Sometimes, due to the transfer of a patient from a medical center to a hospital or even another city, the payment flow is interrupted. With Blockchain's help, the insurance industry and healthcare providers' differences will be eliminated, and when it comes to patient care and treatment, there will be no waste of money.

Effective exchange of research results, which facilitates the production and production of new drugs, is the other of the advantages of using Blockchain in Healthcare.

Smart hospitals, built on an ICT environment of interconnected assets, particularly based on the Internet of things (IoT) [30], are the other field of the healthcare industry that can use the benefits of using Blockchain.

References

1. Siyal, A.A., et al.: Applications of blockchain technology in medicine and healthcare: challenges and future perspectives. Cryptography **3**, 3 (2019)
2. Ekblawand, A., Azaria, A., Halamka, J., Lippman, A.: A Case Study for Blockchain in Healthcare:"MedRec" prototype for electronic health records and medical research data
3. Angeletti, F., Chatzigiannakis, I., Vitaletti, A.: The role of blockchain and IoT in recruiting participants for digital clinical trials. In: 2017 25th International Conference on Software, Telecommunications and Computer Networks (SoftCOM), pp. 1–5 (2017)

4. Shen, B., Guo, J., Yang, Y.: MedChain: efficient healthcare data sharing via blockchain. Appl. Sci. 49–56 (2019). https://doi.org/10.3390/app9061207
5. Blackstone, E.A., Jr. Fuhr, J.P., Pociask, S.: The health and economic effects of counterfeit drugs. Am. Health Drug Benefits **7**, 216–224 (2014)
6. Bocek, T., et al.: Blockchains everywhere - a use-case of blockchains in the pharma supply-chain. In: 2017 IFIP/IEEE Symposium on Integrated Network and Service Management (IM), pp. 772–777 (2017)
7. Lapointe, C., Fishbane, L.: The blockchain ethical design framework. Technology, Governance, Globalization, Innovations (2019)
8. Combating Counterfeit Medicine; WHO Drug Information: Geneva, Switzerland, vol. 20, pp. 3–4 (2006)
9. da Conceição, A.F., et al.: Eletronic Health Records using Blockchain Technology. CoRR abs/1804.10078 (2018). arXiv: 1804.10078. http://arxiv.org/abs/1804.10078
10. de Oliveira, M.T., et al.: Towards a blockchain-based secure electronic medical record for healthcare applications. In: ICC 2019 - 2019 IEEE International Conference on Communications (ICC), pp. 1–6 (2019)
11. Deloitte. Blockchain-opportunities-for-health-care
12. Dias, J.P., Ferreira, H.S., Martins, A.: A blockchain- based scheme for access control in e-health scenarios. In: SoCPaR (2018)
13. Figueroa, S., Añorga, J., Arrizabalaga, S.: An attribute-based access control model in RFID systems based on blockchain decentralized applications for healthcare environments. Computers (2019). https://doi.org/10.3390/computers8030057
14. Dagher, G.G., et al.: Ancile: privacy-preserving framework for access control and interoperability of electronic health records using blockchain technology. Sustain. Cities Soci. (2018). https://doi.org/10.1016/j.scs.2018.02.014
15. Gross, M.S., Miller, R.C.: Blockchain in healthcare today. In: Ethical implementation of the learning healthcare system with blockchain technology (2019)
16. Haber, S., Scott Stornetta, W.: How to time-stamp a digital document. J. Cryptol. **3**(2), 99–111 (1991). https://doi.org/10.1007/bf00196791
17. Mitchell, I., Hara, S.: BMAR - blockchain for medication administration records. Adv. Sci. Technol. Secur. Appl. 231–248 (2019). https://doi.org/10.1007/978-3-030-11289-9_10
18. Jiang, S., et al.: BlocHIE: a BLOCkchain-based platform for healthcare information exchange. In: 2018 IEEE International Conference on Smart Computing (SMART-COMP), pp. 49–56 (2018)
19. Jones, S.S., et al.: Health information technology: an updated systematic review with a focus on meaningful use. https://doi.org/10.7326/M13-1531
20. Juneja, A., Marefat, M.: Leveraging blockchain for retraining deep learning architecture in patient-specific arrhythmia classification. In: 2018 IEEE EMBS International Conference on Biomedical Health Informatics (BHI), pp. 393–397 (2018)
21. Khatoon, A.: A blockchain-based smart contract system for healthcare management. Electronics (2020). https://doi.org/10.3390/electronics9010094
22. Klein, D.M., et al.: Use of the blue button online tool for sharing health information: qualitative interviews with patients and providers. J. Med. Internet Res. (2015)
23. Dimitropoulos, L., Patel, V., Scheffler, S.A., Posnack, S.: Public attitudes toward health information exchange: perceived benefits and concerns. Am. J. Manag. Care (2011)
24. LaKeisha, M., Ellen, W.: The real impact of counterfeit medications. US Pharm. **39**, 44–46 (2014)

25. Liang, X., et al.: Integrating blockchain for data sharing and collaboration in mobile healthcare applications. In: 2017 IEEE 28th Annual International Symposium on Personal, Indoor, and Mobile Radio Communications (PIMRC), pp. 1–5 (2017)
26. Tate, M., Johnstone, D., Fielt, E.: Ethical issues around crowdwork: How can blockchain technology help? (2017)
27. McGhin, T., et al.: Blockchain in healthcare applications: research challenges and opportunities. J. Netw. Comput. Appl. **135**, 62–75 (2015). https://doi.org/10.1016/j.jnca.2019.02.027. ISSN 1084-8045. http://www.sciencedirect.com/science/article/pii/S1084804519300864
28. Mikula, T., Jacobsen, R.H.: Identity and access management with blockchain in electronic healthcare records. In: 2018 21st Euromicro Conference on Digital System Design (DSD), pp. 699–706 (2018)
29. Nakamoto, S.: Bitcoin: A peer-to-peer electronic cash system (2008)
30. European Union Agency for Network and Information Security: Smart Hospitals Security and Resilience for Smart Health Service and Infrastructures (2016). https://doi.org/10.2824/28801
31. Nguyen, D.C., et al.: Blockchain for secure EHRs sharing of mobile cloud based e-health systems. IEEE Access **7**, 66792–66806 (2019)
32. Nichol, P., Brandt, J.: Co-Creation of Trust for Healthcare: The Cryptocitizen Framework for Interoperability with Blockchain (2016). https://doi.org/10.13140/RG.2.1.1545.4963
33. Peterson, K., et al.: A blockchainbased approach to health information exchange networks. In: NIST Workshop Blockchain Healthcare, vol. 1, pp. 1–10 (2016)
34. Kuo, T.-T., Kim, H.-E., Ohno-Machado, L.: Blockchain distributed ledger technologies for biomedical and health care applications. J. Am. Med. Inf. Assoc. **1**, 1–10 (2017). https://doi.org/10.1093/jamia/ocx068
35. Tseng, J.-H., et al.: Governance on the Drug Supply Chain via Gcoin Blockchain. International J. Environ. Res. Public Health **15**(6), 1055 (2018)
36. Weinfurt, K.P., Bollinger, J.M., Brelsford, K.M., et al.: AJOB Empir Bioeth. In: Patients' views concerning research on medical practices: Implications for consent (2016)
37. WHO. CONSTITUTION OF THE WORLD HEALTH ORGANIZATION (1946). https://apps.who.int/gb/bd/PDF/bd47/EN/constitution-en.pdf. Accessed 12 Oct 2020
38. Xia, Q., et al.: MeDShare: trust-less medical data sharing among cloud service providers via blockchain. IEEE Access **5**, 14757–14767 (2017)
39. Tang, Y., et al.: Information Technology & People. In: Ethics of blockchain (2019)
40. Yang, B., Yang, H.: A blockchain-based approach to the secure sharing of healthcare data. In: Proceedings of the Norwegian Information Security Conference 2017, Oslo, Norway (2017)
41. Zhang, P., et al.: Metrics for assessing blockchain-based healthcare decentralized apps. In: 2017 IEEE 19th International Conference on e-Health Networking, Applications and Services (Healthcom), pp. 1–4 (2017)
42. Zhang, P., et al.: Applying software patterns to address interoperability in blockchain-based healthcare apps. ArXiv abs/1706.03700 (2017)
43. Zhao, H., et al.: Lightweight backup and efficient recovery scheme for health blockchain keys. In: 2017 IEEE 13th International Symposium on Autonomous Decentralized System (ISADS), pp. 229–234 (2017)

Innovative Blockchain-Based Applications - State of the Art and Future Directions

Hada Alsobhi$^{(\boxtimes)}$, Abeer Mirdad, Suhair Alotaibi, Mwaheb Almadani,
Inam Alanazi, Mohrah Alalyan, Wafa Alharbi, Rania Alhazmi,
and Farookh Khadeer Hussain

Faculty of Engineering and Information Technology, University of Technology
Sydney, 15 Broadway, Ultimo, Sydney, NSW 2007, Australia
{Hada.Alsobhi,Abeer.Mirdad,Suhair.Alotaibi,Mwaheb.Almadani,Inam.Alanazi,
Mohrah.S.AlAlyan,WafaMatarA.Alharbi,Rania.Alhazmi}@student.uts.edu.au,
Farrokh.Hussain@uts.edu.au

Abstract. Recently, blockchain technology has increasingly being used to provide a secure environment that is immutable, consensus-based and transparent in the finance technology world. However, significant efforts have been made to use blockchain in other fields where trust and transparency are required. The distributed power and embedded security of blockchain leverage the operational efficiency of other domains to be immutable, transparent, and trustworthy. The trust of the published literature in blockchain technology is centered on crypto-currencies. Therefore, this paper addresses this gap and presents to the user several applications in many fields, including education, health, carbon credits, robotics, energy, pharmaceutical supply chains, identity management, and crypto-currency wallets. This paper overviews the knowledge on blockchain technology, discusses the innovation of blockchain technology based on the number of applications which have been introduced, describes the challenges associated with blockchain technology, and makes suggestions for future work.

Keywords: Blockchain · Applications · Blockchain research challenges

1 Introduction and Background

In 2008, blockchain technology was first introduced to the public, along with the emergence of Bitcoin. Blockchain is considered a technological innovation that has broad implications and applications after Ethereum launched the smart contracts [1]. As technological communities and industrial giants are rapidly moving towards service-directed frameworks with distinct application specifications, blockchain applications have come under comprehensive investigation. These systems will allow users to obtain services with more sophisticated features. Many blockchain applications are utilized as an immutable ledger where there is a lack of trust between entities so that a trustworthy third party is no longer needed in

L. Barolli et al. (Eds.): AINA 2021, LNNS 227, pp. 323–335, 2021.
https://doi.org/10.1007/978-3-030-75078-7_33

such systems. Given that centralized storage technologies have become controversial and inadequate, a blockchain-based application can increase transparency, provide immutability, and create trust between organizations. Thus, blockchain technology has attracted the attention of many researchers and practitioners due to blockchain's more advanced functionalities and features over traditional technology systems. In addition to the range of new blockchain use cases occurring every day, including supply chains, healthcare, digital identities, energy, and intellectual property systems. Blockchain technology has recently been used in a number of different applications as a way to maintain trustworthy records and information in a distributed and reliable manner [2].

Researchers have defined blockchain as a distributed ledger technology (DLT) that contains a sequence of blocks to store valuable data and information that can be shared over the network between parties [1]. Blockchain has the following key characteristics: decentralization, persistency, auditability, and anonymity. Blockchain is a technological solution to solve security issues, such as ensuring all transactions are secured with cryptographic hashes and asymmetric-key pairs to sign and verify any type of transaction. Therefore, once the data is recorded in a blockchain, it becomes difficult to change.

Within the blockchain network, all historical transactions between nodes are recorded in distributed ledgers. Nodes or peers are connected to a P2P network and hold a copy of the shared ledger. In order to add and verify the transactions within the system, the miners or crypto miners need to agree on whether any new blocks are valid or not by applying consensus algorithms. The essential operation of the blockchain rely on the consensus process. Since blockchain features work without a leader or a trusted third party, it uses consensus mechanisms to make a decision and guarantee ledgers are compatible in various nodes. There are advantages and disadvantages for each consensus mechanism, but it is important to sustain a system state agreement by following the guidelines of consensus mechanisms. Also, there are several consensus models such as the Proof of Work Consensus Model, Proof of Stake Consensus Model and Proof of Authority Consensus Model. The Consensus Comparison Matrix in [1]. compares the different consensus models. Cryptography and hashing mechanisms are used to construct a chain of data blocks to ensure immutability in the append-only distributed ledgers in the blockchain network. Thus, all data blocks contain a series of transactions appended to the previous blocks using the hash values as part of the stored data.

Smart contracts are a set of codes that are executed by miners. The smart contract was defined in 1994 by Nick Szabo as "a computerized transaction protocol that executes the terms of a contract" [3]. The overall goals of smart contracts are designed to satisfy common contractual conditions, such as payment terms, to minimize both malicious and occasional exceptions and to reduce the need for reliable intermediaries. The smart contract is executed by nodes through the blockchain network; all the executed nodes must derive the same results from the execution, and the results of the execution are recorded on the blockchain. However, not every blockchain can run smart contracts [1].

This objective of this paper is to expose the reader to range of innovation blockchain applications in many sectors such as education, identity management and carbon credit management. A significant amount of the extant literature on blockchain technology in the field of computer science is focused on its use in the finance sector in the form of crypto-currencies. However, there are numerous use cases of blockchain outside crypto-currencies. The purpose of this paper is to enlighten the reader to a number of innovative applications of blockchain technology across a number of domains, each of which has the potential to have a high impact in the respective sector of the application. It explains how integrating such technology can significantly reshape advanced industries of blockchain technology in diverse disciplines, such as education, carbon credits, robotics, energy, pharmaceutical supply chains, identity management, and crypto-currency wallets.

2   Innovative Blockchain-Based Applications

In this section, we outline and discuss a number of blockchain applications across a number of sectors.

2.1   Blockchain in Education

A recent development in the education sector has been the emergence of micro-credentials [4]. Micro-credentials give students the ability to undertake 'micro' offerings such as short courses from different universities. These short courses could be of various lengths - ranging from a few hours to a few weeks. Students also have the flexibility to select bespoke offerings from different universities. The micro-credential content is delivered online as opposed to face- to-face content delivery. This gives students complete flexibility to select micro-credentials from multiple universities, and importantly tailor their learning experience depending on their availability. A significant issue facing the emergence of micro-credentials is the lack of fool-proof mechanisms to verify on-the-fly a micro-credential issued by another institution. Furthermore, technological solutions for managing micro-credentials are in their infancy, so proposing blockchain-based solutions can help not just in the authentication of micro-credentials, but using some of the emerging artificial intelligence approaches to manage and recommend micro-credentials for students.

Over the last few years, the higher education sector worldwide has grown which has resulted in the emergence of new educational offerings, including online courses, online certificates, online degrees, and more recently micro-credentials. The growth of these educational offerings is evidence of the trend towards continuous lifelong learning. A new ecosystem of educational providers, including technology companies, is beginning to emerge with a view to imparting the 'on-demand' and 'just-in-time' skills (using emerging skills) that are required by employers and which are advantageous for employees to have in their portfolio of skills [5].

The higher education (HE) domain is a suitable domain for adopting blockchain technology. Blockchain can make administrative procedures faster and easier when credentials need to be verified. A small number of higher education institutions (HEIs) have started to use blockchain in several applications to share, verify, and validate learning outcomes and certificates [6]. Blockchain technology can help in managing the micro-credentials of academic achievements that a learner accrues over a lifetime. The development of such a reliable, trustworthy, robust, and scalable approach including new technologies will be crucial in supporting community acceptance and the uptake of micro-credentials [5].

Universities around the world are finding ways to recognize micro-credits officially. The Australian Qualification Framework has now opened the door for micro-credentials and most Australian universities, such as the University of Technology Sydney (UTS), RMIT, the University of Melbourne, Griffith University, the University of New South Wales, and the University of Western Sydney recognize short courses, known as micro-credentials. UTS launched its micro-credentials in 2020 [7].

2.2   Blockchain in Managing Carbon Credits

There is a strong movement around the world to reduce carbon emissions. Many countries such as New Zealand, Finland and Australia have plans to entirely cut their carbon emissions in the next few decades. New Zealand, U.K. and Scotland have plans for net zero carbon emissions by 2050 [8]. Carbon credits is a term used for any type of permit that gives the holder the right to produce one to One of carbon dioxide or the same amount of other greenhouse gases [9]. There will be an urgent need for the carbon credits market by 2021, and systems will be needed to allow individuals to trade in carbon credits. Hence, strategies, technologies and the corresponding exchange of carbon credits are needed to measure the carbon credit trade [10].

Currently, carbon credits are used as a mechanism to reward companies and countries that are emitting less CO_2. However, the technology around digital management and digital trading is suffering from many shortcomings such as double counting and a lack of transference in exchange for carbon credits. Hence, it is essential to introduce improvements to the current system. One such solution is blockchain. Using blockchain to manage and trade carbon credits is one of its many uses that has received relatively very little research attention so far. Proposing blockchain technology as a solution will help to create a market that enables the trade of carbon credits as well as to solve the problem of double counting. Blockchain technology can provide a secure, trustworthy and transparent system in which all the information of carbon credits can be managed in a reliable manner.

2.3   Blockchain in Robotics

Over the last few decades, the demand to investigate robotics technology in other industries, such as logistics services, education, healthcare, agriculture and many more has increased [11]. This has led to a dramatic growth in robotics manufacturing [12]. A large number of robotics providers offer robots with a wide variety of capabilities to perform a wide range of functions. Some of these capabilities are combined and integrated in a multipurpose robot that is engineered to provide similar or overlapping services [13]. Integrating blockchain technology in the robotics field has attracted researchers' attention over the last few years. The robotics research areas have a number of research gaps or issues that need to be solved. Blockchain technology presents a promising next-generation approach to solve these pressing research issues. In the remainder of this sub-section, we discuss some of these issues and potential solutions involving blockchain.

Due to the variety of robotic alternatives and the advanced features of robots, the identification and selection of the most suitable robot for a particular task is one of the challenges that face robot service requesters, especially non-expert requesters. So, establishing a reputation system that manages data on robots and robot providers could address the robot selection problem and help robot requesters to make a decision. All reputation systems are prone to manipulation by malicious users which include robot providers who may give high ratings to themselves and/or give their competitors low ratings [14]. So, there is a strong argument for using tamper- proof mechanisms to store reputation values. The secure and verifiable blockchain structure may be used to store the reputation values and prevent data manipulation since the blocks are created and stored in a distributed manner by different peers.

In the case of complex tasks, the robot requesters can now build a cooperation mechanism to formulate a robot swarm from homogeneous or heterogeneous robots to perform the required tasks [15]. Such a swarm structure is highly vulnerable to attack by malicious or byzantine robots. Thus, a number of blockchain-based methods have been proposed to secure the swarm and validate the robots' behaviour [16,17]. Obviously, these blockchain-based methods could bring other great benefits to the robotics swarm such as rewarding valid robots, selecting swarm leaders and allocating tasks to a robot based on its behaviour history.

2.4 Blockchain in Energy Markets and Trading

In the energy sector, large centralized energy providers, who primarily produce electricity (e.g., power grid) using non-renewable energy fossil fuels, are the primary sources for energy production to support the huge energy demand. This traditional power supply has a severe impact on both the environment (e.g., environmental pollution and climate change) and consumers (e.g., energy shortage and increased electricity cost). Driven by these impacts, new clean renewable energy sources (RES), such as solar energy, wind energy and water energy have been developed [18].

The distribution and integration of RES into the energy system requires a new energy marketplace schema to manage and maintain the increasing RES production and consumption and intelligently connect both power providers and consumers in an effective way [19]. This is one of the greatest challenges in the energy sector because, to date, the existing marketplace is unable to handle the dramatic increase of RES in real-time. Furthermore, energy prices in these markets are determined on a national level without considering local energy demand and supply. Therefore, several local energy marketplace (LEM) approaches have been proposed to successfully integrate distributed renewable resources for the local community into the energy system [18].

This local energy market mechanism provides a market that facilitates local trade and access to the local RES for a practical community. A community is a group of energy consumers and prosumers (e.g., a homeowner, a commercial company, an industrial factory or a farm) who have a smart meter and battery storage. There are three main participants in the LEM: an energy producer, an energy consumer and a prosumer, who consumes and produces energy as well.

To ensure the long-term sustainability of this LEM, an innovative, secure and decentralized information technology using blockchain, is needed. Blockchain creates a safe, decentralized, transparent and trustworthy network that empowers the energy trading marketplace and makes it affordable for local residents. Blockchain, as a distributed computing technology, opens a new future direction for the decentralized energy trading market by providing a secure, transparent and trustworthy network trading environment for communities [20].

The LEM is also known as a peer-to-peer (P2P) online trading marketplace. The business model for this blockchain marketplace is based on an interaction between sellers and buyers via an intelligent platform. This trustworthy blockchain platform will allow energy producers to meet and sell their excess electricity at a desired price to consumers directly without an intermediate party. For instance, a producer will create a smart contract on the blockchain through the platform to sell his surplus energy and when a consumer is willing to purchase that electricity, the smart contract will be executed. In practice, the purchased energy will be transferred from the producer's smart meter to the consumer's smart meter and when the consumer receives all the agreed energy, then the money will be transferred from the consumer's wallet on the blockchain to the producer's wallet.

2.5   Blockchain in the Pharmaceutical Sector

Implementing an effective strategy to ensure the existence of reliable and equitable pharmaceutical supply chains is one of the highest priorities in healthcare systems. In developing countries, the distribution of medicines is an urgent issue due to the complexity of managing the pharmaceutical supply chain as there are many stakeholders and it also involves human wellbeing [21].

A pharmaceutical supply chain consists of a set of players, processes, information, and resources which transfers raw materials, and components to finished products or services and delivers them to the customers [22]. The drug industry is facing several problems such as finding a balance in terms of producing drugs to meet consumer demand, tracking every entity in the supply chain, and reducing the number of counterfeit drugs. Blockchain (in conjunction with other approaches such as artificial intelligence) offers a powerful means to solve these problems.

Optimizing the process and management of the pharmaceutical supply chain and sub- standard drug proliferation is a major research challenge. An imperfect supply chain system and the lack of traceability are the two main reasons for the appearance of counterfeit drugs. This problem is due to the lack of shared information between nodes in the supply chain system [22,23]; for instance, manufacturers are not aware of the location of their products after exporting their drugs. Hence, better tools for managing information on drugs are urgently required. The risk to human life due to counterfeit medicines is becoming a global issue. The World Health Organization (WHO) reported increased sales of counterfeit drugs around the world, and it is expected to increase by 35% over the next five years [24]. Developing countries in Africa and Asia are the main areas suffering from counterfeit drugs, which represent around 30% of total medicine sales.

The security and traceability of pharmaceutical companies and pharmacy distributors is becoming extremely highly important to control this problem [25]. Many technologies have been used to improve medicine supply chain systems, but there are still many issues. The introduction of blockchain technology can reap several benefits. Drugs, for example, can be tagged and scanned to be stored securely in a distributed ledger; this ledger will be updated in real-time as the drugs are transferred from one entity to another in the supply chain [26].

2.6   Blockchain in Identity Management

Identity (ID) management is a critical topic that focuses on managing individual characteristics. An ID management system is paramount in personal and organisational communication to provide health, education, business, transportation and government services amongst others. People aim to obtain a secure and reliable ID given its significant value. Thus, blockchain is an ideal technology that securely stores data.

Many institutions use the traditional ID system for data storage in a centralised database that relies on a third party, leading to a lack of interoperability and many security and privacy issues. Moreover, a number of privacy acts outline the ownership of personal ID information (PII) and grant legal rights to certain organisations to use individuals' PII. In 2019, Malaysia was ranked fifth among the nations with the worst personal data security system. Its number of reported data breach cases increased significantly from 64 in 2018 to 178 in 2019 [27].

To address some of the above emergent challenges in ID management, Blockchain ID management studies have recently grown. Different authentication mechanisms were applied, including smart cards and passwords and biometric methods that use fingerprint, facial recognition etc. For example, Estonia is one of the first countries that uses a public-permissioned network in blockchain and digitised their ID system since 2007 through the citizen digital ID. They also started signing documents digitally through a combination of a citizen ID and PIN and have been providing non-citizens digital IDs since December 2014 to become e-residents for commercial activities [29].

A self-sovereign ID facilitates individuals to fully access their IDs without third parties. Blockchain uses a high-security system in ID management to enable users to authenticate and authorise resource access through a digital signature in a digital ID. Furthermore, blockchain can solve data storage problems associated with most ID management systems and provide a secure decentralised network and minimise ID theft [30]. It is a smart way to reduce fraud by increasing transparency and an ideal solution to protect and recover communities' ID in case of disasters, such as fire and earthquake. On the other hand, a federated ID has valuable benefits that allow users to use one authentication in accessing different services while minimising digital ID management and its cost. The most important solutions currently existing are SAML, Shibboleth, Liberty Alliance framework, openID and WS [31]. Blockchain in ID not only focuses on data storage but also offers economic ID to save individuals' reputation data. Storing transactions between different users in blockchain builds a trust relationship.

Blockchain based approaches for managing user identities have universal applications that allows users to own and control their IDs and easily integrate and manage them. Data will be encrypted and stored securely to enhance transparency, privacy and security. Research in the area of Blockchain-based approaches for identity management is in its infancy; hence, there are a number of opportunities and research challenges to be addressed in this space.

2.7   Blockchain-Based Wallet

Payment systems have developed from a traditional payment system that utilizes physical or pocket wallets to a digital payment system that utilizes digital or E-wallets. Digital wallets have become extremely prevalent because large companies such as Google, PayPal, and Facebook have embraced the trend of global payments and offer more e-wallet payment services. The benefits of adopting digital wallets include a convenient and secure payment process, efficient funds transfer, and utility costs reduction. However, when applying traditional digital transactions that leave digital footprints, certain data security risks arise, and entities' sensitive information is available to banks and service providers [32]. On the other hand, blockchain as a payment system is a relatively recent trend driven by the success of Bitcoin and its ability to build a trusted ecosystem due to its decentralization and immutability features. By implementing blockchain technology as a payment medium, a financial business can significantly reduce transaction time and decrease operational transaction costs as it removes many intermediaries [33].

The cryptocurrency system requires a blockchain-based wallet to send and receive cryptocurrencies within the network. Blockchain-based wallets are software or hardware devices that allow individuals to send and receive digital currency within the blockchain network. It contains no currency, and it is more like a keychain that holds private keys with each key being associated with a public key address. Digital wallets in blockchain technology play a significant role in securely maintaining user's public-private key pairs. Deciding on the appropriate and secure blockchain-based wallets is critical when joining the blockchain network since losing private keys means losing the cryptocurrencies associated with that private key [34].

Although blockchain is a promising technology and blockchain-based wallets are here to stay, there is a lack of studies which investigate the security mechanisms that should be used on top of blockchain-based wallets, such as two-factor and biometric authentication. A limited number of studies have introduced a risk-based mechanism for users to determine what security mechanisms they should utilize to secure their wallets based on their requirements. Furthermore, there is a lack of a decision-making model to make trust- based assessments when interacting with other entities. Therefore, there is a critical dilemma in relation to which procedures provide a secure and more trustworthy management scheme for the blockchain-based wallet so that it becomes more reliable, efficient, and stable.

3   Blockchain Adoption Challenges

The emergence of blockchain has had a profound effect on how information is stored and processed securely. However, blockchain adoption in public sectors is still very limited [35]. This section provides an in-depth study on the challenges of the adoption of blockchain in real-world contexts.

One of the biggest challenges in implementing blockchain is the lack of a clear understanding of this technology. This lack of information has led to insufficient recognition of technical compliance, subsequent benefits of acquisition, and potential use cases [36]. Moreover, there is inadequate awareness of the level of organization, where managers have shown unwillingness to adopt novel technology because of their failure to recognize the significant returns on investment and the lack of effective technology use cases. Confusion between Bitcoin and blockchain has also led to this misconception [36]. Technology should isolate itself from the negative connotations of cryptocurrency and the concerted efforts of educated consumers are needed to spread greater awareness of blockchain and its benefits beyond cryptocurrencies alone.

The initial cost of adopting blockchain is very high owing to its complexity. Moreover, the adoption of blockchain technology currently requires either the development of a proprietary solution or soliciting the services of blockchain technology providers, which are very scarce [37]. Because of the technical infancy of blockchain, there is a shortage of skilled professionals in this field. In addition, the new system may require hardware resources owing to the computation-intensive nature of the technology [38]. Therefore, the adoption of a blockchain-based solution currently exceeds the limited financial and operational limits for medium and small enterprises and generally speaking is only being used by large organizations.

Blockchain technology has attracted the attention of many organizations due to its prospects of multi-sector applications outside the finance field where it first began. However, the adoption of blockchain technology experiences challenges such as insufficient knowledge of blockchain and initial costs.

4 Discussion

The above sections in this paper, have outlined a number of innovative blockchain applications across a number of different sectors. Although the above applications of lockchain is not exhaustive, it is extensive in the sense that a smorgasbord of blockchain applications have been outlined and discussed. Each of the blockchain applications presented in paper opens and presents a unique set of research challenges that require further investigation. For example, considering the Carbon Credit Blockchain application presented in Sect. 2.2, a niche research problem specific to this application is that of the development of automated trading algorithms to be able to autonomously transact on behalf of the end users. Another example of a unique research issue is that of Machine Learning methods needed to be able to rank the blockchain wallets in a personalized manner. Discussing each and every research problem emanating from the above mentioned blockchain application is not in the scope of this paper. However, we would like to outline and discuss a few research challenges that are common across most, it not all of the aforesaid mentioned blockchain application.

- **Approaches for ascertaining veracity of the subjective information**
  Most Blockchain applications in the literature assume that the originating source of information is from a reputable or objective source. The existing literature lacks approaches for verifying the validity of subjective information sources prior to it being mined. If the blockchain is taking into account subjective information or information from subjective sources (such as social media), then its legitimacy needs to be ascertained, before it is added as a Block in the blockchain.
- **Novel consensus approaches for mining information from subjective sources**
  Various protocols such as PoW, PoS and PoC or PoA are used by the miners during the mining process. While the working of each of these protocols vary, they are consistent in checking the legitimacy of the source information prior to the mining process. These consensus approaches have been designed to work well when the input information source/s are from objectives sources; however, this observation does not hold true when the input information source is a subjective source. To cater for scenarios wherein the input information source is subjective, there is need for further research to develop mining protocols that can take into account subjective sources of information.
- **Blockchain adoption methodologies**
  With each of the above mentioned blockchain applications, there is a pressing need to develop informed methodologies with a view to maximise the adoption of the niche blockchain application in the respective domain. While there could be some learning derived from the development and use of blockchain adoption strategies from one domain to another domain (for example there could be some commonality between the factors impacting Micro-credential blockchain adoption and Identity management blockchain; the overall developed methodology would have to be bespoke to the specific blockchain application.

5   Conclusion and Future Work

Blockchain technology plays a critical role in the digital transformation of centralized susceptible data structures into a distributed, trackable and auditable data structure. Moreover, blockchain technology is new, and it is continuing to be utilized in new applications and to be developed globally. Therefore, it is necessary to keep developing blockchain applications in many different areas for a future which is free of fraud and deception. This paper highlighted the significance of blockchain technology in different disciplines, including education, carbon credits, robotics, energy, pharmaceutical supply chains, identity management, and cryptocurrency wallets. This could have broader implications to leverage the applications' operational efficiencies to be more secure, transparent, and trustworthy. Also, this is the first research study that provides a collection of applications that introduce the technology of blockchain to the field.

Our future work involves tackling a number of research issues and challenges. One line of enquiry that we intend to keep on pursuing is that of developing innovative blockchain applications in various sectors/realms. The second line of enquiry that we intend to work on is to develop solutions to pressing Blockchain problems such as the ones outlined in Sect. 4.

References

1. Yaga, D., Mell, P., Roby, N., Scarfone, K.: Blockchain technology overview arXiv preprint arXiv:1906.11078 (2019)
2. Woodside, J.M., Augustine, F.K., Jr., Giberson, W.: Blockchain technology adoption status and strategies. J. Int. Technol. Inf. Manag. **26**(2), 65–93 (2017)
3. Zheng, Z., Xie, S., Dai, H., Chen, X., Wang, H.: An overview of blockchain technology: architecture, consensus, and future trends. In: 2017 IEEE International Congress on Big Data (BigData Congress), , pp. 557–564. IEEE (2017)
4. Ralston, S.J.: Higher education's microcredentialing craze: a postdigital-deweyan critique. Postdigital Science and Education, p. 1
5. Turkanović, M., Hölbl, M., Košič, K., Heričko, M., Kamišalić, A.: Eductx: a blockchain-based higher education credit platform. IEEE Access **6**, 5112–5127 (2018)
6. Jirgensons, M., Kapenieks, J.: Blockchain and the future of digital learning credential assessment and management. J. Teach. Educ. Sustain. **20**(1), 145–156 (2018)
7. Gallagher, S.: A new era of micro-credentials and experiential learning. University world news, the global window on higher education (2019)
8. Patel, D., Britto, B., Sharma, S., Gaikwad, K., Dusing, Y., Gupta, M.: Carbon credits on blockchain. In: 2020 International Conference on Innovative Trends in Information Technology (ICITIIT), pp. 1–5. IEEE (2020)
9. Kenton, W.: How does the carbon credit work? December 2020. https://www.investopedia.com/terms/c/carbon_credit.asp
10. Kim, S.-K., Huh, J.-H.: Blockchain of carbon trading for un sustainable development goals. Sustainability **12**(10), 4021 (2020)
11. Atkinson, R.D.: Robotics and the future of production and work. Tech. Rep, Information Technology and Innovation Foundation (2019)

12. Vavra, C.: Electronics manufacturing driving industrial robotics market growth. Control Eng. **64**(12), 14 (2017). Name - Robotic Industries Association; Copyright - Copyright CFE Media Dec 2017; Last updated - 2020-11-18. [Online]. Available: http://ezproxy.lib.uts.edu.au/login?url=https://www-proquest-com.ezproxy.lib.uts.edu.au/trade-journals/electronics-manufacturing-driving-industrial/docview/2130711949/se-2?accountid=17095

13. Hsu, K., Tsai, W., Yang, H., Huang, L., Zhuang, W.: On the design of cross-platform social robots: a multi-purpose reminder robot as an example. In: International Conference on Applied System Innovation (ICASI) 2017, pp. 256–259 (2017)

14. Ahn, J., Park, M., Shin, H., Paek, J.: A model for deriving trust and reputation on blockchain-based e-payment system. Appl. Sci. **9**(24), 5362 (2019)

15. Le Thi Thuy, N., Nguyen Trong, T.: The multitasking system of swarm robot based on null-space-behavioral control combined with fuzzy logic. Micromachines **8**(12), 357 (2017)

16. Strobel, V., Castelló Ferrer, E., Dorigo, M.: Blockchain technology secures robot swarms: a comparison of consensus protocols and their resilience to byzantine robots. Front. Robot. AI **7**, 54 (2020)

17. Strobel, V., Castello Ferrer, E., Dorigo, M.: Managing byzantine robots via blockchain technology in a swarm robotics collective decision making scenario (2018)

18. Mengelkamp, E., Gärttner, J., Rock, K., Kessler, S., Orsini, L., Weinhardt, C.: Designing microgrid energy markets: a case study: the brooklyn microgrid. Appl. Energy **210**, 870–880 (2018)

19. Schleicher-Tappeser, R.: How renewables will change electricity markets in the next five years. Energy Policy **48**, 64–75 (2012)

20. Gencer, A.E., Basu, S., Eyal, I., Van Renesse, R., Sirer, E.G.: Decentralization in bitcoin and ethereum networks. In: International Conference on Financial Cryptography and Data Security, pp. 439–457. Springer (2018)

21. Dowling, P.: Healthcare supply chains in developing countries: situational analysis. Arlington, Va.: USAID— DELIVER PROJECT (2011)

22. Barchetti, U., Bucciero, A., De Blasi, M., Mainetti, L., Patrono, L.: Rfid, epc and b2b convergence towards an item-level traceability in the pharmaceutical supply chain. In: 2010 IEEE International Conference on RFID-Technology and Applications, pp. 194–199. IEEE (2010)

23. Rossetti, C.L., Handfield, R., Dooley, K.J.: Forces, trends, and decisions in pharmaceutical supply chain management. Int. J. Phys. Distrib. Logist, Manag (2011)

24. ten Ham, M.: Health risks of counterfeit pharmaceuticals. Drug Saf. **26**(14), 991–997 (2003)

25. Mackey, T.K., Liang, B.A.: The global counterfeit drug trade: patient safety and public health risks. J. Pharm. Sci. **100**(11), 4571–4579 (2011)

26. Yu, X., Li, C., Shi, Y., Yu, M.: Pharmaceutical supply chain in china: current issues and implications for health system reform. Health Policy **97**(1), 8–15 (2010)

27. Ghani, F.A., Shabri, S.M., Rasli, M.A.M., Razali, N.A., Shuffri, E.H.A.: An overview of the personal data protection act 2010 (pdpa): Problems and solutions

28. Lim, S.Y., Fotsing, P.T., Almasri, A., Musa, O., Kiah, M.L.M., Ang, T.F., Ismail, R.: Blockchain technology the identity management and authentication service disruptor: a survey. Int. J. Adv. Sci. Eng. Inf. Technol. **8**(4–2), 1735–1745 (2018)

29. Sullivan, C., Burger, E.: E-residency and blockchain. Comput. Law Secur. Rev. **33**(4), 470–481 (2017)

30. Panait, A.-E., Olimid, R.F., Stefanescu, A.: Identity management on blockchain–privacy and security aspects. arXiv preprint arXiv:2004.13107 (2020)

31. Kallela, J., et al.: Federated identity management solutions. In: TKK T-110.5190 Seminar on Internetworking (2008)
32. Biryukov, A., Tikhomirov, S.: Security and privacy of mobile wallet users in bitcoin, dash, monero, and zcash. Pervas. Mobile Comput. **59**, 101030 (2019)
33. Zaghloul, E., Li, T., Mutka, M.W., Ren, J.: Bitcoin and blockchain: security and privacy. IEEE Internet of Things J. **7**(10), 10 288–10 313 (2020)
34. Sethy, A., Ray, A.: Leveraging blockchain as a solution for security issues and challenges of paperless e-governance application. Progress in Computing, Analytics and Networkingk, pp. 651–658. Springer (2020)
35. Stratopoulos, T.C., Wang, V.X., Ye, J.: Blockchain technology adoption. Available at SSRN 3188470 (2020)
36. Janssen, M., Weerakkody, V., Ismagilova, E., Sivarajah, U., Irani, Z.: A framework for analysing blockchain technology adoption: Integrating institutional, market and technical factors. Int. J. Inf. Manag. **50**, 302–309 (2020)
37. Koens, T., Poll, E.: The drivers behind blockchain adoption: The rationality of irrational choices. In: European Conference on Parallel Processing, pp. 535–546. Springer (2018)
38. Sadhya, V., Sadhya, H.: Barriers to adoption of blockchain technology (2018)

A Lightweight Authentication Scheme for SDN-Based Architecture in IoT

Nadia Kammoun[1], Ryma Abassi[1], Sihem Guemara El Fatmi[1],
and Mohamed Mosbah[2(✉)]

[1] Digital Security Research Lab, Higher School of Communication of Tunis
(Sup'Com), University of Carthage, Tunis, Tunisia
{Nadia.Kammoun,Ryma.Abassi,Sihem.Guemara}@supcom.tn
[2] Univ. Bordeaux, CNRS, Bordeaux INP, LaBRI, UMR 5800, 33400 Talence, France
mohamed.mosbah@u-bordeaux.fr

Abstract. With the widespread usage of Internet-enabled devices, the Internet of things (IoT) has become popular and well integrated in different countries nowadays. However, the huge amount of data generated from several smart devices in the same IoT network, security of the network and the limited devices capacities of storage and process are a very big concern. To treat such a large database generated from all devices in the IoT network and to handle their security issues, Software Defined Network (SDN) has emerged as a key approach. Keeping focus on security challenges, several entities must authenticate each other before any communication to ensure a trusted network. We adopt in this paper the SDN architecture for an IoT environment, including authentication schemes for things and users based on Elliptic Curve cryptography (ECC).

1 Introduction

The number of connected objects is growing exponentially, making the Internet of Things (IoT). It is a wide network of networks composed by connected smart devices like sensors and actuators. These devices are adopted in several domains such as healthcare, smart transportation, smart cities, waste management... Consequently, industry and individuals are attempting to get in the fast flow of commercialization without paying attention to the safety of IoT networks and devices. Such a neglect exposes these latters to different attacks and endanger IoT users and the whole eco-system. For this reason, every IoT device should be well secured throughout its lifecycle by affording secure booting, authentication, access control... [3].

L. Barolli et al. (Eds.): AINA 2021, LNNS 227, pp. 336–345, 2021.
https://doi.org/10.1007/978-3-030-75078-7_34

However, the security protocols supported by IoT devices must be both robust to vicious attacks and lightweight which is not an easy mission. With the rapid development of IoT, Software-Defined Networking (SDN) has captured attention by offering a programmable low-cost alternative to traditional routers owners. It alleviates nodes workload by restricting their task to forward data between each other through controllers. These latters process network states and security protocols after interpreting application policies into low-level rules. A fundamental concept of SDN divides the network into 3 layers. This structure provides the network with more flexibility; nodes join and leave it many times during their lifetime. On the other hand, this elasticity poses a big challenge regarding authentication and access control of nodes and users which should be taken into consideration.

Access control is a crucial security measures in IoT. It is the way of determining whether a demand should be granted or denied. Users have limited access to different services in the network depending to predefined policies. Otherwise, smart objects form an heterogeneous and highly dynamic environment which increase the risk of infiltrators incoming. Trust management represents an effective solution to limit this risk by attributing a trust level to each device in the network. This value describes the trustworthiness of IoT nodes towards each other and depends of different metrics according to the requirements of networks. In this context, we have previously proposed an SDN architecture based on trust management and access control for IoT ensuring malicious nodes elimination and users access limitation.

However, for such a distributed and decentralized kind of the network, it is crucial to control who are getting at sensors services. In IoT applications, multiple entities like sensing nodes, users and service providers have to authenticate to establish trusted communications by avoiding some attacks like man-in-the-middle, denial-of service (DoS) and cloning.

This paper has five other segments. Section 2, gives an overview of some IoT authentication related works. Our authentication proposition for both of things and users to enhance an IoT network security is presented in Sect. 3. Finally, Sect. 4 is dedicated to the conclusion.

2   Related Works

We refer to some related works based on authentication. Just few of them integrate authentication in SDN architecture for IoT environments. These related works are shown in Table 1.

Table 1. Table of Related works

Approach	Description	Limitation
Identity-based authentication scheme for the Internet of Things [1]	Authors propose an identity-based authentication scheme towards heterogeneous IoT and tested by AVISPA	– No performance analysis for communication overheads or memory [4] – Simulated in a none SDN environment [4]
A Lightweight Authentication Protocol for Internet of Things[6]	Lee et al. implemented a lightweight authentication Protocol based on XOR encryption method instead of other complicated encryption techniques like hash functions	– No consideration of location privacy [4]
A multi-attack resilient lightweight IoT authentication scheme [2]	Propose a secure and lightweight mutual authentication scheme for constrained IoT devices. Its security robustness was tested by AVISPA tool	– many vulnerabilities have been detected in the protocol [8]
An Efficient Authentication and Access Control Scheme for perception layer of Internet of Things [10]	Authors in this paper propose a lightweight authentication mechanism based on (ECC) and an access control scheme based on (ABAC)	– requires complex management and restricts its deployment to limited devices. – provides only theoretical results [11]

3 Authentication Proposition

One of the most critical requirements for IoT is to take into consideration security challenges under limited capacities. Our present contribution completes our previous one which is baptised A New SDN Architecture based on Trust Management and Access Control for IoT. This paper applies an SDN architecture on an IoT network, which aims to improve its security by employing Attribute-based access control (ABAC) to control the access and trust management to exclude malicious nodes. In fact, SDN architecture is based on 3 layers: SDN applications are set by network administrators (application layer). Also in SDN, all control is relayed to a central(s) controller(s) (control layer), whereas data forwarding is ensured by the network devices (infrastructure layer) including IoT nodes. Otherwise, implementing access control in our security model limits users and devices access according to predefined attributes in ABAC paradigm. In addition, for excluding malicious nodes from a network, the best way is to use trust management. A variable trust level is attributed to each node to prove its trustworthiness. Figure 1 illustrates our previous SDN architecture. It shows an infrastructure layer which contains 3 under-layers:

– Cluster Members (CMs): represent most of network's IoT objects and divided into clusters according to our clustering algorithm defined in a previous work [9].

– Cluster Heads (CHs): are some IoT objects with high Trust and Energy levels calculated in [9].
– Bases Stations (BSs), routers and switches.

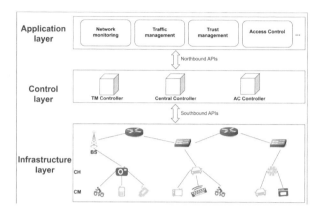

Fig. 1. SDN architecture based on access control and trust management

Unfortunately, these measures are insufficient to secure an IoT network. Users and things authentication before communications is a real necessity. In fact, IoT devices can be accessed by an illegitimate user due to the lack of an authentication mechanism which will lead to a data suppression or manipulation [5]. Traditional authentication mechanisms can't be applicable on IoT devices with constrained resource [5]. The work done by [7] enquired the threats of DoS attacks on SDN control channel and proved that if an authentication measure is missing, controller resources will be exhausted rendering them incapable of affording the intended services.

We propose in this paper an authentication mechanism supported by an SDN controller to alleviate devices workload. We enhance our proposed SDN architecture by implementing an Authentication Controller (AC), which is responsible for the whole authentication process and stores its algorithms. This controller is directly connected to the other controllers to exchange key information about things and users. Although SDN reduces objects batteries consumption by displacing process tasks to controllers, exchanged authentication messages shouldn't be so heavy. In fact, we propose a lightweight authentication scheme based on Elliptic-curve cryptography (ECC). It is a public-key encryption technique that provides smaller key sizes compared to the other techniques while ensuring the same security level. ECC ensures lower storage requirements and communication overheads what makes it the most widely used technique in IoT [1].

The proposed authentication model for IoT networks concerns users and things. It encompasses three phases. The first one, called Registration Phase, consists on gathering security credentials of requestors (users and things) by

AC as described in Sect. 4.2.1 and 4.3.1. The second one, called Authentication Phase, permits to start a mutual communication between requestor and AC, according to the obtained security credentials. Specifications are presented in Sect. 4.2.2 and 4.3.2. The last one, baptised re-authentication, obliges users and things in standby mode for a long time to authenticate again as described in 41.2.3.

3.1   Notations

The set of notations in Table 2 are used in forthcoming algorithms.

Table 2. Table of notation

BS:	Base Station
AC:	Authentication Controller
U:	User
T:	Thing
ID_U:	The User's ID
ID_T:	The Thing's ID
E:	The Thing's residual Energy value
t:	The Thing's trust value
Req:	Requested service of the user from a thing
ID_{AU}:	The user's ID imputed during registration phase by the User
ID_{AT}:	The thing's ID imputed during registration phase by the Thing
F_q:	A finite field
EC:	An elliptic curve defined F_q over a finite field
P:	A point on EC
$||$:	Concatenation operation
$h()$:	Hush function

3.2   Thing's Authentication Solution

According to our predefined SDN architecture, each thing's request passes by BSs before attending its final destination. BSs are responsible for forwarding messages between things from different clusters, things and users as well as things and controllers. Thus, authentication requests sent by sensor nodes are received by AC through BSs. To exchange services within the network, things just need to authenticate with the AC through BSs. We assumed that there are n neighborhood nodes $T_i \in \{T_1, T_2, ...T_n\}$ i n the subnet area. Our proposed key establishment method is based on ECC. AC generates a couple of public keys for each sensor node when it joins the network; one unique public key Pub_{T_i} for each node and AC's public key Pub_{AC}. AC is also responsible for generating the hash function as well as the elliptic curve and its parameters. We suppose that all BSs are already authenticated.

Initially AC selects an elliptic curve EC over finite field F_q and opts for a base point P with large order p (where q and p are primes). Then AC makes it public to the different BSs which in turn issue it to all sensor nodes and users.

3.2.1 Registration Phase

Before starting the authentication protocol between two entities of the network, it is necessary that every communication party pass by a registration process to provide security credentials that will be used for the authentication phase. In the following, we detail the message flow of registration and authentication phases (Fig. 2).

STEP 1: The thing sends a registration request to its directly attached BS containing its public key Pub_{T_i} encrypted using Pub_{AC}, its ID ID_{AT} and a nonce N1 to prevent replay attacks.

STEP 2: BS verifies thing's membership by checking the conformity between ID_{AT} and ID_T which is already registered during the clustering process. In case of compliance, BS transfers the encrypted request using Pub_{AC} to AC accompanied by its authentication certificate $Certif_{BS}$, the trust level t and the energy level E of the thing. In fact, t and E are already calculated during the clustering process [9]

STEP 3: Upon receiving the request from BS, AC first checks $Certif_{BS}$ then it decrypts sensor node's request using its private key Sec_{AC}. AC verifies thing's trust level with the trust management controller to prove that it's not a malicious node and it's worth the authentication. AC calculates the hash of Pub_{T_i} then it sends a request to BS encrypted by Pub_{BS} and which contains; Pub_{T_i} for both BS and the node and Pub_{BS} and $h(Pub_{T_i})$ encrypted by Pub_{T_i} for the node.

STEP 4: BS receives the message, decrypts it and stores a copy of Pub_{T_i}. Then it forwards $E(Pub_{T_i}, h(Pub_{T_i})||Pub_{BS})$ to the node.

The thing ensures that it has been registered by receiving the BS's public key Pub_{BS} encrypted by its own public key Pub_{T_i}. So, it stores it after decryption. It also uses $h(Pub_{T_i})$ to calculate its own private key $Sec_i = a * h(Pub_{T_i})$ such that a $\in F_q$.

3.2.2 Authentication Phase

STEP 5: When the thing needs to authenticate itself, it sends a random number N2 $\in F_q$ encrypted by Pub_{AC} along with ID_{AT} both signed by Pub_{BS} to authenticate AC.

STEP 6: Upon receiving ID_{AT}, BS search in its local storage thing's public key. Then, BS sends the encrypted message by Pub_{AC} and the Pub_{T_i} to specify for AC which thing is requesting the authentication.

STEP 7: AC decrypts the request and replies by sending back N2 with a new N3 randomly selected from $\in F_q$ to attest the thing's identity, both encrypted by Pub_{T_i}.

STEP 8: BS forwards directly the request to the thing.

STEP 9: The thing has to send back N3 encrypted by Pub_{AC} to be authenticated.

STEP 10: BS forwards the request to AC to complete the authentication process.

By receiving N3, AC becomes certain that the thing is not a malicious node and it can exchange services with nodes and users which are already authenticated.

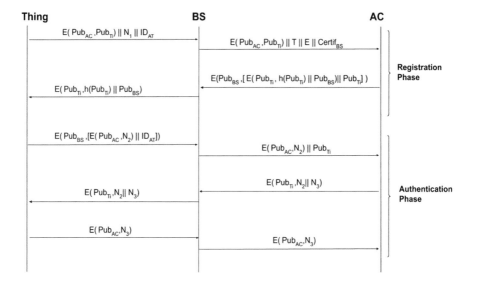

Fig. 2. Registration and Authentication message flow of things

3.2.3   Re-authentication Phase

After things authentication, a session is attributed to each node when it get connected. This session denotes a timer has been launched since the node was authenticated. After 24 h of being *Not_Active*, the thing passes from *Logged_On* to *Logged_off* state, it should request to authenticate again if it still needing a service.

Algorithm 1. Thing's and User's Re-Authentication

$Session\_state = Logged\_On$;

$Start\_time = Current\_time()$;

while ($Session\_state = Not\_Active$) **and** ($Current\_time - Start\_time >= 24hours$)
do

   $Session\_state = Logged\_Off$;

   The thing (user) must be logged on another time;
end while

3.3   User's Authentication Solution

In addition to things, users in the network should authenticate themselves with AC before getting access. The network can support more than one user; $U_i \in \{U_1, U_2, ...U_n\}$. User's authentication is established directly with AC. Once users are authenticated, they can exchange data with things already authenticated.

3.3.1   Registration Phase

Before users authentication, registration phase provides a first filtration of users. Just like for sensor nodes, AC generates a couple of public keys for each user when he joins the network and after obtaining its ID ID_U; one unique public key Pub_{U_i} for each user and AC's public key Pub_{AC}. AC registers all users identities in its database for further use as for authentication.

STEP 1: The user sends a registration request to AC containing its public key Pub_{U_i} encrypted using Pub_{AC}, its ID ID_{AU} and a nonce N4 $\in F_q$ to prevent replay attacks.

STEP 2: AC verifies user's identity by checking the conformity between ID_{AU} and ID_U. At this stage, Algorithm 2 is applied. Each user who enters an ID_{AU} different than its ID_U more than 3 times, its authentication request will be cancelled and he should wait 60 s to request again. Once $ID_{AU} = ID_U$, AC calculates the hush of Pub_{U_i} and sends it to the user with a random number N5 $\in F_q$ encrypted by Pub_{U_i}.

Algorithm 2. User Filtration

---

set $N = 0$;

while $N < 4$ **do**
   if $ID_{AU} = ID_U$ **then**
      The authentication process can be executed
   else if $N = 4$ **then**
      Wait 60 seconds
   else
      $N = N + 1$
   end if
end while

3.3.2 Authentication Phase

STEP 3: The user ensures that it has been registered by receiving $h(Pub_{U_i})$. He first calculates its private key $Sec_{U_i} = b * h(Pub_{U_i})$ such that $b \in F_q$ then it decrypts $E(Pub_{U_i}, N5)$ using its private key. He replies AC by sending back N5 to prove his authenticity and another random N6 $\in F_q$ for the mutual authentication with AC.

STEP 4: Upon receiving the message from the user, AC decrypts it and conclude the authentication process by sending back N6.

By receiving N6, the user is allowed to obtain nodes services in the network. Figure 3 illustrates the message flow of user's authentication phase.

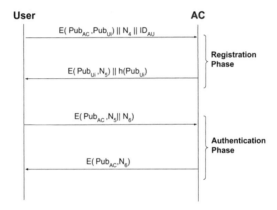

Fig. 3. Registration and Authentication message flow of users

3.3.3 Re-authentication Phase

After user authentication, a session is attributed to each connected user when it gets connection just like sensor nodes. We adopt the same algorithm (Algorithm 1) of thing's re-authentication in Sect. 4.2.3.

4   Conclusion

With the rapid development of IoT, SDN has captured more and more attention. Its centralized control improves the efficiency of devices management and their security. Otherwise, authentication in distributed networks is one of the main security requirements. We implemented in this paper an authentication mechanism based on Elliptic Curves Cryptography (ECC). It consists on defining message flow of registration and authentication phases for both things and users. It also contain an algorithm for things and user re-authentication and another algorithm that filtrates users before authentication. In this way, we theoretically ensured a high level of security within an IoT environment by combining

Authentication to access control and trust management implemented in previous works. Experimentation results will be afforded in future works to validate our mechanism.

References

1. Salman, O., Abdallah, Sarah, Elhajj, I.H.: Identity-based authentication scheme for the internet of things. In: IEEE Symposium on Computers and Communication (ISCC), pp. 1109–1111. IEEE (2016)
2. Adeel, A., Ali, M., Khan, A.N., et al.: A multiattack resilient lightweight IoT authentication scheme. Trans. Emerg. Telecommun. Technol.e3676 (2019)
3. Mohana, S., Priya, T.K.S.L.: Identity-based Encryption for device-to-device Security in IOT Environments. World Sci. News **41**, 120 (2016)
4. Kalkan, K., Zeadally, S.: Securing internet of things with software defined networking. IEEE Commun. Mag. **56**(9), 186–192 (2017)
5. Abbou, A.N., Baddi, Y., Hasbi, A.: Software defined networks in internet of things integration security: challenges and solutions. In: 2018 6th International Conference on Wireless Networks and Mobile Communications (WINCOM), pp. 1–6. IEEE (2018)
6. Lee, J.-Y., Lin, W.-C., Huang, Y.-H.: A lightweight authentication protocol for internet of things. In: 2014 International Symposium on Next-Generation Electronics (ISNE), pp. 1–2. IEEE (2014)
7. Abdullaziz, O.I., Chen, Y.-J., Wang, L.-C.: Lightweight authentication mechanism for software defined network using information hiding. In: 2016 IEEE Global Communications Conference (GLOBECOM), pp. 1–6. IEEE (2016)
8. Lara, E., Aguilar, L., Sanchez, M.A., et al.: Lightweight authentication protocol for M2M communications of resource-constrained devices in industrial internet of things. Sensors **20**(2), 501 (2020)
9. Kammoun, N., Abassi, R., Guemara, S.: Towards a new clustering algorithm based on trust management and edge computing for IoT. In: 15th International Wireless Communications & Mobile Computing Conference (IWCMC). IEEE 2019, pp. 1570–1575 (2019)
10. Ye, N., Zhu, Y., Wang, R. C., Malekian, R., Lin, Q.M.: An efficient authentication and access control scheme for perception layer of internet of things (2014)
11. Ouaddah, A., Mousannif, H., Abou Elkalam, A., et al.: Access control in the internet of things: big challenges and new opportunities. Comput. Netw. **112**, 237–262 (2017)

SVM-Based Ensemble Classifiers
to Detect Android Malware

Md Faiz Iqbal Faiz$^{(\boxtimes)}$

North Eastern Regional Institute of Science and Technology, Itanagar, India

Abstract. Android malware is a significant threat to Android-based devices. Various single-app analysis tools are developed to detect these threats. In this regard, ML-based tools are more effective in detecting single malicious applications due to their robustness and capability to detect zero-day malware. In this work, we propose ensemble classifiers composed of support vector machines (SVMs) to detect Android malware. We discuss the proposed classifiers' effectiveness and do a comparative analysis with existing single SVM-based classifiers, a mixture of SVM and other classifiers, and multi-stage classifiers based on SVM. We also argue that the proposed classifiers can detect app-collusion, a special kind of threat.

1 Introduction

Smartphones are an integral part of our daily activities nowadays. Smartphones have a lot of usage, from smart banking to smart health. Smartphones come up with different mobile operating systems installed in it. Among them, Android-based smartphones are the most popular. One reason for their popularity is that applications (apps) for various utilities are easily available and easy to use on Android. However, the primary reason is that Android is open-source software, so developers find it easy to develop applications that can run on Android.

Nevertheless, with openness comes hazard. Attackers use this property to design malicious applications to attack the Android operating system. Attackers take advantage of the inherent vulnerabilities in Android. To understand the various attacks launched by attackers through malicious applications, we need to understand the structure of the Android operating system and the essential components of the system. In the next section, we briefly discuss Android's structure and communication between the components. Also, we briefly discuss the Android security model.

Google Android is a modern operating system for smartphones with a rapidly expanding market share. The popularity and open-source character of Android facilitate its deployment on other hardware platforms, e.g., netbooks [1] and tablet P.C.s. Android's core security mechanisms are application sandboxing, application signing, and a permission framework to control access to (sensitive) resources. The standard Android permission system limits access to sensitive data (SMS, contacts), resources (battery or log_les), and system interfaces (Internet connection, G.P.S., GSM). Once granted (by the end-user), the assigned permissions cannot be changed afterward, and the Android's reference monitor checks them at runtime. This approach restricts the potential damage imposed by compromised applications. However, Android's security framework exhibits serious shortcomings: On the one hand, the burden of approving application permissions is delegated to the end-user who, in general, does not have the appropriate skills. Hence, malware and Trojans can be installed on end-user devices, as shown by very recent Android Trojans: such as unauthorized sending of text messages, malicious game updates, or location tracking and leaking of sensitive data in the background of running games. On the other hand, Android's security framework is vulnerable to privilege escalation attacks at the application-level.

The contributions of this paper are summarized as follows:

- Development of SVM-based ensemble classifiers to detect Android malware.
- Comparative analysis of SVM-based ensemble classifiers with non-ensemble methods.

2 Related Work

Using machine learning techniques to detect zero-day malware is not new. Recently, there is an upsurge in the number of papers proposing different classification models for Android malware. They employ various static analysis and dynamic analysis tools to devise the feature set to train the classification models accurately. Most methods use static analysis tools due to their efficiency. The machine learning models are built using a large corpus of Android benign and malicious applications. The authors in [2] proposed an ensemble classifier to detect Android malware. They took advantage of the static analysis's efficiency and ensemble classifier's performance to detect Android malware with high accuracy. They employed a random forest ensemble classifier for the detection purpose. They compared their model with other machine learning models such as decision trees, naïve Bayes, and simple logistic. They achieved a detection accuracy of 97–99% with low false positives.

The authors in [3] developed a malware detection system called SiGPID. SiGPID detects Android malware and its families by using significant permissions. The authors utilized the idea of the usage of permissions by Android malware. They introduced three pruning levels to finally select a set of permissions that are the most significant in determining a given application is malware or not. They identified 22 significant permissions. In this, the authors removed the need for using the full permission set as features. The proposed method achieved an accuracy of 93.62% in detecting malware from the data set and 91.4% in detecting unknown/new malware.

In [4], the authors proposed a lightweight classification model, DREBIN, to detect Android malware. The proposed model uses the SVM classification model. The model is lightweight because it can detect Android malware directly on the smartphone. The model uses static analysis to collect many features and store them as a joint vector. The SVM is trained offline outside the smartphone device, and only the learned model is transferred to the device. DREBIN used a dataset of 123,453 applications and 5,560 malware samples. DREBIN outperformed existing detection models by achieving an accuracy of 94%. On popular smartphones, DREBIN took only 10 s to analyze a single app, making it suitable for online detection. The authors in [5] proposed a classification model based on SVM and achieved a detection accuracy of 99% on the given data set.

3   Proposed Methodology

The proposed methodology used here consists of ensemble classifiers. The ensemble classifiers consist of a set of SVMs. We use various kernel functions for our base estimator, which is SVM here. The ensemble method we use here is the bagging method. The bagging method consists of algorithms that build several instances of a base estimator on random subsets of the original training set and then combine their classifications to output a final classification [6]. These algorithms are used to reduce the variance of a base estimator by introducing randomness in the construction procedure and then making an ensemble. Bagging methods are used as a way to reduce overfitting and work best with strong and complex models.

Our proposed models have the following advantages:

1. Our models are more resistant to overfitting than single SVM classifiers.
2. Our models have the flexibility to incorporate different classifiers in addition to SVM classifiers.
3. Our models can achieve good generalization performance.
4. Our models are multi-purpose. We can use it to detect both colluding app-pairs and single malicious apps as explained later.
5. Our proposed models can learn both linear and non-linear datasets.

4   Results and Discussions

4.1   Datasets and Features

In this proposed work, we use a dataset that consists of 15,036 Android applications, obtained from [7]. The Android applications are of two kinds benign or malicious (malware). The full feature set size is 215. The feature types include permissions requested/used, API calls, intent fields. We use the recursive feature elimination technique with logistic regression as the base estimator to select "K" best features. Here, K's value is 178, i.e., we use 178 features for training ensemble classifiers. The ensemble classifiers consist of SVMs as the base estimator. We use different types of kernel functions in our experiment. The various kernel functions are gaussian/RBF, linear, polynomial, and sigmoid. The degree of the polynomial kernel here is 3. We use grid search techniques to optimize the hyperparameters of the SVMs. The ensemble method used here is the bagging method—the number of base estimators in the ensemble classifier range from 10 to 100, respectively. We use 9,954 benign and malicious for training purposes and 5,082 benign and malicious samples for testing.

4.2 Evaluation Metrics

We use different evaluation metrics to evaluate the performance of our ensemble classifiers. We use the following evaluation metrics:

1. True Positive Rate (TPR): TPR is defined as the ratio of correctly classified malicious apps to the total no of malicious apps, given by

$$TPR = \frac{TP}{TP + FN} \tag{1}$$

   TP is the number of correctly classified malicious apps, and FN is the number of malicious apps incorrectly classified as benign. TPR is also known as sensitivity.

2. False Positive Rate (FPR): FPR is defined as the ratio of benign apps incorrectly classified as malicious to the total number of benign apps, given by

$$FPR = \frac{FP}{FP + TN} \tag{2}$$

   FP is the number of benign apps classified as malicious, and TN is the number of apps correctly classified as benign.

3. Accuracy (ACC): ACC is defined as the ratio of the total number of apps correctly classified as benign and malicious to the total number of apps given by

$$ACC = \frac{TP + TN}{P + N} \tag{3}$$

   TP is the number of correctly classified malicious apps, TN is the number of correctly classified benign apps, P is the total number of malicious applications, and N is the total number of benign applications.

4. Precision or Positive Predictive Value (PPV): PPV is defined as the ratio of the number of correctly classified malicious apps to the number of apps predicted as malicious, given by

$$PPV = \frac{TP}{TP + FP} \tag{4}$$

   TP is the number of correctly classified malicious apps, and FP is the number of benign apps incorrectly classified as malicious.

5. F1 score: F1 score is defined as the harmonic mean of precision and sensitivity, given by

$$F1 = 2.\frac{PPV.TPR}{PPV + TPR} \tag{5}$$

4.3 Evaluation of Ensemble Classifiers

Table 1 shows the single SVM classifier evaluation metrics, i.e. when we use an SVM with a specific kernel function. The kernel functions used are RBF,

Table 1. Evaluation metrics of single SVM classifiers for different kernels

Kernel	Accuracy	F1 score	Precision
RBF	0.980	0.972	0.984
Linear	0.975	0.965	0.974
Poly-3	0.960	0.944	0.987
Sigmoid	0.902	0.865	0.869

Table 2. Confusion matrix of single SVM classifiers for different kernels

Kernel	True negatives	False positives	False negatives	True positives
RBF	3194	28	72	1788
Linear	3175	47	80	1780
Poly-3	3201	21	178	1682
Sigmoid	2981	241	257	1603

Table 3. Evaluation metrics of ensemble classifiers for different kernels with E = 10

Kernel	Accuracy	F1 score	Precision
RBF	0.979	0.971	0.982
Linear	0.974	0.964	0.974
Poly-3	0.958	0.940	0.984
Sigmoid	0.904	0.868	0.875

Table 4. Confusion matrix of ensemble classifiers for different kernels with E = 10

Kernel	True negatives	False positives	False negatives	True positives
RBF	3191	31	74	1786
Linear	3175	47	83	1777
Poly-3	3196	26	187	1673
Sigmoid	2994	228	258	1602

Table 5. Evaluation metrics of ensemble classifiers for different kernels with E = 100

Kernel	Accuracy	F1 score	Precision
RBF	0.980	0.972	0.984
Linear	0.975	0.966	0.974

Table 6. Confusion matrix of ensemble classifiers for different kernels with E = 100

Kernel	True negatives	False positives	False negatives	True positives
RBF	3194	28	73	1787
Linear	3175	47	78	1782

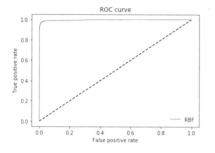

Fig. 1. ROC curve of SVM with RBF kernel

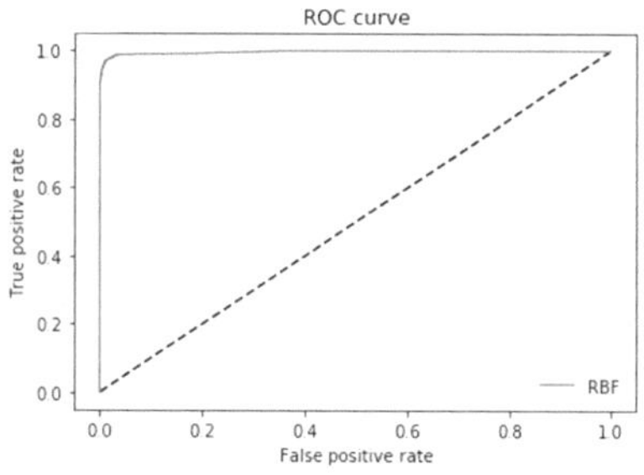

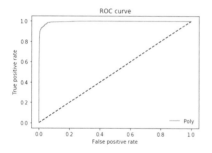

Fig. 3. ROC curve of SVM with polynomial kernel

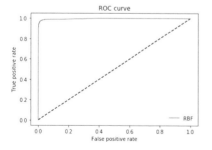

Fig. 5. ROC curve of ensemble classifier with RBF kernel and E =10

Fig. 2. ROC curve of SVM with linear kernel

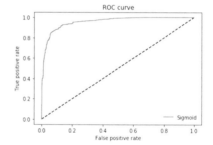

Fig. 4. ROC curve of SVM with sigmoid kernel

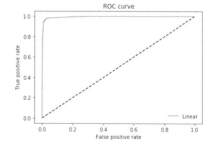

Fig. 6. ROC curve of ensemble classifier with linear kernel and E =10

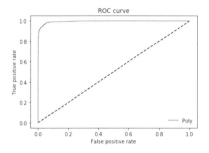

Fig. 7. ROC curve of ensemble classifier with polynomial kernel and E =10

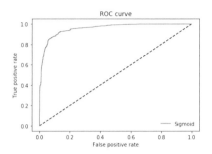

Fig. 8. ROC curve of ensemble classifier with sigmoid kernel and E =10

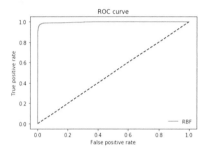

Fig. 9. ROC curve of ensemble classifier with RBF kernel and E =100

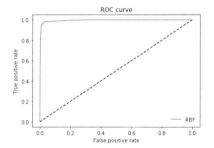

Fig. 10. ROC curve of ensemble classifier with linear kernel and E =100

linear, polynomial, and sigmoid. Poly-3 is the polynomial kernel of degree 3. The RBF-based SVM performs the best as compared to other kernels. Linear-based SVM is the second-best classifier. Table 2 shows the confusion metrics of the given classifiers. Table 3 shows the evaluation metrics of the SVM-based ensemble classifiers. We can see that here also RBF-based ensemble classifier performs the best as compared to other classifiers. The performance difference between single SVM classifiers and SVM-based ensemble classifiers is minute. Table 4 shows the confusion matrix of the ensemble classifiers. Here, "E" is the number of base estimators in the ensemble classifier.

To check the SVM-based ensemble classifiers' performance, when we increase the number of estimators, we increase the number of base estimators to 100. We can see that the RBF-based ensemble classifier's performance remains the same, but the linear-based ensemble classifier's performance increases slightly, as shown in Table 5. Table 6 shows the confusion matrix of the corresponding classifiers. Figure 1, 2, 3 and 4 shows the receiver operating characteristic (ROC) curve of the single SVM classifiers. Figure 5, 6, 7 and 8 shows the ROC curve of the SVM-based ensemble classifiers. Figure 9 and 10 shows the ROC curve of the ensemble classifiers for when we increase the number of estimators to 100.

5 Discussion

The models proposed in this paper can detect Android colluding app-pairs if they use the lightweight framework model as mentioned in [8]. The only thing we need to do is to separate the final prediction function from the training phase. Collect the parameters of the bagging classifier in a separate parameter vector and used them separately. The SVM-based ensemble classifiers are used earlier for detecting breast cancer [9], but a detailed study on their ability to detect Android malware is not studied before. We believe that ensemble classifiers can be a viable option when the dataset is huge, much larger than the dataset used in this experiment. In place of usual soft-margin SVMs, we could use other SVMs such as LINEX SVM [10] or least square SVMs [11] as base estimators.

6 Conclusion

We proposed several SVM-based classifiers. From our detailed experimental analysis, we showed that SVM-based ensemble classifiers could detect Android malware effectively. The proposed models use bagging as the ensemble method. We use majority voting to get the final prediction. Our proposed models are as effective as the single SVM classifiers and are resistant to overfitting due to the inherent constriction procedure. The performance of some SVM-based ensemble classifiers is slightly better than single SVM classifiers.

References

1. Enck, W., Ongtang, M., McDaniel, P.: Understanding android security. IEEE Secur. Priv. **7**(1), 50–57 (2009). https://doi.org/10.1109/MSP.2009.26
2. Yerima, S.Y., Sezer, S., Muttik, I.: High accuracy android malware detection using ensemble learning. IET Inf. Secur. **9**(6), 313–320 (2015). https://doi.org/10.1049/iet-ifs.2014.0099. https://digital-library.theiet.org/content/journals/10.1049/iet-ifs.2014.0099. IET Digital Library
3. Li, J., Sun, L., Yan, Q., Li, Z., Srisa-an, W., Ye, H.: Significant permission identification for machine-learning-based android malware detection. IEEE Trans. Industr. Inf. **14**(7), 3216–3225 (2018)
4. Arp, D.S., Hubner, M., Gascon, M., Hugo Rieck, K.: DREBIN: effective and explainable detection of android malware in your pocket. In: Symposium on Network and Distributed System Security (NDSS) (2014). https://doi.org/10.14722/ndss.2014.23247
5. Lu, Y., Kuo, C., Chen, H., Chen, C., Chou, S.: A SVM-based malware detection mechanism for android devices. In: 2018 International Conference on System Science and Engineering (ICSSE), New Taipei, pp. 1–6 (2018)
6. Breiman, L.: Bagging predictors? Mach. Learn. **24**(2), 123–140 (1996)
7. Yerima, S.Y., Sezer, S.: DroidFusion: a novel multilevel classifier fusion approach for android malware detection. IEEE Trans. Cybern. **49**(2), 453–466 (2019)
8. Faiz, M.F.I., Hussain, M.A., Marchang, N.: Android malware detection using multistage classification models. In: Barolli, L., Poniszewska-Maranda, A., Enokido, T. (eds.) Complex, Intelligent and Software Intensive Systems. CISIS 2020. Advances in Intelligent Systems and Computing, vol. 1194. Springer, Cham (2021)

9. Huang, M.-W., Chen, C.-W., Lin, W.-C., Ke, S.-W., Tsai, C.-F.: SVM and SVM Ensembles in Breast Cancer Prediction. PLoS ONE **12**(1), e0161501 (2017). https://doi.org/10.1371/journal.pone.0161501

10. Ma, Y., Zhang, Q., Li, D., Tian, Y.: LINEX support vector machine for large-scale classification. IEEE Access **7**, 70319–70331 (2019)

11. Suykens, J., Vandewalle, J.: Least squares support vector machine classifiers. Neural Process. Lett. **9**, 293–300 (1999). https://doi.org/10.1023/A:1018628609742

Sorting Algorithms on ARM Cortex A9 Processor

Yomna Ben Jmaa[1](✉), David Duvivier[2], and Mohamed Abid[3]

[1] REDCAD Laboratory, University of Sfax, Sfax, Tunisia
yomna.benjmaa@redcad.org
[2] LAMIH UMR CNRS 8201, Polytechnic University of Hauts-de-France,
Valenciennes, France
David.Duvivier@uphf.fr
[3] CES Laboratory, University of Sfax, Sfax, Tunisia

Abstract. Sorting is considered as one of the most well-known problems in the computer world. It is a common process among several application areas, such as real time decision support systems and intelligent transport applications. In this paper, we propose a software implementation for different sorting algorithms, such as InsertionSort, QuickSort, HeapSort, ShellSort, MergeSort and TimSort on the Zynq Zedboard platform. In addition, the performance of the different algorithms are compared in terms of averages and standard-deviation of computational time, energy consumption and stability. As demonstrated by the experimental results, the ShellSort is 42.1% faster and can even reach 72% when running on the ARM Cortex A9 processor mainly if the number of elements (n) to be sorted is greater than 64. Otherwise, TimSort is the best algorithm. Also, ShellSort is the best algorithm in terms of standard-deviation of computational times and energy consumption.

1 Introduction

Nowadays, embedded systems integrate specific functionalities requiring computing power that a single processor cannot provide. They must respect more or less severe constraints about consumption, speed of operation, performance... with applications in, for instance, Intelligent Transportation Systems (ITS) that play a significant role to minimize accidents, trafic jams and air pollution. These systems are exploited in several domains such as avionics, railway and automotive. Moreover, ITS require decision-making systems that help analyze the environment and the state of the system, to compare several possible solutions. In such systems, many applications use sorting algorithms at different steps. For a given amount of data issuing from ITS systems, the performance of sorting algorithms is affected not only by their complexity, but also by the targeted execution platform(s). Among these platforms, reconfigurable systems on programmable chip (SOPC) architectures [1,2], with the evolution of technologies on the FPGAs provide a significant improvement of the flexible use and minimization of the systems. In fact, a SoPC integrates all the components on the same FPGA. Moreover, the main advantages of a SoPC are the programming speed and the reconfiguration simplicity. Actually, there are two different types of processors for the SoPCs:

- Softcore processor (reconfigurable or virtual processors): can be implemented on a reconfigurable circuit (FPGA). We cite for example ARM processors that are softcore versions of the ARM family. The main advantages of these types of processors are the integration of flexibility into their systems compared to classical processors and the ease of integration into reconfigurable technology.
- Hardcore processor: can be an integrated circuit comprising a reconfigurable part and a non-configurable part or else an integrated circuit in an FPGA/SoPC. For this, the use of hard processors in integrated circuits is a bit flexible. Hard processors must be more powerful, larger and consume more power.

Considering these target execution platforms, sorting algorithms are crucial operations for many applications. Consequently, excluding "basic" but inefficient sorting algorithms, such as SelectionSort, InsertionSort and BubbleSort (see Table 1), it is logically advisable to switch to sorting algorithms of maximum complexities of O(n log(n)) [11, 13] such as MergeSort, HeapSort and QuickSort. Moreover, there exists "enhanced versions" as, for example, the ShellSort, which appears to be as an improvement of the InsertionSort. Whereas TimSort, one example of hybrid sorting algorithms, is a combination of the InsertionSort and MergeSort. The objective of this work is to propose an optimized software implementation of sorting algorithms, such as the InsertionSort, HeapSort, ShellSort, QuickSort, TimSort and MergeSort, on the zynq platform using compilation and synthesis guidelines. In addition, the important objective of this paper is to compare these algorithms on the ARM Cortex A9 processor in terms of performance criteria (averages and standard-deviation of computational time and energy consumption) using limited data size ranging from 8 to 4096 elements to be compatible with the field of real-time decision support as solutions to choose the best trajectory while avoiding obstacles and also reduce the number of accidents for avionic systems.

Table 1. Complexity of sorting algorithms

	Best time complexity	Average time complexity	Worst time complexity
BubbleSort	O(n)	$O(n^2)$	$O(n^2)$
InsertionSort	O(n)	$O(n^2)$	$O(n^2)$
SelectionSort	$O(n^2)$	$O(n^2)$	$O(n^2)$
Odd-EvenSort	O(n)	$O(n^2)$	$O(n^2)$
ShellSort	O(n)	$O(n \log(n)^2)$	$O(n \log(n)^2)$
QuickSort	O(nlog (n))	O(nlog (n)	$O(n^2)$
HeapSort	$O(n \log(n))$	$O(n \log(n))$	$O(n \log(n))$
MergeSort	$O(n \log(n))$	$O(n \log(n))$	$O(n \log(n))$
TimSort	O(n)	$O(n \log(n))$	$O(n \log(n))$

This paper is organized as follows: Sect. 2 presents a state of the art on different sorting algorithms using different platforms. Section 3 gives a brief description of a specific architecture. Section 4 compares software implementations of sorting algorithms. Then, in last section, we conclude our paper.

2   Related Works

As previously mentioned, sorting is commonly used in software applications [4,5], such as decision-making tools, avionics systems, etc. For this reason, researchers are still trying to improve the performance of sorting algorithms [6].

The authors in [7] proposed three different approaches for data sorting acceleration of large data streams: sorting the set of data processing totally on the software side (sorting and merging on PC), a combination of PC side and field programmable gate array (FPGA) platform (hardware sorting with software merging), and fully hardware side (sorting and merging on FPGA), whereas the work proposed by [8] presents two new algorithms; the quick-merge parallel algorithm, which uses a Quicksort algorithm to sort each subset of data on a CPU core and MergeSort algorithm to merge the results obtained for each CPU core and the hybrid algorithm (parallel bitonic algorithm on GPU graphics processor and sequence Mergesort on CPU) using an openMP and CUDA framework. The result shows that if the number of elements is small, then sorting algorithms running on GPUs are faster than those running on multi-core CPU. In contrast, sorting algorithms running on multicore are more efficient for many elements and the hybrid algorithm is a little slower than the efficient quick-merge parallel algorithm on CPU. The authors in [9] proposed an efficient implementation and detailed analysis of MergeSort on current CPU architectures. This implementation exploits the principle architectural characteristics of modern processors to offer meaningful performance advantages. These features integrate cache blocking to reduce access latency, vectorizing for IMD to improve compute density, partitioning work and load balancing between several cores, and multi-way merging to remove the bandwidth bound stages for large input sizes. The result shows that the SIMD implementation with 128-bit SSE is $3.3\times$ faster than the scalar version and the excellent scalability of this implementation with SIMD width scaling up to 16X wider than current SSE width of 128-bits, and CMP core-count scaling well beyond 32 cores.

In the following sections, we develop our proposed implementation of sorting algorithms on an Advanced RISC Machine (ARM) Cortex A9. Then, we compare the obtained performance to the different sorting algorithms in terms of averages and standard-deviation of computational time and energy consumption.

3   A Specific Architecture of the System on Programmable Chip (SoPC)

To perform complex processing on a chip, SoPC devices such as ARM processors contain IP hardware accelerators which are input/output control functions (UART, VGA, PS2, etc.) or intensive processing (FFT, avionics application, sorting algorithm, etc.). These components can be designed as hardware components or reusable intellectual properties (IPs) which refers to a product invented by a company or person for which exclusive rights are distributed. There are 2 types of IP:

- Software IP (Soft core) is a programmable processor or circuit. It is synthesizable and cannot be predictive in terms of consumption and time. This IP is portable and its source code can be modified.
- Hardware IP (Hard core) is a component defined in the generation of Bitstream. Indeed, these components are not flexible and use only one technology. They increase performance while reducing consumption and running time.

Xilinx is providing the Zedboard as part of the System On a Chip (SoC) evaluation and development [10]. It is divided into two combinatorial parts, a Processing System (PS) and a Programmable Logic (PL) as detailed in following paragraphs.

- Field programmable Gate Arrays (FPGA) Part: The Zynq-7000 family is based on the Xilinx All Programmable SoC architecture. It is characterized by 85,000 Slices, 560 KB RAM block, with 220 DSP Slices. This family also integrates a feature-rich dual-core or single-core ARM Cortex A9-based processing system (PS) with programmable logic from a FPGA in a single chip. Moreover, it offers high performance, flexibility, scalability, power consumption and easy use.
- Microprocessor system (PS): The zynq FPGA family consists of one or more ARM cores that also incorporate memory and a set of I/O devices. Hence, the ARM Cortex A9 has a L1 cache memory of 32 KB on addresses and data and a L2 cache memory of 512 KB. Xilinx has adopted the AXI (Advanced eXtensible Interface) protocol for intellectual property (IP) cores starting with the Spartan-6 and Virtex-6 devices.

4   Software Implementation of Sorting Algorithms

Via a set of experiments, we estimate the complexity of six sorting algorithms (InsertionSort, QuickSort, TimSort, HeapSort, ShellSort, MergeSort) on the ARM Cortex A9 processor in terms of average and standard-deviation of computational times, energy consumption and stability.

4.1   Development Environments

First of all, we detail the two different existing development environments:

- Embedded Linux: An Operating System (OS) is a software that manages the hardware devices and processes complex instructions requested by the user or other applications. The development and implementation of a Linux system is characterized by several advantages which are: Making it easier to test new features, Low cost, Reuse of components and codes, Real Time Operating System (RTOS). In addition, Embedded Linux has several disadvantages: Lack of drivers for specific devices, Slow Linux kernel compilation and high learning complexity for development, Unstable system in case of wrong code generation.
- Standalone mode: it allows users to simply compile and run their programs on hardware with no OS. This mode is supported for the following platforms: Zc702, Zc706, Zedboard, microzed, Zybo. Furthermore, in recent versions of Vivado, standalone mode is available.

4.2   Study of the Complexity of Sorting Algorithms

In this section, we focus on InsertionSort, HeapSort, ShellSort, QuickSort, MergeSort, and TimSort using input data encoded in 32 bits and the standalone mode. In this study, we compare the average and standard-deviation of computational time, energy consumption and the stability of the algorithms on the Zedboard platform in order to integrate the "best" algorithm into avionics applications. For the consumption measurement, we connect the Texas Instruments "USB Interface Adapter" module to the Zedboard. For the calculation of computational time, we execute each algorithm on the software part more precisely on the ARM Cortex A9 processor by considering 8, 16, 32, 64, 128, 256, 512, 1024, 2048 and 4096 elements. Finally, in order to evaluate the stability, we calculate the variation coefficients of the sorting algorithms and graphically illustrate the results via boxplots. In this paper, we consider 47 permutations that are generated using Lehmer's method [12] as input data. Therefore, we describe in this section the notions of permutation, statistical dispersion, assumptions and constraints, and explain the different results obtained when running the sorting algorithms on the ARM Cortex A9 processor.

4.2.1   Permutations

Permutations [15] are used in various combinatorial optimization problems. Usually, a permutation is a disposition of a set of n objects 1, 2, 3, ..., n in a specific order where each element occurs just once. For example, there are six permutations of the set 1, 2, 3, (1, 2, 3), (1, 3, 2), (2, 1, 3), (2, 3, 1), (3, 1, 2) and (3, 2, 1). There are exactly n! permutations of n objects and the Lehmer code [12,14] is a specific method for encoding each possible permutation.

4.2.2   Assumptions and Constraints

Based on the work of [3], we assume that the maximum number of elements is 4096 because our objective is to use the "best" sorting algorithm for real-time decision support systems in a specific avionic application, which repeatedly sorts at most 4096 elements. In this case, we have thought about implementing sorting algorithms because, as mentioned in Table 1 their complexity is an approximation for large values of n. In addition to the computational times, we considered the energy consumption for the ARM processor as well as the resources for the implementation on FPGA (LUT: 53200, RAM: 4.9 Mb). Moreover, we calculated the standard-deviation of computational times to assess the temporal variations of each sort. Indeed, this is a very important result for a real-time application. In addition, we calculated the statistical dispersion. Statistical dispersion occurs when measuring the sorting computational times on permutations of the same size due to transmission errors or the noise generated by the OS. The related error is then estimated using standard-deviation (of computational times...). Considering each set of data to sort, the observed margin of error is calculated as the upper and lower limits using the following formulas:

$UBound = Time + standard\_deviation$
$LBound = Time - standard\_deviation$

After that, we calculate the coefficient of variation which is a relative measure of the dispersion of the data around the mean. The coefficient of variation is calculated as the ratio of the average and standard-deviation of computational times, which is expressed as a percentage. This coefficient allows us to compare the degree of variation from one sample to another, even if the means are different. It is therefore an important measure of reliability. However, a straightforward calculation of the standard-deviation of computational times alone does not usually allow us to estimate the relative dispersion of values around the mean. For this reason, if the value of the coefficient of variation is high then the dispersion around the mean is greater. In fact, a coefficient of variation is less than 1% is generally considered sufficient for the mean to be representative, provided that the number of elements in the population is sufficient (i.e. empirically >=30) and that the distribution of values in the population follows a Gaussian distribution. In contrast, if the coefficient is greater than 1 then it is advantageous to represent the results graphically with boxplots which allow to give an idea of the dispersion. We can therefore conclude that the stability of sorting algorithms in terms of computational times is significant if the coefficient of variation less than 1%.

4.2.3   Experimental Results

In this section, we present our results concerning the average and standard-deviation of computational times, energy consumption and stability for sorting algorithms. Firstly, we compare their average and standard-deviation of computational times (in microseconds) on a Zebboard platform (processor ARM Cortex A9) in Table 2, using a number of different elements ranging from 8 to 4096 and many replications (R = 1000). For example, if $n = 4096$, the average computational times are 194007.48 us, 11843.88 us, 6451.24 us, 5752 us, 5684.51 us and 3291.33us for InsertionSort, QuickSort, HeapSort, TimSort, MergeSort and ShellSort, respectively. Finally, we can conclude that if the amount of data is large then the InsertionSort algorithm consumes a lot of time even if $n <= 64$ which is rather small compared to usual applications of the sorting algorithms with huge values of n. For this reason, we did not take these algorithms into consideration. We were rather interested only in the other five algorithms if the number of data is greater than 64. In fact, Fig. 1 shows that TimSort and Merge-Sort algorithms give almost the same value of time if the size of the array is large. In addition, the QuickSort algorithm has a very long computational times compared to others. Subsequently, we notice that ShellSort is 42.1% faster up to 72% when running on the ARM Cortex A9 processor if $n > 64$ otherwise, the TimSort algorithm is the best. In order to assess the temporal stability of the algorithms, Table 2 also presents the standard-deviation of computational times for the sorting algorithms in both cases ($n <= 64$ and $n > 64$). We then observe that the value of the standard-deviation varies between 0 and 79,514. Therefore, it is clear that the ShellSort algorithm has a better average and standard-deviation than others if $n > 64$. For example, if n = 2048, the standard-deviation of computational times is equal to 62.612 for HeapSort, 71.61 for MergeSort, 50.121 for TimSort and 49.94 for ShellSort.

Table 2. Average and standard-deviation of computational times of sorting algorithms

	InsertionSort	HeapSort	ShellSort	QuickSort	MergeSort	TimSort
8	2.28 (0)	2.72 (0)	2.13 (0.369)	3.87 (0.05)	3.29 (0)	1.18 (0.05)
16	7.33 (0.202)	6.71 (0.307)	5.63 (0.625)	10.41 (0.06)	7.79 (0.021)	4.09 (0.08)
32	25.33 (0.339)	17.36 (0.196)	13.43 (0.671)	31.94 (0.175)	17.94 (1.65)	14.37 (0.619)
64	98.83 (0)	44.4 (0.107)	31.55 (2.194)	105.76 (0.18)	40.67 (2.55)	46.32 (3.28)
128	280.33 (0.027)	107.96 (4.28)	76.59 (7.15)	164.24	93.5 (4.069)	103.04 (7.4815)
256	930.55 (0.42)	255.76 (6.404)	168.58 (11.05)	266.49	207.84 (8.1)	220.07 (14.388)
512	3320.21 (0.57)	583.33 (14.18)	437.38 (11.083)	500.2	536.57 (36.26)	553.82 (21.7)
1024	12735	1323.75 (35.88)	1029.52 (28.86)	1214.4	1198.48 (32.83)	1205.97 (30.69)
2048	49314.71	2949.01 (62.612)	2116.45 (49.94)	3499.45	2629.1 (71.61)	2628.01 (50.121)
4096	194007.48	6451.2 (79.514)	3291.38 (59.6)	11843.88	5684.51 (60.08)	5741.88 (67.45)

In term of energy consumption, the measured values of voltage (V) and current (A) for the sorting algorithms remain almost constant when switching from one sorting algorithm to another and varying the size of the array. Indeed, the power value is almost constant for all sorting algorithms (P = 0.12 W). We can therefore conclude that if the algorithm has the best performance in terms of average and standard-deviation of computational times then, it will be considered as the best in terms of power consumption. In fact, Table 3 shows the energy consumption for the different sorting algorithms. As a matter of fact, the ShellSort consumes less energy when the number of elements is greater than 64. Moreover, as shown by Tables 3 and 4, if the number of data $n > 64$, we

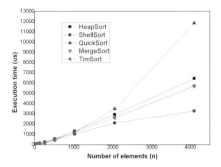

Fig. 1. Computational times of sorting algorithms

can conclude that ShellSort is the best performing algorithm in terms of average and standard-deviation of computational times and power consumption when running on the ARM Cortex A9 processor, otherwise, TimSort is chosen.

Table 3. Energy consumption of sorting algorithms (uJ)

	InsertionSort	HeapSort	ShellSort	QuickSort	MergeSort	TimSort
8	0.27	0.3264	0.2556	0.4644	0.394	0.1416
16	0.88	0.8052	0.6756	1.2492	0.9348	0.4908
32	3.0396	2.0832	1.7316	3.8328	2.1528	1.724
64	11.86	5.328	3.786	12.6912	4.884	5.558
128	33.639	12.955	9.1908	19.706	11.22	12.364
256	111.666	30.6912	20.229	31.978	24.84	26.408
512	398.425	70	52.4856	60.024	64.388	66.458
1024	1528.2	158.85	123.542	145.728	143.8176	144.716
2048	5917.95	353.88	253.97	419.934	315.492	315.36
4096	23282.45	774.14	394.96	1421.2656	682.1412	689.025

From Table 4, we notice that the coefficient of variation for the ShellSort, Heapsort, MergeSort and TimSort is greater than 1% if $n > 64$. As a consequence, we can experimentally conclude that there is a significant variation, which implies that the Boxplots are useful to give an idea about the dispersion of the data. In view of the obtained results, we focus on the Boxplots of the Shell-Sort and TimSort algorithms when $n <= 64$ and $n > 64$. Figure 2 shows that the boxplots for the ShellSort algorithm are almost lines. Moreover, as expected this figure illustrates a significant number of outliers, in general less than 3,74% of sorted data, due to transmission errors or the noise generated by the OS. More specifically, in addition to the percentage of total outliers (lower and upper outliers), the upper outliers (outliers greater than the value of the 3rd quartile (Q3)

Table 4. Coefficient of variation for sorting algorithms

	HeapSort	MergeSort	ShellSort	TimSort
8	0	0	13.566	1.77
16	2.94	0.256	9.31	1.026
32	0.613	9.16	3.86	1.33
64	0.1	6.26	4.94	8.064
128	3.96	4.35	9.335	7.26
256	2.5	3.897	1.25	6.534
512	2.43	6.758	2.533	3.91
1024	2.71	2.73	2.803	2.544
2048	2.2	2.723	2.35	1.9
4096	1.23	1.05	1.8	1.17

of Boxplot) are in general less than 0.8% of sorted data. We also observe that the percentage of outliers for the shellSort algorithm is negligible for n = 128, 256, 512, 1024 and 4096, but we notice that the number of outliers is important for n = 2048. In addition, Fig. 3 shows the Boxplots of the TimSort algorithm on the ARM processor. We notice from this figure a significant number of outliers, less than 7.22% of sorted data. The percentage of upper outliers for the TimSort algorithm on the ARM processor are less than 5% of sorted data.

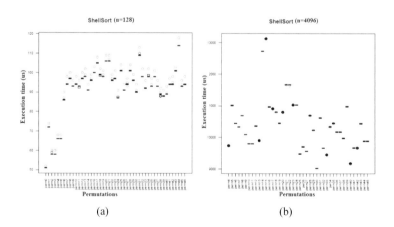

Fig. 2. Boxplots of the ShellSort algorithm on the ARM processor

Therefore, in order to compare the experimental measure of the time complexity of the sorting algorithms, we calculated their average computational times as a function of the array size to assess their temporal variations when n varies. For this purpose, we calculate the dispersion of the computational

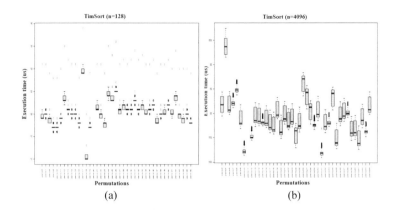

Fig. 3. Boxplots of the TimSort algorithm on the ARM processor

times. Figure 4 shows the execution time (ET) of the ShellSort and TimSort algorithms, LBound and UBound as a function of n. We can see that the two curves are almost linear. In this case and based on the Boxplots, we can give an idea about the stability of the algorithms.

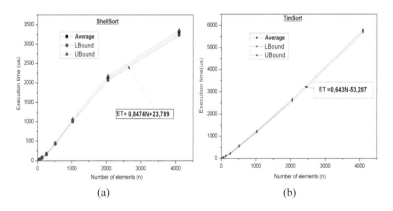

Fig. 4. Temporal variation of the ShellSort and TimSort algorithms as a function of n

5   Conclusion

In this paper, we have presented a software implementation of sorting algorithms to improve their performance in terms of four criteria (Average and standard-deviation of computational times, energy consumption and stability) on a Zynq-7000 platform using different numbers of elements ranging from 8 to 4096 encoded in 4 bytes.

Based on these results, we can conclude that BubbleSort, InsertionSort and QuickSort have a high computational times. In addition, ShellSort is 42.1% faster up to 72% running on the ARM Cortex A9 processor if $n > 64$, otherwise the TimSort algorithm is the best. To conclude, the ShellSort algorithm is the best trade-off between considered criteria.

As future work, we plan to embed the ShellSort, as the winner of the tournament between sorting algorithms in the context of our Intelligent Transportation Systems in decision-making avionic functions, such as path planning algorithms.

References

1. Baklouti, M., Aydi, Y., Marquet, P., Dekeyser, J., Abid, M.: Scalable MPNOC for massively parallel systems - design and implementation on FPGA. J. Syst. Architect. Embedded Syst. Des. **56**, 278–292 (2010)
2. Ben Jmaa, Y.: Implémentation temps réel des algorithmes de tri dans les applications de transports intelligents en se basant sur l'outil de synthèse haut niveau HLS. Ph.D dissertation, University of Valenciennes and Hainaut-Cambresis (UVHC), France (2019)
3. Nikolajevic, K.: Dynamic autonomous decision-support function for piloting a helicopter in emergency situations. Ph.D dissertation, UVHC, France (2016)
4. Miao, M., Jianfeng, W., Sheng, W., Jianfeng, M.: Publicly verifiable database scheme with efficient keyword search. Int. J. Inf. Sci. **475** (2019)
5. Aronovich, L., Ron, A., Eitan, B., Haim, B., Michael, H., Shmuel, T.K.: Systems and methods for efficient data searching, storage and reduction, U.S (2019)
6. Usmani, A.R.: A novel time and space complexity efficient variant of counting-sort algorithm. In: An International Conference on Innovative Computing (ICIC), Lahore, Pakistan (2019)
7. Boyan, L.: A Data Sorting Hardware Accelerator on FPGA. KTH, School of Electrical Engineering and Computer Science (EECS) (2020)
8. Dominik, Z., Marcin, P., Maciej, W., Kazimierz, W.: The comparison of parallel sorting algorithms implemented on different hardware platforms. Comput. Sci. **14** (2013)
9. Chhugani, J., Macy, W., Baransi, A., Nguyen, A.D., Hagog, M., Kumar, S., Dubey, P.: Efficient implementation of sorting on multi-core SIMD CPU architecture. Proc. VLDB Endow. (2008)
10. Patti, R.S.: Three-dimensional integrated circuits and the future of system-on-chip designs. Proc. IEEE **94** (2006)
11. Jmaa, Y.B., Atitallah, R.B., Duvivier, D., Jemaa, M.B.: A comparative study of sorting algorithms with FPGA acceleration by high level synthesis. Computación y Sistemas **23**(1), 213–230 (2019)
12. Diallo, A., Zopf, M., Johannes, F.: Permutation Learning via Lehmer Codes. In: 24th European Conference on Artificial Intelligence (ECAI). Santiago de Compostela, Spain (2020)
13. Jmaa, Y.B., Ali, K.M., Duvivier, D., Jemaa, M.B., Atitallah, R.B.: An efficient hardware implementation of timsort and mergesort algorithms using high level synthesis. In: International Conference on High Performance Computing and Simulation (HPCS), Genoa, Italy, pp. 580–587. IEEE (2017)

14. Magis, A.T., Price, N.D.: The top-scoring 'N' algorithm: a generalized relative expression classification method from small numbers of biomolecules. BMC Bioinf. **13** (2012)

15. Mehdi, M.: Parallel hybrid optimization methods for permutation based problems. Ph.D. dissertation, Lille University of Science and Technology, France (2011)

Cryptanalysis of a Privacy Preserving Ranked Multi-keyword Search Scheme in Cloud Computing

Xu An Wang[(✉)]

Key Laboratory of Information and Network Security, Engineeering University of PAP, Xi'an, China

Abstract. Nowadays outsourcing data to cloud servers is a popular data management way for many users. But how to ensure the security and the usability of the outsourced data is a very challenge problem. Usually encryption of the outsourced data is a good way to protect the privacy. But after encryption, the usability always loses, such as how to search over the outsourced data. Aiming at solving this problem, recently Zhang et al. proposed a privacy preserving ranked multi-keyword search scheme for multiple data owners in cloud computing [IEEE Transactions on Computers, 65(5), 1566–1577, 2018]. However, in this short paper we find their scheme is not secure. Concrete, the adversary can forge the search trapdoor for any keyword, which invalidates the security goal of their original design.

1 Introduction

Outsourcing the data to the cloud servers for management is a good choice for many data owners nowadays. However a very challenge problem is how to ensure the security of the outsourced data while achieve the usability of it. Especially the privacy of the outsourced data should be protected. A natural way is first encrypting the data and then outsourced the ciphertexts to the cloud servers. However such solutions usually lost the usability of the outsourced data. Search on the encrypted outsourced data is the basic requirement for cloud storage. Thus how to efficient retrieve the encrypted data is first needed before privacy preserving cloud storage being used practically. Searchable encryption is such a cryptographic primitive used to solve this problem, until now there are already existing some work [1, 3–12] on this topic.

Recently Zhang et al. [2] proposed a privacy preserving ranked multi-keyword search scheme for multiple data owners in cloud computing [IEEE Transactions on Computers, 65(5), 1566–1577, 2018]. However we find their scheme is not secure, in particular the cloud servers can forge trapdoors for any keyword, which obviously invalidates their scheme's security.

2 Review of Zhang et al.'s Proposal

Here we only review the Construction Initialization algorithm, the Keyword Encryption algorithm and Trapdoor Generation algorithm, we omit review the other algorithms for they are not directly with our attack, interested readers can refer them in [2].

L. Barolli et al. (Eds.): AINA 2021, LNNS 227, pp. 367–370, 2021.
https://doi.org/10.1007/978-3-030-75078-7_37

- Construction Initialization. Let g_1 and g_2 denote the generator of two cyclic groups G and G_1 with order p. Let e be a bilinear map $e: G \times G \to G_1$. Given different secret parameters as input, a randomized key generation algorithm will output the private key used in the system. $k_{a1} \in Z_p^+$, $k_{a2} \in Z_p^+$, $k_{i,f} \in Z_p^+$, $k_{i,w} \in Z_p^+$ where k_{a1} and k_{a2} are the private keys of the administration server, $k_{i,w}$ and $k_{i,f}$ are the private keys of the administration server, $k_{i,w}$ and $k_{i,f}$ are the private keys used to encrypt keywords and files of data owner O_i, respectively. Let $H(\cdot)$ be a public hash function, its output locates in Z_p^+.
- Keyword Encryption. Given the hth keyword of data owner O_i, i.e. $w_{i,h}$, they encrypt $w_{i,h}$ as follows.

$$\widetilde{w_{i,h}} = \left(g^{k_{i,w} \cdot r_O \cdot H(w_{i,h})}, g^{k_{i,w} \cdot r_O}\right)$$

where r_O is a randomly generated number each time, which helps enhance the security of $\widetilde{w_{i,h}}$. For easy description and understanding, they let $E_a' = g^{k_{i,w} \cdot r_O \cdot H(w_{i,h})}$ and $E_o = g^{k_{i,w} \cdot r_O}$.

The data owner delivers E_a' and E_o to the administration server, and the administration server further re-encrypts $E_{a'}$ with his secret keys k_{a1} and k_{a2} and gets E_a.

$$E_a = (E_{a'} \cdot g^{k_{a1}})^{k_{a2}}$$

Therefore $\widetilde{w_{i,h}} = (E_a, E_o)$. The administrative server further submits $\widetilde{w_{i,h}}$ to the cloud server.

- Trapdoor Generation. Assume a data user wants to search keyword $w_{h'}$, so he encrypts it as follows:

$$T'_{w_{h'}} = \left(g^{H(w'_h) \cdot r_u}, g^{r_u}\right)$$

where r_u is a randomly generated number each time.

Upon receiving $T'_{w_{h'}}$, the administration server first generates a random number r_a, and then re-encrypts $T'_{w_{h'}}$ as follows:

$$T_{w_{h'}} = \left(g^{H(w_{h'}) \cdot r_u \cdot k_{a1} \cdot k_{a2} \cdot r_a}, g^{r_u \cdot k_{a1}}, g^{r_u \cdot k_{a1} \cdot r_a}\right)$$

For easy description and understanding, they let $T_1 = g^{H(w_{h'}) \cdot r_u \cdot k_{a1} \cdot k_{a2} \cdot r_a}$, $T_2 = g^{r_u \cdot k_{a1}}$, $T_3 = g^{r_u \cdot k_{a1} \cdot r_a}$, and send $T_{w'_h} = (T_1, T_2, T_3)$ to the cloud server.

3   Our Attack

In this attack, assume the cloud server is malicious, our attack is the following:

1. The malicious cloud server queries the Trapdoor Generation oracle by a random keyword w_h, the Trapdoor Generation oracle should return the trapdoor to it.
2. According to the Trapdoor Generation algorithm, the cloud servers can obtain $T_1 = g^{H(w_h) \cdot r_u \cdot k_{a1} \cdot k_{a2} \cdot r_a}$, $T_2 = g^{r_u \cdot k_{a1}}$, $T_3 = g^{g^{r_u \cdot k_{a1} \cdot r_a}}$ as the trapdoor.
3. The malicious cloud server computes

$$\bar{T}_1 = T_1^{\frac{1}{H(w_h)}}$$

4. Then the malicious cloud server can forge the trapdoor for any keyword w'_h, which is the following

$$T'_1 = \bar{T}_1^{H(w'_h) \cdot r'_u} = g^{H(w_{h'}) \cdot r_u \cdot k_{a1} \cdot k_{a2} \cdot r_a \cdot r'_u}$$

$$T'_2 = T_2^{r'_u} = g^{r_u \cdot k_{a1} \cdot r'_u}, T'_3 = T_3^{r'_u} = g^{r_u \cdot k_{a1} \cdot r_a \cdot r'_u}$$

where r'_u is a randomly generated number from Z_p^+.

5. We can verify $T_{w'_h} = (T'_1, T'_2, T'_3)$ is a valid trapdoor for keyword w'_h for the cloud server, because it follows the structure of valid trapdoor.

Actually the attacker can be anyone, for the query to the Trapdoor Generation oracle is allowed in the security model of their scheme.

4   Conclusion

In this short paper, we show one recently proposed privacy preserving ranked multi-keyword search scheme for multiple data owners is not as secure as the authors claimed. The main reason is that, in their scheme the cloud servers can forge trapdoors for any keyword, which obviously invalidates their scheme's security. Thus we conclude this scheme needs further consideration before being used in practical applications.

Acknowledgements. This work is supported by National Key Research and Development Program of China Under Grants No. 2017YFB0802000, National Nature Science Foundation of China (Grant Nos. 61572521, U1636114), National Cryptography Development Fund of China Under Grants No. MMJJ20170112, Natural Science Basic Research Plan in Shaanxi Province of china (Grant Nos. 2018JM6028), and an Open Project from Guizhou Provincial Key Laboratory of Public Big Data under Grant No. 2019BDKFJJ008. This work is also supported by Engineering University of PAP's Funding for Scientific Research Innovation Team (No. KYTD201805), Engineering University of PAP's Funding for Key Researcher (No. KYGG202011).

References

1. Wang, X.A., Huang, X., Yang, X., Liu, L., Wu, X.: Further observation on proxy re-encryption with keyword search. J. Syst. Softw. **85**, 643–654 (2011). https://doi.org/10.1016/j.jss.2011.09.035
2. Zhang, W., Lin, Y., Xiao, S., Wu, J., Zhou, S.: Privacy preserving ranked multi-keyword search for multiple data owners in cloud computing. IEEE Trans. Comput. **65**, 1566–1577 (2015). https://doi.org/10.1109/TC.2015.2448099
3. Katz, J., Sahai, A., Waters, B.: Predicate encryption supporting disjunctions, polynomial equations, and inner products. In: EUROCRYPT 2008. LNCS 4965, pp. 146–162 (2008)
4. Boneh, D., Waters, B.: Conjunctive, subset, and range queries on encrypted data. In: TCC 2007. LNCS 4392, pp. 535–554 (2007)
5. Boneh, D., Crescenzo, G.D., Ostrovsky, R., Persiano, G.: Public key encryption with keyword search. In: Eurocrypt 2004. LNCS 3027. Springer, pp. 506–522 (2004)
6. Abdalla, M., Bellare, M., Catalano, D., Kiltz, E., Kohno, T., Lange, T., Malone-Lee, J., Neven, G., Paillier, P., Shi, H.: Searchable encryption revisited: consistency properties, relation to anonymous IBE, and extensions. In: Crypto 2005. LNCS 3621. Springer, pp. 205–222 (2005)

7. Golle, P., Staddon, J., Waters, B.: Secure conjunctive search over encrypted data. In: ACNS 2004. LNCS 3089, pp. 31–45 (2004)
8. Baek, J., Safiavi-Naini, R., Susilo, W.: Public key encryption with keyword search revisited. Cryptology ePrint Archive, Report 2005/119 (2005)
9. Li, J., Wang, Q., Wang, C., Cao, N., Ren, K., Lou, W.: Enabling efficient fuzzy keyword search over encrypted data in cloud computating. In: Infocom 2010, pp. 441–445. Cryptology ePrint Achieve. https://eprint.iacr.org/2009/593.pdf
10. Cao, N., Wang, C., Li, M., Ren, K., Lou, W.: Privacy preserving multi-keyword text search in the cloud supporting similarly-based ranking. IEEE Trans. Parallel Distrib. Syst. **25**(1), 222–233 (2014)
11. Zheng, Q., Xu, S., Ateniese, G.: VABKS: verifiable attribute-based keyword search over outsourced encrypted data. Cryptology ePrint Achieve. https://eprint.iacr.org/2013/462.pdf
12. Xia, Z., Wang, X., Sun, X., Wang, Q.: A secure and dynamic multi-keyword ranked search scheme over encrypted cloud data. IEEE Trans. Parallel Distrib. Syst. (2015). https://doi.org/10.1109/TPDS.2015.2401003

Attack of Two Certificateless Aggregate Signature Schemes Based on VANET

Shi Lin[1](✉) and Fu Pu[2](✉)

[1] PAP Engineering University, Xi'an, Shan Xi, China
[2] National University of Defense Technology, Xi'an, Shan Xi, China

Abstract. The Vehicular Ad-hoc Network is the fundamental of smart transportation system in the future. The rapid development of wireless communication technology makes it possible to communicate between vehicles and vehicles, between vehicles and roadside infrastructures. Wang et al. and Hong et al. proposed a certificateless aggregate signature algorithm for Vehicular Ad-hoc Network respectively, and proved the algorithm to be safe under the random oracle model. In this paper, these two programs are analyzed in depth, and it is found that the two solutions both have security problems. The scheme of Wang et al. cannot resist the attack of type II attacker, while the scheme of Hong et al. can neither resist the attack of type I attacker nor the attack of type II attacker. This paper details the attack process of the two schemes, analyzes reasons for the signature forgery of each algorithm and gives improvement measures. This work provides an idea for designing a more secure and efficient certificateless aggregate signature scheme based on Vehicular Ad-hoc Network.

1 Introduction

In 1976, Diffie and Hellman proposed the idea of public key cryptography, which is the most important invention and progress of modern cryptography, and is of great significance for protecting information security in networks. In the traditional public key cryptography, a trusted authentication center is required to issue a certificate to each user. When the number of users is large, the management and maintenance of the certificate is very complicated, which greatly reduces the performance of the system. In order to reduce the management of certificates, Shamir proposed the idea of identity-based public key cryptography in 1984 [1]. Users choose their own identity information as public key, which can effectively solve the problem of certificate management in traditional public key cryptosystem. However, the identity-based public key cryptosystem requires a trusted private key generation center. The private key generation center can obtain the private key of all users and can forge the signature of any user. Therefore, the identity-based public key cryptosystem suffers the key escrow problem.

In 2003, Al-Riyami and Paterson proposed the concept of Certificateless Public Key Cryptography (CL-PKC) [2]. In this system, the private key generation center only generates part of the private key of the user, the user then generates its own private key independently according to the partial private key generated by the private key generation center and the secret value selected by itself, thus solving the certificate management and key escrow problem. So far, a number of certificateless cryptography schemes have been proposed to meet different needs.

Aggregate signature was first proposed by Boneh et al. in 2003 [3], which is a research hotspot in recent years and often appears in top-level conference papers. In an aggregate signature scheme, different users sign different messages separately, and these signatures can be combined into one signature, and the verifier can verify whether the signature is from a specified user by simply verifying the synthesized signature, thereby reducing the signature verification workload and signature storage space. Aggregate signature can be said to be "batch processing" and "compression" technology in the field of digital signatures. Gong et al. first combined aggregate signature and certificateless public key cryptography [4], proposed a certificateless aggregate signature scheme, whose security was proved under the random oracle model. Since then, a large number of certificateless aggregate signatures have been proposed [5–8].

Vehicular Ad-hoc Network (VANET) is a mobile ad-hoc network formed by communications between vehicles and vehicles, between vehicles and roadside infrastructures [9]. The VANET is mainly composed of OnBoard Units (OBU) installed on the vehicles and Road Side Units (RSU) deployed on the infrastructure around the roads. Users share information and access various services provided by adjacent infrastructure through communications between vehicles and vehicles, and between vehicles and infrastructures. While improving efficiency and the riding environment, the network also brings many security threats such as eavesdropping, tampering, and tracking user privacy. In order to realize the safety certification of information transmission in VANET, many scholars have studied the certificateless aggregate signature algorithm based on VANET.

Wang et al. designed a certificateless aggregate signature algorithm based on VANET [10], but the algorithm has many shortcomings in terms of security and efficiency. The solution cannot resist the attack of type II attacker. Once the attacker intercepts the user's one valid signature on a message, the attacker can forge a legitimate valid signature of the user for any message, that is, the algorithm does not satisfy unforgeability. Recently, Hong et al. proposed a privacy-preserving authentication scheme with full aggregation in VANET [11], but the scheme can neither resist the attack of type I attacker nor the attack of type II attacker. This paper gives the detailed attack steps of the two schemes in references [10] and [11], analyzes the reasons for the attack, and gives the improvement measures. In this paper, two typical VANET-based certificateless aggregate signature schemes are deeply analyzed, which provides ideas for designing a more secure and efficient certificateless aggregate signature scheme based on VANET, thus avoiding the similar errors in the protocol designing.

2 Analysis of Wang et al.'s Scheme [10]

2.1 Scheme Review in [10]

Generally, aggregate signature scheme consists of one Key Generation Center (KGC), n signers, one signature aggregator, and one signature verifier. It includes seven algorithms, such as system setup algorithm, partial key extraction algorithm, user key generation algorithm, signature algorithm, aggregate signature algorithm and the aggregate signature verification algorithm. In addition to the above algorithms, the scheme of Wang et al. also includes vehicle registration algorithm, pseudonym generation algorithm and message verification algorithm. The specific algorithm is described as follows:

(1) System setup: The key generation center KGC selects a cyclic additive group G_1 and a cyclic multiplicative group G_2 with the same prime order q, P is a generator of G_1. KGC chooses a random number $s \in Z_q^*$ as the system master key, calculates $P_k = sP$ as the system public key, and selects two secure hash functions H_1, H_2: $\{0, 1\}^* \rightarrow Z_q^*$. Each RSU chooses its secret value y_i and calculates the public key $\overline{P_i} = y_i P$. The system public key and public parameters are disclosed, and the master key s is secretly saved.

(2) Vehicle registration: This algorithm is executed by the Road and Transport Authority (RTA). For a vehicle whose identity is ID_i, RTA selects a hash function: H_3: $\{0, 1\}^* \rightarrow G_1$, calculates $ID_i' = H_3(ID_i)$, ID_i' is the user's fake identity.

(3) Partial Private Key Generation: KGC executes this algorithm. Enter the user's identity ID_i', KGC calculates the vehicle's partial private key $p_i = sID_i'$, and transmits the partial private key to the user through a secure channel.

(4) User key generation: The user randomly selects $x_i \in Z_q^*$ as its secret value and calculates its public key $P_i = x_i P$.

(5) Pseudonym generation: Executed by the RSU. Enter the identity ID_i' of the vehicle U_i, RSU randomly selects $a_i \in Z_q^*$, calculates $F1_i = a_i ID_i'$, $W_i = H_2(F1_i)$, $F2_i = a_i W_i$, and outputs the pseudo identity $F_i = F1_i + F2_i$.

(6) Sign: The specific signing process of vehicle U_i to message m_i is: The vehicle randomly selects $r_i \in Z_q^*$, calculates $U_i = r_i P$, $h_i = H_1(m_i, F1_i, P_i, U_i)$, $V_i = p_i F2_i + h_i r_i P_K + h_i x_i \overline{P_i}$. $\sigma_i = (U_i, V_i)$ is the signature of the vehicle U_i to the message m_i.

(7) Verify: Input the message-signature pair (m_i, σ_i), public key P_i, partial pseudonym $F1_i$ and other parameters. The signature verification process is: Calculate $h_i = H_1(m_i, F1_i, P_i, U_i)$, $W_i = H_2(F1_i)$ and check whether the equation $e(V_i, P) = e(F1_i W_i + h_i U_i, P_K)e(h_i P_i, \overline{P_i})$ holds or not. If it holds, accept the signed message; otherwise reject.

(8) Aggregate: Input the signatures (m_i, σ_i) of n vehicles, the aggregator calculates $m = \{m_1, m_2, \cdots, m_n\}$, $\sigma = (U_1, U_2, \cdots, U_n, V)$, $V = \sum_{i=1}^{n} V_i$ and outputs the aggregate signature as (m, σ).

(9) Aggregate verify: RSU receives the aggregated signatures of n users, and verifies the signature as: Calculate $h_i = H_1(m_i, F1_i, P_i, U_i)$, $W_i = H_2(F1_i)$ and check whether the equation $e(V, P) = e(\sum_{i=1}^{n} (F1_i W_i + h_i U_i), P_K)e(\sum_{i=1}^{n} (h_i P_i), \overline{P_i})$ holds or not; if it holds, accept the aggregated signature, otherwise reject.

2.2 Attack of Wang et al.'s Scheme

After careful analysis, this paper finds that Wang et al.'s scheme is not safe under the attack of type II attacker. This section constructs a specific attack algorithm to prove that Wang et al.'s solution cannot satisfy its claimed unforgeability.

Assume there is a vehicle A, whose real identity is ID_A and fake identity generated by RTA is ID_A', partial private key is $p_A = sID_A'$, secret value is X_A, the corresponding public key is $P_A = x_A P$, pseudonym identity is $F_A = F1_A + F2_A$ ($F1_A = a_A ID_A'$, $W_A = H_2(F1_A)$, $F2_A = a_A W_A$).

The signing process of vehicle A to message m_i is as follows: A randomly selects $r \in Z_q^*$, calculates $U_i = r_i P$, $h_i = H_1(m_i, F1_A, P_A, U_i)$, $V_i = p_A F2_A + h_i r_i P_K + h_i x_A \overline{P_i}$, $\sigma_i = (U_i, V_i)$ is the valid signature of vehicle A to message m_i.

Let Q be the type II attacker (the malicious KGC, knowing the system's master key, but can not replace the user's public key). After attacker Q intercepts the legal signature $\sigma_i = (U_i, V_i)$ of user A on the message m_i, since Q knows the system master key, Q knows the partial private key of user A, Q calculates $h_i = H_1(m_i, F1_A, P_A, U_i)$, $V_i = p_A F2_A + h_i(r_i P_K + x_A \overline{P_i})$, and then $r_i P_K + x_A \overline{P_i} = (V_i - p_A F2_A)/h_i$.

Then, the attacker Q forges a signature of user A to the message m_i' as follows: Let $U_i' = U_i$, calculate $h_i' = H_1(m_i', F1_A, P_A, U_i')$, $V_i' = p_A F2_A + h_i'(V_i - p_A F2_A)/h_i$. $\sigma_i' = (U_i', V_i')$ is the forged signature of user A to message m_i'.

Theorem 1. The signature generated by Q through the above method is legal.

Proof. The verifier verifies the forged signature $\sigma_i' = (U_i', V_i')$ generated by the attacker Q. If the forged signature can pass the verification, then the forged signature is legal.

Verifier calculates $h_i' = H_1(m_i', F1_A, P_A, U_i')$, then

$$
\begin{aligned}
e(V_i', P) &= e(P_A F2_A + h_i'(v_i - P_A F2_A)/h_i, P) \\
&= e(p_A F2_A, P)e(h_i'(V_i - p_A F2_A)/h_i, P) \\
&= e(p_A F2_A, P)e(V_i - p_A F2_A, P)^{\frac{h_i'}{h_i}} \\
&= e(p_A F2_A, P)e(V_i, P)^{\frac{h_i'}{h_i}} e(p_A F2_A, P)^{-\frac{h_i'}{h_i}} \\
&= e(p_A F2_A, P)^{1-\frac{h_i'}{h_i}} e(p_A F2_A + h_i r_i P_K + h_i x_A \overline{P_i}, P)^{\frac{h_i'}{h_i}} \\
&= e(p_A F2_A, P)^{1-\frac{h_i'}{h_i}} e(p_A F2_A, P)^{\frac{h_i'}{h_i}} e(h_i r_i P_K + h_i x_A \overline{P_i}, P)^{\frac{h_i'}{h_i}} \\
&= e(p_A F2_A, P)e(h_i'/h_i(h_i r_i P_K + h_i x_A \overline{P_i}), P) \\
&= e(sID_A' a_A W_A, P)e(h_i' r_i P_K + h_i' x_A \overline{P_i}, P) \\
&= e(F1_A W_A, P_K)e(h_i' P_K, U_i)e(h_i' \overline{P_i}, P_A) \\
&= e(F1_A W_A, P_K)e(h_i' U_i, P_K)e(h_i' P_A, \overline{P_i}) \\
&= e(F1_A W_A + h_i' U_i', P_K)e(h_i' P_A, \overline{P_i})(U_i' = U_i) \quad (1)
\end{aligned}
$$

Since the equation $e(V_i', P) = e(F1_A W_A + h_i' U_i', P_K) e(h_i' P_A, \overline{P_i})$ holds, so the signature verification succeeded. Through the above attack steps, it can be known that the single signature of Wang et al.'s scheme can be forged, thus the aggregate signature can also be forged.

In summary, the VANET-based certificateless aggregate signature scheme proposed by Wang et al. is not safe under the attack of type II attacker.

2.3 Improvement Measures

In the signature phase of Wang et al.'s scheme, the algorithm $V_i = p_A F2_A + h_i r_i P_K + h_i x_A \overline{P_i}$ is not suitable. The second and third part of the right side of the equation both contain h_i, so that the attacker can get the value of $r_i P_K + x_A \overline{P_i}$ under the circumstances of known p_A, thus achieving the forgery purpose. If we change the equation to $V_i = p_A F2_A + r_i P_K + h_i x_A \overline{P_i}$, we can effectively avoid the attack of type II attacker.

3 Analysis of Hong et al.'s Scheme [11]

The scheme proposed by Hong et al. is the latest literature of VANET-based certificateless aggregate signature algorithm. By analyzing the algorithm of Hong et al., it is found that the algorithm is insecure, and it can neither resist the attack of type I attacker nor the attack of type II attacker. The detailed attack steps of the Hong et al. algorithm are given below.

3.1 Scheme Review in [10]

Hong's scheme can be divided into seven algorithms:

(1) System setup: Trusted authority(TA) generates two groups G_1, G_2 with the same prime order q. P is a generator of G_1, PKG chooses a random number $s \in Z_q^*$ and calculates $P_{pub} = sP$, where s is used for partial private key generation and is only known to PKG. TRA chooses a random number $\alpha \in Z_q^*$ and calculates $T_{pub} = \alpha P$, where α is used for pseudo identity generation and is only known to TRA. TAs choose four cryptographic hash functions: H_0, H_1, H_2, H_3.

(2) Pseudonym generation: Before joining the VANET, the vehicle should obtain the pseudonyms generated by TRA. A vehicle V_i chooses a random number $k_i \in Z_q^*$ and calculates $PID_{i,1} = k_i P$, then the vehicle sends $(RID_i, PID_{i,1})$ to TRA in a secure way. After receiving $(RID_i, PID_{i,1})$, TRA first checks whether the RID_i exists in its local database, and then calculates $PID_{i,2} = RID_i \oplus H_0(\alpha PID_{i,1}, VP_i)$ where VP_i is the valid period of PID_i. Then $PID_i = (PID_{i,1}, PID_{i,2}, VP_i)$ is transmitted to PKG via a secure channel.

(3) Partial key generation: Given a pseudo identity PID_i, PKG calculates $Q_i = H_3(PID_i)$, $psk_i = sQ_i$ and sets psk_i as a partial private key. Then PKG transmits (PID_i, psk_i) to the vehicle.

(4) Vehicle key generation: The vehicle V_i chooses a random number $x_i \in Z_q^*$ as its secret key vsk_i and calculates the vehicle public key $vpk_i = x_i P$.

(5) Sign:

1) When a vehicle V_i enters a new RSU's area, it first calculates $H_j = H_1(ID_{Rj})$, $S_i = psk_i + vsk_i H_j$ and stores it in TPD. Note that, H_j and S_i only need to be calculated once if vehicle V_i is under the R_j's coverage. When the vehicle leaves the current area and gets into a new area, they need to be recalculated.

2) When a vehicle V_i needs to sign a message m_i, it randomly picks a pseudo identity PID_i and chooses the current time as the timestamp t_i. Where t_i gives the freshness of the signed message to against reply attack. The vehicle chooses a random number $r_i \in Z_q^*$ and calculates $R_i = r_iP$. Then calculate $h_i = H_2(m_i, PID_i, vpk_i, ID_{Rj})$, $T_i = r_iH_j + h_iS_i$. Finally, $\sigma_i = (R_i, T_i)$ is a signature on $m_i \| t_i$ of PID_i. Then, V_i sends $\{PID_i, m_i, vpk_i, t_i, \sigma_i\}$ to the nearby RSU.

(6) Verify: Once a RSU receives the signed message $\{PID_i, m_i, vpk_i, t_i, \sigma_i\}$, it first checks the freshness of t_i. if t_i is fresh, RSU continues the verification procedure. The RSU R_j calculates $H_j = H_1(ID_{Rj})$ and stores it in its storage. Then R_j calculates $h_i = H_2(m_i, PID_i, vpk_i, ID_{Rj})$, $Q_i = H_3(PID_i)$ and checks whether $e(P, T_i) = e(P_{pub}, h_iQ_i)e(H_j, R_i + h_ivpk_i)$ holds or not. If it holds, accept the signed message; otherwise reject.

(7) Aggregate: assume a set of vehicles $\{V_1, V_2, \cdots, V_n\}$ with pseudo identities $\{PID_1, PID_2, \cdots, PID_n\}$, vehicle public keys $\{vpk_1, vpk_2, \cdots, vpk_n\}$ and corresponding message-signature pairs $\{(m_1\|t_1, \sigma_1 = (R_1, T_1)), \cdots, (m_n\|t_n, \sigma_n = (R_n, T_n))\}$. The RSU calculates $R = \sum_{i=1}^{n} R_i, T = \sum_{i=1}^{n} T_i$ and outputs the aggregated signature $\sigma = (R, T)$.

(8) Aggregate verify: Once an application server receives the certificateless aggregate signature $\sigma = (R, T)$ and corresponding messages, pseudo identities, vehicle public keys. The application will check the freshness of $t_i (i = 1, 2, \cdots, n)$, if t_i is fresh, then the application server calculates $Q_i = H_3(PID_i)$, $h_i = H_2(m_i, PID_i, vpk_i, ID_{Rj})$ and checks whether $e(P, T) = e(P_{pub}, \sum_{i-1}^{n} h_iQ_i)e(H_j, R + \sum_{i=1}^{n} h_ivpk_i)$ holds or not. If it holds, accept the signed message; otherwise reject.

3.2 Attack of Hong et al.'s Scheme

After careful analysis, it is found that Hong et al.'s scheme is not safe under the attack of type I and type II attacker. The following specific attack algorithm is constructed to prove that Hong et al.'s scheme does not satisfy its claimed unforgeability.

3.2.1 Replace Public Key Attack of Type I Attacker

Assume there is a vehicle A, whose pseudo identity is PID_a, partial private key is $psk_a = sQ_a = sH_3(PID_a)$, secret value is x_a, and the corresponding public key is $vpk_a = x_aP$. Now, the vehicle A is in the area of RSU ID_{Rj}, A calculates $H_j = H_1(ID_{Rj})$, $S_a = psk_a + vsk_aH_j$.

Let Q be a type I attacker, who don't know the system master key, but could replace the user's public key. Q randomly selects $b \in Z_q^*$, calculates $vpk_a' = (bP - P_{pub}Q_a)/H_j$, and use vpk_a' to replace vpk_a.

Q forges the signature of vehicle A to m_i as follows:

Q randomly selects $r_i \in Z_q^*$, calculates $R_i = r_iP$, $h_i = H_2(m_i, PID_a, vpk_a', ID_{Rj})$, $T_i = r_iH_j + h_ib$. The attacker outputs $\sigma_a = (R_i, T_i)$ as the forged signature of A to m_i.

Theorem 2. The signature Q generated through the above method is legal.

Proof. The verifier verifies the forged signature $\sigma_a = (R_i, T_i)$ generated by Q. As long as the signature can be verified, that is, $e(P, T_i)$ is equal to $e(P_{pub}, h_i Q_a)e(H_j, R_i + h_i vpk_a')$ (since the attacker Q has replaced user A's public key), then the signature is legal.

The verifier calculates $Q_a = H_3(PID_a)$, $h_i = H_2(m_i, PID_a, vpk_a', ID_{Rj})$, then

$$e(P, T_i) = e(P, r_i H_j + h_i b)$$
$$= e(P, r_i H_j)e(P, h_i b)$$
$$= e(R_i, H_j)e(bP, h_i) \tag{2}$$

And then

$$e(P_{pub}, h_i Q_a)e(H_j, R_i + h_i vpk_a')$$
$$= e(P_{pub}, h_i Q_a)e(H_j, R_i)e(H_j, h_i vpk_a')$$
$$= e(P_{pub} Q_a, h_i)e(H_j, R_i)e(H_j vpk_a', h_i)$$
$$= e(P_{pub} Q_a + H_j vpk_a', h_i)e(R_i, H_j)$$
$$= e\left(P_{pub} Q_a + H_j \frac{bP - P_{pub} Q_a}{H_j}, h_i\right)e(R_i, H_j)$$
$$= e(bP, h_i)e(R_i, H_j) \tag{3}$$

Since the equation $e(P, T_i) = e(P_{pub}, h_i Q_a)e(H_j, R_i + h_i vpk_a')$ holds, so the signature verification is successful. Through the above steps, it can be known that the single signature of Hong et al.'s scheme can be forged, thus the aggregate signature can also be forged.

In summary, the VANET-based certificateless aggregate signature scheme proposed by Hong et al. is not safe under the replace public key attack of type I attacker.

3.2.2 The Attack of Type II Attacker

Assume there is a vehicle A, whose pseudo identity is PID_a, partial private key is $pska = sQ_a = sH_3(PID_a)$, secret value is x_a, and the corresponding public key is $vpk_a = x_a P$. Now, the vehicle A is in the area of RSU ID_{Rj}, A calculates $H_j = H_1(ID_{Rj})$, $S_a = pska + vsk_a H_j$.

The signature phase of vehicle A to message m_i is: A randomly selects $r_i \in Z_q^*$, calculates $R_i = r_i P$, $h_i = H_2(m_i, PID_a, vpk_a, ID_{Rj})$, $T_i = r_i H_j + h_i S_a$. $\sigma_a = (R_i, T_i)$ is the signature of vehicle A to message m_i.

Let Q be a type II attacker (the malicious KGC, who knows the system's master key, but could not replace the user's public key). Since Q knows the system's master key, Q knows the partial private key of user A. After Q intercepts the legal signature $\sigma_a = (R_i, T_i)$ of A to m_i, Q calculates $h_i = H_2(m_i, PID_a, vpk_a, ID_{Rj})$, since $T_i = r_i H_j + h_i S_a$, then $T_i = r_i H_j + h_i(pska + vsk_a H_j) = r_i H_j + h_i pska + h_i x_a H_j$. Q calculates $T_i - h_i pska = H_j(r_i + x_a h_i)$, then $r_i + x_a h_i = (T_i - h_i pska)/H_j$.

Q forges the signature of vehicle A to another message m_i' as follows: Q calculates $h_i' = H_2(m_i', PID_a, vpk_a, ID_{Rj})$, let $f = h_i'/h_i$, $R_i' = fR_i$, $T_i' = h_i' pska + fH_j \frac{T_i - h_i pska}{H_j} =$

$h_i' pska + f(T_i - h_i pska) = fT_i, \sigma_i' = (R_i', T_i')$ is the forged signature of user A to message m_i'.

Theorem 3. The signature Q generated through the above method is legal.

Proof. The verifier verifies the forged signature $\sigma_i' = (R_i', T_i')$ generated by Q. As long as the signature can be verified, that is, the equation $e(P, T_i') = e(P_{pub}, h_i' Q_a) e(H_j, R_i' + h_i' vpk_a)$ holds, then the signature is legal.

First, verifier calculates $h_i' = H_2(m_i', PID_a, vpk_a, ID_{Rj})$, then

$$
\begin{aligned}
e(P, T_i') = e(P, fT_i) &= e(P, T_i)^f \\
&= e(P_{pub}, h_i Q_a)^f e(H_j, R_i + h_i vpk_a)^f \\
&= e(P_{pub}, h_i f Q_a) e(H_j, f(R_i + h_i vpk_a)) \\
&= e(P_{pub}, h_i' Q_a) e(H_j, fR_i + h_i' vpk_a) \\
&= e(P_{pub}, h_i' Q_a) e(H_j, R_i' + h_i' vpk_a)
\end{aligned}
\tag{4}
$$

Since the equation $e(P, T_i') = e(P_{pub}, h_i' Q_a) e(H_j, R_i' + h_i' vpk_a)$ holds, so the signature verification phase succeeded. Through the above attack steps, it can be known that the single signature of the Hong et al.'s scheme can be forged, thus the aggregate signature can also be forged.

In summary, the VANET-based certificateless aggregate signature scheme prposed by Hong et al. is not safe under the attack of type II attacker.

3.3 Improvement Measures

For the type I attacker, change $H_j = H_1(ID_{Rj})$ to $H_j = H_1(ID_{Rj}, vpk_i)$, add vpk_i to hash function to bind the value of vpk_i and H_j together, then it can resist the replace public key attack of type I attacker. For the type II attacker, change $h_i = H_2(m_i, PID_i, vpk_i, ID_{Rj})$ to $h_i = H_2(m_i, PID_i, vpk_i, ID_{Rj}, R_i)$, add R_i to hash functions h_i to increase the correlation between them, and then it can resist the attack of type II attacker.

4 Summary

In VANET, the issue of secure authentication of information transmitted between vehicles and vehicles, between vehicles and roadside infrastructures has been a research hotspot in the field of information security. In this paper, two typical certificateless aggregate signature schemes based on VANET are analyzed. It is found that the schemes can not meet the corresponding security requirements and cannot resist the attacks of type I or II attacker. In this paper, the detailed attack steps of the two schemes are given. The causes of attacks are analyzed respectively, and the improvement measures are suggested. The work in this paper provides ideas to design a safer and more efficient certificateless aggregate signature scheme based on VANET in the next step.

References

1. Shamir, A.: Identity-based cryptosystems and signature schemes. IEEE Trans. Inf. Theory. **22**(6), 644–654 (1976)
2. Al-Riyami, S.S., Paterson, K.G.: Certificateless public key cryptography. In: Advances in Cryptology-Asiacrypt 2003. LNCS, vol. 2894, pp. 452–473. Springer, Berlin (2003)
3. Boneh, D., Gentry, C., Lynn, B., et al.: Aggregate and verifiably encrypted signatures from bilinear maps. In: Advances in Cryptology-Eurocrypt 2003. LNCS, vol. 2656, pp. 416–432. Springer, Berlin (2003)
4. Gong, Z., Long, Y., Hong, X., et al.: Two certificateless aggregate signatures from bilinear maps. In: Proceedings of the IEEE SNPD 2007, pp. 188–193. IEEE (2007)
5. Shen, H., Jianhua, C.H., Jian, S.H., et al.: Cryptanalysis of a certificateless aggregate signature scheme with efficient verification. Secur. Commun. Netw. **9**, 2217–2221. (2016). https://doi.org/10.1002/sec.1480
6. Hui, Z.H.: Insecurity of a certificateless aggregate signature scheme. Secur. Commun. Netw. Secur. Commun. Netw. **9**, 1547–1552 (2016). https://doi.org/10.1002/sec.1447
7. Xiaodong, Y., Jinli, W., Tingchun, M., et al.: A short certificateless aggregate signature against coalition attacks. PLoS ONE **13**(12), e0205453 (2018). https://doi.org/10.1371/journal.pone.0205453
8. Libing, W., Zhiyan, X., Debiao, H., et al.: New certificateless aggregate signature scheme for healthcare multimedia social network on cloud environment. Secur. Commun. Netw. Article ID 2595273, 13 pages (2018). https://doi.org/10.1155/2018/2595273
9. Yongchan, K., Jongkun, L.: A secure analysis of vehicular authentication security scheme of RSUs in VANET. J. Comput. Virol. Hack Tech. https://doi.org/10.1007/s11416-016-0269-z
10. Daxing, W., Jikai, T.: Probably secure certificateless aggregate signature algorithm for vehicular ad hoc network. J. Electron. Inf. Technol. **40**(1), 11–17 (2018)
11. Zhong, H., Shunshun, H., Cui, J., et al.: Privacy-preserving authentication scheme with full aggregation in VANET. Inf. Sci. **476**(2019), 211–221 (2019)

Reduction of Packet Duplication in Broadcasting Inter Vehicular Information Using 5G Network

Mezbaul Islam[✉], Fahmida Akter, Ruhul Amin, Ruhul Amin, and Raqeebir Rab

Ahsanullah University of Science and Technology, Dhaka 1208, Bangladesh
{160104076,160104078,160104070,160104074,raqeebir.cse}@aust.edu

Abstract. With the increased number of vehicles on the roads, the generation of duplicate packets is a common problem in the Vehicular Ad hoc Network (VANET). Duplicate packets increase the overhead of packets in a network. Thus resulting in an increased delay and an unnecessary flow of packets. In a dense network, the problem is more severe. In this paper, we have proposed an effective *Conflict Region Control Flooding (CFCR)* algorithm to reduce the overhead of duplicate packets transmission. Furthermore, we have provided our simulation results depending on three different performance metrics- the accuracy of the proposed method, redundant packets removal percentage, and delay- to show that the approach is effective enough in reducing the overhead of packets over the regular flooding approach.

Keywords: Vehicular Ad hoc Network · 5G · Controlled flooding · Computer network · Delay

1 Introduction

The numbers of vehicles around us are increasing day by day. After houses and offices, it is the vehicles where people spend most of their time. Computer networking experts have suggested a new approach to the idea of wireless networking called Vehicular Ad hoc Networks (VANET) to improve safety and sustain highly effective traffic control systems. VANETs have a potential role in Intelligence Transport Systems (ITS). The main trade in VANETs is the changes of topology, low link lifetime, and the high number of vehicles in the networks. The primary challenge in VANETs is to ensure reliable and efficient packet transmission among the vehicles. In most of the cases in VANETs, information is shared among more than two vehicles, which is broadcast in nature. One of the drawbacks of broadcasting is flooding which in turn causes an increased number of duplicate packets and delay of data transmission in VANETs. In a dense network, the problem is more severe because of the large number of vehicles. Many flooding techniques such as- Simple Flooding, Probabilistic Flooding, Neighbor knowledge Flooding, Area-based Flooding, and Location-based Flooding are introduced for broadcasting information among the vehicles. [1–3] provide their own controlled flooding approaches. In this research, we have proposed an effective *Conflict Region Controlled Flooding* algorithm to prevent the flow of redundant data packets in VANETs over 5G networks. As VANET involves packet flooding, duplication of packets is a

L. Barolli et al. (Eds.): AINA 2021, LNNS 227, pp. 380–394, 2021.
https://doi.org/10.1007/978-3-030-75078-7_39

major issue here. In our proposed method, we chose to approach our goal by the implementation of *Conflict Region* and *Controlled Flooding*. The *Conflict Region* concept helps to reduce the burden of duplication-checking to some selected vehicles while the *Controlled Flooding* approach proposes a protocol to implement duplication-checking. V2V (vehicle-to-vehicle), V2I (vehicle-to-infrastructure), D2D (device-to-device)–are three major methods for constructing vehicular ad hoc networks. A good number of protocols have been proposed over Bluetooth, 5G, and Wifi on VANETs [4]. While [5] shows some drawbacks of Wifi and the effectiveness of using BLE (Bluetooth Low Energy) in V2V.

The rest of the paper is organized as follows: Sect. 2 summarizes the papers we went through for establishing the structure of the research, Sect. 3 describes our proposed method with the help of three algorithms, Sect. 4 presents the performance analysis of the proposed algorithm measured by three different performance metrics, and Sect. 5 contains future work and conclusion.

2   Related Work

A good number of works have been done on VANET for different communication methods (IEEE 802.11, Bluetooth) using the controlled flooding approach. G. Ciccarese et al. [1] introduce an intelligent flooding system to reduce packet duplication in broadcasting safety messages in VANET. According to their method, before sharing a safety message, a vehicle will generate a controlled packet named RTB (Request To Broadcast). Upon receiving, the neighbors will generate a CTB (Clear To Broadcast) packet. The receiver whose CTB will arrive at the sender first will be selected as the forwarder. T. Taleb et al. [2] propose a method where the vehicles that are moving in the same direction are considered in the same group. They suggest a routing algorithm where a path that involves vehicles from the same category is searched by the algorithm. As the control packets are shared among the vehicles of the same group, unnecessary flooding is reduced. While S. Pareek et al. [3] provide a geometric approach to reduce the number of broadcast packets while sharing with the vehicles in different lanes. Shinde et al. [6] suggest a novel approach for the allocation of cognitive radio-based resources while implementing D2D techniques to control interference between cellular devices and D2D vehicles. Naoki et al. [7] propose a system for allowing cars to autonomously collect and exchange data using autonomous knowledge IEEE 802.11-based inter-vehicle connectivity. S. Mumtaz et al. [8] suggest a novel resource allocation policy based on cognitive radio when applying D2D strategies in V2V. This strategy of allocation should monitor the interaction between mobile devices and D2D vehicles. J. Barrachina et al. [5] suggest a density-based approach to urban scenarios for RSU implementation. S. Shah et al. [9] provide an overview of the related 5G vehicle connectivity building blocks. They show the benefit of 5G communication over other existing systems in vehicular communication. It also paves the way for challenges in this sector. O. Lesse et al. [10] propose a controlled flooding scheme that retains the desired flooding properties while limiting the extent of the flooding message across the network. T. Kitani et al. [11] propose a method for collecting, retaining, and disseminating traffic information efficiently using inter-vehicle communication by introducing buses into the traffic information sharing system. R. Frank et al. [12] introduce V2V applications using Bluetooth communication technology.

3   Proposed Methodology

This section explains our proposed methodology with the help of three algorithms where the first algorithm explains the selection of neighbours, the second algorithm shows the steps of the duplication checking procedure, and the third algorithm presents the protocol of reducing the transmission of duplicate packets.

3.1   Description of Scenario

Determination of Source, Target, and Neighbour: In Fig. 1, we have deployed vehicles randomly to show how a vehicle determines its target(s). Here, V_i denotes a vehicle and the circular area determines the transmission range of that vehicle. Two or more vehicles are considered neighbours if they stay within the transmission regions of one another. In Fig. 1, we assume V_1 is the vehicle that is initiating the packet transfer. V_2 is inside the transmission range of V_1. So V_1 is the source and V_2 is the target. Again, V_3 is within the range of V_2. So V_2 is at the same time a source (for V_3) and a target (for V_1). Thus we can say a vehicle can be a source and a target at the same time.

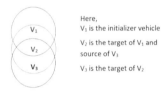

Fig. 1. Determination of a source and a target

Determination of Orientation of a Vehicle: From [13], the transmission range of 5G signal (same as the millimeter wave) is maximum 200 m. However, [14] shows that millimeter wave cannot penetrate several types of objects thus result in a lower propagation distance. Considering the best scenario, we are limiting the coverage for a vehicle up to 200 m in this research with an omnidirectional coverage as antennas provide omnidirectional coverage.

3.2   Explanation of the Proposed Methodology

To explain the proposed methodology, we have provided three algorithms along with their descriptions.

Determination of Conflict Region: Algorithm-1 deals with the approach of how a vehicle determines whether it is in a conflict region or not. Besides, we explain different aspects of the algorithm.

Algorithm 1: Determination of Conflict Region

---

1   $V \leftarrow$ currently checking vehicle
2   $V_{N_{i_{ad}}} \leftarrow$ unique address of a neighbouring vehicle of V
3   $V_{list_{N_{ad}}} \leftarrow$ set of unique addresses of neighboring vehicles of vehicle V
4   **for** each V **do**
5   $\quad$ add $V_{N_{i_{ad}}}$ to $V_{list_{N_{ad}}}$
6   $\quad$ **if** $|V_{list_{N_{ad}}}| > 1$ **then**
7   $\quad\quad$ | V is in conflict region
8   $\quad$ **else**
9   $\quad\quad$ V is not in conflict region

Description of Algorithm-1

Assigning Unique Address: In our proposed method, we are assigning a unique address to each vehicle. This address is called *vehicle_ id*. This address could be implemented in different ways. For example, the MAC address of an antenna can be used as a unique address as a MAC address is unique to every single antenna.

Determination of Conflict Region: In Fig. 2, we have implemented two vehicles, V_1 and V_2. The transmission ranges of V_1 and V_2 are represented with a red circle and a black circle respectively.

Scenario 1: In Fig. 2, V_1 and V_2 are inside each other's transmission regions. So we can say that the two vehicles are neighbours. V_1 adds the unique address of V_2 to its own storage $V_{1list_{N_{ad}}}$. At the same time, V_2 adds the unique address of V_1 to its storage $V_{2list_{N_{ad}}}$. As the sizes of the storage of both vehicles are 1, from Algorithm-1 we can say none of the vehicles are in any conflict region.

Fig. 2. Determining conflict region

Scenario 2: From Fig. 3, we can see that V_1 has two vehicles inside of its transmission region, V_2 and V_3. Hence V_1 will add the unique addresses of the two vehicles to its storage, $V_{1list_{N_{ad}}}$ and the size of the storage will be 2. As the size of the storage is greater than 1, from Algorithm-1, we can say that V_1 is in a conflict region. On the other hand, both V_2 and V_3 have one vehicle inside their transmission ranges which is

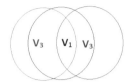

Fig. 3. Conflict region scenario

V_1. So both V_2 and V_3 will have the unique address of V_1 and the sizes of $V_{2_{list_{N_{ad}}}}$ and $V_{3_{list_{N_{ad}}}}$ will be 1. From Algorithm 1, we can say that V_2 and V_3 are not in any conflict region.

An Approach to Check Packet Duplication: Algorithm-2 describes how duplicate packets are checked for flooding and explains various aspects of the Algorithm 2. In our simulation, a packet contains five fields. They are - *vehicle_id*, *packet_id*, *X-coordinate*, *Y-coordinate*, *speed* of the vehicle.

vehicle_ id: $vehicle\_id$ is an auto-generated unique address for each vehicle to prevent the vehicles from processing their own generated packets.

packet_ id: It is the second field of a packet in our proposed method. A vehicle calculates the average of the three other fields- X-coordinate, Y-coordinate, speed and assigns the value in the $packet\_id$ field. We used the average of three fields to make sure that we can provide the maximum possibility to generate a unique $packet\_id$ for a packet.

X-coordinate and Y-coordinate: In our simulation, *X-coordinate* and *Y-coordinate* contain values of the position of a vehicle in X-axis and Y-axis respectively at any given time. A vehicle adds information of its position to these fields.

Speed: It is the last field of a packet. A vehicle adds its current speed to this field.

Description of Algorithm-2: The algorithm runs for vehicles that are in conflict regions to perform duplication checking. According to Algorithm-2, for duplication-checking, packets are stored in the buffer storage at first. Then after the checking is completed, packets are stored in the final storage. Figure 4 represents a buffer and a final storage where the buffer has one packet and the final storage has two packets. The fields that are currently being checked are marked with red color.

Checking of the first field: When a packet arrives at the V_{bs} -

– If V_{fs} is not empty, the system checks if the $vehicle\_id$ of the packet is the same as the currently checking vehicle. If the system finds a match, it deletes the packet from V_{bs} as the packet is its own. Otherwise, it increases i and j by 1 to go to the next stage of checking.

Algorithm 2: Packet Duplication Checking

---

  1 $V_{fs} \leftarrow$ final storage of vehicle V
  2 $V_{bs} \leftarrow$ buffer storage of vehicle V
  3 $V_{u\_id} \leftarrow$ unique id of vehicle V
  4 $BP_p \leftarrow$ one packet stored in V_{bs}
  5 $BP_{p_i} \leftarrow$ i-th field in BP_p
  6 $SP_p \leftarrow$ one packet stored in V_{fs}
  7 $SP_{p_j} \leftarrow$ j-th field in SP_p
  8 $i \leftarrow 0, j \leftarrow 0$
  9 **while** V_{fs} is not empty **do**
 10   **while** j $< V_{fs}$.size() **do**
 11   **if** $BP_{p_i} != V_{u\_id}$ **then**
 12              $i \leftarrow$ i+1, $j \leftarrow$ j+1
 13   **if** $BP_{p_i} == SP_{p_j}$ **then**
 14                  $i \leftarrow$ i+1, $j \leftarrow$ j+1
 15   **if** $BP_{p_i} == SP_{p_j}$ **then**
 16                      $i \leftarrow$ i+1, $j \leftarrow$ j+1
 17   **if** $BP_{p_i} == SP_{p_j}$ **then**
 18                          $i \leftarrow$ i+1, $j \leftarrow$ j+1
 19   **if** $BP_{p_i} == SP_{p_j}$ **then**
 20   delete BP_p from V_{bp}
 21                              $i \leftarrow 0, j \leftarrow 0$
 22   **else**
 23                              $i \leftarrow$ i-4, $j \leftarrow$ j+1
 24   start from 10
 25   **else**
 26                        $i \leftarrow$ i-3, $j \leftarrow$ j+2
 27   start from 10
 28   **else**
 29                    $i \leftarrow$ i-2, $j \leftarrow$ j+3
 30   start from 10
 31   **else**
 32                  $i \leftarrow$ i-1, $j \leftarrow$ j+4
 33   start from 10
 34   **else**
 35   delete BP_p from V_{bp}
 36   add BP_p to V_{fs}
 37   delete BP_p from V_{bs}
 38 **while** V_{fs} is empty **do**
 39   **if** $BP_{p_i} == V_{u\_id}$ **then**
 40   delete BP_p from V_{bs}
 41   **else**
 42   add BP_p to V_{fs}
 43   delete BP_p from V_{bs}

- If V_{fs} is empty, the vehicle checks if the *vehicle_id* of the packet is the same as the currently checking vehicle. If the system finds a match it deletes the packet from V_{bs}. Otherwise, it adds the packet to V_{fs} without any checking and deletes it from V_{bs}.

,

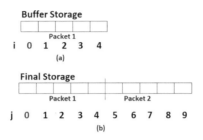

Fig. 4. Techniques to check duplication

Checking of the second field:

- In Fig. 5(a), if the system finds a match in the second field of both packets in V_{bs} and V_{fs}, the system increases the value of i and j by 1 to check the next fields of both packets (Fig. 5(b)).
- If the system does not find a match in Fig. 5(a), it increases j by 4 to jump to the first field of the second packet in V_{fs} and decreases i by 1 to jump to the first field of the packet in V_{bs} (Fig. 5(c)).

,

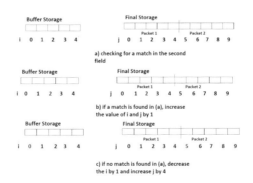

Fig. 5. Techniques to check duplication

Checking of the third field:

- In Fig. 6(a), if the system finds a match in the third field of both packets in V_{bs} and V_{fs}, it increases the value of i and j by 1 (Fig. 6(b)) to check the next field of both packets.

- If the system does not find a match in Fig. 6(a), it increases j by 3 to jump to the first field of the second packet in V_{fs} and decreases i by 2 to jump to the first field of the packet in V_{bs} (Fig. 6(c)).

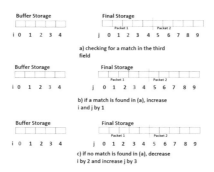

Fig. 6. Techniques to check duplication

Checking of the fourth field:

- If the system finds a match in the fourth field of both packets in V_{bs} and V_{fs}, the system increases the value of i and j by 1 to check the next field of both packets.
- If the system does not find a match, it will increase j by 2 to jump to the fifth field of the second packet in V_{fs} and decreases i by 3 to jump to the first field of the packet in V_{bs}.

Checking of the fifth field:

- If the system finds a match in the fifth field, it means the packet in the V_{bs} is duplicate packet. So the system will delete the packet from the buffer.
- If the system does not find a match, then the system will increase j by 1 to go to the first field of the second packet in V_{fs} and decrease i by 4 to jump to the first field of the packet in the V_{bs}.

3.3   Controlled Flooding

The *Conflict Region Controlled Flooding (CRCF)* (Algorithm 3) algorithm determines how a vehicle will act when it is in a conflict region or not in a conflict region.

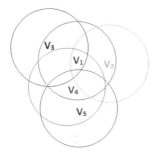

Fig. 7. How algorithm-3 works for conflict regions

Description of Algorithm-3: In Fig. 7, we have deployed vehicles randomly which are represented by V_i sign. In Fig. 7, we have a total of five vehicles. We denote the set of neighbours for a vehicle V_i with the notation V_{N_i}. The sets of neighbours for the vehicles V_1, V_2, V_3, V_4, and V_5 are- $V_{N_1} = \{V_2, V_3, V_4\}$, $V_{N_2} = \{V_1\}$, $V_{N_3} = \{V_1\}$, $V_{N_4} = \{V_1, V_5\}$, $V_{N_5} = \{V_4\}$ respectively. From Algorithm-1, V_1 and V_4 are the two conflicted vehicles while V_2, V_3, V_5 are the three non-conflicted vehicles. We assume V_3 is initiating a packet transfer. So it will share a packet p with its only neighbour, V_1. As packet p has newly arrived and the buffer is empty(considering no packets were in the network before), V_1 will add the packet p to its final storage and will share the packet p to V_2, V_3, and V_4. As V_2 is not a conflicted vehicle, it will add the packet to its final storage without any checking and will send the packet to V_1. V_1 has already received packet p, so it will delete the packet from its buffer. V_3 will delete the packet it has got from V_1 as the packet is its own. Thus packet p no longer circulates in the network consisting V_1, V_2, and V_3.

In the case of V_4, as the packet p is new to V_4, it will add it to its final storage from the buffer. Now V_4 will share the packet with V_1 and V_5. V_1 has a record of the packet p already. So it will discard the packet p. As the packet p is new to V_5, it will add the packet to its final storage and will share the packet with its only neighbour V_4. As V_4 has previously received the packet p, it will not add the packet to its final storage and will delete the packet from its buffer. Hence packet p no longer circulates in the mini-network of V_1, V_4, and V_5. As a result, packet p circulates within the networks of five vehicles without the transmission of any duplicate packets. Thus two conflicted vehicles V_1 and V_4 play a significant role to stop the circulation of duplicate packets in the given network.

Algorithm 3: Conflict Region Controlled Flooding Algorithm

---

1 $V_N \leftarrow$ set of neighbouring vehicles of vehicle V
2 $V_{N_i} \leftarrow$ i-th neighboring vehicle of vehicle V
3 $V_{N_{i_{fs}}} \leftarrow$ final storage of vehicle V_{N_i}
4 $V_{N_{i_{bs}}} \leftarrow$ buffer storage of vehicle V_{N_i}
5 **for** a vehicle V **do**
6 $\quad$ add vehicle_ id
7 $\quad$ add packet_ id
8 $\quad$ add X-coordinate
9 $\quad$ add Y-coordinate
10 $\quad$ add speed
11 $\quad$ make packet and send to all V_N
12 **for** each V_{N_i} **do**
13 $\quad$ **if** V_{N_i} *is in conflict region* **then**
14 $\quad\quad$ store packet p in $V_{N_{i_{bs}}}$
15 $\quad\quad$ **for** each packet p **do**
16 $\quad\quad\quad$ perform duplication checking
17 $\quad\quad\quad$ send all p in $V_{N_{i_{fs}}}$ to all V_N
18 $\quad$ **else**
19 $\quad\quad$ store packet in V_{fs}
20 $\quad\quad$ send all p in V_{fs} to all V_N
21 $\quad$ repeat from 5

---

4   Performance

In this section, we have provided the simulation settings and the performance analysis of our proposed algorithms concerning three different performance metrics - accuracy, redundant packet removal percentage, and average delay of the proposed method.

4.1   Simulation Scenario

As no maps were available, to perform the simulation we made a simple UI (Fig. 8) which represents a two-lane road. For the traffic density, we used sparse and dense networks which are represented in the Figs. 9a and 9b respectively. In Table 1, we have provided a summary of the simulation settings.

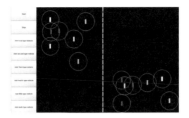

Fig. 8. Simulation scenario

Table 1. Simulation summary

Parameter	Value
Number of vehicles	20, 40, 80, 100, 150
Types of vehicles	12
Signal characteristics	5G
Transmission radius of each vehicle	200 m (120 pixels)
Flooding protocol	CRCF
Traffic density	Sparse and dense

4.2 Performance Analysis

We have performed performance analysis of our proposed method concerning three different performance metrics as mentioned earlier. For the three metrics, we provided comparison between the regular system and the proposed method.

Accuracy: It is the percentage of how accurately the system has filtered out the duplicate packets. Let, *total number of duplicate packets in the final storage* = A and *total number of packets in the final storage* = B. The equation is:

$$Accuracy(\%) = (1 - \frac{A}{B}) \times 100\% \tag{1}$$

We have performed the simulation on both sparse and dense networks. The results are presented in Fig. 10.

Redundant Packet Removal Percentage (RPRP): Let, *number of packets added to the final storage* = A and *number of packets arrived at the buffer* = B The equation of Redundant Packet Removal Percentage is:

$$RPRP(\%) = (1 - \frac{A}{B}) \times 100\% \tag{2}$$

The redundant packet removal percentage indicates the overhead reduced by the flooding approach. We performed the metric for both dense and sparse networks. The achieved data are given in Table 2a and Table 2b.

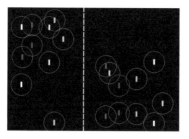

(a) Sparse Network Scenario

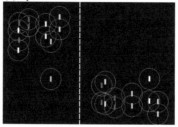

(b) Dense Network Scenario

Fig. 9. Two different traffic densities

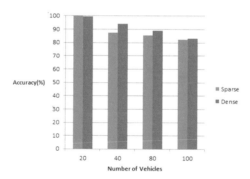

Fig. 10. Accuracy for sparse and dense networks

Average Delay: This is the average time between the creation of a packet and the arrival of the destination vehicle. From the concept of Networking, we know that the average delay of a packet consists of four different types of delays, they are - Queuing delay, Processing delay, Transmission delay, and Propagation delay.

Queuing Delay: It is the time a packet needs to wait in a queue before being processed. The equation of queuing delay is:

$$Queuing\ Delay = \frac{(number\ of\ packets - 1) \times size\ of\ a\ packet}{bandwidth} \tag{3}$$

Table 2. Redundant packet removal percentage

(a) Sparse networks

No. of vehicles	No. of packets received in the final storage (regular flooding)	No. of packets received in the final storage (proposed flooding)	Percentage of removal of duplicate packets
20	34244	16773	52%
40	487335	239053	51%
80	5527959	2757568	51%
100	8136194	4077532	50%
150	21105243	10571693	50%

(b) Dense networks

No. of vehicles	No. of packets received in the final storage (regular flooding)	No. of packets received in the final storage (proposed flooding)	Percentage of removal of redundant packets
20	21591	10757	51%
40	509992	239581	54%
80	5722118	2733676	53%
100	8318768	3988550	53%
150	18494179	8909692	52%

In our method, the queuing delay is divided into two parts. One part is the time a packet needs to wait in a queue and the other part is the time a packet needs to be being checked for duplication. From the data, we achieved in the Tables 2a and 2b, we calculated the average duplication-checking delay for a packet. We provide the data for the processing delay in the Tables 3a and 3b.

Transmission Delay: It is the time needed to put a packet from the storage to the transmission medium. The equation is:

$$Transmission\ Delay = \frac{size\ of\ packet}{bandwidth} \qquad (4)$$

In our performed simulation, each packet has a size of 100 bits. While 5G signal has bandwidth ranging from 25.26 GHz to 52.6 GHz.

Propagation Delay: From the law of Physics, we know that the equation of propagation speed of a signal is:

$$speed = wavelength \times frequency \qquad (5)$$

From [15], we have found that for 5G communication the millimeter wave is used. The spectrum of millimeter waves ranges from 30 GHz to 300 GHz. On the other hand, the wavelength of a millimeter-wave ranges from 1 mm to 10 mm. So the minimum propagation speed of the 5G signal can be 3×10^7 ms^{-1} which makes the propagation delay a negligible one. Thus we can omit propagation delay from the calculation of average end-to-end delay.

Table 3. Average processing delay

(a) Sparse networks

Number of vehicles	Delay (without flooding approach) ms	Delay (with flooding approach) ms
20	0.00003	0.00064
40	0.000012	0.0004
80	0.000017	0.0007
100	0.000017	0.0007
150	0.000017	0.0007

(b) Dense networks

Number of vehicles	Delay (without flooding approach) ms	Delay (with flooding approach) ms
20	0.0	0.0014
40	0.0000002	0.005
80	0.0000005	0.0022
100	0.0000001	0.002
150	0.0000012	0.0032

Except for the processing time in duplication-checking, all other delays are same for the regular flooding and the proposed flooding. From the Tables 2a, 2b, 3a, and 3b we can see that for a small penalty in the delay we can reduce the overhead of packet processing by more than 50%.

5 Conclusion

L. Urquiza-Aguiar et al. [16] provided a packet duplication-checking method which was applied on MAC layers using IEEE 802.11p standard. They achieved 48% and 60% repeated packets for 100 and 150 vehicles respectively whereas our proposed CRCF algorithm has 50% of repeated packets in sparse networks for both 100 and 150 vehicles respectively. Besides, our proposed method has 53% and 52% of repeated packets in dense networks for 100 and 150 vehicles respectively. In the Sect. 4, the provided simulation results show that the proposed flooding approach has significantly increased the performance of flooding. We have achieved a satisfactory accuracy and have been able to reduce the percentage of redundant packets at a substantial amount, although the delay is slightly higher than the normal flooding method. The delay can be reduced with the improvement of the duplication-checking algorithm.

References

1. Ciccarese, G., De Blasi, M., Marra, P., Palazzo, C., Patrono, L.: On the use of control packets for intelligent flooding in VANETs. In: 2009 IEEE Wireless Communications and Networking Conference, pp. 1–6 (2009)

2. Taleb, T., Sakhaee, E., Jamalipour, A., Hashimoto, K., Kato, N., Nemoto, Y.: A stable routing protocol to support its services in VANET networks. IEEE Trans. Veh. Technol. **56**(6), 3337–3347 (2007)
3. Pareek, S., Shanmughasundaram, R.: Implementation of broadcasting protocol for emergency notification in vehicular ad hoc network (VANET). In: 2018 Second International Conference on Intelligent Computing and Control Systems (ICICCS), pp. 1032–1037 (2018)
4. Cheng, X., Yang, L., Shen, X.: D2d for intelligent transportation systems: a feasibility study. IEEE Trans. Intell. Transp. Syst. **16**(4), 1784–1793 (2015)
5. Barrachina, J., Garrido, P., Fogue, M., Martinez, F.J., Cano, J.-C., Calafate, C.T., Manzoni, P.: Road side unit deployment: a density-based approach. IEEE Intell. Transp. Syst. Mag. **5**(3), 30–39 (2013)
6. Shinde, S.S., Yadahalli, R.M., Tamboli, A.S.: Vehicular ad-hoc network localization techniques: a review. Int. J. Electron. Commun. Technol. **3**(2), 82–86 (2012)
7. Shibata, N., Terauchi, T., Kitani, T., Yasumoto, K., Ito, M., Higashino, T.: A method for sharing traffic jam information using inter-vehicle communication. In: 2006 3rd Annual International Conference on Mobile and Ubiquitous Systems-Workshops, pp. 1–7. IEEE (2006)
8. Mumtaz, S., Huq, K.M.S., Ashraf, M.I., Rodriguez, J., Monteiro, V., Politis, C.: Cognitive vehicular communication for 5g. IEEE Commun. Mag. **53**(7), 109–117 (2015)
9. Shah, S.A.A., Ahmed, E., Imran, M., Zeadally, S.: 5g for vehicular communications. IEEE Commun. Mag. **56**(1), 111–117 (2018)
10. Lesser, O., Rom, R.: Routing by controlled flooding in communication networks. In: Proceedings. IEEE INFOCOM'90: Ninth Annual Joint Conference of the IEEE Computer and Communications Societies@ m_The Multiple Facets of Integration, pp. 910–917. IEEE (1990)
11. Kitani, T., Shinkawa, T., Shibata, N., Yasumoto, K., Ito, M., Higashino, T.: Efficient VANET-based traffic information sharing using buses on regular routes. In: VTC Spring 2008-IEEE Vehicular Technology Conference, pp. 3031–3036. IEEE (2008)
12. Frank, R., Bronzi, W., Castignani, G., Engel, T.: Bluetooth low energy: an alternative technology for VANET applications. In: 2014 11th Annual Conference on Wireless On-Demand Network Systems and Services (WONS), pp. 104–107. IEEE (2014)
13. Roh, W., Seol, J., Park, J., Lee, B., Lee, J., Kim, Y., Cho, J., Cheun, K., Aryanfar, F.: Millimeter-wave beamforming as an enabling technology for 5g cellular communications: theoretical feasibility and prototype results. IEEE Commun. Mag. **52**(2), 106–113 (2014)
14. 5G propagation. https://transition.fcc.gov/Bureaus/Engineering_Technology/Documents/bulletins/oet70/oet70a.pdf
15. Frenzel, L.: Millimeter waves will expand the wireless future. Electron. Des. **6**, 30–36 (2013)
16. Urquiza-Aguiar, L., Tripp-Barba, C., Romero, A.: Reducing duplicate packets in unicast VANET communications. In: Proceedings of the 12th ACM Symposium on Performance Evaluation of Wireless Ad Hoc, Sensor, & Ubiquitous Networks, pp. 1–8 (2015)

Speed Control for Autonomous Vehicular in Platoon

Anis Boubakri[✉] and Sonia Metteli Gammar[✉]

CRISTAL research Lab, ENSI, Manouba, Tunisia
anis.boubakri@esprit.tn, sonia.gammar@ensi.rnu.tn

Abstract. As promising applications, AV require road security which could be achieved by controlling the speed of each vehicle. The adaptive control is used for this goal, however, it can cause a speed exceeding and it is not stable for disturbances. To resolve these issues, we propose a speed management application by combining the predictive control and flatness. Simulations shows that our application provides stable speed values while minimizing energy consumption.

Keywords: Autonomous vehicles · Platoon · V2V · Speed control

1 Introduction

An autonomous vehicle (AV) is a vehicle capable of supporting partially or no human interaction, inter-vehicle interaction as well as interacting with its environment (e.g. sensors) in order to improve the road security. In level 0, there is no assistance, the driver has a full control of the vehicle. Level 1 refers as the driving assistance where the driver is the main responsible of driving and the vehicle system is in charge of assisting the driver by supporting either longitudinal control (e.g. speed control) or lateral control (e.g. white line tracking) but not both at the same time. Level 2, known as partial automation, consists of delegating the longitudinal and lateral control of the vehicle to the system while the driver supervises the vehicle system and monitors the environment. For level 3 called the conditional automation, the vehicle system takes in charge the driving operation and the environment monitoring while the driver supervises the vehicle system and does not have all permissions on the vehicles, i.e. he is not allowed to exceed the authorized maximum speed value, while he is allowed to drive his vehicle. The third level vehicles are called semi-AV. In level 4, the vehicle is capable to perform some actions (e.g. parking) in well defined context independently of the driver. Level 5 is called AV. In this case, the driven system can perform all the drive actions in all situations independently of the driver (i.e. the driver has no intervention on the vehicle) [1]. In this paper, we are limited to the AV of the third level.

In the Vehicle-to-Vehicle (V2V) network, vehicles can form a centralized or decentralized platoon. The centralized platoon size is limited by the scope coverage of the Platoon header [2–4], while the size of the decentralized platoon is

L. Barolli et al. (Eds.): AINA 2021, LNNS 227, pp. 395–405, 2021.
https://doi.org/10.1007/978-3-030-75078-7_40

illimited [5,6]. The platoon management is handled by the platoon header, called the leader vehicle [2,6]. One of the most challenging aspect in the platoon is the choice of a suitable speed circulation, minimizing the number of unnecessary acceleration and deceleration. This speed control impacts the energy consumption in a Platoon. The speed control process is comprised of three phases. The first one consists of collecting data, through the network, from the antecedent vehicle and from the local sensors. This phase is crucial for the control process. If a vehicle can not receive reliable real time data, it will not be able to proceed to the next steps. In the second phase, collected data are used to choose the suitable running speed. Finally, the real-time decision is executed.

In this paper, we will propose an autonomous driving application while managing the speed and reducing the energy consumption, and accordingly reducing the number of accidents. This application makes the external environment discovery through exchanging the beacon messages with the V2V vehicles and through exchanging data with the infrastructure (V2I). Based on these data, our application can choose the traffic speed value for each vehicle in the platoon. The chosen speed value is used to generate, in real time, a predictive speed tracking trajectory in order to provide stable and robust speed control in the platoon, minimize the fuel consumption and increase the traffic flow.

The organization of this paper is as following. We will start by presenting related works about speed management for AV in Sect. 2. Section 3 presents a short overview about predictive and flat controls that we will integrate in our approach. In Sect. 4, we will present our application which can help to choose and manage the vehicle speed. Section 5, we will illustrate the simulation of our proposed application and its performance.

2   Related Works

Lad et al. [7] aimed to control the speed value for small electric vehicles based on an existing database which contains information about the external environment of vehicles. They built an experimental platform by using small simulation vehicles which form a centralized Platoon with fixed size and managed by a control computer. The communication is between the control computer and vehicles and there is no communication between vehicles. The control application, implemented in Matlab in the computer, sends speed control signals to every the platoon's vehicle via a WiFi network. The vehicles use STM cards which collect data from the sensors and then send them to the Raspberry card. This latter sends received data via the standard 802.11g WiFi network. The control computer processes these data and sends the suitable control signal to each vehicle. This proposition has some gaps which are: There is no communication between vehicles in the Platoon; The simulation is applied for low speed values; The WiFi coverage is limited by 140 m; The latency delay is a critical factor in the DSRC communication; If the central computer loses the connectivity or is damaged, the Platoon management can not be performed.

Lazar and Tiganasu [8] proposed an approach to regulate the speed value by adjusting the safety distance between vehicles in order to increase the road traffic flow in a centralized homogeneous Platoon. This approach gives a stable speed regulation but not enough robust. However, the authors ignored the transmission delay which represents an imperative requirement in the real-time communication phase for AV.

Kerner [9] aimed to regulate the speed by proposing a new approach of real situation learning in vehicles. He proposed an autonomous driving framework based on the Three-Traffic-Phase approach to overcome the problem of an abrupt deceleration in the Platoon. Based on the classic Adaptive Cruise Control (ACC) approach, an abrupt reduction in the speed can cause traffic problems and reduce traffic flow since the response time required for the AV to react is ignored. The Three-Traffic-Phase Adaptive Cruise Control (TPACC) approach proposed by Kerner solves this problem (there is no fixed time for the previous vehicle in TPACC). The Kerner work ignores the communication between vehicles and focuses on the autonomous driving system. However, ignoring the communication between vehicles can not provide an appropriate driving application since driving decisions are based on the data exchanged between vehicles.

These works handle the speed control by using the adaptive command. However, this command is not stable for disturbances [9]. In addition, these works do not take into account the transmission time and the processing time.

3   Backgrounds

In this section, we will give an overview about the V2V communication in a platoon, the predictive control and the flatness control wich present a background for our proposed application.

3.1   V2V Communication in a Platoon

In a V2V Platoon [10], the communication between vehicles can be carried out via DSRC (Dedicated short-range communications) technology [2,5], VLC technology, a hybrid approach based on both DSRC and VLC (Visible Light Communication) technologies, or even based on cellular network. The DSRC technology is widely used in the AV domain as well as in the vehicular network despite of its limitations. These limits are mainly the packet loss during transmission, the coverage limitation (maximum of 300 m for urban area and a maximum of 1 km outdoor) security and the routing fees [11].

3.2   Predictive Control

Predictive control is robust even when a disturbance occurs [12]. This command is mainly used in slow systems and is difficult to be applied for fast systems such as vehicles. Predictive control is based on discrete incremental model of the system, and by adding an integral action in the model, we obtain the following CARIMA model:

$$A(q^{-1})y(k) = B(q^{-1})u(k) + \frac{C(q^{-1})}{H_S(q^{-1})}w(k) \tag{1}$$

This model allows to eliminate the static errors compared to the reference variable or any constant disturbance. We denote by $u(k)$ and $y(k)$ the command sequences and the output respectively, $w(k)$ is the white noise sequence with zero average, and A, B and C are polynomials parameterized according to a delay operator q^{-1}. These polynomials can be written as following:

$A(q^{-1}) = 1 + a_1 q^{-1} + \ldots + a_{n_A} q^{-n_A}$: integrates the effects of the current exit and previous exit values.

$B(q^{-1}) = b_0 + b_1 q^{-1} + \ldots + b_{n_B} q^{-n_B}$: integrates the effects of the current control and previous controls.

$C(q^{-1}) = 1 + c_1 q^{-1} + \ldots + c_{n_C} q^{-n_C}$: integrates the effects of the current noise and previous noise values.

$H_S(q^{-1}) = 1 - q^{-1}$:Makes it possible to take account not of the output but of its incrementation

The GPC control law is obtained by minimizing a quadratic criterion given by the Eq. (3). The constraints on the signal control or outputs can be integrated in the criterion. The generalized predictive Control GPC algorithm aims to apply a sequence of controls obtained by minimizing of the following cost function [12,13]:

$$\bar{J} = \sum_{j=h_i}^{h_p} [\varepsilon_{y,k+j}]^2 + \lambda \sum_{j=1}^{h_c} [\varepsilon_{u,k+j-h_i}]^2 \tag{2}$$

We denote:$E\{.\}$ is the expectation, $\hat{y}(k+j\,|k)$ the output prediction at time $(k+j)$ while its value is known at time k, hi is the initial costing horizon, h_c is the control horizon, and $\lambda(j)$ is control weighting sequence.

Despite advanced command techniques mentioned on previews related works, a new study is required to produce more effective controls and provide higher degree of stability and robustness. In this section, we describe a approach to control the speed of AV based on predictive control [12,13]. We resort to adopt the predictive command for its several advantages which are: It respects the constraints of the controlled and manipulated variables; Avoid excessive variations on manipulated variables. This allows the use of the actuators such as verins and motor with a long lifetime; it allows the system to perform automatic adaptation when measurable disturbances happen; It provides an ease adjustment of physical parameters which can help us to predict the variation effect; In order to get a robust and accurate control, predictive control is based on the optimization of performance criterion. This latter integrates a sequence of predicted output and helps to anticipate the system evolution. It requires on both future system output values and future reference variable values (in our case traffic speed value).

3.3   Flatness Control

The flatness control [14] is used for dynamic real-time systems with high mobility. In 1992, the flatness control concept was introduced by Fliess et al. [15]. It is characterized by the flat output variable which is used to configure all the variables in the system in real time. The major contribution of flatness control is the planning and monitoring of the trajectory with simple implementation and high specification performance. The flatness control allows tracking the trajectory as well as the regulation of the desired output of the generated trajectory in real time. The reference trajectory can be interpreted as predicted trajectory of the relevant controlled variables. In the case of process described by the model in the general case by Eq. (1). As the systems to be studied are linear and controllable, they are flat [13]. In general case, the process is described by the model in Eq. 1. AV are linear and controllable, therefore, they are flat [14]. We define based on the flat output Z_k, the tracking trajectory related to the control and the output denoted respectively by ε_u and ε_y which are given by the following relations:

$$\varepsilon_{y,k} = y_k - \tilde{B}(q^{-1})Z^d_{k+n_{\tilde{A}}} \tag{3}$$

$$\varepsilon_{\tilde{u},k} = \tilde{u}_k - \tilde{A}(q^{-1})Z^d_{k+n_{\tilde{A}}} \tag{4}$$

where the Z^d_k is the trajectory samples $Z^d(t)$ defined in continuous time. These samples correspond to the desired discrete trajectory for the flat output. Therefore, the flat output trajectory can be written as following:

$$\tilde{A}(q^{-1})\varepsilon_{y,k} = \tilde{B}(q^{-1})\varepsilon_{\tilde{u},k} + C(q^{-1})V_k \tag{5}$$

For this model, we apply the quadratic cost as following:

$$J = \sum_{j=hi}^{hp} \left(\hat{y}(k+j) - y_c(k+j)\right)^2 + \lambda \sum_{j=0}^{hc-1} \left(\Delta u(k+j)\right)^2 \tag{6}$$

The criterion minimizing leads to the following optimal solution:

$$\varepsilon_{u-opt} = -\sum_{j=h_i}^{h_p} m^0_{1,j}\hat{\varepsilon}^0_{y,k+j/k} \tag{7}$$

Knowing that $\hat{\varepsilon}^0_{y,k+j/k}$ satisfies the following equation:

$$C(q^{-1})\hat{\varepsilon}^0_{y,k+j/k} = F_j(q^{-1})\varepsilon_{y,k} + H_j(q^{-1})\varepsilon_{\tilde{u},k-1} \tag{8}$$

4   Proposed Solution

In this section, we will present our application which can help to manage the vehicle speed. Our application is composed of 3 phases. Phase 1 consists of collecting data from the platoon network. We make a decision about the most appropriate speed value in phase 2 and phase 3 is the execution.

4.1 Phase 1: Data Collection

To collect data, we propose a beacon message transmission algorithm as shown in Fig. 2. This algorithm, implemented in all vehicles, aims to transmit data periodically to all platoon vehicles. Once joining the network, a vehicle retrieves the traffic rules, if they exist, from the RSU. Then, it initializes the reception time *Trecep*. When it receives a beacon message, the vehicle updates its positions and traffic conditions (i.e. speed, red lights), and then, broadcasts the speed value. Once the reception time *Trecep* is over, a new cycle begins.

4.2 Phase 2: Speed Value Choice Decision

Management the speed for AV is an imperative and sensitive task. It helps to reduce the energy consumption and reduce the road accidents. Based on information collected from the beacon messages via the V2V and V2I networks, the vehicle On Board Unit (OBU) will decide about the suitable traffic speed in order to avoid unnecessary acceleration and deceleration. Figure 1 illustrates an example of the speed management at traffic lights. We suppose that the vehicle is running at speed value VA. Once the vehicle detects the traffic lights, two cases are possible (Fig. 2):

– Detection of a green light: based on the communication between the vehicle and the RSU, the OBU calculates the required time to switch to a new light color and the distance to pass through this light. Then, it calculates its new speed value. If this speed exceeds the maximum permissible speed, it decreases its speed value so that it passes through the next green light without stopping the vehicle. Otherwise, it increases its speed value to reach the calculated speed value;
– Detection of a red light: Based on collected data, the OBU calculates its new speed value. It can either decreases or increases its speed in order to reach the traffic light switching to the green color.

4.3 Phase 3: Execution

In this section, we aim to present our by proposition a speed management application for AV. We will benefit from the robustness of the predictive control and the fastness of the flatness control. predictive control and flatness algorithm. We suppose that A, B and C are the matrix, Y_c is the speed reference variable, h_p, h_c and λ are the parameters of our algorithm. We start by initializing the value k by 1. For every iteration, a sampling period Ts is chosen. The OBU acquires the converted data from analog to numeric. Then, it updates the vectors and the matrix (i.e. G, f, dx) and lastly, it calculates and applies the conversion control from analog to numeric. Once Ts elapses, a new iteration begins.

The predictive control and flatness, is given based on Eqs. (4) and (5), we deduce that the control structure has the following RTS polynomial form:

$$R(q^{-1})u_k + S(q^{-1})y_k = K(q^{-1})Z^d_{k+n_{\tilde{A}}} \tag{9}$$

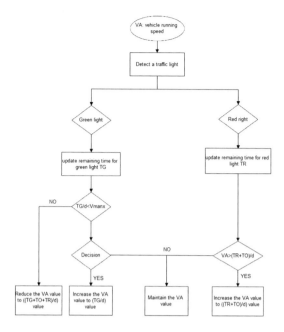

Fig. 1. Speed management at traffic lights

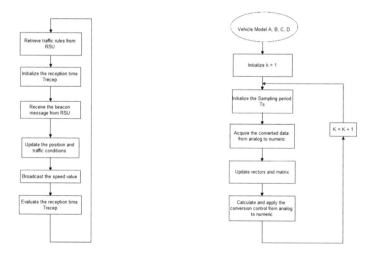

Fig. 2. Beacon message transmission algorithm

Fig. 3. Predictive control and flatness algorithm

while the polynomial K is defined by:

$$K(q^{-1}) = A(q^{-1})R(q^{-1}) + B(q^{-1})S(q^{-1}) \tag{10}$$

This control structure will allow tracking a given trajectory with quasi-null error while ensuring the stability and robustness of the obtained regulator. In a platoon, the speed of the leader vehicle is determined based on the Road Side Units (RSU) rules. This speed is then sent, every 100 ms, to other vehicles in the platoon via a beacon message which contains the Id, speed, direction and position [2,5].

The predictive control and flatness will allow reducing the braking and acceleration time as well as the braking damping based on data collected from the network. In addition, it allows to gain in terms of process control (precision and stability) which will subsequently reduce the consumption of circulation energy. To make further reduction in the energy consumption, unnecessary acceleration and deceleration must be minimized. Acceleration occurs where the is a need to reach the maximum speed allowed by the road rules, such as when the vehicle joins/leaves the platoon, or after speed reduction caused by the presence of traffic lights or obstacles. Deceleration occurs when there is a need to reduce the speed caused by the presence traffic lights or obstacles, or even when destination is reached.

Figure 3 illustrates different steps of the predictive control and flatness algorithm. We suppose that A, B and C are the matrix, Y_c is the speed reference variable, h_p, h_c and λ are the parameters of our algorithm. We start by initializing the value k by 1. For every iteration, a sampling period Ts is chosen. The OBU acquires the converted data from analog to numeric. Then, it updates the vectors and the matrix (i.e. G, f, dx) and lastly, it calculates and applies the conversion control from analog to numeric. Once Ts elapses, a new iteration begins.

5   Simulation and Performance

In our simulation, we consider 10 AV which form a centralized platoon having a distance of 3 km. We suppose that two speed values are predefined by the RSU: from 0 to 2400 m and from 2600 m to 3200 m, the speed value is 30 m/s, while from 2400 m to 2600 m, the speed value is equal to 10 m/s.

Our simulation is implemented using Matlab and OMNET++ software. OMNET++ simulates the exchange of beacon messages between vehicles in order to be insured about the existence of vehicles in the network. The beacon message contains the ID, the speed value, the position and the direction related to the vehicle. Periodically, OMNET++ transmits the chosen speed value to Matlab interface. Matlab generates the speed trajectory in real-time and the application, implemented on the OBU, allows tracking the generated trajectory with quasi-null error.

Figure 4 shows the variation of the speed over time. The leader vehicle V1 receives the maximum speed value predefined by the RSU rules. It broadcasts this speed value, using IEEE802.11p, to all the vehicles belonging to the platoon. These vehicles in turn handle their own speed values. Based on our proposed approach, the speed regulator management gives accurate speed values. Figure 6 shows that it is required approximately 8 s for a vehicle to reach 30 m/s with a reduced energy consumption and a minimized number of acceleration and deceleration.

The simulation results show that by applying the predictive control and flatness, the chosen speed value can not be exceeded because this control can predict the new speed value, while the ACC control adapts the speed value and Consequently traffic problems can occur. The predictive control and flatness ensures the reference trajectory tracking in real-time.

Figure 5 shows that energy consumption is important when switching between different speed values. It is imperative to minimize unnecessary acceleration and deceleration.

To minimize the fuel consumption, we propose to integrate a hysteresis into our solution. The hysteresis ignores the insignificant speed value in order to reduce the unnecessary acceleration and deceleration.

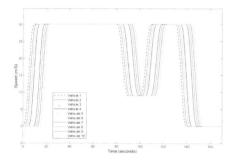 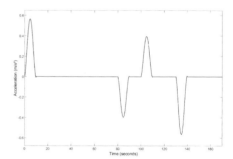

Fig. 4. Traffic speed of vehicles   **Fig. 5.** Leader vehicle acceleration

During driving, an AV may meet an obstacle or a traffic light which requires a speed management. Figure 6 illustrates the speed management performed by the leader vehicle at a traffic light. In this case, the leader vehicle handles its speed according to the situation: reducing the speed in order to reach the green traffic light without having to stop, increasing the speed before the red traffic light goes on, or keeping the same speed value if the traffic light is green. All vehicles follow the same behavior while keeping a safe distance with the antecedent vehicle.

Figure 7 shows the leader vehicle energy consumption over time. The hysteresis value is calculated to be sensitive to the speed value change caused by the meeting obstacles and traffic lights. This reduces the number of unnecessary acceleration and deceleration and increases the road traffic flow.

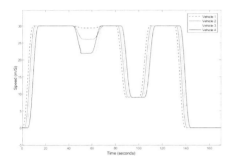
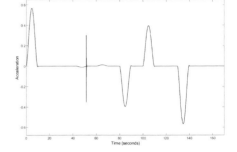

Fig. 6. Speed management at a traffic light

Fig. 7. Leader vehicle energy consumption over time

6   Conclusion

In this paper, we are interested in AV of level 3/4. We proposed a new approach based on predictive and flatness control in order to manage automatically the traffic speed. We started by collecting data from the RSU, and then choosing and applying the most suitable speed in real time. Our proposition shows a huge reduction in the energy consumption while reducing the number of unnecessary acceleration and deceleration. In future work, we aim to implement our solution in real V2V infrastructure.

References

1. Gruyer, D., Magnier, V., Hamdi, K., Claussmann, L., Orfila, O., Rakotonirainy, A.: Perception, information processing and modeling: Critical stages for autonomous driving applications. Annu. Rev. Control. **44**, 10 (2017)
2. Li, Y., Chen, W., Zhang, K., Zheng, T., Feng, H.: DSRC based vehicular platoon control considering the realistic v2v/v2i communications, February 2017
3. Kwon, J.-W., Chwa, D.: Adaptive bidirectional platoon control using a coupled sliding mode control method. IEEE Trans. Intell. Transp. Syst. **15**, 2040–2048 (2014)
4. Jia, D., Ngoduy, D.: Platoon based cooperative driving model with consideration of realistic inter-vehicle communication. Transp. Res. Part C Emer. Technol **68**, 04 (2016)
5. Ucar, S., Ergen, S.C., Ozkasap, O.: Security vulnerabilities of IEEE 802.11p and visible light communication based platoon. In: 2016 IEEE Vehicular Networking Conference (VNC), pp. 1–4 (2016)
6. Segata, M., Lo Cigno, R., Tsai, H.M., Dressler, F.: On platooning control using IEEE 802.11p in conjunction with visible light communications. In: 2016 12th Annual Conference on Wireless On-demand Network Systems and Services (WONS), pp. 1–4 (2016)

7. Lád, M., Herman, I., Hurák, Z.: Vehicular platooning experiments using autonomous slot cars**i.h. was supported by the Czech science foundation within the project GACR 16-19526s. IFAC-PapersOnLine **50**(1), 12 596–12 603 (2017). 20th IFAC World Congress. http://www.sciencedirect.com/science/article/pii/S2405896317328719

8. Lazar, C., Tiganasu, A.: String stable vehicle platooning using adaptive cruise controlled vehicles. IFAC-PapersOnLine **52**(5), 1–6 (2019). 9th IFAC Symposium on Advances in Automotive Control AAC 2019. http://www.sciencedirect.com/science/article/pii/S2405896319306202

9. Kerner, B.S.: Autonomous driving in framework of three-phase traffic theory. Procedia Comput. Sci. **130**, 785–790 (2018). The 9th International Conference on Ambient Systems, Networks and Technologies (ANT 2018)/The 8th International Conference on Sustainable Energy Information Technology (SEIT-2018)/Affiliated Workshops. http://www.sciencedirect.com/science/article/pii/S1877050918304988

10. Boubakri, A., Gammar, S.M.: Intra-platoon communication in autonomous vehicle: a survey. In: 9th IFIP International Conference on Performance Evaluation and Modeling in Wireless Networks, PEMWN 2020, Berlin, Germany, 1–3 December 2020, pp. 1–6. IEEE (2020). https://doi.org/10.23919/PEMWN50727.2020.9293086

11. Yin, J., Elbatt, T., Yeung, G., Ryu, B., Habermas, S., Krishnan, H., Talty, T.: Performance evaluation of safety applications over DSRC vehicular ad hoc networks, pp. 1–9, January 2004

12. Ramos, C., Martínez, M., Sanchis, J., Herrero, J.: Robust and stable predictive control with bounded uncertainties. J. Math. Anal. Appl. **342**(2), 1003–1014 (2008). http://www.sciencedirect.com/science/article/pii/S0022247X07014564

13. Ouammi, A., Achour, Y., Zejli, D., Dagdougui, H.: Supervisory model predictive control for optimal energy management of networked smart greenhouses integrated microgrid. IEEE Trans. Autom. Sci. Eng. **17**(1), 117–128 (2020)

14. Ayadi, M., Haggege, J., Bouallegue, S., Benrejeb, M.: A digital flatness-based control system of a DC motor. Stud. Inf. Control **17**, 201–214 (2008)

15. Fliess, M., Levine, J., Martin, P., Rouchon, P.: A lie-backlund approach to equivalence and flatness of nonlinear systems. IEEE Trans. Autom. Control **44**(5), 922–937 (1999)

Multiple RPL Objective Functions for Heterogeneous IoT Networks

Bishmita Hazarika[1]([✉]), Rakesh Matam[1], and Somanath Tripathy[2]

[1] Indian Institute of Information Technology, Guwahati, India
rakesh@iiitg.ac.in
[2] Indian Institute of Technology, Patna, India
som@iitp.ac.in

Abstract. RPL (Routing Protocol for low power and lossy networks) is an IETF standardized and widely used routing protocol for low-power lossy networks. It selects routes based on an objective function that defines the routing metrics and constraints (if any). The standard specifies multiple routing metrics like expected transmission count (ETX), latency, hop-count, remaining power in the device, etc. to select routes, but none of these metrics singularly are suited for heterogeneous networks. In a power constrained network, several factors effect the quality of service (QoS) and therefore to achieve an optimal routing solution, combination of several metrics is to be considered based on the network. Hence, in this paper multiple objective functions are designed considering the a constrained network with heterogeneous nature, similar to that of a smart grid during both low and high traffic load. The multi-objective metrics consider energy, link quality, delay, and hop-count as parameters to compute the metrics. Simulations show that the proposed multi-objective function(s) increase network lifetime, throughput, and also lowers the packet loss in the network.

1 Introduction

Internet of Things (IoT) is a collection of inter connected physical objects equipped with sensors and embedded computing capabilities, which are in turn connected to the Internet. Technologies empowering the IoT ecosystem have evolved over-the-years starting from wireless sensor networks (WSNs). The major applications of IoT are comprised of tens to thousands of devices that are usually constrained in terms of processing power, memory and energy. In most of the use-case scenarios these devices are typically battery operated. The wireless links interconnecting these devices are typically lossy in nature and support only low-data rates. In other words, the wireless links are unstable and packet deliver ratio is poor. Therefore, these networks are also referred to low-power lossy networks or LLNs. These aforementioned network and device characteristics motivated the IETF to constitute the ROLL (routing over low-power lossy networks) working group to design IPV6 routing protocol for LLNs (RPL). The detailed description and explanation of RPL protocol can be found in [15]. In this protocol the metrics and constraints that define the network path selection process is defined by Objective Function (OF). The choice of the metric(s) is dependent on the network requirements and goals. The choice and design of objective function is based on network and application requirements. Therefore, in a large wireless sensor network, as devices are heterogeneous in terms of physical and technical characteristics, the design of an OF is crucial as there are multiple factors effecting the quality of the network. In this paper we identify the factors responsible for network performance degradation in a heterogeneous network, and further design multiple RPL objective functions tailored to the needs of such heterogeneous networks.

1.1 Motivation and Contributions

The standard RPL objective functions do not consider many parameters that affect the performance of a large heterogeneous and constrained LLN. So, the main goal of this work is to design an OF for heterogeneous networks with device constraints. The main contributions of this paper are:

- Enumerate the factors affecting the network lifetime and network quality of service.
- Design an OF considering device heterogeneity and constraints to boost network lifetime and packet delivery ratio of the LLN.
- The OFs are designed to switch dynamically based on criticality of the network to provide efficient route selection, during both high and low traffic loads.

The rest of the paper is organized as follows. The related work is presented in Sect. 2. The design of proposed objective functions and their working is discussed in Sect. 3 and Sect. 4, respectively. The experimental analysis and simulation results are presented in Sect. 5. Finally, we conclude our work in Sect. 6.

2 Related Work

Routing in large low-power lossy networks is a challenging task. The two widely used and standardized OFs for the RPL routing protocol are MRHOF, where ETX is considered for route selection and hop-count. This OF considers the route with least number of transmissions and re-transmissions Even though ETX is an effective metric it is not suited for all types of LLN. The other standardized OF is called OF0, where the metric is Hop Count. This function only considers the number of hop in a path to calculate the best route. Other proposed metrics like the one presented in [12] and [9] proposes the use of fuzzy logic for path selection. In [11]. Although, this method increases the lifetime of the network than ETX, the QoS parameters like packet delivery ratio decreases. The approaches presented in [10,13,14] are based on RPL multi-instance, but do not consider the energy and delay parameters. The approach used in [8] also proposes an multi instance method and considers ETX, energy and delay metrics as parameters.

The authors in [1–4] also introduce new objective functions with packet-delivery as the primary metric. On the other hand, work presented in [5] and [6] designed objective function for RPL protocol by considering energy and lifetime as the primary metric. In this paper, we aim to design a metric that can dynamically choose an OF based on either network lifetime or QoS requirements.

3   Design of Proposed Objective Functions

The lifetime of the network is based on the routing choices and the balance in selection of nodes. Node energy levels is one of the primary metric that has to be considered to increase network lifetime. Similarly, the physical characteristics of a node highly impact the quality of link. Other metrics like delay and hop-count also contribute to the QoS of the network. The choice of parameters and their combination is crucial for meeting lifetime and QoS requirements. In this paper, two Objective Functions are proposed. The first proposed Objective Function is named POF_1 and second proposed Objective Function is named POF_2. OF1 and POF_1 should not be considered as same. OF1 is the standard OF also known as MRHOF and POF_1 is the first proposed objective function in this paper.

3.1   Design of First Proposed of (POF_1)

(POF_1) aims to increase the network lifetime and decrease the packet loss of the network. It selects the routes by considering the product of link quality of the route combined with the remaining energy/power of nodes on a route in order to include links with higher quality and nodes with higher energy. Delay is another important factor that is shown to affect the network performance therefore is part of the objective function, which gives us the POF_1.

$$POF_1 = \frac{\texttt{Final link quality} \times \texttt{Power}}{\texttt{Delay}} \tag{1}$$

where,

$$\texttt{Final-link-quality} = \frac{\texttt{LQ of sender} + \texttt{LQ of receiver}}{2}$$

$$\texttt{Link Quality (LQ)} = \frac{\texttt{1-receiver power}}{\texttt{receiver-sensitivity}}$$

$$\texttt{Power} = \texttt{remaining energy of the nodes}$$

$$\texttt{Delay} = \texttt{node time to process a packet} \tag{2}$$

3.2   Limitations of POF_1 and Design of Second Proposed of (POF_2)

Although POF_1 increases the lifetime and lowers the packet loss, this OF does not consider the number of hops on a route, which might lead to selection of longer routes. Even if this does not cause any problem during normal scenario but when network traffic is high or application is critical (i.e. where latency is an issue), selection of a longer route over a shorter route is not desirable irrespective of the condition of link quality in the longer route since a route with more hops are formed as a conjunction of multiple and more number of links. Each of these links are of different quality since resulting in different bandwidth and throughput of each individual link. Since the physical properties of the nodes are heterogeneous, the link quality is less likely to be uniform for all the links in a route. Hence, the more number of link conjunction, higher the chance of bottle neck. If in a route packet is moving from a link with higher bandwidth towards a link with lower bandwidth, during high traffic the occurrence of bottle neck at such points will lead to packet loss. Longer the route, higher is the point of conjunctions and more packet loss. Hence during high traffic condition it is more efficient to chose a route with lesser number of link conjunctions.

Hence, POF_2 is proposed for high traffic conditions to avoid more number of hops in order to prevent more link conjunctions and select the best possible route with least hop-count. It is designed to support higher data-rates considering routes with fewer number of hops. POF_2 is computed as the product of link quality and power while considering the delay and hop-count. It implies that during higher traffic the route with better link quality, more power, less Delay and less hop-count will be preferred. All the metrics are additive metrics. They are the summation of values of all the nodes on a route.

$$POF_2 = \frac{\text{Final link quality} \times \text{Power}}{\text{Delay} \times \text{Hop Count}} \tag{3}$$

where,

$$\text{Final-link-quality} = \frac{\text{LQ of sender} + \text{LQ of receiver}}{2}$$

$$\text{Link Quality (LQ)} = \frac{\text{1-receiver power}}{\text{receiver-sensitivity}}$$

$$\text{Power} = \text{remaining energy of the nodes}$$

$$\text{Delay} = \text{node time to process a packet}$$

$$\text{Hop Count} = \text{number of nodes in a route} \tag{4}$$

Therefore, in this paper two objective functions are designed, POF_1 is for regular traffic condition and POF_2 is for critical/high traffic condition. This multi-objective approach balances the load and optimizes the overall performance in case of lifetime and packet loss.

4   Working of Objective Functions

The design of the proposed OFs are illustrated with the help of a small network. The network is comprised of six routes between root node S and node X namely A, B, C, D, E, F. All the routes vary in the number of nodes and properties which effects the link quality, delay and remaining power. The total link quality, total remaining power and total delay in a route can be calculated by adding the individual values of the respective metrics of all the links in the route. For example, the total remaining power factor can be obtained by adding the remaining power of all the nodes in that path. Similarly, the other metrics are also additive. In Fig. 1, the rectangles on the right hand side contains the values of all the six routes where LQ stands for Link Quality, p is the overall remaining power of the route and d stands for delay.

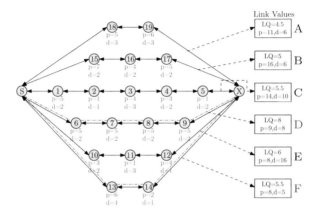

Fig. 1. Network nodes with different values for power, delay, link quality and hop count. LQ = total link quality of the route, p = remaining power, d = delay, h = hop count.

Table 1. Values for different routes between Source and Destination

Objective Function	A	B	C	D	E	F
$POF_1 = \frac{LQ \times p}{d}$	8.25	13.3	7.7	9	8	8.8
$POF_2 = \frac{LQ \times p}{d \times h}$	4.12	4.43	1.54	2.25	2.6	4.4

Figure 1 shows a small network where S is the root and six routes are available between node S and node X namely A, B, C, D, E and F. Routing metrics for both the proposed OFs is presented in Table 1, computed as per the respective formulas for the defined OFs. The link which will have highest value will be considered as the best link between node S and node X. From Table 1, it can be observed that in case of POF_1 route B has the highest value of 13.3 using the OF. The second best route is D where the value is 9. Therefore the best link between node S and node F during normal traffic condition is B and the second best link is D. On the other hand, employing the second proposed OF also results in link

B with the metric value of 4.43, but the choice of second best route is link F with value of 4.4. Therefore, from Fig. 1 although the best route is route B in both the case of high and low traffic, the second best route is not the same. Route D and F (marked in red) stand out as second preferred best route for first and second proposed OFs respectively. Thus, this method of multi-objective optimization approach provides load balancing to the network.

5   Comparison and Results

The performance is evaluated on the NetSim simulator. The network set-up consists of 65 nodes, each configured individually to resemble a network with heterogeneous devices.

Table 2. Simulation setup

Parameters	Values
Simulator	Netsim Standard v12
OF	OF0 (Hop Count), OF1 (ETX), Proposed OF
Number of nodes	65 (3 main powered, 63 battery powered)
Initial Energy of nodes	in the range [0.1 mAh–0.7 mAh]
Delay in nodes	in the range [1 s–8 s]
Data rate in nodes	[150, 350, 500, 650, 800] kbps

The lifetime of the two standard OF of RPL protocol is compared in Fig. 2. The comparison shows that for the lifetime of OF0 (Hop Count) and OF1 (ETX) is almost similar for the network configuration used for the simulation in this paper. Therefore, to evaluate our proposed OFs, we can just compare the output of the proposed solution with any one of the two standard OFs hence the lifetime of the network using OF0 is considered to compare with the proposed OFs.

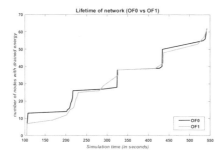

Fig. 2. Difference in network lifetime between OF0 and OF1

B. Hazarika et al.

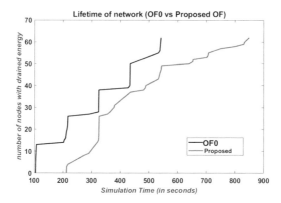

Fig. 3. Difference in network lifetime between OF0 and Proposed OF

Figure 3 shows comparison in terms of overall lifetime of the network between OF0 and the proposed OF. The number of nodes drained off energy is compared against the simulation time. For OF0, nodes start completely draining out after 106 s of simulation. In case of the proposed OF, energy of the nodes are conserved and load is balanced in the links with lower energy hence nodes in the proposed OF only start draining out only after 213 s of running the simulation. Meanwhile for OF0 by the time the simulation reaches 213 s, 15 nodes have already drained out of energy. Hence, it can be clearly seen that the proposed OF increases the lifetime of the network by a significant amount.

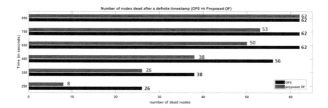

Fig. 4. Number of dead nodes at certain time

Figure 4 shows the number of nodes that are drained of energy after running the simulation for 250 s, 350 s, 450 s, 650 s and 850 s. We can see after 250 s of simulation 26 nodes are drained out of energy for OF0 while for the proposed OF only 8 nodes are drained out of energy. By 650 s, all the nodes are inactive and drained out of energy for OF0 while the network with the proposed OF run up to 850 s before draining out energy from all the nodes.

Now in order to calculate the Packet Delivery Ratio, every network has their own rates of generating data and different traffic conditions at individual point of time. In order to evaluate the overall packet delivery ratio of the network, we need to find the point of traffic which can be considered as the threshold to differentiate between low traffic and high traffic so that POF_1 and POF_2 can be switched when the traffic exceeds that rate. In order to find the optimal switching data rate point for the network used in the simulation for this paper, a trial and error method is used. The highest traffic of the network is 800 kbps so the switching point trial and method is done from 490 kbps onward to find the point from which it can be considered as high traffic.

Table 3. Data rate threshold point at which POF_1 can be switched to POF_2 for best result

Average Data rate	Success packet delivery%
490 kbps	76.2%
550 kbps	76.9%
650 kbps	78.2%
750 kbps	77.6%

In our statistical methodology, if the network switches from POF_1 (OF excluding hop count) to POF_2 (OF including hop count) when average data rate is 490 kbps then the success rate of packet delivery is 76.2%. Similarly, in case of 550 kbps, the success rate is 76.9% and at 650 kbps the success rate is 78.2%. But after 650 kbps, when average data rate is considered as 750 kbps the success rate reduces to 77.6%. Hence 650 kbps is the best optimal threshold value for considering traffic to be high and low for differentiating between the instances. When the average data rate is below 650 kbps the network is considered to be low traffic or normal and when the average data rate exceeds 650 kbps, the network is considered to have high traffic.

To note that, this analysis in Table 3 is purely restricted to the network used for the simulation in this paper. Also, since the considered network is small, the difference in the packet delivery rate is also small but in real test bed these values will be of significantly higher. Hence, Table 4 shows that the proposed OFs results in increase of success rate of packets delivered to 78.2%.

Table 4. Comparison of success rate of delivered packets in OF0, OF1 and Proposed OFs.

Objective function	Packet delivery success %
OF0	32.7%
OF1/MRHOF	61%
Proposed OFs	78.2%

6   Conclusion

In this paper, a combination of multiple objective functions is proposed for a constrained network with heterogeneous devices. The proposed OFs consider the remaining energy of the nodes, delay, link quality, hop-count and criticality of the network. The multi-objective approach optimizes the network lifetime and lower the packet loss in the network. The objective functions are differentiated by a point which can be considered as the threshold for considering the traffic as normal or low. If the average data rate of the network exceeds the threshold then the network is considered to be critical or high traffic and the POF_1 is switched to the POF_2. Evaluation is done to compare the proposed OF with that of the standardized OFs OF1/MRHOF and OF0 and the results show significant improvement in terms of lifetime and packet loss of the network when the proposed OFs are used.

References

1. Boualam, S.R., Ezzouhairi, A.: New objective function for RPL protocol. In: Embedded Systems and Artificial Intelligence, pp. 681–690. Springer, Singapore (2020)
2. Lamaazi, H., Benamar, N.: A comprehensive survey on enhancements and limitations of the RPL protocol: a focus on the objective function. Ad Hoc Netw. **96**, 102001 (2020)
3. Bouzebiba, H., Lehsaini, M.: FreeBW-RPL: a new RPL protocol objective function for internet of multimedia things. Wirel. Pers. Commun. 1–21 (2020)
4. Solapure, S.S., Kenchannavar, H.H.: Design and analysis of RPL objective functions using variant routing metrics for IoT applications. Wirel. Netw. **26**, 4637–4656 (2020)
5. Safaei, B., Monazzah, A.M.H., Ejlali, A.: ELITE: an elaborated cross-layer rpl objective function to achieve energy efficiency in internet-of-things devices. IEEE Internet Things J. **8**(2), 1169–1182 (2020)
6. Moradi, S., Javidan, R.: A new objective function for RPL routing protocol in IoT to increase network lifetime. Int. J. Wireless Mobile Comput. **19**(1), 73–79 (2020)
7. Hassani, A.E., Sahel, A., Badri, A.: A new objective function based on additive combination of node and link metrics as a mechanism path selection for RPL protocol. Int. J. Commun. Networks Inf. Secur **12**(1), 63–68 (2020)
8. Nassar, J., Gouvy, N., Mitton, N.: Towards multi-instances QoS efficient RPL for smart grids. In: Proceedings of the 14th ACM Symposium on Performance Evaluation of Wireless Ad Hoc, Sensor, & Ubiquitous Networks, pp. 85–92, November 2017
9. Lamaazi, H., Benamar, N.: RPL enhancement using a new objective function based on combined metrics. In: 2017 13th International Wireless Communications and Mobile Computing Conference (IWCMC), pp. 1459–1464. IEEE, June 2017
10. Banh, M., Mac, H., Nguyen, N., Phung, K.H., Thanh, N.H., Steenhaut, K.: Performance evaluation of multiple RPL routing tree instances for Internet of Things applications. In: 2015 International Conference on Advanced Technologies for Communications (ATC), pp. 206–211. IEEE, October 2015

11. Todolí-Ferrandis, D., Santonja-Climent, S., Sempere-Payá, V., Silvestre-Blanes, J.: RPL routing in a real life scenario with an energy efficient objective function. In: 2015 23rd Telecommunications Forum Telfor (TELFOR), pp. 285–288. IEEE, November 2015

12. Gaddour, O., Koubâa, A., Baccour, N., Abid, M.: OF-FL: QoS-aware fuzzy logic objective function for the RPL routing protocol. In: 2014 12th International Symposium on Modeling and Optimization in Mobile, ad Hoc, and Wireless Networks (WiOpt), pp. 365–372. IEEE, May 2014

13. Parnian, A.R., Kharazmi, M.R., Javidan, R.: RPL Routing Protocol in Smart Grid Communication 1 (2014)

14. Rajalingham, G., Gao, Y., Ho, Q.D., Le-Ngoc, T.: Quality of service differentiation for smart grid neighbor area networks through multiple RPL instances. In: Proceedings of the 10th ACM Symposium on QoS and Security for Wireless and Mobile Networks, pp. 17–24, September 2014

15. Gaddour, O., Koubâa, A.: RPL in a nutshell: a survey. Comput. Netw. 56(14), 3163–3178 (2012)

A Combination of K-means Algorithm and Optimal Path Selection Method for Lifetime Extension in Wireless Sensor Networks

Wadii Jlassi[1]([✉]), Rim Haddad[2], Ridha Bouallegue[3], and Raed Shubair[4]

[1] Innov'COM Lab/Sup'Com, National Engineering School of Tunis, University Tunis El Manar, Tunis, Tunisia
[2] Laval University, Quebec, Canada
Rim.haddad@eti.ulaval.ca
[3] Innov'COM Lab/Sup'Com, University of Carthage, Tunis, Tunisia
Ridha.bouallegue@supcom.tn
[4] New York University, Abu Dhabi, UAE
raed.shubair@nyu.edu

Abstract. Wireless Sensor Networks (WSN) are special types of wireless networks where hundreds or thousands of sensor nodes are working together. Since the lifetime of each sensor is equivalent to a battery, the energy issue is considered a major challenge. Clustering has been proposed as a strategy to extend the lifetime of wireless sensor networks. Many clustering algorithms consider the residual energy and distance between the nodes in the selection of cluster heads and others rotate the selection of cluster heads periodically. We propose in this article a CH selection followed by making clusters using the K-means algorithm and we present the PRIM algorithm to transmit the packets in multi-hop transmission between CHs and BS and choose the optimal path. The clustering scheme allows to decrease intra-cluster communications and to gain energy efficiency for sensor nodes. Computer simulation results show that our method aims to extend the lifetime of the wireless sensor network efficiently compared to other existing methods.

Keywords: Wireless Sensor Networks · Clustering · K-means · Optimal path

1 Introduction

Wireless Sensor Network (WSN) consists of a large number of sensor nodes deployed in a hostile environment and self-organized to gather and transmit data from the field to a Base Station (BS).

In a WSN, nodes are grouped into partitions called clusters. Various WSNs are employing the cluster structure, which efficiently allocates the resource and energy and thereby maximizes the network lifetime [1]. Clustering algorithms are energy-efficient methods to collect data to BS [2, 3]. The network is divided into clusters. In every cluster, a sensor node is selected as a cluster head. The other members transmit their collected information to the cluster head in a single-hope or multi-hope manner. After gathering information, which are distance separating the nodes and energy for cluster head selection and the optimal path between CH and BS for packets transmission. The cluster head transmits to the base station through the single-stage or multi-stage line [4].

The selection of CH is one of the most critical tasks in the management of WSNs since CHs consume much more energy than other nodes in the network [5]. To elect the cluster heads some protocols use a random number to balance the energy consumption of the sensor nodes over the networks [6]. One of the most recognized protocols in this regard is LEACH [7], it can reduce energy overconsumption by grouping the sensor nodes into clusters to decrease the number of transmitted packets and restrict the direct single-hop communication between the nodes and the base station. Leach-C [8] proposes a centralized clustering algorithm. The K-means algorithm [9] is used to form the clusters such that the distance between the nodes and the CH becomes minimal. The CHs transmit their data to the base station in one-hope. The Hierarchical Agglomerative Clustering (HAC) [10] proposed to group nodes into clusters, then elect CH on a formula that takes into account the position of the nodes and their remaining energy for the CHs packets transmission to the BS. HACMH proposed multi-hop transmission between CHs and BS by applying spanning tree algorithm for choosing the optimal path.

These protocols aim to extend the network lifetime, decrease the number of transmitted packets and reduce direct communication between CHs and BS. This paper introduces the energy usage of sensor networks and how to extend the network lifetime of the sensor nodes by using important criteria.

As an expansion of some previous works in this area, the contributions of this work could be summarized as follows:

- We select CH and optimal path to manage wireless sensor networks efficiently in energy.
- We use K-means algorithm to elect the center of each cluster as a data-gathering point. This algorithm allows to decrease intra-cluster communications.
- We propose a PRIM algorithm for multi-hop transmission between CHs and CH and BS by applying a spanning tree algorithm (PRIM) for choosing the optimal path to avoid the direct transmission between CHs and BS and minimize data transmission time between CHs.

The paper is organized as follows. Section 2 presents the related research, and Sect. 3 introduces the proposed scheme. The simulation results and discussions are presented in Sect. 4. Finally, Sect. 5 concludes the paper and suggests future work.

2 Related Works

Wireless sensor nodes perform the operation of transmitting the data from the source to the destination which should be made in an efficient way so that the data transmission between the sender and the receiver will be in an effective manner. The sensor nodes have limitations in storage, power, latency, constraint bandwidth, and reduced corporal size [11].

In this section, we briefly review some recent researches proposed for clustering in WSNs that are relevant to our approach. Leach protocol [7] is a typical routing scheme based on clustering. The algorithm consists of electing the CHs to distribute the energy consumption evenly among the nodes. The CHs aggregate the data collected from the member nodes and transmit it to the BS. The protocol is based on a probabilistic model that rotates CHs selection in order to balance the energy consumption of nodes in the network. However, Because of the probability model, the Leach protocol has some drawbacks. CHs could be situated very close to each other or in a place where the number of nodes is small. In this case, the protocol doesn't consider the remaining energy of each node. To avoid these drawbacks, the LEACH-C protocol employs a centralized control approach using the location information of the sensor nodes [8]. During the set-up phase, each member node transmits the current location information and energy level to the BS. Then the BS calculates the average energy of the member nodes in the current round and divides the network into a number of clusters. After the selection of a CH in each cluster, the member nodes send the data to the closest CH. Finally, the set-up phase is completed after deciding the routing path. The energy consumption of CH can be reduced by the efficient selection of the CH using the location information of the member nodes. Although, the requirement of getting exact location information of all sensor nodes causes additional energy consumption and the routing path is not always shorter than LEACH protocol.

The protocol HEED [12] chooses CH by referring to the residual energy of each member node. It also considers the inter-cluster communication cost as a secondary clustering parameter. This protocol extends the network lifetime and creates well-distributed clusters. But, the random selection of the cluster heads may cause higher communication between a cluster head and a base station. The ER-HEED [12] (Energy-based Rotated HEED) is considered an improvement of the HEED protocol with the introduction of rotation in equal-sized clusters. This protocol consists of three phases as follows:

– The cluster formation and cluster head selection are performed according to the HEED protocol.
– The CH election is the member of the cluster with the highest energy without the need to perform an election protocol.
– If any cluster member dies, re-clustering is performed by repeating cluster formation and cluster head election step.

The HACSH [11] uses hierarchical agglomerative clustering (HAC) for the formation of clusters from the nodes based on the Euclidean distance (nearest neighbor) between nodes. The HAC algorithm is executed to form k clusters from n nodes with K <= n. After running the HAC algorithm, the distance between node i and the base station (BS) is calculated in Eq. (1):

$$d_i = \sqrt{(x_i - x_{sb})^2 - (y_i - y_{sb})^2)} \qquad (1)$$

After that, we calculate the distance between node i and the base station (BS) in the cluster and the centroid, the nearest node will be designated as CH. Now, each node is associated with one of the k clusters, we calculate the opti point for each cluster which is given by the following formula opti (xopti; yopti) (2, 3) with:

$$X_{opti} = \frac{1}{\sum_{i=1}^{S} E_i} * \sum_{j=1}^{s} E_j x_j \tag{2}$$

$$Y_{opti} = \frac{1}{\sum_{i=1}^{S} E_i} * \sum_{j=1}^{s} E_j x_j \tag{3}$$

Finally, we choose the nearest node of the opti mark (xopti; yopti) as the new CH, always based on the Euclidean distance between the nodes of each cluster and opti point is illustrated in Eq. (4):

$$d_i = \sqrt{(x_i - x_{opti})^2 + (y_i - y_{opti})^2} \tag{4}$$

All nodes of the cluster transmit packets to CH, which will handle to transmit them to the base station (BS).

3   Proposed Work

In this section we propose a data gathering point selection method which employs the K-means algorithm for CHs selection. To transmit the packets from CHs to BS. We propose a minimum spanning tree.

3.1   Network Model

The sensor nodes are randomly distributed in the target area and have the same amount of energy when they are initially deployed. Each node has an ID number and is placed stationary after the deployment. The Base Station can be placed anywhere in the area. Sensor nodes can send the data to it and The CHs are aware of their remaining energy.

3.2   Energy Consumption Model

To evaluate the performance of the proposed method, we will compare its energy model to LEACH. As defined in this model, there are two cases: the free space (fs) and the multi-path (MP). When the distance between a sending node and a receiving node is less than threshold value d0, the free space model, d^2 powerless, is used. Otherwise, multipath model, d^4 powerless, is used. From the two models above, the energy consumption for transmitting a k-bit packet over a distance d is given by Eq. (5):

$$\begin{aligned} E_{Tx}(K, d) &= E_{elec} * K + \varepsilon_{fs} * k * d^2, d < d_0 \\ E_{Tx}(K, d) &= E_{elec} * K + \varepsilon_{mp} * k * d^4, d \geq d_0 \end{aligned} \tag{5}$$

Where E_{elec} is the energy required for processing 1-bit data with the electronic circuits. εfs and εmp is the energy taken for transmitting 1-bit data to achieve an acceptable bit error rate in the case of the free space model and multipath model, respectively.

They are dependent on the distance of transmission. Note that energy dissipation of free space and multipath is proportional to d^2 and d^4, respectively. The threshold, d_0, can be obtained by Eq. (6):

$$d_0 = \sqrt{\frac{\mathcal{E}_{fs}}{\mathcal{E}_{mp}}} \tag{6}$$

The energy taken to receive a k-bit message is calculated in Eq. (7):

$$E_{Rx=E_{elec}} * K \tag{7}$$

3.3 Proposed Scheme

After deployment of sensor nodes in the sensed area, we divide the area into clusters. The clustering decreases redundant data, reduce the number of inter-node communication by localizing data transmission within the clusters and decrease the overall amount of transmission to the base station.

In this section, we will present our approach. The proposed scheme uses K-means algorithm for clusters creation from the nodes based on the Euclidean distance between them.

The algorithm accepts two inputs: $S = \{s1, s2, . sn\}$ sensor nodes with location information and k number of clusters. The output is $C = \{C1, C2 ... Ck\}$ set of clusters with input data partitioned among them.

K-means consists of three steps as follows.

– Step 1: Initial clustering

K-means algorithm is executed clusters creation in wireless sensor networks. First, k out of n nodes are randomly selected as the CHs. Each of the remaining nodes decides its nearest CH according to Euclidean distance.

– Step 2: Re-clustering

After that, each node in the network is assigned to a cluster. The centroid of each cluster is illustrated in Eq. (8).

$$Centroid\,(X, Y) = (\frac{1}{s}\sum_{i=1}^{s} x_i, \frac{1}{s}\sum_{i=1}^{s} y_i) \tag{8}$$

Note that the centroid of a cluster is a virtual node located at the center position of the cluster. In this step, the center of each cluster is updated and the new center will be the average location point of all the sensor nodes locations in the cluster. For each node, recalculate the distance between the node and all cluster centers using Euclidean distance and allocate the closest one with the new center (CH) in each cluster, Step 2 is recursively executed until no point switches clusters and the CH is not changed anymore.

– Step 3: Choosing the CH

As soon as clusters are formed, an ID number is assigned to each node of a cluster according to the distance from the centroid, assigning a smaller number to the closer one. The ID number of a node indicates its order to be chosen as a CH. Therefore, the ID number plays an important role in the selection of a node as CH. The residual energy of the CH is checked every round to retain the connectivity of the network. If the energy of the CH is smaller than the preset threshold, the node in the next order is selected as a new CH. The newly elected CH informs the other nodes of the change of the CH. After clustering and CHs selection, in each round, the proposed scheme adopts multi-hop transmission between CHs and BS by applying a spanning tree algorithm (PRIM) for choosing the optimal path.

3.4 PRIM Algorithm

The algorithm starts with a tree consisting of a single vertex, and continuously increases its size one edge at a time. It halts when all the vertices have been reached.

– Input: A non-empty connected weighted graph with vertices V and edges E (the weights can be negative).
– Initialize: $V_{new} = \{x\}$, where x is an arbitrary node (starting point) from V, $E_{new} = \{\}$ and Repeat until $V_{new} = V$. Then, choose an edge $\{u, v\}$ with minimal weight such that u is in V_{new} and v is not (if there are multiple edges with the same weight, any of them may be picked). Finally, add v to Vnew, and $\{u, v\}$ to E_{new}.
– Output: V_{new} and E_{new} describe a minimal spanning Tree.

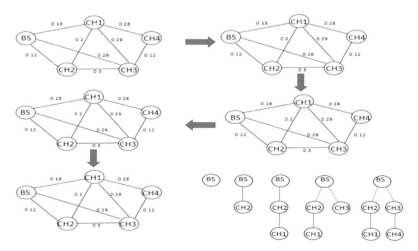

Fig. 1. Steps of the algorithm

In this example, we present and suppose 4 clusters with one BS in Fig. 1. We calculated the distance between all CHs and BS. Initially, the root of the tree is always BS.

The weight between nodes in the graph is represented by the distance between CHs, CH, and BS. After building the minimum spanning tree using the PRIM algorithm, each CH sent packets to the CH or directly to the BS.

4 Simulations Results

In the previous section, we analyzed and evaluated the simulation results with different numbers of clusters. Then, we compared the results with other algorithms namely LEACH-C, ER-HEED, HACMH, and k-means under different scenarios to validate the network performance. The proposed method has been implemented in MATLAB R2015a. Here, we consider a network of 200 sensor nodes randomly distributed in a field of 200 m * 200 m, and the values used in the first model are described in Table 1.

Table 1. Simulation parameters

Parameter	Value
Size of the network	200 m * 200 m
N (Number of deployed nodes)	200
E_0 (initial energy of nodes)	1 J
Eelec	50 nJ/bit
Ecpu	7 nJ/bit
ε_{fs}	10 pj/bit/m^2
ε_{mp}	0.0013 pj/bit/m^4
Position of base station	0 m*0 m
Packet size	4000 bits

The proposed method is analyzed with the different number of clusters in the network. In Fig. 2, First, when k = 3 clusters, the network lifetime is estimated as 1020 rounds. Then, when the number of clusters increases to 10 clusters the network lifetime increases to 1376 rounds. As the cluster number increases, the size of the cluster decreases. So, the energy consumption decreases when the network is divided from 3 clusters to 10 clusters, and the simulation results show the same performance when the number of clusters is 15, 20, and 30.

Next, the proposed scheme is compared with related works in terms of network lifetime and remaining energy. Network lifetime is defined as the number of rounds when all the nodes run out of energy. In Table 2, We compare the behavior of the network in terms of the First Node Dies FND, the Half of the Nodes Alive HNA and the Last Node Dies LN. The configuration parameters for the simulation are presented in Table 1 with the number of clusters is k = 15.

The number of alive nodes is checked to assess the lifetime of the network. In Fig. 3, we note that the proposed scheme has the highest alive node from the beginning

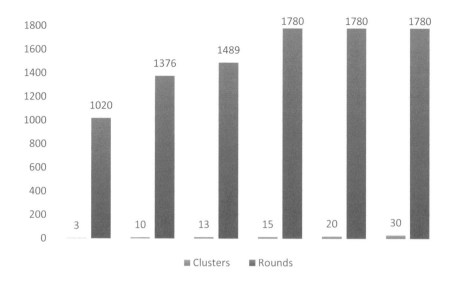

Fig. 2. Number of clusters and rounds

to the total depletion of nodes. After 1500 rounds, almost no alive node is left with Leach-C, ER-HEED, HACMH, and k-means but about 25 nodes are still alive with the proposed scheme. This is since communication overhead is lower than that with compared protocols. Note that the proposed scheme forms the clusters so that the distance between the CH and the member's nodes are minimized also the communication between CHs and CH and BS is optimal.

Table 2. Values of FND, HNA and LND metrics for each algorithm

Algorithm	FND	HNA	LND
LEACH-C	149	488	1032
ER-HEED	183	599	1120
HACMH	191	655	1185
K-MEANS	300	742	1495
Proposed Work	500	1050	1780

Figure 4 shows the remaining energy of the network as the round proceeds. The proposed scheme is always better than Leach-C, ER-HEED, K-means, and HACMH. We observe in Fig. 4 that little energy is left after 1050 rounds with Leach-C and ER-HEED and after 1200 rounds with HACM and K-means. However, the network still has some residual energy until 1780 rounds with the proposed scheme.

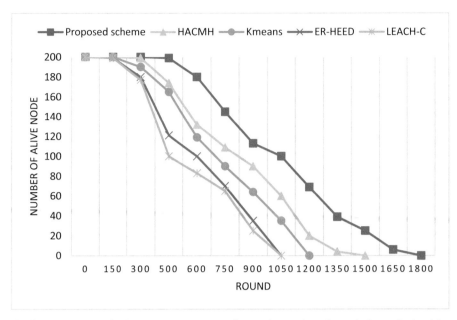

Fig. 3. Distribution of alive sensor nodes according to the number of rounds for each algorithm

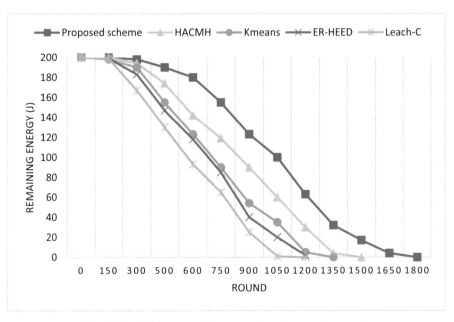

Fig. 4. Residual energy of the network as the round proceeds

5   Conclusion

One of the most challenging issues in Wireless Sensors Networks (WSN) is severe energy restrictions. In this study, we use the center of each cluster as a data gathering point. K-means algorithm is used to find the optimal data gathering point and make clusters. It allows to decrease intra-cluster communications and to increase energy efficiency for sensor nodes. To choose the optimal path, we proposed the PRIM algorithm. The simulation results show that the proposed scheme provides better energy efficiency and a longer network lifetime than the other exits algorithms.

Acknowledgments. We are grateful for the support of the Department of Electrical and Computer Engineering at the New York University of Abu Dhabi (NYU).

References

1. Zytoune, Q., El Aroussi, M., Rziza, M., Aboutajdine, D.: Stochastic Low Energy Adaptive Clustering Hierarchy 47–51 (2008)
2. Ma, J., Shi, S., Gu, X., Wang, F.: Heuristic mobile data gathering for wireless sensor networks via trajectory control. Int. J. Distrib. Sensor Netw. **16**(5), 1550147720907052 (2020)
3. Tatarian, F., Moghaddam, M.H.Y., Sohraby, K., Efati, S.: On maximizing the lifetime of wireless sensor networks in event-driven applications with mobile sinks. IEEE Trans. Veh. Technol. **64**(7), 3177–3189 (2015)
4. Gajjar, S., Sarkar, M., Dasgupta, K.: FAMACROW: fuzzy and ant colony optimization based combined mac, routing, and unequal clustering cross-layer protocol for wireless sensor networks. Appl. Soft Comput. **43**, 235–247 (2016)
5. Thein, M.C.M., Thein, T.: An energy efficient cluster-head selection for wireless sensor networks. ISMS, 287–291 (2010)
6. Sert, S.A., Bagci, H., Yazici, A.: MOFCA: multi-objective fuzzy clustering algorithm for wireless sensor networks. Appl. Soft Comput. **30**, 151–165 (2015)
7. Ghosh, N., Banerjee, I., Sherratt, R.S: On-demand fuzzy clustering and ant-colony optimization-based mobile data collection in wireless sensor network. Wireless Netw. **25**, 1829–1845 (2019)
8. Xiangning, F., Yulin, S.: Improvement on LEACH protocol of wireless sensor network. In: International Conference on Sensor Technologies and Applications, pp. 260–264 (2007)
9. Raval, G., Bhavsar, M., Patel, N.: Enhancing data delivery with density-controlled clustering in wireless sensor networks. Microsyst. Technol. **23**, 613–631 (2017)
10. Abidoye, A.P., Kabaso, B.: Energy-efficient hierarchical routing in wireless sensor networks based on Fog Computing (2020)
11. Abirami, K.S.: Sybil attack in Wireless Sensor Network (2013)
12. Ullah, Z., Mostarda, L.: A comparison of HEED based clustering algorithms - introducing ER-HEED. In: International Conference on Advanced Information Networking and Applications, pp. 339–345 (2016)

Performance Comparison of IEEE802.11ac vs IEEE 802.11n WLAN in IPv6

Samad S. Kolahi[(✉)], Ahmad Khalid Sooran, Faroq Nasim,
and Muhammad Mazhar U. Khan

Unitec Unstitute of Technology, Auckland, New Zealand
skolahi@unitec.ac.nz

Abstract. In this paper, performance of WLAN (Wireless LAN) for peer-peer and client-server networks with WPA-2 security is investigated for both IEEE802.11ac and IEEE802.11n standard. Parameters studied include throughput, and RTT (round trip time) of IPv4 and IPv6 protocols for both TCP and UDP. IPv4 outperform IPv6 in terms of bandwidth and delay due its lower overhead size. IEEE 802.11ac had much better performance than IEEE 802.11n.

1 Introduction

Wireless connectivity and Wi-Fi are terms most people know about today, and is a service most of us take for granted, in that how important it is and how much easier it has made our daily tasks. From our computers, mobile devices, video game consoles, wireless connectivity allows us to communicate with each other, share resources, and browse the world wide web, all without needing to use cluttered wires, and having the flexibility to freely move around is a bonus. With Wi-Fi, we can connect our devices to an access point and then connect to IoT network, through the use of the 802.11 wireless standards which have helped to bring this medium to homes and businesses all around the world.

This paper aims to present real life performance comparison of the two standards, and identify the significant speed advantage IEEE802.11ac offers over IEEE802.11n in IPv6 environment, for both peer-peer and client networks. The operating system used is Windows 10 and Windows Server 2016. Collected data includes throughout, and round trip time, for TCP (Transmission Control Protocol), UDP (User Datagram Protocol), IPv4 and IPv6. Such data was collected with the aid of a traffic management and data generation tool.

With the many advantages that wireless networking offers, one main disadvantage is being less secure than wired networks, due to devices in a wireless local area network being connected wirelessly over the air, and with this, wireless networks are also slower as there is no direct connection between devices [1].

IEEE802.11n was developed in 2009 to replace IEEE802.11g, offering significantly higher throughput than IEEE802.11g standard. In addition, it has also doubled the connection distance from 450 ft. to 820 ft. Until recently, it was the most widely used standard in modern times, and is still adequate in supplying networks with reliable speeds and security. This standard allows connectivity in the 20 and 40 MHz channel sizes in addition to offering dual band connection, meaning it can operate in both the 2.4 GHz and

L. Barolli et al. (Eds.): AINA 2021, LNNS 227, pp. 426–435, 2021.
https://doi.org/10.1007/978-3-030-75078-7_43

5 GHz bands [2]. To produce higher throughput than its predecessors, this standard is the first to introduce MIMO (multiple input, multiple output) system so that signal strength is boosted, and connections between a client and wireless access point become stronger, leading to better performance. Before the development of IEEE802.11n, a single transmitter and receiver was used to transmit signals between devices in a wireless network, known as a radio chain or SISO (single in, single out) systems [3]. It will also allow multiple channels to be aggregated into a single stream, for example allowing two channels of 20 MHz to be aggregated into a single 40 MHz channel, theoretically doubling throughput [2].

Released commercially in 2013, IEEE802.11ac is the latest wireless standard used in home and business networks around the world, and significantly increase wireless performance both theoretically and practically. Basically, IEEE802.11ac is a super charged version of IEEE802.11n and takes the best features and improves upon them further so that a maximum theoretical throughput of 1.7 Gbps is possible. This is the first wireless standard to take throughput up to gigabit speeds [3]. Unlike IEEE802.11n, IEEE802.11ac only operates at the 5 GHz band and can use 40, 80 and 160 MHz bandwidth sizes, whereas IEEE802.11n only supports 20 and 40 MHz channel sizes. It also uses enhanced MIMO, OFDM (Orthogonal Frequency Division Multiplexing) and beam forming technology to provide higher data rate [2].

IPv6 is the latest version of the Internet Protocol (IP), expected to replace IPv4 in the future and addition to providing backwards compatibility, no need for DHCP server, built in encryption, more efficient routing, elimination of NAT (network address translation), better mobility support, better QoS (Quality of Service) and will address IPv4 address shortage [4].

Some related works regarding performance of IPv6 and WLAN IEEE802.11n and 11ac are in [5–12].

To the authors knowledge, there is no research to date in the literature on studying performance comparison of IEEE802.11n peer-peer or client-server networks comparing them to IEEE802.11ac. The motivation behind this study therefore is to produce new results to see how much better the new 802.11ac protocol is, for IPv4, IPv6, TCP, UDP, and for both peer-peer and client server networks.

2 Network Setup

To measure the wireless performances, two networks were set up, a peer-peer and a client server network.

For the client-server network, a Windows Server machine was connected to the WAP4410N IEEE802.11n business.

Wireless access point via a Cat 6 crossover cable for IEEE802.11n testing, and Ubiquiti UniFi HD IEEE802.11ac enterprise access point for IEEE802.11ac testing. The Windows 10 Client machine was connected to the server through the access points wirelessly. The distance between the access points and the client machine was well within few meters, to maintain the maximum signal strength, so that practical results obtained are reliable and consistent. The channel bandwidth of the WAP4410N access point was tested using 40 MHz, and Ubiquiti UniFi HD 11ac Access point using 80 MHz. The hardware specifications for both client and server machines consist of an Intel Core I7 Duo 6300 2.87 GHz, a western digital caviar 160 GB hard drive, 16.00 GB or RAM. The client machine was installed with a TP-Link TL-WN951N wireless N network interface card and Asus AC-68 IEEE802.11ac wireless network interface card. the test bed setup remained consistent for all testing conducted in the client-server section. The test bed diagram is shown in Fig. 1.

Fig. 1. Client-server test bed

For the peer-peer network, two Windows 10 operated machines were connected to the WAP4410N IEEE802.11n business wireless access point for IEEE802.11n testing, and Ubiquiti UniFi HD for IEEE802.11ac testing wirelessly and a workgroup created to allow them to share resources and communicate with each other. The distance between the access point and the clients were well within few meter, to maintain the maximum signal strength, so that practical results obtained are reliable and consistent. The channel bandwidth for WAP4410N 11n access point was 40 MHz, and 80 MHz channel used for the Ubiquiti IEEE802.11ac access point. The hardware specifications for both client machines consist of an Intel Core i7 Duo 6300 2.87 GHz, a Western Digital Caviar 160 GB hard drive, 16.00 GB of RAM. The client machines were installed with a TP-Link TL-WN951N wireless N network interface card and Asus AC-68 IEEE802.11ac wireless network interface card. The test bed setup remained consistent for all testing conducted in the client-server section. The test bed diagram is shown in Fig. 2.

Fig. 2. Peer-peer test bed

3   Data Generation and Traffic Measurement Tools

D-ITG 2.8.1 [13] was the tool used to generate traffic and measure the throughput, and RTT (Round Trip Time). This tool is widely used to evaluate the network performance.

4   Practical Results

This section presents data on the throughput, and RTT of TCP and UDP for both IPv4 and IPv6 on an IEEE802.11ac and IEEE802.11n peer-peer and client server networks. IEEE802.11ac will be compared to IEEE802.11n. Both networks are set up with WPA-2 security implementation for all the networks. For each packet size, a total of 15 runs are carried out, and the results averaged out with the standard deviation calculated. The standard deviation over the average results of 15 runs was less than 0.5%. As there is a very significant gap in performance between IEEE802.11ac and IEEE802.11n, results will not be directly compared in a single graph, so results are presented in separate graphs for each IEEE standard.

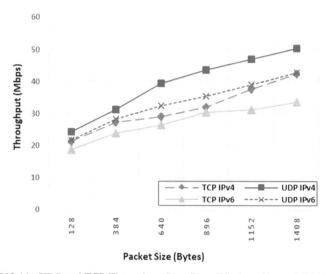

Fig. 3. 802.11n UDP and TCP Throughput Peer-Peer, Wireless Channel Width 40 MHz

Figure 3 presents throughput data on 802.11n and Fig. 4 presents throughput data on 802.11ac on peer-peer WLAN networks. Both wireless networks are tested using TCP and UDP protocols for both IPv4 and IPv6. In most scenarios as the packet size increases, TCP and UDP throughout consistently increases alongside the packets. The throughput on 11ac is significantly greater in all packet sizes tested for both TCP and UDP.

IEEE802.11ac IPv4 UDP outperforms every other packet sizes achieving the highest throughput for all packets except packet 128 Bytes, reaching a maximum throughput of 440 Mbps for packet size of 1408 Bytes. The maximum difference in throughput between 11n and 1ac is at packet 1408 Byte for UDP IPv4 802.11ac having 440 Mbps and UDP IPv4 802.11n having 50 Mbps, where 802.11ac has an advantage of 390 Mbps. For both 802.11n and 802.11ac, IPv4 TCP and UDP outperform IPv6 TCP and UDP, respectively. This is because IPv6 packet has higher overhead than IPv4, and overheads need processing and that makes IPv6 slower.

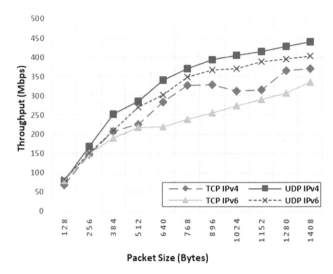

Fig. 4. 802.11ac UDP and TCP Throughput Peer-Peer, Wireless Channel Width 80 MHz

Both networks have UDP IPv4 as the highest performing protocol, followed by IPv6 UDP, IPv4 TCP, and then the least performing is IPv6 TCP in both networks. In all scenarios throughput gradually increase as packet size gets higher, except for IPv4 TCP, which drops in speed in packet 896 Bytes to 329 Mbps, going down to 313 Mbps in packet 1024 Bytes, and then in packet 1152 Bytes increasing again at 316 Mbps, this is the only case where throughput had decreased in all scenarios. This was also observed in other research work that graphs are not smooth and could rise and fall a bit as packet size increase [5, 7, 9], but the overall trend is increasing throughput with packet size. The lowest throughput for 802.11n was 18 Mbps (at packet 128 Byte) for TCP IPv6, while 802.11ac lowest throughput was 68 Mbps (at packet 128 Byte), a significant increase compared to 802.11n.

Figure 5 presents throughput data on 802.11n and Fig. 6 is throughput data on 802.11ac on client-server WLAN networks. Both networks are tested using TCP and UDP protocols for both IPv4 and IPv6. 802.11ac using IPv4 and UDP gave the highest throughput of 808 Mbps. When comparing 802.11ac and 802.11n, the maximum difference in throughput difference is for packet 1408 Byte, UDP IPv4 802.11ac having 808 Mbps and IPv4 UDP 802.11n having 102 Mbps, which makes a huge difference of 706 Mbps, showing just how fast 802.11ac is over its predecessor. IPv4 outperforms IPv6 for TCP and UDP as it has less overhead than IPv6. Both networks have UDP IPv4 as the highest performing protocol, followed by IPv6 UDP, IPv4 TCP, and then the least performing is IPv6 TCP in both networks. One point of deviation is at Fig. 5 (802.11n) where TCP outperforms UDP on packets 128 and 384 for both IPv4 and IPv6. At low packet sizes, the data obtained had some randomness and this was observed in other studies too [5, 7, 9].

TCP has lower throughput as it sends acknowledgements, and that it has higher overhead in its packet. Generally, both networks have UDP IPv4 as the highest performing protocol, followed by IPv6 UDP, IPv4 TCP, and then the least performing is IPv6 TCP

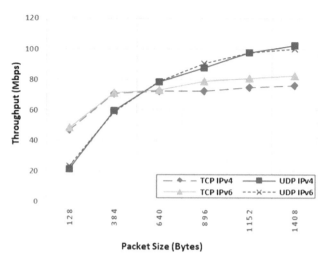

Fig. 5. 802.11n UDP and TCP Throughput Client-Server, Wireless Channel Width 40 MHz.

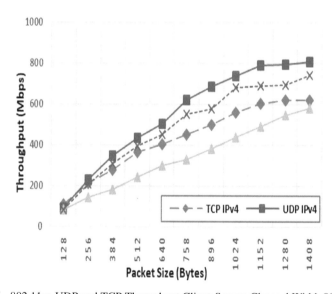

Fig. 6. 802.11ac UDP and TCP Throughput Client-Server, Channel Width 80 MHz.

in both networks. The lowest throughput is in 802.11n at packet size 128 Byte for UDP IPv6 achieving 20 Mbps, while 802.11ac TCP IPv4 at packet 128 Byte achieve 105 Mbps, which is a significant increase.

Figure 7 presents the RTT (round trip time) of the 802.11n peer-peer WLAN network. UDP IPv4 had the least round trip time for each packet, steadily rising as the packet size rises. It achieved a round trip time of 0.07 ms at Packet 128 Bytes, While having 0.18 ms delay at packet 1408 Bytes. The slowest is TCP IPv6, with 0.23 ms at Packet 1408 Bytes. Overall, UDP outperforms TCP for both IPv4 and IPv6. As discussed earlier. The

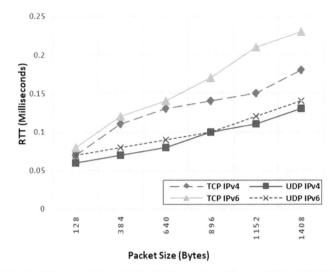

Fig. 7. 802.11n UDP and TCP RTT Peer-Peer, Channel Width 40 MHz.

reason for above is that because UDP packet has less overhead than TCP, TCP sends acknowledgements being a connection oriented protocol, and that IPv4 packet has less overhead than IPv6.

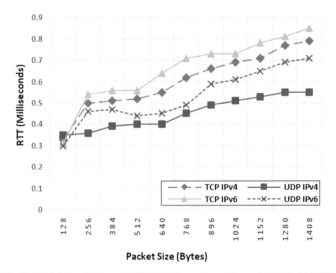

Fig. 8. 802.11ac UDP and TCP RTT Peer-Peer Channel Width 80 MHz.

Figure 8 presents the RTT of the 802.11ac peer-peer WLAN network. IPv4 UDP achieved lower RTT than all other scenarios. It achieved a round trip time of 0.35 ms at packet 128 Byte, while having 0.5 ms at packet 1408 Byte. The highest RTT was for TCP IPv6, with 0.85 ms at packet 1408 Byte. Overall, UDP outperforms TCP for both IPv4 and IPv6. Both networks have UDP IPv4 as the best performing protocol, and having IPv6 TCP as the protocol with the highest latency, due to the IPv6 TCP header being larger than all of them. All protocols had gradually increased in latency as packet sizes increased, however only UDP IPv6 had a drop in packet 512 bytes going to 0.44 ms, while at packet size 384 Bytes had achieved 0.47 ms RTT. After this packet, latency gradually increases again, with packet size 640 Bytes having 0.45 ms, going to 0.50 ms for packet size 768 Bytes.

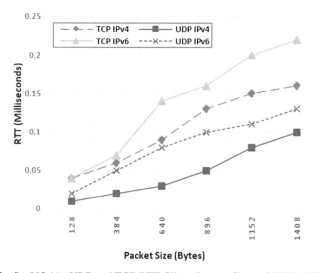

Fig. 9. 802.11n UDP and TCP RTT Client-Server, Channel Width 40MHz.

Figure 9 presents the RTT of the 802.11n client-server WLAN network. UDP IPv4 had the least round trip time for each packet, steadily rising as the packet rises. It achieved a round trip time of 0.01 ms at packet 128 Bytes, this could be because the client-server connection through a wire is much better than that of peer-peer wireless connection over the air. It had the latency of 0.13 ms at packet 1408 Bytes. The slowest is TCP IPv6, taking 0.22 ms at packet 1408 Bytes. Overall client server had better latency than peer-peer counterpart. UDP outperforms TCP for both IPv4 and IPv6. IPv4 have better RTT performance than IPv6. Note these numbers are quite low ranging in the milliseconds, so performance in the real world would be hardly noticeable, and would basically be the same.

Figure 10 is the RTT data as collected from 802.11ac client-server network. UDP IPv4 having 0.19 s round trip time at packet 128 Bytes, while achieving 0.4 ms at packet 1408 Bytes. This delay is significantly better than peer-peer wireless network, due to a server connected to the access point through a wire. TCP IPv6, as expected, had the highest latency, due to the large header in IPv6 and higher overhead in TCP, and connection oriented nature of TCP.

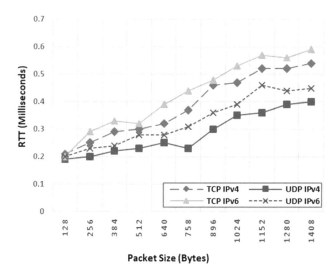

Fig. 10. 802.11ac UDP and TCP RTT Client-Server, Channel Width 80MHz.

5   Conclusions

In this paper, due to the advantages and features of IEEE802.11ac over IEEE802.11n, it gave the highest throughput for both peer-peer and client-server networks at 444Mbps (IPv4 UDP) and 808Mbps (IPv4 UDP), respectively. This throughput is lower than the 1.7 Gbps theoretical bandwidth of 802.11ac. Peer-peer network bandwidth was much less than client server. UDP performed better (higher bandwidth) than TCP, and IPv4 provided higher bandwidth than IPv6. The comparison results from above protocols was presented in the paper.

6   Future Work

The future work includes testing more operating systems such as Linux, using multiple client machines for multi machine throughput testing, and also open system testing to compare with results of this study.

References

1. Zeng, Y., Pathak, Mohapatra, P.: Throughput, energy efficiency and interference char-acterisation of 802.11ac. Transactions on Emerging Telecommunications Technologies (2015)
2. CISCO. CCNA Wireless - Study Notes Part 2: 802.11 standards and WLAN Terminology. Learning Network Cisco: https://learningnetwork.cisco.com/docs/DOC-31396 (2016)
3. Bradley, M., What Is Multiple-In Multiple-Out (MIMO) Technology? https://www.lifewire.com/mimo-wifi-routers-818332
4. Microsoft, "Explore Windows 10 New Features and Update". Retrieved from Microsoft: https://www.microsoft.com/en-nz/windows/features. Accessed April 2018

5. Kolahi, S.S., Almatrook, A.A.: Impact of security on bandwidth and latency in IEEE IEEE802.11ac client-to-server WLAN. In: 2017 Ninth International Conference on Ubiquitous and Future Networks (ICUFN), pp. 893–897 (2017)

6. Newell. P.: Comparison of theoretical and practical performances with 802.11n and 802.11ac wireless networking. In: 2017 31st International Conference on Advanced Information Networking and Applications Workshops (WAINA), pp. 710–715 (2017)

7. Kolahi, S.S. Cao, Y., Chen, H.: Impact of SSL security on bandwidth and delay in IEEE IEEE802.11n WLAN using Windows 7. In: 2016 10th International Symposium on Communication Systems, Networks and Digital Signal Processing (CSNDSP) (2016)

8. Shah, Z.: Throughput comparison of IEEE 802.11ac and 802.11n in an Indoor environment with interference. In: 2015 International Telecommunication Networks and Applications Conference (ITNAC), pp. 196–201 (2015)

9. Kolahi, S.S., Cao, Y.R., Chen H.: Bandwidth-IPSec security trade-off in IPv4 and IPv6 in Windows 7 environment. In: Second International Conference on Future Generation Communication Technologies (FGCT) (2013)

10. Dianu, M.D., Riihijarvi, J., Petrova, M.: Measurement-based study of the performance of IEEE 802.11ac in an indoor environment. In: IEEE International Conference on Communications (ICC'14), Sydney, pp. 5771–5776 (2014)

11. Park. M.: IEEE 802.11ac: Dynamic bandwidth channel access. In: IEEE International Conference on Communications (ICC'11), Kyoto, Japan, 5–9 June 2011, p. 1-5 (2011)

12. Siddiqui, F., Zeadally, S., Salah, K.: Gigabit wireless networking with IEEE 802.11ac: technical overview and challenges. J. Netw. **10**(3), 164–171 (2015)

13. GRID (n.d.) D-ITG, Distributed Internet Traffic Generator. Retrieved March 28, 2018, from GRID: https://www.grid.unina.it/software/ITG/

A Real-Time Intrusion Detection System for Software Defined 5G Networks

Razvan Bocu[1]([✉]), Maksim Iavich[2], and Sabin Tabirca[3]

[1] Transilvania University of Brasov, Brasov, Romania
`razvan.bocu@unitbv.ro`
[2] Caucasus University, Tbilisi, Georgia
`miavich@cu.edu.ge`
[3] National University of Ireland, Cork, Ireland
`tabirca@cs.ucc.ie`

Abstract. The philosophy that founds the world of the Internet of Things apparently becomes essential for the projected permanently connected world. The 5G data networks are supposed to dramatically improve the actual 4G networks' real world significance, which makes them fundamental for the next generation networks of IoT devices. The academic and industrial effort to improve the 5G technological standards considers various routes. Thus, this paper presents the state-of-the-art concerning the development of the standards that model the 5G networks. It values the authors' experience that was gathered during the implementation of the Vodafone Romania 5G networked services. It puts this acquired experience in context by reviewing the relevant similar work, the relevant technologies, and it describes the research directions and difficulties that will probably influence the design and implementation of large 5G data networks. Consequently, the paper presents a machine learning-based real time intrusion detection system, which has been effectively tested in the context of a 5G data network. The intelligent intrusion detection system considers the creation of software defined networks, and it uses artificial intelligence based models. It is able to detect unknown intrusions through the usage of machine learning-based software components. The system has been assessed and the results prove that it achieves superior performance with a lower overhead in comparison to similar approaches, which allows it to be deployed on real-time 5G networks.

1 Introduction

The global Internet network is currently composed of several billion devices, and it is continuously expanding. This trend is determined by the increased usage of consumer electronics, which incorporate a wide array of sensors. It is interesting to note that the limited computational resources of these devices often imply that the data processing is delegated to external third party devices. Furthermore, these devices establish communication links between them, which are used in order to transmit synchronization and state data. The current 4G mobile data networks cannot offer the development possibilities that are necessary for the long term evolution of the IoT networks. Consequently, it is immediate to infer that the 5G networks will form the backbone of the future high speed data networks. The authors have contributed to the design and deployment of Vodafone Romania 5G networked services. Thus, this paper values the experience that was gathered, and consequently puts it in context relative to relevant contributions. Furthermore, the paper describes an intelligent intrusion detection system that considers the creation of software defined networks, and uses artificial intelligence-based models. It is able to detect unknown intrusions by using machine learning algorithms.

The rest of this paper is organized considering the following sections. Section 2 discusses on the technical requirements, which make the 5G Internet of Things (IoT) networks feasible in practice. Section 3 presents the related relevant technologies. Following, the intrusion detection system is described, and its performance is assessed. Consequently, the last section concludes the paper.

2   Technical Requirements

The 5G data networks implement high throughput data links. Nevertheless, the added value that these novel data networks generate is mostly determined by the extensive variety of smart devices, which are supported, and the related compatible applications.

- High throughput data channels – the deployment of the smart applications requires data links that offer transfer speeds of, at least, 25 Mbps, which are intended to support the high definition data containers, the virtual reality (VR) or augmented reality (AR) applications [1].
- Networks that are scalable and structurally flexible – this is determined by the consideration of the mechanism of network functions virtualization (NFV) in order to build the required fronthaul data networks.
- Very low latency – the 5G IoT networks are intended to support smart applications that should send and receive real-time data, which require communication channels with latencies that are no greater than 5 ms [2].
- Reliability and resilience – the existence of sensibly more small network cells in a 5G data network involves that the handover should be conducted in an efficient way, while the network coverage is kept at the optimal levels.
- Data privacy and security – the deployment of applications that process highly sensitive data, such as personal health data, implies that the proper mechanisms should be designed and implemented in order to prevent any illegitimate access attempt.
- Long battery lifetime – the mobility is a central concept in the realm of 5G data networks, and consequently the energy efficiency should be considered.
- Connection density – the 5G data networks are expected to offer concurrent reliable access for a large number of devices, which implies that proper design and implementation decisions are made.

- Mobility – this technical requirement complements the necessity to offer proper conditions for the deployment of many devices, which require reliable mobile intercommunication data links.

It is interesting to note that while the smart devices that are supposed to operate on the 5G networks should process large amounts of data, they often lack the hardware resources that would allow them to accomplish this efficiently. Therefore, the actual data processing is often delegated to cloud-based software systems, which extract useful knowledge out of the raw data through the consideration of data analysis methods [3,4].

3   Fundamental Technologies

The determination of the optimal architectural models for the design of the 5G data networks constitutes the object of sustained research efforts. Nevertheless, any architectural design should consider two perspectives [5,6]. Thus, the data perspective considers the real-time analysis of the data using software defined fronthaul data links, while the control perspective is concerned with the proper management of the network components and the related services, which they determine.

Let us recall that the architecture of a 5G data network should consider significant technical requirements, such as scalability, the ability to implement networked resources using the mechanism of network functions virtualization, which should provide the necessary network function virtualization capabilities [7]. Consequently, extensive functional requirements that support the efficient management of the network should be available. Thus, these should include the efficient configuration of the policies that allow for the mobile devices to behave optimally, the definition of access control policies to the networked resources, and the possibility to fully and efficiently virtualize the existing physical networked resources (Fig. 1).

3.1   Virtualized Wireless Network Function

The virtualized wireless network function (VWNF) is a fundamental process in the realm of the 5G networks design and implementation. It has been effectively used in order to deploy the core of the mentioned 5G network. This process allows for the logical specification of a self-sufficient 5G network by using the network functions virtualization (NFV) on the proper hardware infrastructures. It is immediate to note that this process is interesting from a research and theoretical perspective. Furthermore, it allows for the specialized 5G networks to be deployed on specific infrastructures, such as cloud infrastructures, or telecommunications service providers networks [8]. We have effectively used this mechanism in order to implement specialized networked services on the 5G data network of the respective telecommunications service provider. We have observed that the

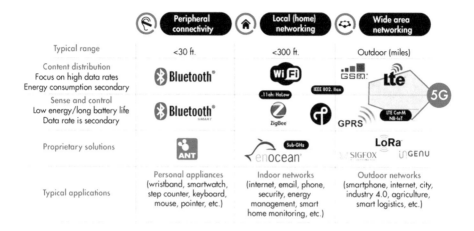

Fig. 1. Essential technologies

virtualized networked environment offers the required logical flexibility and scalability, which allowed us to efficiently deploy the real-time intrusion detection system [9]. This networking virtualization model is visually described in Fig. 2.

We have effectively determined that this virtualization mechanism allows us to properly process the data traffic that flows through the provider's 5G network in order to detect known or potential threat patterns. The contribution that is described in [1] presents an application of this virtualized networking mechanism. We were able to demonstrate, through the design, implementation and deployment of the real-time intrusion detection system, that the mechanism is suitable for the proper creation of the required specialized virtual network, which also supports the findings that are reported in the papers [10].

We have also determined that the logical specification of the 5G networks optimizes the allocation and usage of the radio resources, as we were able to define logical sub-networks that are separately analyzing the 5G data traffic through individual instances of the real-time intrusion detection system. Thus, our findings extend and refine the work that is reported in [11]. The experimental work that we conducted confirms that properly defined and sized virtual 5G networks are able to support even applications that process large amounts of real-time data, like the intrusion detection system.

3.2 Heterogeneous Networks

The concept of heterogeneous networks (HNet) describes a complementary paradigm that is designed to efficiently support the design and implementation of the 5G logical networks that host services. We have used this mechanism in order to efficiently design the runtime environment, which allows the intrusion detection system to properly process all the data traffic on the 5G network [12,13]. Moreover, this complementary mechanism allows us to properly configure the virtualized network environment, so that certain types of data traffic are

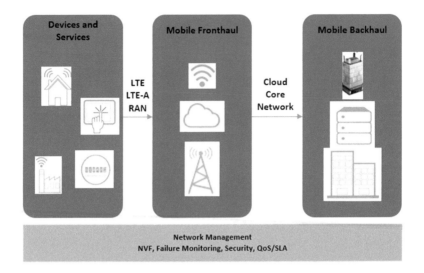

Fig. 2. The virtualized network separation

processed with a higher priority than other types of data traffic, through the specification of proper quality of service policies [14]. The following paragraphs review some interesting scientific contributions that pertain to the scope of the heterogeneous networks, which we have studied during the research process that was pursued.

The work that is reported in the papers [14,15] analyzes the problematic of machine to machine communication (M2M). Thus, the European Telecommunications Standards Institute (ETSI) define the protocols that should be considered during the usage of IoT devices, which lack ample computational power [16,17]. Additionally, the machine to machine applications may be deployed considering the guidelines that are presented in [18] and [19].

The peoples' everyday lives may be potentially influenced by the machine type communications (MTC), which allow the heterogeneous networks to be regarded as a reliable technical variant for the implementation of the 5G IoT networks [20]. The machine type communications is concerned with the continuously increasing amounts of the transferred data, and these challenges are discussed in [21]. Furthermore, the relevant quality of service (QoS) policies are described in [22].

The contribution that is described in [23] suggests that it is possible to deploy a significant number of 5G IoT devices, which will ensure that specific services will be provided through their interaction that also ensures that the data traffic is properly balanced.

Additionally, the design and field deployment of large 5G networked topologies, such as millimeter Wave (mmWave) and Narrowband IoT (NB-IoT) is important [25]. Thus, the system that is reported in this paper actively processes the data traffic that is generated by the enrolled NB-IoT devices using the same machine learning-based intrusion detection core. The contribution that is described in [24] include additional relevant information.

4   The Intrusion Detection System

The architecture of the proposed intelligent intrusion detection system is illustrated in Fig. 3. The architecture of the system consists of three layers: the data traffic forwarding layer, the data management and control layer, and the machine learning-based data analysis layer. The data forwarding layer is responsible for the data traffic monitoring and capturing. It can collect and send the suspect data streams to the control layer, and it also blocks the malicious data traffic according to the instructions of the controller. The data management and control layer identifies the suspicious data patterns, and detects anomalies using the analyzed intercepted data. It also takes proper protection measures according to the decisions made by the data analysis layer, and it consequently instructs the data forwarding layer.

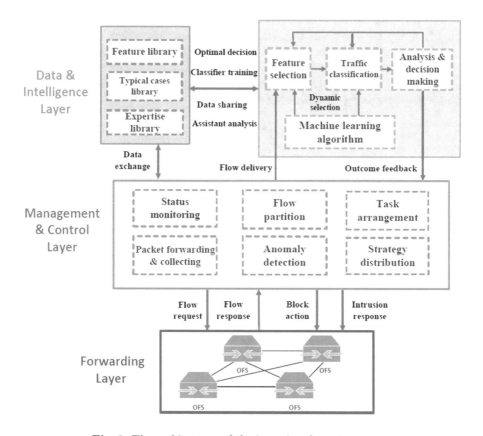

Fig. 3. The architecture of the intrusion detection system

The data forwarding layer provides the data management and control layer with real-time network status information through the real-time collection of suspect data patterns. Furthermore, intrusions are immediately blocked by dropping the malicious packets under the supervision of the other system layers.

The packet collection and data flow partitioning layer provides a more global view of the entire 5G network. The status monitoring module supervises the data network status and continuously analyzes the data packets that it receives in order to analyze them. The data management and control layer processes and parses the received data traffic. Furthermore, it creates relevant clusters of data packets and generates a data fingerprint, which keeps track of the following logical network parameters: the source IP address, the destination IP address, the source port, the destination port, the session duration, and the considered network protocol.

The data fingerprints are used in order to define and label different data flow records, which represent specific network connections and activities. The packet collection and inspection is performed continuously. The data collection and inspection time interval is optimized in order to avoid possible undesirable delays concerning the real-time data analysis process.

The anomaly detection considers some basic flow statistics, which are used to roughly recognize abnormal behaviors and potential anomalies. The particular intrusion detection system's module applies an entropy-based analysis, which is based on the Shannon's theory in order to detect the distribution variations of the analysed data packet samples. The entropy of a random variable x is computed considering the following formula:

$$H(x) = -\sum_{i=1}^{n} p(x_i) log(p(x_i))$$

Here, $p(x_i)$ designates the probability for x to take the value x_i considering all the already detected values. The equation considers four fundamental parameters: the source IP address, the source port, the destination IP address, and the destination port. The values of these parameters are gathered by the real-time traffic analysis component of the system. Thus, considering a particular moment in time, the continuously updated value that the entropy function $H(x)$ provides helps to detect possible malicious data traffic patterns. Considering that E stands for the mean entropy, and S represents the corresponding standard deviation, a possible suspect pattern involves that the value of $H(X)$ is outside the interval $[(E-S),(E+S)]$. Consequently, the suspect data packets are sent over to the proactive data analysis layer for supplementary analysis.

The feature selection component is designed in order to construct and update the features set, which is specific to the detected malicious data patterns. This component is capable to process large amounts of data in a real-time fashion, while removing the data features that are irrelevant to the machine learning core of the system's proactive data analysis layer. Consequently, the data is partitioned into relevant categories, so that malicious data traffic patterns are clearly separated from the benign traffic patterns.

The feature selection component is designed in order to construct and update the features set, which is specific to the detected malicious data patterns. This component is capable to process large amounts of data in a real-time fashion, while removing the data features that are irrelevant to the machine learning core of the system's proactive data analysis layer. Consequently, the data is partitioned into relevant categories, so that malicious data traffic patterns are clearly separated from the benign traffic patterns. The data that is presented in Table 1 considers the performance metrics, which determine five of the table's columns. Furthermore, the performance metrics are calculated considering several fractions of the input data set, which are mentioned in the first column of the table. Let us recall that the data set that is considered for the performance assessment contains 32,000,000 network connections that were analyzed by the intrusion detection system. Furthermore, each connection entity consists of 39 features that are analyzed by the machine learning core of the intrusion detection system. The values of the performance assessment metrics prove that the system scales well with the size of the analyzed data set. Furthermore, the system is able to accurately determine the malicious traffic patterns, while reducing to the minimum the incidence of the false positives. The practical behaviour of the system is especially important in the case of commercial 5G data networks, which transport and process a large number of data transfer sessions that have to be analyzed in a proactive manner.

4.1   The Real World Performance Assessment

The system was installed on the infrastructure of a significant telecommunications services provider. The performance analysis considers the data that was effectively gathered during the real-time intrusion detection process on the provider's 5G data network. The dataset that was considered for the performance assessment contains 32,000,000 analyzed network connections. Each individual connection entity consists of 39 features that are separated into three categories. Thus, the system considers network connections-based features, content-based features, and data traffic-based features. Furthermore, each data traffic entity is marked either as a normal traffic entity, or as a suspicious traffic entity. The latter ones are grouped into four distinct categories: remote to local, probe, user to root, and denial of service.

The performance assessment considers the following metrics: precision (P), reliability (R), tradeoff (T), accuracy (A), and the false positives rate (FP). The precision is defined as the percentage of valid malicious data traffic predictions relative to the total number of predictions that the intrusion detection system makes. The reliability is calculated as the total number of accurately determined intrusion attempts relative to the total number of intrusions. Furthermore, the tradeoff represents a hybrid performance metric between the precision and the reliability, which has the role to provide a better accuracy of the data classification through the following formula:

$$T = 2/((1/P) + (1/R))$$

The accuracy is a ratio that is determined by the sum of the number of legitimate packets and malicious packets properly detected at the numerator, while the denominator is the sum of the accurately detected legitimate and malicious packets plus the incorrectly detected legitimate and malicious packets. Moreover, the false positives rate is determined by the number of legitimate packets that are incorrectly classified over the sum between properly classified legitimate packets and incorrectly classified legitimate packets. The values of the performance metrics, which were obtained, are displayed in Table 1.

Table 1. The values of the performance assessment metrics

Data size	P	R	T	A	FP
20%	97.21%	96.87%	94.24%	94.13%	0.86%
40%	97.02%	96.53%	94.05%	93.92%	1.03%
60%	96.13%	95.89%	93.68%	93.39%	0.93%
80%	96.04%	95.81%	93.48%	93.18%	0.91%
100%	95.47%	95.12%	93.04%	92.07%	1.06%

The data that is presented in Table 1 considers the performance metrics, which determine five of the table's columns. Furthermore, the performance metrics are calculated considering several fractions of the input data set, which are mentioned in the first column of the table. Let us recall that the data set that is considered for the performance assessment contains 32,000,000 network connections that were analyzed by the intrusion detection system. Furthermore, each connection entity consists of 39 features that are analyzed by the machine learning core of the intrusion detection system. The values of the performance assessment metrics prove that the system scales well with the size of the analyzed data set. Furthermore, the system is able to accurately determine the malicious traffic patterns, while reducing to the minimum the incidence of the false positives. The practical behaviour of the system is especially important in the case of commercial 5G data networks, which transport and process a large number of data transfer sessions that have to be analyzed in a proactive manner.

5 Conclusion

The 5G data networks already support relevant real-world applications, and they have the potential to become the backbone of the future always connected human society. Consequently, there are rather difficult design, implementation and deployment problems, which concern all aspects of the 5G networks. Among them, the timely detection of any illegitimate access attempt is essential, especially in the context of a commercial data network. Therefore, this paper presents the state-of-the-art concerning the research that has been made on this very important topic. Furthermore, a real-time intrusion detection system, which is based on the utilization of machine learning techniques, is described. The performance of the system has been tested using real-world data, which has been obtained through the real-time monitoring of the 5G data traffic on the network of a significant Romanian telecommunications services provider. This assessment demonstrates that it is possible to design a software system that blocks most of the illegitimate traffic, which occurs on a high-traffic 5G data network, in a real-time fashion. Moreover, the various existing contributions, which are relevant to the approached topic, are presented in a constructive analytical manner, while the problems that have to be addressed are analyzed, and possible solutions are suggested for their resolution.

References

1. Egham UK: The development of connected things. https://www.gartner.com/newsroom/id/3598917. Accessed 15 Jan 2018
2. Simsek, M., Aijaz, A., Dohler, M., Sachs, J., Fettweis, G.: 5G-enabled tactile internet. IEEE J. Sel. Areas Commun. **34**(3), 460–473 (2016)
3. Akpakwu, G.A., Silva, B.J., Hancke, G.P., Abu-Mahfouz, A.M.: A survey on 5G networks for the internet of things: communication technologies and challenges. IEEE Access **6**, 3619–3647 (2017)
4. Parvez, I., Rahmati, A., Guvenc, I., Sarvat, A.I., Dai, H.: A survey on low latency towards 5G: RAN, core network and caching solutions. IEEE Commun. Surv. Tutor. **20**(4), 3098–3130 (2018)
5. Akyildiz, I.F., Wang, P., Lin, S.C.: SoftAir: a software defined networking architecture for 5G wireless systems. Comput. Netw. **85**, 1–18 (2015)
6. Xia, X., Xu, K., Wang, Y., Xu, Y.: A 5G-enabling technology: benefits, feasibility, and limitations of in-band full-duplex mMIMO. IEEE Veh. Technol. Mag. **13**(3), 81–90 (2018)
7. Chen, M., Qian, Y., Hao, Y., Li, Y., Song, J.: Data-driven computing and caching in 5G networks: architecture and delay analysis. IEEE Wirel. Commun. **25**(1), 70–75 (2018)
8. Boulogeorgos, A.-A.A., et al.: Terahertz technologies to deliver optical network quality of experience in wireless systems beyond 5G. IEEE Commun. Mag. **56**(6), 144–151 (2018)
9. Khal, B., Hamdaoui, B., Guizani, M.: Extracting and exploiting inherent sparsity for efficient IoT support in 5G: challenges and potential solutions. IEEE Wirel. Commun. **24**(5), 68–73 (2017)
10. Xu, L., Collier, R., O'Hare, G.M.: A survey of clustering techniques in WSNs and consideration of the challenges of applying such to 5G IoT scenarios. IEEE Internet Things J. **4**(5), 1229–1249 (2017)
11. Sekander, S., Tabassum, H., Hossain, E.: Multi-tier drone architecture for 5G/B5G cellular networks: challenges, trends, and prospects. IEEE Commun. Mag. **56**(3), 96–103 (2018)
12. Duan, P., et al.: Space-reserved cooperative caching in 5G heterogeneous networks for industrial IoT. IEEE Trans. Industr. Inf. **14**(6), 2715–2724 (2018)
13. Condoluci, M., Araniti, G., Mahmoodi, T., Dohler, M.: Enabling the IoT machine age with 5G: machine-type multicast services for innovative real-time applications. IEEE Access **4**, 5555–5569 (2016)

14. Vilalta, R., Mayoral, A., Casellas, R., Martinez, R., Verikoukis, C., Munoz, R.: TelcoFog: a unified flexible fog and cloud computing architecture for 5G networks. IEEE Commun. Mag. **55**(8), 36–43 (2017)
15. Hasan, M., Hossain, E.: Random access for machine-to-machine communication in LTE advanced networks: issues and approaches. IEEE Commun. Mag. **51**, 86–93 (2013)
16. Lei, K., Zhong, S., Zhu, F., Kuai, X., Zhang, H.: An NDN IoT content distribution model with network coding enhanced forwarding strategy for 5G. IEEE Trans. Industr. Inf. **14**(6), 2725–2735 (2017)
17. Morgado, A., Huq, K.M.S., Mumtaz, S., Rodriguez, J.: A survey of 5G technologies: regulatory, standardization and industrial perspectives. Digit. Commun. Netw. **4**(2), 87–97 (2018)
18. Ndiaye, M., Hancke, G.P., Abu-Mahfouz, A.M.: Software defined networking for improved wireless sensor network management: a survey. Sensors **17**(5), 1–32 (2017)
19. Gringoli, F., et al.: Performance assessment of open software platforms for 5G prototyping. IEEE Wirel. Commun. **25**(5), 10–15 (2018)
20. Palattella, M., Dohler, M., Grieco, A., et al.: Internet of things in the 5G era: enablers, architecture and business models. IEEE J. Sel. Areas Commun. **34**(3), 510–527 (2016)
21. Linge, N., Odum, R., Hill, S., Von-Hunerbein, S., Linnebank, P., Sutton, A., Townend, D.: The impact of atmospheric pressure on the performance of 60GHz point to point links within 5G networks. In: Loughborough Antennas and Propagation Conference (2018)
22. Habiba, U., Hossain, E.: Auction mechanisms for virtualization in 5G cellular networks: basics, trends, and open challenges. IEEE Commun. Surv. Tutor. **20**(3), 2264–2293 (2018)
23. Bocu, R., Costache, C.: A homomorphic encryption-based system for securely managing personal health metrics data. IBM J. Res. Dev. **62**(1), 1–10 (2018)
24. Narayanan, A., et al.: Key advances in pervasive edge computing for industrial internet of things in 5G and beyond. IEEE Access **8**, 206734–206754 (2020)
25. Suomalainen, J., Shahabuddin, S., Mammela, A., Ahmad, I.: Machine learning threatens 5G security. IEEE Access **8**, 190822–190842 (2020)

On the Structure and Assessment of Trust Models in Attribute Assurance

Andreas Grüner[✉] and Christoph Meinel

Hasso Plattner Institute (HPI), University of Potsdam, 14482 Potsdam, Germany
{andreas.gruener,christoph.meinel}@hpi.uni-potsdam.de

Abstract. Online services fundamentally rely on identity management to secure and personalize their presence. Within identity management, attribute assurance techniques target correctness and validity of attributes. These properties are an essential foundation for service provisioning in digital businesses. A myriad of attribute assurance trust models has been published. However, a superior trust model from the various proposals has not been discriminated. Additionally, a profound assessment is challenging due to a missing general notation and approach. In this paper, we work towards the structural characteristics of a secure trust model. To achieve this, we analyze common elements of attribute assurance trust models and outline differentiating factors compared to other domains. Based on the key components, we propose a formal meta-framework to depict existing trust models. Using the framework, characteristics and security attacks of these trust schemes are elaborated. As an outcome, we can conclude that a secure trust model depends on an attack-resistant trust function that considers high trust values and several attestation issuers.

1 Introduction

Identity management plays a significant role at virtually all online services that provide user-specific offerings in digital businesses. An identity is a set of information that characterizes a physical entity and enables an online service to recognize a user. On the same lines, user-specific offerings or benefits at an online service are bound to its particular identity. Therefore, identity management is at the forefront of the online service's security design. Generally, an Identity Provider (IdP) implements identity management processes such as authentication, credential and attribute management. An attribute defines a characteristic of an entity. The IdP is responsible to provide correct facts that reflect the reality. Additionally, the IdP must revoke it when they are not valid anymore.

In open domain identity management models, e.g. federation topologies, the IdP is a trusted third party towards the user and the Service Provider (SP). The IdP, the SP and the user belong to different trust domains and therefore, must trust each other. In a wide range of different scientific subjects, trust is considered a subjective phenomenon that is meaningful in personal relationships. In our opinion, one of the most applicable denotations is the definition of decision

L. Barolli et al. (Eds.): AINA 2021, LNNS 227, pp. 447–458, 2021.
https://doi.org/10.1007/978-3-030-75078-7_45

trust from Jøsang et al. [1] based on the work of McKnight and Chervany [2]. It characterizes trust as *"the extent to which one party is willing to depend on something or somebody in a given situation with a feeling of relative security, even though negative consequences are possible."*

Concerning attribute management, the user and the SP are willing to depend on the IdP for the process of attribute attestations. User and SP rely on transferred attributes that are authentic. The user intends to consume a service, and the SP offers the respective service. Correct and valid attributes are required to provision and potentially invoice the offering accurately. Otherwise, either the service is not usable or the usage might not get invoiced as negative consequence. Both factors restrain the relationship between the user and the SP.

In related research, trust in identity management is holistically referred to as identity trust [1] or trust management in authentication systems [3]. In contrast, the latter one limits the trust context to the public key to identity binding. Nonetheless, Gomi [4] proposes a separation between identity and attestation trust.

We conform to this separation and concentrate our work on trust models in attribute assurance for specifying trust in the correctness of an identity's properties (attestation trust). This research focus is also motivated by the development of a decentralized IdP based on blockchain. This advancement facilitates the separation of the identifier from the actual attributes of an identity. Furthermore, a decentralized IdP fosters the reduction of the traditional IdP to a mere Attribute Provider (AP) [5]. Besides that, a decentralized IdP resolves the IdP as a trusted third party and lets the AP be the last central authority in identity management. Overall, an attribute ecosystem is established to combine properties for a single identity from distinct providers.

Trust models in attribute assurance have been mainly proposed in reference to a specific implementation of a trust management system or authentication scheme. The web of trust and, on the opposite, the chain of trust are the two main directions of trust model development. A web of trust describes the mutual verification of properties by equitable peers. The PGP [6] trust model is one of the popular representatives. On the other side, a chain of trust reflects hierarchical trust models where specific entities confirm properties to other participants. Public-Key Infrastructures (PKI) based on the X.509 [7] standard apply a hierarchical trust model.

These dedicated entities represent trusted third parties. Both models have their distinct advantages and disadvantages. Besides these edge cases, there are many intermediate schemes [8] with differences in the underlying trust modelling. Based on the number of different trust models in attribute assurance and the emerging possibilities of a decentralized IdP, we formulate our research question: *Is there a secure trust model in attribute assurance and how is its structure?*

To address this topic, we analyze the structure of attribute assurance trust models to outline major components and differentiating factors to trust schemes apart from attribute assurance. The structure forms a meta-framework to depict such trust models. Furthermore, we study characteristics and security-related attacks. Moreover, we conclude on characteristics of an attack-resistant attribute assurance trust model based on the previous analysis.

The remainder of this paper is organized as follows. In Sect. 2, related work in this area is described. Subsequently, in Sect. 3 we analyze trust modelling in attribute assurance and provide a structure of respective trust models and desired properties. Finally, we conclude the paper in Sect. 4.

2   Related Work

In 1999, Jøsang [9] proposed an algebra for assessing trust in certification chains. The work's objective is to decide on trust between peers for communication in open networks without having previous interactions. Furthermore, in 2009, Yang et al. [10] published a trust algebra as the foundation for a general trust model. The trust algebra comprises trust evaluation and propagation algorithms for communication partners. Huang and Nicol [11] created a formal semantics based calculus of trust. The calculus provides means to logically model trust relationships and derive trust decisions.

Further research work is done to compare general trust models. Carbone et al. [12] focuses on trust modelling and comparison in dynamic and peer to peer networks. Kinateder et al. [13] concentrate on the comparative study of trust update algorithms of trust models. Fragkakis and Alexandris [14] differentiate trust models for mobile agents. Additionally, Moyano et al. [15] proposed a general conceptual framework for trust models.

Moreover, the PKI domain is an in detail analyzed research field. The main focus of the contributions is on trust in the public key to identity binding. Bakkali and Kaituni [16,17] as well as Haibo [18] propose a logical model to reason about trust in PKI. Huang and Nicol [19] published a general calculus of trust and applied it to identity management. The work enables conclusions about trust-worthiness and risks of certification paths. Furthermore, research concentrates on structures and trust distribution within different types of PKI. For instance, narrative comparisons of different models for PKIs are conducted [20]. Besides that, Maurer [21], Marchesini and Smith [22] and Henderson et al. [23] published various trust models for PKI. Additionally, Ulrich et al. [24] examined an instance of the OpenPGP web of trust. The data is evaluated with regard to network structure and security-relevant criteria. Alexopoulos et al. [3] studied the benefits of using blockchain for trust management in authentication. The authors created a formal model for blockchain-based authentication and studied attacks against the model.

3   Trust Modelling in Attribute Assurance

In this section, we study the structure of trust models in attribute assurance. Thereby, we start with common elements and outline differentiating factors to trust models in outside attribute assurance. Subsequently, we elaborate on attestation and trust networks as well as the trust composition. Finally, security attacks for these trust models are considered and desired properties are presented.

3.1   Common Elements of Trust Models

A trust model environment is comprised of a set of common elements. Primarily, different **entities** are part of the model. These entities rely on each other and reflect the trustors and the trustees. For instance, an entity can be a person, an organization or a company. Furthermore, the **relationships** between the participants are important. From certain entities, trust originates to other entities based on neighbourhood, previous interactions or other important criteria for the trust modelling in the respective domain. Entities that provide trust for other entities are related. These relationships build the foundation for the **trust** evaluation **function**. The trust evaluation function specifies the composition of the trust value in an entity. As a last point, the trust value is used to determine if the interaction with this entity is continued or terminated.

3.2   Distinction to Trust Models in Other Domains

As trust is omnipresent in various domains, manifold trust models have been proposed [1]. Trust models can be based on reputation. Reputation considers previous experiences between the peers. These schemas are applied in agent systems, for instance in peer to peer file-sharing models or in evaluation patterns for market places to judge buyers and sellers. Besides common components, we see direct feedback and trust ageing as specific differences to trust models outside the attribute assurance domain.

No Direct Feedback: Reputation-based trust models retrieve trust from previous experience [25]. Therefore, the prior experiences need to be classified into the categories positive or negative. Positive feedback increases trust while negative feedback decreases trust. This decision must happen on time to influence the trustworthiness of an entity. In the file sharing scenario, the received file can be directly tested for validity. In attribute assurance, a direct decision on the correctness of an attribute, after the SP has received it, might not be possible. Logic checks can superficially verify the attribute, but not conclusively validate it. If the first name or the user's last name is wrong, it might be solely uncovered if an ordered shipment is returned by the logistics company to the seller.

No Trust Ageing: Trust models that specify trust into entities based on prior interactions usually include elements of trust ageing [1]. Interactions that lie further back in the past contribute less to the overall trust score. Recent experience or contact influence the trust level in a significant higher manner. Trust ageing is not formally incorporated into a trust model within attribute assurance, but it can be practically addressed. A property attestation has a limited validity period or can be revoked on demand. For instance, a certificate issued by an authority has an expiration time. Additionally, a revocation mechanism may exist.

3.3   From an Attestation Network to a Trust Network

We separate the structure of trust models in attribute assurance into a graph-based network to depict the relations. An additional set of functional elements refer to the composition of trust. Related trust modelling activities use a graph-based approach [24] or a formal logic [26]. However, a directed graph naturally reflects the relations between the entities. Furthermore, a calculus or logic can determine the actual trust value. We focus on an abstract model and omit peculiarities of an implementation. In particular, we assume the existence of cryptographic measures to secure communication and verify the attestations' origin.

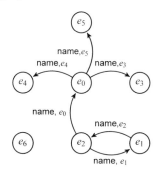

Fig. 1. Sample attestation network

IdPs or APs attest properties for a user and transfer them to the SP. In PKI systems a Certificate Authority (CA) issues certificates for entities to assert a public key to identity binding whereas properties of a user characterize the identity. In the PGP setting, the confirmation that an email address belongs to a public key is also referred to as a certification. Thereby, the email address is the attested attribute. In the Self-Sovereign identity (SSI) ecosystem, attested attributes are referenced by the term verifiable claim or credential. Nonetheless, such an attribute attestation is a confirmation of an entity, e.g. a CA or a user, about a characteristic of another entity.

Definition 1 (Attestation network). An attestation network AN is a directed graph $AN = (E, A)$ that expresses attribute attestations as relations A between the nodes E whereas:

- Nodes E represent all the entities, e.g. IdPs, APs, SPs, CAs or users
- Attestations A constitute asserted attributes by one entity to another. An attestation $a \in A$ is an attribute relation tuple $\langle attribute\ class, attribute\ value \rangle$.

Figure 1 shows a sample graph of an attestation network. The entities attest each other their names whereas node e_0 issues the most assertions. In a PKI environment, the node e_0 can be seen as a CA. In contrast, the entities e_3, e_4 and e_5 constitute regular users. The attestations between entities e_1 and e_2 reflect paradigmatically a web of trust where nodes attest each other their properties. Entity e_6 neither attest nor receives properties. We call a node an AP if it issues at least one property assertion. A user receives at least a single attribute attestation.

An attestation does not directly reflect a trust relation. However, it builds the foundation to assess trust relationships. Concerning the asserted attribute, the issuer may trust the receiver that verification procedures are not deliberately circumvented. Furthermore, the receiver can trust the issuer that delivered private information is adequately protected. Nonetheless, from an outside perspective, the major trust relation exists in case the attestations are presented to a SP for service consumption. The SP or generally any Relying Party (RP) validates the attribute attestations. The RP trusts the issuer of the provided attestations that they are authentic. The attributes of a user must reflect reality. This trust relation between the entities can be depicted in a directed graph as a trust network. The illustrated trust relation in a trust network is context-specific to an attribute class. An AP might be eligible to attest an email address, but it cannot sufficiently verify the name of another entity. Therefore, the trust is dependent on the context of the asserted characteristic.

Definition 2 (Trust network). A trust network TN is a directed graph $TN = (E, R)$ that expresses trust relations R between the nodes E whereas:

- Nodes E reflect all relying parties, e.g. IdPs, APs, SPs, CAs and users
- Trust relations R illustrate a dependency between an entity that acts as trustor and the trustee for attribute assurance. A trust relation $r \in R$ is a tuple $\langle attribute\ class, trust\ rating \rangle$.

The trust rating of a relation within the trust network is a value that belongs to the trust space. The trust space of a model comprises individual trust values that express trustworthiness. These figures are ordered to state comparable differences in trust. For instance, node e_i is more trusted than node e_j by entity e_k. Discrete numeric trust values can be assigned to verbal expressions to drive the understanding of a certain trust level. In Fig. 2 a sample trust network is shown. It is aligned to the previously depicted attestation network. However, it shows only a potential trust situation. The node e_0 receive the most trust for attribute class *name*. Entities e_1 and e_2 only trust each other.

As the attestation network and the trust network are tightly coupled, we investigate both structures' relationship to each other. IdPs, APs, SPs and users can issue claims towards other entities or may receive assertions. Each of the nodes can also act as a RP to accept attestations. Therefore, it expresses a certain level of trust in the attestation issuer. Furthermore, a node may not receive or issue any attestation or trusts respectively is trusted. Thus, both networks encompass the same entities. Having the same nodes on either network, we can investigate a connection between an attribute attestation and a trust relation.

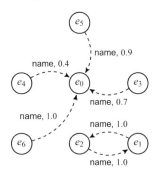

In case a certain node issues a large number of attributes it is likely that these properties are also accepted by other entities. Thus, trust exists in the originating entity. The creation of a large number of assertions that can not be used at any RP is unlikely. On the contrary, if an entity does not assert any characteristics it cannot be deduced that this entity is not trusted at all. For instance, the node may issue attestations in future. Examining the transformation from a single attestation into trust relations, there is a myriad of potential relations that comprises the issuer, receiver or any third party node as RP.

Fig. 2. Sample trust network

Besides this complexity, a detailed derivation of the trust rating in a complex trust space is not feasible. Thus, from a superficial view, there seems to be a connection between an attestation and a trust relation. However, an unambiguous transformation of the attestation network to a trust network is not possible. There might be exceptions in a significantly limited trust model where for instance only one CA exists that must be trusted by definition.

3.4 Making a Trust Decision

We see the attestation and trust network as the foundation for deriving trust. These networks enable a RP to conduct a final trust decision concerning the usage of a supplied attribute. Basically, a trust decision considers all elements related to trust and judges the supplied attributes for acceptance.

Definition 3 (Trust decision). A trust decision D is a tuple $\langle T, B, V, S \rangle$ that is self-evaluating to a binary result either indicating *trust* or *no trust*. A trust decision D_e is made from the perspective of an entity e of the trust network. The tuple elements are:

- Trust function T computes a trust score for an attribute
- Trust base B represents trust ratings towards other entities
- Attestations base V comprise the attribute assertions
- Acceptance rules S defines the acceptance or rejection condition for an attribute

The trust function T is the main component of the decision that describes the aggregation of trust in an attribute. It considers the relationships of the underlying attestation and trust network. The trust base B is a partial graph of the trust network that is reduced to the trust relations originating from the entity e that conducts the trust decision. Furthermore, the attestation base V is a partial graph of the attestation network. This subgraph is a reduction to the attestations of the property for that the trust decision should be conducted. The result of the trust function is a trust score. In a simple case, a list of attestation issuers is accepted. A more complex function may mathematically aggregate trust values of different attestations to an overall score. The final result is matched against the acceptance rules. In general, a rule defines a threshold. If the computed trust score is higher than the threshold, the attribute is accepted for further processing. Otherwise, the characteristics are rejected.

3.5 Characteristics

In the previous section, we defined fundamental elements to depict an attribute assurance trust model. Based on these components, we can derive specific evaluation properties of the networks and the decision process.

Degree of Centralization (DoC): The degree of centralization in a network measures the concentration of relations towards a set of entities. In the attestation network, we focus on the originating entities of assertions. In the trust network, we concentrate on the trust receiving (TR) entities that obtain at least one trust rating. A minor number of attestation issuers respectively trusted nodes in relation to the overall set of entities refer to a chain of trust model. Hence, the attestation issuers or the trusted nodes are seen as a trusted third party. This is reflected by a DoC score that approaches 1. In contrast, a high proportion of attestation issuers or trusted nodes in the trust network reduce centralization and indicate a web of trust model. In this case, the DoC value is close to 0.

$$DoC_{AN} = 1 - \frac{|AP|}{|E|} \qquad DoC_{TN} = 1 - \frac{|TR|}{|E|}$$

Degree of Interconnection (DoI): The degree of interconnection measures the quantity of separated subgraphs (SG) within the attestation or trust network. DoI is related to the strongly connected component measure that is proposed by Ulrich et al. [24]. If solely one graph exists the whole network is interconnected. In case each node is a separate graph, the network is least interconnected possible. A subgraph reflects a trust community where entities rely on each other for the correctness of attestations. A DoI score of 1 describes a highly interconnected network. The metric applies for both the attestation and the trust network.

$$DoI = 1 - \frac{1 - |SG|}{|E|}$$

Issued (IA) and Received (RA) Attestations: The number of attestations issued by a specific entity provides a measure of how active an entity is by providing attestations. The number of attestations that are received by a specific entity reflects its shape. Weakly and strongly attested properties build the foundation for its interaction with RPs.

Attestations for Acceptance (AfA): The metric reflects the minimum quantity of distinct attestations required for acceptance of an attribute at a RP under the condition of default acceptance rules. This measurement is used to evaluate the robustness of the trust function. For instance, AfA is 1 for PKI based on X.509 because one certificate is sufficient for acceptance.

Trust for Acceptance (TfA): The figure indicates the minimum required trust score for an attestation issuer to contribute to the calculated trust score of an attribute and towards its acceptance. Comparable to the previous metric, we use this measurement to assess the security of the trust function. A normalization of the trust model's trust ratings into an interval from [0,1] might be necessary to achieve comparability towards other schemes.

3.6 Security and Attacks

In this section, we interpret generic security objectives towards trust models in attribute assurance and describes attacks against them.

3.6.1   Security Objectives

The triad of availability, integrity and confidentiality reflect the main security objectives in information security. In attribute assurance, availability refers to obtainable and verifiable attributes. A regular user must be able to retrieve properties for its identity from any AP. Additionally, the attributes must be provided to the RP and the RP must be able to verify their origin if necessary. Concerning integrity, the user and the SP expects that attributes are authentic when they are issued. They must reflect reality. Furthermore, if the underlying properties get invalid also the attribute of the identity must be revoked promptly. Besides that, attributes should not be manipulated during the transfer between the entities. Confidentiality references the protection of private data. In particular, it should only be disclosed to authorized entities. Attributes of an identity may represent personal identifiable information of a user that require extraordinary protection. Attack vectors against confidentiality usually comprise transmission and storage protection of attribute data.

3.6.2   Attacks

In this paragraph, we elaborate on attacks on the attribute assurance trust model level. Therefore, we focus on attacks against availability and integrity because their underlying factors are captured in the abstract scheme. We omit the objective confidentiality due to its concentration on the implementation aspects of a trust management pattern.

Censorship: The censorship attack [3] targets the exclusion of a node from the service of an AP. Thus, the AP does not issue attribute attestations towards this entity. The entity is censored. The censorship attack can be motivated by the AP itself or it might be externally enforced on the AP. As a result, the attacked node is not able anymore to participate in interactions with RPs because required attributes are missing. Having several APs is a counter-strategy because the attack effort rises significantly to censor a node at all APs.

Denial of Service: In this context, the denial of service attack targets the AP and tries to prevent completely its attribute attestation service. The AP cannot issue or revoke attestations for any user. The attack affects the AP and its activities, but also restrict regular users to obtain assertions. In contrast to the censorship attack, a large number of nodes are affected. Furthermore, the attack is externally enforced on the AP because the AP has no interest to stop its complete service. As counter-strategy, the trust model must support plenty of APs to avoid a strong dependency on a single AP.

Attribute Forgery: The attribute forgery attack targets to deceive the AP into attesting a wrong property. This behaviour originates from a user that intends to obtain service from a RP under false pretences. The user achieves to circumvent verification procedures of the AP to get the false attribute value attested. This attack may have an impact on the RP and other entities concerning service consumption. As counter-strategy, the RP should not rely only on a single AP. Executing the attack against several APs increases the effort.

Rogue Attribute Provider: An adversary can set up one or more rogue APs to wrongly attest attributes. With this attack, a dedicated subgraph in the attestation and trust network can be built. A RP that falsely trusts rogue APs or applies a generic trust function that considers all APs might be prone to this attack. As a defence mechanism, highly trusted APs should only be considered.

Stale Information: The stale information attack [3] uses outdated information to obtain a service illegitimately. Within this attack, the perpetrator tries to circumvent the revocation mechanism of the AP. Thus, an attribute does not expire or will not get updated in case of changes. As consequence, a RP might still serve a user although the conditions do not hold anymore. Relying on several APs at the same time is a counter-strategy. Thus, the attacker must circumvent revocation mechanism at plenty APs.

Trust Base Manipulation: The trust base manipulation attack targets the trust information of the RP to influence a trust decision about attributes. The adversary increases the trust rating towards an AP or adds new APs with higher trust ratings. Evaluating a characteristic of a user, a property is accepted although it might be wrong. Thus, the RP would be deceived into providing service. Defence strategies can be found on the implementation level, e.g. client hardening.

3.7   Properties of a Secure Trust Model

The major components of an attribute assurance trust model are the attestation and the trust network as well as the trust decision process. The attestation network is solely a result of interaction between entities. Additionally, the trust network relies on subjective trust ratings between the nodes. These two components are an integral part of a trust scheme. However, they can hardly be influenced by modelling activities towards the security of the trust pattern. Studying the factors of the trust decision process, trust and attestation base are reductions of the respective network. Therefore, they are also not significant to determine a secure trust model. Trust function and acceptance rules remain. As the acceptance rules implement a threshold-based approval or rejection, they can be omitted in favour of the trust function. We can normalize the output of the trust function to incorporate differences in the meaning of a rule. Therefore, the trust function is the most significant component of an attribute assurance trust model. To obtain a secure trust model, the trust function must be attack-resistant against the outlined attacks against integrity and availability. Thus, the following properties are of high importance:

1. **High assurance for attribute authenticity:** Important for the RP and the user are correct and valid attributes to consume services. Therefore, attacks against integrity must be mitigated and highest trust on APs must be enforced.

2. **Low dependency on an AP:** The dependency towards one AP or a small number of APs facilitates attacks against integrity and availability. It is easier to execute the attacks against one AP in contrast to several APs to achieve a malicious goal.

4   Conclusion

We analyzed the structure of trust schemes in attribute assurance by formally specifying the attestation and trust network as the foundation. Subsequently, we studied the trust decision process that comprises the trust function, trust base, attestation base and acceptance rules. Based on this framework, we defined important characteristics and elaborated on security attacks against these trust models. As a conclusion, we determined that a secure trust model depends on the security of the trust function. The trust function must incorporate a high assurance that the attributes are authentic and a low dependency towards one attribute provider.

References

1. Jøsang, A., Ismail, R., Boyd, C.: A survey of trust and reputation systems for online service provision. Decis. Support Syst. **43**(2), 618–644 (2007)
2. McKnight, D.H., Chervany, N.L.: The meanings of trust. University of Minnesota, Technical report. MISRC 9604 (1996)
3. Alexopoulos, N., Daubert, J., Mühlhäuser, M., Habib, S.M.: Beyond the hype: on using blockchains in trust management for authentication. In: 2017 IEEE International Conference on Trust, Security and Privacy in Computing and Communications (Trustcom), pp. 546–553 (2017)
4. Gomi, H.: Authentication trust metric and assessment for federated identity management systems. IEICE Trans. Inf. Syst. **95**(1), 29–37 (2012)
5. Grüner, A., Mühle, A., Gayvoronskaya, T., Meinel, C.: A comparative analysis of trust requirements in decentralized identity management. In: 2019 International Conference on Advanced Information Networking and Applications (AINA), pp. 200–213 (2019)
6. Zimmermann, P.R.: The Official PGP User's Guide. MIT Press, Cambridge (1995)
7. Internet Engineering Task Force. RFC 5280. internet x.509 public key infrastructure certificate and certificate revocation list (CRL) profile (2008). https://tools.ietf.org/html/rfc5280. Accessed 30 Dec 2020
8. Grüner, A., Mühle, A., Meinig, M., Meinel, C.: A taxonomy of trust models for attribute assurance in identity management. In: 2020 Workshops of the 34th International Conference on Advanced Information Networking and Applications (WAINA), pp. 65–76 (2020)
9. Jøsang, A.: An algebra for assessing trust in certification chains. In: 1999 Network and Distributed Systems Symposium (NDSS) (1999)
10. Yang, W., Huang, C., Wang, B., Wang, T., Zhang, Z.: A general trust model based on trust algebra. In: 2009 International Conference on Multimedia Information Networking and Security (MINES), pp. 125–129 (2009)

11. Huang, J., Nicol, D.: A formal-semantics-based calculus of trust. IEEE Internet Comput. **14**, 38–46 (2010)
12. Carbone, M., Nielsen, M., Sassone, V.: A formal model for trust in dynamic networks. In: First International Conference on Software Engineering and Formal Methods (SEFM), pp. 54–61 (2003)
13. Kinateder, M., Baschny, E., Rothermel, K.: Towards a generic trust model - comparison of various trust update algorithms. In: Third International Conference on Trust Management (iTrust), pp. 177–192 (2005)
14. Fragkakis, M., Alexandris, N.: Comparing the trust and security models of mobile agents. In: Third International Symposium on Information Assurance and Security (IAS), pp. 363–368 (2007)
15. Moyano, F., Fernandez-Gago, C., Lopez, J.: A conceptual framework for trust models. In: International Conference on Trust, Privacy and Security in Digital Business (TrustBus) 2012, pp. 93–104 (2012)
16. El Bakkali, H., Kaitouni, B.I.: A logic-based reasoning about PKI trust model. In: 6th IEEE International Symposium on Computers and Communications (ISCC), pp. 42–48 (2001)
17. El Bakkali, H., Kaitouni, B.I.: A predicate calculus logic for the PKI trust model analysis. In: IEEE International Symposium on Network Computing and Applications (NCA) 2001, pp. 368–371 (2001)
18. Yu, H., Jin, C., Che, H.: A description logic for PKI trust domain modeling. In: 3rd International Conference on Information Technology and Applications (ICITA), pp. 524–528 (2005)
19. Huang, J., Nicol, D.: A calculus of trust and its application to PKI and identity management. In: 8th International Symposium on Identity and Trust on the Internet (IDtrust), pp. 23–37 (2009)
20. Uahhabi, Z.E., Bakkali, H.E.: A comparative study of PKI trust models. In: 2014 IEEE International Conference on Next Generation Networks and Services (NGNS), pp. 255–261 (2014)
21. Maurer, U.: Modelling a public-key infrastructure. In: European Symposium on Research in Computer Security (ESORICS) 1996, pp. 325–350 (1996)
22. Marchesini, J., Smith, S.: Modeling public key infrastructures in the real world. In: European Public Key Infrastructure Workshop (EuroPKI) 2005, pp. 118–134 (2005)
23. Henderson, M., Coulter, R., Dawson, E., Okamoto, E.: Modelling trust structures for public key infrastructures. In: Australasian Conference on Information Security and Privacy (ACISP) 2002, pp. 56–70 (2002)
24. Ulrich, A., Holz, R., Hauck, P., Carle, G.: Investigating the OpenPGP web of trust. In: European Symposium on Research in Computer Security (ESORICS) 2011, pp. 489–507 (2011)
25. Chirita, P.-A., Nejdl, W., Schlosser, M., Scurtu, O.: Personalized reputation management in P2P networks. In: 2004 International Conference on Trust, Security, and Reputation on the Semantic Web (ISWC), pp. 32–41 (2004)
26. Ries, S., Habib, S.M., Mühlhäuser, M., Varadharajan, V.: Certainlogic: a logic for modeling trust and uncertainty. In: 4th International Conference on Trust and Trustworthy Computing (Trust), pp. 254–261 (2011)

Virtual Environment for Analysis and Evaluation of DDoS Attacks

Ryo Tokuyama[1]([✉]), Yuichi Futa[2], Hikofumi Suzuki[3], and Hiroyuki Okazaki[4]

[1] Graduate School of Bionics, Computer and Media Sciences,
Tokyo University of Technology, 1404-1, Katakuramachi,
Hachioji, Tokyo 192-0982, Japan
g2120022fb@edu.teu.ac.jp
[2] Department of Computer Science, Tokyo University of Technology,
1404-1, Katakuramachi, Hachioji, Tokyo 192-0982, Japan
futayi@stf.teu.ac.jp
[3] Integrated Intelligence Center, Shinshu University,
4-17-1, Wakasato, Nagano 380-8553, Japan
h-suzuki@shinshu-u.ac.jp
[4] Faculty of Engineering, Shinshu University,
4-17-1, Wakasato, Nagano 380-8553, Japan
okazaki@cs.shinshu-u.ac.jp

Abstract. Recently, Distributed Denial-of-Service (DDoS) attacks have been increasing and reports show that the number of DDoS attacks in 2020 increased three times compared to the same quarter of the previous year. In this paper, we develop an attack test environment for evaluation of DDoS attacks, which also includes a basic Distributed Reflection Denial of Service (DRDoS) attack environment. The environment uses virtualization technology in order to adopt the environment easily in various systems. In the attack test environment, the attacker network configurations, reflector, and target are separated. We adjust both the attacker and the reflection server in order to increase the power of the attack. In the study by Suzuki et al. [1], a Unified Threat Management (UTM) was first stopped by external DDoS attacks. In this paper, we confirm that the UTM in our environment stops intermittently under TCP Flood attacks, thus, we consider that our environment succeeded in reproducing virtually the study by Suzuki et al. [1]. We also confirm that in our DRDoS attack test environment, the http communication to the target server is disturbed.

1 Introduction

Recently, Distributed Denial-of-Service (DDoS) attacks have been increasing. The DDoS attack is an attack technique in which multiple machines apply processing loads to a target computer to stop its service. In a DDoS attack, malicious programs are often embedded in multiple machines. By sending and embedding the programs remotely, these multiple machines can attack the target computer all at once.

© The Author(s), under exclusive license to Springer Nature Switzerland AG 2021
L. Barolli et al. (Eds.): AINA 2021, LNNS 227, pp. 459–468, 2021.
https://doi.org/10.1007/978-3-030-75078-7_46

The Distributed Reflection Denial of Service (DRDoS) attack is a technique in which the attacker sends spoof packets from a target computer to a large number of reflection servers and concentrates the responses on the target computer. The reflection servers provide services such as Domain Name System (DNS), Network Time Protocol (NTP), and others.

One of the most well-known DDoS attacks, whose target was Spamhaus.org, a non-profit anti-spam organization in the UK, occured in March 2013. The transmission rate of the attack was 300 Gbit/s. The DNS servers used in the attack were open resolvers. As of October 27, 2013, there were 32 million DNS servers of the open-resolver type, while by December 8, 2019, their number had decreased significantly to 5.46 million, but many still remained in operation [2]. Report [3] mentions that the number of DDoS attacks in 2020 were three times that of the same quarter of the previous year. It is difficult to reduce reflectors because there are 15,000 reflectors in more than 1000 AS [6]. In reference of [7], which proposed DRDoS environment, evaluation of its impact on a target was not described.

Our Contribution
In this paper, we describe a method for developing a virtual environment in which DDoS and DRDoS attacks are generated on purpose. In our attack test environment, the network configurations of the attacker, the reflection server, and the target are separated. We also evaluate the attack test environment. As shown in [1], a Unified Threat Management (UTM) was first stopped by external DDoS attacks. In this paper, we confirm that the UTM in our environment stops intermittently under TCP Flood attacks, therefore, we consider that our environment succeeded in reproducing virtually the study by Suzuki et al. [1]. We also develop a DRDoS attack test environment using ten reflection servers and confirm that the attack succeed in blocking the http connection from the outside. The attack test environment makes it possible to increase the power of the attack.

2   Constituent Technologies

2.1   DDoS Attack

The DDoS attack is an attack technique in which multiple machines apply processing loads to a target computer to stop its services. On the other hand, the DRDoS attack is a technique in which the attacker sends spoof packets from a target computer to a large number of reflection servers and then concentrates the responses on the target computer. The reflection servers provide services such as DNS, NTP, and others. Saddam [5] is one of the DDoS attack tools available in the Python language and includes the following four types of attacks:

- DNS amplification
- NTP amplification

- Simple Network Management Protocol amplification
- Simple Service Discovery Protocol amplification

In this paper, we use the DNS amplification feature of Saddam in DRDoS attacks.

2.2   Intrusion Detection System

A UTM is a single hardware which combines multiple security functions in a network system, including firewall, anti-virus, intrusion prevention system, intrusion detection system (IDS), and web filtering. The IDS is a system which detects, and notifies about access and intrusion from malicious third parties. There are two types of the IDSs. The first type is the network-type IDS (NIDS) and the second type is the host-type IDS (HIDS). A NIDS acquires packets flowing on a network, and analyzes data and protocol header parts of the packets. The snort is the most widely used IDS and is provided in an open source. The snort can analyze real-time traffic and log packets. In this paper, we adopt the snort as NIDS.

3   Previous Research

Suzuki et al. [1] constructed an attack test environment to detect DoS/DDoS attacks by analyzing traffic logs. The DNS Flood was used in the environment. The evaluation of the environment required careful preparation to reduce the impact of an DDoS attack on the other network. In the evaluation, DNS Flood attacks were only executed five times. The attacks used IP addresses randomly spoofed or randomly selected from four fixed addresses. As a result of the attacks, the UTM on the outermost side of the targeted network went down even though the targeted web server remained available.

Issue

Because the attack test environment of Suzuki et al. needs the aforementioned preparation, only limited DDoS attack tests can be performed in the environment. However, the network system is modified many times until its practical operation. Thus, we need to be able to adopt the attack test environment easily in various modified systems. In the actual environment, consideration of the impact on other networks is required. However, we cannot evaluate easily and adaptively that network systems have the resistance to DDoS attacks.

4   Our System

In this paper, we develop an attack test environment using virtualization technology in order to be able to adopt the environment easily in various systems. In our system, we can generate DDoS and DRDoS attacks on purpose. The virtual test environment is equivalent to an actual test environment for DDoS attacks. In our evaluation of the virtual test environment, we also develop a DRDoS attack test environment that uses ten reflection servers.

4.1   Features of Our System

- We use a hypervisor-type virtualization technology for the attack test environment. Task allocation to the CPU in a hypervisor-type virtualization is smoother than that in a host-type virtualization, because the hypervisor-type CPU allocation is performed at a lower level.
- The operating systems of the attacker, target, and reflection servers are CentOS. In all of these servers, Large Receive Offload and Generic Receive Offload are disabled to avoid offloading problems.
- In the DNS, we use TXT records to improve the amplification rate; however, since each TXT record is limited to only 255 characters, we divide each TXT record into 255 character blocks with single quotes and register, so that it does not exceed the 4096 bytes that can be returned by BIND.
- We prepare two attacker servers, each configured to send attack packets to five reflection servers, because we confirmed that the amount of data that can be observed at the target side is twice as large as in the case of sending attack packets from one attack server to ten reflection servers.
- In order to reproduce the communication from the outside, we locate Watch PCs in the attacker section and the snort section of the environment. We use the Watch PCs to check the flow rate of the attack and to establish http connections to the target server.

In the DDoS attack, we use hping3 to perform TCP SYN Flood [4], while in the DRDoS attack, we use the Saddam to attack the target server. Both of these attacks can be executed from the attack server, but in the DRDoS attack, we use a DNS adjusted amplification.

4.2   Network Configuration

The network configuration of the environment consist of the attacker, reflection server, snort, and target sections, as shown in Fig. 1.

Snort monitors network flows to the target section and is corresponding to UTM in the actual environment. To develop an effective attack environment, we add "Watch PC 1" to the attacker section to check the flow rate of the attack. Based on the check results, we evaluate the attack and adjust the DNS server. We also locate "Watch PC 2" for the purpose of monitoring snort. We locate ten reflection servers in the attacker section for the DRDoS attacks.

We use two actual servers, each connects 10 GbE. The attacker and reflector sections are located on actual server 1. The target section and snort are located on actual server 2. We prepare two virtual machines for the attacker servers, ten for routers, ten for reflection servers, one for snort, and one for target server, respectively. We also prepare ten virtual NICs for the attacker server, and connect them to the routers from 192.168.20.0/24 to 192.168.29.0/24, to which the reflection servers belong. We use ten VyOS routers. We develop three virtual NICs for each VyOS router and connect them to the attacker, reflection, and target server, respectively. Table 1 shows the IP addresses of attacker, reflection, snort, and Target servers, while Table 2 the connection segments of attacker, reflection, snort, and Target servers virtual NICs. The reflection server is assigned the IP address shown in Table 1 and is connected only to a VyOS router, which belongs to the reflection server network. The target server is connected to the snort section of the actual server 2 via a LAN segment.

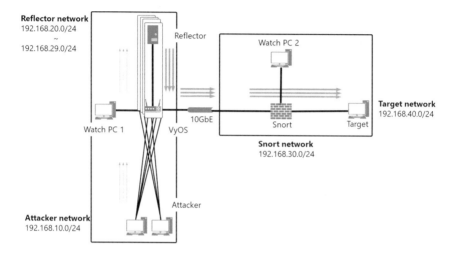

Fig. 1. Network structure

Table 1. IP address of network section

192.168.10.0/24–192.168.11.0/24	Attacker section
192.168.20.0/24–192.168.29.0/24	Reflector section
192.168.30.0/24	Snort section
192.168.40.0/24	Target section

As virtualization technology, we use Hyper-V in actual server 1 and VMware in actual server 2. Table 3 shows the number of CPU cores, memory, and virtual NICs allocate to the attacker, reflection, and target servers. The actual machines are connected to each other using 10GbE.

4.3 DNS Specifications

In this paper, BIND is used for DNS. Here, we explain values in the configuration file of BIND, which are set for the options in named.conf. We set option "recursive" to "no" (disable) and "allow-query" to IP addresses 192.168.10.0/24 and

Table 2. Connected switch of virtual NICs

192.168.10.0/24	Internal virtual switch 1–5
192.168.11.0/24	Internal virtual switch 6–10
192.168.20.0/24–192.168.29.0/24	Internal virtual switch 1–10
192.168.30.0/24	Segment1
192.168.40.0/24	Segment1

Table 3. Allocation for each server

	Attack server	VyOS router	Reflection server	Target server	snort
Number of CPU cores	4	1	8	16	1
Memory allocation	2 GB	2 GB	2 GB	2 GB	16 GB
Number of virtual NICs	5	3	1	1	2

192.168.11.0/24. We prepare a zone named test.local and develop TXT records from host0 to host99. Each TXT record contains 3962 characters, while each TXT record are divided into 255-character blocks by single quotes and register, so that it does not exceed 4096 bytes that can be returned by BIND.

4.4 Snort Specifications

In this study, the snort operates as a network IDS to detect packets transmitted to 192.168.30.0/24. The internal network is defined as 192.168.30.0/24, while the external network as an IP address out of the range mentioned above. We use the rules published in the "Registered Rules" section of the snort official website and configure our own rules to issue an alert when snort detects more than 100 packets in 10 s from port 53, except the internal network.

5 Evaluation

In this section, we describe the evaluation results of each attack simulated in our environment. We also describe the behavior of the DDoS and DRDoS attacks, since we use snort as a UTM to evaluate the attacks.

5.1 DDoS Attack

Figure 2 shows the resource monitor in the attacker servers and the results of execution of a top command when attacks were performed by hping3. Figure 2 shows that hping3 occupies 99.7% of the CPU and transmits data at 22.4 Mbps.

Figure 3 also shows that the snort resource monitor went down, which happened several times. Figure 3 shows that the snort occupies 21.3% of the CPU and receives data at 46.4 Mbps.

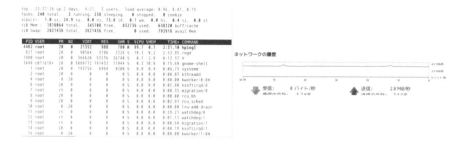

Fig. 2. Attacker's resource monitor

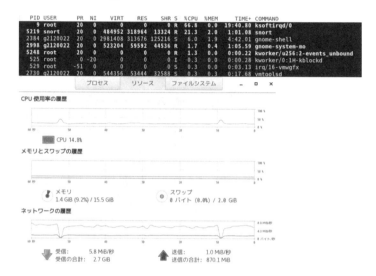

Fig. 3. Snort's resource monitor

5.2 DRDoS Attack

Figure 4 shows the attacker resource monitor in the attacker servers and the
results of execution of a top command when the attacks were performed by
Saddam. Figure 4 shows that Saddam (Python) occupies 100% of the CPU and
transmits data at 14.4 Mbps.

Figure 5 shows the reflection server resource monitor when attacking the
target using Saddam. It can be seen from Fig. 5 that reflection server receives
916 Kbps of data, and transmits the data at 39.2 Mbps. Figure 6 shows that
named occupied 21.9% of the CPU.

Figure 7 shows snort resource monitor and an output of ping to the target
when attacking the target using Saddam. Figure 7 shows that snort occupies
80.7% of the CPU and receives 1.4 Gbps of data.

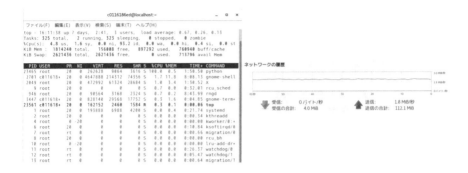

Fig. 4. Attacker's resource monitor

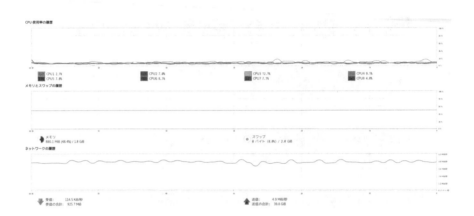

Fig. 5. Reflection server resource monitor

```
top - 21:47:13 up  2:33,  2 users,  load average: 0.01, 0.09, 0.13
Tasks: 258 total,   1 running, 257 sleeping,   0 stopped,   0 zombie
%Cpu(s):  3.7 us,  6.6 sy,  0.0 ni, 88.1 id,  0.0 wa,  0.0 hi,  1.6 si,  0.0 st
KiB Mem : 1863252 total,  104768 free,  871488 used,  886996 buff/cache
KiB Swap: 2097148 total, 2097148 free,       0 used.  577600 avail Mem

  PID USER      PR  NI    VIRT    RES    SHR S  %CPU %MEM     TIME+ COMMAND
 8147 c011618+  20   0 4252628 267324  80788 S  40.7 14.3   0:27.99 gnome-shell
 6714 named     20   0  695228  70360   3628 S  21.9  3.8  23:20.85 named
 7411 root      20   0  450576 107700  67752 S  16.6  5.8   0:03.93 X
    3 root      20   0       0      0      0 S   4.0  0.0   1:26.12 ksoftirqd/0
 3081 root      20   0   40548   4092   3376 S   3.3  0.2   0:59.34 systemd-journal
 6654 root      20   0  306520   9700   4056 S   2.3  0.5   0:27.74 rsyslogd
 9868 c011618+  20   0  832728  34020  18556 S   1.7  1.8   0:01.53 gnome-terminal-
   19 root      20   0       0      0      0 S   1.3  0.0   0:57.56 ksoftirqd/2
 8476 c011618+  20   0  573352  32760  19656 S   1.3  1.8   0:23.65 vmtoolsd
   44 root      20   0       0      0      0 S   1.0  0.0   0:00.25 ksoftirqd/7
 8206 c011618+  20   0  465152  13860   8948 S   1.0  0.7   0:00.25 ibus-x11
```

Fig. 6. Reflection server top command

Fig. 7. Watch PC 2 resource monitor and top command

6 Discussion

In this paper, we locates a Watch PC in the attacker section to reproduce communication from the outside. Based on the results obtained from the Watch PC, we adjust the DNS server to increase the attack flow rate, evaluate the attack, and develop an effective attack test environment.

In this paper, we use the TCP SYN Flood attack as a DDoS attack. In the TCP SYN Flood attack, 30,000,000 packets are sent to the target server, and we confirm that the snort corresponding to UTM, prepared as a defense mechanism, went down several times. As a result of a detailed comparison with the study by Suzuki et al. [1], we confirm that our DDoS attack test environment succeeded in reproducing the same situation, where the UTM goes down after receiving 30,000,000 packets from the start of the attack due to processing overload. Therefore, our environment is considered to match the performance as the actual attack environment. The DNS amplification attack, which as a DRDoS attack generates an attack of 1.4 Gbps, confirm it succeed in blocking the http connection from the Watch PC installed in the attacker section. We thus conclude that the environment is effective.

In Figs. 3, 4, 5, 6 and 7, it can be seen that the DNS amplification attack occupies more bandwidth. On the other hand, the SYN Flood attack cause the network to go down several times. As a result, the snort counted the TCP sessions, and thus the UDP attack did not shut down the snort. The present environment is developed using virtualization technology, therefore, it can easily be applied in various environments. We confirm that our environment can be configured and evaluate appropriate UTMs.

We suppose how to apply our environment to a network system as follows:

- We provide the actual server 1 (the attacker network configurations) and a virtualization image of the Watch PC 2 to developers of the network system.
- Developers try evaluation of the resistance to DDoS attacks.
- Developers check the Watch PC 2 in order to evaluate the influence of DDoS attacks on network flows.

7   Conclusion

In this paper, we described a method for developing a virtual environment in which DDoS and DRDoS attacks are intentionally generated. In the TCP Flood attack, we confirmed the attack test environment to be equivalent to the actual environment, because we recreated the same situation as in [1]. We also developed a DRDoS attack test environment. The DRDoS attack test environment confirmed blocking of the http connection from the outside. We concluded that our virtual attack test environment can be easily applied in various systems and attack tests can be executed at any time.

In future, we will develop an effective defense mechanism against DDoS and DRDoS attacks. In this paper, we use DNS, but we will also verify amplification attacks using different services.

Appendix

This study was supported in part by JSPS KAKENHI Grant Number JP17K00182 and JP18K02917.

References

1. Suzuki, H., Ui, T., Furukawa, S., Yuhara, D., Naruse, S., Asakawa, Y., Nagai, K., Hasegawa, O.: Consideration on implementation of pseudo attack by distributed denial-of-service attack and verification of UTM. JIPS Jpn **21**, 21–28 (2017)
2. Open Resolver Project. http://www.openresolverproject.org/. Accessed 19 Dec 2019
3. DDoS attacks in Q2 2020. https://securelist.com/ddos-attacks-in-q2-2020/98077/. Accessed 1 Jan 2021
4. hping3 - Linux man page. https://linux.die.net/man/8/hping3. Accessed 1 Jan 2021
5. Offensive python saddam. https://github.com/OffensivePython/Saddam. Accessed 1 Jan 2021
6. Kondo, M., Shintani, N., Makita, D., Yoshioka, K., Matsumoto, T.: Study on analyzing Memcached DRDoS attacks and their infrastructures. IEICE Technical report, pp. 114–119
7. Bekeneva, Y.A., Shorov, A.V.: Simulation of DRDoS-attacks and protection systems against them. In: IEEE SCM (2017)

Towards a Secure Smart Parking Solution for Business Entities

Gerald B. Imbugwa[✉] and Manuel Mazzara

Innopolis University, Innopolis, Russia
g.imbugwa@innopolis.university, m.mazzara@innopolis.ru

Abstract. This paper proposes a parking management system that is geared towards business entities. The proposed system will focus on privacy for the different entities who use the system. In a business setup, performance, scalability, and privacy should be at the centre stage. The paper aims to improve on already existing research on smart parking using blockchain. Most of the research done focuses on single entity privacy, single point of failure for the systems deployed on cloud databases, and geographical censorship. This paper proposes a parking management system that will be based on JPMorgan Quorum. The system will focus on privacy, performance, scalability and multiple entities interaction.

1 Introduction

Parking is at the center of the problems of new mobility in urban areas. With urbanization, there has been a surge of movement of people moving from rural to urban or one country to another in search of a greener pasture and better life [17,20]. Therefore, the urban area has expanded its borders to be able to accommodate the high population [10]. With the increasing population, new infrastructure has to be established from housing to roads for the high traffic. According to USToday, motorists spend between 17 to 24 h each year in the search for a parking space in cities. The increase in population has led to 30% of traffic-related issues in the cities [11].

Different parking solutions have been created using, central and decentralized networks to resolve the issues. All solutions have been revolving around:

- Single point of failure which is experienced by the central database. An example of one big technology company going offline for some hours, leading to loss of resources by businesses which uses their services [15].
- Data breach in businesses is experienced when attackers infiltrate the servers to steal valuable information.
- Internet censorship by a government/authority that might prevent a certain application to be used in their geographical area.

It is worth mentioning research done previously [1,16,23] only focuses on how single entities can connect to find parking, as they all do not address how business can utilize parking as a service. A system that has different entities ranging from

L. Barolli et al. (Eds.): AINA 2021, LNNS 227, pp. 469–478, 2021.
https://doi.org/10.1007/978-3-030-75078-7_47

the government, property managers, tenants, service provider and driver can have a full system that can house them in one system on the blockchain.

A system where each of the concerned parties will not have to worry about privacy-related issues regarding their data, ranging from personal information to trade secrets.

We propose a permission blockchain system based on Quorum [4] which will allow different users in the platform to transact with ease, without worrying about privacy, data-breach or loss of their private data or trade secrets.

The subsequent sections are divided into four, where the paper delve into more detail in the subject matter. The paper start by the literature review looking at history of parking, and finish with related smart parking system that are geared towards user privacy using blockchain. Secondly, the system for smart parking by analyzing the proposed system from functions, workflow and technology. Thirdly, a briefly talk on the privacy aspect and last the conclusion on future work.

2   Literature Review

In this section, the paper starts by understanding the underlying history of car parking over the years. Smart parking applications are built on blockchain to tackle the issue of privacy of its users and their limitations.

2.1   History of Parking

In Britain, in 1901, a new parking space located at 6 Denman Street, Central London, was launched by Suburban Electric Carriage Company in conjunction with the city administration of Central London. This multistory car park was the first in the UK and argued to be the first in the world. The parking featured seven floors with an area of 19000 square feet. A fleet of one hundred vehicles could be hosted by the space. Furthermore, the parking also had an elevator which was used to aid the transfer of vehicles to different floors in the multi-storey car park. In the subsequent year, the city administration of Central London & Suburban converted a building in Westminster, which could hold a fleet of 230 cars [8].

In the 1920s, Oklahama, United States, had no regulated parking since there were many lands in the residential areas. Residents could park their cars anywhere they deemed safe to them. However, there was a problem when it came to the shopping areas which lacked sufficient parking space since all parking lots were occupied by the employees. A competition was organized at Oklahoma University to find a solution to the parking problem, and this led to the development and introduction of the park meter. Parking meter was designed and developed by Holger George Thuessen and Gerald A. Hale and was installed in 1935. The installation came with benefits including ease of traffic congestion and increase in business revenue. This new development made customers of the shopping mall happy despite it being the first introduction of parking lot payment [9].

In the 1930s, hotels and apartments in London introduced parking spaces which were built in the basement of the building. This brought about enhanced security and car protection from bad weather. This kind of parking is still used at this moment in most of the buildings and shopping malls in the urban areas, it helps solve the problem of small spaces [14].

In 1974, Oxford introduced pay-on-foot system which was provided by the Swiss and installed at Westgate car park. The drivers on their way in to the parking space were required to collect a ticket and pay for it as they leave. The system is still used to this moment, especially in airports and other recreation facilities around the world [14].

In the 1980s park and ride was introduced, but it might be argued, it came early as of the 1960s. Park and ride concept where commuters would park their cars and use a bus or train for the rest of the journey. The reason behind this was to help decongest the urban areas. Park and ride concept was first seen in Oxford and were used as part time service for a motel [14].

2.2   Related Project

In [6], discusses the limitations of the current parking system regarding user privacy. In addition, the paper examines how the developers of these systems do not put security and privacy at the center stage. The authors propose a framework based on blockchain which will preserve user privacy without relying on third party entities.

In [1], the paper discusses the limitations of the previous research work conducted, which suffers from the single point of failure and data breach. The authors propose a smart parking system that focuses on security, transparency, and availability of parking with the use of consortium blockchain. In addition, the cloaking technique is used to hide a driver's location, thereby preserving their privacy.

In [22], the authors have pinpointed the problem of finding a public parking spot in urban areas can be a challenge. Moreover, another imminent issue that arises is when private entities share their own parking spots and their risk of data privacy. Existing sharing parking system, on the other hand, does not take privacy into consideration. [22] proposes a privacy-enhanced private parking slot sharing scheme that will focus on privacy and their will be no need to use any trusted third party.

3   A System for Smart Parking

In this section, we will go through the parking system overview, as shown in Figure 1 and discuss the entities shown in Fig. 2. This will help to illustrate how the system is supposed to work to achieve its main objective of privacy and prevent data loss or breach.

The system is divided into Parking platform, Network and Blockchain (Quorum):

- The platform consists of a web application which allows different entities to interact with the blockchain.
- Network is a bridge that uses web3.js [5] library which is used to send and get a request from Quorum.
- Quorum, which is a soft fork of Etherium, allows creating private nodes. The private nodes are used to encrypt the data from entities that are not allowed to see or modify the data and transaction.

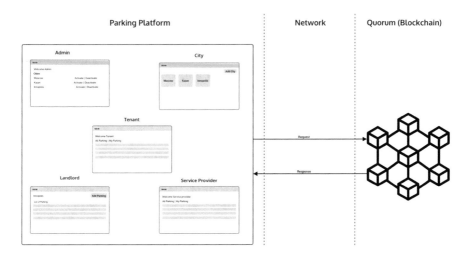

Fig. 1. Parking system with overview

The platform has six entities who have different responsibilities in the system. In Fig. 2, only five entities are shown, since those are the only ones who are part of the web application and the sixth user is not, we will discuss more when discussing each user. The following is a description of each user in the system:

- Admin is a developer of the system and the person who helps to maintain the system. Secondly, the admin (he/she) is responsible for ensuring that all cities registered on the platform are the actual representatives of the government. This process is done to cross-reference with the official government office of the respective city. Lastly, the admin is also responsible for activating and removing system users who violate the terms and policies set by the platform.
- City can be a region/town which is headed by the government. The City is responsible for verifying all the users who register in the system as landlords are the actual owners and the properties registered exists.
- Landlord is anyone who has a property which can be rented for a longer period of time.
- Tenant is anyone who can rent a property for personal use for a period of time.

- Service provider is anyone who rents a property from the landlord to rent it out in smaller quantities or timely base. Service provider is the one responsible for creating a mobile or web application to resell the parking slots to the sixth user who is the driver.

3.1 Smart Contract

The proposed system is to be developed using Etherium smart contracts (Quorum), which will have six major contracts based on the entities. Each smart contract has its set of functions it can perform. We will examine each of the functions and what they do.

Admin contract has three functions. The first is deploying the system for the first time on the decentralized network. Second, the function to activate cities for the first time and deactivate those that have violated the set terms. Last, a kill function which deletes all the contracts on all nodes.

City Contract has four functions. The first function is policies which help the city to implement its own policy for the city, which includes the amount of tax that needs to be paid, whenever any transaction occurs in the city and update them when needed. Secondly, statistic analysis in the city which includes a number of properties but not private data that might include trade secret. Third, the function which helps in the management of landlords in a city, which includes activating tenants, sending warnings, and deactivating users. The fourth function is registration of new cities and putting them in waiting mode.

Landlord Contract also has four functions. The first function is the registration of a new property into the system. Second function is approve or decline a renting request from the tenant or service provider. Third function is setting policies to govern the property, which includes the amount needed to be paid, duration of rent, and any other important policies. Fourth function offers and discount based on the landlord.

Tenant Contract has five functions. First function is a request for renting out a property. Second function is make a payment for rent based on the agreed terms. Third function is management of the property to see available slots and occupied slots. The fourth function is terminate a contract. Last function extending a contract.

Service Provider Contract has seven functions. First function is requesting to rent a property from the landlord. Second function is making payments based on the agreed terms. Third function is terminating a contact. Fourth function is extending a contact. The fifth function is managing of parking slots. Sixth function to allow the service provider view settings to connect to driver contract which helps him build an application to rent out parking spaces. The last function is the registration of service provider.

Driver Contract This contract can only be accessed through service providers. Driver contract has seven functionalities as listed bellow.

- Register a new driver and book a parking slot,
- search for available parking slots around him or another destination,

- see the prices being offered by the service provider,
- set parking duration,
- extend the parking duration,
- make payment for parking,
- and terminate parking.

3.2   Proposed System Workflow

This section illustrates how the system will work after deployment as illustrated in Fig. 3.

The different users will be able to interact with the system once being deployed on the decentralized network through a web application and mobile application which will be created using react [7] and native iOS applications.

A new user will navigate to the web application based on the assumption that a user has a metamask account sign-in [12]. Metamask wallet will be used to authenticate the user with the system and make payments to the system. Accessing the web application for the first time will show a dashboard with a list of cities and a button to add a city. The assumption is that the user might be a city that wants to add a city, landlord, tenant, or service provider. Irrespective of the user, the process is the same since any option selected will present a form where the user will need to fill in details for them to be registered in to the system.

In the scenario where the user is a city, the add city button will be pressed which will present a form. The user will be required to provide details which will be used to crosscheck with the government officials if the people who are trying to be registered are the actual appointed users to manage the city account. Happy ending if they are official personnel, the account will be activated if the account is not will deleted by the admin. The account is deleted to avoid the record of the city in the system and prevent the actual city from registering since the data already exist in the system. Once the account is activated, the user will be directed to a dashboard, where he/she can set their policy for the city. The city is now able to activate and deactivate the landlords.

In the case when user is a landlord, the user will select his or her/their city from the registered one. A new dashboard will be presented listing properties and an option to add a new property. Since he or she is a new landlord, they will select and add a new property. The user fills the form that is presented with the required details and waits for activation from the city. Once activated, the user can add policies and accept or reject proposals from tenants and service providers from their dashboard. Opportunity to manage the service providers, tenants, and multiple properties the user might have added to the system and linked to their account.

In the case the user is a tenant or service provider from the landlord dashboard, he can select the property he likes from the city of his choice. Fill a form with detail and choose a user type between tenant or service provider. Once the landlord accepts a request the user will be navigated to his dashboard. On the dashboard, it will have two lists to shows a list where he can add more properties and my properties which show all the properties with active contract. The dashboard will have other functionalities depending on the account for example, a service provider has settings with an endpoint to connect to his object of the driver's contract to rent the parking.

Depending on the service provider, they can consume the endpoint of the driver object how they need. The functions listed in the previous subsection under driver will be available to create a web or mobile application for their customers.

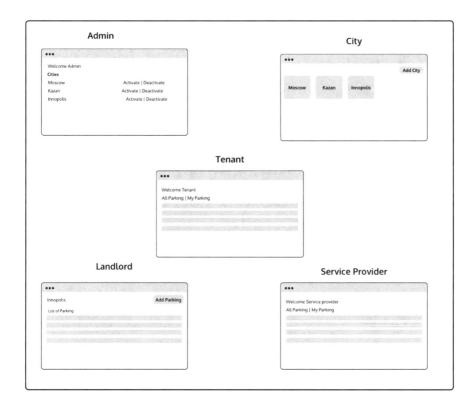

Fig. 2. Entities

3.3 Technological Stack

Truffle suite will be used for development, it offers an effective integration with quorum [18] and react framework [7] which can be added to the box [19]. Quorum is used as it allows deploying private contracts where only the entities that have permission can view and interact with the contracts. Truffle suit offers us with ease of deployment, migration, and writing tests for validating the operation of the contracts, performing the task as expected in the stipulated time before deployment to the decentralized network. Users, on the other hand, can interact with the deployed contracts, using front-end clients that are easy to use.

4 Privacy

The proposed system will ensure the entities in the system will not be able to see the transactions of other members in the system unless they are public. Furthermore, solving a problem experienced in cloud computing of data breach [3, 21]. Service provider will be assured of the privacy of their data, such as the business model.

The encryption of the data in the system is based on Quorum constellation which implements the privacy feature. Constellation is divided as follows:

• Transaction Manager - responsible for transaction privacy. The manager stores and allows access to encrypted transaction data, which are exchanged in encrypted payloads with other participants by transaction managers; however, they do not have access to the sensitive private key.
• Enclave which works in conjunction with the transaction manager to strengthen the privacy by managing the encryption and decryption in an isolated way. Private keys are stored in encalve which is essentially a virtual Hardware security module(HSM), an encryption method.

Quorum will not only attain privacy, the system will also be able to benefit from the gas-less transaction, which will make it cheaper for all users who transact using the system. Last, scalability and performance, which are important quality attributes for enterprise applications [13].

5 Conclusion

In this paper, we proposed a parking management system for business entities focusing on offering parking services, while the systems focus on privacy, performance and scalability [2]. The aforementioned features are achieved by JPMorgan Quorum which is based on Etherium blockhain decentralised network [4].

The parking system will allow the six entities to transact with ease without worrying about the cost and security of their data. Service provider will not have concerns about their business strategies being exposed to the public blockchain network for competitive advantage. Since all data are encrypted before being transmitted on the network, and only the entities with access can use their private key to decrypt and view their data. Future research will include the implementation of the system and identifying ways on how IoT devices can be configured to send information considering that there will be more entities involved in using different devices and manufacturers.

Appendix

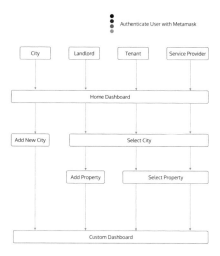

Fig. 3. Parking system workflow

References

1. Al Amiri, W., et al.: Towards secure smart parking system using blockchain technology. In: 2020 IEEE 17th Annual Consumer Communications and Networking Conference (CCNC), pp. 1–2. IEEE (2020). https://doi.org/10.1109/CCNC46108.2020.9045674
2. Baliga, A., et al.: Performance evaluation of the quorum blockchain platform. arXiv preprint arXiv:1809.03421 (2018)
3. Cao, N., et al.: Privacy-preserving multi-keyword ranked search over encrypted cloud data. IEEE Trans. Parallel Distrib. Syst. **25**(1), 222–233 (2013)
4. Morgan Chase, J.P.: Quorum whitepaper (2016)
5. Web3 Foundation: Web3 foundation — web3 foundation nurtures and stewards technologies and applications in the fields of decentralized web software protocols. https://web3.foundation/
6. Hu, J., et al.: Parking management: a blockchain-based privacy-preserving system. IEEE Consum. Electron. Mag. **8**(4), 45–49 (2019). https://doi.org/10.1109/MCE.2019.2905490
7. Facebook Inc.: React – a JavaScript library for building user interfaces. https://reactjs.org/
8. Ison, S., Budd, L.: Parking. In: Handbook on Transport and Urban Planning in the Developed World. Edward Elgar Publishing (2016)
9. Magee, C.C.: Design fob a parking meter device (1936)
10. Marshall, J.D.: Urban land area and population growth: a new scaling relationship for metropolitan expansion. Urban Stud. **44**(10), 1889–1904 (2007)

11. McCoy, K.: Drivers spend an average of 17 hours a year searching for parking spots. https://www.usatoday.com/story/money/2017/07/12/parking-pain-causes-financial-and-personal-strain/467637001/
12. Metamask: A crypto wallet & gateway to blockchain apps. https://metamask.io/
13. Niemi, E.: Quality attributes for enterprise architecture processes. Enterp. Architect. **8** (2013)
14. ParkMark. Brief history of car parks. http://www.parkmark.co.uk/brief-history-of-car-parks
15. Porter, J.: Gmail, Youtube, Google Docs, and other Google services hit by massive outage - the verge. https://www.theverge.com/2020/12/14/22173803/gmail-youtube-google-assistant-docs-down-outage
16. Singh, P.K., et al.: Smart contract based decentralized parking management in its. In: Lüke, K.-H., et al. (eds.) Innovations for Community Services, pp. 66–77. Springer, Cham (2019). ISBN 978-3-030-22482-0
17. Skeldon, R.: Rural-to-urban migration and its implications for poverty alleviation. Asia-Pac. Popul. J. **12**(1), 3 (1997)
18. Suite, T.: Truffle — working with quorum — documentation — truffle suite. https://www.trufflesuite.com/docs/truffle/distributed-ledger-support/working-with-quorum
19. Suites, T.: Boxes — truffle suite. https://www.trufflesuite.com/boxes
20. Tacoli, C., McGranahan, G., Satterthwaite, D.: Urbanization, poverty and inequity: is rural-urban migration a poverty problem or part of the solution. In: The New Global Frontier: Urbanization, Poverty and Environment in the 21st Century, vol. 43 (2008)
21. Wang, C., et al.: Privacy-preserving public auditing for data storage security in cloud computing. In: 2010 Proceedings IEEE INFOCOM, pp. 1–9. IEEE (2010)
22. Wang, L., et al.: Towards Airbnb-like privacy-enhanced private parking spot sharing based on blockchain. IEEE Trans. Veh. Technol. **69**(3), 2411–2423 (2020). https://doi.org/10.1109/TVT.2020.2964526
23. Zinonos, Z., et al.: Parkchain: an IoT parking service based on blockchain. In: 2019 15th International Conference on Distributed Computing in Sensor Systems (DCOSS), pp. 687–693 (2019). https://doi.org/10.1109/DCOSS.2019.00123

The General Data Protection Regulation and Log Pseudonymization

Artur Varanda[1], Leonel Santos[2], Rogério Luís de C. Costa[3(✉)],
Adail Oliveira[3], and Carlos Rabadão[2]

[1] ESTG, Polytechnic of Leiria, Leiria, Portugal
`artur.varanda@pj.pt`
[2] CIIC, ESTG, Polytechnic of Leiria, Leiria, Portugal
`{leonel.santos,carlos.rabadao}@ipleiria.pt`
[3] CIIC, Polytechnic of Leiria, Leiria, Portugal
`{rogerio.l.costa,adail.oliveira}@ipleiria.pt`

Abstract. The General Data Protection Regulation (GDPR) demands the use of various protection levels to ensure that personal data meet information security requirements. One of the techniques the GDPR recommends to protect personal data is pseudonymization, which consists of replacing real data with pseudonyms. Although databases and documents contain much of the personal data that should be protected, several types of log files contain data (like IP addresses, e-mails, and usernames) that may lead to the (direct or indirect) identification of a person. Therefore, log files must also be processed to achieve regulatory compliance with the principle of accountability imposed by the regulation. In this work, we deal with the pseudonymization of log data. We identify and discuss pseudonymization strategies in terms of the log processing phase and management architecture. We experimentally evaluate such strategies using three implementation alternatives, providing conclusions and helpful insights on their usage.

Keywords: Log pseudonymization · Privacy · GDPR

1 Introduction

The General Data Protection Regulation (GDPR) is directly applicable in each EU Member State and ensures a high degree of harmonization of data protection between countries. It requires the implementation of measures to ensure the principle of accountability, which makes the data controller responsible for what he does with "personal data". Personal data is mostly stored in documents and databases. Nevertheless, it is commonly possible to identify files at the system administration level that store data that may lead to the (direct or indirect) identification of a person. For instance, firewalls, routers, VPN servers, operating systems, and applications usually create log files that may store Internet Protocols (IP) addresses, e-mail addresses, and usernames. Hence, as system logs

L. Barolli et al. (Eds.): AINA 2021, LNNS 227, pp. 479–490, 2021.
https://doi.org/10.1007/978-3-030-75078-7_48

commonly store several personal identifiers, they must be processed to achieve regulatory compliance. Some approaches balance privacy and data usability to meet the original purpose of log collection and analysis. These approaches consist of irreversible encoding methods through collision-resistant fingerprint functions to ensure privacy [4].

The GDPR explicitly recommends pseudonymization for assuring the privacy of critical data while enabling the easy processing of personal data beyond its original purposes. This technique considerably reduces the likelihood of identifying an individual from the logs by replacing the original identifiers with masked identifiers.

In this paper, we deal with log pseudonymization. We identify the main phases of the log processing pipeline, discuss the pros and cons of pseudonymizing in each of the identified pipeline phases, describe a generic architecture for log pseudonymization, and experimentally evaluate implementation alternatives.

The next section reviews background and related work. Then, Sect. 3 discusses log pseudonymization strategies. Section 4 presents the experimental evaluation. Finally, Sect. 5 concludes the paper.

2   Background and Related Work

The GDPR specifies rules regarding personal data processing and requires that personal data controllers implement appropriate measures to comply with data protection principles [16]. It also obliges entities handling personal data to make a major operational reform. Therefore, controllers must apply effective technical and organizational measures to ensure a risk-appropriate level of security, which include:

- pseudonymization and/or encryption of personal data;
- ability to ensure the ongoing confidentiality, integrity, availability, and resilience of treatment systems and services;
- ability to re-establish availability and access to personal data promptly in the event of a physical or technical incident;
- processes for regularly testing, assessing, and evaluating the effectiveness of the measures taken to ensure the secure processing of information.

Systems are typically configured to collect and store access, error, and security audit logs. All these logs contain personal data. Although personal data cannot be collected or stored without the consent of the holders, logs may be collected for the legitimate purpose of detecting and preventing fraud and unauthorized access and to ensure system security (GDPR recital 49).

According to the GDPR (Article 4(1)(a)), "personal data" is any information related to an identified or identifiable living individual. Also, information collected that may lead to the (direct or indirect) identification of a person is considered personal data. Personal data that is pseudonymized or encrypted but that can be used to re-identify a person remains personal data. On the other hand, personal data that is anonymized to make the holder unidentifiable is not considered personal data. For data to be truly anonymized, the process must be irreversible.

An individual would be identified within a group when someone could distinguish him from all other members of the group [1]. Hence, distinct data may be used together to turn an identifiable person into an identified person [1]. The GDPR, in recital 30, exemplifies that IP addresses may be combined with other information received by the servers and used as part of a process of profile creation and natural person identification. In such context, IP addresses would be "personal data".

2.1 Log Format and Management

Systems and applications generate logs of various types, including those originated by e-mail servers, web servers, and database servers. These logs can provide useful information to troubleshoot equipment problems and about individuals who perform attacks on system components. The main logging sources are [9] security software (e.g., anti-malware, IDS, firewalls, routers, VPN servers), operating systems (system events and audit logs), and applications (e.g., e-mail servers, web servers, and databases). But log files can also provide information to hackers and cybercriminals who can use them to compromise the system and data security.

Log Format. Log format and syntax define how these messages are formed, transported, stored, monitored, and parsed. In the simplest case, logged events can be treated as text strings, and log analysis can be done by simple text search. In automatic event analysis processes, log generators and users must agree on the syntax used. Examples of such protocols are the Syslog [3] or the W3C Extended Log Format [5]. Each field of a formatted event log contains the representation of certain information. Unfortunately, many logs are not (totally) conform to any specific or predetermined standard, so they are considered free text [2]. Log data may be in plain text or binary files (e.g., Windows Event Log file [6]). In the latter format, conversion tools (e.g., Windows Event Viewer tool) must be used to access the information, converting binary data into readable events.

Log Management. A few years ago, centralized log aggregation simply consisted of a storage host running a service (e.g. *syslogd*) that collects environment logs into a folder [17]. If there is a local system vulnerable, these files are exposed and may be corrupted. Nowadays, Security Information and Event Management (SIEM) is a security solution for log management with additional functionalities [7]. Currently, a typical log management infrastructure consists of hardware, software, networks, and devices that generate, transmit, store, analyze, and discard data. Usually, the infrastructure is divided into three levels: generation, analysis, and presentation. The first level contains the hosts that generate the logs. Agents running on such hosts collect logs and forward them using messages via TCP or UDP. The second level consists of at least one log server that receives and stores logs from first level hosts. Second level servers contain a log management platform that collects logs from their operating system and also from client hosts. The third level consists of applications used to view or analyze the logs and/or statistics. Such architecture brings several benefits in terms of record aggregation, indexing and efficient searching, data loss prevention, data analysis, and monitoring and alerting.

2.2   Related Work

Sonntag [14] uses sequential hashing, combining a *forward chain* and a *backward chain* of keys to create pseudonymized data from log files. First, a key K is used to encrypt all considered data entries in a certain period. In the next period, the hash of the original key K is used as the new key to encrypt data. Then, in a third time slot, the new key is the hash of the one used in the second period (i.e. the hash of the original key K), and so on. The author states that a new encryption key $K2$ should be chosen for the last time slot and a *backward chain* of keys should be used (i.e. the last but one time slot would use the hash of $k2$, and the key of the last but two slots would be formed by the hash of $k2$, and so on). This approach has usability and performance limitations, as the number of entries to be encrypted must be known in advance and the consecutive use of hash functions may degrade performance.

Portillo-Dominguez and Ayala-Rivera [13] describe SafeLog, which is a component used in cloud-based log management environments. SafeLog is used (as a proxy) to filter and anonymize the log data that is sent to the cloud. Its main components include (i) a catalog of keywords and attributes (based on laws and other legal instruments), (ii) a pattern matcher (over user-specified rules), and (ii) a module that applies some de-identification techniques, like tokenization, hashing, and aggregation. In [15], authors evaluate, in terms of a *risk score*, the use of tokenization, suppression, and generalization to transform personal data. Kasem-Madani et al. [8] propose a pseudonymization toolkit for semi-structured data. It takes input data and a set of pseudonymization rules (specified in an XML-based language) and generates a pseudonymized representation of input data in XML format. Neumann et al. [12] claim that it is not trivial for a non-expert to choose the best pseudonymization techniques and parameters to use, and propose a software framework that measures privacy risk in multi-actor systems to support the selection of pseudonymization strategies by non-expert users.

3   Log Processing, Pseudonymization and Management

The centralized log management architecture has several benefits regarding log aggregation, searching, accessing, security, monitoring and alerting, and analysis. In such architecture, the typical phases of the log processing pipeline are *generation*, *ingestion*, *indexation*, *storage*, and *presentation*. To use a log pseudonymization solution in such architecture, one must intervene in one or more of the processing phases, and the most important aspects to consider when choosing the phase on which to use pseudonymization are:

1 - Log Generation. Processing pseudonymization in the log generation phase (by applications or systems) involves developing independent solutions for each application. On the other hand, there is no need to take additional measures at the pseudonymization level to ensure compliance with the regulation, which frees the infrastructure manager from liability.

2 - Log Ingestion. Pseudonymization processing in the ingestion phase (i.e., before the logs are indexed and stored) is done by a transformation module internal to the log management infrastructure. Usually, a user-developed transformation module can be used, enabling the adoption of a wide range of design options and pseudonymization functions. On the other hand, it requires specific code to process inputs of different types, mainly by defining field extraction patterns.

3 - Index duplication. The use of index duplication for pseudonymization implies a duplication of archived messages, one indexing the original values and the other the pseudonymized values. If messages are replicated on the hosts that generate them, network traffic will also double in the transport between clients and the log management server. The index search over original data or pseudonymized data is equally fast and efficient, which is the great advantage of this method.

4 - Log presentation. Pseudonymization in this phase is done by applying appropriate filters to the page returned by the search engine. Messages aren't pseudonymized before archiving, leading to a much faster and efficient ingestion process. Also, for a user with privileges to access identifying data, searching is much more efficient because there is no need for re-identification. A limit on the queries available for users that haven't access to identifying data must be imposed. Otherwise, performance may degrade during search results exhibition, as information has to be pseudonymized before it is presented.

Due to the complexity of generating specific code to process each type of input, the use of pseudonymization during log generation is unfeasible in real-world large infrastructures. Hence, our centralized GDPR-compliant architecture for log management, storage, and analysis, receives logs of various formats and uses pseudonymization in the management server. Data containing identifiable values are replaced by pseudonyms using summary functions. A rainbow table is used for eventual re-identification.

Figure 1 presents some common log sources (e.g., web servers and operating systems) and the main components of the considered architecture, including (i) a log collection tool which collects logs from several sources and clients and transfers them to the log management server, (ii) local log collection and forwarding agents and tools collect logs from the management server host and also forward incoming messages to the central management platform, and (iii) a log management platform, which does field configuring and extraction, messages transformation, data pseudonymization and second index creation, and also provides searching, browsing and reporting capabilities. In the log management module, logs with identified personal data are pseudonymized and sent to an index. At the same time, a second index may be created with the original values to allow re-identification.

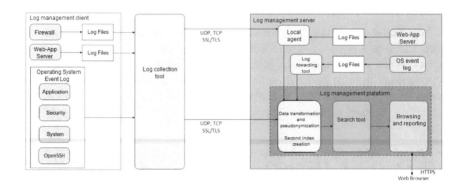

Fig. 1. Main components of the log management and pseudonymization plataform

4　Experimental Evaluation

We experimentally evaluated pseudonymization alternatives using three management platforms: Graylog, Splunk, and ELK Stack. The log management server runs Linux Ubuntu 16.04 (64 bits) and client machines run Windows 7. As log sources, we used the operating systems' events logs (including the SSH server log), web and application servers (i.e. Microsoft IIS and Apache HTTP Server) access logs, and firewall logs for the Linux server (i.e. UFW firewall). We used NXLog [11] as the log collection tool and Linux *Syslog*[1] to access logs on Ubuntu host.

4.1　Graylog-Based Management

Graylog is an open-source log management platform that uses the MongoDB database management system to store its configurations. It uses *Elasticsearch* to index, store and search log messages.

In this scenario, NXLog collects the messages from remote log sources and forwards them to Graylog Input Manager. Messages from the local operating system are collected and forwarded to the input manager using *Syslog*. The Graylog Input Manager must be configured to receive messages of distinct types. For messages from Windows-based hosts, we used GELF TCP, while for Syslog messages we used the Syslog UDP protocol.

Messages are analyzed by the field extractors (we used **grok** patterns) just after passing through the input manager (and before being indexed). In Graylog,

---

[1] https://tools.ietf.org/html/rfc5424.

all received messages are part of a *root stream* and may also be organized as part of other streams based on user-specified rules. Then, messages that belong to a certain stream may be processed in *pipelines*. The pseudonymization of log messages using pipelines can be done using one of the strategies represented in Fig. 2: (i) duplication of indexes and (ii) creation of re-identification messages.

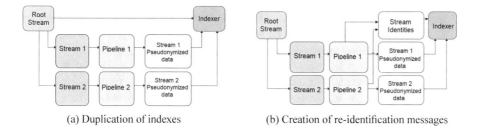

(a) Duplication of indexes   (b) Creation of re-identification messages

Fig. 2. Pseudonymization and multiple streams

Duplication of Indexes - Messages from different sources are sent by NXLog in duplicate, to allow pseudonymization by duplicating indexes. The root stream is made up of duplicate sets of messages. One of them goes directly to the indexer. The other is classified into several streams that will be the inputs of distinct pipelines. In each pipeline, data is transformed according to its own rules. The identifying data is pseudonymized using hash functions and sent to outer streams, which are then ingested by the indexer.

Creation of Re-identification Messages - Re-identification messages (composed of pairs corresponding to the identifying values and the respective summaries) are created and are forwarded to identity streams. These streams function as identity files and are only accessible to users with a privileged profile, to allow re-identification under the terms stipulated by the data protection regulation.

The Duplication of indexes strategy duplicates messages at the outgoing hosts, doubling the volume of data in transport and storage, but allows searching for identities in a non-pseudonymized index, whose access is exclusive to a privileged profile. On the other hand, the creation of re-identification messages and their forwarding to an identity file involves fewer resources, both in terms of traffic and storage, but prevents identity searches.

4.2 ELK Stack-Based Management

The ELK Stack is a set of open source modules consisting essentially of *Elasticsearch*, *Logstash* and *Kibana*. *Elasticsearch* allows to perform complex analyzes in real-time and functions as an indexer, a search engine and event analyzer. Kibana is the web interface of ELK Stack. This module connects to *Elasticsearch* and allows to create real-time dashboards, as well as configure resources and manage the stack as a whole. Logstash is the module responsible for receiving messages from collection and forwarding agents (e.g. *NXLog*), extracting their fields, transforming them and forwarding them to *Elasticsearch*. Beats are message collection modules from various sources: files (*Filebeat*); network elements (*Packetbeat*); Metrics (*Metricbeat*); Windows events (*Winlogbeat*); and more. These modules forward messages to *Logstash* for transformation or send them directly to *Elasticsearch* without being processed.

All the messages from windows-based systems are sent to the log server via NXLog [11] agent to server-side *Logstash*. It also receives logs from Linux-based systems and from the system itself.

Logstash receives the messages and transforms them using the module functions. Logs with identifying data are pseudonymized and sent to an index. At the same time, a second index is created with the original values to allow re-identification. Indexed logs can then be searched and viewed through *Kibana* using a Web Browser.

Input Configuration. The *Logstash* pipeline has two required blocks, the *Input* block and the *Output* block; and an optional block, which is the *Filter* block. The Input block receive messages from a variety of sources (e.g. Filebeats or TCP and UDP ports). The *Filter* block transform messages according to a set of rules. Finally, the *Output* block send messages to a destination, usually an index from *Elasticsearch*. These blocks are configured by editing a configuration file (i.e. *logstash.conf*).

Although it may be necessary to process several types of entries according to specific rules, only one pipeline can be used in *Logstash*. Each message can then be identified with a unique type through the type field. In the *filter* block, each filter set to apply to each message type can be defined using flow control operators. Similarly, in the *output* block, flow control operators can be used to forward each type of message to a specific destination.

Field Extraction. Field extraction is done by using the **grok** filter in the *Logstash* pipeline. In case of failure when extracting a message, the **grok** filter automatically adds a tag with the identifier *_grokparsefailure*. If we want to discard these messages from the *Logstash* pipeline, we can use the **drop** filter. This message purge is especially useful when we want to pseudonymize some information. A message that is badly handled by the **grok** filter may send personal data to the indexer in the *_raw* field since pseudonymization filters are only applied to extracted fields.

Pseudonymization with the ELK Stack. We now exemplify the pseudonymization processes of Apache web server access messages in the *Logstash* pipeline. We configure inputs (as described above) and label them as *win_apache*. Then, we create copies of input messages (labeled as *win_apache_pairs*). Consider we want to pseudonymize the *c_ip* field, corresponding to the IP address of the client accessing the web server. Original messages will be pseudonymized and cloned messages (*win_apache_pairs*) will lead to re-identification messages.

The **fingerprint** filter replaces the contents of the *c_ip* field with its summary resulting from the *HMAC SHA256* function using the *HMAC_SECRET_KEY* key. The **mutate** filter uses the replace option to replace the contents of the *ShortMesage* field with a similar string, but with an IP corresponding to the *c_ip* field, replaced by the field name. In other words, the client IP address in the full message is replaced by the string "c_ip". The **fingerprint** filter is also applied to pseudonymized messages. The only difference is that the original value is kept in the *c_ip* field. On the other hand, the corresponding summary is stored in the new *hash* field.

As the index contains values from many sources other than the web server, the field with the original data should have a generic name instead of "c_ip". Thus, the new *value* field is created with the mutate filter with the *add_field* option and stores the contents of the *c_ip* field. Since the re-identification message is intended to contain only the *value* and *hash* fields, the **prune** filter is used to discard the remaining fields. The *type* field is kept because it is used in the output configuration.

Finally, a second **fingerprint** filter is used to label the message with a *fingerprint_id* resulting from the *MurmurHash* summary function applied to the value contained in the *hash* field. Messages that have the same *fingerprint_id* correspond to identical re-identification messages, with the same contents in the *value* and *hash* fields. *Win_apache_pairs* messages are sent to the *logstash-pairs* index and stored at the address corresponding to the contents of the *fingerprint_id* field. This prevents duplication of re-identification messages in the index. The contents of the *logstash-pairs* index are only accessible to users with a restricted profile. Finally, pseudonymized messages are sent to the *logstash-main* index.

4.3   Splunk-Based Management

Splunk uses its indexing and searching engine. Its web interface also provides the searching and reporting capabilities and may be used to extract fields from messages (using regular expressions or delimiters). NXLog messages sent to the server are not formatted. TCP ports are used for messages incoming from Windows systems, while UDP is used for messages from the server's Syslog. Using Splunk, we evaluated the use of pseudonymization in log ingestion and visualization phases.

Pseudonymization in the Log Presentation Phase - The Splunk search language consists of five components: search terms, commands, functions, arguments, and clauses. We created code for querying expressions to identify and pseudonymize fields of interest. The message displayed in the search result is identical to the original, but with the values of personal data fields replaced by pseudonymized data.

Pseudonymization of data in the presentation phase uses pre-defined search expressions. Also, it makes necessary to create a profile limited to predefined search expressions. Alternatively, a profile can be configured with exclusive access to monitoring panels (dashboards) that present results of scheduled queries periodically or in real-time. In any case, the pseudonymisation solutions in the presentation phase imply a limitation for the user who is limited in terms of search options.

Pseudonymization in the Log Ingestion Phase - In this method, the records are indexed and archived with pseudonymized identification values. The duplication of the indexes (maintaining the original values in one of them) is required to enable the searching and analysis of data. Alternatively, a table of correspondence can be created between identities and their summaries. In any case, access to data must be restricted to met information security requirements imposed by the regulation.

The transformation of the records in the ingestion phase is done through the configuration of four files (*inputs.conf, props.conf, transforms.conf* and *fields.conf*). The first file contains configuration related to incoming messages. In the second, the transformation steps performed in the ingestion phase are declared. The operations performed in each of the transformation steps are defined in the third file. The fourth file contains the declaration of the new fields to be indexed.

The pseudonymization in the ingestion phase can make the ingestion process time-consuming since messages have to be processed before indexing, and increases the delay to making messages available. However, it makes queries much more efficient, since the data have already been transformed in the ingestion phase. Ingestion processing can be accelerated in Splunk using customized modules based on user-developed routines. The use of such modules also allows the use of complex functions that integrate, for example, HMAC and symmetric encryption functions.

4.4   Discussion

Evaluated applications have distinct characteristics and each of them is more suitable for using a certain pseudonymization method or implementing pseudonymization at a certain phase of the log processing pipeline. In Graylog, field data transformations are possible by configuring pipelines, but the hashing functions available are limited. For instance, the HMAC function is not available but it is possible to implement this algorithm using the existing functions. In Splunk, the functions can only be applied to internal fields of the server, which causes delays in the ingestion process. The ELK Stack solution integrates a large number of processing filters and was the only one that made it possible to simultaneously implement the pseudonymization of data through the use of an HMAC function and the use of an identity file in a separate index with restricted access.

Using data pseudonymization in the presentation phase leads to an efficient ingestion process, but requires a significant effort in configuring pre-defined search expressions and profiles. On the other hand, pseudonymization in the ingestion phase increases the delay in making messages available but significantly improves queries' performance. Pseudonymization in the duplication of indexes increases resource consumption but enables queries for identities in a non-pseudonymized index, whose access should be exclusive to privileged profiles. Resource consumption can be reduced through the creation of re-identification messages and identity files, but it prevents identity searches.

5 Conclusions and Future Work

The GDPR imposes the principle of accountability to data controllers, making them responsible for what they do with "personal data". Due to the presence in log files of a large amount of data that can be used to identify persons (directly or indirectly), log data must be processed in a way to achieve regulatory compliance.

This work deals with log pseudonymization. We identified log pseudonymization strategies in terms of their use on the log processing pipeline (i.e., record generation, log ingestion, duplication of indexes, and log presentation) and discussed the pros and cons of each strategy. A generic architecture for centralized log management and pseudonymization was described and experimentally evaluated using three log management solutions. Such architecture functions as an aggregator of log messages from various sources, where pseudonymization is done through transformation functions applied to the log processing pipelines. Documents composed of original values and their corresponding pseudonyms are sent to a restricted access index, to allow re-identification. We demonstrated the proposed architecture for the pseudonymization of event logs is quite versatile. The prototype developed using ELK Stack enables the use of two important features: pseudonymization of data using an HMAC function [10] and re-identification messages stored in a separate restricted index (which may be used in exceptional cases, duly authorized, to comply with the regulation).

The event log generating sources and their data structures are immense and very diverse. As future work, we intend to refine the parsing and field extraction processes for other log sources, developing ways to transform the original messages into easy to process messages.

Acknowledgements. This work is partially funded by National Funds through the FCT (Foundation for Science and Technology) in the context of the project UIDB/04524/2020.

References

1. Bolognini, L., Bistolfi, C.: Pseudonymization and impacts of big (personal/anonymous) data processing in the transition from the directive 95/46/EC to the new EU general data protection regulation. Comput. Law Secur. Rev. **33**(2), 171–181 (2017)
2. Chuvakin, A., Schmidt, K., Phillips, C.: Logging and Log Management: The Authoritative Guide to Understanding the Concepts Surrounding Logging and Log Management. Syngress Publishing, Rockland (2012)
3. Gerhards, R.: The syslog protocol. RFC 5424, pp. 1–38 (2009)

4. Ghiasvand, S., Ciorba, F.M.: Assessing data usefulness for failure analysis in anonymized system logs. In: 17th International Symposium on Parallel and Distributed Computing (ISPDC-2018) (2018)

5. Hallam-Baker, P.M., Behlendorf, B.: Extended log file format: W3C working draft WD-logfile-960323 (1996). https://www.w3.org/TR/WD-logfile.html. Accessed 04 Apr 2021

6. Ham, J., Davidoff, S.: Network Forensics: Tracking Hackers Through Cyberspace. Prentice Hall, Hoboken (2012)

7. Harper, A., VanDyke, S., Blask, C., Harris, S., Miller, D.: Security Information and Event Management (SIEM) Implementation. McGraw-Hill, New York (2010). https://doi.org/10.1036/9780071701082

8. Kasem-Madani, S., Meier, M., Wehner, M.: Towards a toolkit for utility and privacy-preserving transformation of semi-structured data using data pseudonymization. In: Data Privacy Management, Cryptocurrencies and Blockchain Technology, pp. 163–179 (2017)

9. Kent, K., Souppaya, M.P.: SP 800-92. guide to computer security log management. Technical report, Gaithersburg, MD, USA (2006)

10. Krawczyk, H., Bellare, M., Canetti, R.: RFC2104: HMAC: keyed-hashing for message authentication (1997)

11. NXLog Ltd. NXLog community edition reference manual (2018). https://nxlog.co/docs/nxlog-ce/nxlog-reference-manual.html

12. Neumann, G.K., Grace, P., Burns, D., Surridge, M.: Pseudonymization risk analysis in distributed systems. J. Internet Serv. Appl. **10**(1), 1–16 (2019)

13. Portillo-Dominguez, A.O., Ayala-Rivera, V.: Towards an efficient log data protection in software systems through data minimization and anonymization. In: Proceedings - 2019 7th International Conference in Software Engineering Research and Innovation. CONISOFT 2019, pp. 107–115 (2019)

14. Sonntag, M.: Pseudonymizing log entries with time-selective disclosure. In: Workshops der INFORMATIK 2018 - Architekturen, Prozesse, Sicherheit und Nachhaltigkeit, Köllen Druck+Verlag GmbH, Bonn, pp. 119–127 (2018)

15. Tachepun, C., Thammaboosadee, S.: A data masking guideline for optimizing insights and privacy under GDPR compliance. In: Proceedings of the 11th International Conference on Advances in Information Technology, pp. 1–9 (2020)

16. The European Parliament and the Council of the European Union: Regulation (EU) 2016/679 of the European parliament and of the council of 27 April 2016 on the protection of natural persons with regard to the processing of personal data and on the free movement of such data, and repealing directive 95/46. The Official Journal of the European Union **59**, 1–88 (2016)

17. Turnbull, J., Matotek, D., Lieverdink, P.: Pro Linux System Administration. Apress, New York (2009). https://doi.org/10.1007/978-1-4302-1913-2

A Review of Scaling Genome Sequencing Data Anonymisation

Nikolai J. Podlesny[(✉)], Anne V. D. M. Kayem, and Christoph Meinel

Hasso-Plattner-Institute, Potsdam, Germany
{Nikolai.Podlesny,Anne.Kayem,Christoph.Meinel}@hpi.de

Abstract. Sequencing genomes and analysing their variations can make an essential contribution to healthcare research on drug discovery and advancing clinical care, for instance. Genome sequencing data, however, presents a special case of highly sparsely populated, multi-attribute, high-dimensional data, in which each record (tuple) can be associated with more than tens of thousands of attributes on average. Since anonymising genome sequencing data is a necessary pre-processing step for privacy-preserving genomic data analysis for personalised care, discovering all the quasi-identifier combinations required to preserve anonymity is essential; This requires verifying an exponential number of quasi-identifier candidates to identify and remove all unique data values, an NP-hard problem for larger datasets. Furthermore, recent work classifies this problem to be at the very least W [2]-complete and not a fixed-parameter tractable problem. Thus, achieving efficient and scalable anonymisation of genome sequence data is a challenging problem. In this paper, we summarise the uniqueness of ensuring privacy in the context of (whole) genome sequencing. Further, we show and compare the latest trends to discover quasi-identifiers (QID) in large-scale genome data and concepts to counter the exponential runtime growth during QID candidate processing in this field. Finally, we present an architecture incorporating previous enhancements to enable near real-time QID discovery in high-dimensional genome data based on vectorised GPU-acceleration. Achieving anonymisation processing in our experiments in just a few seconds, which corresponds to speedups by factor 100, can be essential in life-or-death situations like triage.

1 Introduction

Sequencing genomes and analysing patients' genomic variations contribute to personalised medicine and research on vaccine discovery, triage, and advancing clinical care. The technical achievement of (whole) genome sequencing was denominated repetitively as an enormous promise for the greater public good [1] and proved its advancements during the past global pandemic of COVID-19 [2]. Simultaneously, the gained insights in individuals' genomics currently come with fear and cost of its patient's privacy. In contrast to standard high-dimensional datasets from commercial and web data, typically used for experimentation

L. Barolli et al. (Eds.): AINA 2021, LNNS 227, pp. 491–501, 2021.
https://doi.org/10.1007/978-3-030-75078-7_49

and testing anonymisation algorithms, genome sequence data presents a worst-case scenario of high-dimensional data. Data gathering techniques often produce highly sparsely populated, multi-attribute, high-dimensional whole-genome sequence data that can have up to thousands of describing attributes per record [3]. Protecting this information from private data exposure when sharing for research, triage, or other means is fundamental and an integral part of our highest ethical standards. Yet, established strategies and algorithmic approaches for privacy-preserving data processing and publishing have been summarised as impractical to ensure privacy for these special case data settings [1]. This classification comes back to the NP-hard nature of syntactic data anonymisation techniques and the necessity to verify the exponential number of quasi-identifier candidates. The situation that tens of thousands of attributes need to be checked links back to the algorithmic enumeration problem, yet necessary to find quasi-identifiers that endanger patient privacy. Achieving efficient and scalable anonymisation of genome sequence data is thus a challenging problem.

Contributions. We pick up the open problem of privacy-preserving publishing of patients genome data and make two contributions in this paper:

- summarise, show, and compare latest trends to discover quasi-identifiers (QID) in large-scale genome data and concepts to counter the exponential runtime growth during QID candidate processing in this field
- present an architecture incorporating previous enhancements to enable near real-time QID discovery in high-dimensional genome data based on vectorised GPU-acceleration

Outline of the Paper. The rest of the paper is structured as follows. We discuss related work in Sect. 2. In Sect. 3, we offer characteristics of genome sequence data and discuss the matter of quasi-identifiers (QID) in genome sequence data. Section 4 addresses scaling patterns for conducting QID discovery on a large-scale and presents the latest concept of accelerating the QID search. We present and compare results from different QID search implementations to enable genome sequence data anonymisation in Section. Section 6 summarises the contributions of this paper and discusses avenues for future research in genome data anonymisation.

2   Related Work

Whole-genome sequencing is the process of determining the order of nucleotides, and their four bases (adenine, guanine, cytosine, and thymine) in a given organism [4]. After comparing individuals' sequence against a standard, variations – known as single-nucleotide polymorphism (SNPs) – are marked to determine the delta. Such processing results in detailed information on the organism's DNA, which can serve as an essential basis for health-related research in diagnostics, prevention, and exploration. However, as McGuire et al. [5] have pointed out, whole-genome sequencing research raises several ethical considerations such as genetic discrimination, psychological impacts due to inadvertent information exposure, and loss of anonymity. Especially in the health sector and on a genome level, privacy compromises might have dramatic consequences for patients like the de-anonymisation of US Governor William Weld's medical information [6], the exposure of tens of thousands of private health data that included patient names, dates of birth, social security numbers, lab results, and diagnostics data [7].

Already after the completion of the first whole-genome sequences, McGuire et al. raise privacy concerns, discuss complexities of informed consent, and outline the need for an empirical study on the effects of data sharing [5]. Naveed et al. offered survey insights on genome data privacy with biomedical specialists, characterise field-specific privacy problems and provide an enumeration of key privacy challenges in genomics like large scale datasets [8]. Based on a case study on Alzheimer's disease, Wagner substantiates the uniqueness of individual genomic specifics, their highly sensitive information and further offers measures for genomic privacy, metric selection, interpretation, and visualisation options [9]. Humbert et al. present reconstruction attacks based on statistical relationships between the genomic variants through graphical models and belief propagation to expose familial correlations [10]. The "US Presidential Commission for the study of bio-ethical issues" identified genome sequencing as clinical care advancement and enormous promise for the greater public good. Simultaneously, they address data privacy challenges and conclude impracticability for absolute privacy in whole-genome sequencing (WGS) [1].

Guaranteeing privacy technically is challenging since anonymising datasets is known to be NP-hard [11], W [2]-complete and not fixed-parameter tractable (FPT) [12] due to the underlying enumerative combinatorial nature [13]. Whole-genome sequencing (WGS) often produces thousands of attributes per record. Because the attribute values differ from per to person, it is possible to obtain highly sparsely populated, multi-attribute, high-dimensional datasets for genomic analysis with at least tens of thousands of describing attributes (columns) [3]. To avoid private data exposure like in the instance of US Governor William Weld's medical information [6], unique attribute values must be removed as part of the anonymisation processing. These unique attribute values are known as quasi-identifiers (QID) serving in combination with auxiliary data attackers as a link to draw a conclusion and derive private information [14]. There is not much research addressing the search for quasi-identifier in the environment of large-scale genome sequencing to the best of our knowledge. Malin et al. presented a method to protect genomic sequences data through generalisation lattices for smaller datasets with less than 400 sequences [15]. Chen et al. offer a scalable approach similar to the "seed-and-extend" method to outsource genome sequence mapping on low-cost cloud platforms [16]. While targeting a secure computation setup based on hashing and fingerprinting for data linkage, the variation mapping outcome may still serve as a quasi-identifier and be used for private data exposure. In the context of genome-wide association studies, Johnson et al. [17] offer privacy-preserving data mining algorithms introducing non-trivial amounts of random noise, which promise more accurate results on correlation analysis (e.g., SNP disease coherence). Yet, open questions remain towards scalability and practicability of sparsely populated, multi-attribute datasets. Kushida et al. offer a systematic literature review of 1798 prospective citations and conclude that current de-identification strategies have their limitations, full anonymisation is challenging, and further work is needed notably to protect genetic information [18].

As part of this work, we will pick up the previously raised concerns and stated problems with scalability and privacy, particularly data anonymisation for large scale genome sequencing data. To this end, we will provide some background on genome sequencing and the requirements for anonymising genome sequencing data successfully in the next section.

3   Genome Sequence Data Anonymisation

In recent years, however, the increased need to share genome sequence data with third-party data analytics service providers, for instance, fortify the issue of privacy. While whole-genome sequencing (WGS) and data anonymisation are well-explored research areas, their combination lacks practical insights.

Genome Sequence Data: A Characterisation. As part of genome sequencing, the order of nucleotides is determined. For this purpose, the whole genome is itemised in its four bases (adenine, guanine, cytosine, and thymine) [4]. Given this classification, a single whole-genome dataset can be coded as a tuple (POSITION, VALUE) for

$$0 < POSITION < 250M \ (INT), \ VALUE \in \{A, C, G, T, a, c, g, t, *\} \quad (1)$$

with CHAR (1). This may be represented in a traditional relational data schema combined with an arbitrary patient identifier. When processing nucleotides' order, the idea is to compare the actual genomic sequence to a reference case to identify substitutions for a single nucleotide at a specific position within the genome. Those substitutions or mutations are known as *single-nucleotide polymorphism (SNP)*. Current research suggests the existence of roughly 4 to 5 million SNPs in a person's whole genome making its combination, and especially its mutation composition unique [9,19]. Even the combination of four or fewer SNPs is often unique and can serve as a quasi-identifier (see Fig. 1a). Such uniqueness combined with auxiliary data might result in individual de-anonymisation, draw conclusions, and derive private information [6,14].

Anonymity Processing of Genomes Sequence Data. The combination of (rare) describing attributes like SNPs, disease history, intake adherence or drug subscriptions can form unique patterns. While anonymity is known as the quality of lacking the characteristic of recognisability or distinction, those unique patterns can be abused to re-identify data records to their original owner. Discovering and dissolving unique patterns through attribute combinations, also known as quasi-identifier, is one of the core principles in data anonymisation.

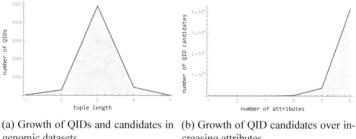

(a) Growth of QIDs and candidates in genomic datasets

(b) Growth of QID candidates over increasing attributes

Fig. 1. Genome sequencing data characteristics

In genomics, one of the most common mutations is SNP *C677T* within the *MTHFR* enzyme. *C677T* represents a single value mutation, where C (cytosine) is replaced by T (thymidine) at position 677 with the effect of decreasing the enzyme activity by up to 70%. While this particular SNP is relatively commonly represented in the population, combining this insight with several other non-related mutations makes the combination unique. Therefore, individuals' anonymity in a dataset shall be granted if no quasi-identifier remains.

Definition 1. Quasi-identifier
Let $F = \{f_1, .., f_n\}$ be a set of all attribute values and $B := \mathscr{P}(F) = \{B_1, .., B_k\}$ its power set, i.e. the set of all possible attribute value combinations. A set of selected attribute values $B_i \in B$, is called a quasi-identifier, if B_i identifies at least one entity uniquely and all attribute values $f_j \in B_i$ are not standalone identifiers.

During the search for quasi-identifiers (QID) all attribute value combinations as QID candidates need to be assessed which sums up to: $C_2(n) = \sum_{r=1}^{n} \binom{n}{r} = \sum_{r=1}^{n} \frac{n!}{(r!(n-r)!)} = 2^n - 1$ where n is the population of attributes (SNPs) and r the subset of n, while r must equal all potential lengths of subsets of attributes. Figure 1b delineates this exponential candidate growth. A proven and subtle strategy to discover quasi-identifiers is by iterating over all QID candidates, grouping by their attribute values set and counting each group size [20]. Standard aggregate functions like SQL92 can accomplish this. As soon as multiple grouped attributes have a group count, it serves as a quasi-identifier.

Hereinafter, architectural considerations required for realising whole genome sequencing will be considered, quasi-identifiers discovery in practice and on a large scale studied and subsequently data anonymisation activities assessed.

4   Scaling QID Search for Genome Sequencing in Practice

Distributing and scaling enumeration problems, like the discovery for quasi-identifiers in large-scale genome sequencing data, is reducible to high-performance computing. Typically, a search scheme can be scaled horizontal or vertical depending on several factors like response time, pricing, scalability demands and network I/O. For the alignment process of genome sequencing data to find mutations, a distributed horizontal architecture approach like Map-Reduce has been implemented in the real-world before [21].

While Map-Reduce is a well-established framework for cheap and easy scaling of computing resources [22], vertical scaling techniques combined with the latest in-memory concepts like dictionary encoding, reverse indices and optimises on L1–L3 cache hits for lightning-fast calculations and reduction of CPU cycles [23,24] proved their applicability in the past as well [25,26]. Figure 2 illustrates such architecture.

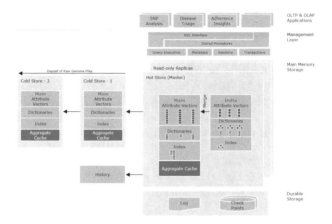

Fig. 2. Vertical scaling architecture

As mentioned before in Sect. 3, the QID search can be implemented relatively efficiently through grouping with standard aggregate functions in SQL92 considering a record *count* c of $c = 1$. As a result of this, any column combination which is unique for at least a single row serves as a quasi-identifier. If a group of attributes has less than two and more than none data records, this attribute combination serves as a quasi-identifier, respectively. As hundreds of millions of attribute combinations and therefore aggregation statements needed to be processed, there are options for optimising its execution to achieve the best query runtime possible. Grouping and aggregation statements are extremely efficient in column-wise storage settings with reversed indices in place [23], as not the entire dataset needs to be loaded, iterated and parsed. Instead, the courser can jump directly to the corresponding row in the virtual-file, appreciating a reverse-index hash table and allocating just the row's content representing the column values. Such mechanic reflects mainly in vertical scaling settings [24] on the query performance, as parallel aggregation, dictionary encoding and particularly the L1–L3 cache hierarchy significantly accelerates cache hits, reducing the cycle latency (see Table 1). These optimisation features not only fall in place for a single aggregation function but also for executing a variety of grouping statements with overlapping input variables. Here, the same dictionary encoding and intelligent sorting of aggregating tuples may significantly increase L1–L3 cache hits minimising CPU cycles for the entire query runtime. The dictionary itself already contains the count in its index table in the best case. Separately, parallel aggregation can also be applied for multiple grouping operations sharing the same (partial) memory, which again improves cache hits, making the main memory I/O the slowest bottleneck in the hardware chain. Similar observations have been explored by Kessler et al. while implementing k-anonymity, and local differential privacy as part of an enterprise database management system view principle [28].

Table 1. Cache hierarchy and CPU cycle latency [27]

Data source	Cycle latency (approx.)
L1 CACHE hit	4 cycles
L2 CACHE hit	10 cycles
L3 CACHE hit	Line unshared 40 cycles
L3 CACHE hit	Shared line in another core 65 cycles
L3 CACHE hit	Modified in another core 75 cycles
remote L3 CACHE	100–300 cycles
Local DRAM	60 ns

Recent technological enhancements offered the new perspective of leveraging GPU hardware to accelerate computationally intensive tasks through massive-parallelisation. GPUs offer a high compute density, with massive computations per memory access, high throughput, and increased latency tolerance. While GPUs memory capacities are still limited, it also lacks a similar large L1–L3 cache as with CPU architecture. Individual CPU cores are faster and smarter than a single GPU core instruction set. The sheer number of GPU cores and the massive parallelism they offer make up the single-core clock speed difference limited instruction sets. For this purpose, Braghin et al. work [29] can be extended to or rendered as vector operation using standard SQL-like aggregation with out-of-the-box libraries like the open data science framework cuDF, a RAPIDS Nvidia initiative that enables GPU acceleration. This way, given a single Tesla V100, the QID search can be parallelised on 5120 CUDA cores promising a massive runtime acceleration.

These trends and concepts will assess different scaling options in the following section.

5   Experiments

Our empirical model aims to compare QID discovery performance on different scaling and optimisation schemes for large high-dimensional genome datasets. To this end, we mimic real-world environments by employing the following hardware for our experiments.

Hardware. Our experiments were conducted on a GPU-accelerated high-performance compute cluster, housing 160 CPU cores (Xeon Gold 6140), 1TB RAM, and 10x Tesla V100 with a combined Tensor performance of 1120 TFlops. The execution environment for GPU related experiments will be restricted to 10x dedicated CPU core and a single, dedicated Tesla V100 GPU.

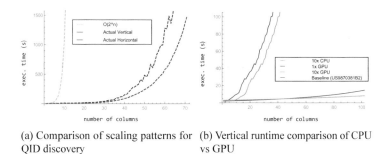

(a) Comparison of scaling patterns for QID discovery

(b) Vertical runtime comparison of CPU vs GPU

Fig. 3. Query runtime complexity for quasi-identifier (QID) discovery

Experiments. The evaluation is based on a synthetic genome dataset publicly available at github.com [30]. A variety of work exists exploring cache optimisation options [31] also particularly for aggregation functions on massively parallel processing systems [32,33]. We are particularly interested in the runtime performance uplift of interacting aggregation functions executed during the QID search scheme. Acknowledging the theoretical time complexity of $O(2^n)$ [13], Fig. 3a depicts the outcome of algorithmic runtimes with different scaling patterns using similar emulated environments. A tiny improvement can be derived from Fig. 3a in favour of vertical scaling. This is not unexpected as with vertical scaling, the network I/O is reduced and the aggregation based quasi-identifier (QID) search scheme benefits (see Sect. 4). A much larger acceleration can be seen in Fig. 3b, where based on the vertical scaling pattern GPU has been employed to execute the same search scheme in a vectorised manner. Using the available GPU resources, a massive decrease in computation is apparent. While the time complexity remains for the enumeration problem, parallelisation alleviates its runtime effects quite massively by factors of 100x in the given scenario.

Subsuming, we observe a light performance uplift as anticipated with vertical scaling patterns over horizontal ones for the same underlying dataset and similar hardware environment. We expect the query runtime uplift to be fortified with an increasing number of describing attributes. Simultaneously, main memory capacity is a natural limitation towards this approach's practicability, since as soon as swapping takes the place of cache performance, uplifts will collapse. Simultaneously, we observe a massive decrease in runtime for GPU accelerated clusters promising near real-time results to discover privacy endangering quasi-identifiers in high-dimensional datasets.

6   Conclusion and Future Work

Understanding the human genome through its sequencing is often referred to as the most valuable insight possible [1]. Doing it on a large-scale is a huge milestone for personalised care and digital health. In this work, the open problem of privacy-preserving publishing of patients' genome data has been considered. We presented and compared the latest trends to discover quasi-identifiers (QID) in large-scale genome data and discussed optimisation concepts to counter the exponential runtime growth during QID candidate processing in this field. Further, we present an architecture incorporating previous enhancements, including vectorised GPU-acceleration and showed that it enables near real-time QID discovery in highly sparsely populated, multi-attribute, high-dimensional genome datasets. The experiments confirm that implementing quasi-identifier discovery via standard aggregate functions in SQL92 is a practical solution for large-scale genome sequencing. Combined with GPU hardware, it permits query runtime improvements by more than 100x. As future work, conducting more extensive experiments would help identify re-identification risks concerning auxiliary data. In particular, real-world evidence on data anonymisation's side-effects on the conducted genome-disease correlation analysis would be exciting.

References

1. Gutmann, A., Wagner, J., Ali, Y., Allen, A.L., Arras, J.D., Atkinson, B.F., Farahany, N.A., Garza, A.G., Grady, C., Hauser, S.L., et al.: Privacy and progress in whole genome sequencing. Presidential Committee for the Study of Bioethical (2012)
2. Paden, C.R., Tao, Y., Queen, K., Zhang, J., Li, Y., Uehara, A., Tong, S.: Rapid, sensitive, full-genome sequencing of severe acute respiratory syndrome coronavirus 2. Emerg. Infect. Dis. **26**(10), 2401 (2020)
3. Sbalzarini, I.: The Algorithms of Life - Scientific Computing for Systems Biology. Keynote talk at ISC High Performance, June 2019
4. International Human Genome Sequencing Consortium: Initial sequencing and analysis of the human genome. Nature **409**(6822), 860 (2001)
5. McGuire, A.L., Caulfield, T., Cho, M.K.: Research ethics and the challenge of whole-genome sequencing. Nat. Rev. Genet. **9**(2), 152 (2008)
6. Barth-Jones, D.: The 're-identification' of governor William Weld's medical information: a critical re-examination of health data identification risks and privacy protections, then and now (2012)
7. Davis, J.: Health data, medical documents exposed by labcorp website error, January 2020
8. Naveed, M., Ayday, E., Clayton, E.W., Fellay, J., Gunter, C.A., Hubaux, J.P., Malin, B.A., Wang, X.: Privacy in the genomic era. ACM Comput. Surv. (CSUR) **48**(1), 1–44 (2015)

9. Wagner, I.: Evaluating the strength of genomic privacy metrics. ACM Trans. Privacy Secur. (TOPS) **20**(1), 1–34 (2017)
10. Humbert, M., Ayday, E., Hubaux, J.-P., Telenti, A.: Quantifying interdependent risks in genomic privacy. ACM Trans. Privacy Secur. (TOPS) **20**(1), 1–31 (2017)
11. Meyerson, A., Williams, R.: On the complexity of optimal k-anonymity. In: Proceedings of the Twenty-Third ACM SIGMOD-SIGACT-SIGART Symposium on Principles of Database Systems, pp. 223–228. ACM (2004)
12. Bläsius, T., Friedrich, T., Schirneck, M.: The parameterized complexity of dependency detection in relational databases. In: Guo, J., Hermelin, D. (eds.) 11th International Symposium on Parameterized and Exact Computation (IPEC 2016). Leibniz International Proceedings in Informatics (LIPIcs), Dagstuhl, Germany, vol. 63, pp. 6:1–6:13. Schloss Dagstuhl–Leibniz-Zentrum fuer Informatik (2017)
13. Podlesny, N.J., Kayem, A.V., Meinel, C.: Attribute compartmentation and greedy UCC discovery for high-dimensional data anonymization. In: Proceedings of the Ninth ACM Conference on Data and Application Security and Privacy, pp. 109–119. ACM (2019)
14. Wong, R.C.-W., Fu, A.W.-C., Wang, K., Pei, J.: Minimality attack in privacy preserving data publishing. In: Proceedings of the 33rd International Conference on Very Large Data Bases, VLDB 2007, pp. 543–554. VLDB Endowment (2007)
15. Malin, B.A.: Protecting genomic sequence anonymity with generalization lattices. Methods Inf. Med. **44**(05), 687–692 (2005)
16. Chen, Y., Peng, B., Wang, X., Tang, H.: Large-scale privacy-preserving mapping of human genomic sequences on hybrid clouds. In: NDSS (2012)
17. Johnson, A., Shmatikov, V.: Privacy-preserving data exploration in genome-wide association studies. In: Proceedings of the 19th ACM SIGKDD International Conference on Knowledge Discovery and Data Mining, pp. 1079–1087 (2013)
18. Kushida, C.A., Nichols, D.A., Jadrnicek, R., Miller, R., Walsh, J.K., Griffin, K.: Strategies for de-identification and anonymization of electronic health record data for use in multicenter research studies. Med. Care **50**(Suppl.), S82 (2012)
19. Lister Hill Center for Biomedical Communications. Genomic Research (2019)
20. Podlesny, N.J., Kayem, A.V., von Schorlemer, S., Uflacker, M.: Minimising information loss on anonymised high dimensional data with greedy in-memory processing. In: International Conference on Database and Expert Systems Applications, pp. 85–100. Springer, Cham (2018)
21. Jespersgaard, C., Syed, A., Chmura, P., Løngreen, P.: Supercomputing and secure cloud infrastructures in biology and medicine. Ann. Rev. Biomed. Data Sci. **3**, 391–410 (2020)
22. Zaharia, M., Chowdhury, M., Franklin, M.J., Shenker, S., Stoica, I.: Spark: cluster computing with working sets. HotCloud **10**(10–10), 95 (2010)
23. Plattner, H., Zeier, A.: In-Memory Data Management: Technology and Applications. Springer, Heidelberg (2012)
24. Färber, F., May, N., Lehner, W., Große, P., Müller, I., Rauhe, H., Dees, J.: The SAP HANA database-an architecture overview. IEEE Data Eng. Bull. **35**(1), 28–33 (2012)
25. Schapranow, M.-P., Häger, F., Plattner, H.: High-performance in-memory genome project: a platform for integrated real-time genome data analysis. In: Proceedings of the 2nd International Conference on Global Health Challenges, pp. 5–10 (2013)
26. Schapranow, M.-P., Plattner, H., Meinel, C.: Applied in-memory technology for high-throughput genome data processing and real-time analysis. In: Proceedings of the XXI Winter Course of the Centro Avanzado Tecnológico de Análisis de Imagen, pp. 35–42 (2013)

27. Levinthal, D.: Performance analysis guide for intel® core™ i7 processor and intel® xeon™ 5500 processors (2009)
28. Kessler, S., Hoff, J., Freytag, J.-C.: SAP HANA goes private: from privacy research to privacy aware enterprise analytics. Proc. VLDB Endow. **12**(12), 1998–2009 (2019)
29. Braghin, S., Gkoulalas-Divanis, A., Wurst, M.: Detecting quasi-identifiers in datasets. US Patent 9,870,381, 16 January 2018
30. Podlesny, N.J.: Synthetic genome data (2021)
31. Pullen, D.M., Sieweke, M.A.: Optimizing cache efficiency within application software. US Patent 7,124,276, 17 October 2006
32. Ramesh, B., Kraus, T.B., Walter, T.A.: Optimization of SQL queries involving aggregate expressions using a plurality of local and global aggregation operations. US Patent 5,884,299, 16 March 1999
33. Plattner, H., Mueller, S., Krueger, J., Mueller, J., Schwarz, C.: Aggregate query-caching in databases architectures with a differential buffer and a main store. US Patent 9,740,741, 22 August 2017

Social Experiment of Road Sensing on Guided Autonomous Driving Vehicle in Snowy Cold Region

Akira Sakuraba[1]($\boxtimes$), Goshi Sato[2], Yoshia Saito[3], Jun Hakura[3], Yoshikazu Arai[3], and Yoshitaka Shibata[1]

[1] Regional Cooperative Research Division, Iwate Prefectural University, Takizawa, Iwate, Japan
`{a_saku,shibata}@iwate-pu.ac.jp`
[2] Resilient ICT Research Center, National Institute of Information and Communications Technology, Sendai, Miyagi, Japan
`sato_g@nict.go.jp`
[3] Faculty of Software and Information Science, Iwate Prefectural University, Takizawa, Iwate, Japan
`{y-saito,hakura,arai}@iwate-pu.ac.jp`

Abstract. Autonomous driving vehicle is one of solutions to provide transportation rights with an alternative modal transportation in the community which is increasing of aging population in rural area. There is technically challenged for determination of driving path of the vehicle to realize autonomous driving in snowy cold region. In order to determine it, sensing of road surface weather which categorizes how the condition of the road in front of the vehicle, is a sort of essential information. This paper reports the social experiment of road surface weather sensing system which intends to realize autonomous vehicle in snowy cold region. The authors developed an onboard environment sensor system, visualization platform, and a vehicle-to-everything (V2X) communication system which consists of multiple wavelength wireless system. The authors had a social experiment using the prototype which is installed on actual guided autonomous driving vehicle system, the proposed system categorized and logged road state with 99.5% accuracy.

1 Introduction

Aging society is a major social interest for many nations. Some researchers pointed out that one thirds population will be aged 65+ by 2030 in Japan [1], the aged society would bring many challenges into current society. One of challenging which is caused by aged society is securing transportation rights for elderly people due to close of public transportation system. For instance, in 2004, number of service route of bus was decreased to 90% which is compared to 1989 [2], it suggests transportation rights for elderly people has been threaten. Residents in countryside depend on private automobiles in order to move, however physically aging complicates them to drive by themselves. Even worse, in countryside both faster aging of community and bus routes closing has growing earlier than urban area, oftentimes extinction of their communities is being concerned.

© The Author(s), under exclusive license to Springer Nature Switzerland AG 2021
L. Barolli et al. (Eds.): AINA 2021, LNNS 227, pp. 502–513, 2021.
https://doi.org/10.1007/978-3-030-75078-7_50

Autonomous driving vehicles could be an effective technology which realizes driver-less transportation. Automotive industries or tech companies are initiatively working on developing of these type of vehicle which is designed for private passenger car. Autonomous vehicle based public road transportation systems such as driverless buses will be more efficient mobility especially operation in countryside area. There is also some demonstration experiments with implementation of autonomous driving and remote monitored public transportation. The practical automated transportation systems could be a solution to maintain bus routes against short of drivers.

On the other hands, autonomous vehicles in cold snowy district have multiple and complicated challenges. Practical implementations of autonomous vehicles detect pedestrians, other vehicles, or obstacles on the road with onboard RGB cameras, millimeter wave sensors, or LiDAR. These sensors often are influenced by noise which is caused snowfalls, fog, and snow drifting.

In addition to the challenge, we consider how the vehicle obtain road surface weather in forward road section as one of the challenges. All-season operable autonomous vehicle strongly requires decision of driving path and speed which depends on road surface weather as the same as manual driving vehicle. Therefore, at the first stage to realize this type of vehicle, there is a demand of sensing system which estimates how road condition of ahead road section.

This paper reports a social field experiment of onboard road weather sensing platform on a guided autonomous driving vehicle system. The authors installed an onboard road surface sensing system on the autonomous vehicle which is operated as public transportation system, and collected road surface information in countryside village where is located in snowy cold district while vehicle is on the move. Installing on guided autonomous vehicle has advantage for use as the moving sensor node that the vehicle travels regularly on scheduled route with programmed velocity and position on the road. This paper also analyzed collected road state information with this type vehicle as a sensor vehicle of floating car data.

2   Related Works

Vehicle onboard sensors for weather related use is an interesting topic today. Huang et al. introduced a method for predicting 24 h later weather using onboard system by multilayer perceptron and long short-term memory (LSTM) model [3]. They use temperature, humidity, ultraviolet, pressure, PM 2.5, and rainfall sensors as onboard node which is installed on the bus. In temperature which is predicted with their proposed method, 3.51–6.06% error and they concluded that system achieves effective weather monitoring and information management, misalignment may be present due to the significant weather changes.

Computer vision based analysis is very popular method for road safety application, weather classification method is also an interesting topic. Iparraguirre et al. proposed a single onboard RGB camera based traffic sign recognition method [4]. They attempt to realize robust recognition against fog, shadows, low contrast image, reflection, and heavy rain with CNN model based classification and image processing in XYZ color space. They achieved real-time processing and overall accuracy of 92% in their evaluation.

We are developing an onboard road surface weather sensing system with multiple type sensors. The system logs outside air/road surface temperature, outside air humidity, qualitative road weather state and friction coefficient using near infrared (NIR) road sensor. These sensor data are associated with geolocation which is positioned with GNSS. We had a social experiment on intercity bus service which contains 200 km round-trip travelling [5] and the system collected road state information with high accuracy.

3   Onboard Road Sensing Platform

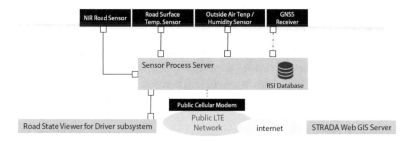

Fig. 1. Onboard Sensor Unit and connected sensor device, subsystems, and GIS server located over internet.

3.1   Onboard Sensor Unit

Table 1. List of onboard sensors connected to Onboard Sensor Unit.

Sensor type	Collectable data
Air temperature	Outside air temperature around the vehicle
Humidity	Outside air humidity around the vehicle
Road temperature	Road surface temperature under the vehicle
NIR road state	Estimated road state and estimated friction coefficient between tires and road surface State := (dry I damp I wet I slash I snow I ice)
GNSS receiver	Positioning of the vehicle and clock synchronization among subsystems

Onboard Sensor Unit is an onboard unit which logs road weather condition related environment data and categorizes road surface weather. Onboard Sensor Unit connected to environment sensors, illustrated in Fig. 1.

The system logs values from these sensors which is described in Table 1. It contains quantitative fields such as outside air temperature, humidity, road surface temperature, and friction coefficient. There is a qualitative field such as road state. These information can describes that when, where, how, and how much the road condition was there.

Onboard Sensor Unit logs averaged value of these sensor output to database at 1 Hz sampling rate. In this system, this information is defined as road state information (RSI).

We also used this system in past work [5], however we limited to the minimal configuration of the system which records just environment data around the vehicle.

In this configuration, Onboard Sensor Unit has a connection to public cellular modem to deliver RSI data and send periodically current working vehicle information to GIS server. The vehicle information contains identification of the vehicle, location of the vehicle, timestamp, sensor values, and vehicle speed. We describe GIS server in Sect. 3.3.

3.2 Road State Viewer for Driver

Road State Viewer for Driver visualizes categorized road state on the offline digital map. This GIS visualization application overlays circles which is different colored depends on categorized road state on associated geolocation in RSI records. The overlaid graphics are based on RSI data which is obtained on own vehicle.

In order to obtain ahead road condition in direction of travel, system attempts to exchange RSI data via vehicle-to-everything (V2X) wireless communication between other vehicles which is moving to reverse direction of travel or roadside unit (RSU). Incoming RSI data is also reflected and overlaid on the map when RSI data is received (Fig. 2).

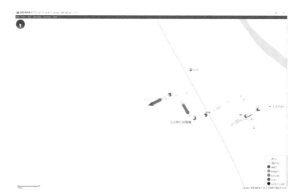

Fig. 2. A screenshot of Road Condition Warning System. The indication of categorized road state by different colors; white: light blue: damp, navy blue: wet, yellow: slash, purple: snow, and scarlet: ice.

3.3 STRADA Web GIS

STRADA (*Surveillance and Tracking with the Road Analyzing in Distributed Architecture*) is a web GIS based visualization application. We intend the users of

STRADA who are not taking the wheel such driver who before travelling by car or operation manager who works for public transport organization.

The primary function of this application is similar to Road State Viewer for Driver, the GIS visualizes to overlay RSI data on associated geolocation on digital map. In addition, this application can display other information in RSI record such outside air/road surface temperature, outside air humidity, friction coefficient on the road, and categorized road state which are illustrated as Fig. 3.

(1) Visualized road weather state from delivered RSI. Colors fill the circle are equivalent to dry: sky blue, damp: light green, snow: amber, and snow: orange. Circles are rendered bigger the state is more hazardous.

(2) Visualized friction coefficient from delivered RSI. Bigger and reddish circles on the map are rendered where low friction was recorded.

Fig. 3. Examples of visualized RSI records by STRADA web GIS.

STRADA displays vehicle location of the onboard system is sensing at the time on the map. When the web server receives current working vehicle information from sensing vehicle, server caches immediately to provide current working vehicle location and sensor information for requesting client which intends to show GIS application.

STRADA is designed for both desktop environments and smart devices, which is based on responsive web design.

3.4 Prototype

We prepared prototype in effort to perform the social evaluation.

We have an implementation of Onboard Sensor Unit is based on Linux based system on Raspberry Pi 3B+ single board computer. We selected Optex CS-30TAC for the road surface temperature sensor, Azbil HTY7843 for sensor both of the temperature and humidity of outside air, and Optical Sensors RoadEye Model SD for NIR road state categorization and friction sensor. These sensors are connected to the Onboard Sensor Unit with wired. Sensor data collecting system is written by Python 3.8 with Nginx 1.14.2 and Flask 1.0.2. The system manages sensor data on SQLite lightweight database.

Implementation of STRADA web GIS is also based on Python 3.8 with Nginx 1.18.0, with digital map is rendered with Leaflet 1.7.1 interactive digital map. Delivering of RSI from vehicle to STRADA server depends on ownCloud [6].

We implemented Road Condition Warning System as a WinForms application with MapFan SDK for Windows which provides offline map function.

4 Social Evaluation in Mountainous Village

This section mentions a social experiment to confirm operability on autonomous driving vehicle. The experiment was held over 2 days in the end of January 2021. The weather of both days was stably cloudy or fine, recorded temperature at Takanosu the nearest surface weather observation station was between $-4.4\,°C$ and $4.9\,°C$.

4.1 Location and Autonomous Vehicle Route

The experiment field is located Kamikoani-village, Akita-prefecture where in northern part of Honshu, Japan. The village has populated about 2,200 people, and 48.7% of residents are aged 65+. This village is categorized a typical mountainous rural area.

A nonprofit is operating autonomous driving electric vehicles and 3 routes in commission from the village, these network of route is intended for transport elderly people attends to the hospital or their shopping. We chose to perform the experiment on the route of Kosawada-Dōkawa which was explained on Fig. 3 and it has 5.2 km round-trip travel at 7.0 km/h at scheduled speed.

4.2 Autonomous Driving Vehicle in Kamokoani-Village

The operator uses Yamaha G30Es electric golf carts based guided autonomous driving vehicle the above described routes. This guided vehicles control angle of steering in order to determine vehicle's travelling path which depends on electromagnetic induction wire placed along to the routes. There are buried magnetic tags under the pavement on the routes to instruct control of vehicle such changing of speed or halt the vehicle.

The vehicle can travel 12 km/h at the maximum speed with 4 passengers.

A driver boards on the autonomous vehicle to monitor traffic condition around the vehicle and switching to manual driving as appropriate. For example, in case of the vehicle is approaching to collapsed snow walls which are blocking the designed path,

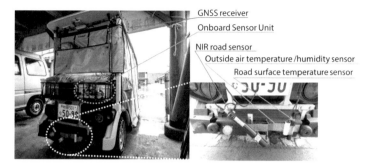

Fig. 4. Installation of Onboard Sensor Unit and connected sensors on the autonomous vehicle

the driver does override automatic steering and switches automatic mode again when the vehicle has been swerved and passed.

We installed an Onboard Sensor Unit on the vehicle. Any sensor heads of outside air temperature/humidity sensor, road surface temperature sensor, and NIR road state sensor are exposed in the rear of body which is shown in Fig. 4.

4.3 Acquired Road State Information Data

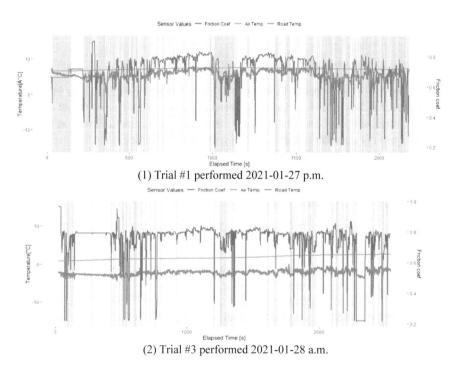

(1) Trial #1 performed 2021-01-27 p.m.

(2) Trial #3 performed 2021-01-28 a.m.

Fig. 5. Varying of friction coefficient, outside air temperature, and road surface temperature values on RSI records by time series. Background colors are filled with the determined road surface state and corresponding to gray: dry, green: damp, yellow: snow, and red: ice.

As described above, we had 4 round-trips which is composed of 2 different timeframes attempting on each different 2 days. The system had obtained 9,139 RSI samples all over the trials.

Figure 5 plots varying of outside air temperature, road surface temperature, and outside air humidity by time series. Background color describes categorized road state by different colors. We investigated that categorization of road state on the whole records had 99.95% accuracy rate.

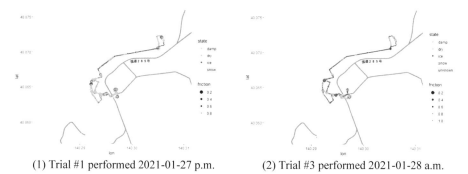

(1) Trial #1 performed 2021-01-27 p.m.   (2) Trial #3 performed 2021-01-28 a.m.

Fig. 6. Visualized road weather state and friction coefficient on the road on digital map

Figure 6 illustrates visualized road state on RSI records on the map by geolocation. Figure 6 (1) describes afternoon in day 1 and Fig. 6 (2) describes around 10 am on day 2. It is confirmed that some points recorded on day 2 disappeared snow or ice state which is categorized on day 1. This change of road state is assumed that high road surface temperature which is increased by sunlight.

4.4   Demonstration of STRADA web GIS

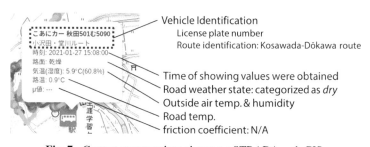

Fig. 7.   Current sensor values shown on STRADA web GIS.

We also demonstrated STRADA web GIS for members of nonprofit. The system overlaid vehicle an icon and a balloon which shows current sensor onboard sensor values on digital map as shown in Fig. 7. The system also rendered successfully temperature, humidity, friction coefficient, and road weather state RSI record uploaded from the autonomous driving vehicle.

5 Analysis and Discussion

5.1 Correlation Analysis on Quantitative Fields

Firstly, we analyzed correlation among acquired sensor output of quantitative values; outside air temperature, road surface temperature, outside humidity, and friction coefficient.

Firstly, the authors attempted overlooking their correlation using heatmap which is illustrated in Fig. 8. The result indicates clearly positive correlation between outside air temperature and road state temperature over different travels.

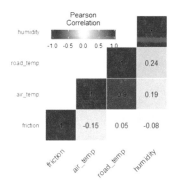

Fig. 8. Correlation heatmap among each sensor values.

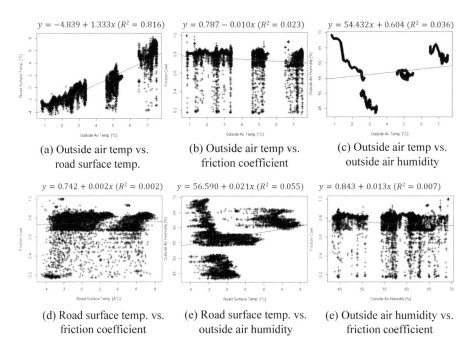

(a) Outside air temp vs. road surface temp.

(b) Outside air temp vs. friction coefficient

(c) Outside air temp vs. outside air humidity

(d) Road surface temp. vs. friction coefficient

(e) Road surface temp. vs. outside air humidity

(e) Outside air humidity vs. friction coefficient

Fig. 9. Scatter plots among each sensor values.

Figure 9 denotes scatter diagrams and regression lines among quantitative fields. Generally, road surface temperature influences friction coefficient, low road surface temperature is regarded as low friction coefficient due to low temperature which is characterized on icy road. However the result indicates no correlation between them, due to friction coefficient has a remarkably with distribution even the same road surface temperature was observed. The characteristics of friction coefficient which is quite varying values regarding the other sensor value was stable, can be seen correlation between other fields such outside air temperature and humidity.

5.2 Analysis on Road Weather State Field

Next, the authors analyzed distribution of the relationship between outside air temperature and road surface temperature and focuses on categorized road weather state. Figure 10 plots the relationship of them. In the same outside air temperature, system observed dry, damp, ice, snow in descending order of road surface temperature in road weather state field. Interestingly, the point of air temperature was above 5 °C, we discovered that there is wider variance of road surface temperature on records which is categorized to ice or snow.

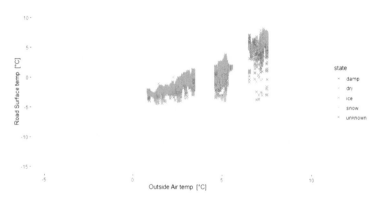

Fig. 10. Relationship among outside air temperature, road surface temperature, categorized road weather state.

System deals road weather state which is a typical qualitative field, it is required to analyze by different method. Therefore we analyzed them with polyserial correlation to measure the correlation between road weather state and other fields which has continuous values. We defined the ordinal scale for road weather state field which is based on the level of hazard in driving as following; dry: 1, damp: 2, snow: 3, ice: 4, and unknown: 0. The authors analyzed the correlation between them as Table 2. There is weak correlation between the state and outside air temperature. Moreover, friction coefficient on the road has strong negative correlation with the road state. Surprisingly, road surface temperature did not have no correlation with the state in our dataset.

Table 2. Correlation coefficient between road surface weather state and other sensor fields. Signs * describe $p < 0.01$ in Chi-square test.

Road Surface Weather State vs.	Polyserial Correlation
Outside Air Temperature	0.2527 *
Road Surface Temperature	0.0127 *
Outside Air Humidity	0.1329 *
Friction Coefficient on the Road	-0.8886 *

5.3 Advantages of Use of Autonomous Vehicle for Public Transport as Sensor Vehicle

In effort to realize predicting or providing precise road condition, especially machine learning based approach requires large scale dataset of RSI. Public transport vehicles travels scheduled time and route in periodically and repeatedly. This characteristics allows to build and to renew the very large dataset.

In addition, autonomous vehicle can drive accurately the position on the lane as well as accurate control of own speed, even if the vehicle is design to be guided by road infrastructure or not. This feature is adequate to collect RSI data in the same travelling condition repeatedly.

Also current implementation of autonomous driving vehicle travels at relatively low speed which compared with full-manual driving vehicle. If the sampling rate of onboard sensor cannot be changed, the sensor on slower vehicle can collect and represent RSI data with more high resolution relatively. For example, autonomous vehicle which is operated in Kamikoani-village can record RSI data every 0.3 m resolution while the vehicle is on the move. This feature is effective to warn the hazardous road section precisely.

These reasons are advantages for use of autonomous driving vehicle which is in charge of public transportation service as the sensor vehicle.

6 Conclusion

This paper reported the social experiment for onboard road weather sensing platform on a guided autonomous driving vehicle based public transportation system. The onboard system equips multiple type sensors which collects environmental parameters such temperature and humidity of outside air, road temperature, friction coefficient on the road, and categorized road surface states. The system associates these sensor values with geolocation in order to use several type of applications. For the platform, system provides road state and other environmental information for non-driver who is not taking the wheel as well as road state warning system for driver of the manual driven vehicle.

The authors had been performed social experiment in the northern part of Japan to collect road state information with the onboard platform installed on practical autonomous vehicle. The result of RSI data analyzing suggests that there is strong positive correlation between outside air temperature and road surface temperature, however, friction has no significant correlation with any other environment sensor value at this time.

The authors investigated the advantage of autonomous vehicle based public transportation system should use as the moving sensor node. Autonomous vehicle in public transport can drive precisely and frequently at low speed moving, it is suitable to obtain road state information and build large road state dataset.

For the future work, authors are now considering to install more multiple type sensors such LiDAR, far-infrared camera, or precipitation sensor in order to investigate which elements contribute and influent to the worse road state. Also authors are attempting to build spatiotemporal large scale dataset of road state information.

Acknowledgments. We applicate The Association of Mobile Service Kamikoani-village Non-profit cooperate to perform the field experiment for sensing. This work was supported by JSPS KAKENHI JP20K19826, JSPS KAKENHI JP20K11773, and Strategic Research Project Grant, Iwate Prefectural University, Japan.

References

1. Muramatsu, N., Akiyama, H.: Japan: super-aging society preparing for the future. Gerontologist **51**(4), 425–432 (2011)
2. Hashimoto, S.: Application and utilization of public transportation in aging society (in Japanese). In: Research Reports of Kobe City College of Technology, No. 50, pp. 137–140 (2012)
3. Huang, Z., Chen, Y., Wen, C.: Real-time weather monitoring and prediction using city buses and machine learning. Sensors **20**(18), 5173 (2020)
4. Iparraguirre, O., Amundarain, A., Brazalez, A., Borro, D.: Sensors on the move: onboard camera-based real-time traffic alerts paving the way for cooperative roads. Sensors **21**, 1254 (2021)
5. Sakuraba, A., et al.: Web, artificial intelligence and network applications. In: Proceedings of the Workshops of the 34th International Conference on Advanced Information Networking and Applications (WAINA 2020), pp. 595–604 (2020)
6. ownCloud: Share files and folders, easy and secure. https://owncloud.com/. Accessed 04 Mar 2021

A Cloud-Based Mobile Application for Women with Gestational Diabetes

Mahmoud Elkhodr[1,3]($\boxtimes$), Shobana Ashokkumar[2,3], Belal Alsinglawi[2,3],
Omar Darwish[2,3], Ola Karajeh[3,4], and Ergun Gide[1,3]

[1] Central Queensland University, Rockhampton, Australia
{m.elkhodr,e.gide1}@cqu.edu.au
[2] Western Sydney University, Penrith, Australia
b.alsinglawi@westernsydney.edu.au
[3] Ferrum College, Ferrum, VA, USA
odarwish@ferrum.edu
[4] Virginia Polytechnic Institute and State University, Blacksburg, VA, USA
okarajeh@vt.edu

Abstract. The widespread presence of Gestational Diabetes Mellitus (GDM) is increasing in recent years. Women with GDM have to continuously monitor their blood glucose levels to reduce pregnancy risks. Existing practices in Australia comprise monitoring the blood glucose levels by taking readings at least three times a day and then manually sending the readings to the healthcare professionals. This research proposes the use of a Mobile application that can streamline communications with healthcare professionals and provide users with health and diet recommendations to regulate blood glucose levels. The application is intended to be used mainly by women with gestational diabetes. It works by providing healthcare professional access to the glucose's readings via the cloud. The app also provides diet and physical activities' recommendations. It also allows users to join community support groups online where they can have access to frequently asked questions and can raise their own queries, which can be answered by healthcare professionals as well.

Keywords: Gestational Diabetes Mellitus · mHealth ·
Self-management of diabetes · Smartphone application · Blood glucose

1 Introduction

Gestational Diabetes is one of the most common chronic and non-communicable diseases. Gestational Diabetes Mellitus (GDM) is caused by glucose intolerance. It is a condition that occurs only during pregnancy. The hormones produced by the placenta are the main cause of gestational diabetes. These hormones resist the insulin produced by the body [1]. This requires intervention such as following a strict diet and, in most cases, it requires medical intervention by way of injecting insulin in the body before each meal. Women with GDM must

L. Barolli et al. (Eds.): AINA 2021, LNNS 227, pp. 514–523, 2021.
https://doi.org/10.1007/978-3-030-75078-7_51

continuously monitor their blood glucose levels to reduce pregnancy risks. By the year 2045, it is estimated that 308 million women will live with diabetes [5].

Management of gestational diabetes requires daily monitoring of blood glucose levels, maintaining a healthy diet, taking medications or insulin and doing physical activities, which are challenging to accomplish especially during pregnancy. This is where Mobile health (mHealth) technologies can alleviate some of the encountered challenges. According to a recent survey [2], the penetration of mobile health app rose from 16% to 48% between the years 2014 and 2018. mHealth allows patients to monitor their condition by themselves. Patients can access their health records through their mobile phones and get assistance such as treatment and advise on the management of the disease [6].

With the evolvement of cloud computing and Internet of Things technologies mHealth users can self-monitor their conditions and feedback the information back to the healthcare providers using technologies such as smart and wearable computing, bio and Internet of Things (IoT) sensors, and smart phones. Many smartphone applications intended for the self-management of diabetes can be found on both Android and iOS platforms. To this end, the main objective of this research is to determine the effectiveness of the assistive technologies used for the management of gestational diabetes and to develop a cloud-based solution that would not only improves the self-monitoring and management of the disease (gestational diabetes) but also enable women to receive community support and exchange experiences.

The structure of this report is organized as follows: Sect. 2 provides some literature review of the related work and their limitations. Section 3 outlines the architecture of the proposed system. In Sect. 4, the features of the proposed framework and the mobile application are presented. Section 5 reports on the implementations of the proposed mobile application. Section 6 provides some evaluation of the developed prototype. Conclusion and future work are provided in Sect. 7.

2   Related Works

Recently, Telemedicine Monitoring systems have been applied and exploited in the context of GDM patients' self-monitoring. In [12] a telemedicine-based system was proposed. Women with gestational diabetes used this system to monitor their glucose level. Women were randomly allocated to two corresponding groups either the internet or the control group. Women in the internet group were given an Accu-Check meter and a smart phone which has the application setup in it. The app transfers the patient's blood glucose readings to the healthcare server through the short message service (SMS). The blood glucose values from the smartphone are sent to the application via infrared transmission. This service reduced the face-to-face visit by 62% compared to regular prenatal visits. There were some limitations faced by some women while using this service, such as data transmission failure and defects in meter. The application also lacked certain self-management functionalities. Other studies such as those reported in [7,8,10] show that using a smartphone for the self-management of the disease has shown significant improvement in controlling the disease. Some of these applications, such as those reported in [3,10], were limited in scope as they targeted only a particular country or set of users. Telemedicine solutions such as the one reported in [8,12], that are based on web applications are less frequently used given the dominance of smartphone in this space. Applications such as [3,4,9,10], lacked a comprehensive functionality as they didn't provide diet recommendation or physical activity modules.

3   The Proposed System Architecture

The proposed system is based on a three-tiers architecture. Figure 1 shows the main components of the proposed system. It consists of three main modules: the patient module, the cloud service provider module and the health care provider module. One advantage of the proposed architecture is usability and portability. The cloud service provider used in this project is Microsoft Azure. The cloud service provider module consists of the web server and the database. The smartphone application is connected to the cloud-based backend service by utilizing the Mobile App service in the Azure portal. The transmitted 'readings' values from the smartphone are stored in the cloud database, which can be accessed by health care professionals through a web interface.

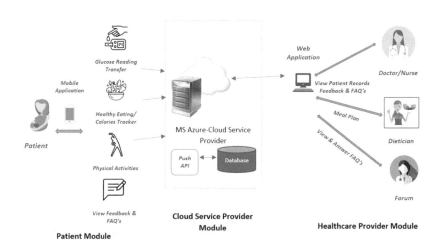

Fig. 1. The proposed system architecture.

3.1   Web Server

In this module the readings from the glucometer are transmitted to the mobile application via Bluetooth, which is then transmitted to the database hosted in the cloud. The web server acts as an interface between the patient and the health care provider. The web server receives the request and sends responses via the corresponding protocol.

3.2   Push Notification

This module is used to send push notifications to the website as well as the mobile application. Whenever a new blood glucose reading is transmitted, this module sends an alert to the health care professionals in the form of push notifications informing them that a new reading has arrived. So, the healthcare professional can view the readings through the web application and provides feedback. This is done via a push notification API.

3.3   Web Application

This module is used by the healthcare professionals. A push notification is provided when the server receives the glucose reading from the patients. The web application shows the records of a patient along with their previous readings in a graph format. The web application is hosted on the cloud and configured with the Azure SQL server, which fetches the data from the backend database.

4   System Features

The followings are the main system features.

4.1   The Mobile Application

The mobile application used by the patient has the following functionalities.

– Allows the patient to transmit their daily blood glucose readings to the cloud
– Provides visualisation of the BG readings
– provides diet recommendation for a healthy lifestyle and suggests suitable foods that have low Glycemic Index (GI).
– Provides information on suitable physical activities during pregnancy and allow patients to track their walking activities
– Allows the patient to View feedback or suggestions left for them by the healthcare professional
– Provides access to a gestational diabetes community support group
– Provide access to a frequently asked questions section. This section lists the most frequently asked queries with answers from the healthcare professionals.

4.2   The Web Application

This application is used by the healthcare professionals to track the patient's health status. It enables them to check the patient's current glucose readings and their past glucose readings in the form of graphs. They can send feedback to the patients and suggest diet plans. They can also encourage them to do more physical activities. The feedback can be viewed by the patient using their application. Healthcare professionals can also send reminder notifications to patients regarding their upcoming appointments.

5   Details of the Implementation

The mobile application is created in Visual Studio 2019 IDE with Xamarin Framework. The coding language used is C. Azure Table Storage was used as the backend. The app is developed to run on Android 9.0 supported devices. The Cloud Service Provider (CSP) utilized for this project is Microsoft Azure. In Azure, a specific Storage Account resource is created for storing the data from the mobile application as well as the website. For trend analysis and to visualize the data over a period, Power BI (Business Intelligence) resource is utilized. The values stored in the Storage Account in Azure are used by Power BI to provide data insights and visualization.

5.1   The Application Screenshots

Connect Device and Log BG Readings: Figure 2 and 3 show the process of logging the BG reading using the app. The values received from the glucometer are directly transferred to the BG transfer module if the device is connected via Bluetooth, which is shown in Fig. 2. The app also allows users to manually enter the BG readings. The patient must specify the meal type (Fasting, Breakfast, Lunch, Dinner, and Bedtime) along with the activity settings (Walking, Yoga, and Exercise), how they had done, and the food which they had eaten. These data help the dietitian to have a detailed analysis of the patient. When the user clicks on the Submit button the BG readings are transmitted to the Azure Table Storage. The readings are then viewed by the healthcare professionals via the Web App which is hosted on the Azure Cloud.

The Visualization Module: Figure 4 shows the visualization of the BG values over a period of time. The user can access this feature by navigating via the Blood Glucose Reading menu ->Trend Analysis. This feature provides the user with a visualization of their BG readings within the selected period.

Diet Recommendations: Figure 5 details the diet recommendation feature of the mobile application. This module provides the user with a search bar, where they can search for a healthy diet. This feature helps gestational diabetes patients to have a healthy and balanced diet and allows them to track the calories consumption.

Fig. 2. Connect Device via Bluetooth.

Fig. 3. Blood Glucose Reading Log.

Community Support Group: Figure 6 shows the screenshot of the Community Support Group and the FAQ module. This feature allows the user to join a Facebook group. Also, it allow them to access a set of frequently asked questions with answers from healthcare professionals.

6   Evaluation

A usability Testing (UT) is conducted on the prototype. Table 1 provides the Test Plan used in this work.

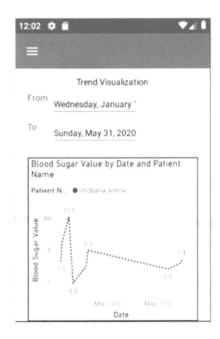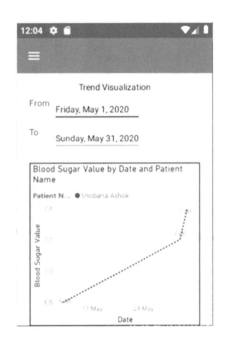

Fig. 4. Trend visualization of BG readings.

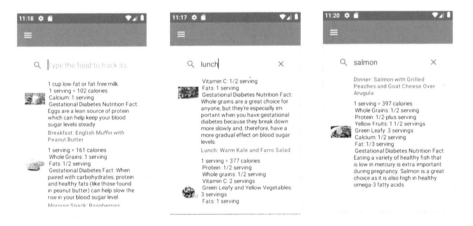

Fig. 5. Health eating diet with calorie tracker.

The usability test was conducted based on the recommendations by Nielsen [11]. According to Nielsen, 85% of the usability flaws can be identified by 3–5 participants. The following three requirements were considered while conducting the test: choosing the right participants for the test, allocating them with tasks to perform, and thirdly allow them to talk about the application. The usability test was conducted with 5 participants. At first, the participants were provided with information about the application and its usage. Then they performed the usability test based on the scenarios presented in Table 1.

Fig. 6. Access to Group & FAQ's.

Table 1. Test plan for usability testing.

App feature	Module	To perform
BG reading log	Scenario 1 - It is after breakfast (10 am)	Action: Open the App and log the BG reading. Information to record: BG – 7.2 mmol/l, Meal Type - After Breakfast, Activity – None, What you had – Oatmeal
Data Visualization	Scenario 2 - Patient wants to view her BG reading Trend	Action: Navigate to Trend Analysis visualize the BG readings for month
Healthy Diet Recommenda-tion	Scenario 3 – A patient has an urge to eat she has never had before; Have a look through the app for suggestions and carbohydrates estimate for this new dish.	Action: Navigate through the app and how do you achieve that
Log Physical Activity	Scenario 4 – See how much calories you have burnt during walking	Action: Go to Physical Activities module and see how much distance you have walked
Access FAQ's	Scenario 5 – Patient is unclear about GDM. She wants to know in detail about GDM	Action: Explore the app to achieve this task

6.1   Evaluation of Results

The participants struggled with Scenario 1 as they found it difficult to connect the device via Bluetooth to the Glucometer. Most of them were able to use the BG log upon their second try. For Scenario 3, the participants suggested to include more diet plans and to enable patients to share their own diet plans. The other scenarios were successfully completed by the participants. Some of the feedback received also pointed to the need to clean and improve the design of the application.

7   Conclusion

This paper proposed the use of a mobile application to manage gestational diabetes. The proposed app provides patients with a solution to self-manage their disease along with obtaining support from their health care professionals. The app also provides diet recommendation, physical activities, and motivation to patients. The community-based support group feature allows patients to obtain peer-support, and provide linkages to available community and government resources. Improving the UI of the app is amongst the planned activities for future works. A trial and a user-based evaluation will then be conducted.

References

1. Gestational diabetes – Diabetes Australia. https://www.diabetesaustralia.com.au/about-diabetes/gestational-diabetes/
2. Accenture: Accenture 2018 Consumer Survey on Digital Health
3. Alotaibi, M., Albalawi, M.: A mobile gestational diabetes management and educational system for gulf countries: system architecture. In: 2018 9th IEEE Control and System Graduate Research Colloquium (ICSGRC), pp. 193–196. IEEE, Shah Alam, Malaysia, August 2018. https://doi.org/10.1109/ICSGRC.2018.8657496, https://ieeexplore.ieee.org/document/8657496/
4. Ambroise, N., Boussonnie, S., Eckmann, A.: A smartphone application for chronic disease self-management. In: Proceedings of the 1st Conference on Mobile and Information Technologies in Medicine (2013)
5. Cho, N., Shaw, J.E., Karuranga, S., Huang, Y., da Rocha Fernandes, J.D., Ohlrogge, A.W., Malanda, B.: IDF Diabetes Atlas: global estimates of diabetes prevalence for 2017 and projections for 2045. Diabetes Res. Clin. Pract. **138**, 271–281 (2018)
6. Free, C., Phillips, G., Felix, L., Galli, L., Patel, V., Edwards, P.: The effectiveness of M-health technologies for improving health and health services: a systematic review protocol. BMC Res. Notes **3**(1), 250 (2010). https://doi.org/10.1186/1756-0500-3-250, https://bmcresnotes.biomedcentral.com/articles/10.1186/1756-0500-3-250
7. Hirst, J.E., Mackillop, L., Loerup, L., Kevat, D.A., Bartlett, K., Gibson, O., Kenworthy, Y., Levy, J.C., Tarassenko, L., Farmer, A.: Acceptability and user satisfaction of a smartphone-based, interactive blood glucose management system in women with gestational diabetes mellitus. J. Diabetes Sci. Technol. **9**(1), 111–115 (2015). https://doi.org/10.1177/1932296814556506
8. Homko, C.J., Santamore, W.P., Whiteman, V., Bower, M., Berger, P., Geifman-Holtzman, O., Bove, A.A.: Use of an internet-based telemedicine system to manage underserved women with gestational diabetes mellitus. Diabetes Technol. Therapeutics **9**(3), 297–306 (2007). https://doi.org/10.1089/dia.2006.0034
9. Mackillop, L., Loerup, L., Bartlett, K., Farmer, A., Gibson, O.J., Hirst, J.E., Kenworthy, Y., Kevat, D.A., Levy, J.C., Tarassenko, L.: Development of a real-time smartphone solution for the management of women with or at high risk of gestational diabetes. J. Diabetes Sci. Technol. **8**(6), 1105–1114 (2014). https://doi.org/10.1177/1932296814542271

10. Mackillop, L.H., Bartlett, K., Birks, J., Farmer, A.J., Gibson, O.J., Kevat, D.A., Kenworthy, Y., Levy, J.C., Loerup, L., Tarassenko, L., Velardo, C., Hirst, J.E.: Trial protocol to compare the efficacy of a smartphone-based blood glucose management system with standard clinic care in the gestational diabetic population. BMJ Open **6**(3) (2016). https://doi.org/10.1136/bmjopen-2015-009702, https://bmjopen.bmj.com/content/6/3/e009702

11. Nielsen, J.: Usability Engineering. Morgan Kaufmann Publishers Inc., San Francisco (1994)

12. Pérez-Ferre, N., Galindo, M., Fernández, M.D., Velasco, V., de la Cruz, M.J., Martín, P., del Valle, L., Calle-Pascual, A.L.: A Telemedicine system based on Internet and short message service as a new approach in the follow-up of patients with gestational diabetes. Diabetes Res. Clin. Pract. **87**(2), e15–e17 (2010)

Genetic Algorithms for Scheduling Examinations

Farshid Hajati[1]($\boxtimes$), Alireza Rezaee[2], and Soheila Gheisari[1]

[1] College of Engineering and Science, Victoria University Sydney, Sydney, Australia
{farshid.hajati,soheila.gheisari}@vu.edu.au
[2] Department of Mechatronics Engineering, Faculty of New Sciences and Technologies, University of Tehran, Tehran, Iran
arrezaee@ut.ac.ir

Abstract. The range of problems to which genetic algorithms have been applied is quite broad. Timetable scheduling is a complex optimization problem. The purpose is to solve this problem by using genetic algorithms. There are two implementations for this problem: the first one solves the complete issue, and the second one divides the problem into two phases and solves the second phase. The results show that the second implementation has better solutions than the first one.

1 Introduction

The machine learning has been used in many applications [3–11]. The genetic algorithm [1, 2] is a method for solving optimization problems that is based on natural selection, the process that drives biological evolution. We can use genetic algorithms to solve a variety of optimization problems that are not well suited for standard optimization algorithms including examinations scheduling. Examinations scheduling is a complex optimization problem. However, it seems that using genetic algorithms for examinations scheduling remains elusive.

This study focuses on the use of genetic algorithms in scheduling the examinations. The objective is to determine the factors which have a potential influence on the result. One of the other purposes is to compare two implementations of this problem. In this paper first we describe the problem specifications and some criteria for the solutions. After that we will describe two implementations for this problem and explain the details of the steps of the implementations. At last we view the results of running the program with different input data and parameters and compare the results of them.

2 Problem Description

In the examinations scheduling problem, we have some rooms and some time-slots. Each time-slot is a period of time in a day. Time-slots are completely separate from each other and do not overlap. For example if we allocate 5 days for the exams and we have 4 exams in each day, the number of time-slots will be 20. We allocate each exam one time-slot and a few numbers of rooms. The number of allocated rooms is between 1 and 4. The rooms are common among examinations that are in the same time-slot.

The examinations scheduling problem has two phases. The first phase is time-slot allocation phase and the second phase is room allocation phase. In the first phase, we determine a time-slot for each examination. In the second phase, we allocate rooms for examinations that are in the same time-slots. The purpose is to schedule examinations according to various criteria, which are listed below:

- The total number seats that are allocated to an examination should be equal with the number of registered students in that exam. When students are allocated to all the rooms for a given examination, the number of reserved seats should not be more or less than the actual number of student.
- When allocating a room or rooms for an exam the most important criteria is to ensure that all student are allocated a seat. When students are allocated to all the rooms for a given examination, there should be sufficient capacity in the rooms for all students.
- For various reasons it is desirable that if a room is open it is as full as possible. This criterion should be considered too.

The importance of these criteria may be different in different situations. But it seems that the number of orphan students is important in most of situations.

3 Implementation

We implement examinations scheduling problem using genetic algorithms. Genetic algorithms are a family of computational models inspired by evolution. These algorithms encode a potential solution to a specific problem on a simple chromosome-like data structure and apply recombination operators to these structures to preserve critical information. An implementation of a genetic algorithm begins with a set of population of typically random chromosomes. One then evaluates these structures and allocates reproductive opportunities in such a way that those chromosomes, which represent a better solution to the target problem, are given more chance to reproduce than those chromosomes, which are poorer solutions.

We can implement the problem in two ways. First we can consider a general implementation which solves the complete problem including phase number one and phase number two synchronously. In this implementation, the genetic algorithm determines a time-slot and one or more rooms for each exam. It also determines the number of students in each room. In this implementation, the search space is extensive.

In the second way, we can split the problem into two separate phases. The first phase is time-slot allocation phase and the second phase is room allocation phase. We can implement and optimize the room allocation phase, considering the first phase is done before. In this implementation, we can reduce the search space and get better results.

Initially we consider the first way of implementation and describe it. After that, we convert the initial implementation to the second implementation with a little change. Then we can compare the results of two implementations.

Now we describe the first way of implementation. First, we should encode the solution on a chromosome. Then we should define a fitness function for evaluating the potential solutions. After that, we must create a population of potential solutions and evolve it. We give the program the information of all of the exams and rooms and the number of time-slots. The information of the exams includes the identifier number of exam, number of registered students and the duration of it. The information of the rooms includes the identifier number of room and the capacity of the room. We should determine the number of initial population and the number of evolve.

3.1   Planning a Chromosome

At the heart of the genetic algorithm is the Chromosome. The Chromosome represents a potential solution and is divided into multiple genes. Step one; therefore, is to decide on the makeup of our chromosomes, which includes how many genes we want and what those genes will represent. In our program, we should determine a time-slot, one or more rooms and the number of students in each room for all of the exams in the solution. Therefore, we define our chromosome as described in Fig. 1.

Time-slot number	Room number	Number of students	Room number	Number of students	Room number	Number of students

Fig. 1. Chromosome definition

In each chromosome, we have the above structure for each exam. We consider that each exam can have at most four rooms. Therefore, for each exam we have nine genes in the chromosome. First gene indicates the time-slot and the other genes indicate the room number and number of students in each room, which means two genes per room in one exam. The number of students in a room can be zero.

In the chromosome, we have the time-slot number, room number and number of students in that room (four pair of them) for the first exam. Then we have the time-slot number, room number and number of students in that room (four pair of them) for the second exam and so on.

3.2   Fitness Function

The role of the fitness function in a genetic algorithm is to determine the merit of each chromosome in a population. For calculating the fitness of a chromosome we assume a number as maximum fitness which means that the fitness function returns that number for a perfect solution. We also define penalties for undesirable events. For each time-slot, the program calculates the penalty for that time-slot and the total penalty will be sum of the penalties in the all of time-slots. Then program calculates fitness from subtracting the total penalty from the maximum fitness.

Now we explain each penalty, which is associated with an undesirable event. The key criterion that needs to be fulfilled is that of eliminating orphaned students. That is wemust ensure that all students in all subjects are allocated to a room. The program calculates the penalty of orphan students by multiplying the number of orphan students by a constant number say REMAINDER_STUDENTS_PENALTY. When students are allocated to all the rooms for a given examination, the number of reserved seats should not be more or less than the actual number of student. The program calculates the penalty for inequality of allocated seats and actual number of students by multiplying the difference of these two numbers by TOTAL_COURSE_CAPACITY_PENALTY and the proportion of these two numbers. For evaluating split exam penalty the program first calculates the proportion of number of students in each split. The penalty is calculated by multiplying this proportion by a constant number SPLIT_EXAM_PENALTY. The penalty for empty seats in a room is calculated by multiplying the number of empty seats by EMPTY_SEATS_PENALTY. If the proportion of the capacity of the room to the number students in that room is greater than two, then the penalty for empty seats will be multiplied by this proportion. The penalty of mixed duration exams is calculated by multiplying the number of mixed duration exams by MIXED_EXAMS_PENALTY.

3.3 Second Implementation

In this implementation, we assume that we know the time-slot for each exam. The program should solve the problem for one time-slot. Actually, the program solves the second phase and optimizes room allocation. We can do this with a little change in the first implementation. The program receives the information of exams in one time-slot instead of the information of all of the exams. We set the number of time-slots in the program to one instead of actual number of time-slots. After these changes the program will optimizes the room allocation phase for a given time-slot. We should run the program for all of the time-slots for a complete result.

This kind of implementation is useful when the time-slot is specified for each exam. For example we may want to specify the time-slot for each exam manually for some reasons such as scheduling according the free times of teachers or allocating hard exams different days.

4 Results

First, we start with a simple example to see the dependency of the result to various parameters. The initial population is 100 and the program evolves the population 100 times. Table 1 shows the amount of each penalty, Table 2 shows the information of the exams and Table 3 shows the information of the rooms.

Table 1. Penalty

REMAINDER_STUDENTS_PENALTY	500
SPLIT_EXAM_PENALTY	200
EMPTY_SEATS_PENALTY	100
MIXED_EXAMS_PENALTY	200
TOTAL_COURSE_CAPACITY_PENALTY	500

Table 2. Exams information

Exam#	Students#	Duration
1	10	60
2	20	100
3	30	60
4	40	100

Table 3. Rooms information

Room#	Capacity
1	20
2	30
3	40

Table 4 shows a sample solution of running the program with above data. Table 5 shows the result of running the program four times and the average of the results. In Table 5, the program calculates the number of orphan students, number of rooms per exam, number of mixed duration exams, number of empty seats in the rooms and number of empty rooms in each time-slot. The numbers in the table are the average of the calculate numbers in the all of time-slots.

Table 4. A sample solution.

Exam#	Time-slot	Room#	Students#	Room#	Students#	Room#	Students#	Room#	Students#	Room#	Students#
1	2	2	10	-	-	-	-	-	-	-	-
2	2	3	20	-	-	-	-	-	-	-	-
3	2	2	11	3	19	-	-	-	-	-	-
4	1	3	40	-	-	-	-	-	-	-	-

Table 5. Results for 100 evolves.

	Run1	Run2	Run3	Run4	Average
Average Number of Orphan Students	0	0	0	1	0.25
Average Number of Rooms per Exam	1.7	2	1.6	1.3	1.6
Average Number of Mixed Exams	0	0	2	0.5	0.6
Percentage of Empty Space in Each Room	25%	24%	30%	15%	23%
Average Number of Empty Rooms	1	0.5	1	1	0.8

If we increase the number of evolve of the population from 100 to 1000, the results will change. Table 6 shows the result of running the program in this situation four times and average of the results. The table shows an improvement in the results. The table shows that the number of rooms per exam decreases and it shows that the number of empty seats in the rooms decreases too. The number of empty rooms increases by increasing the number of evolves. The number of mixed duration exams decreases by increasing the number of evolves. However, the number of orphan students does not change.

Table 6. Results for 1000 evolves.

	Run1	Run2	Run3	Run4	Average
Average Number of Orphan Students	0	0	0	1	0.25
Average Number of Rooms per Exam	1.1	1.3	1	1	1.1
Average Number of Mixed Exams	0.5	0.5	0.5	0.5	0.5
Percentage of Empty Space in Each Room	7%	10%	6%	0%	5%
Average Number of Empty Rooms	1.5	1	1.5	1.5	1.3

Table 7. Exams information.

Exam#	Students#	Duration
1	10	60
2	20	100
3	30	60
4	40	100
5	50	60
6	60	100
7	70	60
8	80	100
9	90	60
10	10	100
11	10	60
12	20	100
13	30	60
14	40	100
15	50	60
16	60	100
17	70	60
18	80	100
19	90	60
20	10	100

After that we run the program for more exams and students. This time the number of time-slots is 5. The initial population is 100 and the program evolves the population 100 times. Table number 1 shows the amount of each penalty, Table 7 shows the information of the exams and Table 8 shows the information of the rooms. In this test the total number of students is 920 and the total capacity is 1000 which can be calculated from multiplying the total capacity of rooms by the number of time-slots.

Table 8. Rooms information.

Room#	Capacity
1	20
2	30
3	40
4	50
5	60

Table 9 shows the result of running the program two times and the average of the results. The average number of orphans is high. Each exam needs approximately 3 rooms and this is not desirable. Average number of empty rooms shows that approximately there is not any empty room in the time-slots.

Table 9. Results for 100 evolves.

	Run1	Run2	Average
Average Number of Orphan Students	31	44	37.75
Average Number of Rooms per Exam	2.8	2.8	2.8
Average Number of Mixed Exams	5.2	4	4.6
Percentage of Empty Space in Each Room	28%	21%	24/5%
Average Number of Empty Rooms	0.6	0.4	0.5

If we increase the number of evolve of the population from 100 to 1000, the results will change. Table 10 shows the result of running the program in this situation two times and average of the results. The number of orphan students decreases and the number of orphan students in previous result is approximately three times greater. It is a great improvement in the results. The other significant changes are in the number of mixed duration exams and the number of empty seats in each room.

As we mentioned earlier, there is another kind of solution for this problem. In this solution the problem is divided into two phases. The first phase is time-slot allocation phase and the second phase is room allocation phase. We show the result of implementing and optimizing the room allocation phase, considering the first phase is done before. We consider the previous example and we assume that in the first phase we choose the exams in time-slot number 1. Table 11 shows these exams. Now we run the program four times. Two times the program evolves the population 100 times and two times it evolves the population 1000 times. Table 12 and Table 13 show the results. It is obvious that the

Table 10. Results for 1000 evolves.

	Run1	Run2	Average
Average Number of Orphan Students	12	14	13
Average Number of Rooms per Exam	2.5	2.7	2.6
Average Number of Mixed Exams	3.6	2	2.8
Percentage of Empty Space in Each Room	13%	13%	13%
Average Number of Empty Rooms	0.6	0.2	0.4

number of orphan students decreases significantly and is zero. Minimizing the number of orphan students is the main goal of the program. Therefore the second solution has more desirable results.

Table 11. Exams information.

Exam#	Students#	Duration
1	10	60
2	20	100
3	30	60
4	40	100

Table 12. Results for 100 evolves.

	Run1	Run2	Average
Average Number of Orphan Students	0	0	0
Average Number of Rooms per Exam	2.2	2	2.1
Average Number of Mixed Exams	8	6	7
Percentage of Empty Space in Each Room	23%	17%	20%
Average Number of Empty Rooms	0	1	0.5

Table 13. Results for 1000 evolves.

	Run1	Run2	Average
Average Number of Orphan Students	0	0	0
Average Number of Rooms per Exam	2.2	2	2.1
Average Number of Mixed Exams	5	3	4
Percentage of Empty Space in Each Room	16%	13%	14.5%
Average Number of Empty Rooms	1	1	1

5 Conclusion

In this paper we described different ways of implementing examinations scheduling problem and the difference between them. We showed the result of each implementation and compared them. The results showed that if we divide the problem into two phases we can get better results. The number of orphan students decreases in this approach. Minimizing the number of orphan students was the most important goal of the implementation so we can conclude that the second implementation is better. The reason is limiting the search space of the problem in the second implementation, so it gives better results in fewer amount of time.

References

1. Whitley, D.: A genetic algorithm tutorial. Stat. Comput. **4**(2), 65–85 (1994)
2. Melanie, M.: An Introduction to Genetic Algorithms. MIT Press, Cambridge (1999)
3. Abdoli, S., Hajati, F.: Offline signature verification using geodesic derivative pattern. In: 22nd Iranian Conference on Electrical Engineering (ICEE), Tehran, pp. 1018–1023 (2014)
4. Barzamini, R., Hajati, F., Gheisari, S., Motamadinejad, M.B.: Short term load forecasting using multi-layer perception and fuzzy inference systems for Islamic countries. J. Appl. Sci. **12**(1), 40–47 (2012)
5. Shojaiee, F., Hajati, F.: Local composition derivative pattern for palmprint recognition. In: 22nd Iranian Conference on Electrical Engineering (ICEE), Tehran, pp. 965–970 (2014)
6. Hajati, F., Raie, AA., Gao, Y.: Pose-invariant 2.5 D face recognition using geodesic texture warping. In: 11th International Conference on Control Automation Robotics and Vision, Singapore, pp. 1837–1841 (2010)
7. Ayatollahi, F., Raie, A.A., Hajati, F.: Expression-invariant face recognition using depth and intensity dual-tree complex wavelet transform features. J. Electron. Imaging **24**(2), 23–31 (2015)
8. Pakazad, S.K., Faez, K., Hajati, F.: Face detection based on central geometrical moments of face components. In: IEEE International Conference on Systems, Man and Cybernetics (SMC 2006), Taiwan (2006)
9. Hajati, F., Cheraghian, A., Gheisari, S., Gao, Y., Mian, A.S.: Surface geodesic pattern for 3D deformable texture matching. Pattern Recogn. **62**, 21–32 (2017)
10. Hajati, F., Faez, K., Pakazad, S.K.: An efficient method for face localization and recognition in color images. In: IEEE International Conference on Systems, Man and Cybernetics (SMC 2006), Taiwan (2006)
11. Hajati, F., Raie, A.A., Gao, Y.: Pose-invariant multimodal (2D+ 3D) face recognition using geodesic distance map. J. Am. Sci. **7**(10), 583–590 (2011)

Solving Job Scheduling Problem Using Genetic Algorithm

Soheila Gheisari[1], Alireza Rezaee[2], and Farshid Hajati[1(✉)]

[1] College of Engineering and Science, Victoria University Sydney, Sydney, Australia
{soheila.gheisari,farshid.hajati}@vu.edu.au
[2] Department of Mechatronics Engineering, Faculty of New Sciences and Technologies, University of Tehran, Tehran, Iran
arrezaee@ut.ac.ir

Abstract. The efficient scheduling of independent computational jobs in a computing environment is an important problem where there are some deadlines for each job to become complete. Finding optimal schedules for such an environment is (in general) an NP-complete problem, and so heuristic approaches must be used. Genetic algorithms are known to give the best solutions to such problems. The purpose of this paper is to propound a solution to a job scheduling problem using genetic algorithms. The experimental results show that the most important factor on the time complexity of the algorithm is the size of the population and the number of generations.

Keyword: Job Scheduling · Genetic Algorithm · Optimal Scheduling

1 Introduction

Machine learning has been used widely in all aspects of the human life [8–16]. Scheduling as a global concept is one of the most important subjects in our everyday life. Another fact is that almost everything is going to be done using computers; and so computers are going to play a vital role in our even simplest ordinary jobs, shopping, banking, communications and lots of such ordinary needs are examples of such simple jobs. Therefore it seems to be very important to schedule jobs on a computer system in the optimum order to satisfy certain characteristics such as least response time, least delay penalty, most performance, and so on.

During the life of genetic algorithms many researchers developed reasonable methods to give a solution for scheduling problem including methods and algorithms for the more special problem, the job scheduling problem on a computer system with different processors. In this project I've focused on the job scheduling problem as well. The aim is to reduce the space complexity and also the time complexity of the algorithm without losing the aspects of the best solution [1].

I have given a full description for the job scheduling problem I've worked on, in the next section. There is an introduction to genetic algorithms in section three. Section four describes the tools I have used to implement my solution. A fully detailed description of my genetic Algorithm solution to this problem is available in section five. Then in the next section I've figured out the obtained results. Finally conclusions are given in section seven.

2   The Job Scheduling Problem

The scheduling problem in its general form is a mapping of a set of jobs or tasks to a set of processors (or machines). The job characteristics (processing time, precedence constraints, deadlines and penalties for certain undesirable properties), machine environment (number of processors, interconnection, power of processors), and performance objectives are the input to the scheduler, and the mapping, or schedule, is the output. The scheduler allocates resources to events. Events are the smallest indivisible schedulable entities, which are given to the scheduler as inputs. Requirements are also input to the scheduler and may range from a real time deadline to the requirement of determining if an improvement in performance is possible. The scheduler takes the events, environment and requirements as input, and produces a schedule [3].

The main problem here is the fact that for a reasonable solution it is expected to meet several requirements, which may conflict or may be in oppositions; therefore an optimum solution have to make a trade off among all its requirements to gain an optimal overall value according to its requirements. I've limited this global definition in my own project, following is the description of the project I worked on: there are a set of tasks to be done, each task has an arrive time, some units of work to be executed this may be thought as the instructions of a program, a deadline at which the task must be completed and if at this time it is not done a penalty will be assigned to that, and also the penalty so called. There are also a set of resources with different powers represented in units of work that resource can do in each unit of time [2].

A solution to this problem has to agree with following requirements. First, the overall response time must have minimum value. The second requirement is concerned with deadlines; the overall deadline has to be minimal as well. Another evident property is that there should not be any overlaps between execution times of different tasks on a single resource. In other words if a task is executing on a certain resource during t1 to t2, no other task can start executing on the same resource at a time between t1 to t2.

Another criterion should be the task's waiting time to get start. The best solution should also have a minimal overall waiting process, and also the minimal execution time. In my project I've taken care of all these requirements, and the project finds the best solution which complies best with all of them. I've demonstrated a whole description of the details of the solution in section five.

3   An Overview of Genetic Algorithms

Genetic algorithms maintain a population of structures, which evolve according to a set of genetic operators, such as selection, crossover and mutation. Each individual in the population is called a chromosome. A chromosome is made up of a list of genes. Genes are the smallest part of the genetic algorithm solutions, they carry units of information with them, and a special set of genes in a special order is treated as solution for the problem, this solution is represented by a chromosome. Each chromosome in the population receives a measure of its fitness according to the values of its genes [4].

Crossover and mutation operations are there to change the values of some genes in the chromosome. The selection operation works on the population and selects a set of chromosomes in the population for the next generation according to a selection method; roulette wheel selection and tournament selection are the most popular methods [5].

4   Tools

I've implemented my project in java; using java genetic algorithm package (jGap), which is a source forge project. This package has implemented all aspects of a genetic algorithm in form of a framework. I've used some built in facilities of this package to develop my own program [6].

The chromosomes of my program are made up of instances of Integer Gene class to represent different values of my solution. I've defined my own fitness function based on requirements of the job scheduling problem I worked on. Then I used a default built in configuration of the package to make a population and evolve it to gain the best solution. This default configuration uses a natural selector as its selection operation called BestChromosomeSelector, which in each generation takes the top n chromosomes into the next generation. The number n is the size of the population (the number of chromosomes in the population) which I've decided to be a constant value through all generations. Which chromosomes are the best is decided by evaluating their fitness value [7].

The configuration also takes the help of a predefined crossover and mutation function as well. The crossover function is defined to make new offspring from two selected parents. The crossover rate in this function has been fixed at the half of the population size which is a constant number in all populations during the evolve process [6].

Another necessary genetic operation is the mutation operator; the mutation operation I used performs with a rate of 1/15, which results in 1/15 of genes being mutated on average.

5   Solution Details

I've described the job scheduling problem I worked on briefly in section two. In this section I'm going to describe it in a more formal way and then I'll give my solution to this problem in details.

The problem is to schedule a set of n registered jobs, called set J on a set of m available resources, called set R. Each job i has three attributes: $arriveTime_i$, $unitsOfWork_i$ Deadline and Penalty. The arrive Time attribute represents the time at which the task (job) is registered. The next attribute, unitOfWork, indicates the amount of work required to complete the task. The deadline attribute is a limit on the time at which the execution of the task has to become complete. If the task is not complete before the specified time a penalty equal to the amount specified in the attribute penalty will be assigned per each unit of time the task completion is delayed.

A solution to this problem, the output of the scheduler, contains a pair of ($startTime_i$, $resource_i$) for each task, which means that the task i is started at $startTime_i$ on resource $resource_i$. For each task i $executionT$ is calculated based on $unitsOfWor$ of the task and the speed of the resource executing the task. The $endTime$ property is then defined to be the sum of the task's $startTime$ and the calculated $executionTime$. According to these definitions for a correct solution it is necessary to hold the following properties:

1. no time conflicts on a resource are allowed:

    $$\forall i \neq j, \; resource_i = resource_j \; srartTime_i > srartTime_j > srartTime_j > endTime$$

2. 2. $\forall i \; srartTime_i > arriveTime_i$

    To simulate such a solution I've defined each gene to represent a ($startTime$, $resource_i$) pair for the corresponding task, therefore each chromosome (representing a solution) is made up of n genes. This representation by itself does not satisfy any of the above necessary conditions, and it's possible to have a chromosome which is not a solution at all. To overcome this problem I've took care about this situation in the fitness function; a solution which does not satisfy these properties will get a very little fitness value, which will result in its extinction very soon. To make it more flexible (especially during the crossover phase) the defined gene above has been split into two simpler genes, the resulting chromosome is shown in Fig. 1.

Fig. 1. The representation of a chromosome

The only remaining part of the solution is to define a fitness function. Such a function should result in larger values for better solution and less values for poorer solution. Therefore to gain a reasonable fitness function for this problem we need to define what a better solution is. To do this some criteria are needed then the solution which best fits those criteria will be the optimal solution. The followings are the criteria I've chosen to select the best solution:

*   The overall conflict among the tasks scheduled on a single resource.
*   The overall penalty
*   The overall difference of arriveTime and startTime of all tasks
*   The overall executionTime

The better the solution the fewer sums of these factors. Each chromosome is evaluated according to these four factors. At the beginning the fitness value is set to the most available integer, and then the value of each factor is calculated using the tasks attributes, resources properties and calculated values such as endTime mentioned above, the value of each factor is then multiplied by a special coefficient, which shows the importance of the factor related to other factors, these scaled values are then subtracted from the initial value of the fitness function to gain the exact fitness value of the chromosome. This value is then used in other genetic algorithm operations, mostly the selection and crossover operations to evolve the population.

6 Results

In this section I have analyzed the results achieved from the algorithm described in the previous section. The last solution of the algorithm results in is affected by the following parameters:

- The size of the population.
- The number of generations.
- The coefficients of the factors described above.

Although the impact of the first two parameters is obvious and is trained several times, during the lifetime of the genetic algorithms, finding an effective and efficient value for these parameters is very important. It's because although a bigger population will result in a more precise, fitter solution but it also takes more space and time to yield to this result and in a similar fashion does the second parameter, the number of generations to be evolved.

According to these all I found that an effective population for this problem should contain 1500 chromosomes and an effective number of generations should be 300.

Besides this is the more important part of dealing with the values of the four coefficients corresponding to the factors mentioned above. Indeed a more relative coefficient for the "Resource usage conflicts" denotes that the algorithm emphasizes to find a correct solution, while a bigger coefficient for the "deadline violation penalty" will result in a solution which tries to complete jobs before the deadlines. And a bigger coefficient for the "overall execution time" will select a solution with less total amount of time spending to execute the tasks, which means the optimum resource allocation in the system. Such a solution might not provide the two necessary properties described before, and so will not be a solution at all. Therefore we need to make a trade off among all these factors to achieve a correct, deadline complier, optimal solution. The affect of these parameters on a specific problem is summarized in Table 1.

The results in Table 1 are the obtained result of the program running to solve a job scheduling problem with the following characteristics:

- there are 14 tasks with overall units of works equal to 161
- 5 resource with an average speed of 2 units of work per time unit
- another limitation is that all the tasks have to be done at time unit 60

According to the first three rows of the table the running time of the program is really proportional to the size of the population and the number of generation evolved. Rows 3, 4, and 5 show how the coefficient of the conflict penalty effects on the given result.

Table 1. The result of running of the program with different values for parameters, the changed values on each running related to the previous are in bold characters.

No.	Pop size	generations	conflict	exec time	penalty	arrive_start time	chromosomes	fitness	running time
1	1000	200	7*1000	78*1000	0.0*2000	95*1	361469	2.147476552E9	35 s
2	**1500**	200	3*1000	67*1000	0.0*2000	118*1	542034	2.147480529E9	58 s
3	1500	**300**	8*1000	62*1000	0.0*2000	82*1	812006	2.147475565E9	89 s
4	1500	300	4*1000	64*1000	0.0*2000	106*1	811888	2.147479541E9	88 s
5	1500	300	14*1000	64*1500	0.0*2000	86*1	811919	2.147469561E9	92 s
6	1500	300	7*1000	65*1500	0.0*2000	99*1	811958	2.147476648E9	96 s
7	1500	300	9*1000	61***2000**	0.0*2000	86*1	811836	2.147474561E9	105 s
8	1500	300	0*1000	63***1000**	120.0*20	121*1	812066	2.147243526E9	98 s
9	1500	300	4*1000	64*1500	0.0***200**	88*1	812236	2.147479559E9	94 s
10	1500	300	1***5000**	67*1000	10.0*200	110*1	812231	2.147481537E9	96 s
11	1500	300	0*5000	71*1000	20.0*200	115*1	811949	2.147481532E9	104 s
12	1500	300	0*5000	63*1000	100.0*200	128*0	811657	2.147473519E9	107 s

Line seven has the most coefficients for the execution time, which truly offers the least execution time.

The impact of the penalty coefficient on the amount of the penalty is shown in line 8; another significant change indicating in this line is in the value of the conflicts which is decreased to zero, which in the two last rows is again reached zero as a result of its very large coefficient. The last row shows that the "distance of the *startTime* to *arriveTime*" does not a major effect on the final result of the problem. Figure 2 shows the output of the program corresponding to the last row of the Table 1.

```
Task0:  arriveTime: 0, deadline: 12, workUnits: 5, startTime: 5, Resource4 (3), endTime: 7

Task1:  arriveTime: 1, deadline: 9, workUnits: 4, startTime: 3, Resource1 (2), endTime: 5

Task2:  arriveTime: 0, deadline: 19, workUnits: 7, startTime: 14, Resource1 (2), endTime: 18

Task3:  arriveTime: 1, deadline: 9, workUnits: 3, startTime: 4, Resource3 (3), endTime: 5

Task4:  arriveTime: 2, deadline: 12, workUnits: 3, startTime: 2, Resource2 (1), endTime: 5

Task5:  arriveTime: 0, deadline: 12, workUnits: 15, startTime: 0, Resource4 (3), endTime: 5

Task6:  arriveTime: 2, deadline: 20, workUnits: 17, startTime: 18, Resource4 (3), endTime: 24

Task7:  arriveTime: 5, deadline: 20, workUnits: 20, startTime: 14, Resource3 (3), endTime: 21

Task8:  arriveTime: 1, deadline: 11, workUnits: 10, startTime: 8, Resource1 (2), endTime: 13

Task9:  arriveTime: 1, deadline: 12, workUnits: 3, startTime: 0, Resource3 (3), endTime: 1

Task10: arriveTime: 1, deadline: 12, workUnits: 13, startTime: 7, Resource4 (3), endTime: 12

Task11: arriveTime: 0, deadline: 50, workUnits: 31, startTime: 35, Resource4 (3), endTime: 46

Task12: arriveTime: 0, deadline: 27, workUnits: 19, startTime: 23, Resource3 (3), endTime: 30

Task13: arriveTime: 0, deadline: 12, workUnits: 11, startTime: 6, Resource3 (3), endTime: 10
results --> aSDiff: 128, execTime: 63, penalty: 100.0, confilict penalty: 0, fitness: 2.147473519E9

process started at: Wed Jan 25 12:14:26 IRST 2006, and completed at: Wed Jan 25 12:16:13 IRST 2006
```

Fig. 2. A sample result of the program.

Table 2 summarizes the effect of the size of the problem on the running time of the project. The given running times show that the size of the problem specially the number of the resources has not a straight effect or significant effect on the time complexity of the algorithm. The dominant factor on the time complexity is the size of the population and also the number of generations.

Table 2. The effect of the size of the problem on the running time.

No	Mumber of tasks	Number of resources	Pop size	Number of generation	Running time
1	2	5	1500	300	79 s
2	30	5	1500	300	216 s
3	30	10	1500	300	239 s
4	30	25	1500	300	220 s

7 Conclusion

In this paper I have presented a genetic algorithm solution to a special job scheduling problem. The given approach tries to make better solutions according to several criteria with different importance and necessity. I have evaluated the performance of the algorithm in job scheduling using real tests, and the experimental results achieved have shown its effectiveness and efficiency in finding an optimal or a near optimal scheduling solution in terms of a large number of jobs and available resources. Further work could experiment using the advantage of the solution of a low cost primitive heuristic job scheduling algorithm such as local search techniques. Another technique to gain a better solution with a better performance is to combine the given genetic algorithm with other artificial intelligence approaches, such as a Tuba search or a simulated annealing approach.

References

1. Abraham, A., Buyya, R., Nath, B.: Nature's heuristics for scheduling jobs on computational grids. In: Proceedings of the 8th International Conference on Advanced Computing and Communications, pp. 45–52. Tata McGraw-Hill, India (2000)
2. Wang, L., Cai, J., Li, M., Liu, Z.: Flexible Job Shop Scheduling Problem Using an Improved Ant Colony Optimization, Scientific Programming, pp. 9016303 (2017)
3. Ritchie, G., Levine, J.: A fast, effective local search for scheduling independent jobs in heterogeneous computing environments. Proceedings of the 22nd Workshop of the UK Planning and Scheduling Special Interest Group (2003)
4. Aarts, E.H.L., Van Laarhoven, P.J.M., Lenstra, J.K., Ulder, N.L.J.: A computational study of local search algorithms for job shop scheduling. ORSA J. Comput. **6**, 118–125 (1994)

5. Rezaee, A., Ajelli, A.: Problem solving of graph correspondence using genetics algorithm and ACO algorithm. Int. J. Innov. Res. Sci. Eng. Technol. **2**(12), 7785–7791 (2013)
6. Jang, W., Jong, D., Lee, D.: Methodology to improve driving habits by optimizing the in-vehicle data extracted from OBDII using genetic algorithm. In: 2016 International Conference on Big Data and Smart Computing (BigComp). pp. 313–316 (2016)
7. Shamsieva, A.M., Arkov, V.U.: On genetic algorithm methodology for robust system design. In: 2011 IEEE 12th International Symposium on Computational Intelligence and Informatics (CINTI), pp. 421–426 (2011)
8. Abdoli, S., Hajati, F.: Offline signature verification using geodesic derivative pattern. In: 22nd Iranian Conference on Electrical Engineering (ICEE), Tehran, pp. 1018–1023 (2014)
9. Barzamini, R., Hajati, F., Gheisari, S., Motamadinejad, M.B.: Short term load forecasting using multi-layer perception and fuzzy inference systems for Islamic countries. J. Appl. Sci. **12**(1), 40–47 (2012)
10. Shojaiee, F., Hajati, F.: Local composition derivative pattern for palmprint recognition. In: 22nd Iranian Conference on Electrical Engineering (ICEE), Tehran, pp. 965–970 (2014)
11. Hajati, F., Raie, A.A., Gao, Y.: Pose-invariant 2.5 D face recognition using geodesic texture warping. In: 11th International Conference on Control Automation Robotics and Vision, Singapore, pp. 1837–1841 (2010)
12. Ayatollahi, F., Raie, A.A., Hajati, F.: Expression-invariant face recognition using depth and intensity dual-tree complex wavelet transform features. J. Electr. Imaging **24**(2), 3–31 (2015)
13. Pakazad, S.K., Faez, K., Hajati, F.: Face detection based on central geometrical moments of face components. In: IEEE International Conference on Systems, Man and Cybernetics (SMC 2006), Taiwan (2006)
14. Hajati, F., Cheraghian, A., Gheisari, S., Gao, Y., Mian, A.S.: Surface geodesic pattern for 3D deformable texture matching. Pattern Recogn. **62**, 21–32 (2017)
15. Hajati, F., Faez, K., Pakazad, S.K.: An efficient method for face localization and recognition in color images. In: IEEE International Conference on Systems, Man and Cybernetics (SMC 2006), Taiwan (2006)
16. Hajati, F., Raie, A.A., Gao, Y.: Pose-invariant multimodal (2D + 3D) face recognition using geodesic distance map. J. Am. Sci. **7**(10), 583–590 (2011)

Functionizer - A Cloud Agnostic Platform for Serverless Computing

Oliviu Matei$^{(\boxtimes)}$, Katarzyna Materka, Paweł Skyscraper, and Rudolf Erdei

Cluj-Napoca, Romania
{oliviu.matei,rudolf.erdei}@holisun.com, {kmaterka,pskrzypek}@7bulls.com

Abstract. The article presents MELODIC/FUNCTIONIZER platform, developed to support serverless deployment. It is a complete cross-cloud, cloud-agnostic deployment platform with advanced, utility-based optimization and multi-layer and real time monitoring features. This platform exploits the modelling language CAMEL, which advances the state-of-the-art through its ability to cover all aspects relevant to the application life-cycle in a rich manner. The platform has been tested with a complex application for face recognition. These platform and the serverless architecture is stable and very scalable, and the results are very good in terms of resources and execution time.

1 Introduction and Motivation

Serverless computing is a dynamic field that is currently in full development and leads to new ways of exploiting of cloud applications. The key benefits delivered by this model include budget savings as well as the flexible scaling of the serverless functions. Based on that, different tools like Fission and Serverless.com, have been introduced to support cloud DevOps specialists in the configuration and operations of serverless components. Major cloud providers like Amazon, Google and Microsoft offered serverless platform, which are able to flexibly deploy and scale of such components.

On the other hand, based on our experience and knowledge, currently there is tools or library able to support the configuration and management of multicloud systems. In this paper we are introducing the FUNCTIONIZER platform which covers this feature by adding these features to the MELODIC multicloud optimization platform. The MELODIC platform, extended in FUNCTIONIZER project is able to deploy and manage serverless components in multicloud environment, as the addition to the already existing support for virtual machines and containers. This extension has the following features: it reuses existing work and requires less development effort, it fulfills requirements related to the effective operations of multicloud applications, it supports and solve some challenges related to the operation and deployment of serverless components in multicloud environment.

L. Barolli et al. (Eds.): AINA 2021, LNNS 227, pp. 541–550, 2021.
https://doi.org/10.1007/978-3-030-75078-7_54

In this paper we present in details MELODIC/FUNCTIONIZER platform [3], which extends the MELODIC platform by providing advanced support for automatic deployment and optimization of the serverless components.

1.1 State of the Art Review and Available Software

Kristina and Skrzypek [2] present a comprehensive survey of the serverless frameworks.

The Fission platform uses Cybernetics to operate a mini-serverless platform over the already existing client-chosen cloud infrastructures. Its most important features include support for all programming languages and also maintains a pool of containers to solve the cold start problem [1].

Kubeless[1], another provisioning framework with Kubernetes under the hood, also supports multiple programming languages and custom run times. Via Kafka, the framework provides its own event triggering. It has a UI where functions that can be called and monitored [12,13].

IronFunctions[2] relies on Kubernetes, Docker Swarm and Mesosphere for resource management [10]. It is also able to import Lambda functions and supports any programming language. IronFunctions also has an UI for calling and updating functions and also solves the cold-start problem by linking hot functions to hot containers.

A provisioning framework for AWS Lambda or Amazon Cloud [2] is Sparta[3]. It is developet in GoLang, so all configuration and infrastructure elements can be defined as types. Via its UI, Grafana, it can monitor the status of functions.

Another provisioning framework, very similar to IronFunctions, is Fn[4]. Functions can be grouped into what is called an app, also variables can be configured in three levels: application, function and route.

For the Python and Java programming languages, we have Snafu[5] [15], a serverless host process. It can import from platforms like AWS Lambda and IBM OpenWhisk and export to other platforms, like Fission and Kubeless. It supports faasification [16] as well as multiple connectors/triggering mechanisms to functions.

An abstraction framework over AWS Lambda, Azure Functions, IBM OpenWhisk, Google Cloud Functions and SpotInst, us Serverless[6] [2]. It can be integrated with Kubeless and Fn. Configuring the deployment of functions is done via CloudFormation modelling language. Monitoring the health of the platform is also implemented in it, as a feature. Also, it can provide support for WSGI apps through Flask, Django or Pyramid.

---

[1] https://kubeless.io/.
[2] https://open.iron.io/.
[3] https://gosparta.io/.
[4] https://fnproject.io/.
[5] https://github.com/serviceprototypinglab/snafu.
[6] https://serverless.com/framework/docs/getting-started/.

2 Requriements, Architecture and Workflow

In this section, the overall architecture of the MELODIC/FUNCTIONIZER platform is presented. The platform was developed under the scope of the H2020 MELODIC project and extended within FUNCTIONIZER project to support serverless deployment. This is a complete cross-cloud, cloud-agnostic deployment platform with advanced, utility-based optimization and multi-layer, real time monitoring features. This platform exploits the modelling language CAMEL, which advances the state-of-the-art through its ability to cover all aspects relevant to the application life-cycle in a rich manner, while catering for the models@run.time paradigm at the deployment and monitoring aspects, enabling the adaptive provisioning of cloud applications. MELODIC also features an advanced Event Management System (EMS), which is able to monitor the cloud application across different clouds and abstraction levels, thus providing the right monitoring feedback for the continuous optimisation loop. The latter is realised through advanced optimisation using various methods in conjunction with constraint programming models.

The application in the platform is described in the CAMEL language [14]. The CAMEL language is the advanced cloud modelling language, which extends the concepts of the Topology and Orchestration Specification for Cloud Applications (TOSCA) standard published by OASIS (The Organization for the Advancement of Structured Information Standards).

The components of the described MELODIC platform uses the same application model. MELODIC can be considered as a implementation of the models@run.time approach for cloud computing applications. The CAMEL model is cloud provider independent and agnostic to any cloud provider. The key advantage of the CAMEL model is ability to create a constraint programming model (CP MODEL). This model is the input to the solvers, which are able to solve cloud optimization problem.

MELODIC/FUNCTIONIZER has been created based on the outcome from previous projects. It has integrated their components s into one platform. The very important challenge was to integrate various components and tools which use various integration approaches (tools, protocols, integration methods). That was the reason that the selection of the integration layer was very important to build the described platform. Finally the Enterprise Service Bus with orchestration based on Business Process Management standard was selected as the most efficient strategy for integration. ESB is very popular integration approach used by many companies and integration. BPM is the most used way of modeling business processes in IT systems. The integration layer of the platform contains two planes: **Control Plane** for business logic integration and controlling, and **Monitoring Plane** for monitoring related activities.

The presented integration methodology has the following benefits: orchestration flow is defined in business process through the UI interface instead of hard coding, r both integration methods are supported (synchronous and asynchronous), high availability, wide support of the integration methods and protocols, easy integration with external applications due to standard protocols and interfaces (Fig. 1).

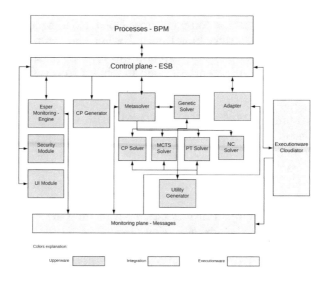

Fig. 1. The architecture of the platform

The sequence of actions performed by the MELODIC/FUNCTIONIZER platform is presented in Fig. 2.

User is modelling the application and requirements in CAMEL language. Based on the prepared model the CP Model is generated. The actual values of metrics are used to create optimization problem. After that, the selection of the most appropriate solver is done by Metasolver. The selected solver starts solving the optimization problem. For each candidate solution the Utility Generator is invoked to calculate utility value.

The best found solution of the optimization problem is to passed to the Adapter component to perform deployment. The Adapter orchestrates deployment of the solution and monitoring component through Cloudiator.

After the deployment, application metrics starts being collected by monitoring component.

Real values of metrics are started to be sent to the Metasolver. The Metasolver decides about starting the reconfiguration process based on the values of these metrics.

After starting the process of reconfiguration the Metasolver enrich the CP model with the latest, real values of metrics. Then again passes the CP Model to solver. The solver starts optimization process again using the Utility Generator to calculate the utility function value. After finding the best current solution, the deployment is executed again, as described above.

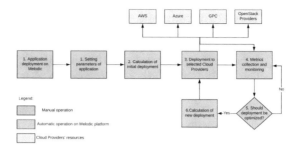

Fig. 2. The workflow of the platform

3   Platform Testing

This section covers a short description of the testing environments and their scope, as well as some particular restrictions and guidelines that apply to them. There are three stages of testing, which are directly related to the respective testing environments: development tests, integration tests and acceptance tests.

In development test environments, the tests will be performed only by developers. In integration test environments, the tests will be executed by developers and testers, and in acceptance test environments, the tests will be executed by testers.

3.1   Development Tests Environment

The purpose of Development Tests environments is to ensure that the code produced by a developer is executable before committing changes to the component's master branch repository. This kind of environment is suitable for performing Unit Tests. There should be at least one development test environment for each developer involved in the project. Each such environment should be created and maintained by the developer himself and software builds should be performed manually on demand. A developer's test environment should contain only platform modules needed for a certain development (as there is no need to maintain the whole platform).

3.2   Integration Tests Environment

The purpose of an Integration Tests environment [3] is to allow developers to integrate the components that they develop with stable releases of the rest of the platform components. There are one integration environment created and maintained including all of the project's application components. It is utilizing Continuous Integration (CI) practice with the use of the Jenkins automation server. Software builds are performed automatically from the master branch after each commit and an additional build is executed on a nightly basis. This type of environment is used to perform Functional Tests [9].

3.3 Acceptance Tests Environment

An Acceptance Tests [11] environment is used by the test team to perform tests over final versions of the platform. Therefore, all the project's components are installed and used. The Acceptance Tests environment is also using of Jenkins CI. The main difference is that builds are performed not from the master branch, but from specific release (or release candidate) branches, when needed. At the beginning, smoke tests are executed in the Acceptance Tests environment. Smoke testing consists of a subset of test cases from functional testing, executed at the beginning of the functional phase to ensure that the system is stable. In the next step, with the use of this test environment all of the following types of tests are performed: Load, Security and Acceptance.

4   Platform Validation

The validation application of the Functionizer platform has been developed as described by Matei et al. in [8].

The business case must validate the Functionizer platform [4], by demonstrating that serverless computing can be effectively incorporated in the multi-cloud domain and demonstrate how Functionizer makes deployment and management of multi-cloud data-intensive applications faster, simpler and cheaper. The business case focuses on a solution, which takes an audio/video stream from wearable devices (e.g. smart glasses) and processes it on a server for face/object recognition. It has applications in several fields, such as: industry, medicine, emergency response and training.

The features employed by the solutions include:

Audio/Video Streaming: There is a two-way audio/video stream, complete with H.264 Encoding.

Chat: The base mean of communication for file transfer and drawings. Can be used when audio/video feed is low quality.

Drawings: Users can draw simple shapes, useful for pointing and explaining.

File Transfer: Files can be transferred between users and viewed. Most useful formats are AutoCAD, text, images (e.g. manuals).

Platform Independency: The solution was tested on: Android 5.1+, ReticleOS, iOS, and Windows. Tested devices include: desktops/notebooks, smart glasses (ODG R7 (HL) and Epson Moverio BT-300), and smartphones with the indicated OS.

Snapshots/Recordings: Users can save snapshots from the live feed, or even record it, for various purposes.

Multi-user Video-Conferencing: Multiple users can be involved in a conference call, with basic information (GPS coordinate and browser).

4.1   Serverless Architecture

The image/face recognition related modules are very complex, therefore we present the second level decomposition of AR Assistance architecture related to them and are depicted in Fig. 3.

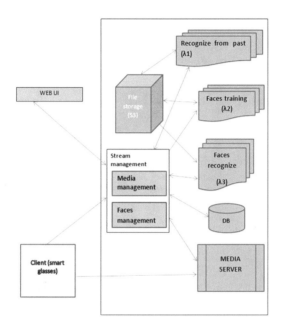

Fig. 3. The serverless architecture of the application

The colours of the components have the following meanings: the **gray** components represent the clients (dark grey is the web client and the light grey depicts the wearable client); The **green** components are already serverfull components; the **ochre** components cloud side components.

The architecture consists of two tiers, the 1st one being the client and the 2nd one being the business logic, consisting of:

- Serverful components are deployed in the cloud and refer to: Stream management, server (Kurento Media Server), Database (DB) (Maria DB)
- Serverless components: 3 lambda functions: Recognize from past, Faces training and Faces recognize, 3rd party component for file storage (e.g. Amazon S3 service).

4.1.1   Sequence Diagram

The sequence diagram for AR Assistance is depicted in Fig. 4. The red life lines are serverful (belonging to components living indeterminable), whereas the green life lines belong to serverless components, with a relatively short lifespan.

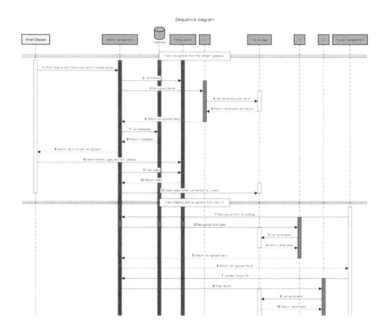

Fig. 4. The serverless sequence diagram

The sequence diagram depicted in Fig. 4 has two lines: on-line (from the wearable client) and off-line (from the web client), split with a horizontal bar. All use cases start from the client - web for training, respectively wearable for production.

The wearable client connects to media manager and establishes a communication session. In turn, media manager captures the frames from the Media Server (which is connected to the wearable via WebRTC) and transfers them to λ_3 (for face recognition), which returns the metadata of the recognized person (if it's the case) to Media Manager and from there to the client. For face recognitions, λ_3 uses images stored in File Storage.

Concurrently, the video stream stored for further detailed processing and historical augmentation. Off-line, the face recognitions can be run on stored videos, an action initiated from the web client.

However, the face recognition algorithm needs to be trained for running properly. This sequence has been depicted as last one because it is less important in the economy of the application (as it runs less than the former ones).

Extensive tests have been performed and reported by Matei et al. [7]. The serverless architecture is stable and very scalable, and the results are quite good (but not yet perfect or usable, from a military point of view). However, a reduction of the computation time is needed if the application is to be used in real life scenarios, such as the ones reported by Matei et al. [5,6].

5   Conclusions and Future Works

The presented MELODIC/FUNCTIONIZER platform introduces support of serverless components management, automatic deployment and optimization. It allows to management the holistic applications which contains different types of components: virtual machines, containers and serverless components. The platform will be further extend towards maturity and commercialization with focus on user interface and platform robustness. Also, the research work on polymorphic adaptation is currently conducted. It should allow to introduce additional level of optimization, due to optimization of application component deployment type. It will be possible to select best component deployment type, i.e. virtual machine, container or serverless, based on the current workload and applications needs.

Acknowledgement. This work has received funding from the Functionizer Eurostars project E!11990.

References

1. Baldini, I., Castro, P., Chang, K., Cheng, P., Fink, S., Ishakian, V., Mitchell, N., Muthusamy, V., Rabbah, R., Slominski, A., et al.: Serverless computing: Current trends and open problems. In: Research Advances in Cloud Computing, pp. 1–20. Springer (2017)
2. Kritikos, K., Skrzypek, P.: A review of serverless frameworks. In: 2018 IEEE/ACM International Conference on Utility and Cloud Computing Companion (UCC Companion), pp. 161–168. IEEE (2018)
3. Kritikos, K., Skrzypek, P.: Simulation-as-a-service with serverless computing. In: 2019 IEEE World Congress on Services (SERVICES), vol. 2642, pp. 200–205. IEEE (2019)
4. Kritikos, K., Skrzypek, P., Moga, A., Matei, O.: Towards the modelling of hybrid cloud applications. In: 2019 IEEE 12th International Conference on Cloud Computing (CLOUD), pp. 291–295. IEEE (2019)
5. Matei, O., Anton, C., Bozga, A., Pop, P.: Multi-layered architecture for soil moisture prediction in agriculture 4.0. In: Proceedings of international conference on computers and industrial engineering, CIE, vol. 2, pp. 39–48 (2017)
6. Matei, O., Anton, C., Scholze, S., Cenedese, C.: Multi-layered data mining architecture in the context of internet of things. In: 2017 IEEE 15th International Conference on Industrial Informatics (INDIN), pp. 1193–1198. IEEE (2017)
7. Matei, O., Erdei, R., Moga, A., Heb, R.: A serverless architecture for a wearable facerecognition application. In: Proceedings of RISS 2021 (2021)
8. Matei, O., Skrzypek, P., Heb, R., oga, A.: Transition from serverfull to serverless architecture in cloud-based software applications. In: 4th Computational Methods in Systems and Software 2020. IEEE (2020)
9. Maxwell, P., Hartanto, I., Bentz, L.: Comparing functional and structural tests. In: Proceedings International Test Conference 2000 (IEEE Cat. No. 00CH37159), pp. 400–407. IEEE (2000)
10. McGrath, G., Brenner, P.R.: Serverless computing: design, implementation, and performance. In: 2017 IEEE 37th International Conference on Distributed Computing Systems Workshops (ICDCSW), pp. 405–410. IEEE (2017)

11. Miller, R., Collins, C.T.: Acceptance testing. In: Proceedings XPUniverse, p. 238 (2001)
12. Mohanty, S., et al.: Evaluation of serverless computing frameworks based on kubernetes (2018)
13. Mohanty, S.K., Premsankar, G., Di Francesco, M., et al.: An evaluation of open source serverless computing frameworks. In: CloudCom, pp. 115–120 (2018)
14. Rossini, Alessandro: Cloud application modelling and execution language (camel) and the passage workflow. Adv. Serv. Oriented Cloud Comput-Workshops ESOCC **567**, 437–439 (2015)
15. Spillner, J.: Snafu: Function-as-a-service (faas) runtime design and implementation. *arXiv preprint* arXiv:1703.07562 (2017)
16. Spillner, J., Dorodko, S.: Java code analysis and transformation into AWS lambda functions. *arXiv preprint* arXiv:1702.05510 (2017)

Interactive Application Deployment Planning for Heterogeneous Computing Continuums

Daniel Hass[1]([⊠]) and Josef Spillner[2]

[1] Endress+Hauser InfoServe, Weil am Rhein, Germany
`daniel.hass@endress.com`
[2] Zurich University of Applied Sciences, Distributed Application Computing
Paradigms, Winterthur, Switzerland
`josef.spillner@zhaw.ch`
`http://blog.zhaw.ch/splab/`

Abstract. Distributed applications in industry are modular compositions involving containers, functions and other executable units. To make them deployable and executable in production, they need to be assigned to heterogeneous resources in computing continuum's, consisting of multiple clouds, devices and other runtime platforms. Existing assignment processes are neither transparent nor interactive. To overcome this limitation, we introduce the Continuum Deployer, a tool capable of reading application descriptions (e.g. Helm charts for Kubernetes), interactively performing and comparing assignment algorithms (e.g. to multiple K8s and K3s cloud/edge/fog/device resources), and exporting deployment files (e.g. Kubernetes manifests). We evaluate the tool with empirical package analysis, exemplary deployments and a synthetic experiment to appropriately address scalability concerns.

1 Introduction and Problem Statement

Computing continuums combine multi-cloud resources with local devices, including resource-constrained (mobile) edges and fogs [1]. They permit the optimal distribution of complex distributed application functionality along the data path. Sensor data processing continuums, for instance, may perform filtering early at the point of sensing, and complex analytics later in appropriate clouds [2]. Building distributed applications for such infrastructures has become feasible due to the proliferation of many microservice and serverless technologies [3] and corresponding packaging formats for compositions and workflows. However, which application part should go to which resource along the continuum is still an open challenge especially in automation-focused industrial domains. We refer to it as *application–continuum resource assignment problem*. Recent research has led to many automated algorithms to solve the assignment problem under

© The Author(s), under exclusive license to Springer Nature Switzerland AG 2021
L. Barolli et al. (Eds.): AINA 2021, LNNS 227, pp. 551–560, 2021.
https://doi.org/10.1007/978-3-030-75078-7_55

constraints and preferences. Proposals encompass the definition of cost/utility functions for the Hungarian algorithm, rule-based matchmaking [4], topology splitting and matching [5], and device-driven adaptive deployment [6]. However, none of these approaches keep the person responsible for deployment in the loop. Many DevOps engineers and administrators would prefer a guided approach where automation is built in but certain decisions can be controlled and followed in a transparent way. Such an anticipated approach is in line with recent trends towards responsible and explainable artificial intelligence, equally applying to automation decisions in upcoming continuums such as decentralised and osmotic computing, cloud robotics, cloud-based manufacturing processes and vehicular clouds [7]. To accommodate this need, this paper introduces the *Continuum Deployer*, a novel interactive tool to solve the assignment problem especially for fast-paced industrial DevOps processes.

2   Solution Space and Method

The solution space to constrained assignment is potentially large in practice due to the following factors:

1. Lack of resource descriptions. Although there are standards like the Common Information Model to describe infrastructure capabilities, they are rarely used in practice. Resources themselves (e.g. mobile phones, smart watches, IoT devices or servers) also do not ship with self-descriptive capabilities.
2. Application description variety. In contrast to resources, engineers have various approaches available to describe their applications and how to deploy them. Modern distributed applications are either deployed through infrastructure as code or through declarative formats, including low-level Docker Compose files and Kubernetes manifests, high-level Helm charts and Cloud-Native Application Bundles (CNAB), Open Application Model (OAM) or Topology and Orchestration Specification for Cloud Applications (TOSCA) files for container-based compositions, and AWS Serverless Application Models (SAM) for function-based compositions [5].
3. Scalability. Each application part may be scaled independently from the others, leading to the need to reserve a multiple of its resource requirements on the same or on different resources.
4. Constraints and preferences. Application descriptions may ship with a-priori constraints on where deployments is allowed. In practice, this information is often absent and may have to be added in an interactive and incremental way, in each iteration followed by a check of the resulting deployment plan. The same process is needed for custom preferences – for instance, to declare that although a database could run on the (mobile) edge, it should instead run in the cloud because a later upgrade of the application needs to access it there.

To address this large solution space, Fig. 1 contains the approach that gradually achieves solutions of different maturity (design, implementation, evaluation) for a subset of industrially relevant cloud application technologies. The remaining sections cover the steps of the proposed method and lead to the presentation of an applied deployment tool which we make available as open source software.

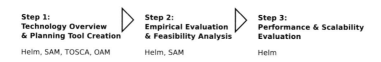

Fig. 1. Solution approach

3   Tool Design

We assume that both resources and application descriptions have matching properties, in particular: CPU cores, memory capacity, and preference labels such as 'private cloud' or 'in-vehicle', and that these properties are accurately described for at least the majority of these entities. Labels are the central mean within the Continuum Deployer for the user to express certain constraints and preferences with regard to the deployment placement. Each node and workload can be assigned with zero to as many labels as the user desires. A suitable node must or should (depending on configuration) possess all of the workloads labels or more to be considered for a deployment. Unlabelled workloads are able to run on any of the available nodes.

The labels are then used to distinguish resource capabilities, ranging from vast and elastically scalable cloud resources to constrained device resources which, due to these constraints, might not always allow for executing all workloads resulting from the heterogeneous packaging formats. Figure 2 shows a subset of the problem space, leading to the research question: How can a complex microservice-based application, packaged as Helm chart, AWS SAM, OAM appfile or any other format, be deployed to resources with different resource constraints while adhering to user preferences?

The anticipated tool design thus needs to fulfil the following requirements:

1. Extensible support for multiple application packaging formats, deployment languages and assignment resolver algorithms (matchers).
2. Consideration of scaling factors and the resulting resource multiplicity.

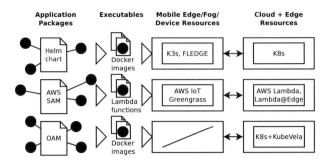

Fig. 2. Schema of deployment planning for various application packaging formats

3. Specification and consideration of user-specific preferences, primarily in the sense of preferring one resource over another (i.e. cloud-first, device-first) but also allowing for more fine-grained trade-offs.
4. Interactive as well as non-interactive use, both with tolerance for incomplete resource specifications.

4   Assignment Algorithms

To ensure the functionality of extensible algorithms, we implemented and integrated two exemplary matchers with vastly different capabilities.

1. Greedy search. The greedy solver asks for a single optimisation target. Implemented are the resource and workload sorting and subsequent greedy matching for CPU and memory, although other metrics can be trivially added. The largest workloads are probed for placement on a sorted list of resources. In this list resources appear in descending order based on the selected optimisation target.
2. SAT search. The constraint satisfaction problem (CP-SAT) solver [8] offers multiple options with regard to the optimisation target. Implemented are six single and multiple targets, including the maximisation of idle CPU, the minimisation of idle memory, or the maximisation of idle combined resources. The implementation documentation gives more details on what is implemented and how to implement further combinations. This solver uses constrained programming to define rules and constrains that describe the resource matching problem in mathematical terms. Afterwards this optimisation is solved as optimal as possible. The results of this solver differ from the greedy ones: if this solver cannot come up with an optimal solution the run will fail and all resources are displayed as unschedulable. This feasibility constraint is enforced on each label group (if labels are defined).

5   Tool Implementation

Continuum Deployer is implemented as command-line tool with several input assistance features including syntax completion and adaptive menus. For the purpose of automation, arguments can be passed to reduce or even avoid any interactive step. A state machine is used to control the internal workflow and any shortcuts through passed arguments, as shown in Fig. 3.

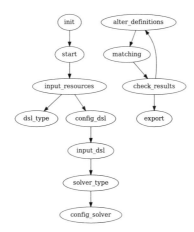

Fig. 3. State machine representing the process of deployment planning

The initial interaction screen implies the parsing of the specified resource definition file and shows the initial resource utilisation. Figure 4 shows the case of three initially 'empty' resources, referring to all resource metrics (processor utilisation, memory allocation) beyond the necessary minimum occupied by the operating system, virtualisation environment and cloud stack (e.g. Kubernetes).

Afterwards, the user selects a domain-specific language, and furthermore selects among given file format options (Fig. 5). Eventually, an application description in the selected domain-specific language (DSL) and format is loaded, and the matchmaking to assign the set of application parts to the set of resources begins. The deployer implementation supports Helm as DSL with local, templated and packaged Helm charts as flavours.

The matchmaking performs the user-selected algorithm (SAT, greedy or user-extensible search). The SAT solver is implemented with the Google OR tools for combinatorial optimisation.

Eventually, the best assignment is visualised. Even the best one may not be sufficient and requires a (minimum) amount of additional resources. In the given scenario (Fig. 6), a basic deployment of the Nginx web server is possible but a replicated deployment is not, as evidenced by the inability to find a suitable spot for the replica `nginx-deployment-1--2`.

```
 / ___|___ _ __ | |_(_)_ __ _ _ _   _ _ __ ___    | _ \ ___ _ __ | | ___  _   _  ___ _ __
| |   / _ \| '_ \| __| | '_ \| | | | | | | '_ ` _ \   | | | |/ _ \ '_ \| |/ _ \| | | |/ _ \ '__|
| |__| (_) | | | | |_| | | | | |_| | |_| | | | | | |  | |_| |  __/ |_) | | (_) | |_| |  __/ |
 _____/|_| |_|\__|_|_| |_|\__,_|\__,_|_| |_| |_|  |____/ \___| .__/|_|\___/ \__, |\___|_|
                                                                 |_|            |___/
```
v0.13.0-dev0 - Author: Daniel Hass

```
Name: node-1
CPU: 2   MEMORY: 8192 MB
CPU |                               | 0%
RAM |                               | 0%
DEPLOYMENTS:
LABEL:
----------------------------------------
Name: node-2
CPU: 3   MEMORY: 2048 MB
CPU |                               | 0%
RAM |                               | 0%
DEPLOYMENTS:
LABEL:
----------------------------------------
Name: node-3
CPU: 4   MEMORY: 4096 MB
CPU |                               | 0%
RAM |                               | 0%
DEPLOYMENTS:
LABEL: cloud:public
----------------------------------------
```

Fig. 4. Start screen of Continuum Deployer after resource parsing

```
Enter DSL type: helm

Configure chart_origin:

[0] chart - Takes a local helm chart or archive as input
[1] yaml - Reads an already templated YAML file
```

Fig. 5. Selection of cloud application DSL and packaging format

The chosen assignment can be refined by interactively tuning the resource definitions, for instance removing a resource that turned out to be under-utilised. Eventually, the resulting deployment plan can be exported into the Kubernetes manifest format. This allows the use of a separate deployment tool, ranging from simple `kubectl` invocations in DevOps scenarios to sophisticated GitOps deployers [9], to turn the plan into an actual deployment.

6   Evaluation

6.1   Empirical Application Evaluation and Feasibility Analysis

We evaluate two popular and industrially relevant packaging formats, Helm charts for container compositions and SAM for serverless applications. We are interested in knowing how many of them specify resource constraints in order to make them eligible for a resource-aware distributed multi-cloud or edge-cloud deployment. All evaluations are conducted with October 8, 2020 data snapshots.

```
Name: node-1
CPU: 2    MEMORY: 8192 MB
CPU |                                   | 0%
RAM |                                   | 0%
DEPLOYMENTS:
LABEL:
-----------------------------------------
Name: node-2
CPU: 3    MEMORY: 2048 MB
CPU |###################                | 66%
RAM |#######                            | 25%
DEPLOYMENTS:
          nginx-deployment-4, cpu=2.0, memory=512, label=[]
LABEL:
-----------------------------------------
Name: node-3
CPU: 4    MEMORY: 4096 MB
CPU |#######################            | 80%
RAM |##################                 | 62%
DEPLOYMENTS:
          nginx-deployment-1-0, cpu=0.4, memory=512, label=[cloud:public]
          nginx-deployment-1-1, cpu=0.4, memory=512, label=[cloud:public]
          nginx-deployment-1-2, cpu=0.4, memory=512, label=[cloud:public]
          nginx-deployment-3, cpu=2.0, memory=1024, label=[]
LABEL: cloud:public
-----------------------------------------

[Error] The following workloads could not be scheduled:
Name: nginx-deployment-2
CPU: 4.0          MEMORY: 512 MB
LABEL:
-----------------------------------------
```

Fig. 6. Best (albeit invalid) greedy match between application and resources

Due to the popularity of Helm, several public marketplaces to share Helm-packaged cloud applications exist, such as KubeApps Hub, Helm Hub and Artifact Hub. We include 306 Helm charts retrieved from KubeApps Hub into our analysis. They reference 459 deployable container images. Out of those, 114 are resource-constrained, with 113 constraining memory and 109 constraining CPU cycles. The physical units differ - 93% use Mi to refer to memory in mebibytes (multiple of 2^{20} bytes), while the remainder uses Gi as well M (multiple of 10^6 bytes). For CPU usage, 97% use m to refer to milli-vCPU, while the remainder uses fractions of vCPUs (1, 0.1). To facilitate the comparison, we unify all units to mebibytes and vCPU fractions, respectively. Figure 7a/b contain the resulting breakdowns of memory and CPU constraints.

Moreover, we compare with 535 SAM files retrieved from the AWS Serverless Application Repository. With 63 of them bundling multiple deployable Lambda functions, they reference a total of 615 Lambda functions. Out of those, 387 are resource-constrained; due to the Lambda execution model that allocates CPU cycles proportionally to memory, only the memory size is specified explicitly. In contrast to helm, the physical units are equalised, and allocations are restricted to a small set of possible values. Figure 8 contains the breakdown of memory allocations, clearly showing a dominance of small functions that can run on resource-constrained devices.

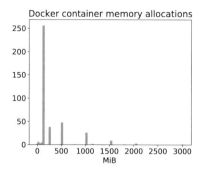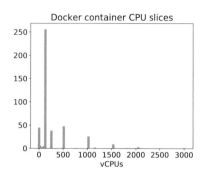

Fig. 7. Frequency of resource allocations for Docker containers in Helm charts

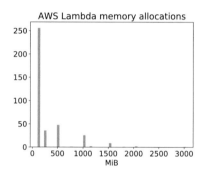

Fig. 8. Frequency of resource allocations for Lambda functions in SAM files

The analysis of Helm charts and SAM files shows that our assumption that at least a significant portion of applications packages ship with maintained resource constraints holds true and therefore the deployer is able to perform useful work.

6.2   Exemplary Performance Evaluation

We measure the placement determination of two popular Helm charts, `wordpress` and `redis`, with the two built-in algorithms, greedy solver and SAT solver, in various configurations to convey the practical feasibility of incorporating a placement decision in real-time as part of deployment workflows. Both charts are modest in size and are fed to the solver in different formats. Redis consists of one secret, two configure maps, three services and two stateful sets, or a total of eight Kubernetes objects in rendered template format. Wordpress consists of a dependency chart, `mariadb`, and seven Kubernetes objects. In compressed chart format.

All measurements are averaged across 100 invocations to reduce the influence of outliers. Figure 9 summarises the results. There are no grave differences between any of the algorithm combinations, although evidently the Redis numbers are lower due to the already rendered charts. Moreover, the performance is worst for the holistic idle resource maximisation for Wordpress, whereas it is worst for idle CPU-only maximisation for Redis. In all cases, the DevOps overhead is less than half a second and thus acceptable to trigger the matching from code commits.

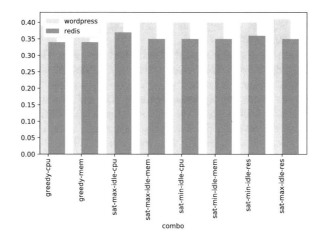

Fig. 9. Performance of placement determination algorithms

6.3 Synthetic Scalability Evaluation

The assignment problem is NP-hard and thus requires specific scalability techniques for interactive matching to ensure that any chosen algorithm will remain controllable by the user. We assign up to 10 synthetically generated multi-container applications to 10 randomly determined resources. While this typically yields a first feasible solution within one second through the SAT solver configured for maximum idle resource optimisation, finding the optimal solution even for only 4 applications may take up almost 3 min on a 2.6 GHz CPU. The Continuum Deployer addresses this problem by (i) showing intermediate results, (ii) letting the user accept those results instead of waiting for the optimum, and (iii) allowing for roundtripping with adjusted configuration and relaxed constraints. The batch mode thus allows for inclusion in latency-critical processes such as CI/CD hooks assuming that non-optimal deployments are acceptable.

7   Conclusions and Material

With this paper, we have introduced the Continuum Deployer as solution for transparent and interactive deployments of composite applications to heterogeneous resources across computing continuums. We followed a systematic approach to contribute a practically useful and extensible tool and to evaluate some of the possible combinations between application packaging formats and solution algorithms. The implementation of Continuum Deployer is available as open source at https://doi.org/10.5281/zenodo.4584220.

References

1. Balouek-Thomert, D., Renart, E.G., Zamani, A.R., Simonet, A., Parashar, M.: Int. J. High Perform. Comput. Appl. **33**, 6 (2019). https://doi.org/10.1177/1094342019877383
2. Donno, M.D., Tange, K., Dragoni, N.: IEEE Access **7**, (2019). https://doi.org/10.1109/ACCESS.2019.2947652
3. Añel, J.A., Montes, D.P., Iglesias, J.R.: Cloud and Serverless Computing for Scientists - A Primer. Springer (2020). https://doi.org/10.1007/978-3-030-41784-0
4. Spillner, J., Gkikopoulos, P., Buzachis, A., Villari, M.: In: 2nd International Workshop on Cloud, IoT and Fog Systems/Security (CIFS)/14th IEEE/ACM International Conference on Utility and Cloud Computing (UCC) (2020)
5. Saatkamp, K., Breitenbücher, U., Kopp, O., Leymann, F.: In: ed. by Ferguson, D., Muñoz, V.M., Cardoso, J.S., Helfert, M., Pahl, C. CLOSER 2017 - Proceedings of the 7th International Conference on Cloud Computing and Services Science, Porto, Portugal, April 24-26, (SciTePress, 2017), pp. 247–258 (2017). https://doi.org/10.5220/0006371002470258
6. Quang, T., Peng, Y.: In: 2020 IEEE International Conference on Pervasive Computing and Communications Workshops, PerCom Workshops 2020, Austin, TX, USA, March 23-27, 2020, pp. 1–6. IEEE (2020). https://doi.org/10.1109/PerComWorkshops48775.2020.9156140
7. Wintersberger, P., Nicklas, H., Martlbauer, T., Hammer, S., Riener, A.: In: AutomotiveUI 2020: 12th International Conference on Automotive User Interfaces and Interactive Vehicular Applications, Virtual Event, Washington, DC, USA, September 21-22, 2020, pp. 252–261. ACM (2020). https://doi.org/10.1145/3409120.3410659
8. Col, G.D., Teppan, E.: In: Bogaerts, B., Erdem, E., Fodor, P., Formisano, A., Ianni, G., Inclezan, D., Vidal, G., Villanueva, A., Vos, M.D., Yang , F. (eds.) Proceedings 35th International Conference on Logic Programming (Technical Communications), ICLP 2019 Technical Communications, Las Cruces, NM, USA, September 20-25, EPTCS, vol. 306. (2019), EPTCS, vol. 306, pp. 259–265 (2019). https://doi.org/10.4204/EPTCS.306.30
9. Spillner, J., Boruta, D., Brunner, T., Gerber, S., Kosmaczewski, A.: Syn: GitOps on stereoids with kubernetes the swiss way. In: 3rd International Conference on Microservices (Microservices) (2020)

MEDAL: An AI-Driven Data Fabric Concept for Elastic Cloud-to-Edge Intelligence

Vasileios Theodorou[1](✉), Ilias Gerostathopoulos[2], Iyad Alshabani[3],
Alberto Abelló[4], and David Breitgand[5]

[1] Intracom Telecom, Peania, Greece
`theovas@intracom-telecom.com`
[2] Vrije Universiteit Amsterdam, Amsterdam, The Netherlands
`i.g.gerostathopoulos@vu.nl`
[3] BitSparkles, Sophia Antipolis, France
`iyad.alshabani@bitsparkles.com`
[4] Universitat Politècnica de Catalunya, Barcelona, Spain
`aabello@essi.upc.edu`
[5] IBM, Haifa, Israel
`davidbr@il.ibm.com`

Abstract. Current Cloud solutions for Edge Computing are inefficient for data-centric applications, as they focus on the IaaS/PaaS level and they miss the data modeling and operations perspective. Consequently, Edge Computing opportunities are lost due to cumbersome and data assets-agnostic processes for end-to-end deployment over the Cloud-to-Edge continuum. In this paper, we introduce MEDAL—an intelligent Cloud-to-Edge Data Fabric to support Data Operations (DataOps) across the continuum and to automate management and orchestration operations over a combined view of the data and the resource layer. MEDAL facilitates building and managing data workflows on top of existing flexible and composable data services, seamlessly exploiting and federating IaaS/PaaS/SaaS resources across different Cloud and Edge environments. We describe the MEDAL Platform as a usable tool for Data Scientists and Engineers, encompassing our concept and we illustrate its application though a connected cars use case.

1 Introduction and Motivation

Modern consumers seek for personalized innovative services and superior user-experience that can only be achieved through novel data-driven technologies. Connected Cars, Smart City and Industry 4.0 are notable examples of domains that are backed by mission-critical applications, fueled by and heavily dependent on data. Such applications typically need to process vast amounts of data at various levels to extract actionable information in a timely, reliable and privacy-preserving manner. The emergence of Cloud computing has been a huge leap

© The Author(s), under exclusive license to Springer Nature Switzerland AG 2021
L. Barolli et al. (Eds.): AINA 2021, LNNS 227, pp. 561–571, 2021.
https://doi.org/10.1007/978-3-030-75078-7_56

forward to effectively host applications, with on-demand resources and pay-as-you-go business model significantly simplifying management and reducing upfront-investment costs.

With the advances in virtualization and cloud-native technologies and the abundance of devices, efficient ways have emerge to store and process data away from centralized data centers and "on-the-Edge", i.e., closer or even right at the data sources. This emerging service delivery paradigm referred to as *Edge Computing*, promises to lead to decreased latency, which is of paramount importance for time-critical applications, e.g., autonomous driving, as well as to more efficient utilization of both the current ubiquitous computation resources (smartphones, telecom servers, cars' on-board units, IoT devices, etc.) and communication bandwidth. Equally importantly, it enables the efficient analysis of data that due to practical, legal, or confidentiality constraints are not allowed to leave the environments in which they were generated, or their transfer entails great performance or other costs. Edge environments are well represented in Cloud offerings of major Cloud providers, usually as enablers for Internet of Things (IoT) scenarios (AWS IoT Greengrass, Azure IoT Edge, Google Cloud IoT Core).

The problem is that existing Edge computing resources are underutilized, while the network is overutilized. The reason is that although modern Edge offerings support some data preparation and pre-processing services at the Edge, data still needs to be transferred to a central location to be properly analysed. Deploying and managing data analytics applications at the Edge is still not straightforward, since existing Cloud solutions for the Edge miss the data modeling and operations perspective. Instead, they focus on an infrastructure and platform level, being oblivious of applications running on top of them. This view encumbers intelligence from Cloud to the Edge, e.g. decision-making processes of when to move data analytics tasks between Cloud and Edge versus when to move data. Thus, existing Cloud solutions for the Edge cannot actively support organizations in the continuous development, operations, and lifecycle management of data analytics applications (DataOps) [3], which is essential for effectively leveraging data for competitive advantage. **Overall, there is no solution yet that supports advanced DataOps on the Cloud-to-Edge continuum, despite the abundance of mature, yet disconnected, Cloud solutions for data analytics at the Edge.**

To illustrate this problem, we consider the scenario of continuously collecting data from a large number of vehicles and combining them with other context data such as that from wearable sensors and smartphones to detect driving behaviors. The data must be analysed so that statistics over large datasets can be calculated and AI/ML prediction models can be trained to identify correlations and mine frequent patterns. This scenario includes performing anomaly detection to identify different safety-related events, e.g. sudden loss of driver's focus and scoring of driver behavior. Nevertheless, driver's sensitive data produced within the car may not be allowed to leave the vehicle, or may entail privacy restrictions on being shared among different service providers. In addition, as the number of

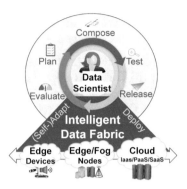

Fig. 1. Continuous data application life-cycle management on the cloud-to-edge continuum.

cars increases, there is a significant rise in the volume of data that needs to be analysed, as well as in the complexity of required data and model management. The challenge then is to deploy, test, execute, and manage service components in the most efficient way, both regarding response time and resource utilization (network bandwidth, compute, storage), from the vehicle to the Cloud, while at the same time respecting data privacy restrictions.

Another challenge is that there is also a methodological gap on how data scientists can deal with the challenges of the Cloud-to-Edge continuum, i.e. the volatility and dynamicity of resources, the varying quality and utility of diverse data sources all along the data path, and the difficulty in discovering and managing relevant data assets [8]. A crucial question is how to abstract and obtain a data-centric view of the underlying infrastructure and assets, while at the same time considering the capabilities and opportunities they offer and avoiding vendor lock-in effects. Essentially, the data scientist should be concerned with the data aspects of the analytics workflows, which in turn poses a requirement for sophisticated automation mechanisms to handle and optimize infrastructure and deployment aspects, as well as datasets and data models management and operation, even across operational domains.

Our proposed solution aims to support data scientists in the DataOps activities of building and maintaining data analytics applications of high flexibility and quality on the Cloud-to-Edge continuum, optimally utilizing Cloud/Edge resources and services (Fig. 1). In particular, it aims to contribute to the evolution of Cloud services for data analytics in the Cloud-to-Edge continuum by:

- Introducing the MEDAL concept—an *Intelligent Data Fabric* as a continuum on the data application layer, formed by the federation of semantically enabled, cloud-native data-centric constructs acting as building blocks.
- Offering a platform for AI-driven Cloud-to-Edge DataOps that provides the data scientist with a comprehensive data-centric view over Edge/Fog/Cloud assets, as well as the ability to manage and automate the lifecycle and

operation of data-intensive analytics and ML workflows, deployed in a distributed fashion that respects data locality and cost models of data operations.

To this end, we introduce innovations in the areas of: (i) Cloud-native Data Fabric across the continuum; (ii) DataOps over the Cloud-to-Edge continuum; (iii) AI-Ops for runtime adaptations over the continuum; and (iv) semantic representation and management of Cloud-to-Edge resources and data assets to ultimately provide a flexible, scalable, and cost-effective platform for Cloud-to-Edge intelligence.

In Sect. 2, we describe the main concepts and methodologies empowering our approach; in Sect. 3, we showcase the application of MEDAL on an illustrative conncerted cars use case; finally, in Sect. 4, we conclude this paper with our remarks.

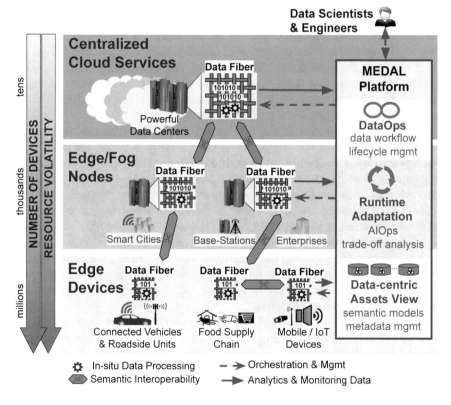

Fig. 2. The MEDAL platform for an intelligent continuum of cloud-native data fibers.

2   An Intelligent Cloud-to-Edge Data Fabric

We adopt a data architectural angle where the main structural component of our data analytics workflow is the *Data Fiber*, which we define as homogeneous wrapper of data assets and services at the data layer. We consider that the *Data Fabric* is formed by the federation of *Data Fibers* of different volumes and capacities on the Cloud-to-Edge continuum (Fig. 2). The Data Fabric facilitates data representation, storage, processing, access and exchange and can be realized using Data Lake technologies [5] in a distributed manner. The high flexibility and configurability provided by Data Fibers as our structural units primarily stems from following cloud-native principles, according to which data ingestion and state is decoupled from data processing and analytics. This allows for paying the effort and cost of data transformation/integration when it is required, on-demand.

At a deployment level, Data Fibers across the continuum are realized as containerized micro-services with a focus on scalability and resilience. Data Fibers are equipped with advanced data profiling and summarization mechanisms, as well as with cloud-native capabilities at the resource layer, fostering rapid instantiation of data workflows over collected data where the data resides—a concept also known as in-situ processing. Data Fibers may belong to one or more administrative domains (e.g., in multi-cloud setups) and need to interconnect and to interoperate. Moreover, Data Fibers are highly dynamic and volatile, making it essential to manage their efficient and automated cloud-native orchestration at the infrastructure layer, including primitives such as dynamic provisioning/decommissioning, auto-scaling and migration, trigger-able via declarative interfaces.

We envision the *MEDAL Platform*, a platform to elastically manage and orchestrate Data Fibers and their federations over the continuum, while offering a unified view over underlying data assets and resources to Data Scientists and Engineers (Fig. 2). Thus, the MEDAL Platform composes an intelligent Data Fabric for managing heterogeneous data and resources adaptively, on demand, facing versatile needs and requirements. To achieve these objectives, MEDAL bases on innovative DataOps principles, tools and techniques for managing the complete lifecycle of data applications; AIOps mechanisms for intelligent response to observed events and evolving requirements; and semantic annotation of data assets and metadata management processes, as we further describe in the following subsections.

2.1   DataOps in the Cloud-to-Edge Continuum

Recently, the DataOps paradigm has emerged as a catalyst towards data workflow automation, aiming at streamlining data operations, accelerating data application development and fostering quality and continuous improvement throughout all phases of data workflow development and operation [1,3]. DataOps combines ideas from agile methodologies, DevOps, and lean manufacturing and tries to deal with changing requirements and accelerate time to market, break the silos between development and operations, and improve quality by reducing non-value-add activities. DataOps views the development of data analytics as a continuous process and focuses on how to make it iterate faster and with higher quality by advocating both following best practices and using the right tools.

We tailor the DataOps paradigm and apply it to the development of data analytics in the Cloud-to-Edge continuum. We adopt the "infinite loop" of DataOps according to which the development of data analytics passes through different phases: Planning, Composition, Testing, and Release of logical data workflows and Orchestration, Adaptation, and Evaluation of deployed workflows on the continuum. Our DataOps framework includes both (i) methodological principles and best practices that guide data scientists and (ii) tools that help speed up the design and automate the testing, quality assurance, and deployment of data analytics workflows in the continuum. Contrary to other Cloud frameworks and platforms for Edge computing, we support the complete development lifecycle, from design to maintenance, and put emphasis on continuous integration and deployment of data analytics workflows. To this end, the MEDAL Platform, incorporates tools for the following features:

Data Workflow Composition. MEDAL Platform provides data scientists with customized access to input their data queries as workflows and compositions of different data processing tasks. It exposes (i) visual editors for highly automated development (akin to mashup tools such as Node-RED), and (ii) script/code editors (akin to Jupyter Notebook) for end-users to directly input their code and define data services. Data Service Composition provides the logical model of a data workflow, which is further mapped by service orchestration tools to a physical model over the available resources. Data workflows can optionally be annotated with requirements of geographical restrictions, resource affinity, capacity (CPU, RAM, throughput), priority and isolation, for optimized mapping.

Monitoring Dashboard. MEDAL Platform provides visualizations for interactive data and analytics exploration, exposing information about the data analytics outputs and quality at the various application ensembles. In addition, it provides visual graphs for the health, status and availability of infrastructure, as well as log monitoring for event management, as exposed by the Cloud-to-Edge resources.

Autonomic Cloud-to-Edge Management and Orchestration. The MEDAL Platform manages flexibility and adaptivity of data workflows as well as provisioning and data asset-aware coordination of Cloud/Fog/Edge services. In this respect, data workflow deployment (including both service binding and job scheduling) and quality control are performed in a resource-aware fashion, matching available resources' characteristics with data workflow requirements. Data service orchestration can be realised using workflow and data pipeline management open source tools.

Continuous Quality Control. The MEDAL Platform provides data workflow testing and optimization environments for data engineers, for continuous improvement of data and infrastructure compositions. They include mechanisms to create staging environments using Cloud and Edge nodes and test data, and to automate the testing of data workflows in those environments. In particular, input data quality is continuously estimated using the Semantic Knowledge Base described below. Once a data workflow passes its prescribed quality tests, it is deployed in production, where its quality and operation continue to be monitored and profiled.

2.2 AIOps for Elastic Cloud to Edge Intelligence

The inclusion of Edge devices, Fog nodes and corresponding services into the pool of Cloud resources introduces new challenges related to volatility, mobility, dynamicity and capacity limitations [8]. Advanced IT operations over such complex and dynamic environments are necessary for maintaining quality of deployed data analytics while minimizing resource usage costs. The challenge here is to introduce Edge Intelligence [2,7] mechanisms both (i) for supporting the elastic lifecycle management and interoperation of Data Fibers (i.e., the data infrastructure layer of our approach) and (ii) for managing the distributed nature of data analytics and ML pipelines spread across the continuum. However, such intelligent mechanisms can only take place with the appropriate visibility and reaction over performance data across all disparate Cloud-to-Edge resources. AIOps [4] has recently been proposed as an effective paradigm to exploit AI/ML techniques towards IT operations automation, by correlating data across different interdependent environments and providing real-time, actionable insights over system behaviors, as well as recommendations and (semi-)automated corrective actions. AIOps services provide timely awareness and proactive actions over service quality degradation, resource utilization changes and system mis-configurations, using event management mechanisms combined with application logic to identify root causes and to trigger appropriate restorative management workflows.

We adopt an AIOps angle of high automation with services of built-in intelligence, where runtime adaptation mechanisms play a central role for closing the loop from issue detection or prediction, to autonomic response. Adaptation mechanisms are crucial for managing unpredictability of resources' and services' availability, as well as for accounting for the varying availability and quality of data along the Cloud-to-Edge continuum, which can also continuously change. Runtime adaptation primitives are instilled into proactive management workflows and include:

Quality-Driven Scheduling: Re-allocation and re-scheduling of data collection and data analytics tasks to sensing/compute nodes based on intelligent monitoring of data and analytics quality;

Flexible Data/ML Model Deployment: Move data models across levels (i.e., closer to Cloud or closer to Edge) to efficiently utilize resources and maintain analytics quality, affecting where data aggregation/model training [9] takes place and thus the necessity of transferring unaggregated/training data across levels;

Elasticity of Data Fibers: Dynamic provisioning, auto-scaling and migration primitives for the Data Fibers across the continuum to respond to detected or predicted over-/under-utilization of resources and to adjust to evolving data analytics requirements (e.g. increase sample size for higher accuracy).

Runtime adaptations follow the Monitor-Analyze-Plan-Execute over Knowledge (MAPE-K) control loop. In the Monitoring phase, data at both the infrastructure and platform level (e.g. CPU load, memory consumption), and at the application level (e.g. application telemetry data on data analytics accuracy and precision, logs). The Analyze phase is responsible for preprocessing, combining, and applying AI/ML techniques for identifying situations that trigger adaptations, also throwing relevant events. Such situations can be both negative (e.g. reduced output quality of a deployed data workflow) and positive (e.g. addition of Edge nodes bringing in opportunity to increase service availability). In the Plan phase, different adaptation actions or plans are determined and compared to each other. If more than one plan is available, a decision is taken either via involving a human operator or (to be fully autonomous) via prioritization based on the contribution of each plan to meeting certain predefined and prioritized goals (e.g. load balancing, increase of output quality). We should note here the importance of cost models [6] for the evaluation of alternative plans, which play the role of a Knowledge Base and are continuously augmented with historical data from monitoring and reaction to past events.

Finally, in the Execute phase, the selected plan is rolled out via the activation of a workflow including a series of concrete changes (e.g. provisioning of a new Data Fiber, decommissioning of another one, and starting a computation on the new Data Fiber).

2.3   Semantic Interoperability in the Cloud-to-Edge Continuum

Semantic interoperability takes place both at the data layer and the infrastructure resource layer to seamlessly manage heterogeneous resources and services across the Cloud-to-Edge continuum. This can only be achieved with semantically rich information about available computing resources and data assets, combined with appropriate mechanisms for persisting and exchanging such information. In this respect, we introduce the concept of a decentralized *Semantic Knowledge Base* that acts as the source of information used by both (i) management entities to monitor Data Fibers and obtain a unified view over the available resource, data, and service assets; (ii) assets to discover and interoperate with each other. Information includes metadata about infrastructure resource characteristics, data assets (data sources, schemas, profiling, data quality, information available, etc.), monitored runtime state (utilization, active sessions). The Semantic Knowledge Base is also enhanced by predictive cost models for future performance estimations that provide recommendations about the deployment of analytics workflows over the available resources.

To be scalable and allow partially autonomous operation, the Semantic Knowledge Base is decentralized, i.e. there is no central node which keeps track of the metadata in the whole continuum. Instead, nodes form metadata exchange clusters dynamically and only share with other clusters in the continuum the metadata necessary for inter-cluster provisioning and management of data workflows. This way, the single point of failure is avoided and a certain degree of autonomicity in resource management and scheduling of operations is retained by each cluster (which could also be a single node). While the discovery and metadata-exchange process can be continuous, the synapsis of federations between Data Fibers at the continuum can take place on-demand and have a temporal nature.

3   Application on the Connected Cars Use Case

Modern cars are equipped with a plethora of sensors, enabling a variety of services in the context of safety, control and entertainment. Insurance companies, as well as city and road safety administrators and fleet owners, are particularly interested in automatized car analytics such as driving behavior analysis (DBA) and predictive maintenance. The value chain ranges from processing units and actuators embedded in the car, to service providers using car data to provide advanced connected services (for the driver, for the manufacturer, for the city, etc.). The execution of analytics over generated data can take place inside the vehicle's Onboard Units (OBUs), at centralized cloud environments or at intermediary nodes along the Edge to Cloud data path (i.e., Edge/Fog Nodes), such as Roadside Units (RSUs) or cellular network infrastructure (Mobile Base Stations acting as MEC points of presence).

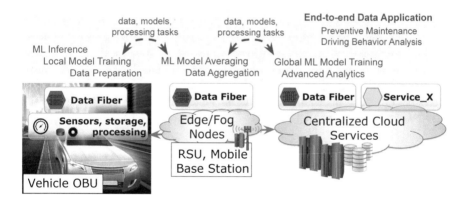

Fig. 3. Intelligent data fabric for connected cars applications.

In Fig. 3, we depict the application of the MEDAL concept on this use case. Data Fibers are instantiated through the MEDAL Platform as interconnected containers at the different levels, forming an intelligent Data Fabric. Onboard the car, at the OBU, the Data Fiber collects data from car sensors and performs local storage and processing (i) for *ML inference* tasks such as the diagnosis of hazardous driving behavior or its prediction due to vital signs; (ii) for *local model training* in case of distributed machine learning data applications, e.g., collection of sensitive (DBA) data from thousands of drivers and federated learning of correlations without any raw data actually leaving any car and (iii) for data preparation so that data can be transformed and cleansed accordingly before being moved to higher level Data Fibers. At the Edge/Fog nodes (RSU, Base Station etc.), the Data Fiber performs *model averaging* over the model parameters received from the Data Fibers on the various cars, or *data aggregation*. On the powerful centralized Cloud, the Data Fiber performs global model training and advanced analytics tasks, possibly interoperating with other available services at the Cloud. The scheduling of tasks between different levels, as well as the activation, termination, scaling and migration of Data Fibers, is managed in an automated fashion by the platform, according to availability and cost of resources (e.g., density of cars over a particular geographical area at a particular time) and changing application requirements which are dependent on situational awareness (e.g., spawn Data Fibers in multiple cars close to a traffic collision).

4   Conclusion

In this work, we have introduced MEDAL—a novel concept for the efficient management of the complete lifecycle of data applications deployed all along the Cloud-to-Edge continuum. We constructed the notions for an intelligent Data Fabric composed of Data Fibers—our semantically-enabled cloud-native distributed building units that can dynamically launch, federate and scale on and across the different levels of the Cloud-to-Edge continuum. We described the DataOps, AIOps and semantic annotation principles underpinning MEDAL and we illustrated our approach through a use case from the connected cars domain. In contrast with existing Cloud solutions, MEDAL fully exploits available knowledge about data assets over the continuum and uses this information to provide a unified data and monitoring view to application developers, as well as to make informed decisions about management, orchestration and adaptation of data workflows.

Acknowledgments. This work is partially supported by the TANDEM project, co-financed by the European Union and Greek national funds through the Operational Program Competitiveness, Entrepreneurship and Innovation, under the call RESEARCH CREATE - INNOVATE (project code: T2EDK-02825).

References

1. Christopher, B., Gil, B., Eran, S.: The dataOps cookbook. In: DataKitchen. 2nd Edition (2019)
2. Deng, S., Zhao, H., Fang, W., Yin, J., Dustdar, S., Zomaya, A.Y.: Edge intelligence: the confluence of edge computing and artificial intelligence. IEEE Internet of Things J. **7**(8), 7457–7469 (2020)
3. Julian, E.: DataOps-towards a definition. In: LWDApp, pp. 104–112 (2018)
4. Masood, A., Hashmi, A.: AIOps: predictive analytics and machine learning in operations. In: Cognitive Computing Recipes: Artificial Intelligence Solutions Using Microsoft Cognitive Services and TensorFlow, pp. 359–382 (2019)

5. Nargesian, F., Zhu, E., Miller, R.J., Pu, K.Q., Arocena, P.C.: Data lake management: challenges and opportunities. Proc. VLDB Endow. **12**(12), 1986–1989 (2019)
6. Stefanidis, V.A., Verginadis, Y., Baur, D., Przezdziek, T., Mentzas, G.: Reconfiguration penalty calculation for cross-cloud application adaptations. In: 10th Int. Conf. on Cloud Computing and Services Science (CLOSER), pp. 355–362 (2020)
7. Theodorou, V., Diamantopoulos, N.: GLT: edge gateway ELT for data-driven intelligence placement. In: IEEE/ACM RCoSE-DDrEE@ICSE 2019, pp. 24–27 (2019)
8. Varghese, B., Wang, N., Barbhuiya, S., Kilpatrick, P., Nikolopoulos, D.S.: Challenges and opportunities in edge computing. In: IEEE International Conference on Smart Cloud (SmartCloud) (2016)
9. Yang, Q., Liu, Y., Chen, T., Tong, Y.: Federated machine learning: concept and applications. ACM Trans. Intell. Syst. Technol. **10**(2) (2019)

Adaptive Container Scheduling in Cloud Data Centers: A Deep Reinforcement Learning Approach

Tania Lorido-Botran[1($\boxtimes$)] and Muhammad Khurram Bhatti[2]

[1] Bilbao, Spain
[2] Information Technology University, Lahore, Pakistan
khurram.bhatti@itu.edu.pk

Abstract. Cloud data centers rely on virtualization to run a diverse set of applications. Container technology allows for a more lightweight execution, in comparison with popular Virtual Machines. Efficient scheduling of containers is still challenging due to varying request arrival patterns, application-specific resource consumption and resource heterogeneity in physical servers. Besides, containers are also more prone to resource contention and performance interference. Cloud providers need to overcome these challenges with a goal in mind: maximize resource utilization to satisfy as many requests as possible. This paper introduces *RLSched*, a deep reinforcement learning-based (DRL) scheduler that is self-adaptive and automatically captures the resource usage dynamics in the data center. The scheduler is based on a decentralized actor-critic multi-agent architecture that enables for parallel execution and faster convergence. *RLSched* relies on an enhanced network model with action shaping, which filters invalid actions and prevents the agent to fall into a sub-optimal policy. The proposed scheduler is compared against other state-of-the-art DRL methods on a simulated data center environment based on real traces from Microsoft Azure. The results show faster convergence and higher number of containers placed per session.

Keywords: Cloud computing · Container scheduling · Deep reinforcement learning

1 Introduction

Over the last decade or so, cloud computing services have replaced private data centers. Many businesses, government organizations and research centers rely on external clouds to run their workloads, that can be as diverse as online websites, to streaming services, large-scale scientific workflows and research workloads. Cloud providers face the challenge to continually optimize the resource usage of the data center, while giving service to dynamic requests from different clients.

Virtualization provides an abstraction layer of the physical resources and eases the continuous allocation and de-allocation of client applications. Virtual Machines became the popular choice, despite their execution overhead and significant boot-up times (minutes). In contrast, container technology provides a more lightweight alternative that is

© The Author(s), under exclusive license to Springer Nature Switzerland AG 2021
L. Barolli et al. (Eds.): AINA 2021, LNNS 227, pp. 572–581, 2021.
https://doi.org/10.1007/978-3-030-75078-7_57

already offered by major cloud providers (e.g. Microsoft Azure Kubernetes Service [19]). However, containers do not offer such strong isolation as VMs do, and they are more prone to resource contention and performance interference. Thus, it is more urgent and necessary to have an effective scheduling algorithm to mitigate such problems and improve resource utilization [9].

(Deep) Reinforcement Learning (DRL) has successfully been used to tackle cloud orchestration problems, including but not limited to container scalability [16], VM auto-scaling [7], and (in particular) VM placement problem [11]. However, due to the large difference between containers and virtual machines, RL related methods that perform well in virtual machine scheduling scenarios cannot be applied to container scheduling scenarios directly [16].

This paper fills the gap and introduces *RLSched*, an effective DRL-based scheduler for the container placement problem. The proposed solution is able to effectively (1) adapt to the continuously evolving conditions of the data center, that varies as requests are allocated and de-allocated, (2) provide the best placement for each incoming request in a timely manner under varying resource utilization levels. Ultimately, *RLSched* reduces the resource fragmentation, enabling a higher number of containers to be placed in the same infrastructure.

The contributions of this paper are as follows:

- Modeling of the container scheduling problem in the form of state definition, action space and reward function (Sect. 4).
- Highly scalable (multi-agent) and self-adaptive scheduler based on a actor-critic RL method (Sect. 5).
- Enhanced model with the use of action shaping (Sect. 5).
- Evaluation of different state-of-art RL approaches used in cloud resource management (Sect. 6).
- Validation of the proposed scheduler in a simulated environment based on real-world traces from a Microsoft Azure data center (Sect. 6).

2   Background: Basics of Deep Reinforcement Learning

This section gives an intuitive description of the different DRL without going heavy into the mathematics. The basic elements in an RL problem are the *agent* and the *environment*. The agent continuously interacts with the environment, observes the current *state* and decides the best *action* to take. After some time, the agent will observe the *reward* obtained after applying that action. The goal is to learn an optimal *policy* $pi_\theta(s|a)$ that maps each state with its optimal action.

Depending on the learning strategy, RL algorithms can be classified into three categories:

- *Value-based*: The policy is implicit. The goal is to learn the action-value function $Q_\theta(s,a)$ and from there infer the policy $\pi(s) = \arg\max_a Q(s,a)$. The value function $Q_\theta(s,a)$ represents the expected reward for each (state,action) pair.
- *Policy-based*: They learn the policy $\pi_\theta(a|s)$, which implicitly maximizes reward over all policies.

- *Actor-based*: They learn both the action-value function and the policy.

As the name suggests, *value-based methods* rely on learning the $Q_\theta(s, a)$ function. Such value-function can be represented with a neural network that takes the state as input, and outputs the expected reward for each action. This is the basis for the widely popular Deep Q-learning [2] method. Double DQN (DDQN) is a further improvement that uses 2 function approximators that are alternatively updated and prevents the bias that DQN might have towards certain actions.

Estimating the Q-function sometimes is more challenging than learning the policy itself. For that reason, *policy-based* methods propose learning the policy directly and usually have better convergence, and are more effective in high dimensional spaces (e.g. large data center). The vanilla *Policy Gradient* method relies on a DNN representation of the policy $\pi_\theta(a|s)$. Learning the policy translates into finding the optimal weights θ by calculating policy gradients. Proximal Policy Optimization (PPO) [6] is a very popular method that typically works well in many scenarios.

Finally, *Actor-Critic methods* combine the best of both worlds. They learn both a policy (actor) and a value function (critic). The actor takes as input the state and outputs the best action. It essentially controls how the agent behaves by learning the optimal policy (policy-based). The critic then evaluates the action by computing the value function (value based). The interaction typically leads to a better policy. *Asynchronous Advantage Actor Critic* A3C algorithm [4] introduces a parallel architecture for faster convergence.

3   Related Work

Container scheduling tools typically rely on heuristics, such as variants of heuristic packing algorithms, such as Best-Fit Decreasing (BFD) and First-Fit Decreasing (FFD). The default scheduler in Google Kubernetes [18] offers a variety of heuristics such as affinity or load level that can be combined to obtain the desired effect. Docker Swarm used to offer a random and binpack policy, that prioritized PMs with the highest resource utilization. However, the most recent version [17] only offers Spread algorithm with the aim to balance the load across PMs. Lu, Ni, and Zhang [3] extend Spread with a linear weighted model that includes other parameters, such us the unused amount of memory. Overall, heuristics can do a fair job for the specific goal they are designed for, but they lack adaptability.

A common strategy in the state-of-art is the use of various optimization techniques. Zhang et al. [8] tackle job scheduling in containers as a multi-objective optimization problem and propose a linear programming model for that purpose. As one can expect, the approach fails to capture the non-linearities associated in the resource usage patterns, that can be captured by a deep neural network (with activation functions). There is a large corpus of publications that rely on genetic algorithms or any kind of evolutionary optimization approach. Kaewkasi and Chuenmuneewong [5] propose a container scheduling algorithm based on ant colony optimization algorithm, with the goal of balancing the resource usage. Zhang et al. [13] rely on genetic algorithms to improve the overall energy utilization in the data center. Mseddi et al. [12] propose using particle swarm optimization. However, we focus on a *cloud provider-centric* approach where the main goal is to reduce the resource wastage. With this goal in mind, Mseddi et al. [12] rely on a metaheuristic based on particle swarm optimization for container placement and task scheduling in fog environments, with the goal to maximize the number of served clients. However, any of the explored optimization methods rely solely on the current state of the data center to make a placement decision. As RL techniques have shown, it is possible to learn the environment dynamics (request arrival pattern, resource usage in PMs) and derive an optimal policy. There is yet to explore the potential of applying DRL to container management, and in particular, more novel and efficient RL methods such us policy gradient or actor-critic methods.

Cloud environments offer a variety of resource management challenges and RL has been successfully applied to a variety of resource management problem in the cloud: such as VM auto-scaling [7], VM placement problem [11] and even container scalability [16]. Value-based methods are the most widely used. Zhang et al. [16] propose the use of SARSA (a variant of Q-learning) to control horizontal scalability of containers. [7] introduce a VM auto-scaling solution based on DQN and D3QN. [10] effectively use DDQN for task management in edge environments. However, as experiments will show, value-iteration strategies might converge to sub-optimal policies. In contrast, Funika and Koperek Pawełand Kitowski [14] propose the use of PPO, a policy-based method, for application auto-scaling.

To the best of our knowledge, *RLSched* is the first attempt to explore the effectiveness of value-based and policy-based for the container scheduling problem and successfully apply an actor-critic based method, that is highly scalable thanks to its multi-agent architecture. In contrast to other proposals, *RLSched* is *cloud provider-centric*, as it focuses on maximizing the number of containers served in the data center, and thus, increase expected revenue.

4   Container Scheduling Problem

The scheduling of containers in data centers can be modeled as an instance of the *online multi-dimensional bin packing problem* [1]. This is a well-known *NP*-hard problem. The environment is comprised of a set of n physical machines (PMs) where container requests must be placed. In order to apply DRL to the problem, this paper introduces the following formalization:

State Space: The state is composed of two parts: the resource usage of each physical server and the next k requests to be placed by the scheduler. The resource usage at a given time t is a multi-dimensional vector $[m,n]$, where m is the number of physical machines, described as a set of resources n. For simplicity, this paper considers a single request per placement ($k = 1$) and focuses on two resources, CPU and memory. Resource utilization is a normalized continuous variable ranged $[0,1]$.

Action Space: The action space is discrete, $[1,\ldots,m]$. Each action a_i indicates the PM number where the container will be placed.

Reward Function: There is a variety of goals that can be used: minimize energy utilization, reduce application latency, improve thermal spread across data center. The reward function can be expressed as a vector of size R corresponding to each of the reward goals to be considered. Given the need of cloud providers to maximize revenue, the chosen goal is to maximize the number of requests that can be accommodated in the data center, by reducing the amount of wasted resources. Thus, the reward function is defined as the negative accumulated resource wastage: $R = -\sum_{1,...,N}(1 - U_i)$, where U_i is a vector representing the resource usage of physical machine i. Additionally, the environment heavily penalizes a bad placement action, where the selected PM that cannot satisfy the current initial request. In that case, the episode ends and the environment returns a very low reward.

5   RLSched: In-Depth

Cloud providers rely on managing the infrastructure efficiently to serve as many clients as possible, which turns into direct revenue increase. A bad placement of the containers leads to resource fragmentation, thus, less requests can be served. Container needs vary over time and overbooking of resources might lead to bad performance, client dissatisfaction and revenue loss. It is imperative for a cloud provider to rely on a solid scheduler, capable of making the right judgement for every request and ultimately reducing the resource wastage.

The self-adaptive nature of DRL makes it ideal for the container scheduling problem, but it has two main associated challenges. One is achieving the optimal policy, that is, to make the best placement possible, and avoid falling into sub-optimal policies (not a *just good enough*). The second challenge in DRL is the training time, during which the algorithm explores different action until it *learns* the best policy.

Core Algorithm: *RLSched* utilizes the *Asynchronous Advantage Actor Critic* A3C algorithm at its core [4]. As described in Sect. 2, AC methods combine the best from value-based and policy-based approaches, in that they learn both a policy (actor) and a value function (critic). The interaction between the actor (policy-based) and the critic (value based) typically leads to a better policy. Thus, this translates to the algorithm of choice as for container scheduling. In order to have a highly scalable scheduler, *RLSched* is based on an asynchronous multi-agent architecture. Several agents collect samples and compute gradients for the policy and value function. Periodically, a main network (function approximator) collects all gradients and performs and update, and in turn, distributes a copy of the updated parameters to each agent. This architecture is highly parallelizable and enables faster convergence.

Enhanced Network Model with Action Shaping: *RLSched* relies on a deep neural network as a function approximator, which takes as input the current state s and outputs the reward r_i associated to each physical machine/slot. In other words, the model will estimate the expected reward for placement the incoming request on each of the physical servers. It is also able to make estimates for unseen cases.

In order to speed-up the convergence and avoid sub-optimal policies, the network model is enhanced with *action shaping* [15]. This extra layer applies a mask or filter that eliminates invalid actions, those PMs where the container resource needs cannot be satisfied.

6   Experimental Evaluation

The goal of the experimental evaluation is double: to test *RLSched* under various scenarios based on real-world traces and analyze its convergence and performance, and to evaluate different state-or-art RL approach used in cloud resource management, including policy gradient and value-based methods.

Testing Environment: A simulated environment of a data center. CPU and memory usage are drawn from probability distributions obtained from real traces from Microsoft Azure data center [20]. All algorithms are implemented on Python v3.8 and models are implemented using Tensorflow v2.5.0, and trained on a GPU. The hardware for the experiments is a machine with Intel Cor i7-10510U, 16GB of RAM, NVIDIA GeForce MX330.

Scenarios: (1) Static: Once a container is placed in the data center, it remains indefinitely. (2) Dynamic: Each container will be de-allocated after some duration, drawn from a probability distribution.

State-of-the-Art RL Methods: *RLSched* is compared against 3 other RL methods used in the literature to tackle resource management problems in the cloud (described in Sect. 3). DQN [7] and DDQN [10] are based Value-based learning, and PPO [14] is based on policy gradient. Finally, A2C is also evaluated: an actor-critic method that lacks the parallel architecture from *RLSched*.

6.1   Convergence Analysis

In order to focus on the convergence of *RLSched* and comparison RL methods, the scenarios chosen have enough capacity to accommodate the series of container requests. Each scenario is comprised of 100 homogeneous PMs (same capacity). In each episode, the scheduler receives a series of container requests, 200 for the static scenario and 500 for the dynamic scenario. For fairness in the results, all tested methods rely on the same network architecture, a fully connected neural network with 2 layers and 256 units per layer.

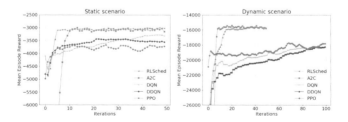

Fig. 1. Convergence for static (left) and dynamic scenario (right)

Figure 1 shows the training process which easily illustrates the convergence speed. The y-axis represents the accumulated reward over an episode. In terms of convergence, RLSched is the clear winner, closely followed by A2C. Looking at the static scenario, RLSched quickly converges (within 5 iterations), followed by A2C that needs 10 iterations, and both achieve the highest episode reward. DQN has largely been used for similar problems, but does not achieve the optimal policy. The same happens with DDQN. Surprisingly, PPO performs worse in the static scenario, but obtains a higher reward in the dynamic scenario.

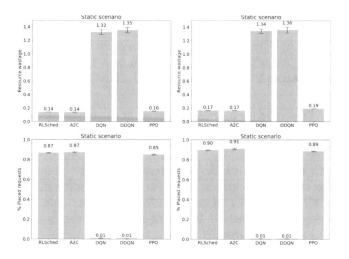

Fig. 2. Resource wastage and request acceptance ratio for static (left) and dynamic scenarios (right)

6.2 Performance Evaluation

The previous experiment looked at how *RLSched* is able to adapt and learn to varying resource utilization and multiple allocations and de-allocation. This next experiment focuses on the real performance from the cloud provider perspective, measured based on two performance metrics. The first one average resource wastage per episode normalized by total capacity of the data center. This metric represents a platform-independent way of quantifying the *goodness* of an scheduler, in contrast to other metrics e.g. energy consumption or cost model, that would be highly dependant on the platform or cloud provider. For each episode, the scheduler has to process a given number of container requests. In a highly utilized data center, a good scheduling setup reduces resource fragmentation and increases the number of requests that can be placed. The second performance metric quantifies the percentage of placed requests per episode (out of the complete series).

Results for both metrics are presented in Fig. 2. *RLSched* and each of the RL counter parts are evaluated under two scenarios composed of 100 homogeneous requests, static and dynamic, with 400 and 500 requests per episode respectively. The clear first conclusion is that value-based approaches, DQN and DDQN, converge to a sub-optimal policy that learns to reject requests to end the episode early, and it translates into very few placed requests. PPO has a slightly worse performance than A2C (as expected). *RLSched* performs very similarly to A2C, as they only different in the parallel architecture.

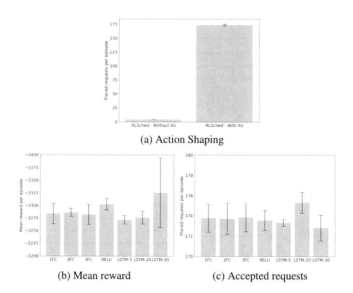

(a) Action Shaping

(b) Mean reward   (c) Accepted requests

Fig. 3. Action shaping analysis (a), model architecture performance in terms of reward (b) and accepted requests (c).

6.3 Model Architecture Exploration

RLSched uses an enhanced model with *Action shaping*, an extra layer that filters invalid actions. This feature plays a major role in the scheduling performance, with two main benefits. The first one, is the reduction of the action space and thus improving the convergence of the scheduler. The second one is the prevention from falling into a sub-optimal policy. Figure 3a shows the results from running *RLSched* on a static scenario with 50 PMs and 200 requests per episode. As clearly shown, without Action Shaping, the scheduler learns a sub-optimal policy that leads to a small number of correctly placed requests. In contrast, *RLSched* with Action Shaping is able to fit up to 175 requests per episode.

The basic model comprises 2 Fully Connected (FC) layers of 256 units. The following experiment explores several variants of the basic architecture: 1, 2 or 3 Fully connected Layers (1FC, 2FC, 3FC) with TanH activation function, 3FC with ReLu as activation function, and a wrapper with (Long short-term memory) LSTM layer and various unit lengths. The environment is a static scenario with 50 homogeneous PMs and results are captured in Figs. 3b and 3c. In this case, the metrics are presented without normalization for clarity purposes (the differences are subtle). Using a single FC layer leads to high variability in the reward, which suggests that it cannot capture the state complexity. Using 2FC has a very desirable stability in the mean reward (thus, the baseline choice for *RLSched*). Replacing tanh with ReLu as activation function leads to worse reward and lower number of placed requests. Finally, it is worth noting that LSTM layers bring an extra benefit in the scheduler performance. This is a promising line of research for future work, in which the use of an LSTM-based model could be used to predict the resource usage dynamics in data centers.

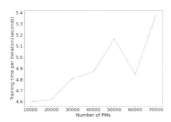

Fig. 4. Mean training time per iteration for different number of PMs

6.4   Scalability

RLSched is highly scalable thanks to its distributed and asynchronous architecture and easily trainable for larger data centers. As Fig. 4 shows, each training iteration takes between 5 and 6 s for data center sizes ranging from 10000 to 70000 PMs (for a static scenario).

7   Conclusions and Future Work

Container scheduling problem imposes additional challenges, due to the tendency of containers to suffer from performance interferences and resource contention. This paper introduces *RLSched*, a self-adaptive solution based on DRL that achieves minimized resource wastage in the data center. It relies on the use of A3C, an actor-critic method, with an Asynchronous architecture that enables for quick converge to optimal policy and high levels of scalability. The scheduler uses the current resource utilization of the data center, plus the request as the input for the network model (a fully connected DNN) and applies action shaping to the output to filter invalid actions. Experiments show that *RLSched* is able to quickly to converge and perform better online scheduling than other popular RL algorithms, shown in a reduced resource wastage and increased number of placed containers in the same episode.

There are a number of optimizations that could be explored. Experiments show that the use of LSTM layers in the network approximator might help in predicting future resource usage and perform a better placement. *RLSched* could be expanded to a multi-cloud environment by taking into account the network latency as part of the reward and also accept requests for multiple containers, including their communication dependencies.

References

1. Csirik, J., Woeginger, G.J.: On-line packing and covering problems. In: Online Algorithms, pp. 147–177 (1998)
2. Mnih, V., et al.: Playing atari with deep reinforcement learning. In: arXiv preprint arXiv:1312.5602 (2013)
3. Shenglin, L., Ni, M., Zhang, H.-B.: The optimization of scheduling strategy based on the Docker swarm cluster. Inf. Technol. **40**(7), 147–151 (2016)
4. Mnih, V., et al.: Asynchronous methods for deep reinforcement learning. In: International Conference on Machine Learning. PMLR, pp. 1928–1937 (2016)
5. Kaewkasi, C., Chuenmuneewong, K.: Improvement of container scheduling for docker using ant colony optimization. In : 2017 9th International Conference on Knowledge and Smart Technology (KST), pp. 254–259. IEEE (2017)
6. Schulman, J., et al.: Proximal policy optimization algorithms. In: arXiv preprint arXiv:17 07.06347 (2017)
7. Wang , Z., et al.: Automated cloud provisioning on aws using deep reinforcement learning. In: arXiv preprint arXiv:1709.04305 (2017)
8. Zhang, D., et al.: Container oriented job scheduling using linear programming model. In: 2017 3rd International Conference on Information Management (ICIM), pp. 174–180. IEEE (2017)
9. Bhimani, J., et al.: Docker container scheduler for I/O intensive applications running on NVMe SSDs. In: IEEE Transactions on Multi-Scale Computing Systems, 4.3, pp. 313–326 (2018)
10. Zhang, Q., et al.: A double deep Q-learning model for energy-efficient edge scheduling. IEEE Trans. Serv. Comput. **12**(5), 739–749 (2018)
11. Bingqian, D., Chuan, W., Huang, Z.: Learning resource allocation and pricing for cloud profit maximization. In: Proceedings of the AAAI Conference on Artificial Intelligence, Vol. 33(01), pp. 7570–7577 (2019)
12. Mseddi, A., et al.: Joint container placement and task provisioning in dynamic fog computing. IEEE Internet of Things J. **6**(6), 10028–10040 (2019)
13. Zhang, R., et al.: A genetic algorithm-based energy-efficient container placement strategy in CaaS. IEEE Access **7**, 121360–121373 (2019)
14. Funika, W., Koperek, P., Kitowski, J.: Automatic management of cloud applications with use of Proximal Policy Optimization. In: International Conference on Computational Science, pp. 73–87 Springer (2020)
15. Kanervisto, A., Scheller, C., Hautamäki, V.: Action space shaping in deep reinforcement learning. In: 2020 IEEE Conference on Games (CoG), pp. 479–486 IEEE (2020)
16. Zhang, S., et al.: A-SARSA: a predictive container auto-scaling algorithm based on reinforcement learning. In: 2020 IEEE International Conference on Web Services (ICWS), pp. 489–497 IEEE (2020)
17. Docker. Docker Swarm. https://docs.docker.com/engine/swarm/how-swarm-mode-works/services/. Accessed Jan 2021
18. Kubernetes. Google Kubernetes. https://kubernetes.io/docs/concepts/scheduling-eviction/kube-scheduler/. Accessed Jan 2021
19. MS. Microsoft Azure Kubernetes Service (AKS). https://azure.microsoft.com/en-us/services/kubernetes-service/. Accessed Jan 2021
20. Trace. Azure Public Dataset. https://github.com/Azure/AzurePublicDa%20taset. Accessed Jan 2021

Reliable Server Pooling Based Workload Offloading with Mobile Edge Computing: A Proof-of-Concept

Thomas Dreibholz[1]([✉]) and Somnath Mazumdar[2]([✉])

[1] Simula Metropolitan Centre for Digital Engineering,
c/o OsloMet – Storbyuniversitetet,
Pilestredet 52, 0167 Oslo, Norway
`dreibh@simula.no`
[2] Department of Digitalization, Copenhagen Business School,
Howitzvej 60, 2000 Frederiksberg, Denmark
`sma.digi@cbs.dk`

Abstract. In recent times, mobile broadband devices have become almost ubiquitous. However, battery-powered devices (such as smartphones), have limitations on energy consumption, computation power and storage space. Cloud computing and Mobile Edge Computing (MEC) can provide low-latency compute and storage services at the vicinity of the user, MEC in particular due to the upcoming 5G networks. However, the complexity lies in how to simply and efficiently realise MEC services, with the auxiliary public (multi-)cloud resources? In this paper, we propose a proof-of-concept for using Reliable Server Pooling (RSerPool) as a light-weight layer of managing resource pools and handling application sessions with these pools. Our approach is simple, efficient, has low overhead and is available as open source. Here, we demonstrate the usefulness of our approach by measuring in a test setup, with a 4G testbed connected to MEC and public multi-cloud resources.

1 Introduction

The current trend of making large powerful mobile devices (especially smartphones) has become increasingly widespread. Interestingly, such devices lack required computational power as well as proper storage capacity and/or I/O speed, apart from suffering energy-related issues. Overall, executing complex applications with large computational/storage requirements can be delegated to cloud services. It is well-known that latency-tolerant applications are well suited for cloud services, while latency-sensitive applications are suffering in the cloud platform.

This work has been supported by the European Community through the 5G-VINNI project (grant no. 815279) within the H2020-ICT-17-2017 research and innovation program. Parts of this work have also been funded by the Research Council of Norway under project number 208798/F50. The authors would like to thank Ann Edith Wulff Armitstead for her comments.

L. Barolli et al. (Eds.): AINA 2021, LNNS 227, pp. 582–593, 2021.
https://doi.org/10.1007/978-3-030-75078-7_58

It is claimed that Mobile Edge Computing (MEC) [1] in upcoming 5G networks is a way to solve the latency issues, by providing cloud services in data centres nearby the mobile user. However, not much work is available on how to actually realise services for MEC. Reliable Server Pooling (RSerPool) [2,3], is an IETF standard for a light-weight server pooling approach, which initially targeted server redundancy in telephone signalling systems. Current existing literature could be divided into two broad categories. One class of works is primarily focusing on proposing efficient strategies to optimise the respective objective functions, such as latency or energy, by using various algorithms or models, like for instance genetic algorithms [4] and Fuzzy Logic [5], to name a couple. Another category proposes architectures and frameworks for improving the QoS via workload offloading. In [6], a model is proposed to reduce the average response time for mobile users in offloading their workloads to the cloudlets. The primary components of the framework are the cloudlet, software-defined network and the cellular network infrastructure. However, we have not found any testbed-related work based on the Reliable Server Pooling mechanism.

In this paper, we introduce a simple approach for service offloading from mobile devices to MEC resources and even public (multi-)cloud resources [7,8], depending on availability and workload. The goal of our approach is to combine existing software systems to provide a solution which meets the following goals: simplicity, efficiency, low-overhead, and open-source. To achieve this, we combined and adjusted these components: RSPLIB [2] for RSerPool-based session handling of MEC services; VNF-based Evolved Packet Core (EPC) for the 4G/5G network based on OPENAIRINTERFACE [9,10]; OPEN SOURCE MANO [11] for service orchestration of EPC and MEC systems; and OPENSTACK for hosting the compute resources for EPC and RSerPool-based MEC services. We have demonstrated the applicability of our proposed approach by a proof-of-concept in a testbed setup. Finally, we also have provided some discussions about this ongoing work related to improving the system.

2   Component Description

In this section, we have provided the required background information of the components which are primarily responsible for managing and orchestrating the MEC as well as network-related services.

2.1   Reliable Server Pooling (RSerPool)

Reliable Server Pooling (RSerPool) [2,12,13] was originally motivated by the need for handling server redundancy in telephone signalling systems. However, the problem of handling server redundancy is generic and regularly triggers "reinventing the wheel". RSerPool is therefore a generic, application-independent framework. A particular property of it is to be simple and light-weight, making it suitable also for devices with very limited resources. RSPLIB[1] [2] is the most widespread open-source implementation of RSerPool.

---

[1] RSPLIB: https://www.uni-due.de/be0001/rserpool/.

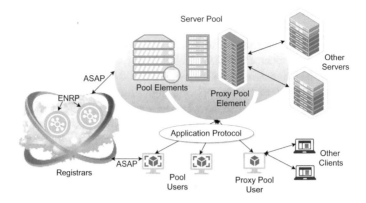

Fig. 1. The RSerPool architecture

The RSerPool architecture, as described in detail in [2,3], is depicted in Fig. 1. In an RSerPool setup, a number of servers, each providing a certain service, form a pool. Servers of a pool are denoted as Pool Elements (PE). Within its operation scope, a pool is identified by its unique Pool Handle (PH, e.g. a string like "Data Analysis Pool"). The handlespace, which is the set of all pools of an operation scope, is managed by Pool Registrars (PR, also denoted as registrars). Since a single PR would be a single point of failure, RSerPool setups should consist of at least two registrars. They synchronise the handlespace by using the Endpoint haNdlespace Redundancy Protocol (ENRP) [2,14]. An operation scope is limited to an organisation or company. Unlike services like the Domain Name System (DNS), RSerPool does not intend to scale to the whole Internet. This significantly simplifies the architecture, making it very light-weight (see also [15]). Nevertheless, pools can be distributed over large geographic areas [2], to achieve a high resilience of services, e.g. to keep a service running in case of an earthquake. Servers can dynamically register to, and deregister from, a pool at a PR of the operation scope, by using the Aggregate Server Access Protocol (ASAP) [2,16]. The ASAP connection is also used for monitoring the availability of the PE by a keep-alive mechanism [12]. Proxy pool elements can connect "legacy" non-RSerPool servers.

ASAP is also used by clients, denoted as Pool Users (PU) in the context of RSerPool, to access the resources of a pool. They can query a PR of the operation scope to select PE(s). This selection is performed by using a pool-specific pool member selection policy [2,12,17], which is usually just denoted as pool policy. Examples of pool policies are Round Robin (RR) and Least Used (LU). ASAP can also be used between PU and PE, then realising a Session Layer functionality between a PU and a pool. ASAP can then also support the actual Application Layer protocol to handle failovers and help with state synchronisation [2,12]. Proxy pool users can connect "legacy" non-RSerPool clients to a pool.

2.2   Open Source MANO and the SimulaMet OAI EPC

Network Function Virtualisation (NFV) is a crucial part of 5G networks: Network functionalities can be realised as Network Services (NS), which are composed of Virtual Network Functions (VNF). NSs can then be instantiated as Virtual Machines (VM) in data centres. This allows for a very high flexibility: NSs can dynamically be instantiated when needed and removed when not needed any more. Furthermore, VNF instances can be scaled as needed. However, managing and orchestrating NFV is a complex task. An increasingly popular framework for this purpose is OPEN SOURCE MANO[2] (OSM) [11]. OSM is the orchestration platform from ETSI. It utilises an underlying Network Function Virtualisation Infrastructure (NFVI) for instantiating the Virtual Deployment Units (VDU) as VMs. A commonly used NVFI is OPENSTACK, but OSM supports other frameworks as well.

Based on OSM, we developed a VNF for the Enhanced Packet Core (EPC) of OPENAIRINTERFACE[3] (OAI), denoted as SIMULAMET OAI VNF[4] [9]. In particular, it can be used to easily realise a tailor-made EPC for custom 4G/5G testbed setups. Our EPC (see also Fig. 2) consists of four VDUs [9]:

1. *Home Subscriber Server* (HSS) is the central database containing the information about users and their subscriptions. The HSS functionalities include mobility management, session establishment, user authentication and access authorisation. It provides its service to the MME via the S6a interface.
2. *Mobility Management Entity* (MME) handles the procedures of attaching and detaching as well as service requests of *User Equipment* (UE) and eNodeBs. It communicates with eNodeBs over the S1-C interface (particularly using SCTP as Transport Layer protocol), with SPGW-C over the S11 interface, and with HSS over the S6a interface.
3. *Control Plane of the Packet Data Network Gateway* (SPGW-C) provides the control part of a combined Serving Gateway (SGW) and Packet Data Network Gateway (PGW). That is, OAI combines SGW and PGW, but uses *Control and User Plane Separation* (CUPS). The SPGW-C handles control requests from the MME via the S11 interface, and communication with the SPGW-U via the SXab interface.
4. *User Plane of the Packet Data Network Gateway* (SPGW-U) handles the forwarding of user traffic between the *Public Data Network* (PDN) at the SGi interface (i.e. usually the public Internet) and the eNodeB over the S1-U interface. User traffic between eNodeB and SPGW-U is tunnelled via GPRS Tunnelling Protocol (GTP). The setup of user traffic tunnels is controlled by the SPGW-C over the SXab interface.

The configuration flexibility of our VNF can be utilised easily by using NSs, e.g. by adding MEC resources. We will explain the details next in Sect. 3.

---

[2] OPEN SOURCE MANO: https://osm.etsi.org.

[3] OPENAIRINTERFACE: https://www.openairinterface.org.

[4] SIMULAMET OAI VNF: https://github.com/simula/5gvinni-oai-ns.

586   T. Dreibholz and S. Mazumdar

3   Proposed Approach

Here, we are going to explain our approach to handle the workload offloading challenge. The basic scenario is illustrated in Fig. 2: the UEs run a certain application, which is demanding when it comes to computation and/or storage. They have very limited resources. MEC resources to support this application are available nearby the user. Furthermore, it may also be possible to have additional resources in public (multi-)clouds [7,8] somewhere in the Internet. Application examples for workload offloading could be:

- Processing measurement data recorded by the UE, e.g. to apply computation-intensive Machine Learning (ML) algorithms on specialised hardware;
- Performing post-processing and advanced compression of real-time recorded video/audio data as well as storage for further usage by the UE;
- Mining crypto currency for payment of other services, e.g. for allowing the user to read pay-per-view articles of an online journal or newspaper.

Our approach to apply RSerPool is illustrated in Fig. 3. The resources are added into a pool, here identified by its PH "Processing Pool". That is, the servers in the MEC, and possibly also auxiliary servers in a public (multi-)cloud, become PEs of this pool. From the implementation perspective, it is easy to also start a server instance on the UE itself if the application allows this, due to resource constraints. RSerPool itself is very light-weight, which means it will not add a large management overhead [2,15]. The PE on the UE would then be part of the pool too. If everything else fails, for instance due to loss of network coverage, it would allow to run the application on the UE device, albeit with reduced performance. The PU-side of the application then just needs to pick a *suitable* PE from the pool (or even use multiple PEs in parallel) to use their service. The choice of a suitable PE is, as described in Subsection 2.1, the goal

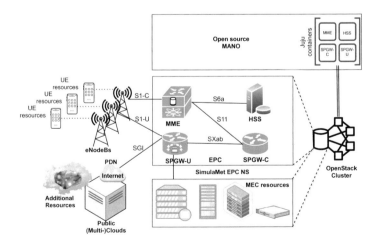

Fig. 2. Testbed setup with EPC, OPEN SOURCE MANO and mobile edge computing

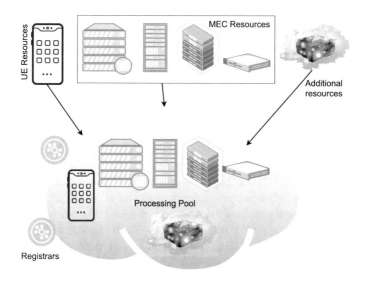

Fig. 3. RSerPool with mobile edge computing and public (multi-)cloud resources

of the pool policy. It is important to use the right pool policy here, so that the following goals are achieved:

1. Only use the PE on the UE if there is no other choice, e.g. no network coverage or MEC/public (multi-)cloud resources available.
2. Use public (multi-)cloud resources only when they are a suitable choice, e.g. when the MEC resources are highly utilised.
3. Otherwise, use the MEC resources.
4. Apply load balancing.

Least Used (LU) [2,17] select the PE p where its load L_p is lowest, and round robin or random among PEs with the same lowest load. It will therefore not differentiate between MEC, public (multi-)cloud and UE resources, i.e. not satisfying the first 3 goals. Priority Least Used (PLU) [2] adds a PE-specific load increment constant I_p to LU. That is, PEs are chosen based on the lowest sum $L_p + I_p$. Setting $I_{p_{\mathrm{MEC}}} < I_{p_{\mathrm{PMC}}}$ for all MEC PEs p_{MEC} and public (multi-)cloud PEs p_{PMC}, as well as $I_{p_{\mathrm{UE}}} = 100\%$ for the UE PE. This PLU pool policy setting achieves our goals as described above.

Finally, PRs are needed to manage the handlespace. Clearly, at least one PR needs to run within the core network (e.g. as part of the MEC setup). Another PR can run locally to the additional PEs in the public (multi-)cloud. To allow the UE to run a local PE without network coverage, the UE would need to run its own PR instance. It is light-weight, i.e. having only low memory and CPU requirements, as well.

4   Testbed Description

For our proof-of-concept evaluation in Sect. 5, we made the following setup: OSM is running "Release EIGHT" in a dedicated VM, connected to an OpenStack setup as NFVI. An NS using the SimulaMet OAI VNF is deployed by OSM in the NFVI. This NFVI is also used for the MEC resources. The User Equipment (UE) is a regular PC. Another regular PC is used as eNodeB. It is running the eNodeB software from OAI, stable version 1.2.2. As Software-Defined Radio (SDR) board, an Ettus B210 connected via USB 3.1 is engaged.

As application, the CalcAppProtocol model from [2, Section 8.3][12] – and provided as part of RSPLIB [2] – is used: a PE has a given request handling *capacity* given in the abstract unit of calculations/s. An arbitrary application-specific metric for capacity may be mapped to this definition, e.g. CPU operations, processing steps, disk space usage, etc. Each request has a request size, which is the number of calculations consumed by the processing of the request. A PE can process multiple requests simultaneously, following the multi-tasking principle. The user-side performance metric is the handling speed. The total time for handling a request d_{Handling} is defined as the sum of queuing time, start-up time (dequeuing until reception of acceptance acknowledgement) and processing time (acceptance until finish). The *handling speed* (in calculations/s) is defined as: $\text{HandlingSpeed} = \frac{\text{RequestSize}}{d_{\text{Handling}}}$.

In our setup at SimulaMet in Oslo, Norway, we use:

- n PU instances may run on the UE. Each PE generates requests with an average size of 1,000,000 calculations at an average frequency of 10 s (negative exponential distribution for both).
- There is 1 PE on the UE, with a capacity of only 200,000 calculations/s.
- 2 PEs are deployed as MEC resources in OpenStack, each with a capacity of 1,000,000 calculations/s.
- In total, 4 PEs as public multi-cloud resources are deployed, each with a capacity of 1,000,000 calculations/s. These PEs are distributed as multi-cloud using the NorNet Core [18,19] infrastructure, with each one PE in Longyearbyen, Gjøvik, Tromsø (all Norway) and Haikou (China). That is, there are significant delay differences (see also [20] for details), between around 20 ms within Norway, and more than 300 ms between Norway and China.
- PEs accept up to 4 requests in parallel. When fully loaded, further requests get rejected, and a new PE has to be selected.
- 1 PR on the UE, 1 PR in the MEC cloud, and 1 PR in Stavanger, Norway.

5   Proof-of-Concept Evaluation

For our proof-of-concept setup, we have chosen the PLU policy (see Sect. 3) with $I_{\text{PU}}{=}100\%$, $I_{\text{MEC}}{=}10\%$ and $I_{\text{PMC}}{=}50\%$. That is, MEC PEs should be preferred, unless highly loaded. Then, PMC PEs should be used instead. The

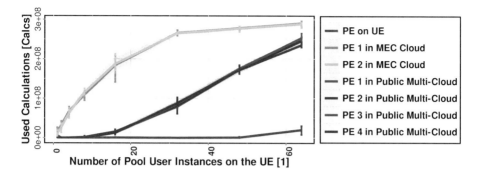

Fig. 4. Used capacity (in calculations) of each PE

PE on the UE itself should only be used in case of severe overload (or loss of network connectivity). Each measurement run has been repeated 5 times, with a measurement duration of 300 s. The results show the average of these runs, together with absolute minimum and maximum (thin error bars) as well as 10% and 90% quantiles (thick error bars).

The results for the used capacity on the PEs when increasing the number of PU instances on the UE from 1 to 64 are shown in Fig. 4. As expected, under low load, mostly the MEC resources are used (green curves). 1 PU generating requests of 1,000,000 calculations every 10 s utilises the pool of 6 PEs (2×MEC+4×PMC, providing 1,000,000 calculations/s per PE) only by around 1.67% on average. It is clearly visible that only the MEC PEs are used. For 16 PUs, the average pool load is already 26.7%. Then, when sometimes a MEC PE already handles 2 or 3 simultaneous requests, it becomes reasonable to use a PMC PE instead. The pool load increases to around 53.3% for 32 PUs, then leading to an increased usage of the 4 PMC PEs. 48 PUs mean 80% average utilisation, while 64 PUs are overload at around 107%.

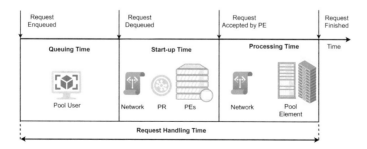

Fig. 5. Queuing time, start-up time and processing time

Figure 1 illustrates the 3 components of the request handling time – queuing, start-up and processing – in detail. Table 1 shows the results for the queuing time,

Table 1. Average per-PU total queuing time

Pool users	Q_{mean}	Q_{min}	Q_{max}	Q_{Q10}	Q_{Q90}
1	37.51	3.79	106.98	4.16	88.29
2	9.88	1.11	39.18	2.68	17.95
4	9.50	0.42	26.44	0.90	21.98
8	10.32	0.64	29.96	3.88	20.72
16	15.57	0.00	78.19	1.47	33.12
32	31.26	0.00	173.75	4.66	65.87
48	37.90	0.60	341.95	5.97	73.62
64	174.33	1.46	1081.93	27.24	435.95

i.e. the total time requests spent in the PU's queue during the measurement time. While it remains low ($\lesssim 38$ s) for decent pool loads, it increases significantly – as expected – when the pool is no longer able to process the overload (64 PUs, i.e. 107% load). In this case, the UE PE also gets used.

The corresponding start-up and processing times are displayed in Fig. 6 and Fig. 7. As expected, the start-up time remains reasonably constant. As long as a PE is not completely full (4 simultaneous requests), it can accept a new request. At 64 PUs, requests get rejected – and another PE selection adds to the start-up

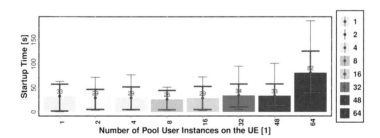

Fig. 6. Average per-PU total start-up time

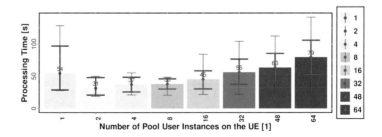

Fig. 7. Average per-PU total processing time

time. The processing time remains at a similar level at low to medium pool loads. That is, the PLU pool policy works as intended, realising a load balancing in the pool. Finally, the processing time increases at high load and overload, when capacity becomes a scarce resource.

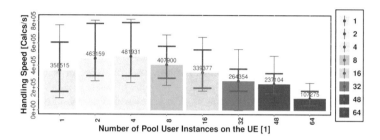

Fig. 8. Average request handling speed (in calculations/s) at the PU

The average handling speed of the pool, as shown in Fig. 8, is reasonable from the user's perspective, as long as the pool is not overloaded. That is, even at 48 PUs running on the UE (which itself would only have a PE capacity of 200,000 calculations/s for *all* requests when disconnected from the pool!), the *per-PU* handling speed is still much more than 218,000 calculations/s.

Hence, our simple proof-of-concept setup achieves the set goals: It is a simple, light-weight, efficient workload offloading from UE into a MEC+PMC system.

6   Conclusions and Future Work

There is a growing demand for offloading workload from mobile devices into (multi-)clouds, because these devices are resource-constrained and battery-powered. The MEC ecosystem aims at providing low-latency communication between user device and cloud service instance or compute device, while using public network services. However, there is a need for a simple, light-weight solution for maintaining sessions with pools of MEC- and (multi-)cloud resources. Reliable Server Pooling (RSerPool) is a standard used for server redundancy and session handling.

In our paper, we propose a simple but efficient approach for using RSerPool as a solution for the workload offloading issue. It is possible to realise an application with light-weight management overhead and low configuration effort by applying useful pool policy configurations. This lightweight property even offers the possibility to run a server instance on the user device itself, to provide local processing – within the device's limits – as a last resort when everything else fails. We presented a simple proof-of-concept evaluation to show the effectiveness of our approach.

As part of future work, it is necessary to evaluate our approach in more detail, in larger setups and different scenarios. It is also useful to integrate the

592 T. Dreibholz and S. Mazumdar

deployment more tightly into OPEN SOURCE MANO (OSM), to provide the pool element and pool registrar functionalities as part of network services.

References

1. Hu, Y.C., Patel, M., Sabella, D., Sprecher, N., Young, V.: Mobile edge computing - a key technology towards 5G. ETSI White Paper **11**(11), 1–16 (2015)
2. Dreibholz, T.: Reliable server pooling – evaluation, optimization and extension of a novel IETF architecture. PhD thesis, University of Duisburg-Essen, Faculty of Economics, Institute for Computer Science and Business Information Systems (2007)
3. Lei, P., Ong, L., Tüxen, M., Dreibholz, T.: An overview of reliable server pooling protocols. In: Informational RFC 5351, IETF (2008)
4. Wang, J., Wu, W., Liao, Z., Sherratt, R.S., Kim, G.J., Alfarraj, O., Alzubi, A., Tolba, A.: A probability preferred priori offloading mechanism in mobile edge computing. IEEE Access **8**, 39758–39767 (2020)
5. Sonmez, C., Ozgovde, A., Ersoy, C.: Fuzzy workload orchestration for edge computing. IEEE Trans. Netw. Serv. Manag. **16**(2), 769–782 (2019)
6. Xiang, S., Nirwan, A.: Latency-aware workload offloading in the cloudlet network. IEEE Commun. Lett **21**(7), 1481–1484 (2017)
7. Dreibholz, T., Mazumdar, S., Zahid, F., Taherkordi, A., Gran, E.G.: Mobile edge as part of the multi-cloud ecosystem: a performance study. In: Proceedings of the 27th Euromicro International Conference on Parallel, Distributed and Network-Based Processing (PDP), Pavia, Lombardia/Italy, pp. 59–66 (2019)
8. Hong, J., Dreibholz, T., Schenkel, J.A., Hu, J.A.: An overview of multi-cloud computing. In: Proceedings of the 1st International Workshop on Recent Advances for Multi-Clouds and Mobile Edge Computing (M2EC) in conjunction with the 33rd International Conference on Advanced Information Networking and Applications (AINA), Matsue, Shimane/Japan, pp. 1055–1068 (2019)
9. Dreibholz, T.: Flexible 4G/5G testbed setup for mobile edge computing using openairinterface and open source MANO. In: Proceedings of the 2nd International Workshop on Recent Advances for Multi-Clouds and Mobile Edge Computing (M2EC) in conjunction with the 34th International Conference on Advanced Information Networking and Applications (AINA), Caserta, Campania/Italy, pp. 1143–1153 (2020)
10. Ocampo, A.F., Dreibholz, T., Fida, M.R., Elmokashfi, A.M., Bryhni, H.: Integrating cloud-RAN with packet core as VNF using open source MANO and OpenAirInterface. In: Proceedings of the 45th IEEE Conference on Local Computer Networks (LCN), Sydney, New South Wales/Australia (2020)
11. Reid, A., González, A., Armengol, A.E., de Blas, G.G., Xie, M., Grønsund, P., Willis, P., Eardley, P., Salguero, F.J.R.: OSM scope, functionality. Operation and Integration Guidelines. White paper, ETSI (2019)
12. Dreibholz, T., Rathgeb, E.P.: Overview and evaluation of the server redundancy and session failover mechanisms in the reliable server pooling framework. Int. J. Adv. Internet Technol. (IJAIT) **2**(1), 1–14 (2009)
13. Dreibholz, T., Zhou, X., Becke, M., Pulinthanath, J., Rathgeb, E.P., Du, W.: On the security of reliable server pooling systems. Int. J. Intell. Inf. Database Syst. (IJIIDS) **4**(6), 552–578 (2010)

14. Xie, Q., Stewart, R.R., Stillman, M., Tüxen, M., Silverton, A.J.: Endpoint handlespace redundancy protocol (ENRP). RFC 5353, IETF (2008)
15. Dreibholz, T., Rathgeb, E.P.: An evaluation of the pool maintenance overhead in reliable server pooling systems. SERSC Int. J. Hybrid Inf. Technol. (IJHIT) **1**(2), 17–32 (2008)
16. Stewart, R.R., Xie, Q., Stillman, M., Tüxen, M.: Aggregate server access protcol (ASAP). RFC 5352, IETF (2008)
17. Dreibholz, T., Tüxen, M.: Reliable server pooling policies. RFC 5356, IETF (2008)
18. Dreibholz, T.: NorNet at hainan university in 2021: from simulations to real-world internet measurements for multi-path transport research - a remote presentation. Keynote Talk at Hainan University, College of Information Science and Technology (CIST) ((2021)
19. Gran, E.G., Dreibholz, T., Kvalbein, A.: NorNet core - a multi-homed research testbed. Comput. Netw. Spec. Issue Future Internet Testbeds **61**, 75–87 (2014)
20. Dreibholz, T.: HiPerConTracer - a versatile tool for IP connectivity tracing in multi-path setups. In: Proceedings of the 28th IEEE International Conference on Software, Telecommunications and Computer Networks (SoftCOM), Hvar, Dalmacija/Croatia (2020)

Acquisition and Modeling of Website Parameters

Krzysztof Zatwarnicki[1]([✉]), Stanislav Barton[1], and Damian Mainka[2]

[1] Department of Computer Science, Opole University of Technology,
76 Prószkowska Street, 45-758 Opole, Poland
s.barton@po.edu.pl
[2] MainkaIT, Opole, Poland

Abstract. During the last 30 years, the web has evolved from simple information HTML pages to complex applications supporting business, television, newspapers, entertainment, and others. While there are many articles on website popularity, there has been little work in understanding the complexity of individual web pages. In the article, we present a measurement-driven study of the complexity of web pages today. We measured 426 866 web pages in about 12 weeks. Our study is devoted to two problems. The first problem was to describe the complexity of a web page with metrics based on the content they included and the kind of service they offered. The second focus of our study was to build probabilistic models of observed distributions. Such models can be used in HTTP request generators modelling the work of modern web systems. Separate models are proposed for each category of web pages and all pages together.

1 Introduction

Internet and especially web services and applications have become part of our daily life. During the last thirty years, web pages have changed significantly. Used initially to host text and images, web pages now include several content types, from images, videos, music to scripts executed in client browsers, which make the page look and behave more like an application than static content. In combination, rendering a single web page often involves fetching several objects with varying characteristics from multiple servers working in a computing cloud.

The very fast development of the Internet and the rapid increase in the number of customers force the need to conduct research aimed at increasing the efficiency and reliability of solutions used on the web. Before new solutions can be widely used for production purposes, they must pass the research phase and testing, thanks to which it is possible to determine the quality of the proposed solutions against other proposals, the scope of applications and scalability. Various techniques are used to evaluate new solutions before they can go into the production phase. These include analytical approach, simulation research and research performed on prototypes loaded with the traffic of artificially generated customers. In all of the techniques, it is required that the nature of user-generated traffic, user requests, and resources retrieved in the requests be consistent with those observed on the Internet. Therefore, there is a need to conduct research on the characteristics that describe the Internet load.

Most prior work on web measurements focuses on characterizing the web graph [6,17], or studying the changes of content on web pages [12]. This leads to a better understanding of web usage but does not provide analysis about websites themselves. Understanding the complexity of web services and individual pages is vital on several fronts. We can start with browser developers who need to identify the aspect of web page complexity that impacts user-perceived performance. Additionally, website providers increasingly incorporate third-party providers such as CDNs into their pages and need techniques to evaluate their impact on users. Engineers developing distribution strategies in web clusters and web cloud systems need models of typical websites and pages to design optimal load distribution algorithms and methods.

In this paper, we present a measurement-driven study of the complexity of web pages today. We measured 426 866 web pages in about 12 weeks, beginning from mid-January 2020. Our study focuses on two problems. First, we want to quantify the complexity of a web page with metrics. The pages are character-ized by the size and the content fetched during the rendering process, i.e. the number of fetched (embedded) objects, their size and type. All of the pages are divided into four categories: Business, Health, Computers and News, according to Alexa—a traffic analysis company [1].

The second focus of our study is to build a probabilistic model that can gen-erate parameters characterizing web pages according to observed distributions. Separate models are proposed for each category of web pages and for all pages together. The proposed models are intended for use in HTTP request genera-tors modelling the work of artificial web clients, especially in simulation-driven research on web cluster systems and web clouds.

The paper is structured as follows. Section 2 contains an overview of publi-cations related to the topic of the article. The specification of the software used to download and analyse web pages is presented in Sect. 3. The results of the analysis are discussed in Sect. 4, while Sect. 5 contains proposed probabilistic models. Section 6 summarizes the article.

2 Related Work

There have been many publications analysing different aspects of web pages. They include works devoted to understanding web page structure, tools improv-ing web performance, and measurements of web applications. Most of the works are focused on web traffic or web protocols.

Early work on web traffic models and metrics describes static properties like distribution of page and image sizes. For example, Williams et al. discuss the results of research of three different data sets collected in 2004, Hernan-dez et al. discuss the evolution of Web traffic from 1995 to 2003 [14]. In 2001, Cherkasova and Karlsson [8] showed several new trends in Web server workloads.

Barford et al. [4] study changes in Web client access patterns between 1995 and 1998. Other important work in the early studies of the web can be found in [2, 21, 27].

More recent work also describes dynamic aspects inherent in Web 2.0 technologies like JavaScript, CSS and AJAX. Ihm [15] and Pries [16] analyse the implications of evolution from static to dynamic contents for web traffic models. On the basis of those works, Butkiewicz et al. [17] proposed metrics more adequate for Web 2.0 traffic. Analysing the characteristics of mobile web traffic is yet another trend, e.g. Johnson [18], Knox [19], and how it differentiates from computer web traffic, e.g. Sanders et al. [20]. The latest articles also describe the characteristics of HTML 5.0 web pages, e.g. Mendes et al. [13] and Arvidsson et al. [3].

In parallel, there are works devoted to the problem of measuring the end-user's Quality of Experience (QoE). Some of them describe how to qualify user-experience [21, 23] while other studies investigate the impact of Internet protocols on web browsing quality [9, 19, 20]. Particular interest is also given to the upload or download link when investigating bottlenecks [25, 30].

It should be noted that as far as there are publications devoted to traffic and web protocols, there is little work on quantifying and understanding website complexity and especially modelling the website parameters like the size of HTML content, the number of objects embedded in the page, the size of embedded objects etc. Among the few studies, we can mention Crovella and Bestavros [10] and Barford et al. [5].

In this work, we describe the complexity of the web pages measured, and we propose statistical models reflecting the characteristics of average websites of different categories. The presented models are designed to build simulation models to assess the quality of operation of web systems, in particular web clouds, cluster-based web systems and Content Delivery Networks (CDN).

3   Acquisition of Website Parameters

3.1   Data Collection

The main aim of conducting our research was to just characterize web pages served on web services and to build models of such web pages. To achieve this, we analysed the pages and their elements that were downloaded from the domain or sub-domains of the main domain of the websites. We did not consider the elements of the web pages which were downloaded but originated from sources other than the selected original ones.

We analysed websites belonging to four different categories: Business, Health, Computers and News. Addresses of the most popular websites and their categorization were taken from Alexa [1] – one of the most popular web services providing web traffic data, global rankings and other information on 30 million websites. About 150 thousand web services were visited, from which 426 thousand individual web pages were downloaded during the experiments. The measurement was conducted from January 16 to April 6, 2020.

Our web page sample includes landing pages, as well as other pages belonging to the services, indicated as links on the websites. Our work involved fetching the pages, rendering and analysing them. Also, embedded objects like css and js files, images, audio and video content, belonging to and indicated on the pages were downloaded to be analysed separately. Table 1 presents the number of the pages analysed and the embedded objects.

Table 1. The number of the analyzed web pages and the embedded objects

Category	Number of pages	Number of embedded objects
Business	163 551	7 261 827
Health	84 001	3 455 883
Computers	154 722	5 207 757
News	24 592	3 456 484
Summary	426 866	19 381 951

The following data were collected for each page:

- sizes of HTML rendered for the page,
- size of embedded objects,
- number of embedded objects in the page,
- size of the page measured as the sum of sizes of HTML rendered for the page and sizes of embedded objects,
- category of embedded objects.

3.2 Tools

We used many different tools in the automated process of collecting the data. Google Chrome v.7.9.317, running in headless mode (command line), was the most important tool. The chrome engine was downloading, rendering, and saving the HTML part of the pages. We used the wget [31] to download and analyse embedded objects. The process of collecting data was controlled by scripts written for research purposes. The results were saved to a database. All the experiments were run concurrently (up to 32 separate processes) and were conducted on the Windows 10 operating system.

4 Characterizing Complexity

The analysis we have made has two main aims. First, in this section, we analyse and discuss the obtained metrics describing web pages. This involves page sizes, the number of embedded objects fetched to load the page, sizes of those objects and their categories, and, in the end, sizes of HTML/XHTML documents. Next, in Sect. 5, we build models of web pages belonging to different categories.

4.1 Page Size

The cumulative distribution of page sizes for different categories of web services is presented in Fig. 1 and Table 2. It contains additional metrics like mean, median and 90th percentile values. As we can see, the median values for Computers and News are the smallest (253 kB and 270 kB). However, the 90th percentile for News is the greatest and equals 2000 kB.

The median and 90th percentile values significantly differ from the mean values, which are much greater due to high standard deviations. The mean values for News amount to 10011 kB, while for Computers only to 550 kB. Additionally, according to [11,28,29] the mean page size varies from 1400 to 3000 kB, depending on the data source. The measured values of page sizes are smaller. This is due to the fact that when calculating the page size we take into account only objects downloaded from the domain of the page. Additionally, most of the results presented in the articles concern research conducted for the landing web pages. In our study, we examine not only landing web pages but also other pages available in the web services.

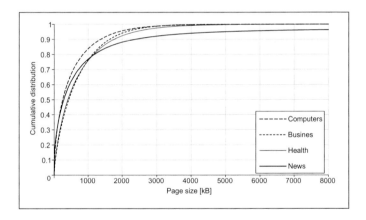

Fig. 1. Cumulative distribution of page sizes for different categories of pages

Table 2. Cumulative distribution of page sizes for different categories of pages

	Business	Computers	Health	News
Mean	718 364	550 895	2 756 672	10 011 833
Median	457 840	253 335	458 140	270 430
90th percentile	1 605 917	1 366 476	1 737 023	2 011 593

4.2   Number of Embedded Objects

Number of embedded objects. The median number of embedded objects varies
from 19 in News pages to 31 in Busines pages. News pages have the biggest
number of objects in the case of the 90th percentile, which is 131, and Computer
pages – the smallest one, i.e. 83, see Table 3. About 5% of News pages have more
than 400 objects, see Fig. 2 and Table 3.

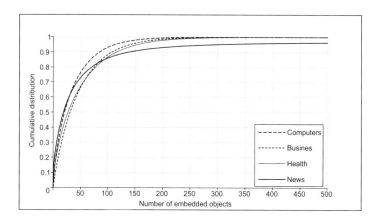

Fig. 2. Cumulative distribution of the number of embedded objects for different categories of pages

Table 3. Cumulative distribution of the number of embedded objects for different
categories of pages

	Business	Computers	Health	News
Mean	46.5	35.8	96.5	273.7
Median	31	21	29	19
90th percentile	83	83	117	131

The mean number of different types of embedded objects per page for different page categories is presented in Fig. 3. The object type was determined from
the HTTP response header when the object was retrieved. There are four types
presented: application representing today javascript files, text that in almost all
cases contains css content, fonts of different kinds, and the most popular type –
image.

Most of the pages contain little more than 10 images. Only News pages
contain an average of 40 pictures. Business, Health and Computer pages use
from 3 to 5 application scripts. The number of fonts used usually does not
exceed 3. Also, in most cases, the number of css files is smaller than 4.

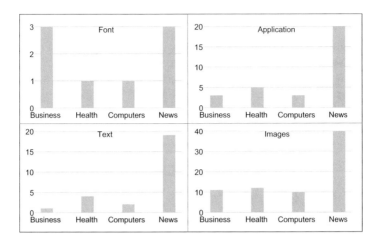

Fig. 3. Mean number of different types of embedded objects per page for different page categories

4.3   HTML Document Sizes

The mean and median values do not vary much in the case of HTML document sizes, see Fig. 4 and Table 4. The biggest documents (with the biggest median) belong to the category of Health pages. The smallest documents belong to Business pages. Pages in Health and News categories have a small number of small HTML documents.

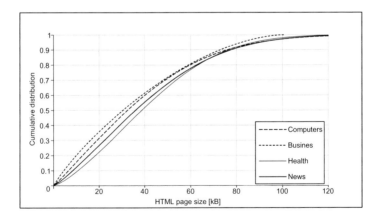

Fig. 4. Cumulative distribution of HTML document sizes for different categories of pages

Table 4. Cumulative distribution of HTML document sizes for different categories of pages

	Business	Computers	Health	News
Mean	36 028	37 984	43 031	42 570
Median	29 845	32 987	39 343	34 460
90th percentile	76 902	78 072	79 678	80 327

4.4   Sizes of Embedded Objects

The embedded objects used on the web page have different structures and purposes. Some of them are text files containing information on formatting the page or are javascript programs run when the page is rendered or when the user works with the page. Pictures and videos are binary data and are part of the website directly viewed and evaluated by users. Figure 5 and Table 5 present results for embedded object sizes. The median value of the sizes does not vary much between categories (from 3.2 kB to 4.2 kB). However, the 90th percentile is almost twice as big for News (74 kB) as for Business (47 kB). Additionally, the mean values are much bigger than the median values.

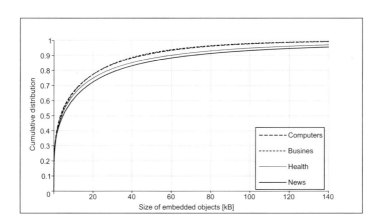

Fig. 5. Cumulative distribution of embedded object sizes for different categories of pages

Table 5. Cumulative distribution of embedded object sizes for different categories of pages

	Business	Computers	Health	News
Mean	16 846	14 697	28 634	36 683
Median	3 495	3 289	3 760	4 232
90th percentile	47 680	49 544	62 317	73 909

5 Modeling Website Parameters

The second focus of our study was to build a probabilistic model of the categories
of pages analysed. Such a model can be useful when building HTTP request
generators modelling the work of artificial web clients, especially in simulation-
driven research.

Table 6. Numerical models of pages of different categories

	Function $\mathbf{F(x)}$	$\mathbf{R^2}$	$\mathbf{p}$	$\mathbf{X_{max}}$
	Busines			
Page Size	$\mathrm{erf}\left(64.963396\,x^{0.679810}\right)$	0.995	0.959	603 573
Number of emb. obj.	$\left(1-(1-x)^{247.626760}\right)^{0.9289606}$	0.999	0.975	11 989
HTML Size	$\left(1-(1-x)^{1.694583}\right)^{0.885441}$	0.992	0.933	100
Size of emb. obj.	$\left(1-(1-x)^{16450.038945}\right)^{0.296008}$	0.934	0.892	596 690
	Computers			
Page Size	$\mathrm{erf}\left(19.545611\,x^{0.535063}\right)$	0.992	0.988	264 594
Number of emb. obj.	$\left(1-(1-x)^{544.585251}\right)^{0.738174}$	0.995	0.907	22 859
HTML Size	$\mathrm{erf}\left(2.786606\,x^{1.0492765}\right)$	0.988	0.967	174
Size of emb. obj.	$\left(1-(1-x)^{64.8063754}\right)^{0.274135}$	0.911	0.915	2 587
	Health			
Page Size	$\left(1-(1-x)^{4830.841387}\right)^{0.691644}$	0.995	0.958	4 336 513
Number of emb. obj.	$\mathrm{erf}\left(62.979693\,x^{0.615792}\right)$	0.995	0.634	75 850
HTML Size	$\mathrm{erf}\left(33.583876\,x^{1.310522}\right)$	0.997	0.992	955
Size of emb. obj.	$\mathrm{erf}\left(15.453365\,x^{0.329546}\right)$	0.972	0.341	150 988
	News			
Page Size	$\dfrac{x^{0.004871}}{1+13.388374\,e^{-56.721173\,x^{0.317079}}}$	0.976	0.957	4 336 513
Number of emb. obj.	$\dfrac{x^{0.006485}}{1+19.703743\,e^{-39.152385\,x^{0.305865}}}$	0.969	0.930	75 850
HTML Size	$\mathrm{erf}\left(22.396271\,x^{0.6264173}\right)$	0.995	0.934	995
Size of emb. obj.	$\dfrac{x^{0.000462}}{1+5.524628\,e^{-41.198557\,x^{0.305431}}}$	0.976	0.957	150 988
	Compilation			
Page Size	$\left(1-(1-x)^{5134.315371}\right)^{0.6264173}$	0.995	0.958	4 336 513
Number of emb. obj.	$\left(1-(1-x)^{1527.319823}\right)^{0.737083}$	0.985	0.658	75 850
HTML Size	$\mathrm{erf}\left(18.031233\,x^{1.062706}\right)$	0.997	0.992	955
Size of emb. obj.	$\mathrm{erf}\left(26.678036\,x^{10.335099}\right)$	0.931	0.584	596 690

To build the model, we used the regression analysis of the cumulative distribution function. Let us denote the sought regression function as $f(x, c_i), i = 1 \ldots n$, where unknown coefficients are marked c_i. Considering graphs of individual cumulative probabilities, the sought regression function must be a nonlinear function of both x and coefficients c_i. We searched for these coefficients using the Least Squares Method. Due to their nonlinearity, we used the Gauss-Newton iteration method. The results of our analysis and calculations are presented in Table 6.

The functions $F(x)$ in Table 6 are regression functions modelling the cumulative distributions of page sizes, numbers of embedded objects, sizes of HTML document and sizes of embedded objects for different categories of pages. The functions $F(x)$ are normalized in both the x-axis and the y-axis, which means that the definition domains are always $0 \leq x \leq 1$ and the function domain is $0 \leq F(x) \leq 1$. The x value is normalized in the following way $x = \frac{X}{X_{max}}$, where X is the value for which the probability is calculated, X_{max} is the biggest value observed (provided in Table 1). The erf is an error function (also called the Gauss error function). It is worth noticing that $F(\frac{X}{X_{max}}) = P(r \leq X)$, where the right-hand side represents the probability P in which the random variable r takes on a value less than or equal to X.

The regression function quality can be assessed according to the coefficient of determination R^2 and a p-value calculated by the method of the analysis of variance (ANOVA). As we can notice in Table 6, both values are very high and often close to 1 in each case, which indicates that the proposed models represent well-observed distributions.

6   Summary

In this paper, we present the results of the measurement-driven study of the complexity of web pages. We characterized the complexity based on both the content they included and the kind of service they offered. We found out that News sites contained significantly more embedded objects, especially images, and much bigger HTML documents than sites of other categories. By contrast, Business and Computer sites are the smallest as regards website sizes, the number of embedded objects and HTML document size.

Based on the collected data, we were able to build models of pages of different categories. To build models, we used the regression analysis of the cumulative distribution functions. To tune, the model method of the least squares, combined with the Gauss-Newton iteration method, was used. The proposed model can be used to build HTTP request generators modelling the work of artificial web clients in simulation-driven research.

References

1. Alexa – traffic analysis company. https://www.alexa.com/. Accessed 10 Oct 2020
2. Arlitt, M.F., Friedrich, R., Jin, T.: Workload characterization of a Web proxy in a cable modem environment. ACM Performance Eval. Rev. **27**(2), 25–36 (1999)
3. Arvidsson, A., Grinnemo, K., Chen, E., Wang, Q., Brunstrom, A.: Web metrics for the next generation performance enhancing proxies. In: 2019 International Conference on Software, Telecommunications and Computer Networks (SoftCOM), Split, Croatia, pp. 1–6 (2019)
4. Barford P., Bestavros, A., Bradley, A., Crovella, M.: Changes in web client access patterns, characteristics and caching implications In: Special Issue on World Wide Web Characterization and Performance Evaluation; World Wide Web Journal, (1998)
5. Barford, P., Misra, V.: Measurement. IMA Workshop on Internet Modeling and Analysis, Minneapolis, MN, January, Modeling and Analysis of the Internet (2004)
6. Broder, A., et al.: Graph structure in the web. Comput. Networks **33**(1), 309–320 (2000)
7. Butkiewicz, M., Madhyastha, H.V., Sekar, V.: Understanding website complexity: measurements, metrics, and implications. In: Proceedings 2011 ACM SIGCOMM Conference on Internet Measurement Conference, pp. 313–328 (2011)
8. Cherkasova, L., Karlsson, M.: Dynamics and evolution of web sites: analysis, metrics and design issues. In: Proceedings of the 6th IEEE Symposium on Computers and Communications, Hammamet, Tunisia, 64–71 (2001)
9. Cook, S., Mathieu, B., Truong, P., Hamchaoui, I.: QUIC: better for what and for whom? In: Proceedings of IEEE International Conference on Communications (ICC) (2017)
10. Crovella, M.E., Bestavros, A.: SelfSimilarity in World Wide Web traffic evidence and possible causes. In: SIGMETRICS 1996, USA, Philadelphia (1996)
11. Everts, T.: The average web page is 3MB. How much should we care? https://speedcurve.com/blog/web-performance-page-bloat/. Accessed 10 Oct 2020
12. Fetterly, D., Manasse, M., Najork, M., Wiener, J.: A large-scale study of the evolution of web pages. Softw. Practice and Experience **34**(2), 213–237 (2004)
13. Mendes, J., Laranjeiro, N., Vieira, M.: Toward characterizing HTML defects on the Web. Software Practice and Experience **48**(1), 750–757 (2018)
14. Hernandez-Campos, F., Jeffay, K., Donelson-Smith F.: Tracking the evolution of web traffic: 1995–2003. In: Proceedings of 11th IEEE/ACM International Symposium on Modeling, Analysis and Simulation of Computer and Telecommunications Systems (MASCOTS), pp. 16–25 (2003)
15. Ihm, S., Pai, V.S.: Towards understanding modern web traffic. In: 2011 ACM SIGCOMM Internet Measurement Conference, pp. 295–312 (2011)
16. Johnson, T., Seeling, P.: Landing page characteristics model for mobile web performance evaluations on object and page levels. In: 2015 IEEE International Conference on Communications (ICC), pp. 3616–3621 (2015)
17. Kleinberg, J.M., Kumar, S.R., Raghavan, P., Rajagopalan, S., Tomkins, A.: The web as a graph: measurements, models and methods. In: Proceedings COCOON (1999)
18. Knox, A., Seeling, P.: Mobile web page characteristics: delivery and stability considerations. In: 2017 14th IEEE Annual Consumer Communications Networking Conference (CCNC), pp. 37–40 (2017)

19. Lychev, R., Jero, S., Boldyreva, A., Nita-Rotaru, C.: How Secure and Quick is QUIC? Provable Security and Performance Analyses. In: Proceedings of the IEEE Symposium on Security and Privacy (SP), (2015)
20. Manzoor, J., Drago, I., Sadre, R.: How HTTP/2 is changing web traffic and how to detect it. In: Network Traffic Measurement and Analysis Conference, TMA 2017, Dublin, Ireland, pp. 21–23 (2017)
21. Pitkow, J.E.: Summary of WWW characterization. World Wide Web **2**(1–2), 3–13 (1999)
22. Pries, R., Magyari, Z., Tran-Gia, P.: An HTTP web traffic model based on the top one million visited web pages. In: Proceedings of the 8th Euro-NF Conference on Next Generation Internet (NGI), pp. 133–139 (2012)
23. Sanders, S., Sanka, G., Aikat, J., Kaur, J.: The influence of client platform on web page content: meas-urements, analysis, and implications. In: Web Information Systems Engineering – WISE 2015. Lecture Notes in Computer Science, Springer, Cham, pp. 1–16 (2015)
24. Sackl, A., Casas, P., Schatz, R., Janowski, L., Irmer, R.: Quantifying the impact of network bandwidth fluctuations and outages on web qoe. In: Seventh International Workshop on Quality of Multimedia Experience, QoMEX 2015, Pilos, Messinia, Greece, pp. 1–6 (2015)
25. Saverimoutou, A., Mathieu, B., Vaton. S.: A 6-month analysis of factors impacting web browsing quality for QoE prediction. In: Computer Networks, Elsevier (2019)
26. Seufert, M., Wehner, N., Casas, P.: Studying the impact of HAS qoe factors on the standardized qoe model P.1203. In: 38th IEEE International Conference on Distributed Computing Systems, Vienna, Austria, ICDCS (2018)
27. Williams, A., Arlitt, M., Williamson, C., Barker, K.: Web workload characterization: ten years later. In: Tang, X., Jianliang, X., Chanson, S.T. (eds.) Publish info Web content, pp. 3–22. Springer, New York (2005)
28. Web Almanac By HTTP Archive: HTTP Archive's annual state of the web report https://almanac.httparchive.org/en/2019/. Accessed 10 Oct 2020
29. Website Performance, Webpages Are Getting Larger Every Year, and Here's Why it Matters https://www.pingdom.com/blog/webpages-are-getting-larger-every-year-and-heres-why-it-matters/. Accessed 10 Oct 2020
30. Yang, Y., Zhang, L., Maheshwari, R., Kahn, Z.A., Agarwal, D., Dubey, S.: A point of presence recommendation system using real user monitoring data. In: Passive and Active Measurement -17th International Conference, PAM 2016, Heraklion, Greece (2016)
31. Wget - package for retrieving files https://www.gnu.org/software/wget/. Accessed 1 Oct 2019

Present Clinicians with the Most Relevant Patient Healthcare Data Through the Integration of Graph DB, Semantic Web and Blockchain Technologies

Steven Delaney[1]([✉]), Doug Schmidt[2], and Christopher Chun Ki Chan[3]

[1] Ryerson University, Toronto, Canada
steven.delaney@ryerson.ca
[2] Capital Blockchain, Toronto, Canada
doug.schmidt@sympatico.ca
[3] Department of Information Management, Chaoyang University of Technology, Taichung, Taiwan
Christopherckchan@cyut.edu.tw

Abstract. Many governments are examining the move to a unified patient centric healthcare system to reduce costs and improve healthcare services. This would provide clinicians (A Clinician is a healthcare professional who deals directly with a patient, such as doctors, nurses, pharmacists, specialists and first responders.) with a holistic view of all patient data. Combined with increases in patient data generated by advances in medical devices, it allows the extract of more information increasing recall, but it requires great cognitive effort to identify relevant information. This additional effort will exasperate the impact of the trend of decreasing time clinicians have to interact with their patients. It raises the risk to clinicians of cognitive overload. This paper proposes a solution that presents clinicians with the most relevant patient healthcare data based on the clinicians role and the healthcare scenario. The goal is to provide the clinician a more efficient method to determine the proper treatment. This proposal is applicable to current scenarios where the key relevant data is readily available.

Keywords: Blockchain · Healthcare · Medical · Privacy · Semantic web · Semantic DB · Graph DB · EHR · EHR relevancy · Clinicians · Shared data · Patient · Interoperability · Ontology

1 Introduction

A patient centric view of healthcare data in a unified healthcare system promises to lower healthcare costs which currently average about 10% of the GDP for most developed countries [1]. A unified healthcare system would provide rapid access to a holistic view of the patients data; allowing for faster determination and application of treatment. [1]. This unified system is of high interest to governments to lower costs and improve healthcare systems. [2].

© The Author(s), under exclusive license to Springer Nature Switzerland AG 2021
L. Barolli et al. (Eds.): AINA 2021, LNNS 227, pp. 606–616, 2021.
https://doi.org/10.1007/978-3-030-75078-7_60

In Canada, patient data resides in the HIMS[1] of multiple healthcare organizations [1]. The heterogeneous nature of the multiple HIMS data silos complicates interoperability in support of a patient centric view of healthcare data [1]. Interoperability issues include linking data to the correct patient, EHR standards, data quality, performance and scalability, data privacy, access and security controls [1–3].

The interoperability issue of data quality with respect to the usability of data directly correlates with clinician interaction time with patients [4] due to an increased cognitive load on clinicians to navigate increased volumes of patient healthcare data [3, 4, 7, 8]. The latter point will be exasperated in a unified healthcare system, due to the combined volume of patient data from multiple EHR's from clinics, IOTs, pharmacies and hospitals.

Addressing the issue of data usability can have a significant impact on the performance of clinicians in patient care [6].

2   Objective

The objective of this paper is to propose a system which synthesizes a patients healthcare information and presents it to a clinician in order of relevancy to the patients current condition. This system combines healthcare data and medical facts using schemas[2] to determine and retrieve information is most relevant to the patients condition and treatment.

3   Approach

Our initial approach was to create schema's that represent Semantic DB query patterns. Clinicians including First Responders could simply select a schema to identify the EHR patient data most relevant to the healthcare scenario. Semantic relationships would be used to expand the query to include additional data determined to be useful.

We started with a Semantic based approach as there is a strong level of interest and success in the application of Semantic Web technology to provide consistency in medical terminology and categorization of clinical and medical data [6, 8].

However, our ability to utilize current Semantic healthcare databases to identify relevant data presented challenges which echoed similar comments in other papers. For example; (1) The interrelationships between medical data is highly complex and not well suited to the implementation of relationships supported by Semantic DBs [9]. (2) There are many healthcare Semantic DBs rich in medical information but often, the design is focused on categorization which was difficult to map to clinical use cases [8]. (3) Often, we found healthcare databases using codified terminology that was unintuitive making the data and structure difficult to understand [8]. (4) The use of custom ontologies to support our solution would have introduced undesired maintenance effort to keep the ontologies up to date. [8].

For these reasons, we took the approach displayed in Fig. 1 and described as follows: (1) A Graph DB is used as an index to data residing in Semantic and EHR databases. (2)

---

[1] Healthcare Information Management System.

[2] A Schema in this context refers to a pre-determined query pattern(s).

EHR databases provide data specific to the patient. (3) Semantic DBs provide scientific medical information such as, known drug side effects, indicators of health, and the known effects of drug combinations.

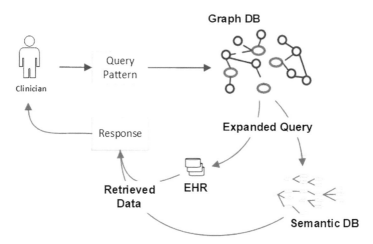

Fig. 1. This solution diagram describes the approach to our proposed solution. The solution employs Graph DB expanded query and Semantic DB and EHR queries

4   Proposed Solution Design

The solution design logical architecture is described in Fig. 2. It represents the solution that will be built for our Proof of Concept (PoC). A description of the various logical architecture components follows the diagram.

4.1   Clinician UI

This component represents the user interface. It handles user authentication, selection and initiation of the healthcare query and presentation of the retrieved data. This component introduces the concept of a schema.

The UI also handles the presentation of the retrieved data. The determination of what data is most relevant is based on (1) Input from the clinician on what information is most essential [10]. (2) The age of the data with priority given to the most recent [7].

4.2   Schema (Graph DB Query Patterns)

Within the context of this solution, a schema consists of one or more Graph DB query patterns to determine what data is to be retrieved from one or more EHR and Semantic Web DB's. The clinicians specialty/role and current healthcare scenario will determine the clinician's selection of schema. For example, our research shows that the data needs of a cardiologist differ from those of a dermatologist. The significant value of a Schema based information retrieval solution is that Schemas can be customized to the more than 60 Specialist categories in North America.

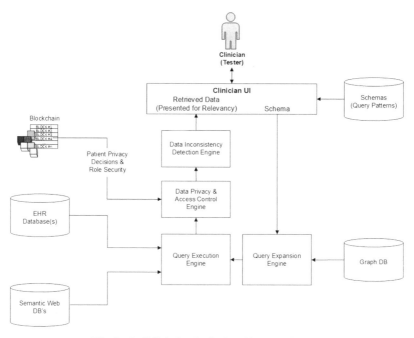

Fig. 2. PoC Solution logical architecture design.

4.3 Query Expansion Engine

The purpose of the Query Expansion Engine is to query the Graph DB using the query patterns contained in the schema passed to it by the UI. The query patterns will be expanded as per instructions in the schema utilizing the data relationships in the Graph DB.

4.4 Graph DB

The Graph DB nodes will represent either a single healthcare data entity or a Semantic DB query. Data entity node properties will contain information on what EHR databases to query, the name of the data entity in each EHR and the requirements to transform the data retrieved in a common format for display. This allows the solution to pull the same type of healthcare data (e.g. patients cholesterol level) from multiple EHR's albeit, taken at different times to form a history of the patients data and changes over time. For nodes representing Semantic DB queries, the Semantic DBs can be used to provide additional non-personal scientific information. For example, a patients prescribed drug could trigger bringing back information on that drugs dosage, side effects and instructions to take if overdosed.

The examples reflect our requirements for the Graph DB nodes and do not necessarily indicate how the nodes will be configured.

In Table 1, the name used internally for the healthcare data is stated as well as the database and field name to be retrieved. Since we cannot guarantee that every EHR database will format common information in the same manner, we have included a provision to transform the retrieved data into a common format for use in this solution. Also included, is a rating of the privacy risk presented by non-authorized access to this information. This will be passed to the *Data Privacy and Access Control Engine* to determine if the user is authorized to view this type of information. It should be noted that for all EHR data retrieved, the *Query Execution Engine* will retrieve data about the data such as, the date the data was collected.

Table 1. Example of the property requirements for an EHR node in the Graph DB.

Node Field	Example	Description
Name	Cholesterol_Level	Name of the Data Entity used in the solution
Property 1	Best EHR Database	The name of the EHR database containing information on cholesterol levels for the patient.
Property 1a	BEHR_cholesterol	The name of the data field in the EHR Database containing the patients cholesterol level.
Property 1b	BEHR_cholesterol_DT	The cholesterol value retrieved will be transformed to conform to the format standards of this solution. This is accomplished by passing the cholesterol value returned to an API using the code in this property.
Property 2	2nd Best EHR Database	same as Property 1
Property 2a	2BEHR_cholesterol	same as Property 1a
Property 2b	2BEHR_cholesterol_DT	same as Property 1b
Property 3	Data Privacy Type	**DR** : This information on it's own can be used to identify the patient. **IRx** : This data when combined with other information could be used to infer the identity of the patient. The value x (1 to 10) indicates the estimated level of risk of this occuring should this information be known. **NONE:** The identity of the patient cannot be inferred by this information.

The schema may or may not require that additional information general to the healthcare data (non-private) is to be retrieved as shown in Table 2. The choice will either be set within the schema or presented as an optional check box for the clinician. Using our cholesterol example, the clinician may want to know what the indicator values are for cholesterol. This information could be pulled from an appropriate Semantic DB.

Table 2. Example of the property requirements for a Semantic Query node.

Node Field	Example	Description
Name	SMDB_cholesterol	The name of the semantic database query.
Property 1	Best SMDB End Point	The URI for the semantic web end-point.
Property 2	SMDB_cholesterol_Q	The semantic database query.

The Graph DB node and relationship model will be created with the assistance of medical practitioners for several healthcare subject areas.

4.5   Query Execution Engine

The purpose of the *Query Execution Engine* is to receive the information passed to it by the *Query Expansion Engine* and create and execute the various queries to the specified EHR and Semantic DBs. The engine will also handle the formatting of the retrieved data and the collection of information about the data such as, the date the data was collected. This engine must have API's or URI's to all the EHR and Semantic Web databases as well as the authority to retrieve the data. Further, the engine must be able to match the patients identification with that in the EHR database to ensure the data retrieved does indeed belong to the patient of interest. Security and patient identification is not an issue with the Semantic DBs as the information is openly known scientific facts not related to any specific patient.

The engine design will incorporate protections for data at rest, data in use and data in transit to better protect patient privacy.

4.6   Data Privacy and Access Control Engine

This purpose of the *Data Privacy and Access Control Engine* is to ensure that patient privacy and application security controls are in place. This engine will receive the data retrieved by the *Query Execution Engine* and will filter out and clear any information that the viewer is not authorized to view. The reasoning for not relegating this function to the systems hosting the EHR data is that it reduces system complexity and reliability risks associated with the requirement to synchronize multiple security administration systems and processes.

4.7   Blockchain

The use of blockchain technology is leveraged to hold the systems security role configuration and, record patient privacy decisions regarding their healthcare data. In this PoC, the *Data Privacy and Access Control Engine* will utilize the blockchain to enforce privacy controls and patient privacy decisions. Our vision of blockchain will be limited in terms of the implementation in the PoC. This is done in order to control the scope of the PoC. Risks will be mitigated by using synthetic and anonymized patient data.

Our vision for blockchain would have the EHR services utilize the same blockchain for applying security controls for external access to data in their systems. This supports all parties having consistent definition of the privacy controls for each patient with minimal lag time for changes to propagate. This would support the ability for patients to record their own privacy decisions to the blockchain. It would ensure that all users of the patients data adhere to the same controls and all access to the patients data is recorded and transparent to the patients.

4.8   Data Inconsistency Detection Engine

The purpose of the *Data Inconsistency Detection Engine* is to examine the data retrieved before it is presented. It compares "like" data to determine if there are any significant discrepancies. For example, is the patients blood type in one EHR different than another?

If yes, the clinician might want to run a blood test to determine the correct blood type for initiating treatment. It may also be an indicator that there is a mismatch in the patients identity between two EHR's meaning that the patients data presented is unreliable.

5 Advantages of This Solution Design

Advantages to this design include (1) EHR and Semantic Web data continue to reside in their respective databases and systems. This means that administration, operations, development, support, hosting and the purpose for which they were designed remain unchanged. Our solution only needs to ensure that the API's, URI's and queries continue to work. (2) The only data our solution needs to maintain is that required by the clinicians using the solution. Our solution does not need to be concerned with all the other data in the EHR and Semantic Web databases. (3) Our design handles the interoperability with multiple EHR data bases at the data entity level. Because we are using only a subset of the data, the interoperability issues are significantly lessened. (4) No private patent information is permanently retained in the solution simplifying risk management. (5) Because the solution only retains information on healthcare data, the solution is largely immune to changes in the structure of the EHR and Semantic Web databases provided the queries are not significantly affected. (6) The solution is agile to the extent that new relationships, data entities and Semantic Web data can be easily incorporated and adjusted in the Graph DB. (7) There is the potential in the future to provide additional value through enhancements to the Graph DB. For example, placing data in the relationships to indicate new relationships and the lessening or removal of existing relationships.

6 Solution Challenges

Solution challenges include (1) The continued reliability and availability of the supporting EHR and Semantic Web databases. (2) Since data is retained in external EHR systems, there is the risk of performance issues with respect to response time. (3) Although the solution has applicability in some areas today, it will not achieve it's full potential until unified patient centric healthcare data is available in some form. (4) The ability for patients to control the privacy of their healthcare data and see consistency of the application of their privacy elections, will not occur until there is a structure, perhaps similar to our thinking on blockchain, in place to support it.

7 Next Steps

Our next step is to prove the core of the solution which is to generate expanded queries for an EHR and Semantic DB using a query pattern for a very simple use case. Our team clinicians will validate the quality of the results. From there we will test using more complex use cases and slowly build out the automation of the process.

8 Test Use Case

The EHR data selected for the initial simple use case was based on discussion with an Addiction Specialist (MD). The Addiction Specialist (MD) may meet virtually with over 100 patients in a single day leaving a limited allocation of time with each patient and little leeway in order to remain on schedule. It is therefore important that the time spent retrieving and viewing the patients healthcare information be as short as possible in order to maximize the time spent engaging with each patient.

The selected EHR data in Table 3 represents the data most relevant to the Addiction Specialist (MD) in order to update themselves on the patients history and for making decisions regarding treatment or changes in treatment.

Table 3. Type of EHR data to be used in our simple test case.

Ref#	Pri	Data Entity	Description	Query
1	1	Demographics	Patients demographics e.g. Age, Sex, Weight	Base
2	7B	ECG	ECG showing the patients QT/QTC value This is important for determining if it is safe to prescribe drugs that might lengthen the QT/QTC value	Base
3	7C	Infectious diseases	A list of the patients current infectious diseases obtained from a lab report	Base
4	4	Medical history	The patients medical history contains diagnosed past infectious & non-infectious diseases, current non-infectious diseases, and conditions (e.g. spina bifida)	Base
5	7A	POC UDS	The results of the most recent Point of Care UDS drug test	Base
6	2	Prescribed medication	Patients prescribed drugs	Base
7		Prescribed drugs surgery	Indicate if any drugs were prescribed due to a surgeries with an emphasis on the most recent surgeries	Expanded
8		Prescribed drugs drug allergies	Indicate if the patient is taking a prescribed drug that is known to cause them an allergic reaction	Expanded
9	5	Psychiatric history	Patients psychiatric history For example, bipolar disorder	Base
10	6	Surgical history	List of all the patients surgical history with an emphasis on the most recent	Base
11	3	Drug allergies	List known drug allergies	Base

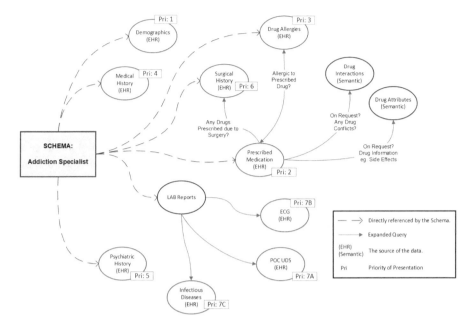

Fig. 3. Addiction Specialist (MD) treatment session – The relationship model on which the test Graph DB will be based.

The meaning of the columns is; Ref#: a reference number to the data entity. Pri: is relative importance of the data to the Addiction Specialist (MD). Data Entity: Is the name of the data entity. Description: The information represented by the data entity. Part of Query: will contain with the terms "Base" or "Expanded". "Base" indicates that this data is referenced by name in the schema. "Expanded" indicates that the data is referenced through a relationship to another data entity that has been retrieved.

The test data in the Semantic DB is defined in the following Table 4. The meaning of the columns is the same as for the EHR data described in Table 3. The only difference is that the data is retrieved only "On Request".

Table 4. Type of Semantic data to be used in our simple test case

Ref#	Pri	Data Entity	Description	Query
1	N/A	Drug attributes	Factual information on the specified patients prescribed drug. For example, side effects, dosage	On Request
2	N/A	Drug interactions	Provides known negative reactions the specified drug has to other drugs	On Request

The relationship model for the identified EHR and Semantic data is displayed in Fig. 3. This will be our starting point for creation of the Graph DB model. The Fig. 3

identifies the relationships between the nodes. The dashed lines do not indicate node relationships. These lines indicate the nodes directly referenced in the selected schema.

9 Conclusion

The ability to present the most relevant patient data to a clinician assists the speed and determination of treatment to the patient. Although the relevancy of data is important, the benefit extends only to the availability of information. Even small amounts of additional healthcare data can make a significant difference in some healthcare scenarios (e.g. the initial use case).

A key benefit of the proposed solution is that it can be implemented with limited technical requirements and will not incur significant interoperability issues to the existing healthcare infrastructure. Furthermore, the solution can be made operational within a healthcare scenario of limited scope and scale over time.

To maximize the benefits, the challenge of data availability and protection of patient privacy must be addressed. It is possible this system combined with the use of blockchain to protect patient privacy may be one way of addressing these challenges. For future work, we hope to demonstrate this system with a proof of concept.

The goal of this solution is to provide the most valuable and useful data so a clinician can make the most "informed" decision possible in the most timely and cost effective manner.

References

1. Sharma, R.: Blockchain in Healthcare, France Canada Chamber Of Commerce Ontario (2018). www.fccco.org. 5 June 2018
2. Special Issue Electronic Solutions for Artificial Intelligence Healthcare, 15 June 2019, A Review on the Role of Blockchain Technology in the Healthcare Domain (2019). https://doi.org/10.3390/electronics8060679
3. WETSEB 2019 Proceedings of the 2nd International Workshop on Emerging Trends in Software Engineering for Blockchain, pp. 52–55, 27 May 2019, Investigating quality requirements for blockchain-based healthcare systems, Montreal, Quebec, Canada, Copyright 2019 IEEE Press (2019). https://doi.org/10.1109/wetseb.2019.00014
4. HIMSS EHR Usability Task Force, June 2009, Defining and Testing EHR Usability: Principles and Proposed Methods of EHR Usability Evaluation and Rating. https://mospace.umsystem.edu/xmlui/handle/10355/3719
5. Tielman T. Van Vleck, Daniel M. Stein, Peter D. Stetson, Stephen B. Johnson, 2007, Assessing Data Relevance For Automated Generation Of A Clinical Summary, Pages 761–765, AMIA 2007 Symposium Proceedings, PMCID: PMC2655814, PMID: 18693939
6. David, R., Francis, R., Joan, A.L.V., Fabio, C., Saram, E., Patrizia, M., Roberta, A., Carlo, C.: An ontology-based personalization of health-care knowledge to support clinical decisions for chronically ill patients. J. Biomed. Inform. **45**, 429–446 (2012). https://doi.org/10.1016/j.jbi.2011.12.008
7. Cheng, T., Hemant, J., Huimin, Z.: Semantic Query Expansion for Effective EHR Retrieval, China Summer Workshop On Information Management - Proceedings, KunMing China, 29–30 July 2008, pp. 6–11 (2008)

8. Xhemal, Z., Bujar, R., Florije, I., Jaumin, A.: State of the Art of Semantic Web for Health-care, 2015, World Conference on Technology, Innovation and Entrepreneurship, Procedia - Social and Behavioral Sciences, **195**, 1990–1998 (2015). https://doi.org/10.1016/j.sbspro.2015.06.213

9. Renzo, A.: A comparison of current graph DB models. In: 2012 IEEE 28th International Conference on Data Engineering Workshops (2012). https://doi.org/10.1109/icdew.2012.31, Electronic ISBN:978-0-7695-4748-0, Print ISBN:978-1-4673-1640-8

10. Jukka, P., Wray, B., Henry, T.: A temporally adaptive content-based relevance ranking algo-rithm. In: Proceedings of the 28th Annual International ACM SIGIR Conference on Research and Development in Information, pp. 647–648. Salvador, Brazil, ACM New York, NY, USA ©2005, ISBN:1-59593-034-5 (2005). https://doi.org/10.1145/1076034.1076171

Automatically Injecting Semantic Annotations into Online Articles

Hamza Salem[1](✉), Manuel Mazzara[1](✉), and Said Elnaffar[2](✉)

[1] Innopolis University, Republic of Tatarstan, Russia
h.salem@innopolis.university, m.mazzara@innopolis.ru
[2] American University in the Emirates, Dubai, United Arab Emirates
said.elnaffar@aue.ae

Abstract. Extracting information from the Web is one of the most trending subjects for data analysts and scientists. Web scraping is one of the prominent means to do so by parsing HTML pages and extracting data from their embedded tags. To that end, software developers write customized scripts for each target Website. Crafting such scripts is a challenging task due to the different structures of Web sites and their dynamic rendering. Yet, a more challenging task is to infer semantic annotations from a news article. It is interesting, for example, to have a word such as the US in an already published article to be automatically annotated semantically by its population and area. We argue that these annotations should not be mere hyperlinks to be consumed by human readers but rather they can be machine readable using the Web standard RDF (Resource Description Framework). Embedding such RDF inside the HTML page of the news article should enrich it with automatically generated semantics that can boost its SEO (Search Engine Optimization) and are ready to be consumed by conventional Web scrapers. As a proof of concept, we built a prototype that focuses on the plain text of an article, without stipulating the existence of structured tags and attributes, to generate a new Web document that is augmented with semantic annotations using the RDF markup language, readable by humans and machine-consumable by Web scrapers.

1 Introduction

The World Wide Web is increasingly becoming the ultimate source of knowledge in almost every field in our life. This opened virtually unlimited opportunities to the Internet users to exploit data in various ways [1]. One of the common ways of searching and browsing the net is for regular users to enter certain keywords or phrases. Another important segment of searchers are conducted by those bots developed by coders who aim to automate the process of data extraction from the Web. The latter category of users, i.e., the search bots, are not after the mere dump of data stored as pages but rather they seek more embedded information that they can use for data analysis, comparisons, surveys, etc. Hence, the concept of Web data extraction is common among software developers nowadays.

L. Barolli et al. (Eds.): AINA 2021, LNNS 227, pp. 617–624, 2021.
https://doi.org/10.1007/978-3-030-75078-7_61

Web data extraction is the process of gleaning useful data from the Web using some techniques such as Web Usage Mining, Web Scraping, and Semantic Annotation [2]. These techniques leverage other technologies such as natural language processing, machine learning, information retrieval, databases, and ontologies [2]. Web scraping, as it is often referred to, is an automation process that developers build by writing customized scripts, called scrapers, that identify data embedded inside Web documents, especially HTML pages, and extract them [3].

The Semantic Web is an extension of the World Wide Web through standards introduced by the World Wide Web Consortium (W3C) [4]. Semantic Web technologies enable us to create data stores on the Web, build vocabularies, and write rules for handling data. Oceans of data can be also linked and related to each other using standards such as RDF, SPARQL, OWL, and SKOS [4–6]. The objective of the Semantic Web is to make Internet data machine readable and reasoned about using metadata technologies such as the Resource Description Framework (RDF) and Web Ontology Language (OWL) [6]. Fortunately, the RDFa Core, is one of the RDF versions whose specification attributes can be expressed using any markup language [7]. This means that embedded data already available in a markup language such as the HTML and its tags can be reused by the RDFa markup, and thus, publishers do not need to inject or repeat a significant amount of data in the content of a document [8]. To that end, we argue in this work that we can transform the mere syntax of a text document (e.g., news article) to a semantically annotated one using the simple markups of HTML and RDFa. This paper is organized as follows. Section 2 is a brief overview of the literature of efforts to derive semantics from online documents. In Sect. 3, we present our methodology to extract semantics, and in Sect. 4 we discuss the implementation details of the prototype. Finally, Sect. 5 concludes the paper and sketches some future work.

2   Literature Review

The literature is rich with efforts that aim at either generating semantic annotation or extracting them. For example, the authors in [9] investigated how to extract semantic annotation using the KIM (Knowledge and Information) technique [10] to build a large knowledge base of annotations in the form of searchable metadata. However, for this approach to work, having structured semantic layer and data schema is mandatory, a condition that many Web sites do not satisfy.

Other works used middleware to extract semantic data from Web sites. For instance, in [11] the authors used Virtuoso Sponger [12] to generate linked data originating from different data sources with a wide range of data representation and formats. Semantic Fire [2] is another middleware that can extract and store semantic data as RDFs. In general, this middleware approach entails prior configuration in order to operate on domain-specific Web sites (e.g., real estates vs. car services).

The authors in [13] developed a method that performs induction of first-order logic rules to extract data from unstructured web resources. This method extracts data from web sites with an unknown DOM tree structure to extract semantic information without external supervision. Nevertheless, this approach is still highly contingent on the existence of content-specific attributes and Natural Language Processing (NLP) patterns.

Beno [14] developed Doc2RDFa, an HTML rich text processor with the ability to automatically and manually annotate content. However, the system is domain-specific and confined to legal documents. In a previous work [15], we argued that we can derive semantic annotations from the plain content (text) without the necessity of having it structured (e.g., using HTML tags and attributes) and we built a prototype using a pattern matching as a proof of concept.

3   Methodology

We argue that our methodology, which is depicted in Fig. 1, can derive semantic annotation from the sole textual content without the necessity of having a prior structure embodied as HTML tags or any markups. Overall, our approach uses the NLP to transform paragraphs to individual sentences, then words, which are augmented with semantics using HTML tags in the RDFa format. As shown in Fig. 1, we initially scrape the HTML of a given news Web site using pattern mining [15]. The pattern miner scrapes the Title and the Body of the news document in order to obtain the plain text content. Subsequently, we generate tokens from words where each word gets tagged by a syntactic category based on its POS (Part-of-speech) such as a verb or a noun. To augment this word with semantics, we connect to external sources of knowledge, such as, but not limited to:

1. **PyDictionary**: PyDictionary is a Python library that provides the meaning, translation, synonym, and antonym of a word. It derives such info from sources such as synonym.com, Google Translate, and WordNet [16].
2. **Search Results returned by Google** (see Fig. 3).

The first source of knowledge, PyDictionary, is an example of how we can get semantics about a word directly by a simple programmatic query submitted to the library. The second source is an example of indirect retrieval of semantics as we ask a scraper bot to extract the info (e.g., place, event, and person) from the result page (Fig. 3) returned by Google.

Finally, and to annotate the original document with the retrieved semantics, we inject RDFa properties into the HTML structure for each token (word) processed. The newly generated document can be used by humans (upon rendering) or by the machine (upon scraping).

4   Design and Implementation

To implement the above methodology, we used our pattern mining tool that we previously built to scrap Web sites [15]. Figure 2 shows how the Tokenizer splits article body into sentences (see Code Snippet 1) using the NLTK library in Python [17].

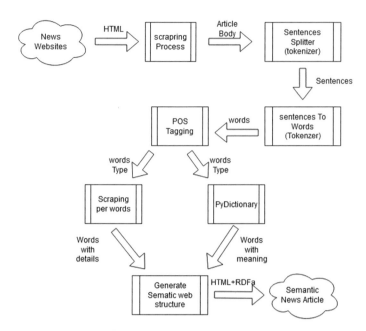

Fig. 1. Generating semantic web structure using scraping and NLP workflow

```
#data is the Article body
import nltk
tokenizer = nltk.data.load('tokenizers/punkt/english.pickle')
arr=tokenizer.tokenize(data)
#arr is array of sentence output
```

Code Snippet 1: The Tokenizer splits article into sentences

Each sentence is fed into the word-tokenize() function of NLTK to produce a list of words or tokens. Code Snippet 2 shows the implementation in Python.

```
#data is the Article body
for i in arr:
    tokens = nltk.word_tokenize(i)
    tokensTags=nltk.pos_tag(tokens)
    print(tokens)
    print(tokens_tags)
# Tokens = ['And', 'the', 'US', 'continues', 'to', 'this',
'day', 'to','end', 'Iranian', 'lives',]
# tokensTags= [('And', 'CC'), ('the', 'DT'), ('US', 'NNP'),
('continues', 'VBZ'), ('to', 'TO'), ('this', 'DT'),
('day', 'NN'), ('to', 'TO'), ('end', 'VB'), ('Iranian', 'JJ')
, ('lives', 'NNS')]
```

Code Snippet 2: The Tokenizer splits sentences into tokens and tags each

By getting tokensTags we had the type of the word if it verb or linking word or noun, as you see token ('US', 'NNP') is a noun it refers to "United States". By Using the tag we can take only nouns to be entered to pyDictionary to get the meaning of the word as descriptions, see example in Code snippet 3.

```
from PyDictionary import PyDictionary
print(PyDictionary(tokensTags).getMeanings()[tokensTags]
#{'Noun': ['North American republic containing 50
states - 48 conterminous states in North America plus
Alaska in northwest North America and the Hawaiian
Islands in the Pacific Ocean; achieved independence in
1776']}
```

Code Snippet 3: PyDictionary getMeanings() function

To overcome this limitation, we resorted to a wider pool of knowledge, namely the result page that we typically get when we Google a certain word, as shown in Fig. 3. We built a scaping bot (Code Snippet 4) using the BeautifulSoup library [18] which provides idiomatic ways of navigating, searching, and modifying the parse tree of HTML and XML files.

```
 import urllib
import requests
from bs4 import BeautifulSoup

# desktop user-agent
USER_AGENT = "Mozilla/5.0 (Macintosh;
Intel Mac OS X 10.14; rv:65.0) Gecko/20100101 Firefox/65.0"

query = tokensTags
query = query.replace(' ', '+')
URL = "https://www.google.com/search?client=firefox-b-d&q="+query

headers = {"user-agent": USER_AGENT}
resp = requests.get(URL, headers=headers)

if resp.status_code == 200:
    soup = BeautifulSoup(resp.content, "html.parser")
    results = []
    res=soup.find('div',{'id':'wp-tabs-container'})
    .find("div",{'class':'kno-rdesc'})
    print(res.text)
```

Code Snippet 4: Scrape Google results description

Figure 3 shows the returned results page by Google for the word US. Our BeautifulSoup-based scraper bot locates the red-framed area, using the find() function, and extracts the information from that part of the HTML structure (Code Snippet 4).

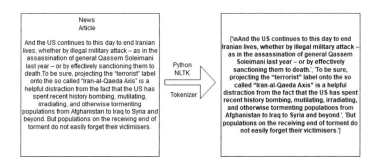

Fig. 2. Tokenizer Splitting article body to sentences

Fig. 3. The retrieved and scraped Google result page upon searching for a word

In the last step, we merge all extracted semantic annotations and inject them as HTML and RDFa attributes in the newly generated document of the news article (Code Snippet 5). These semantic annotations can be expressed as visual RDFa graphs [8] to be viewed by the site visitors or consumed by machines using scraper bots (Fig. 4).

```
<div vocab="https://schema.org/" typeof="Noun">
  <p> And the
    <span property="description" >
    <meta property="description" content="is a country of 50 states
    covering a vast swath of North America, with
    Alaska in the northwest and Hawaii extending the nations
```

```
    presence into the Pacific Ocean.
    Major Atlantic Coast cities are New York,
    a global finance and culture center,
    and capital Washington, DC." />
    US
    </span> continues to this day to end Iranian lives
  </p>
</div>
```

Code Snippet 5: Example HTML + RDFa

Fig. 4. The semantically annotated RDFa pages can be visualized and machine-read

5   Conclusion and Future Work

In this work, we showed the possibility of augmenting raw online content, such as news articles, with semantic annotations that are automatically derived from different sources of knowledge. It is a transformation from sematic-less Web document into another one that is enriched with semantics that can be manifested as hyperlinks or floating tips to be consumed by the visiting reader. Semantic annotations can even be expressed as a pictorial graph that explains the meanings of terms and their relationships as a Web ontology [6]. The benefits are not limited to humans, machines can also tap into such embedded RDFs inside the HTML using Web scrapers and SEO tools. As a future work, we plan to apply our approach on more case studies and investigate the possibility of extending it to photos and videos embedded in online articles in order to identify the main objects (or concepts) inside them using machine learning.

References

1. Brin, S., Motwani, R., Page, L., Winograd, T.: What can you do with a web in your pocket? IEEE Data Eng. Bull. **21**(2), 37–47 (1998). http://dblp.uni-trier.de/db/journals/debu/debu21.html#BrinMPW98

2. Laender, A.H.F., Ribeiro-Neto, B.A., da Silva, A.S., Teixeira, J.S.: A brief survey of web data extraction tools. SIGMOD Rec. **31**(2), 8493 (2002). https://doi.org/10.1145/565117.565137

3. Malik, S.K., Rizvi, S.: Information extraction using web usage mining, web scrapping and semantic annotation. In: 2011 International Conference on Computational Intelligence and Communication Networks, pp. 465–469 (2011)

4. Berners-Lee, T., Hendler, J., Lassila, O.: The semantic web. Sci. Am. **284**(5), 34–43 (2001)

5. Pérez, J., Arenas, M., Gutierrez, C.: Semantics and complexity of SPARQL. ACM Trans. Database Syst. (TODS) **34**(3), 1–45 (2009)

6. Gómez-Pérez, A., Corcho, O.: Ontology languages for the semantic web. IEEE Intell. Syst. **17**(1), 54–60 (2002)

7. Dolog, P., Nejdl, W.: Challenges and benefits of the semantic web for user modelling. In: Proceedings of the Workshop on Adaptive Hypermedia and Adaptive Web-Based Systems (AH2003) at 12th International World Wide Web Conference. Citeseer, Budapest (2003)

8. Adida, B., Birbeck, M.: RDFA core 1.1 (2007)

9. Malik, S.K., Rizvi, S.A.: Information extraction using web usage mining, web scrapping and semantic annotation. In: 2011 International Conference on Computational Intelligence and Communication Networks, pp. 465–469. IEEE (2011)

10. Ontotext, K.: Platform (2011)

11. Ferrara, E., De Meo, P., Fiumara, G., Baumgartner, R.: Web data extraction, applications and techniques: A survey. Knowledge-based systems **70**, 301–323 (2014)

12. Chen, H., Chau, M., Zeng, D.: Ci spider: a tool for competitive intelligence on the web. Decision Support Syst. **34**(1), 1–17 (2002)

13. Fernández-Villamor, J.I., Iglesias, C.A., Garijo, M.: First-order logic rule induction for information extraction in web resources. Int. J. Artif. Intell. Tools **21**(06), 1250032 (2012)

14. Beno, M., Filtz, E., Kirrane, S., Polleres, A.: Doc2RDFa: semantic annotation for web documents (2019)

15. Salem, H., Mazzara, M.: Pattern matching-based scraping of news websites. J. Phys. Conf. Ser. 1694, 012011 (2020). https://doi.org/10.1088/1742-6596/1694/1/012011

16. Karkar, R., Nagdev, S., Gangrade, P., Gatade, D.D.: Transformation of sentimental impact for documents. Transformation **5**(04) (2018)

17. Perkins, J.: Python text processing with NLTK 2.0 cookbook. Packt Publishing Ltd. (2010)

18. Richardson, L.: Beautiful soup documentation. Dosegljivo (2007). https://www.crummy.com/software/BeautifulSoup/bs4/doc/. Dostopano 7 July 2018

A Color Adaptation Method in Picture Story with Emotional Expression of Body Motion

Huynh Thi Kim Chi[1], Kosuke Takano[2(✉)], and Kin Fun Li[3]

[1] Course of Information and Computer Sciences, Graduate School of Engineering, Kanagawa Institute of Technology, 1030 Shimo-ogino, Atsugi, Kanagawa, Japan
s1985021@cco.kanagawa-it.ac.jp
[2] Department of Information and Computer Sciences, Kanagawa Institute of Technology, 1030 Shimo-ogino, Atsugi, Kanagawa, Japan
takano@ic.kanagawa-it.ac.jp
[3] Department of Electrical and Computer Engineering, University of Victoria, Victoria, BC V8W 3P6, Canada
kinli@uvic.ca

Abstract. In this paper, we present color adaptation of an image in a picture story with emotional expression of body motion. In the modern education, it has become increasingly important to cultivate aesthetic sensibility of children for enriching their open-minded and creative personality. Current advanced computer-aided education systems for sensibility education such as the eurhythmics approach can support emotional education and brings new experiences to both teachers and students. During the class of aesthetic sensibility learning using such systems, it is important for students to share emotions, feelings, and intentions for activating student's collaboration; however, the conventional systems have not focused on amplifying and conveying an atmosphere generated by emotions and feelings of students. For this purpose, we develop a system that can amplify and convey the atmosphere created by the student's physical motions by dynamically changing pictures for proceeding picture stories. By several experiments using our prototype, we evaluate the feasibility of the proposed system.

1 Introduction

Emotional development is vital in helping children grow into well-adjusted adults. Being able to express feelings through words and pictures for sharing personal emotions allows children to be healthy emotionally and psychologically.

Current advanced computer-aided education systems for sensibility education such as the eurhythmics approach can support emotional education and brings new experiences to both teachers and students. During the class of aesthetic sensibility learning using such systems, it is important for students to share emotions, feelings, and intentions for activating student's collaboration. There are many researches of extracting and recognizing emotions [1–3, 6], and these methods regarding emotion recognition can be applied for implementing such sensibility education systems; however, the conventional systems have not focused on amplifying and conveying an atmosphere generated by emotions and feelings of students.

L. Barolli et al. (Eds.): AINA 2021, LNNS 227, pp. 625–634, 2021.
https://doi.org/10.1007/978-3-030-75078-7_62

In this paper, we present a color adaptation method in picture story with emotional expression of body motion. The feature of our system is that our system can amplify and convey the atmosphere created by students' physical motions by dynamically changing the color of pictures based on the relationship between emotion and color [8–10] through a collaborative physical movement. Our system helps children recognize feelings, at an early age to primary school age, based on emotional expressions such as fear, surprise, anger, sorrow, happiness, and disgust with gestures, and encourage them to express and share the individual emotions by making visual stories.

By initial experiments using our prototype, we evaluate the feasibility of the proposed system for discussing educational implications using our system.

2 A Collaborative Picture Story Generation System with Emotional Expression

2.1 Collaborative Picture Story Generation

The purpose of our proposed system is to support aesthetic sensibility learning, where students play together with pictures.

Figure 1 shows the overview of our system that supports aesthetic sensibility learning, where students play together with nature sounds and pictures. During the students' ensemble performance, their facial expressions and body movements are continuously captured by sensor devices and analyzed, and the emotion associated with the facial expressions and body movements are extracted based on the analysis.

Here, the performance elements of student x consist of facial expression f_x and body motion m_x. Therefore, suppose that a picture story includes s pictures, $p_1, p_2, \ldots, p_s$, the performance of student x is represented as a sequence of performance elements: $(f_x, m_x)_1$, $(f_x, m_x)_2$, $(f_x, m_x)_3$, $\ldots$, $(f_x, m_x)_s$ (Fig. 2).

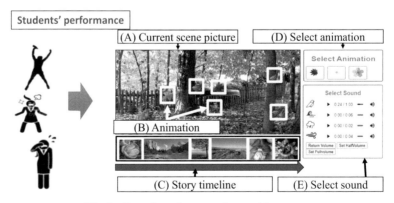

Fig. 1. Overview of proposed ensemble system

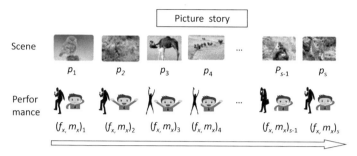

Fig. 2. Student's performance along picture story

2.2 Emotion Judgement Model

The Emotion judgement model extracts emotion e_x of a student x from the student's performance elements $(f_x, m_x)_s$ (Fig. 3). As for the definition of emotion e_k, in this study, six basic emotions, anger, disgust, fear, happiness, sorrow, and surprise are defined for construction of the emotion model, since it is deemed that these six emotions are universal over the world [1].

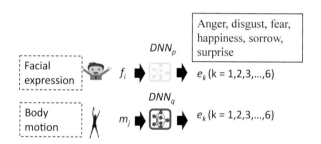

Fig. 3. Emotion judgement model

For predicting emotions from student's physical motions, we construct several types of deep neural networks (DNNs) using Convolutional Neural Network (CNN) and Long Short-Term Memory (LSTM). The DNNs predict two kinds of emotions from student's body movements. One is the emotion that is classified into six emotions, sorrow, happiness, surprise, disgust, anger, fear, and another is the positiveness that has two attributes, positive and negative (Fig. 4). For example, the six emotions are used to select the natural sounds and change the color of images that matches student's emotions, and the positiveness is used to set the volume of sound and the speed of animation, respectively.

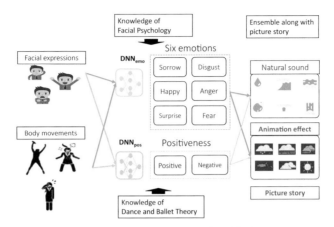

Fig. 4. Emotional models for ensemble with picture story

2.3 Scene Conversion and Animation Effect According to Emotion

In the picture story, each scene p_s consists of animation AN_s, picture PI_s, and sound SO_s, that is, $p_s = (AN_s, PI_s, SO_s)$. A scene converter converts a p_s to $p^*_s = (AN^*_s, PI^*_s, SO^*_s)$ according to emotion predicted by the emotion judgement model (Fig. 5). In order to implement scene conversion, Generative Adversarial Networks (GANs) [7, 8] is leveraged to automatically generate a new scene p^*_s according to student's emotion e_x.

Figure 6 shows an example of animation effects for leaves fall, flowers fall, and snow fall. In addition, Fig. 7 shows an example of scene conversion along picture story by adopting contents to student's emotion. Thus, emotions are propagated and amplified by changing the colors, sounds, and animations of the content in the picture story.

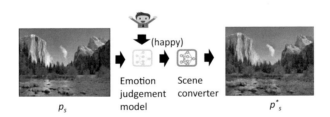

Fig. 5. Scene conversion according to predicted emotion

(a) Leaves fall   **(b)** Flowers fall   **(c)** Snow fall

Fig. 6. Example of animation effect

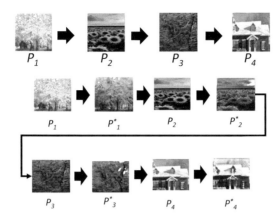

Fig. 7. Example of scene conversion according to emotion along picture story

2.4 Propagation and Amplification of Emotion

The extracted emotions are amplified and shared between students playing together by dynamically changing pictures and sounds in the picture story. Besides, as shown in Fig. 8, our system allows remote classes, even in foreign countries, to connect to an ensemble session and play together at the same time so that students in each class can share the emotions and the atmosphere created by students' performance.

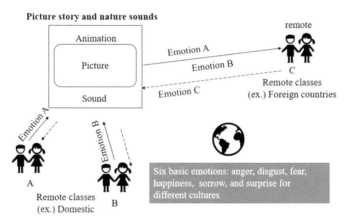

Fig. 8. Playing ensembles between remote classes

3 Experiment

3.1 Experiment 1

In this experiment, we confirm that a body movement within an image can be recognized and two emotions, positive and negative, can be estimated.

We collect dance videos on YouTube where six kinds of labels, "happiness", "surprise", "fear", "sadness", "anger", and "disgust", are attached for each movement. In addition, OpenPose is used for making a posture link image by extracting a posture link information of a human in an image. Then, we labeled each posture link image as "positive" for "happiness" and "surprise" motions, and "negative" for "fear", "sadness", "anger", and "disgust" motions. Table 1 shows an overview of image classification by positive and negative, and Fig. 9 shows an example of posture link images. We applied pre-trained Convolutional Neural Network (CNN), VGG-16 for training posture link images with positive and negative labels. As for test data, we collected 17 images by ourselves.

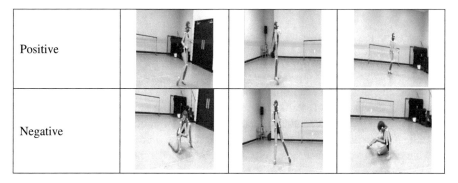

Fig. 9. Example of posture link images (dance video)

Table 1. Image classification by positive and negative

Emotions	Number	Size
Positive	1,680	368 × 368
Negative	1,680	368 × 368

Figure 10 shows a recognition accuracy of classification, where 88% for positive posture link images, 67% for negative posture link images, and the average rate are 74%. In addition, Fig. 11 shows an example of classification result. From these results, we can confirm that a human posture within an image can be recognized and two emotions, positive and negative, can be predicted by learning posture link images.

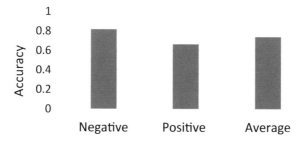

Fig. 10. Recognition accuracy of classification

Emotion	Images		
Positive			
Negative			

Fig. 11. Example of classification result

3.2   Experiment 2

In this experiment, we confirm the image color conversion is possible using six colors corresponding to emotions by Generative Adversarial Networks (GANs). We use StarGAN [7] which is one of implementation of GANs.

In the experiment, we focus on three categories of images, (1) flowers, (2) butterflies, and (3) fish, and collected 500 images for each category of 6 colors, and 3000 images in total. When we apply GANs, each image is resized to 178px × 218px or 128px × 128px.

Figures 12, 13, and 14 show examples of the results of color conversion. From these results, we can confirm that the color of the selected image can be converted to other colors associated with six emotions.

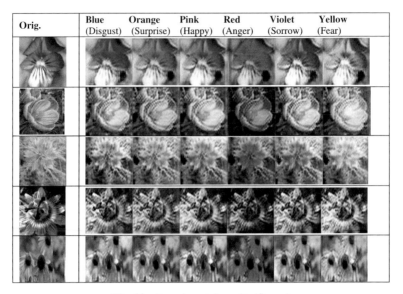

Fig. 12. Experimental results (flower)

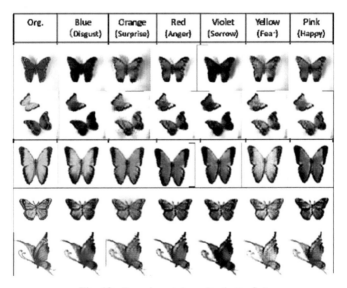

Fig. 13. Experimental results (butterfly)

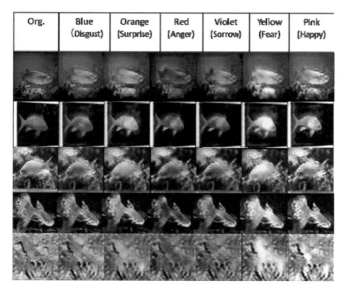

Org.	Blue (Disgust)	Orange (Surprise)	Red (Anger)	Violet (Sorrow)	Yellow (Fear)	Pink (Happy)

Fig. 14. Experimental results (fish)

4   Conclusion

In this paper, we have proposed a color adaptation method in picture story with emotional expression of body motion.

In the experiment, we evaluated the feasibility of our proposed system and showed the results as an initial step that our system can extract emotions from body motion and convert the color of flower, butterfly, and fish images to a new color based on the extracted emotions.

In our future work, first, we will implement the basic functions to extract emotions from facial expression, body action and to adjust sounds and animations in each scene in the picture story according to students' emotions. Then, we will evaluate how our system works for supporting sensibility education.

References

1. Ekman, P.: Universal facial expressions of emotion, vol. 8. The University of California, Autumn (1970)
2. Berkeley Keltner, D., Cowen, A.S.: Self-report captures 27 distinct categories of emotion bridged by continuous gradients. https://doi.org/10.1073/pnas.1702247114. Accessed 5 Sept 2017
3. Plutchik, R.: The nature of emotions. Am. Sci. **89**, 344–355 (2001)
4. Hanada, M.: Correspondence analysis of color–emotion. Associations 30 August 2017, pp. 224–237 (2017)
5. D'Andrade, R.G.: The colors of emotion. Am. Ethnol. **1**, 49–63 (2009)
6. Ullah, A., Ahmad, J., Muhammad, K., Sajjad, M., Wook, S.: Action recognition in video sequences using deep bi-directional LSTM with CNN features. Vis. Surveill. Biometr.: Pract. Challenges Possibil. 1155–1166 (2017)

7. Choi, Y., Choi, M., Kim, M., Ha, J.-W., Kim, S., Choo, J.: StarGAN: unified generative adversarial networks for multi-domain image-to-image translation. In: IEEE Conference on Computer Vision and Pattern Recognition (CVPR), pp. 8789–8797 (2018)
8. Kaneko, T., Kameoka, H., Tanaka, K., Hojo, N.: Cyclegan-VC2: Improved cycleGAN-based non-parallel voice conversion. In: 2019 IEEE International Conference on Acoustics Speech and Signal Processing (ICASSP) ICASSP 2019, pp. 6820–6824 (2019)
9. Wu, H., Zheng, S., Zhang, J., Huang, K.: GP-GAN: towards realistic high-resolution image blending. In: MM 2019: Proceedings of the 27th ACM International Conference on Multimedia, pp. 2487–2495, October 2019
10. Tsai, Y.-H., Shen, X., Lin, Z., Sunkavalli, K., Lu, X., Yang, M.-H.: Deep image harmonization. In: Proceedings of the IEEE Conference on Computer Vision and Pattern Recognition (CVPR), pp. 3789–3797 (2017)

First-Impression-Based Unreliable Web Pages Detection – Does First Impression Work?

Kenta Yamada[(✉)] and Hayato Yamana

Department of Computer Science and Communications Engineering, Waseda University, Tokyo, Japan
{yamada,yamana}@yama.info.waseda.ac.jp

Abstract. Considering the continuous increase in the number of web pages world-wide, detecting unreliable pages, such as those containing fake news, is indispensable. Natural language processing and social-information-based methods have been proposed for web page credibility evaluation. However, the applicability of the former to web pages is limited because a model is required for each language, while the latter is poorly adapted to changes, owing to its dependence on external services that can be discontinued. To solve these problems, herein we propose a first-impression-based web credibility evaluation method. Our experimental evaluation of a fake news corpus gave an accuracy of 0.898, which is superior to those of existing methods.

1 Introduction

Owing to the widespread availability of personal computers and smartphones, children and adults can nowadays easily access the internet. Thus, distinguishing unreliable web pages, such as those containing rumors, misleading content, and fake news, is indispensable [1]. The features employed for web page credibility evaluation are categorized into the following: content [1–4], involving statistics on texts and elements, linguistic [4, 5], concerning natural language processing output, and social [1, 2], such as the web page reputation on SNS. Among these features, the linguistic and social features exhibit severe limitations. The linguistic features are language-dependent, and thus, require a specific model for each language. Conversely, social features depend on external services that can be terminated, and the reputations of these services are commonly inaccurate.

To overcome these challenges, herein a first impression-based web credibility evaluation method is proposed. This method integrates the web page usability and performance, which represent content features, thereby eliminating limitations of existing methods. The method relies on the prominence–interpretation [6] and dual-processing [7] theories, assuming that users adopt a heuristic evaluation that initially considers the web page appearance based on fast or low-cost thinking. We adopted the Google lighthouse, a tool for scoring web page quality, to assess the first impressions of users. A fake news corpus was then evaluated, and the results were compared with those of existing methods. Finally, we examined the difference between the evaluation based on the first impression of a web page by a user and that associated with deep consideration.

© The Author(s), under exclusive license to Springer Nature Switzerland AG 2021
L. Barolli et al. (Eds.): AINA 2021, LNNS 227, pp. 635–641, 2021.
https://doi.org/10.1007/978-3-030-75078-7_63

Related studies are presented in Sect. 2, while the first impression-based web credibility evaluation method is described in Sect. 3. In Sect. 4, the proposed method is examined. Finally, the conclusions of the paper are provided in Sect. 5.

2   Related Studies

2.1   Prominence–Interpretation Theory

Fogg's prominence–interpretation theory (PI theory) [6] originated from one of the earliest studies on web page credibility evaluation. This theory is based on a four-year quantitative study involving 6,500 participants, and it assumes that users rely on two steps (prominence and interpretation) for decision-making. Prominence can be linked to the perception of a web page, and it is influenced by the following five factors: user involvement, web page topic, user task, user experience, and differences between individuals. Among these factors, involvement, that is, the motivation and ability to scrutinize a web page dominates. Fogg et al., therefore, concluded that each user evaluates the credibility of a web page through prominence, followed by interpretation within a short period. According to Fogg et al., 46.1% of users first consider the appearance of a web page [8].

2.2   Dual Processing Model-Based Theory

Metzger [7] explained web credibility evaluation using the dual-processing theory. In this theory, two modes of human thinking including the fast (*system 1*) and slow (*system 2*) are proposed, and these are responsible for human decision-making and reasoning [9]. Although system 1 is low-cost because it involves rapid decision-making, it is associated with systematic reasoning flaws, whereas system 2, characterized as high-cost, is linked to systematic reasoning. Metzger concluded that a user utilizes system 2 for evaluation when motivated; otherwise, the user evaluates the web page design heuristically.

2.3   Web Page Credibility Evaluation Features

The features used for evaluating web page credibility are classified into content, linguistic, and social. Content features involve statistics on words, characters, morphemes, and sentences as well as the web page structure. Conversely, linguistic features represent the output of natural language processing of a web page, while social features incorporate the citations and reputation of a web page on SNS such as Twitter and Facebook.

Olteanu et al. [2] proposed a web page credibility evaluation method comprising 25 content and 12 social features. The evaluation was performed using a support vector machine (SVM), decision trees, and extremely randomized trees (ERT) on the Microsoft credibility dataset [10], and this produced an F-score of 0.75. The social features exploited included Facebook, Twitter, Bitly, Delicious, Alexa Rank, and Google PageRank; however, because some features were discontinued, information on the method is unavailable.

Wawer et al. [5] then introduced 183 linguistic features using the general inquirer (GI) [11] (https://www.wjh.harvard.edu/~inquirer/) [2]. The GI is a popular tool for content analysis that contains a dictionary based on key psychosocial and psycholinguistic theories. Using logistic regression (LR), they achieved an F-score of 0.83 with the Microsoft credibility corpus. However, although linguistic features improve the classifier, the need for a language-specific morphological analyzer and dictionary limits their application to the more than 100 languages in the world.

As an extension of content features, Otani et al. [3] added the placement, size, and domain of the external resources embedded in a web page. This produced a classification of external resources into ads and others termed AD/non-AD. Evaluation using the gradient boosting decision tree produced an F-score higher than that of Wawer et al., thereby highlighting the importance of the element features of a web page for credibility evaluation.

In 2017, Kakol et al. [1] conducted a credibility assessment of 5,000 web pages involving 20,000 participants. User criteria such as informativity, objectivity, design esthetics, and references were employed for the evaluation. In addition, in 2021, Song et al. [4] proposed a fake news detection model by using news articles and the associated images for crossmodal attention residual and multichannel networks (CARMN). The model, which produced an accuracy of 0.922 on Weibo posts, is based on CNNs, and implicitly adopts content and linguistic features.

3   First Impression-Based Credibility Detection

In the present study, we propose a web credibility evaluation method based on user first impression, without need for linguistic or social features. The method exhibits robustness for web pages and eliminates third-party services such as SNS. It simulates the prominence step of the prominence–interpretation theory and system 1 of the dual-processing-based theory in a user, generating results with accuracies comparable to those of existing methods.

Owing to the assumption that web credibility evaluation through first impression is controlled by *the appearance and the structure* of a web page, we adopted usability and performance features. To obtain these features, we used the NodeCLI version of Google lighthouse (https://developers.google.com/web/tools/lighthouse). Google lighthouse was developed for scoring the quality of a web page by examining the following five parameters: performance, accessibility, best practices, search engine optimization (SEO), and progressive web apps (PWA), and then assigning scores varying from 0 to 100. Among these, we utilized the scores of performance (41 features), accessibility (35 features), and best practices (14 features) because these parameters are related to the usability and performance of a web page. The scores were used as inputs for the extreme gradient boosting (XGboost) machine learning model [12]. This model was trained to classify the input of a web page as reliable or unreliable. To the best of our knowledge, no study has employed the Google lighthouse for web page credibility evaluation.

4   Experimental Evaluation

4.1   Comparison with Previous Methods

As baseline methods, we selected those of Olteanu et al. [2], Ootani et al. [3], and Wawer et al. [5]. Content and linguistic features were used, while social features were excluded because some APIs of social services such as the Google PageRank employed in [2, 3] no longer exist.

We used a fake news corpus (https://github.com/several27/FakeNewsCorpus) and compared the classification accuracies of the methods. The fake news corpus is an English language dataset involving ten classes of labels. As presented in Table 1, the web pages labeled *reliable* were considered reliable samples, while those labeled *fake, unreliable,* and *conspiracy* were utilized as unreliable samples. The number of URLs from each domain was limited to 20 pages, and these were selected randomly to minimize bias. In addition, inaccessible web pages were excluded. We introduced a 5-fold cross-validation to compare the accuracy of the methods with parameters of the XGBoost optimized by grid search.

According to the results presented in Table 2, the proposed method produces a higher accuracy than the existing methods. This demonstrates that usability and performance features are adequate for web page credibility evaluation. We then investigated the contribution of each feature using an interpretation method known as SHAP [13]. Consequently the following features were identified as the major contributors: security-related features such as redirects-http (whether http websites are redirected to https), completeness-related features such as aria-valid-attr-value (whether valid values are specified for area attributes), and performance-related features including the speed index (time for a page to load).

Table 1. Summary of the data used for comparison with previous studies

Label		Num. of URLs	Num. of domains
Ground Truth (Label in this experiment)	Labels in Fake News Corpus		
Reliable	Reliable	1,533	77
Unreliable	Fake, Unreliable, Conspiracy	1,683	112

Table 2. Summarized data highlighting the comparison with previous studies

Method	Features				Accuracy
	Content	Linguistic	AD/ Non-AD	Usability/ Performance	
Olteanu et al. [2]	✓				0.737
Wawer et al. [5]	✓	✓			0.797
Ootani et al. [3]	✓	✓	✓		0.811
Proposed				✓	**0.898**

4.2   Evaluation of the Proposed Method on Web Page Content

This experiment is used to evaluate the applicability of the proposed method to web page content in languages other than English. We constructed a dataset by inviting 48 students to evaluate the credibility of selected web pages. The procedure, which was adopted from [1] is as follows:

1) Set the following six keywords in Japanese: money, economy, medical care, health, politics, and hobbies. Then, expand each keyword to include 100 words by randomly using the word2vec trained by the Japanese Wikipedia.
2) Search for each word in step 1 using the Google search engine, and record the top 10 ranked and 90th to 100th ranked URLs, thereby producing 12,000 URLs.
3) Randomly select three web pages for each domain from the URLs in step 2.

Next, labels were assigned to the web pages using the following procedure:

a) Each student evaluated 400 web pages in step 3 according to first impression using a five-point scale.
b) Same web page will be evaluated by 3 students, with a rating of ≥ 4 considered reliable, while ≤ 2 was unreliable, and 3 was neutral.
c) Web pages with more than two reliable (unreliable) rating were classified as reliable (unreliable), while those with fewer than two rating were excluded.

The final dataset comprised 584 web pages (175 unreliable and 409 reliable). The trained model in Sect. 4.1 classified the web pages in the dataset as unreliable and reliable, as presented in Table 3. Although the accuracy of 0.812 is lower than that for the fake news corpus, the method still produced good results.

Table 3. Confusion matrix from the evaluation of content from different web pages

| | | Proposed method | |
		Unreliable	Reliable
Ground Truth	Unreliable	**112**	63
	Reliable	47	**362**

4.3   Comparison of the First Impression- and System-2 Based Evaluations

We examined the discrepancy between credibility evaluation results of the first impression and deep consideration (corresponding to System 2 in Sect. 2.2) approaches. Twenty-four students were invited to evaluate the credibility of the web pages used in Sect. 4.2 through deep consideration on a five-point scale. In this evaluation, the students were asked to document reasons for their decisions. Any reason considered illogical was excluded, and after applying steps b and c presented in Sect. 4.2, a system 2-based evaluation dataset was obtained for 279 web pages. Web pages that failed to satisfy step c in Sect. 4.2 were excluded. Data highlighting differences between the two evaluation methods are presented in Table 4. The results suggests that ground truthing of the web credibility evaluation dataset is required for deep consideration, that is, the system 2-based evaluation. This is because of gaps in the first impression-based evaluation.

Table 4. Summary of differences in evaluations based on first impression and system 2

#web pages		First impression-based evaluation	
		Reliable	Unreliable
System 2-based evaluation	Reliable	**26**	16
	Neutral	37	**18**
	Unreliable	10	19

5 Conclusion

In the present study, a first impression-based method suitable for credibility evaluation of diverse web pages was developed. The results of an experiment on a fake news corpus was characterized by an accuracy of 0.898, which was superior to those of existing methods. The ratio of flipped evaluation between first impression and deep consideration was 20%. This revealed that evaluating the credibility of some web pages using first impression was challenging. Future work will improve the proposed method to handle web pages that require deep consideration for credibility evaluation. The dataset used in Sect. 4.2 can be obtained at https://github.com/yamanalab/WebCorpusJP.

Acknowledgment. This work was supported by JSPS KAKENHI (Grant Number 17KT0085).

References

1. Kakol, M., Nielek, R., Wierzbicki, A.: Understanding and predicting web content credibility using the content credibility corpus. Inf. Proc. Mang. **53**, 1043–1061 (2017)
2. Olteanu, A., Peshterliev, S., Liu, X., Aberer, K.: Web credibility: features exploration and credibility prediction. In: Proceedings of the European Conference on Information Retrieval, ECIR 2013. LNCS, vol. 7814, pp. 557–568 (2013)
3. Ootani, K., Yamana, H.: External content-dependent features for web credibility evaluation. In: Proceedings of the 2018 International Conference on Big Data, pp. 5414–5416 (2018)
4. Song, C., Ning, N., Zhang, Y., Wu, B.: A multimodal fake news detection model based on crossmodal attention residual and multichannel convolutional neural networks. Inf. Proc. Mang. **58**, 102437 (2021)
5. Wawer, A., Nielek, R., Wierzbicki, A.: Predicting web page credibility using linguistic features. In: Proceedings of the 23rd International Conference on WWW, pp. 1135–1140 (2014)

6. Fogg, B.J.: Prominence-interpretation theory: explaining how people assess credibility online. In: Proceedings of the CHI 2003 Extended Abstracts on Human Factors in Computing Systems, pp. 722–723 (2003)
7. Metzger, M.J.: Making sense of credibility on the Web: models for evaluating online information and recommendations for future research. J. Am. Soc. Inform. Sci. Tech. **58**, 2078–2091 (2007)
8. Fogg, B.J., Soohoo, C., Danielsen, D.R., Marable, L., Stanford, J., Tauber, E.R.: How do users evaluate the credibility of Web sites?: A study with over 2,500 participant. In: Proceedings of the DUX 2003, pp. 1–15 (2003)
9. Stanovich, K.E.: Who is Rational? Studies of Individual Differences in Reasoning? Psychology Press, London (1999)
10. Schwarz, J., Morris, M.: Augmenting web pages and search results to support credibility assessment. In: Proceedings of the SIGCHI Conference on Human Factors in Computing Systems, pp. 1245–1254 (2011)
11. Stone, P.J., Dunphy, D.C., Ogilvie, D.M., Smith, M.S.: The General Inquirer: A Computer Approach to Content Analysis. MIT Press, Cambridge (1966)
12. Chen, T., Guestrin, C.: XGBoost: a scalable tree boosting system. In: Proceedings of the 22nd ACM SIGKDD, pp. 784–794 (2016)
13. Lundberg, S.M., Lee, S.I.: A unified approach to interpreting model predictions. In: Proceedings of the 31st Conference on NIPS, pp. 4768–4777 (2017)

Gender and Academic Performance: A Case Study in Electrical Engineering

Linlin Zhang and Kin Fun Li[✉]

University of Victoria, Victoria, Canada
{linlinz,kinli}@uvic.ca

Abstract. Education institutions are promoting diversity and inclusivity these days. Traditionally, engineering schools are under-represented by certain groups, especially females. In this work, a case study is presented on academic performance of males and females in the Department of Electrical and Computer Engineering at the University of Victoria. The analytics performed is based on 1840 students over a period of 12 years. The overall composition of the student population is presented. The characteristics of successful graduates are explored, including gender correlation to program duration and GPA. Similarly, issues with failed male and female students are examined with respect to academic probation, failed courses, GPA, and program duration. Insights on gender's influence on academic performance and advices for at-risk students are presented.

1 Education Analytics Introduction

As an active research area in the past decade, education analytics involves converting raw education-related data into useful information and deeper knowledge, for academic advisors and administrators, with the aim to improve a student's academic performance as well as the institution's retention rate [4].

In many countries, there is a big gender gap in the fields of STEM, even though academic institutions and industries have advocated diversity and inclusivity. This paper focuses on education analytics related to gender in an electrical and computer engineering program. Section 2 gives a brief survey of recent literature on gender analytics. Our case study is detailed in Sect. 3 including the program background and the data used. Section 4 examines some characteristics of successful graduates of the program while Sect. 5 investigates issues related to failure.

2 Survey on Education Analytics Focusing on Gender

There is a considerable number of existing papers on education analytics with a direct or indirect focus on gender. A search on IEEE Xplore for related publication from 2016 to 2020 found a total of ten relevant papers. Four of the papers have a major emphasize on gender in the analysis while the other six have gender as one of the influential factors in their studies.

2.1   Major Focus on Gender

Alhusban et al. [1] examine the effect of gender on specializations as some have more females than males. Álvarez et al. explore whether there are significant differences between the gender in planning, learning, and self-evaluation in an introductory computer programming course [2]. Piad et al. [7] use data mining to identify patterns and find gender is one of the major factors that are most influential on IT employability.

2.2   Gender is One of the Influential Factors

Using the educational datasets xAPI from Kaggle and rules generated by the apriori algorithm, Rahman et al. [9] try to find which factor, such as gender and absenteeism, has the most impact on a student's academic success at all levels of education. In order to identify at-risk students and to decrease late dropout rate, Salazar-Fernandez et al. [10] use process mining discovery techniques to build prediction models. Factors, such as gender and income, are examined and evaluated on their impact on failure rate.

To predict whether a student is to graduate on time, Gunawan et al. [3] investigate which attributes have bigger influence on the prediction. Gender is the second most influential factor, after GPA. Patil et al. [6] develop a performance prediction model based on classification of data from prior classes. The trained model is applied to new students, considering their gender, marks, and rank in entrance examinations.

In [8], Putpuek et al. are interested in predicting students' final GPA. The features used in their prediction models include gender and previous educational background. Focusing on distance learning, Liu et al. [5] use age, gender, and previous education level as factors to plan online course curriculum and to advice course selection.

3   Institution and Dataset

This paper focuses on students in the Department of Electrical and Computer Engineering (ECE) at the University of Victoria (UVic). ECE offers courses in electrical, computer, software and general engineering, in a tri-semester format. Students must complete eight academic terms, interleaved with co-op work terms, which they have to complete at least four in order to graduate. A student normally requires five years of studies and work experience to graduate.

Similar to many other ECE programs in North America, it has an unbalanced student population with respect to various under-represented groups, especially females. In order to better understand the gender issues within ECE, we investigate gender related parameters and their impacts on performance, program duration, failures, and program change.

3.1   The Dataset

The raw dataset was extracted from university database. It consists of 1898 undergraduate students in ECE who have taken courses between May 2008 to August 2019. The cleaned dataset consists of 1840 students, with 199 females and 1641 males, or 10.82% and 89.18% of the total population, respectively.

There are over 100 features in the dataset, including student characteristic features like gender, citizenship, and previous institutions. Each student record shows the course taken and grades obtained. In this study, we concentrate on the following features:

- Graduation status: graduated or not
- Academic standing: good standing, probation, failed
- Program year: year 1, year 2, year 3, year 4, year 5
- Gender: male, female
- Course registered: designation (e.g., SE Software Engineering) and number
- Registered term: 2009 Summer, 2018 Fall, etc.
- Course grade: letter grades from F to A+ and the corresponding grade point (e.g., F = 0, A = 8, etc.)

3.2 Definitions and Terminologies

To facilitate presentation and discussion, the following defined terms are used throughout this paper:

- Dept-Course: courses with designation a student takes in the program, such as ECE electrical/computer engineering, SE software engineering, etc.
- Non-Dept-Course: courses other than Dept-Course
- Registered-Course: a course registered by a student, which outcome could be passed, failed, and dropped
- Registered-Terms: a term in which a student has at least one Registered-Course
- Good-Standing: evaluation of a student's performance showing satisfactory progress and the student is allowed to continue in the program
- Probation: evaluation of a student's performance showing unsatisfactory progress and the student must satisfy some conditions, before being re-admitted to the program
- Graduated: a student who completed the program and obtained the degree
- Failed: a student who left the program after Probation
- Switch-Dept: a student who switched to another department
- In-Progress: a student is in Good-Standing and has not Graduated yet

3.3 Overview of Student Status Based on Gender

An overview of the status of all students in the dataset, based on gender, is shown in Table 1:

From Table 1, one can observe that in the ECE program:

- Males outnumber females: females only constitute 10.8% of the total student population while the national ECE average in Canada is about 15%. This translates to additional effort needed in female student recruitment
- Females and males have similar percentage of failing: this makes sense assuming a uniform distribution among the population

Table 1. Student status distribution

Category	Female (% among females)	Male (% among males)	Subtotal number
Population	199	1641	1840
Graduated	94 (47.2%)	909 (55.4%)	1003
Failed	9 (4.5%)	65 (4.0%)	74
Switch-Dept	22 (11.1%)	146 (8.9%)	168
In-Progress	74 (36.7%)	521 (31.8%)	595

- Females have a higher percentage of switching to another department: it is of interest to examine the reasons why they switch and which department they are switching to. This may help future design of the curriculum as student retention is a major goal of many schools
- Males have a higher percentage of graduating: this is an interesting phenomenon that also exists in other Canada engineering schools. Since degree completion is the major goal of all current and prospective students, the patterns of successfully graduated (and failed) students need further investigation, and are examined in the next section. Findings would be very helpful to identify at-risk students and enable advisors to provide recommendations in a timely fashion

4   Characteristics of Successful Graduates

As the grade point average (GPA) is a universal evaluation barometer for academic performance, it is a determinant factor in comparing performance. At UVic, GPA is calculated based on a 9.0 scale with 5.0 roughly translated into a B grade.

4.1   Gender and GPA

Figure 1 shows the GPA range of graduated male and female students. Female graduates perform better at the higher range GPA (6 to 9) but not as good as males at the lower GPA range (below 5), while they have the same percentage in the middle of the scale. This is a rather interesting pattern that is worthy of further investigation by reviewing other factors.

Table 2 shows the overall summary statistics of the GPA among the two groups. The females indeed perform slightly better with a higher mean and median, though within each group there are no big fluctuation with both having similar standard deviation. One can conclude that most of the graduates (over 65%) in both groups fall in the GPA range of 4 to 7, and there is no significant difference in terms of their GPA.

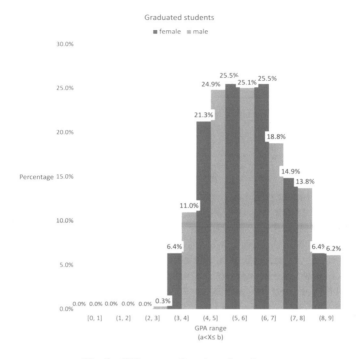

Fig. 1. GPA range of graduated students.

Table 2. Overall summary statistics.

Gender	Mean	Median	Standard deviation
Female	5.94	5.91	1.31
Male	5.67	5.53	1.37

4.2 Gender and Program Duration

Normally, a student is expected to graduate in 5 years or 15 terms with both academic and co-op components completed. However, this is rarely the case due to various reasons such as leave of absence, failing courses that are pre-requisites to subsequent courses, extended co-op employment instead of the standard four months, etc.

It is of interest to see what are the typical number of terms that students take to graduate. Figure 2 depicts the program duration of the successful graduates, while Table 3 shows the summary statistics. It can be observed that females tend to take slightly longer time to graduate as compared to males. This could be attributed to females fail early-year courses more often, or they tend to take less courses per term so they can learn better with more focus. In the next sections, these hypotheses will be explored.

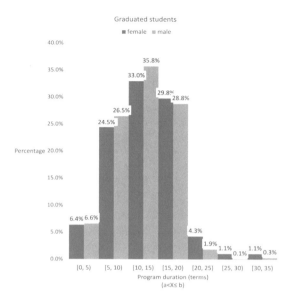

Fig. 2. Program duration of graduated students.

Table 3. Program length (terms) of graduates.

Gender	Mean	Median	Standard deviation
Female	12.63	13	5.30
Male	11.82	12	4.57

4.3 Program Duration, GPA, and Gender

To investigate the correlation between program duration and GPA among the genders, a scatter plot is shown in Fig. 3. The linear regression slope of the males is negative, which is as expected. The weaker students, with lower GPAs, may have failed courses or taken fewer courses per term, resulting in a longer time frame to graduate. For females, the regression slope is almost zero. This indicates that the females perform more consistently independent of the number of terms to graduate.

An interesting part of Fig. 3 is at the right side of the plot. It seems that quite a few students having high GPAs but take much longer time to graduate. Upon further investigation, it is found that one female student took three courses on average per term thus making her program longer. The other twelve students, 9 males and 3 females, have continued their studies at the graduate level pursuing higher degrees in both master and doctoral. In this small sample, 2% of the graduated females and 0.55% of the graduated males continue with higher degree studies at the same institution.

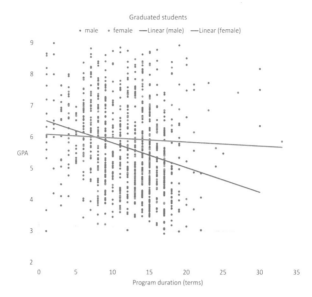

Fig. 3. Program duration and GPA of graduates.

5   Issues with Failed Students

For every institution and program, it is vital to review the issues and progress of failed students. One cannot simply leave it to the students but rather needs to intervene to improve their chance of successful completion of the program. By examining these issues, educators may be able to give advance warning and even pre-emptive actions to assist the at-risk students.

5.1   Year of Probation

It seems to be common wisdom that if an engineering student can survive the first and second year of their studies, then she could make it through the program. A student under probation could be re-admitted to the program if certain conditions are satisfied, mostly by repeating and passing some failed courses; though quite a few students get a second probation. The academic year that a student gets their last probation would show when the student could not satisfy the program's academic regulation and is required to withdraw or change program.

It is found that most failed students get their last probation in their second year, which is compatible with the results in Sect. 5.3 where the program duration and gender of failed students are explored. Moreover, over 90% of the failed students get their last probation before or during their third year of studies.

5.2   GPA of Failed Students

Another issue of interest among failed students is their GPA as shown in the summary statistics in Table 4. Failed females and males have similar mean and median GPAs, but the females have a wider range with a standard deviation of 2.34. The implication of wide distribution is the difficulty in predicting the GPA of failed female students.

Table 4. GPA of failed students.

Gender	Mean	Median	Standard deviation
Female	3.91	2.52	2.34
Male	3.31	3.00	1.66

5.3 Program Duration of Failed Students

Examining the average time failed students stay in the program is a useful exercise to aid resource planning for administrators. Figure 4 shows the number of terms that failed students stay in the program while Table 5 illustrates the corresponding summary statistics. All female students failed within 9 terms while 18.5% of the males failed from the program beyond 10 terms. This information can be used to advice proactively the failed students in their third year (terms 5 and 6). It also gives planners an idea of resources needed for the third- and fourth- year of the program.

5.4 Number of Failed Students

The statistics of students who failed at least one course versus those who have never failed are shown in Table 6. Females have a smaller probability of failing at least one course, as compared to males. This difference is too small, though, to be useful in student advising and resource planning.

Often when a student fails one or more courses, the student is still able to complete the program and graduate. Therefore, it is of interest to see how these students perform as compared to students who never fails. As shown in Table 7, students of both genders who have failed at least one course, tend to have lower GPA than those who have not failed any courses. The same conclusion can also be made for compulsory courses only.

The relationship between total number of failed courses and GPA is shown in Fig. 5. For both groups, the regression slope is negative indicating a falling trend, and the greater number of failed courses, the lower the GPA. Though, the regression slope of females who failed courses shows a faster decreasing trend that the males.

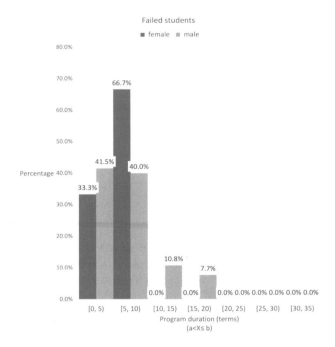

Fig. 4. Program duration of failed students.

Table 5. Program duration of failed students.

Gender	Mean	Median	Standard deviation
Female	5.30	5	2.29
Male	6.09	5	4.31

Table 6. Number of failed students.

Gender	At least one failed course	No failed course
Female	94 (47.24%)	105 (52.76%)
Male	839 (51.13%)	802 (48.87%)

Table 7. GPA of students with failed courses.

	At least one failed course		No failed course	
	Female	Male	Female	Male
Mean	4.21	3.97	6.35	6.24
Standard deviation	1.25	1.25	1.31	1.31

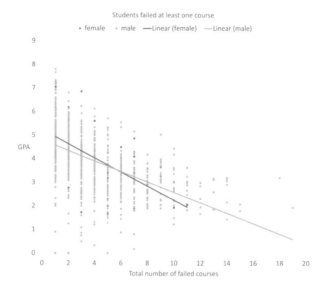

Fig. 5. Number of failed courses versus GPA.

6 Conclusion and Future Work

In this work, we examined gender related academic performance in electrical and computer engineering. We presented a brief survey showing related work in the literature, most of which have based their findings and recommendations on facts and statistics. Our overall approach focuses on quantitative analysis but at the same time tries to find the qualitative reasons to explain certain trends and various scenarios. In addition to the general statistics of the student population in the dataset, we also explored the characteristics of successful students (those who graduated) and the issues of the weak students who failed in the program. We plan to explore other factors, such as demographics and family affluence, on academic performance. Also, we have started an investigation to compare various academic aspects of domestic and international students.

References

1. Alhusban, S., Shatnawi, M., Yasin, M., Hmeidi, I.: Measuring and enhancing the performance of undergraduate student using machine learning tools. In: 11th International Conference on Information and Communication Systems, pp. 261–265, April 2020
2. Álvarez, C., Fajardo, C., Meza, F., Vásquez, A.: An exploration of STEM freshmen's attitudes, engagement and autonomous learning in introductory computer programming. In: 38th International Conference of the Chilean Computer Science Society, pp. 1–8, November 2019
3. Gunawan, Hanes, Catherine: Information systems students' study performance prediction using data mining approach. In: Fourth International Conference on Informatics and Computing, pp. 1–8, October 2019
4. Microsoft White Paper on Education Analytics. https://pulse.microsoft.com/uploads/prod/2018/07/MicrosoftEducationAnalytics.pdf. Accessed 2 July 2020
5. Liu, S., d'Aquin, M.: Unsupervised learning for understanding student achievement in a distance learning setting. In: IEEE Global Engineering Education Conference, pp. 1373–1377 (2017)
6. Patil, R., Salunke, S., Kalbhor, M., Lomte, R.: Prediction system for student performance using data mining classification. In: Fourth International Conference on Computing Communication Control and Automation (2018)
7. Piad, K.C., Dumlao, M., Ballera, M.A., Ambat, S.C.: Predicting IT employability using data mining techniques. In: Third International Conference on Digital Information Processing, Data Mining, and Wireless Communications, pp. 26–30 (2016)
8. Putpuek, N., Rojanaprasert, N., Atchariyachanvanich, K., Thamrongthanyawong, T.: Comparative study of prediction models for final GPA score: a case study of Rajabhat Rajanagarindra University. In: IEEE/ACIS 17th International Conference on Computer and Information Science, pp. 92–97 (2018)
9. Rahman, A., Mutiarawan, R., Darmawan, A., Rianto, Y., Syafrullah, M.: Prediction of students academic success using case based reasoning. In: 6th International Conference on Electrical Engineering, Computer Science and Informatics, pp. 171–176, September 2019
10. Salazar-Fernandez, J., Sepúlveda, M., Munoz-Gama, J.: Influence of student diversity on educational trajectories in engineering high-failure rate courses that lead to late dropout. In: IEEE Global Engineering Education Conference, pp. 607–616, April 2019
11. Supianto, A.A., Julisar Dwitama, A., Hafis, M.: Decision tree usage for student graduation classification: a comparative case study in Faculty of Computer Science Brawijaya University. In: International Conference on Sustainable Information Engineering and Technology, pp. 308–311 (2018)

Evaluation Approach for Smart Charging Ecosystem – with Focus on Automated Data Collection and Indicator Calculations

Marit K. Natvig[1($\boxtimes$)], Shanshan Jiang[1], Svein Hallsteinsen[1], Salvatore Venticinque[2], and Regina Enrich Sard[3]

[1] SINTEF, Strindveien 4, Trondheim, Norway
{marit.k.natvig,shanshan.jiang,svein.hallsteinsen}@sintef.no
[2] Department of Engineering, University of Campania "Luigi Vanvitelli", Aversa, Italy
salvatore.venticinque@unicampania.it
[3] Eurecat Technology Centre, Barcelona, Spain
regina.enrich@eurecat.org

Abstract. Access to charging is a prerequisite for the transition to electric mobility. There are however challenges related to charging and charging infrastructures, e.g., charging availability, grid capacity during peak hours, and the CO_2 intensity of the energy mix provided. This paper suggests measures to be taken in a smart charging ecosystem to mitigate the challenges. The impact of the measures must however be evaluated. The objective of the paper is to suggest an evaluation approach, with focus on quantitative aspects. The measures of relevance, the associated indicators for the impact evaluation, and an overview of the research data needed is provided. In addition, data content examples and calculation details are described for two indicators – the charging flexibility provided by the EV users and the peak to average ratio characterising the load balancing. Scenarios to be evaluated and how simulations are used to complement the evaluation of the demonstrators are addressed.

1 Introduction

The communication from the European Commission on "Sustainable and Smart Mobility Strategy – putting European transport on track for the future" [1] states that the uptake of zero emission vehicles must be boosted. However, citizens cannot be expected to replace their fossil cars with electric vehicles (EVs) unless they have easy and predictable access to charging. Thus, the strategy recognizes the need for charging infrastructures and the importance of the ongoing revisions of the alternative fuel directive (current version [2]) and the building directive (current version [3]). The revisions will address a smooth integration of charging infrastructures into the electricity grid as well as charge points in buildings.

The establishment of charging infrastructure must be sustainable from a societal, economic, and environmental point of view. The infrastructures themselves require investments, and with an electrified transport sector, the power demand will increase, and power grids may get overloaded. Thus, grid investments are required, and the energy provided must be as green as possible [4].

The European Horizon 2020 project GreenCharge builds on previous work on energy smart neighbourhoods [5] and addresses the above challenges. In the GreenCharge concept, the charge management and smart local energy management work together to facilitate a transport system running on green energy. EV users get charging support, and peaks in the power grid and grid investments are reduced through load balancing. When many vehicles are plugged into the grid around the same time (e.g., on returning home from work), the energy management balances demand with available supplies, supplies from renewable energy sources included.

The concept aims for smart charging ecosystems where actors, devices, infrastructures, and software systems provide services to each other [6]. The ecosystems are cross sectorial, involving the building, charging, and energy sectors.

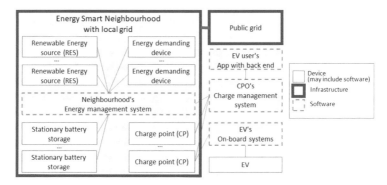

Fig. 1. Smart charging ecosystem components

Figure 1 illustrates the ecosystem components. The building sector is within an energy smart neighbourhood (ESN) encompassing one or more buildings. It may have local renewable energy sources (RES) like solar plants, stationary batteries for energy storage, and devices like heating/cooling devices, washing machines and dishwashers.

The charging sector is represented by charge points (CP) in the premises of the ESN, the charge management system of the charge point operator (CPO), the EVs charging at the CPs, on-board systems for charging, and systems supporting the EV user regarding charging, e.g., an App used by the user or the navigation system.

The energy sector is represented by the power grids and associated systems. The local power grid has a local energy management system ensuring the best possible use of energy across the ESN. Devices and EV charging sessions are started and stopped, according to rules and the energy availability. The use of the energy from local RES is also managed, and if surplus, the energy may be stored in stationary batteries.

This paper describes a method for smart charging ecosystems evaluations. The quantitative aspects are emphasized since these are customized to the concept. Section 2 presents related work on evaluation frameworks. Section 3 defines relevant measures and an indicator framework for impact evaluations. Section 4 and 5 address the research data needed for indicator calculations, and calculation strategies for two indicators. Section 6 addresses how demonstrators and simulations will contribute to the evaluation and exemplifies the use of indicators. Finally, Sect. 7 concludes and describes the remaining work.

2 Related Work on Evaluation Approach

According to Lervåg [7], impact studies in the transport sector traditionally are performed by ex post evaluations of implemented services, field operational tests, and simulations. They all may build on classic evaluation strategies involving a comparison of the before and after situations, or goal-oriented evaluation approaches where the results are compared with predefined criteria.

Several evaluation frameworks for the transport sector address impact assessments through the comparison of before and after situations. The FESTA methodology [8] provides guidelines for the evaluation of intelligent transport systems with focus on driver behaviour in field operational tests. The CIVITAS evaluation framework [9] defines an impact evaluation approach as well as guidelines for implementation process evaluations. The framework offers a common approach with pre-defined indicators to urban mobility projects funded by the European Commission, the GreenCharge project included. Electric mobility issues are however not addressed.

Lervåg [7] states that current evaluation strategies have shortcomings caused by the rapid development of technology and limited access to empirical knowledge due to the complexity of full scale implementations in real life situations. It may also be difficult to establish baseline data. To cope with the shortcomings, the use of program theory is suggested to find dependencies, and to support the development of policies.

3 Indicators Needed for Impact Evaluation

The impact evaluation approach described by this paper builds upon the CIVITAS evaluation framework [9] and supports the assessment of the impact of certain measures through use of indicators. The same indicators are established before and after the introduction of measures, and the differences are analysed.

Table 1. Measures in measure groups and related indicators

Groups	Measure (S) or (B)	Description	Indicators (C) or (GC)
Charging	Public CP (S)	CP can be used by the public	I1 Awareness level (C) I2 Acceptance level (C) I3 Perception level of physical accessibility (C) I4 Operational barriers (C) I5 Number of EVs (GC) I6 Number of CPs (GC) I7 Utilisation of CPs (GC) I8 Charging availability(GC) I9 Charging flexibility (GC)
	Private CP (S)	CP is owned and used by one EV user	
	Shared CP (S)	CP is shared when not used by the owner	
	Booking of CP (B)	Time slot for charging is booked in advance	
	Flexible charging (B)	EV user allows charging at any time before a latest finish time	
	Priority charging (B)	EV user requests charging as fast as possible with priority over non prioritised users if not enough energy for all	
Smart energy manage-ment	Local RES (S)	Energy from local RES is exploited	I10 Peak to average ratio (GC) I11 Self-consumption (GC) I12 Energy mix (GC) I13 CO2 emissions (C) I14 Share of battery capacity for V2G (GC)
	Stationary battery storage (S)	Energy is stored locally for later use when it is advantageous	
	Exploiting V2G (Vehicle-to-Grid) (B)	Energy from connected EVs is used when it is advantageous	
	Optimal and coordinated use of energy (B)	Energy use (charging included) is coordinated with energy availability and optimised to maximise the use of green energy and to reduce peak loads	

Table 1 lists and groups the measures to be implemented through the software systems in Fig. 1 to facilitate smart charging ecosystems. Some measures are state-of-the-art (S), while others go beyond state-of-the-art (B). Both types are included since the first facilitates the implementation of the second, and since combinations (e.g., shared CPs and booking of CPs) need to be evaluated. It may be difficult to assess the effect of individual measures. Thus, all measures in one group are evaluated as a whole, and Table 1 lists the indicators of relevance for each group. The indicators are either adopted from the CIVITAS framework (C) and adapted to the smart charging ecosystem, or they are defined by the GreenCharge project (GC).

3.1 Charging Measures

The objective of the charging measures is to provide better and more predictable access to charging services to the EV users. In addition, the measures should arrange for good utilisation of existing CPs to limit the need for additional CPs. It is also crucial that the charging services are designed to arrange for optimal use of energy.

The public/private/shared CP measures are about how EV users get access to charging, whether they have dedicated, private CPs, or must share CPs with others. A sharing of private CPs arranges for better availability and utilisation of the CPs. Bookings of CPs arrange for predictable access to charging services and may mitigate the so-called charging anxiety.

The priority and flexible charging measures are about when EVs are to be charged. With priority charging, the charging starts immediately and is accomplished as fast as possible. With flexible charging, the EV user accepts that the charging can be done at any time before a deadline. Flexible charging is most relevant when EVs are connected for a longer period. In such cases and when supported by the EV, the charging can be started and stopped several times according to what is optimal with respect to the grid capacity, other energy demands, the energy price, the availability of renewable energy, etc.

For all types of charging, a charging request should be provided. It defines the energy demand and the time slot in which the charging should take place. The latter may be hours or days ahead. Today, the EV user may have to provide the charging request manually, e.g., through an App. In the future, an integration with vehicle on-board systems (for access to the current state of charge), travel planners or navigation systems (for charge planning and scheduling support) and decision support systems (for adaption to habits, plans, etc.) may support the EV user and automatically make suggestions for the information needed.

The indicators selected for the charging measures cover several aspects. Indicator I1 – I4 (on awareness, acceptance, accessibility, and operational barriers) are evaluated by means of qualitative data collected from EV users on the awareness and perception of the charging services, as defined by the CIVITAS framework. These indicators are to a large extend about how successful the implementation of the measures is from the EV user's point of view and provide an important context for the analysis of the other indicators.

Indicator I5 and I6 are about the diffusion of eMobility by addressing the number of charge points and the number or share of EVs.

Indicator I7 – I9 (utilisation, availability, and flexibility) are about the charging behaviour and flexibility. The charge point utilisation addresses connection times, charging time, and use of energy. The charging availability is about the fulfilment of charging demands and about how booked time slots are used (e.g., delays in plug-in time and blocking after booked time slot). Charging flexibility addresses the flexibility of the EV user with respect to when the charging can take place.

Demonstrators and simulated scenarios will address sub-sets of the charging mea-sures, and the measures will be evaluated as one package by means of a selection of the indicators. An indicator may also constitute a context for other indicators, and they may influence each other. A high number of charge points may for example give a higher number of EVs, and high awareness and acceptance about the need for flexibility may increase the charging flexibility provided by the EV users.

3.2   Smart Energy Management Measures

The smart energy management measures aim to fulfil individual energy demands while ensuring optimal use of energy to minimise both the CO_2 emissions and the peaks in the electricity grid. The latter will reduce the need for costly grid investments and may also reduce the energy costs if the tariff rewards a reduction of peaks.

The local RES measure is about local production of green energy, e.g., by means of PV panels. The measures on stationary battery storage and exploiting Vehicle-to-Grid (V2G) arrange for flexibility with respect to when energy is used. In case of surplus or if it is not optimal to use the green energy from RES immediately (e.g., due to high availability of cheap energy from the distribution grid), the energy can be stored in the batteries of connected EVs or in stationary batteries. When energy costs are high, or when the energy demand exceeds availability, the stored energy can be used.

As mentioned, the aim is optimal use of energy, and the measures on optimal and coordinated use of energy do an optimisation across energy demands in the ecosystem through an integration and control of energy sources and energy demanding equipment (RES, stationary batteries, charging infrastructure, heating and cooling devices, washing machines, etc.). The optimisation should be based on the current situation as well as on prognosis on both future energy demands and energy production from RES (i.e., prediction-based energy optimisation). The input to the prognosis will be historical data, charging requests received ahead of the actual arrival of the EVs, and weather forecasts.

Indicator I10 is about the peaks in the energy use compared to the average value. Ideally, there should be no peaks in the energy use from the public grid, just a flat curve. The optimisation mentioned above aims to flatten the curve, and this is also supported using RES and stationary batteries.

I11 and I14 (self-consumption and V2G) is about the share of energy produced locally that is consumed locally and the share of the battery capacity in connected EVs that is available for use. In general, it is advantageous that both are high. Locally produced energy should be prioritised since prices for energy bought are higher than those for energy sold. Access to energy in EV batteries increases the flexibility.

I12 – I13 are about the energy mix and the CO2 emissions. The energy mix is the share of different energy sources in the energy provided, and the mix in the local grid improves if local RES is used. The CO2 emissions in eMobility depends on the energy mix in the energy which the EVs are charged with.

As for the charging measures, the smart energy management indicators are influenced by each other. Increased self-consumption due to local RES may for example give a greener energy mix and reduce CO2 emissions. There are also dependencies between the charging measures and the smart energy management measures. A high charging flexibility will arrange for a reduction of the peaks since the charging can be accomplished when energy is available. High flexibility may also increase the self-consumption and thereby decrease the CO_2 emissions.

4 Research Data Needed to Calculate Indicators

Research data are collected in three ways: 1) Manually through surveys and interviews; 2) semi-automatically or automatically by means of the software systems running at demonstrators; and 3) through simulations. In the following, we focus on the quantitative data from 2) and 3). These datasets are designed in collaborations with experts on electric mobility and energy management to facilitate automated calculations of the quantitative indicators described in Sect. 3 (I5 – I14).

4.1 Dataset Entity Types

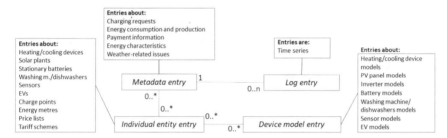

Fig. 2. Files with research data entries

The datasets with quantitative data are organized into the entity types in Fig. 2. Device model entries define among others EV models with properties like battery capacity, charging/discharging efficiency, maximum charging/discharging power, etc.

Individual entity entries define the physical entities involved in a demonstrator or simulation scenario - one entry for each EV, solar plant, stationary battery, etc. These will refer to their respective device model entries when this is relevant.

Metadata and log entries describe dynamic events or situations, such as charge requests, charge sessions, weather conditions, energy import/export and energy mix. A metadata entry provides overall information and refers to the individual entities involved as well as log entries that describe the situation over time by means of time series (i.e., timestamps with related values).

4.2 Research Data for Quantitative Analysis

Figure 3 shows some of the classes in the research data information model. The orange classes are related to charging. The EV model class defines the properties of the EV being charged, among others the AC and DC charging efficiency and the maximum power for AC and DC charging and discharging. For each charging session, a charging request defines the requested connector type, the location where charging is requested, and the charging constraints. The latter supports the charging measures in Sect. 3.1. Earliest start and latest finish time (EST and LFT) define the booked time slot and the potential flexibility. Initial and target state-of-charge (SoC) define the energy demand. Priority can be requested, and a minimum energy content must be provided in case the energy availability is limited.

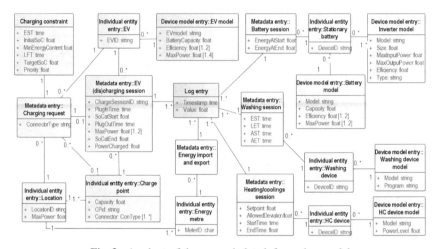

Fig. 3. A subset of the research data information model

An actual charging/discharging session is linked to its charging request and to a charge point (CP). The data of relevance are plug in and plug out time, SoC at start and end, the maximum power for charging/discharging, and the power charged. The associated log entries provides the actual charging/discharging profile.

The blue classes are about other aspect affecting the load such as energy import/export, battery sessions, and other types of sessions. Metadata and log entries define the characteristics and the energy import/export/charging/discharging/use over time. Battery sessions have start and end time, and the battery model has charging/discharging efficiency and maximum power. Heating/cooling sessions also have start and end times, a setpoint and allowed deviations. Washing sessions are carried out within a timeslot defined by the EST and LFT and have actual start and end times (AST/AET). All sessions are linked to individual entities, and these are linked to the respective device models.

The research data on device models and individual entities are mainly created manually. The metadata and log entries are however established automatically by software systems (see Fig. 1). The implementation must be integrated in and followed up as a part of the software development. On-board systems in general do not provide data on EV battery's SoC (not supported by the current standards but might be so in the future). Thus, the EV user's App must collect the SoC from the user.

5 Calculation of Selected Indicators

In the following, we describe the calculation of two indicators – I9 Charging flexibility and I10 Peak to average ratio. These indicators are chosen since they are crucial to the understanding of the success of smart charging ecosystems. A high degree of charging flexibility facilitates load balancing, and the peak to average ratio is about the success of the load balancing. The data elements from the information model in Fig. 3 used in the calculations are listed in Table 2.

Table 2. Symbols used in calculations

Symbol	Description	Research data element (see Fig. 3)
B_cap	Battery capacity	EV_model.BatteryCapacity
EV_{max_cp}	Max charging/discharging power for AC/DC	EV_model.MaxPower
CP_{chrg_cap}	Max charging/discharging power (if such a limit exists)	Charge_point.Capacity
I_{soc}	Initial SoC	Charging_constraint.InitialSoC
T_{soc}	Target SoC	Charging_constraint.Target_SoC
Min_{ec}	Min. energy content the user can cope with	Charging_constraint.MinEnergyContent
T_{dep}	Expected departure (latest finish time)	Charging_constraint.LFT
T_{arv}	Expected arrival time (earliest start time)	Charging_constraint.EST
T_{pin}	EV plugin time	EV_charging/discharging_session.PlugInTime
T_{pout}	EV plugout time	EV_charging/discharging_session.PlugOutTime

Symbol	Description	Symbol	Description
O_{eng}	Offered energy for V2G	R_{eng}	Requested energy
EV_{chrgp}	EV charging power	A_{flexT}	Actual flexibility time
$EV_{dischrgp}$	EV discharging power	O_{flexT}	Offered flexibility time specified in the request
P_{max}	Max power over a period T	P_{avg}	Average power over a period T

5.1 Charging Flexibility

A charging request specifies the charging constraints in terms of when the charging must be finalised and how much the EV must be charged. To allow for prediction-based energy optimisation, requests need to be received prior to arrival of the EVs.

EV charging is considered as flexible loads because the charging can be interrupted and the speed can be regulated, and a charge session may be spread in time as long as the charging constraints can be met. In general, the longer an EV is connected in portion to the needed time to charge to the desired energy level (depending on the energy required for charging and the charge speed), the more flexibility it provides.

If the EV is connected longer than the period specified in the request, the actual flexibility is higher than the flexibility provided by the user. If V2G is enabled, there is additional flexibility when the EV battery can be used as an energy source. Thus, there are three flexibility indicators. Each has a value within the range [0,1], where the value closer to 1 represents a better flexibility:

1. Offered flexibility: The flexibility the EV user provides with respect to when the charging can be accomplished as determined from the charging constraints.
2. Actual flexibility: The actual flexibility that the system could have utilised based on when the EVs are actually plugged in and out.
3. V2G flexibility: The flexibility the EV user is willing to provide through V2G.

The indicators are defined and calculated as following (see symbols in Table 2):

$$I_{offered\_flexibility} = 1 - \frac{R_{eng}}{EV_{chrgp}} \cdot \frac{1}{O_{flexT}} = 1 - \frac{B_{cap}(T_{soc} - I_{soc})}{EV_{chrgp}} \cdot \frac{1}{T_{dep} - T_{arv}}$$

$$I_{actual\_flexibility} = 1 - \frac{R_{eng}}{EV_{chrgp}} \cdot \frac{1}{A_{flexT}} = 1 - \frac{B_{cap}(T_{soc} - I_{soc})}{EV_{chrgp}} \cdot \frac{1}{T_{pout} - T_{pin}}$$

$$I_{V2G\_flexibility} = 1 - \frac{O_{eng}}{EV_{dischrgp}} \cdot \frac{1}{O_{flexT}} = 1 - \frac{B_{cap}(1 - Min_{ec})}{EV_{dischrgp}} \cdot \frac{1}{T_{dep} - T_{arv}}$$

where

$$EV_{chrgp} = min(EV_{max\_cp}, \ CP_{chrg\_cap}), \ EV_{dischrgp} = min(EV_{max\_cp}, \ CP_{chrg\_cap})$$

5.2   Peak to Average Ratio

The peak to average ratio (PAR) indicator is meant to determine how flat the load curve is. It can be calculated as P_{max}/P_{avg} (see Table 2) within a time period T, assuming: i) there are multiple power samples within this period; ii) all of them greater or equal to 0 (only one direction for electricity flow is considered, i.e.: consumption); and iii) PAR is 1 if P_{max} equals 0. The minimum value for this ratio is 1, and it indicates the power in the installation or subsystem is constant, while bigger values indicate the power occasionally reaches high values and the rest of the time is much lower. It may be calculated for a CP, a charging infrastructure, a sub-network in the ESN or the feed-in line supplying the ESN. An objective of the ESN system is to keep it as close to 1 as possible to make the most use of the physical or logical (by contract) power capacity in the infrastructure.

6   Scenarios to be Evaluated

Scenarios are defined to investigate how the technology would work. Aspects such as charge planning and booking, charging in different types of neighbourhoods, and V2G are addressed in real life demonstrations and simulations.

6.1   Demonstrations and Simulations

Scenarios are adapted to local needs and contexts and implemented in real life ESN demos in Barcelona, Bremen, and Oslo by means of the measures in Sect. 3. The demos address, among others, home charging in a housing cooperative with a common garage, charge at work for office buildings, and the sharing of CPs. Measures for smart energy management like optimal and coordinated use of energy, local RES, and stationary battery storage are combined with measures for charging to study the effects on the energy demand. EVs that are connected over a longer period, typically overnight or the whole working day, may offer flexible charging but may also request priority charging. The combination of shared CPs and booking of CPs may give predictable charging, but the frequency of blockings must be investigated.

In line with Lervåg [7], the traditional impact evaluation of the demos have shortcomings. Implemented instances of the measures are rather few and in small scale due to budgetary constraints and the limited duration of the project. To overcome these constraints and broaden the basis for the evaluation of future scenarios with much higher density of EVs than we see today, we apply simulations. Based on the collected data we can simulate the impact of the GreenCharge concept in a more diverse set of scenarios, both with respect to size of the ESN and diversity and dimensioning of included measures.

Three scenarios proposed for simulation corresponds to the implemented demos and may provide interesting feedback to demo owners about how the installations will behave in possible future scenarios. Other artificial scenarios are created by combining and/or replicating elements from the demonstrators and representing ESNs of varying size, complexity, and context, closer to the project vision than the implemented ones, and thus allowing to investigate the impact of more full-fledged deployment of the GreenCharge concept.

For each scenario we will run simulations varying systematically one characteristic of the scenario at a time and computing the indicators. In this way we will investigate to which extent and in which way the varying characteristics impact the indicators. The varying characteristics are listed in Table 3. Mostly they correspond to the presence and dimensioning of the measures implemented in the demonstrators. The dimensioning in some cases corresponds to the indicator framework in Table 1.

Table 3. Varying characteristics

Varying characteristic	Variation
Local energy management	Optimisation method (centralised or distributed) and optimisation criteria (minimise energy cost or maximise energy greenness)
Number of EVs	EV penetration (e.g., 25%, 50%, … 100%)
Local RES	Percentage of total consumption, e.g., 0.25, 0.5, …
Stationary battery storage	Capacity as % of total consumption
Grid connection capacity	% of average consumption
Internal transfer capacity	With and without constraints, gradual removal of bottlenecks
Price model for the calculation and sharing of energy cost	E.g., energy only or mixed energy/power, fixed or Time of Use (ToU) or spot
Share of booking	Depends on available data
Share of battery capacity for V2G	Select EVs arbitrarily. Share of capacity drawn from distribution

6.2 Example on Use of Indicators

To demonstrate the impact of charge flexibility on loads, we use research data from the CoSSMic research project [10] and present three examples with charging at a charge station with a maximum capacity of 6 kW. The distribution of power among the CPs is controlled, and the charging of EVs is started and stopped accordingly.

Table 4 lists the indicators used. Each EV will request an amount of energy (E_{req}) and can charge at a maximum speed P_{chg}. Based on the optimisation of the charge station, each EV will get an amount of energy delivered (E_{del}). Assuming the EV charging constraints given in the charge requests correspond to the actual connection periods, the Energy Management Systems can charge the EVs from the arrival time (T_{arv}) to the departure time (T_{dep}). The actual time of charge completion (T_{chc}) depends on the optimisation policy.

The baseline example in Fig. 4a) includes a washing machine starting at 10:00, a dishwasher starting at 15:00, a freezer continuously running, and three EVs plugged in at 7:00, 10:00 and 9:30. The energy demand is in total 29.4 kWh. Figure 4b) shows the time-series of the total demanded power. With no charging flexibility, the peak to average ratio (PAR) is 7.62, as P_{max} equals 9.34 kW, and P_{avg} is 1.2 kW. With a power limitation of 6 kW, the total power consumption exceeds the threshold from 10:00 to 12:12. In this interval, 3.4 kWh cannot be delivered to charge the EVs.

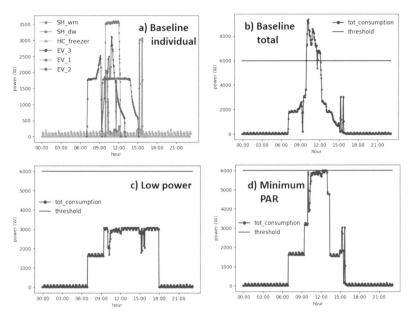

Fig. 4. Time-series examples – baseline, low power, minimum PAR

If the users driving the EVs will leave at 19:00, the charging flexibility can be exploited to reduce the peak demand. In Fig. 4c) and d), the optimisation aims to keep the power peak below 6 kW, and to minimise the PAR, giving priority to the EVs that arrived earlier and satisfying all energy demands.

Table 4. Indicators for example scenario

EV	P_{chg} kW	E_{req} kWh	T_{arv}	Baseline		Low Power		Min PAR	
				T_{chc}	Flex	T_{chc}	Flex	T_{chc}	Flex
EV1	3.6	8.2	07:00	12:33	0	13:30	0.86	18:05	0.86
EV2	1.8	10	09:30	15:18	0	15:37	0.69	16:37	0.69
EV3	3.1	9.8	10:00	13:13	0	13:05	0.68	13:20	0.68
SUM_{req_eng} (kWh)				29.4		29.4		29.4	
SUM_{del_eng} (kWh)				26		29.4		29.4	
PAR				7.62		5		2.5	
Charging Flexibility				0		0.53		0.53	

7 Conclusion and Further Work

This paper presents a method for evaluation of measures in a smart charging ecosystem. The measures, the indicator framework, the research data needed, and examples for indicator calculations have been described. The approach is hybrid, targeting both real life demos and simulation scenarios. Thus, we can also investigate scale ups and varying factors that could not be realised in real life due to limited demo size, capabilities and complexity, time, and budget. The approach covers a variety of aspects of relevance to cross sectorial smart charging ecosystems, as defined by the GreenCharge concept.

Further work is to be done in GreenCharge project regarding analysis of economic measures and its impact in the charging and smart energy management results. Among these measures, the impact of rewarding and penalising certain behaviours, such as incentivising charging flexibility or assigning penalties to users blocking booked CPs after expected departure time, will be investigated.

The evaluation process in the GreenCharge has already started by collecting research data from the demos, a process that will last about seven months. Baseline data will be further complemented through simulations, by disabling the smart energy management features.

Acknowledgments. Authors of this paper, on behalf of GreenCharge consortium, acknowledge the European Union and the Horizon 2020 Research and Innovation Framework Programme for funding the project (grant agreement no. 769016). The authors would like to thank all project partners for technical work related to evaluation and project management.

References

1. European Commission: Sustainable and Smart Mobility Strategy – putting European transport on track for the future, in COM (2020) 789, European Commission, Editor. Brussels (2020)
2. European Parliament: Directive on on the deployment of alternative fuels infrastructure, in DIRECTIVE 2014/94/EU, E. Parliament, Editor (2014)
3. European Parliament, Directive on on the energy performance of buildings, in DIRECTIVE 2010/31/EU E. Parliament, Editor (2010)
4. Sørensen, Å.L., et al.: Smart EV charging systems for zero emission neighbourhoods. ZEN Report No. 5 (2018)
5. Jiang, S., et al.: A distributed agent-based system for coordinating smart solar-powered microgrids. In: 2016 SAI Computing Conference (SAI). IEEE (2016)
6. Natvig, M., Jiang, S., Hallsteinsen, S.: Stakeholder motivation analysis for smart and green charging for electric mobility. In: Workshops of the International Conference on Advanced Information Networking and Applications. Springer (2020)
7. Lervåg, L.-E.: Evaluation of intelligent mobility services-connecting research and policy by using program theory. In: European Transport Conference 2016 Association for European Transport (AET) (2016)
8. FOT-Net & CARTRE: FESTA Handbook, Version 7 (2018). https://connectedautomated driving.eu/wp-content/uploads/2019/01/FESTA-Handbook-Version-7.pdf. Accessed 19 Feb 2021
9. Engels, D.: Refined CIVITAS process and impact evaluation framework (2017)
10. Amato, A., et al.: Software agents for collaborating smart solar-powered micro-grids. In: Smart Organizations and Smart Artifacts, pp. 125–133. Springer (2014)

The Use of Automatic Vehicle Location (AVL) Data for Improving Public Transport Service Regularity

Benedetta Argenzio, Nicola Amatucci, Marilisa Botte, Luca D'Acierno$^{(\boxtimes)}$,
Luca Di Costanzo, and Luigi Pariota

Department of Civil, Architectural and Environmental Engineering,
Federico II University of Naples, 80125 Naples, Italy
{nicola.amatucci,marilisa.botte,luca.dacierno,luca.dicostanzo,
luigi.pariota}@unina.it

Abstract. Smart and sustainable mobility systems are crucial elements for tomorrow's cities, whose facilities will be increasingly connected automated and environmentally friendly. Therefore, many strategies based on alternatively-powered vehicles, shared transport modes and micromobility solutions are being developed. However, the promotion of traditional public transport solutions still remains a key factor. Therefore, increasing their attractiveness, thus positively influencing the modal split, turns out to be an important goal to be achieved. In this context, we proposed a methodology, based on the use of Automatic Vehicle Location (AVL) data, for improving transit service performance. Finally, in order to show the feasibility of the proposed approach, it has been applied to a real bus line operating in the south of Italy.

1 Introduction

Nowadays the necessity of relying on smart and environmentally friendly transport systems represents a crucial factor for assuring a proper level of sustainability of our cities. Within the European context, the goal is to reach a climate-neutral status by 2050, as stated by the European Parliament in [1]. Clearly, this requires a multidisciplinary approach; however, from a mobility perspective, several strategies are being investigated for this purpose.

Firstly, alternatively-powered vehicles are being tested (e.g. electric and biogas) [2,3]. Concerning electric power supply, such vehicles are characterised by a local emission equal to zero; moreover, the production of electrical power by means of renewable sources (such as solar or wind) increases their sustainability [4]. Obviously, the bottlenecks for the complete development of such systems still remain the autonomy of batteries, the speed of re-charging and the capillarity of charging stations on the territory [5].

Further, alternative transport modes are being promoted, such as scooters, hoverboards or bikes, mostly as shared options, which stands alongside to the traditional car-sharing services which, however, require a more impactful infrastructure [6–8].

L. Barolli et al. (Eds.): AINA 2021, LNNS 227, pp. 667–676, 2021.
https://doi.org/10.1007/978-3-030-75078-7_66

Together with what above, a key factor is represented by the optimisation and sustainable management of traditional public transport modes, i.e. rail and transit systems. To this purpose, many strategies have been developed for increasing service performance [9–13], as well as reducing related energy consumption and environmental impact [14–19]. The attractiveness of these services also depends on their capacity in satisfying demand features. In order to make such modes demand-responsive, a crucial role is played by estimation demand flow techniques and passenger behaviours modelling tasks [20–25].

Finally, also the use of IoT and big data applications are becoming increasingly adopted and represents an important driver for moving towards smart and autonomous mobility frameworks [26–31]. In this context, our aim is to develop a methodology, based on the use of Automatic Vehicle Location (AVL) data, for optimising regularity of a bus line, thus making such a transport mode more attractive and affecting positively the modal split.

The rest of the paper is thus organised: Sect. 2 presents the proposed methodology, Sect. 3, applies it to a real context and Sect. 4 illustrates conclusions and further developments.

2   The Proposed Methodology

Service regularity represents one of the main factors affecting the attractiveness of transit service. However, ensuring it is very difficult due to the numerous delays affecting the service. In particular, we can have primary delays (i.e. delays arising firstly due to deviations from planned conditions) and secondary delays (i.e. due to the interaction of the considered vehicle with other vehicles on the same network). The latters are more serious in the case of rail systems, where in many cases the constrained driving makes impossible to overtaken or dodge a slower train.

Specifically, primary delays are addressing by adding a certain time rate, i.e. the so-called extension time, to the minimum travel time of a bus; while, secondary delays are addressing by adding a further time aliquot, i.e. the so-called buffer time, which is aimed to prevent knock-on effects and preserve service stability.

Such time rates affects the cycle time of the line, i.e. the time required by a vehicle to achieve the initial condition. Clearly, the estimation process is different in the case of round trip lines, which have 2 terminus stops (Fig. 1) and circular lines, which have just 1 terminus stop (Fig. 2).

In particular, in the case of round trip lines, we have:

$$CT_{rtl}^{min} = RT_{ot} + DT_{ot} + IT_{ot} + RT_{rt} + DT_{rt} + IT_{rt} \qquad (1)$$

where CT_{rtl}^{min} is the minimum cycle time of a round trip line; RT_{ot} and RT_{rt} are running times, respectively, in the case of outward and return trip; DT_{ot} and DT_{rt} are dwell times, respectively, in the case of outward and return trip; IT_{ot} and IT_{rt} are inversion times, respectively, in the case of outward and return trip.

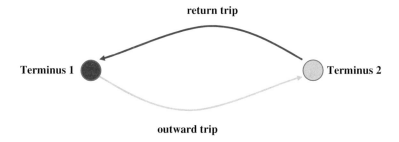

Fig. 1. Framework of a round trip line.

It is worth noting that, the inversion time of the outward trip has to be computed at Terminus 2; while, the inversion time related to the return trip has to be computed at Terminus 1.

By adding extension times in the case of outward trip, ET_{ot}, and return trip, ET_{rt}, as well as buffer times in the case of outward trip, BT_{ot}, and return trip, BT_{rt}, to the minimum cycle time, the planned cycle time of a round trip line, CT_{rtl}^{plan}, can be obtained as follows:

$$CT_{rtl}^{plan} = CT_{rtl}^{min} + ET_{ot} + BT_{ot} + ET_{rt} + BT_{rt} \tag{2}$$

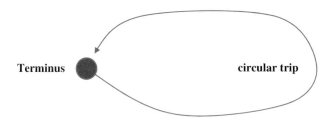

Fig. 2. Framework of a circular line.

Likewise, in the case of circular lines, we have:

$$CT_{cl}^{min} = RT_{cl} + DT_{cl} + IT_{cl} \tag{3}$$

where CT_{cl}^{min} is the minimum cycle time of a circular line; RT_{cl} is the related running time; DT_{cl} is the related dwell time; IT_{cl} is the related inversion time.

The planned cycle time of a circular line, CT_{cl}^{plan}, similarly to the above, can be computed by adding the extension time, ET_{cl}, and the buffer time, BT_{cl}, to the minimum cycle time, that is:

$$CT_{cl}^{plan} = CT_{cl}^{min} + ET_{cl} + BT_{cl} \tag{4}$$

As already mentioned, a proper quantification of the above-described time rates is fundamental for addressing services delays and, therefore, for ensuring a certain degree of service regularity, as will be shown in the application section.

3   Application to a Real Bus Line

In order to show the effectiveness of the proposed approach, it has been applied to a real context, i.e. a circular transit line within the Naples Bus System in the south of Italy. It covers about 22 km, through 31 stops.

Data provided by the on-board location devices are represented by the travel times of the circular line (TT_{cl}), i.e. the sum of running (RT_{cl}) and dwell times (DT_{cl}), as shown by Eq. 5.

$$TT_{cl} = RT_{cl} + DT_{cl} \tag{5}$$

Detected data have been analysed by testing two statistical distributions, i.e. Gaussian and Erlang functions (see Eq. 6 and 7).

$$F_G\left(x\right) = \int_{-\infty}^{x} \frac{1}{\sqrt{2 \cdot \pi \cdot \sigma^2}} \cdot e^{-\frac{(x-\mu)^2}{2 \cdot \sigma^2}} \quad \text{with } \mu \in R, \ \sigma^2 \in R^+, \ x \in R \tag{6}$$

$$F_E\left(x\right) = \int_{0}^{x} \frac{\lambda^k \cdot x^{k-1} \cdot e^{-\lambda \cdot x}}{(k-1)!} \quad \text{with } k \in N^+, \ \lambda \in R^+ \cup \{0\}, \ x \in R^+ \cup \{0\} \tag{7}$$

where μ and σ^2 are, respectively, the mean and the variance of the Gaussian function; λ and k are, respectively, the rate and the shape parameters of the Erlang function.

In particular, for both of them, function parameters have been calibrated and related goodness of fit has been identified by means of the computation of R^2 value. Results are shown in Table 1 and 2, respectively for Gaussian and Erlang distributions. While, Fig. 3 presents a graphical comparison between the calibrated functions and surveyed data.

Tables 3 and 4 show, respectively, travel times and extension times for different confidence levels (i.e. 80%, 85%, 90%, 95%) in the case of surveyed data and calibrated functions.

Table 1. Gaussian function parameters.

μ parameter	σ^2 parameter	R^2 value
56.285	14.920	0.8088

Table 2. Erlang function parameters.

k parameter	λ parameter	R^2 value
20	0.3434	0.8378

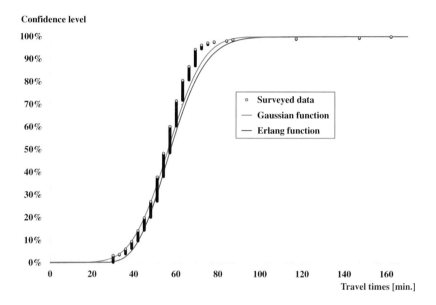

Fig. 3. Cumulative distributions.

In order to assess how the estimated times affect service performance, it is necessary to evaluate the cycle time. In particular, by considering an average travel time equal to 56.3 min and an inversion time equal to 10 min, the minimum cycle time results to be equal to 66.3 min. Then, according to the adopted confidence level, the planned cycle time can be computed (see Table 5).

As can be seen, as the confidence level increases, outcomes of surveyed and calibrated data become increasingly different; however, the difference between the two calibrated distributions remains low for justifying the use of a statistical distribution different from the Gaussian function, which allows to obtain feasible results with a low computational effort.

Further, in order to quantify the investment, in terms of number of buses, required for ensuring a certain level of service, the following analysis has been carried out.

The service headway (H_{serv}) between two subsequent buses may be calculated as a function depending on service frequency (φ_{serv}), that is:

$$H_{serv} = \frac{1}{\varphi_{serv}} \qquad (8)$$

Table 3. Travel time evaluations [minutes].

Confidence level	Surveyed data	Gaussian function	Erlang function
80%	63.0	67.3	68.8
85%	66.0	69.8	71.7
90%	69.0	73.1	75.4
95%	72.0	77.9	81.2

Table 4. Extension time evaluations [minutes].

Confidence level	Surveyed data	Gaussian function	Erlang function
80%	6.7	11.0	12.5
85%	9.7	13.5	15.4
90%	12.7	16.8	19.1
95%	15.7	21.6	24.9

Table 5. Planned cycle times [minutes].

Confidence level	Surveyed data	Gaussian function	Erlang function
80%	73.0	77.3	78.8
85%	76.0	79.9	81.7
90%	79.0	83.1	85.4
95%	82.0	87.9	91.2

The minimum number of buses (NB_{min}) required to perform the service may be calculated as a function depending on the adopted cycle time (CT_{cl}^{ad}) and the service frequency (φ_{serv}), that is:

$$NB_{min} = int\left(CT_{cl}^{ad} \cdot \varphi_{serv}\right) + 1 \qquad (9)$$

Table 6. Service parameters in the case of minimum cycle time adoption.

φ_{serv} [buses/h]	H_{serv} [min]	NB_{min} [#]
2.0	30.0	3
3.0	20.0	4
4.0	15.0	5
5.0	12.0	6
6.0	10.0	7
7.5	8.0	9
8.0	7.5	9
10.0	6.0	12
12.0	5.0	14

Therefore, according to Eq. 8 and Eq. 9, service parameters in the case of minimum and planned cycle times have been computed for different values of service frequency (Tables 6 and 7).

Results confirm the importance of relying on a robust function for the estimation of extension times and, therefore, of the cycle time to be planned for ensuring a certain regularity of the service. Obviously, as the confidence level

Table 7. Service parameters in the case of planned cycle time adoption.

φ_{serv} [buses/h]	H_{serv} [min]	Confidence level [%]	NB_{min} [#]		
			Surveyed data	Gaussian funct.	Erlang funct.
2.0	30.0	80%	3	3	3
		85%	3	3	3
		90%	3	3	3
		95%	3	3	4
3.0	20.0	80%	4	4	4
		85%	4	4	5
		90%	4	5	5
		95%	5	5	5
4.0	15.0	80%	5	6	6
		85%	6	6	6
		90%	6	6	6
		95%	6	6	7
5.0	12.0	80%	7	7	7
		85%	7	7	7
		90%	7	7	8
		95%	7	8	8
6.0	10.0	80%	8	8	8
		85%	8	8	9
		90%	8	9	9
		95%	9	9	10
7.5	8.0	80%	10	10	10
		85%	10	10	11
		90%	10	11	11
		95%	11	11	12
8.0	7.5	80%	10	11	11
		85%	11	11	11
		90%	11	12	12
		95%	11	12	13
10.0	6.0	80%	13	13	14
		85%	13	14	14
		90%	14	14	15
		95%	14	15	16
12.0	5.0	80%	15	16	16
		85%	16	16	17
		90%	16	17	18
		95%	17	18	19

and service frequency increase, also the difference between outcomes increases. Indeed, for instance, in the case of a frequency equal to 12 buses/h, the minimum cycle time reveals the need of 14 buses for ensuring this level of service. However, this value arises up to 19 by considering a planned cycle time with a confidence level equal to 95%. This validates the effectiveness of the proposed approach also for evaluating the fleet investment required in the planning stage for a certain service. Likewise, the same analysis could be performed for identifying the maximum service regularity allowed with respect to the available number of buses.

4   Conclusions and Research Prospects

In this paper, a methodology based on the use of Automatic Vehicle Location (AVL) data, for improving transit service performance, has been proposed. In particular, the goal was to properly quantify system parameters in correspondence of different confidence levels, thus suitably supporting the planning of service regularity, as well as helping dispatchers in acting specific control strategies.

It is worth noting that, with suitable onboard devices, the proposed methodology can be easily applied also to the case of a hybrid fleet, as well as e-fleet contexts.

As research prospects, we propose to test the developed approach in the case of different network contexts, as well as to customize it to different time slots as, for instance, within the day (i.e. peak and off-peak hours) and the week (i.e. weekdays and holidays).

References

1. European Parliament: The European Green Deal: European Parliament resolution of 15 January 2020 on the European Green Deal (2019/2956(RSP)), pp. 1–24 (2020). https://www.europarl.europa.eu/doceo/document/TA-9-2020-0005_EN.html. Accessed Feb 2021
2. Dahlgren, S.: Biogas-based fuels as renewable energy in the transport sector: An overview of the potential of using CBG, LBG and other vehicle fuels produced from biogas. Biofuels (in press). https://doi.org/10.1080/17597269.2020.1821571
3. Rizopoulos, D., Esztergár-Kiss, D.: A method for the optimization of daily activity chains including electric vehicles. Energies **13**(4), 1–21 (2020). https://doi.org/10.3390/en13040906
4. Turki, F., Guetif, A., Sourkounis, C.: Contactless charging electric vehicles with renewable energy. In: Proceedings of the 5th International Renewable Energy Congress (IREC 2014), Hammamet, Tunisia (2014). https://doi.org/10.1109/IREC.2014.6826998
5. Adhikari, M., Ghimire, L.P., Kim, Y., Aryal, P., Khadka, S.B.: Identification and analysis of barriers against electric vehicle use. Sustainability **12**(12), 1–20 (2020). https://doi.org/10.3390/su12124850

6. Ratinho, T.: E-scooters, bikes and urban mobility: lessons from the streets of Paris. The Conversation (on line journal) (2019). https://theconversation.com/e-scooters-bikes-and-urban-mobility-lessons-from-the-streets-of-paris-125619. Accessed Feb 2021

7. Bieliński, T., Ważna, A.: Electric scooter sharing and bike sharing user behaviour and characteristics. Sustainability **12**(22), 1–13 (2020). https://doi.org/10.3390/su12229640

8. Turoń, K., Kubuc, A., Chen, F., Wang, H., Lazarz, B.: A holistic approach to electric shared mobility systems development - modelling and optimization aspects. Energies **13**(21), 1–19 (2020). https://doi.org/10.3390/en13215810

9. Gallo, M., D'Acierno, L., Montella, B.: A multimodal approach to bus frequency design. WIT Trans. Built Environ. **116**, 193–204 (2011). https://doi.org/10.2495/UT110171

10. Cascetta, E., Cartenì, A., Carbone, A.: The quality in public transportation: the Campania regional metro system. Ing. Ferrov. **68**(3), 241–261 (2013). http://www.ingegneriaferroviaria.it/web/it/node/526

11. Di Maio, A., Botte, M., Montella, B., D'Acierno, L.: The definition of bus fleet operational parameters: the dwell time estimation. In: Proceedings of the 20th IEEE International Conference on Environment and Electrical Engineering (IEEE EEEIC 2020) and 4th Industrial and Commercial Power Systems Europe (I&CPS 2020), Madrid, Spain (2020). https://doi.org/10.1109/EEEIC/ICPSEurope49358.2020.9160592

12. Esztergar-Kiss, D.: Trip chaining model with classification and optimization parameters. Sustainability **12**(16), 1–15 (2020). https://doi.org/10.3390/su12166422

13. D'Acierno, L., Botte, M.: Railway system design by adopting the Merry-Go-Round (MGR) paradigm. Sustainability **13**(4), 1–21 (2021). https://doi.org/10.3390/su13042033

14. De Martinis, V., Gallo, M., D'Acierno, L.: Estimating the benefits of energy-efficient train driving strategies: a model calibration with real data. WIT Trans. Built Environ. **130**, 201–211 (2013). https://doi.org/10.2495/UT130161

15. Cartenì, A., Henke, I.: External costs estimation in a cost-benefit analysis: the new Formia-Gaeta tourist railway line in Italy. In: Proceedings of the 17th IEEE International Conference on Environment and Electrical Engineering (IEEE EEEIC 2017) and 1st Industrial and Commercial Power Systems Europe (I&CPS 2017), Milan, Italy (2017). https://doi.org/10.1109/EEEIC.2017.7977614

16. D'Acierno, L., Botte, M.: Passengers' satisfaction in the case of energy-saving strategies: a rail system application. In: Proceedings of the 18th IEEE International Conference on Environment and Electrical Engineering (IEEE EEEIC 2018) and 2nd Industrial and Commercial Power Systems Europe (I&CPS 2018), Palermo, Italy (2018). https://doi.org/10.1109/EEEIC.2018.8494575

17. Calise, F., Cappiello, F.L., Cartenì, A., Dentice d'Accadia, M., Vicidomini, M.: A novel paradigm for a sustainable mobility based on electric vehicles, photovoltaic panels and electric energy storage systems: case studies for Naples and Salerno (Italy). Renew. Sustain. Energy Rev. **111**, 97–114 (2019). https://doi.org/10.1016/j.rser.2019.05.022

18. Esztergar-Kiss, D., Braga Zagabria, C.: Method development for workplaces using mobility plans to select suitable and sustainable measures. Res. Transp. Bus. Manag. (in press). https://doi.org/10.1016/j.rtbm.2020.1005

19. Gallo, M., Marinelli, M.: Sustainable mobility: a review of possible actions and policies. Sustainability **12**(18), 1–39 (2020). https://doi.org/10.3390/su12187499

20. Bifulco, G.N., Cartenì, A., Papola, A.: An activity-based approach for complex travel behaviour modelling. Eur. Transp. Res. Rev. **2**(4), 209–221 (2010). https://doi.org/10.1007/s12544-010-0040-3

21. Cartenì, A., Galante, G., Henke, I.: An assessment of models accuracy in predicting railways traffic flows: a before and after study in Naples. WIT Trans. Ecol. Environ. **191**, 783–794 (2014). https://doi.org/10.2495/SC140661

22. Ercolani, M., Placido, A., D'Acierno, L., Montella, B.: The use of microsimulation models for the planning and management of metro systems. WIT Trans. Built Environ. **135**, 509–521 (2014). https://doi.org/10.2495/CR140421

23. Caropreso, C., Di Salvo, C., Botte, M., D'Acierno, L.: A long-term analysis of passenger flows on a regional rail line. Int. J. Transp. Dev. Integr. **1**(3), 329–338 (2017). https://doi.org/10.2495/TDI-V1-N3-329-338

24. Di Mauro, R., Botte, M., D'Acierno, L.: An analytical methodology for extending passenger counts in a metro system. Int. J. Transp. Dev. Integr. **1**(3), 589–600 (2017). https://doi.org/10.2495/TDI-V1-N3-589-600

25. D'Acierno, L., Botte, M., Montella, B.: Assumptions and simulation of passenger behaviour on rail platforms. Int. J. Transp. Dev. Integr. **2**(2), 123–135 (2018). https://doi.org/10.2495/TDI-V2-N2-123-135

26. D'Acierno, L., Cartenì, A., Montella, B.: Estimation of urban traffic conditions using an Automatic Vehicle Location (AVL) System. Eur. J. Oper. Res. **196**(2), 719–736 (2009). https://doi.org/10.1016/j.ejor.2007.12.053

27. Dureja, A., Suman, S.: Efficient transportation: future aspects of IoV. Int. J. Veh. Inf. Commun. Syst. **5**(3), 290–308 (2020). https://doi.org/10.1504/IJVICS.2020.110994

28. Esztergar-Kiss, D., Kerenyi, T., Matrai, T., Aba, A.: Exploring the MaaS market with systematic analysis. Eur. Transp. Res. Rev. **12**, 1–16 (2020). https://doi.org/10.1186/s12544-020-00465-z

29. Etzioni, S., Hamadneh, J., Elvarsson, A., Esztergar-Kiss, D., Djukanovic, M., Neophytou, S., Sodnik, J., Polydoropoulou, A., Tsouros, I., Pronello, C., Thomopoulos, N., Shiftan, Y.: Modeling cross-national differences in automated vehicle acceptance. Sustainability **12**(22), 1–22 (2020). https://doi.org/10.3390/su1222976

30. Jung, I.: An IoT-based smart parking management system. Int. J. Comput. Vis. Robot. **10**(2), 122–132 (2020). https://doi.org/10.1504/IJCVR.2020.105680

31. Lytras, M.D., Chui, K.T., Liu, R.W.: Moving towards intelligent transportation via Artificial Intelligence and Internet-of-Things. Sensors **20**(23), 1–4 (2020). https://doi.org/10.3390/s20236945

Modelling Behavior in a Route Choice Driving Simulation Experiment in Presence of Information

Roberta Di Pace[1]([✉]), Stefano de Luca[1], Francesco Galante[2], and Luigi Pariota[2]

[1] Department of Civil Engineering, University of Salerno, Fisciano, SA, Italy
{rdipace,sedeluca}@unisa.it
[2] Department of Civil and Environmental Engineering, University of Naples 'Federico II',
Naples, Italy
{francesco.galante,luigi.pariota}@unina.it

Abstract. Modelling route choice decision making in Advanced Traveler information System (ATIS) contexts is still a crucial task. In particular, two main categories of variables can be identified in order to model travelers' behaviors: the former may be defined as endogenous and are related to the experiment environment; the latter may be defined as exogenous (referring to the respondents involved in the experiment). This paper focuses on the analysis of exogenous variables. An experiment is carried out using a driving simulator, on a real route choice context (a sub-area of the urban network in the city of Naples, in the Campania Region) reproduced in a virtual reality. All data are analyzed by aggregate and statistical approaches to preliminarily investigate the correlations between some exogenous variables and the collected choices of drivers. Furthermore, collected observations have been modelled by applying the Structural Equation Model (SEM) approach to model the effect of information on switching behaviors.

1 Introduction and Motivations

Analyzing and modelling route choice under real-time information provision is a fundamental issue for both transportation planners and decision makers. Indeed, the proper design of Advanced Traveler Information Systems (ATIS) requires a realistic, reliable and efficient modelling framework (e.g., [1, 5, 17]). In general, two different kinds of variables may influence the route choice making mechanism with ATIS (Advanced Traveler Information Systems): the endogenous variables and the exogenous variables. Most of the literature has mainly investigated endogenous variables, such as travel time variability and traffic congestion driver socioeconomic characteristics and trip purpose [33, 34, 41–43], route complexity [35] and situational factors such as weather conditions. Several other studies have focused the attention on to the impacts that different kinds of information may have on route choice: information quality [5, 36], information contents [36, 38–40], amount and source. Concerning the exogenous variables, several researchers have illustrated the role of drivers 'attitudes towards real-time information [32], with particular attention to information acquisition and usage, trust (34), usefulness of information [33] and risk-prone tendency. In conclusion, it is common opinion that route choice modelling must consider traditional determinants (e.g., travel time, driving environment, network complexity, drivers' socioeconomic characteristics, traffic congestion) but also drivers' attitudes, beliefs, and cognitive workload [9, 23, 36]. Hence, there is an increasing attention on the role of human factors and cognitive aspects in route choice modelling and an increasing need for specific simulation environments able to reproduce realistic driving contexts and to collect realistic behavior. Based on previous research this paper focuses on exogenous variables and how to observe such variables in a virtual environment. To investigate travelers' behavior in route choice contexts with information systems, the Stated Preference (SP) approach is usually adopted. Several technological based approaches were adopted for survey collection to analyze travelers' behavior under ATIS (see [18]) the most straightforward are the technological platforms (see [1, 10]). These platforms make it possible to observe travelers' behavior in a more realistic way compared to traditional tools (such as. telephone interviews), which are applied to SP techniques, even if the nature of the experiments is assisted by these technological platforms (see [2, 5, 14, 27, 28, 36, 39]) In general, two different kinds of platforms may be adopted: the driving-simulator (DS) and the travel simulator (TS). When using the DS, the respondent must experience the chosen route (by means of a computer-based virtual-reality simulation), while with the TS the respondent is notified by feedback about the actual travel time of all routes chosen and unchosen. A limited number of studies have been carried out by adopting DSs. However, one of the recent improvements to the web-based survey is related to the integration of environmental experiments by incorporating some components of virtual reality such as pictures, graphics, photos etc. (see [29]). In fact, the main weaknesses of SP experiments are related to the burden and the complexity of the experiment. As shown in previous research (see [16]) assisting respondents with appropriate technological tools allows for quicker learning of how the system performs and consequently how to react to its stimuli. In particular, the main effects are related to information acceleration and to risk perception; in fact, driving in a virtual reality situation directly affects the penalty perception of travelers by inducing a more realistic risk aversion tendency mainly in the case of inaccurate (and for that reason, more uncertain) conditions of choice (i.e., inaccurate information). Within the previous cited context, the paper's aim is analyzing travelers' behaviors in a virtual reality (realistic) route choice context, by accounting some latent variables related to the attitudes of information users. A real urban network scenario in the presence of ATIS in DS (validated in a previous stage of a research project) was implemented. Furthermore, collected data were analyzed by statistical approaches to identify significant variables affecting travelers' behavior [37]. The paper is organized as follows: Sect. 2, provides a brief introduction to the methodology in terms of experiment design and scenario implementation in a Route Choice Driving Simulator (RCDrivingSim); in Sect. 3, the details of the collected survey are shown in terms of aggregate and statistical analysis and in terms of modelling approaches (i.e., the structural equation model); in Sect. 4, conclusions and further research perspectives are discussed in last section.

2   Methodology

To collect drivers' choices in a route choice context an experiment in a virtual reality was done. The experiment was carried out by using the RCDrivingSim and in terms of the network, the ring road "est-ovest" and a sub-area of the urban network of the city of Naples, in Campania Region, was considered during the experiment. Many traveler behavior investigations have adopted the cross-sectional approach, in particular, to estimate the "before-and-after" behaviors, however, it has been shown that a panel survey approach is more suitable in the case in which the same people are involved in the experiment, even if, it may be difficult to detect the effect of "treatment" on the people. Based on previous consideration, drivers are asked to repeat the experiment in a context "with" and "without" information and they are asked to make their choices to arrive on time at the destination.

2.1   The Survey Instrument: The RCDrivingSim

The DS which was adopted in this experiment here discussed is the same as that used for the results from previous research (see [16]). In the DS, respondents drive in a virtual reality environment and route choices are made by driving and steering with respect to the provided information by VMSs. RCDrivingSim uses the same software of the immersive dynamic DS at Naples University, VERA, (see [3, 19]). Like VERA the software used for driving simulation is SCANeR®II r2.22 from the Oktal company, however it is a pc-based DS, with a desktop support which induces less realism for the analysis of drivers' behavior, but sufficient for the applications related to route choice behavior. The driver controls the driving tasks using accelerator and brake pedals and a steering wheel (Logitech™ MOMO Racing Force Feedback Wheel), able to provide force feedback, as well as six programmable buttons (ignition, horn, turn signals, etc.), sequential stick shifters and paddle shifters. The graphic computer renders, antialiased at 60 Hz, a full 19" monitor projection of the view from the driver's seat in the car. The graphics laptop also generates sound effects consistent with the roadway environment, including, but not limited to engine, turn signal, horn, road/tire, wind, tire squeals, and other traffic sounds. The panel includes in terms of instruments, a tachometer and a speedometer as well as turn signals and gear settings. Finally, the (virtual clock) is not applied in the RCDrivingSim, in fact, the characteristics of the chosen alternative are experienced by the respondent by driving.

2.2   The Experiment

The origin-destination pair considered is the connection between Capodimonte (point A in Figs. 1 and 2) and Vomero, around the Collana Stadium (point B in Figs. 1 and 2). Depending on the congestion degree, they can make three different alternatives route choices: the off ramp Vomero (route1, see Fig. 1 a), the off ramp Camaldoli (route 2, see Fig. 1 b) and the off ramp Arenella (route 3, see Fig. 2 a). The three alternative routes are composed by a part of the urban network (where the mean speed limit is equal to 50 km/h) and a part of the ring road "est-ovest" of the city of Naples (where the mean speed limit is equal to 80 km/h). Furthermore, the routes have different characteristics, in particular: Route 1, (the mean speed limit is 80 km/h) comprised of 2/3 of the ring road "est-ovest" and 1/3 of the urban network; Route 2, is comprised of 1/2 of the ring road "est-ovest" and 1/2 of the urban network; Route 3, is comprised of 1/3 of the ring road "est-ovest" and 2/3 of the urban network. The experiment here described, has been preceded by another pilot experiment to validate the implemented scenario. Furthermore, before starting the experiment, respondents are asked to answer to a questionnaire related to their own sociodemographic characteristics and to their attitudes with respect to the risk perception in order to classify them in terms of risk aversion and risk propensity.

Respondents are asked to repeat their driving sessions, 6 times without and 6 times with information; at each driving simulation over 6 trials, they are asked to drive one time without information and successively with information (for each respondent 12 observations are collected in all). To reduce the degree of fatigue in drivers, they are asked to drive each day for three successive days to collect the effect of information on the switching behaviors. At the end of the experiment respondents are required to answer an ex post questionnaire in order to collect some information about the quality of the experiment (also important in order to validate an ex post in the prototype) and some further information about the reason for their choices (the most chosen route can be motivated by the shortest one, for distance).

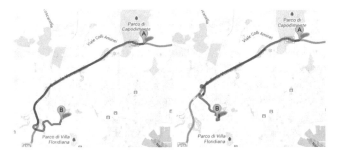

Fig. 1. a) Route 1 b) Route 2

Fig. 2. a) Route 3 b) snapshot of the experiment

In the scenario with information, respondents are provided via VMS with an alert on the degree of traffic congestion (e.g., queue) and waiting time in queue ("the duration of queue is … min"; see Fig. 2 b). Routes 2 and 3 are characterized by the same traffic conditions, whereas for route 1 three different traffic conditions are considered in a context without (low for 12% of the time, intermediate for 30% of the time and high for 58% of the time) and with information (low for 43% of the time, intermediate for 40% of the time and high for 17% of the time). Based on the distributions of the degree of congestion, in the absence of information, route 1 was the shortest one 35% of the times, route 2 was the shortest one 65% of the times and, thus, route 3 was never the shortest one. On the other hand, in the presence of information systems, route 1 was the shortest one 36% of the times, route 2 was the shortest one 64% of the times and route 3 was the shortest one 2% of the times. The mean and the standard deviation of instantaneous travel times of each route are shown in Table1.

Table 1. Instantaneous travel times

distr. parameters	Route 1	Route 2	Route 3
mean [min] (s.d. [min])	10 (6.8)	8 (0.72)	13 (0.08)

3   Survey Results

In this section the results of the aggregate and statistical analyses are shown. For this experiment, 40 respondents were involved and were randomly assigned to each scenario. The sample was composed of 47% females; 65% of the sample had possessed a driving license for less than ten years; half the sample was aged between 20 to 30 years old, 18% between 31 to 40. The educational attainment was 6% secondary school, 35% bachelor's degrees; 41% master's degrees; 18% PhDs. Most of the respondents were familiar with technologies: 35% of them sometimes use GPS devices and navigators, 97% of them often use internet, 67% of them sometimes buy on-line, 80% are used to office productivity software.

3.1   Aggregate and Statistical Analysis

Table 2 shows the distribution parameters of each route whereas in Table 3 the actual degrees of congestion are shown; finally, the aggregate routes' share is reported in Table 4.

Table 2. Actual travel times

Scenario	distr. Parameters (min)	Route 1	Route 2	Route 3
No Information	mean (s.d.)	11 (7.37)	9 (0.45)	12 (0.01)
Information	mean (s.d.)	10 (6.20)	9 (0.59)	12 (0.06)

Table 3. Actual degrees of congestion

Scenario	distr. Parameters (min)	High	Intermediate	Low
No Information	mean (s.d.)	6 (0.0)	8 (0.57)	12 (2.41)
	frequency	12%	30%	**58%**
Information	mean (s.d.)	6 (0.002)	9 (0.11)	12 (1.08)
	frequency	17%	40%	**43%**

Table 4. Aggregate routes' share

Scenario	Route 1	Route 2	Route 3
Without Information	**58%** *(35%)*	42% *(65%)*	0% *(0%)*
With Information	15% *(36%)*	**84%** *(64%)*	1% *(0%)*

*In the brackets: the percentage in which the route has been the shortest one

In accordance with the literature, two different drivers' behaviors may be observed in a route choice context: uncertain (or ambiguous) and based on risk perception (see [15]) behaviors. In the former case, the probability of the outcomes is unknown whereas in the latter case these are known. With respect to our experiment, the ambiguity and risk perception can be referred to the different contexts of choice: without information and with information. In fact, in the case of context of choice, the absence of information can induce confused behaviors (less or more) depending on the different degrees of congestion of route 1, whereas in the presence of information choices can be affected by both degrees of the congestion of route 1 and the presence of the information system, which assists drivers in their choices. Furthermore, it can be observed that the choice context may be classified as uncertain by the effect of degrees of traffic congestion and by degrees of accuracy in particular with respect to route 1, but this is not our case because the information in the ATIS scenario is considered accurate [6–8]. In a scenario without information, the ambiguous drivers' behaviors can be defined in accordance with confused reactions consecutively inducing a higher percentage of choices of route 1; whereas in the case of presence of information, the risk perception behaviors induce a higher percentage of choices of longer and most reliable routes, as with route 2. In fact, in this last case, respondents can estimate the outcomes probability by the help of (accurate) information systems, and as expected, they show a more risk averse propensity. To preliminary investigate the relation between the risk aversion propensity and the observed switching behaviors, some statistical nonparametric tests (the Kruskall Wallis test) are carried out. The answers collected in a preliminary questionnaire (ex-ante questionnaire) are considered and analyzed. Statistical tests show the "significant effect" of the question "Is is very neg. if arriving 5/10 min. late" (Chi square = 0.848; Asymp. Sig. = 0.027) on switching choices.

4 Modelling Approach

Preliminary investigations expected that some (latent, non-observable) attitudes may be revealed about choices on the effect of the context perception and these may influence drivers' behavior. Several works in literature have emphasized the importance of including some psychological factors in modelling decision making, thus the necessity of enriching the adaptability of discrete choice modelling in these cases [4, 30, 31] including some perceptions and attitudes related to the respondents by introducing the latent variables. In this paper the model parameters are estimated by the application of Structural Equations Modelling approach [24, 25].

4.1 Structural Equations Model: Framework Formulation

The SEM econometric modelling approach is adopted to handle endogenous and exogenous variables and, in the meantime, some unobserved variables (i.e., the latent variables) related to the observed variables by linear combination (see [20]). The main variables considered in the mathematical formulation are the exogenous variables (x), the exogenous latent variables (ξ), the endogenous latent variables (η) and the endogenous variables (y); furthermore, the terms ζ, δ and ε are the error terms. The Structural equation model (SEM) is.

$$\eta = \alpha + \mathbf{B} \cdot \eta + \Gamma \cdot \xi + \zeta \tag{1}$$

The Measurement models (MMs) are (already based on some simplification by fixing and constraining some parameters; see [26])

$$\mathbf{x} = \Lambda_x \cdot \xi + \delta, \mathrm{E}(\delta) = 0, \ Cov(\delta) = \theta_\delta \tag{2}$$

$$\mathbf{y} = \Lambda_\mathbf{y} \cdot \eta + \varepsilon, E(\varepsilon) = \mathbf{0}, \ Cov(\varepsilon) = \theta_\varepsilon \tag{3}$$

The above introduced variables are related, as shown in Fig. 3, to the structural equation and the measurement models where:

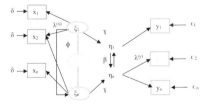

Fig. 3. SEM equations framework

4.2   Estimation Results

The model is estimated by the software LISREL 8.8 (SSI - Scientific Software International, International). Before estimation, some statistical nonparametric tests (results are not shown for the sake of brevity) were carried out to identify the significant effect of variables on observed (switching) choices and then classify the main variables for modelling estimation.

Table 5. Exogenous variables

Grouping variable (w.r.t. exogenous latent variables)	Variable	Value
Utility	Compliance	Binary variable (1 id compliant)
	Gender	Binary variable (1 if male)
	Job	Discrete variables
Risk Aversion	Age $_{20-25}$	Continuous variables
	Age $_{25-30}$	
	Age $_{30-35}$	
	Familiarity with technology (TecFam)	Discrete variables

In fact, the variables considered in the modelling approach are the exogenous variables (see Table 5 below for a more detailed description), the endogenous latent variables, utility (Utility) and the risk aversion (RiskAv), and, finally, the endogenous variables which are the switching observed choices and the perception indicators (i.e., the answers obtained in the ex-ante questionnaire with respect to the risk aversion). For the perception indicators, the Exploratory factor analysis is made (i.e. the Varimax-Rotated Factor Loadings) to relate these to the "RiskAv" attitude.

The detailed description of parameters estimation is shown in Table 6 and by the path diagram in Fig. 4.

The convergence of the model is reached after 39 iterations. In terms of goodness of fit statistics, from the obtained results the $\chi 2$ test statistic value (0.422) shows that the model fits data significantly (level of significance is 1%). Furthermore, other indicators (RMSEA = 0.012, the Non-Normed Fit Index, NNFI = 0.852, the Comparative Fit Index, CFI = 0.945) indicate an acceptable fit.

Considered attributes confirmed that in the presence of information, assuming that in the case of accurate information (as in the herein explained experiment) a high value of compliance is induced, this (represented by binary variable, 1 if compliant with information system, 0 otherwise) may be considered as an explaining variable. Furthermore, two sociodemographic variables are shown to significantly affect the travelers' switching behaviors: the gender (male positively influence the propensity to switching) and

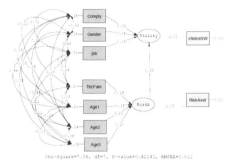

Fig. 4. Path diagram of the structural equation model (LISREL output)

the job. Furthermore, as shown in the model, the utility is affected by the influence of latent attitudes to the risk (aversion); this variable is related to the declared propensity, in the ex-ante questionnaire, at risk aversion. Finally, in accordance with the literature (see Tsirimpa et al., 2009) different classes of age are shown to be strictly related to risk aversion (RiskA), however, lower classes negatively affect the risk aversion in the sense that younger drivers usually shown more risk propensity attitudes. Risk aversion is, also positively affected by the binary attributes which explains the high degree of interaction and technological confidence of users (this is also information collected through an ex-ante questionnaire).

Table 6. Estimation results

Measured variables	Endogenous variables	
	Utility	Risk aversion
Choice	1.00[a]	-
Λ_y	–	-
RiskA	–	1.00[a]
Beta		
Utility	–	–
Risk aversion	+ 0.33 (+1.92)	–
Λx	*Uility*	*RiskAv*
Compliance	+ 0.03 (+1.25)	
Gender	+ 0.04 (+1.23)	
Job		+ 0.05 (+0.75)

(continued)

Table 6. (*continued*)

Measured variables	Endogenous variables	
	Utility	Risk aversion
Age $_{20-25}$		−0.15 (*−0.98*)
Age $_{25-30}$		+ 0.10 (*+1.77*)
Age $_{30-35}$		+ 0.10 (*+0.88*)
Familiarity with technology		−0.25 (*−1.98*)

[a]Fixed parameters

5   Conclusions and Research Perspectives

In order to support the development of sustainable mobility [11, 12, 21, 22] in the era of Cooperative Intelligent Transportation Systems and Connected and Automated Vehicles [13], modelling users' behavior in route choice contexts is still are relevant issue. The paper aims at analyzing travelers' behaviors in a virtual reality route choice context, by accounting some latent variables related to the attitudes of information users. In fact, as shown, some risk aversion attitudes may directly affect the users' behaviors in a route choice context. The collected survey is related to the context in which respondents are provided without and with information to observe the switching behaviors through the effect of information provided. The information is introduced to reduce the degree of uncertainty which usually influences travelers' choices and as well known, to be effective, only accurate information needs to be introduced. The results here shown are considered a continuation of a research project characterized by threefold aims: i) Set-up and validation of some technological tools for Stated preference experiments; ii) Set-up and validation of a Travel Simulator (TS) web-based platform (see [6]); iii) Set-up and validation of a Route choice Driving Simulator (DS; see [16]); iv) Carry out a stated preference experiment in the case of ATIS route choice contexts by using both tools; v) Developing a data fusion procedure for data TSs and matching DSs surveys. In particular, the focus of the paper is related to point (ii) and the experiment is carried out using a driving simulator. The story board experiment may be summarized in the beginning step and ex ante questionnaire, during which sociodemographic information and some perception indicators are collected regarding the respondents in the experiment, the intermediate step during which respondents may reproduce their choices in a route choice driving simulation environment, and the ending step during which users are requested to answer to an ex-post questionnaire in order to collect some further information on the scenario perceptions (these are also used for scenario validations) and some information on the reason for their choices. Data are analyzed through aggregate and statistical approaches, and by modelling approaches (the Structural equation model is applied). Based on preliminary investigation, two endogenous latent variables are considered: the utility, related to the observed switching behaviors and the attitudes to the risk aversion related to some perception indicators (obtained by the declaration collected during the ex-ante questionnaire). The utility function is explained by compliance, gender, and job exogenous attributes, whereas the risk aversion is explained by classes of age and trusting information declared by users during the ex-ante questionnaire. Finally, as expected the modelling approach has shown the "effect" of risk aversion on utility function estimation. The main fields of investigation for further research will be finalized halfway through point (ii) and when developing the data fusion procedure as described in step (iii) of the research project.

References

1. Adler, J.L.: Investigating the learning effects of route guidance and traffic advisories on route choice behaviour. Transp. Res. Part C **9**(1), 1–14 (2001)
2. Avineri, E., Prashker, J. N.: The impact of travel time information on travellers' learning under uncertainty. In: 10th International on Travel Behaviour Research, 10–15. August 2003 in Luzern Swiss (2003)
3. Baldoni, F., Galante, F., Pernetti, M., Russo, M., Terzo, M., Toscano, M.: Tuning and objective performance evaluation of a driving simulator to investigate tyre behaviour in on-center handling manoeuvres. Veh. Syst. Dyn. **49**, 1423–1440 (2011)
4. Bekhor, S., Albert, G.: Accounting for sensation seeking in route choice behavior with travel time information. Transp. Res. Part F: Traffic Psychol. Behav. **22**, 39–49 (2014)
5. Ben-Elia, E., Di Pace, R., Bifulco, G.N., Shiftan, Y.: The Impact of travel information's accuracy on route-choice. Transp. Res. Part C: Emerg. Technol. **26**, 146–159 (2013)
6. Bifulco, G.N., Di Pace, R., Viti, F.: Evaluating the effects of information reliability on travellers' route choice. Eur. Transp. Res. Rev. **6**(1), 61–70 (2014)
7. Bifulco, G.N., Simonelli, F., Di Pace, R.: Endogenous driver compliance and network performances under ATIS. In: Proceedings of IEEE Intelligent Transportation Systems Conference (ITSC'07), Seattle, Washington (USA), pp. 1028–1033 (2007)
8. Bifulco, G.N., Simonelli, F., Di Pace, R.: The role of the uncertainty in ATIS applications. In: Application of Soft Computing – Updating the State of Art, vol. 52, pp. 230–239 (2009)
9. Birrell, S.A., Young, M.S.: The impact of smart driving aids on driving performance and driver distraction. Transp. Res. Part F: Traffic Psychol. Behav. **14**(6), 484–493 (2011)
10. Bonsall, P., Firmin, P., Anderson, M., Palmer, I., Balmforth, P.: Validating the results of a route choice simulator. Transp. Res. Part C **5**(6), 371–387 (1997)
11. Calise, F., Cappiello, F.L., Cartenì, A., Dentice d'Accadia, M., Vicidomini, M.: A novel paradigm for a sustainable mobility based on electric vehicles, photovoltaic panels and electric energy storage systems: case studies for Naples and Salerno (Italy). Renew. Sustain. Energy Rev. **111**, 97–114 (2020)
12. Cartenì, A.: Urban sustainable mobility. Part 2: Simulation models and impacts estimation. Transp. Probl. **10**(1), 5–16 (2015)
13. Cartenì, A.: The acceptability value of autonomous vehicles: A quantitative analysis of the willingness to pay for shared autonomous vehicles (SAVs) mobility services. Transp. Res. Interdiscip. Perspect. **8**, 100224 (2020)
14. Chang, H.L., Chen, P.C.: Impact of uncertain travel information on drivers' route choice behaviour. In: Transportation Research Board 88th Annual Meeting, Washington, D.C., USA, January 2009
15. de Palma, A., Ben-Akiva, M., Brownstone, D., Holt, C., Magnac, T., McFadden, D., Moffatt, P., Picard, N., Train, K., Wakker, P., Walker, J.: Risk, uncertainty and discrete choice models. Mark. Lett. **19**, 269–285 (2008)
16. Di Pace, R., Galante, F., Pariota, L., Bifulco, G.N., Pernetti, M.: Collecting data in ATIS context: travel simulator platform vs. route choice driving simulator. In: Proceedings of the 90th TRB Annual Meeting, Washington U.S.A., 23–27 January 2011, pp. 1–16 (2011)

17. Di Pace, R.: Analytical Tools for ATIS (Strumenti Analitici per Applicazioni ATIS), Ph.D. thesis, Università degli Studi di Napoli "Federico II"- Facoltà di Ingegneria (2008). https://www.fedoa.unina.it/view/people/Di_Pace,_Roberta.html
18. Papinski, D., Scott, D.M., Doherty, S.T.: Exploring the route choice decision-making process: a comparison of planned and observed routes obtained using person-based GPS. Transp. Res. Part F 12, 347–358 (2009)
19. Galante, F., Mauriello, F., Montella, A., Pernetti, M., Aria, M., D'Ambrosio, A.: Traffic calming along rural highways crossing small urban communities: driving simulator experiment. Accid. Anal. Prev. 42(6), 1585–1594 (2011)
20. Golob, T.F.: Structural equation modeling for travel behavior research. Transp. Res. Part B: Methodol. 37(1), 1–25 (2003)
21. Henke, I., Cartenì, A., Molitierno, C., Errico, A.: Decision-making in the transport sector: a sustainable evaluation method for road infrastructure. Sustainability 12(3), 764 (2020)
22. Henke, I., Cartenì, A., Francesco, L.D.: A Sustainable evaluation processes for investments in the transport sector: a combined multi-criteria and cost-benefit analysis for a new highway in Italy. Sustainability 12(23), 9854 (2020)
23. Jamson, A.H., Merat, N.: Surrogate in-vehicle information systems and driver behaviour: effects of visual and cognitive load in simulated rural driving. Transp. Res. Part F: Traffic Psychol. Behav. 8(2), 79–96 (2005)
24. Jöreskog, K.G., Sörbom, D.: LISREL 8 User's Reference Guide. Lincolnwood, Scientific Software International (1999b)
25. Jöreskog, K.G., Sörbom, D., Du Toit, S., Du Toit, M.: LISREL 8: New StatisticalFeatures. Third printing with revisions. Scientific Software International, Lincolnwood (2003)
26. Kaplan, D.: Structural Equation Modeling: Foundations and Extensions. Sage Publications, Newbury Park (2000)
27. Katsikopoulos, K.V., Duse-Anthony, Y., Fisher, D.L., Duffy, S.A.: The framing of drivers' route choice when travel time information is provided under varying degrees of cognitive load. Hum. Factors Ergon. Soc. 42(3), 470–481 (2000)
28. Katsikopoulos, K.V., Duse-Anthony, Y., Fisher, D.L., Duffy, S.: A risk attitude reversals in drivers' route choice when range of travel time information is provided. Hum. Factors Ergon. Soc. 43(3), 466–473 (2002)
29. Lapietra, M., Pronello, C.: Experimental results from the test on the new Stated Preferences valuation method, the Continuous Attribute-Based Stated Choice, in the web based surveys. In: The 8th International Conference on Survey Methods in Transport, Annecy, France, 25–31 May 2008 (2008)
30. McFadden, D.: The choice theory approach to market research. Mark. Sci. 5(4), 275–297 (1986)
31. de Dios Ortúzar, J., Willumsen, L.G.: Modelling Transport, 4th edn. Wiley, Publication Prentice Hall, Englewood Cliffs (1988)
32. Paz, A., Peeta, S.: Information-based network control strategies consistent with estimated driver behavior. Transp. Res. Part B: Methodol. 43(1), 73–96 (2009)
33. Peeta, S., Paz, A., DeLaurentis, D.: Stated preference analysis of a new very light jet based on-demand air service. Transp. Res. Part A: Policy Pract. 42(4), 629–645 (2008)
34. Peeta, S., Yu, J.W.: Adaptability of a hybrid route choice model to incorporating driver behavior dynamics under information provision. IEEE Trans. Syst. Man Cybern.-Part A: Syst. Hum. 34(2), 243–256 (2004)
35. Peeta, S., Yu, J.W.: A hybrid model for driver route choice incorporating en-route attributes and real-time information effects. Netw. Spat. Econ. 5(1), 21–40 (2005)
36. Ranney, T.A., Baldwin, G.H., Smith, L.A., Martin, J., Mazzae, E.N.: Driver behavior during visual-manual secondary task performance: occlusion method versus simulated driving (No. DOT HS 811 726) (2013)

37. Song, D., Peeta, S., Hsu, Y.T.: Psychological effects of real-time travel information on traveler route choice decision-making process (No. 17–01162). (2017).
38. Tian, H., Gao, S., Fisher, D. L., Post, B.: A mixed-logit latent-class model of strategic route choice behavior with real-time information. In: Transportation Research Board 91th Annual Meeting, Washington, D.C., USA, January 2012.
39. Tian, H., Gao, S., Fisher, D.L., Post, B.: Route choice behavior in a driving simulator with real-time information. pp. Transportation Research Board 90th Annual Meeting, Washington, D.C., USA, January 2011
40. Tsirimpa, A., Polydoropoulou, A., Antoniou, C.: Development of a latent variable model to capture the impact of risk aversion on travelers' switching behavior. J. Choice Model. **3**(1), 127–148
41. Tversky, A., Kahneman, D.: Judgment under uncertainty: heuristics and biases. Science **185**, 1124–1130 (1974)
42. Yu, J.W., Peeta, S.: Experimental analysis of a hybrid route choice model to capture dynamic behavioral phenomena under advanced information systems. KSCE J. Civil Eng. **15**(1), 175–185 (2011)
43. Zhang, L., Levinson, D.: Determinants of route choice and value of traveler information: a field experiment. Transp. Res. Rec. **2086**(1), 81–92 (2008)

A Smart Road Application: The A2 Mediterranean Highway Project in Italy

Ilaria Henke[1(✉)], Gennaro Nicola Bifulco[1], Armando Carteni[2], Luigi Di Francesco[2], and Antonio Di Stasio[2]

[1] Department of Civil, Construction and Environmental Engineering, University of Naples "Federico II", 80125 Napoli, Italy
{ilaria.henke,gennaronicola.bifulco}@unina.it
[2] Department of Engineering, University of Campania "L. Vanvitelli", 81031 Aversa, Italy
{armando.carteni,luigi.difrancesco}@unicampania.it,
antonio.distasio@studenti.unicampania.it

Abstract. Today technological innovation and digital transformation represent a concrete challenge for the transport sector, toward high quality and sustainable infrastructures and services. Within this issue, road mobility is undergoing a digital transformation based on both automation and connectivity, which promises to improve road safety, energy consumption and driving comfort, through vehicles to vehicles and vehicles to infrastructure communication and interaction (devices). The process of digital transformation for road infrastructures (and services) involves many aspects which concerns not only technological problems but also regulatory and legal aspects, both for the construction of new infrastructures (or their revamping) and for the management. Starting from these considerations, the aim of the paper was twofold: i) critically analyze the current state of practice of the Italian smart roads legislation; ii) describe the first smart road pilot project in south of Italy, the A2 Mediterranean Highway, in term of both smart technologies considered and corresponding investment costs.

Keywords: Smart road · Digital transformation · New technologies · Cost-benefit analyses · Transportation planning · Sustainable mobility

1 Introduction

In recent years, technological innovation and digital transformation represent a concrete challenge for the transport sector, toward high quality and sustainable infrastructures and services (e.g. [1]). Road mobility is undergoing a digital transformation based on both automation and connectivity, which promises to improve road safety, energy consumption and driving comfort, through vehicles to vehicles and vehicles to infrastructure communication and interaction (devices) (e.g. [2, 3]), thus aiming at the design of so-called Smart roads. Smart roads could be defined as "*roads, coupled with digital information, able to provide capabilities for advanced applications such as traffic monitoring, real-time data delivery to drivers, improved safety conditions, or support to the development of connected and autonomous vehicles (CAVs), among others*" [4]. Smart roads are configured as the set of road infrastructures that aim at a more efficient and effective management of the infrastructures improving benefits both for the users (e.g. [5–7]) and for the infrastructure operators (e.g. [8]). The focus of smart road is the driver and his safety, enabling real-time information, for example, about alternative paths, environmental and driving conditions. To achieve these aims, more than still talking the Intelligent Transport System (ITS), it is more appropriate to talking about the Cooperative Intelligent Transport System (C-ITS) (e.g. [9–11]). C-ITS allows connectivity of driver with the IT (Information Technology) tools used by them (for example mobile tools) and connectivity of vehicles traveling on roads. The connectivity types of vehicles traveling on roads can be classified into: *i)* V2V (vehicle to vehicle) connectivity, i.e. the transfer of information between two and more vehicles traveling on the same road; *ii)* V2I (vehicle to infrastructure) connectivity, i.e. the transfer of information from a vehicle to road infrastructure on which circulating*; iii)* connectivity of I2V infrastructures (Infrastructure to vehicle), i.e. the transfer of information from a road infrastructure to the vehicles traveling on it. The European Commission in the White Paper of the 2011 identifies the development of the Cooperative Intelligent Transportation System as a valid tool for achieving the goals of zero deaths on the road and zero impact on the climate by 2050 [12]. The performances of digital transformation of infrastructures are strongly influenced by the technological development of vehicles, currently equipped with an increasing number of driving assistance systems and which soon will evolve towards automatic driving (e.g. [13]). The process of digital transformation for road infrastructures (and services) involves many aspects which concerns not only technological problems but also regulatory and legal aspects, both for the construction of new infrastructures (or their revamping) and for the management (security, use of data, protection of privacy, responsibility of the various actors involved, use of the radio spectrum). The steps have to be taken in this context are many and at a European level different directive have been enhanced in order to define these aspects and encourage the development of C-ITS technologies on road infrastructures. Precisely, in 2010, the Directive 2010/40 /EU defines the ITS Action Plan with the aim of promoting the development of ITS technologies in Europe. In this Directive, the priority actions are defined, such as: a) the organization throughout the European Union of information services on multimodal mobility; b) the organization throughout the European Union of traffic information services in real time; c) data and procedures for free communication to users, where possible, of universal minimum information on traffic related to road safety; d) the harmonized organization throughout the European Union of an interoperable electronic emergency call service (eCall); e) the organization of information services for safe parking areas for heavy goods vehicles and commercial vehicles; f) the organization of reservation services for secure parking areas for heavy vehicles and commercial vehicles [14]. In addition to this EU directive, different delegated regulations are issued by the European Commission, such as: i) Delegated Regulation (EU) No. 305/2013 concerning the harmonized preparation throughout the territory of European Union of an interoperable electronic emergency call (eCall) service; ii) Delegated Regulation (EU) No. 885/2013, regarding the provision of information services on safe parking areas for heavy vehicles and commercial vehicles; iii) Delegated Regulation (EU) No. 886/2013 concerning data and procedures for the free communication to users, where possible, of universal minimum information on traffic related to road safety; iv) Delegated Regulation (EU) No. 962/2015 relating to the provision throughout the European Union of real-time traffic information services; v) Delegated Regulation (EU) No. 1926/2017 concerning the provision throughout the European Union of information services on multimodal mobility. In this context, pilot projects have been proposed in Europe for the construction and/or test of smart roads facilities. Among these it is possible to mention the pilot project C-ROADS in Austria, launched in February 2016 and concluded becoming operational in 2019 which affects the motorways connecting Vienna and Salzburg, the Brenner corridor and the surroundings of Graz. The project consists of equipping 300 km of highways with C-ITS [15]. In Belgium/Flanders, the C-Roads has been designed with the main objective of operating and evaluating the implementation of a cloud-based "virtual infrastructure" for an effective distribution of C-ITS services that connect road users with the Traffic Management Center (TMC), allowing the TMC to interact directly with end users. This pilot project also offers the opportunity to upgrade traffic information services and traffic management services. It, started in 2018, ended and became operational in 2020 [16]. With respect to the Italian context, in 2016 the government identified in the digital transformation of the road infrastructures one of the main aims for transport infrastructures development [17]. Furthermore, this action allows to pursue also one of the main aims of the national transportation planning based on the useful, streamlined, and shared development of infrastructures, in addition to the enhancing of the existing infrastructural heritage. In this context, countless pilot projects have been promoted in Italy within the most strategic national road axes: the "E45-E55 Orte-Mestre" route, the "Alemagna road 51" in Veneto region, the "Catania ring road" and the "A19 Palermo-Catania" highway in Sicily region, the "A90 highways", the "Rome ring road", the "A91 Rome-Fiumicino" airport motorway in Lazio region, the A2 Mediterranean Highway [18]. From a regulatory point of view, Italy is lagging behind the European directives, and recently has issued different of documents (discussed in the Sect. 2.1) in order to promote and regulate the development of smart roads in the country. The national road transport operator in Italy, ANAS S.p.A., in 2018, promotes the "Smart Road program" for a total investment program of about 1 billion euros with an initial funding of about 250 million euros, also thanks to European contributions (as part of the PON Infrastructure and Networks Operational Program 2014–2020 of the Italian Ministry of Infrastructure and Transport) starting with a project on the A2 Mediterranean Highway. Starting from these considerations, the aim of the paper was twofold: *i)* critically analyze the current state of practice of the Italian smart roads legislation; *ii)* describe the first smart road pilot project in south of Italy, the A2 Mediterranean Highway, in term of both the main technologies considered and the corresponding implementation costs. The paper is structured into three sections: Sect. 2 illustrates the current state of practice of the Italian smart roads' legislation; Sect. 3 describe the A2 Mediterranean Highway smart road project; Sect. 4 reports the main conclusions and the research perspectives.

2 The Current State of Practice of the Italian Smart Roads' Legislation

The reference regulatory framework is extremely recent, in fact, despite the fact that ITS were already included in the General Transport and Logistics Plan in 2001 as a key measure for the implementation of the sustainable mobility goals (e.g. [19]), it was only in 2012 (Law no.221 of 2012),11 years later, that Directive 2010/40/ EU is implemented within the Italian regulatory paradigm. The EU Directive, as illustrated in paragraph 1, defines the priority actions to be taken throughout the territory of the European Union. In 2013 with the Decree for the Diffusion of Intelligent Transport Systems in Italy, the requirements for the dissemination, design and implementation of ITS were defined. In 2014, with Decree 446, the Ministry of Infrastructure and Transport was identified as the national responsible for the adoption of the National Plan for the development of ITS systems. Despite what is defined by current legislation, the objectives and targets set (such as the formulation of data on infrastructures and traffic flows as a free open-data source or the Public Index of information on infrastructures and traffic) have not yet been achieved. In 2016 in the preliminary study *"Functional Standards"* for Smart Roads [20] the performance and technical-functional aspect of smart roads are specified.

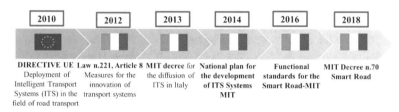

Fig. 1. The main stages of the Italian regulatory framework regarding Smart Roads

In 2018, Ministerial Decree No. 70 establishes that the digital transformation process is initially applied to the road infrastructures of the TEN-T, core and comprehensive, as well as to new connecting infrastructures between elements of the TEN-T network, and subsequently to all infrastructures belonging to the INT (Integrated National System). Furthermore, the minimum functional standards of smart roads are defined (not entering into the merits of specific technologies) and regulates the experimentation of automatic and connected vehicles on public roads (Fig. 1). According to Ministerial Decree 70/2018, Smart Roads are classified in to two types: "Type I Smart Road" and "Type II Smart Road". Type I Smart Roads belong to the Infrastructures belonging to the TEN-T network, core and comprehensive, and, in any case, the entire motorway network where functional standards are envisaged that allow both the monitoring of infrastructures with communications between vehicles and infrastructure and communication between vehicles. Type II Smart Roads are the infrastructures belonging to the INT (Integrated National System) not classified as type I smart roads where communication systems between vehicles and infrastructures are envisaged for monitoring the infrastructures.

3   The A2 Mediterranean Highway Smart Road Project

The A2 highway, also known as the "Mediterranean Highway" is a highway located in south of Italy that involves the Campania, Basilicata and Calabria regions (Fig. 2) with a total extension of 442.2 km, an integral part of the European Helsinki-Valletta corridor. The history for the build of this highway was long and tortuous, which began on December 14, 1961 when the Board of Directors of ANAS, chaired by the Minister of Public Works, approves the preliminary design of the Mediterranean Highway. On 21 January 1962, in the presence of the President of the Council of Ministers, the works began, declaring that they will be completed within two years, they were completed in 1964. In 1972 a landslide near Lagonegro damaged the Taggine and Sirino viaducts. This caused the closure of part of the motorway and the opening of a detour (a new street diversion, a bottleneck between km 131 and km 132), which will last for over 40 years. Between 1996 and 1997, the government approved a modernization program based on the traffic flow examination, the motorway envisages the construction of three lanes in each direction only for the first 53 km (from Salerno to Sicignano degli Alburni), while the rest of the motorway will be adapted and made safe, with the elimination of dangerous curves, the reduction of slopes and, where necessary, the construction of new tunnels and the demolition of viaducts. The completion of the works took place on 22 December 2016, opening at traffic more than one year in advance of the project delivery date scheduled for 2017. In 2018 the Italian government in agreement with ANAS has decided to improve this highway developing the first smart road project in Italy. Thus, a pilot project was launched, where a total investment of € 100 million is expected, of which € 21 million allocated to MIT's PON and Networks 2014–2020 [21]. In 2016, a call for tenders was launched for the purpose of ASR 35/2016, concerned about the supply and installation of systems and workstations for the implementation of the advanced Smart Road technological infrastructure, for the connectivity of ANAS users and operators on the A2 and on the "R.A.2 Salerno – Avellino" motorway junction for a total amount of € 20 million. In 2018, work began which aims, through the application of digital transformation processes, to increase road safety aimed above all at reducing accidents, greater travel comfort, management and the improvement in ordinary traffic conditions and extraordinary critical events, the resilience of transport networks, the digital interaction with the territories crossed, together with the interoperability with connected vehicles, until to reach the gradual circulation of self-driving vehicles. Consistently with Ministerial Decree 70/2018 and with the "position paper" of the Ministry of Infrastructures and Transport, the Mediterranean Highway smart road will perfume the aims of: i) a safe travel: roads with adequate levels of maintenance and driving support solutions also to the future implementation of autonomous driving (e.g. [13]); ii) an efficient infrastructure: increasing the operating factors of the infrastructure, promptly intervening in emergencies by providing a preferential channel and rapid management of the emergency and rescue; iii) an effective infrastructure: traffic monitoring and management of growing traffic volumes with dynamic lanes, vehicle-infrastructure communication (V2I). To meet this aim, ANAS in a pre-feasibility study [22] has decided to adopt a series of technologies to make the highway a Smart Road. Among these, the main ones are described below. A data communication network with a high bit-rate optical fiber has been arranged to guarantee the connectivity of users and IT tools as well as that between vehicles and infrastructures.

The management and control in real time of the conditions of the infrastructures will take place through an IoT (Internet of Things) technology. IoT sensors, intelligent

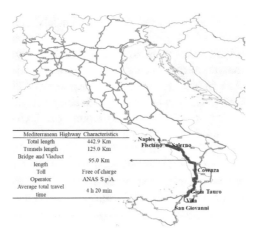

Mediterranean Highway Characteristics	
Total length	442.9 Km
Tunnels length	125.0 Km
Bridge and Viaduct length	95.0 Km
Toll	Free of charge
Operator	ANAS S.p.A.
Average total travel time	4 h 20 min

Fig. 2. The application case study: the A2 Mediterranean Highway (source: processing starting from https://www.stradeanas.it/it)

devices, will be set up along the infrastructure, capable of providing information to the system controller on the status of the main infrastructures (road surface; road barriers, Guard Rail, New Jersey and Mobile Barriers, Bridges/Viaducts; tunnels) climatic conditions and operating conditions concerning the traffic and transport of goods. In fact, through the use of sensors for the road surface and with the installation of multifunctional stations, it is possible to communicate to the users who travel the road information on weather conditions, possible queues, presence of construction sites, etc. Furthermore, a Smart Camera will be installed along the Smart Road, which will be managed and controlled by a specialized software, which will allow to detect: dangerous conditions (such as a stationary vehicle, vehicle moving in the opposite direction of travel, presence of pedestrians, presence of debris on the roadway); traffic conditions in real time (average speed, traffic volume, traffic density, lane occupation [%], vehicle length [m]); vehicle characteristics (it will be able to recognize vehicle license plates and therefore recognize police vehicles, test plates, etc. as well as automatically detect the orange ADR dangerous goods warning panels with decoding of the Kemler and UN Codes). Through the use of the smart camera, by sending the video signal for video analysis it will be possible to analyze the road runoff in the medium and short term. The Smart road will be equipped with adaptive traffic management systems capable of dynamically creating the best possible outflow conditions and implementing the actions and traffic control plans envisaged at a strategic level. The information concern about the deviation of vehicle flows on alternative routes, with a view to suggesting the optimal path, in the event of traffic events or adverse weather situations based on the results of the estimation / forecast models [23–26]. The information to the vehicles will be provided through PMV (Variable Message Panel) or through On Board Unit (OBU) (devices installed on vehicles). To ensure communication between vehicles (V2V) and between vehicles and infrastructure (V2I), the highway will be prepared for a Wi-Fi in motion system, used for C-ITS applications. These systems will allow the connection of the user's mobile device in order to receive the appropriate information ensuring the continuous exchange of information to and from the customer, in complete safety and without distractions. Other functions, usable from the Wi-Fi network, will be allowed only when the vehicle is stationary. The Wi-Fi in motion technology allows a continuity of the signal to moving vehicles even at maximum speeds such as 150 km/h, with a large signal capacity designed for Mobile APP in continuous data exchange even in hands-free mode, with fast and safe "Handover", such as to keep communication always active along the road section when passing from one "access point" to another. In addition, Road-side unit and radio modules design on industrial WLAN technology for V2I communication, will be located in a way to allow connection to vehicles equipped with On Board Unit (OBU), that are devices installed on board-unit v2x vehicles that meet the industry standard [22]. Starting from the technological systems described above, the main infrastructure characteristics (e.g. number of tunnels, road length) of the highway and the list of unit price of the considered technologies [27], it was possible to estimate the investment cost for the smart road (Table 1).

Table 1. The construction cost of Smart Road Mediterranean Highway

Costs	Economic value*km (thousand euro/km)	Total High Scenario (Mln euro)	Total Low Scenario (Mln euro)
High bit-rate communication network	€ 11.8	€ 5.2	€ 3.6
IoT technology	€ 62.3	€ 27.6	€ 20.0
Wi-Fi in motion	€ 3.0	€ 1.3	€ 1.0
Road monitoring system	€ 14.0	€ 6.2	€ 6.2
Total Investment costs for one roadway	€ 91.1	€ 40.3	€ 30.8
Total Investment costs for two roadways	€ 182.2	€ 80.6	€ 61.6
Other costs		€ 15.7	€ 12.0
Total investment costs		**€ 96.3**	**€ 73.6**

Precisely, two scenarios were considered: an "*high scenario*", which provides the use of the total amount of technological systems considered, and a "*low scenario*", which provides only some of the technological systems starting from the ones already implemented on the highway (e.g. optical fiber and Wi-Fi systems). For these two scenarios, the investment cost was estimated taking into account both the devices costs and their installation ones. The overall investment cost range between 62 to 81 million euros. In addition to the investment cost, the following were also considered: i) unexpected costs equal to 6.5% of the investment cost for dual carriageways; ii) general expenditure costs equal to 13% of the total investment cost in line with the provisions of the new procurement code (D.L. 18.04.2017, n. 56). General expenses take into account, for example, contract costs and related taxes, construction costs, costs of deposits. In total it is estimated that the cost of carrying out the work varies between 74–96 million euros. The estimation of the overall investment costs does not mean that this technological upgrade should be implemented. Against these costs, the benefits that this investment will produce for users and non-users (externalities) will be estimated to evaluate if the overall balance between costs and benefits is positive for the community and for the environment. Among the quantitative techniques commonly used for these evaluations' analyses [28–34] there are both cost benefit (e.g. [35–38]) and multi-criteria analyses (e.g. [39, 40]). Recently, the Italian government approved the "Guidelines for assessment of Investment Projects" that, jointly with the EU guidelines, defines the methodologies to develop both ex-ante and ex-post evaluation analyses, and these could be applied also for smart road applications. Actually, no evaluation analysis has yet been conducted in Italy on a smart road, also due to the difficulty of estimating the benefits produced by these new technologies, which are probably significantly different (e.g. reduction in traffic accidents, increased in driving comfort), to those commonly more relevant, for example, for a new road/railway (e.g. average time saved, reduction in greenhouse gases emission), and this will be a future challenge for research on these issues.

4   Conclusions and Developments

Technological innovation and digital transformation will increasingly represent an enabling factor capable of helping to create modern, lean, and quality infrastructures offering new services to citizens, greater effectiveness and efficiency for the transport of passengers. Smart Roads therefore represent a technological development of the transport sector through Smart Mobility, ITS Systems – Intelligent Transport System, CAV – Connected and Automated Vehicles. The process of digital transformation for road infrastructures (and services) involves many aspects which concerns not only technological problems but also regulatory and legal aspects, both for the construction of new infrastructures (or their revamping) and for the management (security, use of data, protection of privacy, responsibility of the various actors involved, use of the radio spectrum). Furthermore, many countries are currently developing smart technologies for road infrastructures and vehicles which, in the near future, will evolve towards automatic driving. Usually, in the evaluation analyses (e.g. Cost-Benefit analysis) related to traditional transport projects (e.g. a new road or transport service), users impact (i.e. mainly generalized transport cost variation) represents about the 80–90% of the total expected benefits. By contrast, investment in smart roads suggest that this kind of intervention will be able to produce benefits different from those of traditional infrastructures, that are more relevant for externalities (impacts for non-users) and less significant for users. These technologies promise to reduce road accidents, fuel consumption, greenhouse gas emissions and increase driving quality and passenger comfort, rather than reducing vehicles' travel times. This may lay the ground for a new season for the evaluation analysis, pointing out the main role of externalities. Within this issue, no one smart road evaluation analysis has been still conducted in Italy, probably due to the difficulty in defining and estimating the benefits produced by these new technologies. This paper has tried to partially fill this gap, analyzing the first smart road pilot project in Italy in term of both smart technologies considered and corresponding investment costs estimation. Among the research perspectives there is the definition and the successive estimation of the main benefits produced by a smart road project, towards the non-users (externalities), the users and the transport operator that manage the read, and aiming in performing the first smart road cost benefits analysis in Italy.

References

1. Shaheen, S.A., Chan, N.: Mobility and the sharing economy: potential to facilitate the first- and last-mile public transit connections Built Environ. **42**, 573–588 (2016)
2. Botte, M., Pariota, L., D'Acierno, L., Bifulco, G.N.: An overview of cooperative driving in the European Union: policies and practices. Electronics **8**(6), 616 (2019)
3. Kuenzel, R., Teizer, J., Mueller, M., Blickle, A.: SmartSite: intelligent and autonomous environments, machinery, and processes to realize smart road construction projects. Autom. Constr. **71**, 21–33 (2016)
4. Barazzetti, L., Previtali, M., Scaioni, M.: Roads detection and parametrization in integrated BIM-GIS using LiDAR. Infrastructures **5**(7), 55 (2020)
5. Papageorgiou, M., Diakaki, C., Dinopoulou, V., Kotsialos, A., Wang, Y.: Review of road traffic control strategies. Proc. IEEE **91**(12), 2043–2067 (2003)
6. Wiegand, G.: Benefits and challenges of smart highways for the user. In: IUI Workshops (2019)
7. Shapiro, J.M.: Smart cities: explaining the relationship between city growth and human capital (2003). Available at SSRN 480172
8. Mangiaracina, R., Tumino, A., Miragliotta, G., Salvadori, G., Perego, A.: Smart parking management in a smart city: costs and benefits. In: 2017 IEEE International Conference on Service Operations and Logistics, and Informatics (SOLI), pp. 27–32. IEEE (2017)
9. Botte, M., D'Acierno, L., Bifulco, G.N.: C-ITS communication: an insight on the current research activities in the European Union. Int. J. Transp. Syst. 3 (2018)
10. Dameri, R.P., Ricciardi, F.: Leveraging smart city projects for benefitting citizens: the role of ICTs, pp. 111–128. Springer, Cham, In Smart City Networks (2017)
11. Toppeta, D.: The smart city vision: how innovation and ICT can build smart, "livable", sustainable cities. Innov. Knowl. Found. **5**, 1–9 (2010)
12. European Commission: WHITE PAPER Roadmap to a Single European Transport Area – Towards a competitive and resource efficient transport system (2011)
13. Cartenì, A.: The acceptability value of autonomous vehicles: a quantitative analysis of the willingness to pay for shared autonomous vehicles (SAVs) mobility services, vol. 8. Transportation Research Interdisciplinary Perspectives (2020a). 100224. ISSN 2590–1982
14. EP and CoEU: Directive 2010/40/Eu of 7 July 2010 on the framework for the deployment of Intelligent Transport Systems in the field of road transport and for interfaces with other modes of transport (2010)
15. Froetscher, A., Monschiebl, B.: C-roads: elements of c-its service evaluation to reach interoperability in europe within a wide stakeholder network: Validation steps and comparative elements used in a living lab environment in austria. In: 2018 IEEE 87th Vehicular Technology Conference (VTC Spring), pp. 1–5. IEEE (2018)
16. C-Roads: Annual pilot overview report 2018 (2019). https://www.c-roads.eu/
17. Ministero dell'economia e delle finanze: Documento di economia e Finanza 2017. Allegato - Connettere l'Italia: fabbisogni e progetti di infrastrutture (2017)
18. ANAS: SMART ROAD: Anas porta l'Italia verso la mobilità del futuro (2021). https://www.stradeanas.it/it/smart-road-anas-porta-l-italia-verso-la-mobilita-del-futuro
19. Calise F.; Cappiello F.L., Cartenì, A., Dentice d'Accadia, M., Vicidomini, M.: A novel paradigm for a sustainable mobility based on electric vehicles, photovoltaic panels and electric energy storage systems: case studies for Naples and Salerno (Italy). Renew. Sustain. Energy Rev. **111**, 97–114 (2019)
20. Ministero Delle Infrastrutture e dei Trasporti: Standard funzionali per le Smart-Road – position paper (2016). https://www.mit.gov.it/

21. Ministero dell'economia e delle finanze: Documento di economia e Finanza 2018 (2018). https://www.rgs.mef.gov.it/
22. ANAS: SMART ROAD - La strada all'avanguardia che corre con il progresso (2018). https://www.stradeanas.it/
23. De Luca, S., Papola, A.: Evaluation of travel demand management policies in the urban area of Naples. WIT Trans. Built Environ. 52 (2001)
24. De Luca, S., Di Pace, R.: Modelling the propensity in adhering to a carsharing system: a behavioral approach. Transp. Res. Procedia **3**, 866–875 (2014)
25. Di Pace, R., Marinelli, M., Bifulco, G.N., Dell, M.: Modeling risk perception in ATIS context through fuzzy logic. Procedia-Soc. Behav. Sci. **20**, 916–926 (2011)
26. Bifulco, G.N., Simonelli, F., Di Pace, R.: Endogenous driver compliance and network performances under ATIS. In: 2007 IEEE Intelligent Transportation Systems Conference, pp. 1028–1033 (2007)
27. ANAS: Elenco prezzi 2017 – Impianti tecnologici (2017). https://www.stradeanas.it/
28. Cartenì, A., Henke, I.: The evaluation of public investments according to the cost-benefit analysis: an application to the formia-gaeta railway line [La valutazione degli investimenti in opere pubbliche attraverso l'analisi costi-benefici: Un'applicazione alla riqualificazione della linea ferroviaria formia-gaeta]. Ingegneria Ferroviaria **74**(9), 651–681 (2019)
29. Cartenì, A., Henke, I., Di Bartolomeo, M.I., Regna, M.: A cost-benefit analysis of a fully-automated driverless metro line in a high-density metropolitan area in Italy. In: Proceedings – 2019 IEEE International Conference on Environment and Electrical Engineering and 2019 IEEE Industrial and Commercial Power Systems Europe, EEEIC/I and CPS Europe (2019), 11–14 June Genova, Italy
30. Gallo, M., D'Acierno, L., Montella, B.: A multimodal approach to bus frequency design. WIT Trans. Built Environ. **116**, 193–204 (2011). https://doi.org/10.2495/UT110171
31. Botte, M., D'Acierno, L.: Dispatching and rescheduling tasks and their interactions with travel demand and the energy domain: models and algorithms. Urban Rail Transit **4**(4), 163–197 (2018)
32. D'Acierno, L., Placido, A., Botte, M., Montella, B.: A methodological approach for managing rail disruptions with different perspectives. Int. J. Math. Models Methods Appl. Sci. **10**, 80–86 (2016). https://naun.org/cms.action?id=12152
33. D'Acierno, L., Gallo, M., Montella, B., Placido, A.: Analysis of the interaction between travel demand and rail capacity constraints. WIT Trans. Built Environ. **128**, 197–207 (2012). https://doi.org/10.2495/UT120181
34. D'Acierno, L., Botte, M.: A passenger-oriented optimization model for implementing energy-saving strategies in railway contexts. Energies **11**(11), 1–25 (2018). https://doi.org/10.3390/en11112946
35. Carteni', A.: A cost-benefit analysis based on the carbon footprint derived from plug-in hybrid electric buses for urban public transport services. WSEAS Trans. Environ. Dev. **14**, 125–135 (2018). ISSN/E-ISSN: 1790–5079/2224–3496
36. Cartenì, A., De Guglielmo, M.L., Pascale, N.: Congested urban areas with high interactions between vehicular and pedestrian flows: a cost-benefit analysis for a sustainable transport policy in Naples, Italy. Open Transp. J. **12**(1), 273–288 (2018a)
37. Carteni', A., Henke, I., Molitierno, C.: A cost-benefit analysis of the metro line 1 in Naples, Italy. WSEAS Trans. Bus. Econ. **15**, 529–538 (2018b). Print ISSN: 1109–9526
38. Carteni, A., Henke, I.: External costs estimation in a cost-benefit analysis: the new Formia-Gaeta tourist railway line in Italy. In: Conference Proceedings - 2017 17th IEEE International Conference on Environment and Electrical Engineering and 2017 1st IEEE Industrial and Commercial Power Systems Europe, EEEIC/I and CPS Europe (2017). art. no. 7977614

39. Cartenì, A., Henke, I., Di Francesco, L.: A sustainable evaluation processes for investments in the transport sector: a combined multi-criteria and cost–benefit analysis for a new highway in Italy. Sustainability **12**(23), 9854, 1–27 (2020a)
40. Cartenì, A., Henke, I., Mallozzi, F., Molitierno, C.: A multi-criteria analysis as a rational evaluation process for building a new highway in Italy. WIT Trans. Ecol. Environ. **217**, 713–723 (2018c)

A Big Data Analysis and Visualization Pipeline for Green and Sustainable Mobility

Dario Branco[1], Beniamino Di Martino[1,2(✉)], and Salvatore Venticinque[1]

[1] Department of Engineering, University of Campania "Luigi Vanvitelli",
Aversa, Italy
{dario.branco,beniamino.dimartino,
salvatore.venticinque}@unicampania.it
[2] Department of Computer Science and Information Engineering, Asia University,
Taichung, Taiwan

Abstract. Big data analysis is a relevant activity to evaluate the impact of innovative technologies leveraging the digital transformation of smart cities. This paper illustrates the definition and realization of a Big Data Pipeline for data curation, analysis and visualization of key performance data and indicators (KPIs), for the evaluation of the impact of innovative technologies on Sustainable Mobility. It has been designed and developed to support the evaluation activities of the GreenCharge project, which provides cities with technological solutions and business models for effective implementation and management of charging infrastructures for electric vehicles. It is currently in operation in three GreenCharge pilots to estimate the impact of the operating technological solutions and business models.

1 Introduction

GreenCharge project aims to innovate cities with technological solutions and business models for cost-effective implementation and control of charging infrastructures for electric powered vehicles. Many solutions have been investigated in literature to leverage the transition to electromobility [5]. Among the other, GreenCharge will also investigate the exploitation of charging scheduling service that includes booking and billing for electric vehicles charging to optimize the utilization of available charging stations, to decrease the grid consumption and improve the user's charging experience. GreenCharge will enable cities and municipalities to make the transition to zero emission/sustainable mobility with innovative model of business [2], technology and instructions for efficient and successful deployment of the EV's charging infrastructure and to improve the efficiency of their operation. The guidelines will integrate the experience of projects pilot, simulations and advice on the location of charging points, on the reduction of investments in the grid and on political and public communication measures to accelerate the adoption of electromobility and will be aligned with the processes of the Sustainable Urban Mobility Plan (SUMP). GreenCharge will examine the effects of these innovations in pilot sites in Barcelona, Bremen and Oslo to evaluate the proposed approach. CIVITAS Evaluation Framework[1] will be the base for the GreenCharge evaluation methodology [7]. GreenCharge has adopted and customized the CIVITAS framework, applying it to e-mobility, using the methodology and procedures results from this project but also making a contribution to it. The core concept of the CIVITAS framework are key performance indicators (KPIs) and measures. A KPI is related to one or to a combination of parameters that permit a quantitative evaluation of the impact of the project on a particular component of transport [6]. The evaluation framework has naturally been adapted to the application context of e-mobility, paying attention to the evaluation technology and to the assessment of stakeholder acceptance.

In this paper, we describe the work performed and results achieved in defining and realizing a Big Data Pipeline for data curation, analysis and visualization of key performance data and indicators related to the evaluation of the impact of GreenCharge project measures on Green and Sustainable mobility in the context of the project pilots. Related work we have performed is related to the definition of an ontology and a semantic framework for evaluation of e-mobility actions [4] and to application of Big Data technology and Machine Learning to perform Tweets analysis in order to evaluate smart and sustainable urban mobility actions in Barcelona pilot site [3].

2   Design of the Big Data Pipeline

The conceptual model of the Big Data Pipeline devised for GreenCharge is shown in Fig. 1. It is composed of the following phases: Data Collection, Data Storage, Data Curation, Data Analysis and Data Visualization.

In Fig. 2 the *Component Diagram* of the Pipeline is illustrated, by which the implemented software components are identified, together with their composition and interfaces.

After the first phase of data collection, the data curation phase continuously checks the data format and consistency in a shared repository by a **datachecker** component. This operation is performed on each loaded and/or modified data by multiple rules according to the type of data to be checked. The other two components of the pipeline are the **KPI Calculator**, which carries out the actual computation of the analytics and the **KPI visualizer**, which instead reads the results of the computation and provides a numerical and/or graphical view of the calculated KPIs values.

In order to perform the calculation, the necessary data, taken from measurements made in the various pilot cities, but also the ones generated by simulated scenarios [1], are stored and organized (after anonymization) in an **Permanent File Storage**, with which the KPI calculator must be able to interoperate.

---

[1] https://civitas.eu/.

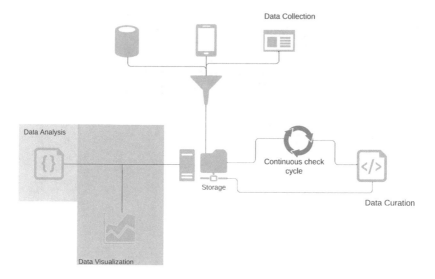

Fig. 1. Conceptual model of Big Data Pipeline Architecture

There was therefore the need not only to define a structure for organizing the files themselves, but also to define the format of the data themselves.

Types of data collected can be divided into four macro-areas: Devices, Individual Entities, Recordings and logs, Other Data.

2.1 DataChecker Component

The **Datachecker** operates in the third phase of the pipeline. As it was already mentioned, it deals with the verification of the data loaded in the **Permanent File Storage** and its task is to ensure compliance with the defined data format. The check is performed both on the file name and on the its content, which are specific of the kind of data. Furthermore, the data providers are notified via email when the software finds any inconsistency in the added or modified files. The software also provides a graphical interface that helps data providers and managers to have an complete overview of the amount of data loaded into the repository and how many of them are correct or incorrect. It also provides information and statistics on how much data has been loaded and correctness, for each data type and for each data provider, proving to be a very useful tool for the monitoring the population of the data lake. Precise indications are reported to the data providers regarding which checks failed and which ones completed successfully. Since the following three phases of data curation, data analysis and data visualization are closely related to each other, a component diagram shows their high-level relationships in Fig. 2.

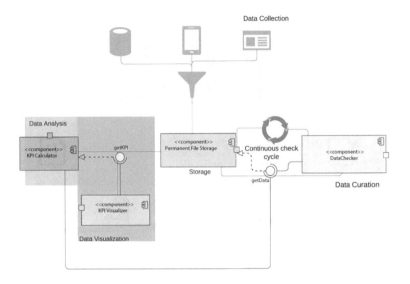

Fig. 2. Component diagram of the pipeline

2.2 KPI Calculator Component

The **KPI Calculator** and the **KPI Visualizer** are two distinct components that interact via bash script. This design choice to split the two components allows to provide the end user with also a command line software that can be executed to perform the calculation. The main components are located on two different servers. The first is responsible for keeping the data and the second for processing it. The temporary archive on the "Web Server" will contain that subset of data that is necessary for calculating the KPIs just in the period of interest. It will be cleared right the execution. Only the results of the calculation are stored in files owned by the user. The *Activity Diagram* in Fig. 3 schematically shows what happens when a user asks to view the KPIs. A dashboard allows for selecting the input parameters and starting the KPI calculation software via a bash command. The calculation software tries to connect to the **Permanent file storage** via *sftp*. The script download only those files that are necessary for the requested calculation. If there are no the required data, an error is returned. The data is extracted into data structures of the *Key-Value* type, where the *key* is the name of the attribute present in the downloaded file and the *value* is the corresponding value within the file. The time-series, on the other hand, are saved in a three-dimensional matrix where the first dimension represents the time-series while the second one represents the time value and the last one represents the measured value. After the data extraction, the actual calculation is performed using the algorithms defined for each individual KPI (some examples of calculation algorithms are presented in the following section). The structure of the output files is also of *key-value*. The computed KPIs are saved in tabular form in *csv* files, which are then read by the **KPI Visualizer**.

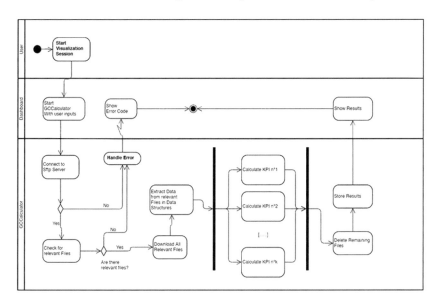

Fig. 3. Activity diagram

2.3 KPI Visualizer Component

The **KPI Visualizer** component aims at providing to the users a graphical presentation of KPIs by a web interface. The interface has been designed to be as simple as possible, but which simultaneously it offers all the necessary functionalities. The main task of this component is to read the output file generated by the **KPI Calculator** in the folder relative to the user who started the evaluation session and to convert the content into a key - value structure. This structure is then used to fill the various components of the web page. As can be seen from the use case diagram of the **KPI Visualizer** there are various types of visualization for the various KPIs which differ in the type of input data and the display method. Business KPIs needs a time period ranging from one year to three years (the calculation of the single value is carried out month by month) as shown in Fig. 7. SelfConsumption KPI that is represented as a value but also as a chart as in Fig. 6. Other numeric KPIs are represented as shown in Fig. 5 and needs the period of interest and the demo site of interest in order to be calculated. The tool is also able to export a summary report which include the output data and allows to download all the produced charts in the different formats. In total, the entire software system is able to calculate and display over 30 KPIs.

3   Implementation and GUI

The **DataChecker** Component has been implemented in Python. The software runs in batches. It is organized into abstract classes, which define the three types of input data, and their implementation that depends on the specify data. Finally, each class has methods that perform the various checks on the data structure and on the values they contain. The work of Datachecker is needful for the correct operation of the other component of the pipeline that deal with the analytics. Some screenshots of the web interface of the Datachecker are shown in Fig. 4. They report the results of the analytics related to the data curation and represent a timely report for the data providers who are informed about any errors in their loaded data.

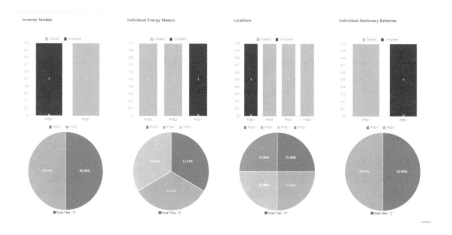

Fig. 4. DataChecker view

The **KPI Visualizer** Component, as well as the web interface of the datachecker, was developed using PHP, Bootstrap and Javascript technologies.

After the log-in, the user can select the period he is interested in and the place of interest. A REST call activates the calculator and loads the results in the dashboard shown in Fig. 5. The self consumption chart is also calculated, as shown in Fig. 6, where in green it is possible to see the solar energy production, in red and in blue the consumer loads and in purple the self consumption value that varies during the period of interest. The total value of self consumption is shown instead in a percentage bar as shown in Fig. 5.

Fig. 5. Calendar selection screenshot

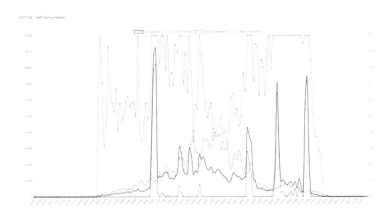

Fig. 6. Self-consumption graph screenshot

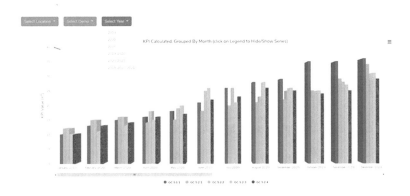

Fig. 7. Business KPI graph screenshot

4   Use Cases

In this section we illustrate the application of the Pipeline for the analysis, computation and visualization of two KPIs specifically relevant for renewable energy *self-consumption* and *sharing of EV batteries* to other appliances, utilizing Vehicle-to-Grid technologies.

4.1   Self-consumption

Self-consumption can be defined as the share of total photo-voltaic production consumed directly by the owner of the plant.

$$Self\text{-}Consumption = \frac{Energy\_Used\_From\_Panel}{Total\_Panel\_Production} \qquad (1)$$

To define self-consumption in a more formal way, we can denote the instantaneous power consumption of the building with $L(t)$ and generation instantaneous power from photo-voltaic system with $P(t)$. The amount of Power generated that is used directly at the site is limited by the smaller profile, be it production or consumption. It can be expressed as follows:

$$M(t) = Min(L(t), P(t)) \qquad (2)$$

where $M(t)$ is the instantly overlapping part of the production profiles and consumption (instant self-consumption). In the event that the energy is stored (either in batteries or thermally), the formula becomes:

$$M(t) = Min(L(t), P(t) + S(t)) \qquad (3)$$

where $S(t)$ is the input or output power from the unit of storage, with $S(t) < 0$ unloading (consumption) $S(t) > 0$ in charging phase (production). This takes into account losses due to charging, storage and unloading. At this point you can redefine self-consumption (SC):

$$\varphi_{SC} = \frac{\int_{t=t_1}^{t_2} M(t)dt}{\int_{t=t_1}^{t_2} P(t)dt} \qquad (4)$$

The visualization of such KPI is shown in Fig. 6.

4.2   Share of Battery Capacity for V2G

The aim of this KPI is to observing the users' behavior from research data and evaluating how much the battery of the EV could be exploited after the agreement of the user in the case V2G would have been supported. The required data for the KPI computation are the *Max Charging power of EV [CP]*, the *Status of charge at arrival [SOC]*, the *Target Status of charge (that means energy demand) [TSOC]*, the *Departure time [DP]*, the *Arrival time [AT]*. Finally we suppose that V2G discharging power is equal to Charging Power and that the charge point can charge/discharge the EV at the maximum power (we want to evaluate the use share, without taking into account the capability by the CP to exploit it).

In Fig. 8(a) the charge point charges the EV at maximum power until the target SoC. After that it can use the remaining time to charge and discharge the remaining capacity. It has room to further charge the battery, but it is usefulness if it cannot be discharged. The amount of energy that can be charged and discharged is:

$$SBC = CP[DT - AT - \frac{(TSOC - SoC)}{CP}] * 0.5 \tag{5}$$

In the Fig. 8(b) there is a reduced capacity, but if the users accepts that the battery can charged and discharge more times the amount of energy can flow is the same. In Fig. 8(c) the CP exploits the V2G capability at beginning, for example according to when a peak of energy demand or a peak of res production is expected. This specific KPI is shown to the user through the dashboard shown in Fig. 5.

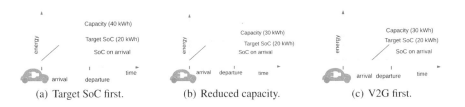

(a) Target SoC first.   (b) Reduced capacity.   (c) V2G first.

Fig. 8. Charging scenarios for V2G

5   Conclusions

The evaluation of operating innovation for the digitalization of smart cities cannot disregards the automation of data processing whose goal is to provide the user with a report on the achieved impact by a set of measured KPIs. To this aim, we presented a big data analysis and visualization pipeline and a software tool that implements such a computational workflow for the evaluation of technology innovation in the context of e-mobility. The proposed approach and the related software can be easily extended with additional analytics and KPIs, and to support wider heterogeneity of data, in the same application context or in any different domains.

Acknowledgments. Authors of this paper, on behalf of GreenCharge consortium, acknowledge the European Union and the Horizon 2020 Research and Innovation Framework Programme for funding the project (grant agreement no. 769016).

References

1. Aversa, R., Branco, D., Di Martino, B., Venticinque, S.: Greencharge simulation tool. In: Advances in Intelligent Systems and Computing. AISC, vol. 1150 , pp. 1343–1351 (2020)
2. Buur, J.: Participatory design of business models. In: Proceedings of the 12th Participatory Design Conference: Exploratory Papers, Workshop Descriptions, Industry Cases, PDC 2012, vol. 2, pp. 147–148. ACM, New York (2012)
3. Di Martino, B., Colucci Cante, L., Graziano, M., Enrich Sard, R.: Tweets analysis with big data technology and machine learning to evaluate smart and sustainable urban mobility actions in Barcelona. In: Advances in Intelligent Systems and Computing. AISC, vol. 1194, pp. 510–519 (2021)
4. Di Martino, B., Colucci Cante, L., Venticinque, S.: An ontology framework for evaluating e-mobility innovation. In: Advances in Intelligent Systems and Computing. AISC, vol. 1194, pp. 520–529 (2021)
5. Dijk, M., Orsato, R.J., Kemp, R.: The emergence of an electric mobility trajectory. Energy Policy **52**, 135–145 (2013)
6. Marijuán, A.G., Etminan, G., Möller, S.: Smart cities information system key performance indicator guid version:2.0. Technical report (2017). ENERC2/2013-463/S12.691121
7. Venticinque, S., Di Martino, B., Aversa, R., Natvig, M., Jiang, S., Enrich Sard, R.: Evaluating technology innovation for e-mobility. In 20:19 IEEE 28th International Conference on Enabling Technologies: Infrastructure for Collaborative Enterprises (WETICE), pp. 76–81. IEEE (2019)

A Simplified Map-Matching Algorithm for Floating Car Data

Federico Karagulian[✉], Gabriella Messina, Gaetano Valenti, Carlo Liberto, and Fabio Carapellucci

Italian National Agency for New Technologies, Energy and Sustainable Economic Development (ENEA), Via Anguillarese 301, 00123 S. Maria di Galeria, Roma, Italy

{federico.karagulian,gabriella.messina,geatano.valenti,
carlo.liberto,fabio.carapellucci}@enea.it

Abstract. We present a simplified map-matching algorithm that could be considered a robust tool to identify the correct path between consecutive GPS traces over a large number of scenarios avoiding ambiguous route assignment consistent with trajectory samples. Our formulation relies on a hidden Markov model (HMM) framework including multiple features such as the travelled distances between consecutive GPS traces, the signal quality and the direction of travel. The accuracy of the algorithm was evaluated using Floating Car Data (FCD) from a large fleet of privately owned cars and commercial vehicles equipped with devices capable of acquiring GPS positions with a sampling period of about 30 s. Experimental results showed an average accuracy of the model of about 85%. Results suggest our model is suitable not only to identify trajectories for specific origins and destinations, but also to extract traffic and travel time patterns.

1 Introduction

Among the products developed with satellite technology, the Global Positioning Systems (GPS) represents the most important for number of applications in all commercial sectors [1]. The increasing diffusion of geo-localization systems based on the Satellite Technology (GPS, GLONASS, BEIDOU, Galileo) installed on vehicles allows gathering information about trips across geographical areas. This information is stored in massive amount of data named Floating Car Data (FCD). FCD is referred to GPS time series collected by moving vehicles through an on-board terminal or a mobile phone [2]. The high spatial and temporal coverage of FCD cannot be achieved by the traditional fixed monitoring stations. Therefore, the use of FCD may represent a valuable solution to remotely monitor travel patterns and to complement traffic detection by fixed sensors, thus reducing the cost of traffic monitoring infrastructure. FCD is also used by applications like Google Navigation [3] and Waze [4] to predict travel times and to collect traffic data from movement profiles of smartphones. Among the advantages offered by FCD, there is the possibility to obtain accurate traffic and travel information at various temporal and spatial scales.

Map matching algorithms combine a digital road map with sequences of positioning data received from the GPS to obtain the routes that represent the sequences of roads travelled by vehicles.

L. Barolli et al. (Eds.): AINA 2021, LNNS 227, pp. 711–720, 2021.
https://doi.org/10.1007/978-3-030-75078-7_70

Although several map-matching methodologies [5–7] have been proposed, current techniques suffer from two main shortcoming: (1) the availability of updated digital maps describing the road network geometry, topology and traffic related map attributes, (2) imprecise location and speed measurements due to GPS errors. Moreover, GPS cannot trace the devices in enclosed spaces, such as buildings and tunnels.

Recently, map-matching has been implemented with several techniques which include geometrical analysis [8], Extended Kalman Filter [9] and Hidden Markov Model (HMM) [2, 10]. The map-matching algorithm proposed in this work combines the standard HMM with two additional constraints to improve the overall accuracy of the model over a large number of trip scenarios: (1) the comparison between every travelled distance, obtained from the GPS device, with the total length of candidate edges; (2) the plausibility of the correct temporal sequence of crossed nodes and the travel direction of trajectories.

Our methodology has been tested over a large road network constituted by urban, extra-urban and rural subnetworks.

2   Data Setups

FCD Data.   Floating Car Data (FCD) used in this work was provided by VEM-Solutions [11], a company, entirely owned by the Viasat Group holding, that integrates the most modern telematics technologies to guarantee complete safety of the vehicle and its occupants. FCD was gathered over a large area covering the province of Salerno (Italy) and part of the city of Naples from 31 August to 2 October 2019. A GPS record includes a set of fields: an anonymous identifier of the vehicle, the position's longitude and latitude, a timestamp of the measurement, a progressive distance between two consecutive traces, a signal 0/1 indicating the status of engine, the instant speed, the type of the vehicle (passenger and commercial vehicles) and the GPS signal quality level.

The average sampling rate was about 30 s per record. For the study area, about 165 million records were collected from about 94k moving vehicles of which 93% were cars and 7% commercial vehicles. FCD was pre-processed to filter outliers due to GPS signal quality or device failures and to build a new dataset organized by trips for each vehicle ID.

The Road Networks.   The road network (thereafter called *Graph*) data was downloaded from OpenStreetMap [12] and pertains to an area of 50 km diameter around the province of Salerno (Italy). For the purpose of this study, we only focused on the "driving" network without considering roads dedicated for cycling and walking. We should note that the geographical extension of the FCD also included the highly urbanized city of Naples. The *Graph* consisted of about 200k edges for a total length of 42000 km. Moreover, the administrative border maps and the road classification attributes were used to identify urban/extra-urban roads in the overall network.

In the following, the geographical position of a node and of a GPS trace in the *Graph* will be indicated as x_i and g_k, respectively.

Hardware and Software. Elaborations for map-matching were carried out with the following hardware: Intel®, Xenon®, Gold 6142 CPU®, 2.59 GHz processor, 64 CPUs, 128 GB RAM. FCD, network data and map-matching outputs were stored into a PostgreSQL data base. The map-matching algorithm was entirely written with the open-source programming language Python.

3 Map-Matching Algorithm

Map-matching is a process to associate a sequence of geographical coordinates to a real path on the road network. In this work, we explore the relation between a series of GPS traces and the edges in an existing *Graph* through an ordered temporal list of nodes identified by the most probable sequence of edges crossed by a moving vehicle. The relationship between traces, edges and nodes is described by the Emission and the Transition probability that will be briefly explained below.

The Emission Probability is referred to the distance between a node x_i and the nearby GPS trace g_k weighted over the standard deviation σ calculated from all distances between each GPS trace and a 20-m buffer built around the GPS trace [13]. Overall, the value of σ associated to our FCD was estimated ranging from 5 to 20 m. We must point out that σ is the important parameter that may strongly affect the performance of the map-matching process especially in urban context. However, the GPS *accuracy* is an intrinsic property of the GPS sensor device that cannot be improved in the map-matching process.

The Emission Probability is assumed following a normal distribution as follow:

$$p_E(x_i|g_k) = \frac{1}{\sqrt{2\pi}\sigma} e^{-\frac{d_{ik}^2}{2\sigma^2}} \tag{1}$$

where $d_{ki} = \{g_k|r_i\}$ $(r_i \in R)$ is the distance between the trace and the part of the nearby edge intersecting the circular buffer. For a node x_i, we can obtain the most probable corresponding GPS trace g_k through the calculation of the emission probability (Fig. 1).

On the other hand, the Transition Probability is the probability the vehicle travels from one node x_i to the next one $x_j = x_{i+1}$ along a certain path. To estimate the Transition Probability we considered the difference between the Euclidean distance $d(x_i, x_{i+1})$ and the shortest path distance $D(x_i, x_{i+1})$ among two consecutive nodes [14]:

$$p_T(x_i|x_{i+1}) = e^{-[d(x_i,x_{i+1})-D(x_i,x_{i+1})]} \tag{2}$$

where $D(x_i, x_{i+1})$ is calculated by the Dijkstra algorithm [15]. Therefore, the transition probability represents the probability that a given path is travelled by a vehicle moving between two consecutives traces (Fig. 1).

3.1 Map-Matching Methodology

At the basis of the map-matching model there is the assumption of a Markov process that consists in the determination of the most probable node sequence associated to GPS traces [16].

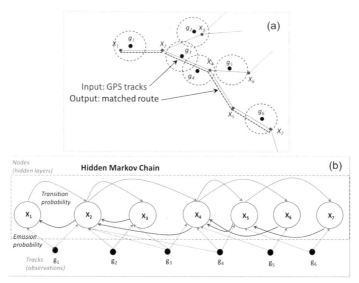

Fig. 1. *(a)* representation of GPS traces g_k in a road network where X_i are the nodes. Each buffer intersects one or more edges determining the probable number of adjacent nodes to be used in the Hidden Markov Model (HMM). *(b)* representation of the Markov Chain where *emission* probabilities are referred to the probability having a GPS trace close to a node, whereas *transition* probabilities are referred to the probability having transitions between nodes.

In order to match a sequence of traces with a probable path in the *Graph*, we use a generalization of the Viterbi decoding over a *Hidden Markov Model (HMM)* [2, 17]. According to the Markov process, the probability having a trace crossing a node at a given time, only depends on the node crossed immediately before. The algorithm choses the path corresponding to the sequence of most probable nodes taking into consideration the emission probability and the transition probability.

These are used in the Viterbi algorithm to find the most probable nodes and to store them in a prioritized list of *maximum probability nodes,* thereafter called *MaxProbNode* list.

As shown in Fig. 1(a), the buffer area of each GPS traces (g_k) intersects the edges whose nodes (x_i) at their extremes are probable candidates to build the correct matched path.

The main steps of the proposed map-matching algorithm can be summarized as follow:

1. for each trace point in a trip $T = [g_1, g_2, g_3, \ldots g_k]$, a 20-m buffer zone is defined to find edges lying in proximity of the trace. This allows dumping all candidates' nodes with very low emission probabilities. The schematic representation of the map-matching process in Fig. 1(a) clearly shows edges are identified through the intersection of the buffer with the road segment crossed by each trace. Therefore, we associate an *"adjacency list"* of neighbor edges r_i to each trace and we indicate this as $E(g) = \{E_{g1}, E_{g2}, \ldots \ldots E_{gk}\}$ $(k = 1, 2 \ldots n)$, where $E_{g_k} = \{g_k | r_1, r_2, \ldots \ldots r_n\}$ $(r_n \in Graph)$ is the sub-list of node-pair/edges $(x_i, x_{i+1}) = r_i$ associated to each trace g_k.

2. the emission probability $p_E(x_i | g_k)$ is computed for each neighbor node x_i, whereas a transition probability $p_T(x_i | x_{i+1})$ is calculated between each node x_i and all its incident nodes x_{i+1} (Fig. 1(b)).

3. for each pair of nodes (x_i, x_{i+1}) in the *adjacency list* $E(g)$ we verify the shortest path distance between two nodes $D(x_i, x_{i+1})$ is within the distance between two temporally consecutive traces:

$$D(x_i, x_{i+1}) \leq \Delta_{FCD}(g_k, g_{k+1}) \tag{3}$$

$D(x_i, x_{i+1})$ is calculated by the Dijkstra algorithm [15], whereas $\Delta_{FCD}(g_k, g_{k+1})$ is the distance travelled between two temporally consecutive traces and it is computed using the progressive distance between traces from the FCD.

Condition (3) is used in order to avoid matching an edge longer than the effective travelled distance.

According to the above constraints, the maximum likelihood node x_i^T over the Markov chain is the one with the highest joint *emission* and *transmission* probabilities:

$$x_i^T = arg\ max(max(p_T(x_i | x_{i+1})) \times p_E(x_i | g_k)) x_i \in E_{g_k} \tag{4}$$

However, if the node x_i^T is not among the candidate node list $E_{g_{k+1}}$ associated to the temporally consecutive trace g_{k+1}, it is removed from the *adjacency list* $E(g)$ and the process is reinitialized. The above process is repeated for the next trace point and, the list of most probable nodes *MaxProbNode* is populated consequently. The algorithm ends when the last point is reached.

Pseudo code of the Map-matching algorithm developed for the case study (see text for more details).

```
Input: trip: T = [g₁, g₂,g₃, .....gₖ], road network: Graph
Output: route = [(x₁,x₂), (x₂,x₃),........,(xᵢ,xᵢ₊₁)]
```

```
For i = 1 to K do
    find edges E(g) = {E_g1, E_g2,......E_gk} crossing each trace gₖ
    Let MaxProbNode[ ] store the max. joint probabilities
    REPEAT
        for each (xᵢ, xᵢ₊₁) = rᵢ in E(g) do
            Compute distance between traces Δ_FCD(gₖ, gₖ₊₁)
            Compute distance between nodes D(xᵢ, xᵢ₊₁)
            if D(xᵢ, xᵢ₊₁) < Δ_FCD(gₖ, gₖ₊₁)
                Compute p_T(xᵢ|xᵢ₊₁)
                Compute p_E(xᵢ|gₖ)
            c=arg max(max(p_T(xᵢ|xᵢ₊₁)) × p_E(xᵢ|gₖ))
            MaxProbNode[xᵢ] = c
    WHILE MaxProbNode[xᵢ] not in E_gk = {gₖ| r₁, r₂, .....rₙ} do
        remove xᵢ from MaxProbNode
        remove xᵢ from E(g)
```

By comparing the position of each node with the sequence of neighbor traces, it is verified that the sequence of the *MaxProbNode* list is associated to edges oriented towards the same direction. This allows the identification of traces that, within a temporal sequence, are related to edges in the opposite direction compared to the edges related to the previous/next trace. Nodes, assigned to edges of opposite direction with respect to adjacent edges are removed from the *MaxProbNode* list.

The node sequence in the *MaxProbNode* list is not necessarily associated to a sequence of connected edges. Therefore, the final *route* $[(x_1, x_2), (x_2, x_3), \ldots, (x_i, x_{i+1})]$ is obtained by computing the shortest path between each consecutive pair of nodes using the Dijkstra algorithm [15]. Subsequently, the *MaxProbNode* list is accordingly updated and the less probable nodes are removed from the *adjacency list*. On the other hand, the prioritized *MaxProbNode* list is maintained during the iterations for each trace to speed up the evaluation of the following nodes and to reduce the number of computations.

3.2 Map-Matching Outputs

Figure 2 shows two examples of outputs obtained with the map-matching process performed in the area of Salerno for the whole month of September 2019. Matched paths are shown for motorways, secondary and tertiary roads in rural and urban areas. As we can see, map-matching worked well on trajectories of different length. Unmatched traces were eventually located at the beginning or at the end of the trajectory (*trip*). This was observed in several circumstances where the trace corresponding to the origin and destination was located in residential/commercial areas of private properties where roads were not reported in the OSM dataset.

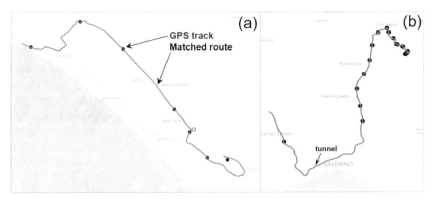

Fig. 2. Map matching outputs for different sequences of GPS traces. *(a)* vehicle travelling along urban roads. *(b)* vehicle travelling along extra-urban roads.

Overall, about 4.5 million of trips were processed with an average rate of 160 *trips* a minute. Fig. 3 shows the traffic volumes from map-matching carried out in our case study. Traffic volumes were estimated considering the total number of paths crossing the edges.

As shown in Fig. 3(a), cars were mostly travelling along primary, secondary, tertiary and residential roads. Instead, commercial vehicles were mainly travelling over motorways and highways (Fig. 3(b)).

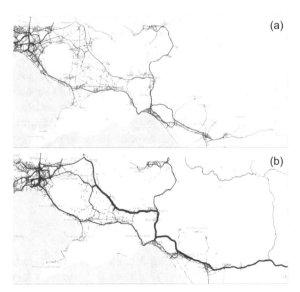

Fig. 3. Traffic count obtained from map-matching of FCD with the road network around the area of Salerno Italy (a) for cars and for (b) commercial vehicles during the month of September 2019. Thickness of roads is proportional to the number of times (counts) each edge was crossed.

As reported in recent works, the accuracy of the map-matching algorithm was defined as the ratio between the length of the matched *route* and the distance travelled by the vehicle within a *trip* [5, 7]. For every *trip*, correct matching was observed when the *route* was mapped on any road segment of the *Graph*. In order to avoid accounting low accuracies derived from map-matching of very short *trips,* the latter were not considered (inset Fig. 4). Compared to the majority of the matched routes, which had length greater than 2.7 km, short trips were identified as the matched route of length less than 700 m. These represented only 5% of the total matched routes.

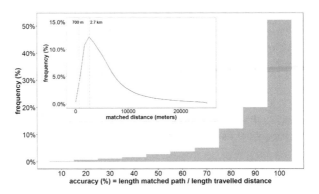

Fig. 4. Distribution of the accuracy obtained from the map-matching process over the area of Salerno (Italy). Inset shows the distribution of the lengths of matched routes.

As reported in other works [5, 7], it is not always possible to determine the road segment crossed by every trace for two main reasons: (a) possible topological error in the *Graph:* the information provided in the digital street map might presents missing roads or incorrectly classified roads; (b) cases when traces are located in the middle of road junctions and therefore, traces could be attributed to either exits of the junctions. Because of that, the matched route could be longer or shorter than the progressive distance travelled by each trace. Overall, within an error of 10–20%, the average map-matching accuracy, was about 85% (Fig. 4). This result was in line with other works [7, 18, 19] which reported map-matching of FCD on urban and extra-urban networks with sampling rates ranging from 5 to 60 s.

4   Conclusions and Future Work

In this work, we presented a map-matching methodology based on a HMM for FCD at an average sampling rate of about 30 s. This algorithm accounts for trace position to calculate emission probabilities and uses the shortest path distance to calculate transition probabilities. In addition, it also compares travelled distances with length of candidate roads and verifies the correct temporal sequence of nodes and direction of candidate roads. The overall average accuracy of the proposed map-matching model was about 85% with an average sampling time of about 30 s. Overall, the proposed algorithm works well to solve some errors such as matching traces onto elevated roads, roundabout or parallel lanes.

Therefore, the proposed methodology is suitable for analyzing large amount of FCD to estimate traffic volumes for several applications in the mobility sector.

Future improvements will explore a methodology to build a dynamic buffer area around each trace that keeps into account a prioritized list of edges depending to different environments such as in urban or extra-urban areas.

Acknowledgments. The research was carried out within the project SENTINEL (Sistema di pEsatura diNamica inTellIgente per la gestioNE deL traffico pesante) funded by the Ministry of Education and Ministry of University and Research.

References

1. Maddison, R., NiMhurchu, C.: Global positioning system: a new opportunity in physical activity measurement. Int. J. Behav. Nut. Phys. Activity **6**, 73 (2009). https://doi.org/10.1186/1479-5868-6-73
2. Thiagarajan, A., Ravindranath, L., LaCurts, K., Madden, S., Balakrishnan, H., Toledo, S., Eriksson, J.: VTrack: accurate, energy-aware road traffic delay estimation using mobile phones. MIT web domain (2009)
3. Li, J., Boonaert, J., Doniec, A., Lozenguez, G.: Traffic flow multi-model with machine learning method based on floating car data. In: 2019 6th International Conference on Control, Decision and Information Technologies (CoDIT), pp. 512–517 (2019). https://doi.org/10.1109/CoDIT.2019.8820434
4. Jeske, T.: Floating Car Data from Smartphones: What Google and Waze Know About You and How Hackers Can Control Traffic. Presented at the (2013)
5. Goh, C.Y., Dauwels, J., Mitrovic, N., Asif, M.T., Oran, A., Jaillet, P.: Online map-matching based on hidden Markov model for real-time traffic sensing applications. Presented at the IEEE Conference on Intelligent Transportation Systems, Proceedings, ITSC (2012). https://doi.org/10.1109/ITSC.2012.6338627
6. Huber, S., Rust, C.: Calculate travel time and distance with Openstreetmap data using the open source routing machine (OSRM). Stata J. **16**, 416–423 (2016). https://doi.org/10.1177/1536867X1601600209
7. Torre, A.D., Gallo, P., Gubiani, D., Marshall, C., Montanari, A., Pittino, F., Viel, A.: A map-matching algorithm dealing with sparse cellular fingerprint observations. Geo-spatial Inf. Sci. **22**, 89–106 (2019). https://doi.org/10.1080/10095020.2019.1616933
8. Chen, D., Driemel, A., Guibas, L.J., Nguyen, A., Wenk, C.: Approximate Map Matching with respect to the Fréchet Distance. In: 2011 Proceedings of the Workshop on Algorithm Engineering and Experiments (ALENEX), pp. 75–83. Society for Industrial and Applied Mathematics (2011). https://doi.org/https://doi.org/10.1137/1.9781611972917.8.
9. Obradovic, D., Lenz, H., Schupfner, M.: Fusion of map and sensor data in a modern car navigation system. J. VLSI Sign. Process. Syst. Sign. Image Video Technol. **45**, 111–122 (2006). https://doi.org/10.1007/s11265-006-9775-4
10. Pink, O., Hummel, B.: A statistical approach to map matching using road network geometry, topology and vehicular motion constraints. In: 2008 11th International IEEE Conference on Intelligent Transportation Systems, pp. 862–867 (2008). https://doi.org/10.1109/ITSC.2008.4732697
11. VEM-Solutions: https://www.vemsolutions.it

12. OpenStreetMap: https://www.openstreetmap.org
13. Qi, W., Lei, W., Kennedy, R.A.: Iterative Viterbi decoding, trellis shaping, and multilevel structure for high-rate parity-concatenated TCM. IEEE Trans. Commun. **50**, 48–55 (2002). https://doi.org/10.1109/26.975743
14. Gao, X., Ramesh, V., Boult, T.E.: Statistical characterization of morphological operator sequences. In: Proceedings of the 7th European Conference on Computer Vision-Part IV, pp. 590–605. Springer-Verlag, Berlin, Heidelberg (2002)
15. Dijkstra, E.: A note on two problems in connexion with graphs. Numerische Mathematik. (1959). https://doi.org/10.1007/BF01386390
16. Rabiner, L.R.: A tutorial on hidden Markov models and selected applications in speech recognition. Proc. IEEE **77**, 257–286 (1989). https://doi.org/10.1109/5.18626
17. Hummel, B.: Map Matching for Vehicle Guidance (Draft), /paper/Map-Matching-for-Vehicle-Guidance-(-Draft-)-Hummel/0cbddf0548ab44d4cc74fa30841eeda31cd51429. Accessed 15 Oct 2020
18. Andersen, O., Torp, K.: Sampling frequency effects on trajectory routes and road network travel time. In: Proceedings of the 25th ACM SIGSPATIAL International Conference on Advances in Geographic Information Systems, pp. 1–10. Association for Computing Machinery, New York, NY, USA (2017). https://doi.org/10.1145/3139958.3140024
19. Hu, Y., Lu, B.: A hidden markov model-based map matching algorithm for low sampling rate trajectory data. IEEE Access. **7**, 178235–178245 (2019). https://doi.org/10.1109/ACCESS.2019.2958982

Correction to: Software-Defined Networking: Open-Source Alternatives for Small to Medium Sized Enterprises

Paul Thornley and Maryam Bagheri

Correction to:
Chapter "Software-Defined Networking: Open-Source
Alternatives for Small to Medium Sized"
in: L. Barolli et al. (Eds.): *Advanced Information*
***Networking and Applications*, LNNS 227,**
https://doi.org/10.1007/978-3-030-75078-7_20

In the original version of the Chapter, the following correction have been incorporated: The chapter title "Software-Defined Networking: Open-Source Alternatives for Small to Medium Sized" has been changed to "Software-Defined Networking: Open-Source Alternatives for Small to Medium Sized Enterprises".

The updated version of this chapter can be found at
https://doi.org/10.1007/978-3-030-75078-7_20

Author Index

Printed in the United States
by Baker & Taylor Publisher Services